中国农业科学院
兰州畜牧与兽药研究所科技论文集
（2015）

中国农业科学院兰州畜牧与兽药研究所　主编

中国农业科学技术出版社

图书在版编目（CIP）数据

中国农业科学院兰州畜牧与兽药研究所科技论文集.2015／中国农业科学院兰州畜牧与兽药研究所主编.—北京：中国农业科学技术出版社，2017.12
ISBN 978-7-5116-2950-0

Ⅰ.①中…　Ⅱ.①中…　Ⅲ.①畜牧学-文集②兽医学-文集　Ⅳ.①S8-53

中国版本图书馆 CIP 数据核字（2017）第 004659 号

责任编辑　闫庆健
文字加工　鲁卫泉
责任校对　贾海霞

出 版 者　中国农业科学技术出版社
北京市中关村南大街 12 号　邮编：100081
电　　话　(010)82106632(编辑室)　(010)82109704(发行部)
(010)82109703(读者服务部)
传　　真　(010) 82106625
网　　址　http://www.castp.cn
经 销 者　各地新华书店
印 刷 者　北京建宏印刷有限公司
开　　本　880 mm×1 230 mm　1/16
印　　张　39.25　彩插 14 面
字　　数　1 030 千字
版　　次　2017 年 12 月第 1 版　2017 年 12 月第 1 次印刷
定　　价　150.00 元

#《中国农业科学院兰州畜牧与兽药研究所科技论文集（2015）》

编写委员会

前　　言

近年来，在中国农业科学院科技创新工程的引领下，研究所的科研水平快速提升。我所科研人员和管理人员不但有工作上的热情，更有对工作认识上的高度和对学科理解上的深度。他们在紧张繁忙的实践活动中，笔耕不辍，将自己的研究成果写成论文。这不单是科研人员和管理人员的工作总结、过程记录，更是他们智慧的结晶，最终成为研究所的一笔宝贵财富。

为了珍惜这笔财富，加强优秀论文的交流与传播，营造更加浓厚的学术氛围，促进科研水平和管理水平的提升，切实推进研究所的科技创新，科技管理处搜集了 2015 年研究所科研人员公开发表的论文编印成《中国农业科学院兰州畜牧与兽药研究所科技论文集》第四卷，共 152 篇。由于时间仓促，可能还有论文未能收录，敬希鉴谅！

编者

2017 年 12 月

目　录

Synthesis and Evaluation of Novel Pleuromutilin Derivatives with a Substituted Pyrimidine Moiety

YI Yun-peng[1], YANG Guan-zhou[2], ZHANG Chao[1], CHEN Jiong-ran[1], LIANG Jian-ping[1*], SHANG Ruo-feng[1]

(1. Key Laboratory of New Animal Drug Project of Gansu Province, Key Laboratory of Veterinary Pharmaceutical Development, Ministry of Agriculture, Lanzhou Institute of Husbandry and Pharmaceutical Sciences of CAAS, Lanzhou 730050, China; 2. School of Pharmacy, Lanzhou University, Lanzhou, 730020, China)

Abstract: A series of novel pleuromutilin derivatives possessing 6−hydroxy pyrimidine moieties were synthesized via acylation reactions under mild conditions. The *in vitro* antibacterial activities of the synthesized derivatives against methicillin − resistant *Staphylococcus aureus* (MRSA), methicillin − resistant *Staphylococcus epidermidis* (MRSE), *Bacillus subtilis* (*B. subtilis*), and *Escherichia coli* (*E.coli*) were tested by the agar dilution method. The majority of the screened compounds displayed potent activities. Compounds 3 and 6a were found to be the most active antibacterial agents against MRSA and MRSE. Moreover, in the vivo experiment, compound 6a showed comparable antibacterial activity to that of tiamulin, with ED^{50} of 5.47mg/kg body weight against MRSA.

Key words: Pleuromutilin derivatives; Synthesis; Antibacterial activity; MRSA

1 INTRODUCTION

Many hospitals lack the effect way to the late phase of infection as a result of the continuing rise in antibiotic resistance among pathogenic bacterial strains. Therefore, development of new antibacterial agents possessing a novel mechanism of action is of great importance to the medical community[1].

Pleuromutilin (1, Fig.1), a natural antibiotic with antibacterial activities, was isolated in 1951 from two basiomycetes species, *Pleurotus* mutilis and *Pleurotus passeckeranius*[2]. The pleuromutilin class has a unique mode of action, which involves inhibition of protein synthesis by primarily inhibiting ribosomal peptidyl transferase center (PTC)[3,4]. Further studies demonstrated that the interactions of tricyclic core of tiamulin are mediated through hydrophobic interactions and hydrogen bonds, which are formed mainly by the nucleotides of domain V[5,6].

---

* Corresponding author. E-mail: yiyp2013@163.com, shangrf1974@163.com

The process of semi-synthesis based on natural products, especially the complex natural products, is to be the predominant avenue to new antibiotics[7].The chemical modifications of pleuromutilin have been made for an attempt to improve the antimicrobial activities and *in vivo* efficacy after the identification of structure of pleuromutilin.The introduction of a thioether at the C-22 position of pleuromutilin analogues and the presence of a basic group enhance antibacterial activity[8]. Thus, the modifications which have focused on their C-14 side chain have led to three drugs: Tiamulin (Fig.1.), valnemulin and retapamulin (Fig.1.).Tiamulin and valnemulin were approved for veterinary, and retapamulin became the first pleuromutilin drug approved for use in humans[8-12].

A range of pleuromutilin derivatives with a heterocyclic ring at the C14 side chain has been recently synthesized and studied[13]. Structure activity relationship (SAR) study showed the synthesized pleuromutilin derivatives with a heterocyclic ring at the C14 position may increase hydrogen bonding and $\pi-\pi$ stacking interactions and thereby have strong antibacterial properties[14-16].In addition, compounds bearing primary amine substituents at pyridine ring incorporated into the C14 side chain exhibited antibacterial activity[16].The molecular docking results revealed the substituents with hydrogen donators, for example, hydroxy and amino group, at the C14 side chain enhance the binding affinities by hydrogen bondings and thus should be introduced to produce new analogues with higher antibacterial activities[17].

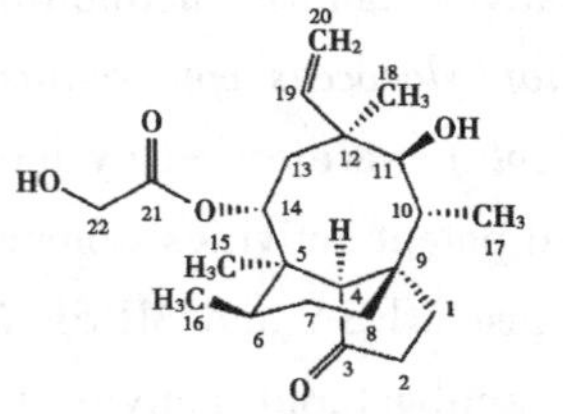

Pleuromutilin, 1

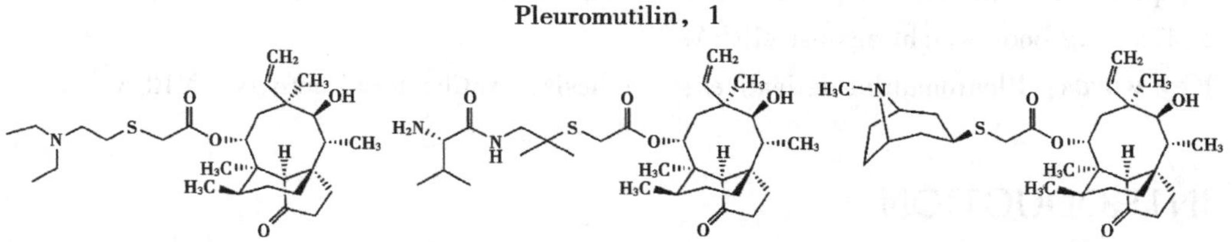

Tiamulin　　Valunemulin　　Retapamulin

Fig.1　Structures of pleuromutilin and its drugs

Previous work in our group has led to the synthesis of pleuromutilin derivatives with a 1,3,4-thiadiazole ring at the C14 position.The antibacterial studies prove that heterocyclic ring bearing polar groups at the C14 side chain of pleuromutilin derivatives may raise their antibacterial activity[17,18]. This conclusion is supported by the work of Ling et al.[16].They designed and synthesized a serial of pleuromutilin with pyridine ring bearing amino groups which exhibit excellent *in vitro* antibacterial activity against both sensitive and resistant Gram-positive bacterial strains.The compounds with pyrimidine ring were wide found in nature products, such as nucleotides, thiamine (vitaminB1) and alloxan[19].The pyrimidine system also turned out to be an important pharmacophore, and is found in many synthetic compounds such as antibacterial drug, trimethoprim, barbiturates and the HIV drug, zidovudine[19].We now report the design, synthesis, and antibacterial studies of novel pleuromutilin derivatives with 6 - hydroxy pyrimidine incorporated into the C14 side chain for identification of a higher soluble and more efficacious drug candidate.

2 RESULTS AND DISCUSSION

2.1 Chemistry

The general synthetic route to build the pleuromutilin derivatives is illustrated in Scheme 1. The lead compound 14-O-[(4-amino-6-hydroxy-pyrimidine-2-yl) thioacetyl] mutilin (3) was prepared by nucleophilic substitution of 22-O-tosylpleuromutilin (2) with 4-amino-6-hydroxy-2-mercaptopyrimidine monohydrateunder basic conditions in 78% yield.Pleuromutilin derivatives 4 and 5 were always simultaneously obtained as isomers from 3 because of the ketoeenol equilibrium of the pyrimidinone scaffold [20].The reaction generally gave 4 and 5 (1.0: 2.1 ratio) in the presence of K_2CO_3 as base.

Compounds 6aeh were directly obtained by condensation reactions between the amino group of compound 3 and the carboxyl group of carboxylic acids or amino acids in which amino groups were protected by tert-butoxycarbonyl (BOC) groups.The reactions were performed at room temperature in the presence of 1-ethyl-3-(3-dimethyllaminopropyl) carbodiimide hydrochloride (EDCI), 1-hydroxybenzotriazole (HOBt) and N, N-Diisopropylethylamine (DIPEA) as base.The protected amino groups were hydrolysed using a solution of TFA in DCM for 30 min and the compounds 6e-h were obtained in a yield of 41-73%.We also extensively synthesized but obtained with trace the isomers of compounds 6b-d: 14-O-[(4-(2-Chlorobenzamide)-6-hydroxypyrimidine-2-yl) thioacetyl] mutilin, 14-O-[(4-(4-Chlorobenzamide)-6-hydroxypyrimidine-2-yl) thioacetyl] mutilin, 14-O-[(4-(2-Methylbenzamide)-6-hydroxypyrimidine-2-yl) thioacetyl] mutilin, 14-O-[(4-(3-Methylbenzamide)-6-hydroxypyrimidine-2-yl) thioacetyl] mutilin, 14-O-[(4-(2-Methoxylbenzamide)-6-hydroxypyrimidine-2-yl) thioacetyl] mutilin, and 14-O-[(4-(3-Methoxylbenzamide) - 6 - hydroxypyrimidine - 2 - yl) thioacetyl] mutilin. The structures of the synthesized derivatives were characterized by IR, 1H NMR, 13C NMR and HRMS spectra (Supplementary data) and further supported by the single crystal X-ray diffraction analysis of compound 4 (Fig.2).

2.2 Antibacterial activity

The minimum inhibitory concentrations (MIC) of the obtained compounds 3, 4, 5 and 6a-h in comparison to tiamulin fumarate were tested against drug-resistant microorganisms, including three Gram (+), methicillin-resistant *Staphylococcus aureus* ATCC 29213 (MRSA), methicillin-resistant *Staphylococcus epidermidis* ATCC 35984 (MRSE) and *Bacillus subtilis* ATCC 11778 (*B. subtilis*), and one Gram (-), *Escherichia coli* ATCC 25922 (*E.coli*) by the agar dilution methods.The results are given in Table 1. Compounds 3 and 6a showed the most antibacterial activities which were superior or comparable antibacterial activities to that of tiamulin fumarate, with MIC values both in the 32-0.125mg/mL range.Compounds 4, 5, 6g and 6h showed slightly less activity against MRSA and MRSE and lower potent against *B.subtilis* when compared to that of tiamulin fumarate.However, all the compounds except 6a displayed lower antibacterial activities against *E.coli* in comparison to that of tiamulin fumarate.

Compounds 3, 4, 5, 6g and 6h with a primary amine on the terminal C-14 glycolic acid side chain presented improved activity against MRSA and MRSE compared with the compounds 6b-f. To our surprise, compound 6a with an acetyl group on the C-14 side chain displayed highest activity than the other compounds. The results of MICs indicated that the introduction of the primary amine into the C-14 glycolic acid side chain could enhance antibacterial activity, which was consistent with previous reported[17,18].

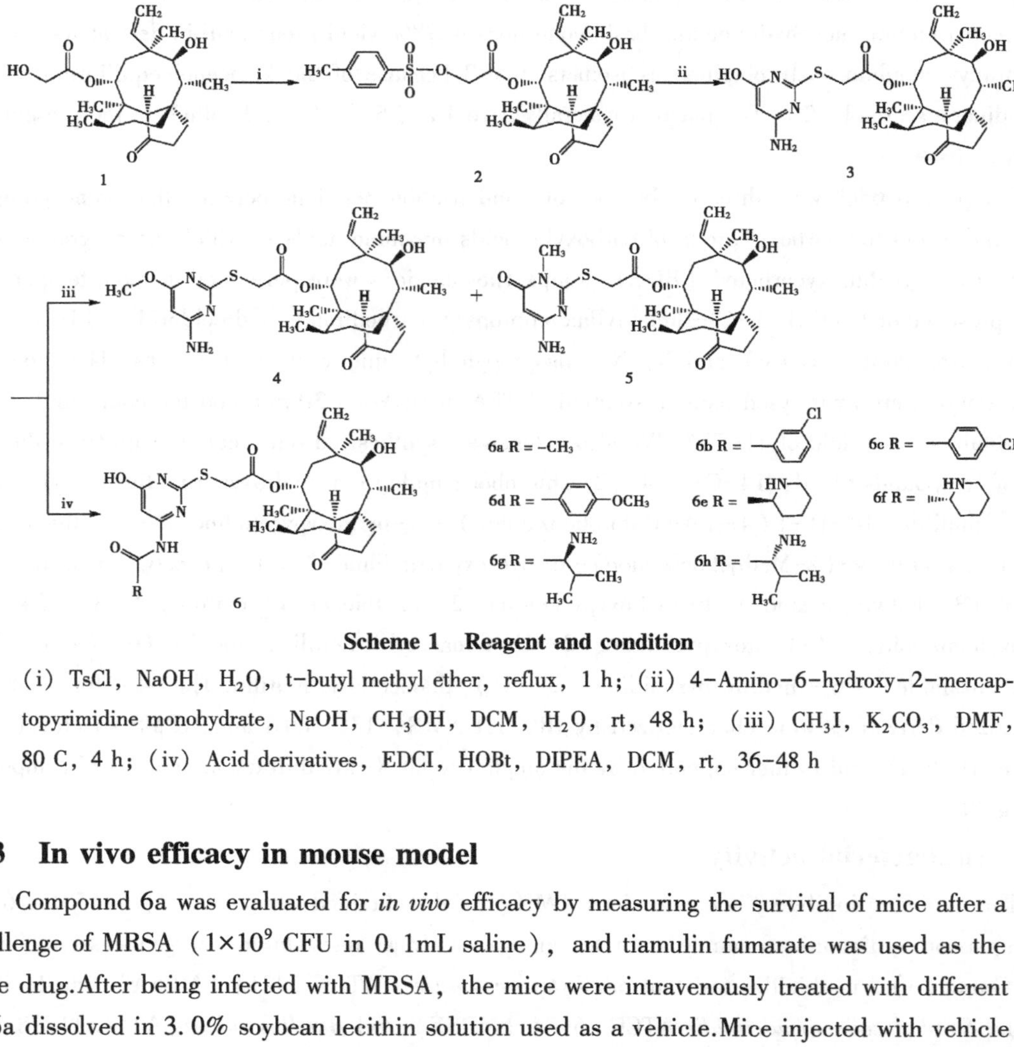

Scheme 1 Reagent and condition

(i) TsCl, NaOH, H_2O, t-butyl methyl ether, reflux, 1 h; (ii) 4-Amino-6-hydroxy-2-mercaptopyrimidine monohydrate, NaOH, CH_3OH, DCM, H_2O, rt, 48 h; (iii) CH_3I, K_2CO_3, DMF, 80 C, 4 h; (iv) Acid derivatives, EDCI, HOBt, DIPEA, DCM, rt, 36-48 h

2.3 In vivo efficacy in mouse model

Compound 6a was evaluated for *in vivo* efficacy by measuring the survival of mice after a lethal challenge of MRSA (1×10^9 CFU in 0.1mL saline), and tiamulin fumarate was used as the reference drug. After being infected with MRSA, the mice were intravenously treated with different doses of 6a dissolved in 3.0% soybean lecithin solution used as a vehicle. Mice injected with vehicle alone showed 100% mortality in this model. Treatment with 6a and tiamulin fumarate displayed dose-dependent protection and led to the survival of the mice infected with MRSA (see Fig.3), with ED^{50} of 5.47 and 5.95mg/kg body weight, respectively. Thus, 6a showed comparable antibacterial activity to that of tiamulin against MRSA in the mouse systemic model. The result of *in vivo* efficacy showed that 6a was able to protect animals in a dose-dependent fashion and might act as a potent antibacterial drug.

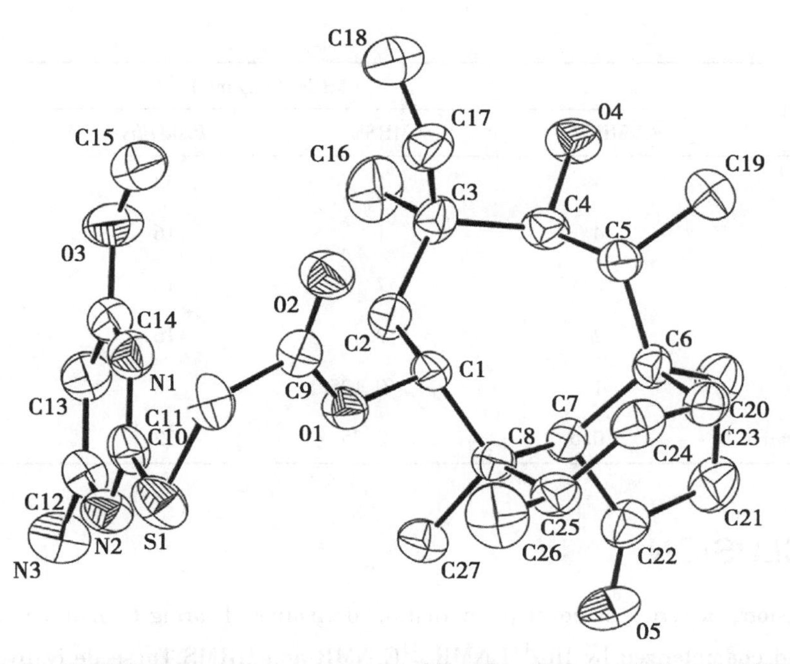

Fig.2 ORTEP diagram for compound 4 with ellipsoids set at 75% probability

(hydrogen atoms were omitted for clarity)

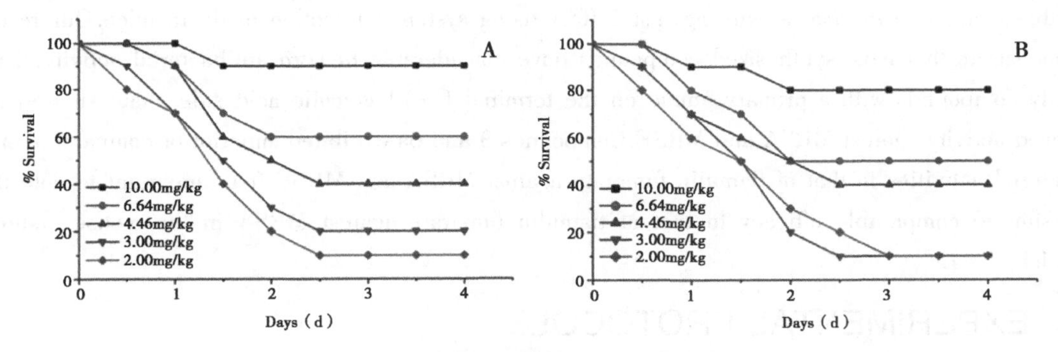

Fig.3 Efficacy of compound 6a

(A) and tiamulin fumarate, (B) in mouse systemic infection model

Table 1 *In vitro* antibacterial activities of pleuromutilin derivatives

Compounds	MICs (mg/mL)			
	MRSA	MRSE	*B.subtilis*	*E.coli*
3	0.25	0.125	32	16
4	1	0.5	16	32
5	2	1	16	32
6a	0.125	0.125	32	8
6b	32	16	64	16
6c	4	2	32	32

(continued)

Compounds	MICs (mg/mL)			
	MRSA	MRSE	*B. subtilis*	*E. coli*
6d	4	1	32	16
6e	4	1	16	16
6f	8	2	16	32
6g	2	0.5	16	32
6h	1	0.125	8	64
Tiamulin fumarate	0.5	0.25	32	8

3 CONCLUSION

In conclusion, a series of novel pleuromutilin derivatives bearing 6-hydroxy pyrimidine were synthesized and characterized by IR, ^{1}H NMR, ^{13}C NMR and HRMS. These derivatives were initially evaluated for their *in vitro* antibacterial activities against three Grampositive strains (MRSA, MRSE and *Bacillus subtilis*) and a Gramnegative strain *Escherichia coli*. Compound 6a was chosen for further evaluation *in vivo* activity against MRSA using systemic infection mode in mice. Our results demonstrate that most synthesized compounds have considerable *in vitro* antibacterial activity. Especially compounds with a primary amine on the terminal C-14 glycolic acid side chain showed improved activity against MRSA and MRSE. Compounds 3 and 6a exhibited superior or comparable antibacterial activities to that of tiamulin fumarate against MRSA and MRSE. It is important to note that 6a showed comparable efficacy to that of tiamulin fumarate against MRSA in the mouse systemic model.

4 EXPERIMENTAL PROTOCOLS

4.1 General

All the starting materials were of reagent grade. The solvents used for the isolation/purification of the compounds were purified prior to use unless noted otherwise. All reactions were monitored by thin-layer chromatography (TLC) using 0.2-mm-thick silica gel GF254 pre-coated plates (Qingdao Haiyang Chemical Co., Ltd, Shandong, China). After elution, the plates were visualized under UV illumination at 254 nm for UV active materials. Further visualization was achieved by staining with a 0.05% $KMnO_4$ aqueous solution. All column chromatography purifications were carried out on a 200-300mesh silica gel (Qingdao Haiyang Chemical Co., Ltd, Shandong, China) with conventional methods. IR spectra were obtained on a NEXUS-670 spectrometer (Nicolet Thermo, Edina, MN, USA) using KBr thin films, and the absorptions are reported in cm^{1}. ^{1}H NMR and ^{13}C NMR spectra were recorded using Bruker-400 MHz spectrometer (Bruker BioSpin, Züich, Züich State, Switzerland). High-resolution mass spectra (HRMS) were obtained with a Bruker Daltonics APEX II 47e mass spectrometer equipped with an electrospray ion source.

4.2 Chemistry

4.2.1 14-O-[(4-Amino-6-hydroxy-pyrimidine-2-yl) thioacetyl] mutilin (3)

To a solution of 4-amino-6-hydroxy-2-mercaptopyrimidine monohydrate (1.65g, 10mmol) in 20mL methanol, 10 M NaOH (1.1mL, 11mmol) was added and stirred for 30 min. A solution of 22-O-tosylpleuromutilin (5.33, 10mmol) in 20mL of DCM was added dropwise to the reaction mixture. The mixture was stirred for 36-40 h at room temperature and evaporated under reduced pressure to dryness. The crude product was extracted with a solution of ethyl acetate (60mL) and water (20mL) and treated with saturated $NaHCO_3$. The target compound 3 was then precipitated from the solution and purified by flash silica column chromatography (ethyl acetate: ethanol 20 : 1 v/v) to yield 3.93 g (78%). IR (KBr) 3368, 2930, 1729, 1629, 1575, 1542, 1457, 1286, 1117, 1019, 981, 809cm^{-1}; ^{1}H NMR (400 MHz, $CDCl_3$) δ 6.47 (dd, J = 17.4, 11.1 Hz, 1H), 5.74 (d, J=8.3 Hz, 1H), 5.32 (d, J=11.0 Hz, 1H), 5.24-5.13 (m, 2H), 4.79 (s, 1H), 4.11 (q, J=7.1 Hz, 1H), 3.85 (d, J=16.5 Hz, 1H), 3.69 (d, J=16.5 Hz, 1H), 3.35 (d, J=6.2 Hz, 1H), 2.29-1.98 (m, 6H), 1.75 (d, J=14.1 Hz, 1H), 1.63 (dd, J=21.3, 11.1 Hz, 2H), 1.56-1.18 (m, 9H), 1.10 (d, J=20.2 Hz, 4H), 0.86 (d, J=6.8 Hz, 3H), 0.71 (d, J=6.9 Hz, 3H); ^{13}C NMR (100 MHz, $CDCl_3$) δ 216.93, 166.90, 165.97, 162.65, 160.33, 139.23, 117.16, 84.04, 74.59, 70.10, 60.39, 58.10, 45.43, 43.94, 41.85, 36.69, 36.00, 33.25, 30.39, 26.88, 26.34, 24.81, 21.03, 16.79, 14.89, 11.45; HRMS (ES) calcd $[M+H]^+$ for $C_{26}H_{37}N_3O_5S$ 504.2526, found 504.2520.

4.2.2 14-O-[(4-Amino-6-methoxyl-pyrimidine-2-yl) thioacetyl] mutilin (4) and 14-O-[(4-Amino-1-methyl-6-oxo-1, 6-dihydropyrimidin-2-yl) thioacetyl] mutilin (5)

Methyl iodide (0.71 g, 5.0 mmol) was added dropwise into a suspension of compound 3 (2.51 g, 5.0 mmol) in anhydrous DMF (20mL) and K_2CO_3 (15 mmol). The reaction mixture was stirred at 80-85 ℃ for 4 h. After the mixture was cooled, the inorganic material was taken off by filtration and the solvent was removed in vacuum. The solid residue, consisting of a mixture of 4 and 5 isomers, was obtained and purified using silica column chromatography (petroleum ether: ethyl acetate 1 : 15 v/v) to yield 0.75 g (29%) and 1.58 g (61%) of compounds 4 and 5, respectively. Compound 4: IR (KBr) 3374, 2928, 1732, 1626, 1585, 1548, 1390, 1307, 1273, 1212, 1152, 1117, 1048, 1017, 982, 917 cm^{-1}; ^{1}H NMR (400 MHz, DMSO) δ 6.67 (s, 2H), 6.11 (dd, J = 17.7, 11.2 Hz, 1H), 5.52 (d, J = 8.2 Hz, 1H), 5.42 (s, 1H), 5.02 (dd, J = 23.6, 14.4 Hz, 2H), 4.49 (d, J = 6.0 Hz, 1H), 3.85 (s, 1H), 3.75 (s, 2H), 3.41 (t, J=5.7 Hz, 1H), 2.39 (s, 1H), 2.32-1.92 (m, 5H), 1.63 (s, 2H), 1.49-1.15 (m, 9H), 1.02 (s, 5H), 0.80 (d, J=6.8 Hz, 3H), 0.58 (d, J=6.8 Hz, 3H); ^{13}C NMR (100 MHz, DMSO) δ 217.59, 169.52, 168.33, 168.09, 165.63, 141.25, 115.71, 82.04, 73.12, 70.20, 60.23, 57.72, 53.71, 45.43, 44.46, 42.01, 36.82, 34.48, 33.69, 30.58, 28.93, 27.10, 24.92, 21.24, 16.56, 14.99, 11.96; HRMS (ES) calcd $[M+H]^+$ for $C_{27}H_{39}N_3O_5S$ 518.2683, found 518.2681.

Compound 5: IR (KBr) 3422, 2929, 1730, 1630, 1509, 1457, 1414, 1282, 1154,

1117, 1094, 807; ^{1}H NMR (400 MHz, DMSO) δ 6.28 (s, 2H), 6.10 (dd, J = 17.8, 11.2 Hz, 1H), 5.52 (d, J = 8.1 Hz, 1H), 5.05 (dd, J = 27.5, 14.5 Hz, 2H), 4.92 (s, 1H), 4.52 (d, J=5.9 Hz, 1H), 4.07–3.96 (m, 2H), 3.42 (d, J=5.4 Hz, 1H), 3.34 (s, 1H), 2.40 (s, 1H), 2.22–1.93 (m, 5H), 1.62 (dd, J = 24.8, 12.6 Hz, 2H), 1.52–1.13 (m, 9H), 1.05 (s, 4H), 0.82 (d, J = 6.8 Hz, 3H), 0.61 (d, J = 6.8 Hz, 3H); ^{13}C NMR (100 MHz, DMSO) δ 217.55, 170.79, 166.83, 161.88, 161.51, 160.51, 141.22, 115.85, 81.23, 73.13, 70.83, 60.22, 57.65, 45.43, 44.63, 43.90, 42.04, 36.82, 34.56, 30.57, 29.07, 27.09, 24.93, 21.23, 16.61, 14.89, 11.98; HRMS (ES) calcd $[M+H]^+$ for $C_{27}H_{39}N_3O_5S$ 518.2683, found 518.2696.

4.2.3 General procedure for the synthesis of compounds 6a–f

A mixture of the carboxylic acids derivative (2.2 mmol), 0.38 g 1-ethyl-3-(3-dimethylaminopropyl) carbodiimide hydrochloride (2.0 mmol), 0.27 g 1-Hydroxybenzotriazole (2.0 mmol) and 0.39 g N, N-diisopropylethylamine (3.0 mmol) were dissolved in DCM (15 ml) and stirred for 15 min. Then, 0.50 g compound 3 (1 mmol) was added in one portion and the reaction was stirred at room temperature for 36–48 h. The mixture was washed with saturated aqueous $NaHCO_3$ and brine, and then dried with Na_2SO_4 overnight. Before the synthesis of compounds 6g and 6h, the amino of valine should be protected by ditertbutyl dicarbonate (BOC_2O). After treatment with $NaHCO_3$ and brine, the residue was treated with a mixture of 10mL trifluoroacetic acid (TFA) and 10mL DCM at room temperature for 30 min. The reaction mixture was quenched with 25% aqueous $NaHCO_3$ (15mL) and washed with water, dried with anhydrous Na_2SO_4 overnight and rotary evaporated to dryness. The crude residue obtained was purified by silica gel column chromatography (petroleum ether: ethyl acetate 1 : 2 v/v) to afford the desired compounds.

4.2.4 14-O-[(4-Acetamido-6-hydroxypyrimidine-2-yl) thioacetyl] mutilin (6a)

White solid, 0.80 g (yield 73%); IR (KBr): 3448, 2931, 1779, 1729, 1627, 1588, 1549, 1459, 1400, 1372, 1299, 1201, 1162, 1117, 1030, 981, 939 cm^{-1}; ^{1}H NMR (400 MHz, $CDCl_3$) δ 6.47 (dd, J = 17.2, 11.0 Hz, 1H), 5.73 (d, J = 7.9 Hz, 1H), 5.32 (d, J = 10.8 Hz, 1H), 5.17 (d, J = 17.5 Hz, 2H), 4.69 (s, 2H), 3.85 (d, J = 16.4 Hz, 1H), 3.78–3.56 (m, 3H), 3.34 (d, J=4.6 Hz, 1H), 2.16 (ddd, J=75.4, 48.8, 32.3 Hz, 5H), 1.75 (d, J = 13.9 Hz, 2H), 1.63 (d, J=9.1 Hz, 3H), 1.28 (dd, J = 45.4, 39.0 Hz, 9H), 1.13 (s, 4H), 0.86 (d, J=6.2 Hz, 3H), 0.71 (d, J=6.4 Hz, 3H); ^{13}C NMR (101 MHz, $CDCl_3$) δ 216.96, 167.08, 166.35, 162.71, 160.59, 139.24, 117.18, 84.18, 74.59, 70.05, 58.41, 58.11, 45.44, 44.32, 43.95, 41.86, 36.70, 36.01, 34.45, 33.29, 30.40, 26.89, 26.34, 24.82, 18.43, 16.81, 14.90, 11.47; HRMS (ES) calcd $[M+H]^+$ for [$C_{29}H_{41}N_3O_5S$ 560.2788, found 560.2784.

4.2.5 14-O-[(4-(3-Chlorobenzamide)-6-hydroxypyrimidine-2-yl) thioacetyl] mutilin (6b)

White solid, 0.53 g (yield 41%); IR (KBr): 3368, 3215, 2931, 1735, 1625, 1586, 1547, 1459, 1283, 1240, 1116, 981, 739 cm^{-1}; ^{1}H NMR (400 MHz, $CDCl_3$) δ 8.20–7.91 (m, 2H), 7.73–7.34 (m, 2H), 6.49 (s, 1H), 6.08 (s, 1H), 5.72 (s, 1H), 5.21 (dd, J= 50.2, 22.6 Hz, 4H), 4.11 (d, J=6.7 Hz, 1H), 3.97–3.57 (m, 2H), 3.34

(d, J=6.1 Hz, 1H), 2.40-1.90 (m, 6H), 1.63 (td, J=47.2, 13.3 Hz, 3H), 1.42 (s, 4H), 1.36-1.18 (m, 3H), 1.10 (s, 4H), 0.85 (d, J=5.5 Hz, 3H), 0.72 (s, 3H); ^{13}C NMR (100 MHz, $CDCl_3$) δ 217.15, 170.39, 167.95, 165.03, 164.49, 162.37, 139.15, 134.89, 134.19, 130.37, 130.00, 129.70, 128.54, 117.12, 91.07, 74.63, 69.84, 60.42, 58.17, 45.46, 43.98, 41.89, 36.77, 36.00, 34.48, 34.05, 30.43, 26.89, 26.37, 21.05, 16.84, 14.90, 11.45; HRMS (ES) calcd $[M+H]^+$ for $C_{33}H_{40}ClN_3O_6$ 642.2399, found 642.2397.

4.2.6 14-O-[(4-(4-Methylbenzamide)-6-hydroxypyrimidine-2-yl) thioacetyl] mutilin (6c)

White solid, 0.53 g (yield 43%); IR (KBr): 3375, 2930, 1735, 1624, 1586, 1458, 1298, 1255, 1166, 1117, 1071, 980, 745 cm^{-1}; ^{1}H NMR (400 MHz, $CDCl_3$) δ8.06-7.97 (m, 2H), 7.29 (s, 1H), 6.59-6.36 (m, 1H), 6.09 (dd, J=6.6, 3.0 Hz, 1H), 5.72 (d, J=6.3 Hz, 1H), 5.36-5.23 (m, 1H), 5.21-5.06 (m, 2H), 3.91-3.67 (m, 2H), 3.33 (s, 1H), 2.44 (d, J=1.1 Hz, 2H), 2.33-2.01 (m, 5H), 1.74 (d, J=14.5 Hz, 2H), 1.66-1.50 (m, 3H), 1.51-1.18 (m, 9H), 1.16-1.04 (m, 4H), 0.85 (d, J=6.4 Hz, 3H), 0.73 (dd, J=4.3, 2.3 Hz, 3H); ^{13}C NMR (101 MHz, $CDCl_3$) δ 217.10, 170.22, 167.98, 164.83, 163.60, 145.16, 139.19, 130.52, 129.39, 125.89, 117.11, 91.10, 74.63, 69.72, 60.40, 58.18, 45.46, 43.98, 41.89, 36.79, 36.01, 34.49, 34.08, 30.45, 26.89, 26.36, 24.84, 21.81, 21.04, 16.84, 14.91, 11.45; HRMS (ES) calcd $[M+H]^+$ for $C_{34}H_{43}N_3O_6S$ 622.2945, found 622.2938.

4.2.7 14-O-[(4-(4-Methoxylbenzamide)-6-hydroxypyrimidine-2-yl) thioacetyl] mutilin (6d)

White solid, 0.59g (yield 46%); IR (KBr): 3369, 2933, 1734, 1605, 1581, 1549, 1511, 1458, 1401, 1299, 1251, 1160, 1116, 1069, 1024, 846 cm^{-1}; ^{1}H NMR (400 MHz, $CDCl_3$) δ 8.07 (d, J=8.8 Hz, 2H), 6.94 (d, J=8.8 Hz, 2H), 6.47 (dd, J=17.4, 11.0 Hz, 1H), 6.07 (s, 1H), 5.70 (d, J=8.4 Hz, 1H), 5.33-5.22 (m, 2H), 5.14 (d, J=17.4 Hz, 1H), 3.86 (d, J=8.8 Hz, 3H), 3.79-3.61 (m, 2H), 3.32 (d, J=6.2 Hz, 1H), 2.36-1.93 (m, 5H), 1.74-1.50 (m, 3H), 1.47-1.14 (m, 9H), 1.15-1.04 (m, 4H), 0.84 (d, J=6.9 Hz, 3H), 0.71 (d, J=6.8 Hz, 3H); ^{13}C NMR (101 MHz, $CDCl_3$) δ 217.16, 170.12, 168.02, 164.87, 164.37, 163.23, 139.16, 132.67, 120.85, 117.09, 113.96, 91.10, 74.61, 69.74, 58.17, 55.57, 45.45, 44.49, 43.97, 41.88, 36.77, 35.99, 34.48, 34.06, 30.43, 26.87, 26.36, 24.82, 18.41, 16.82, 14.90, 11.44; HRMS (ES) calcd $[M+H]^+$ for $C_{34}H_{43}N_3O_7S$ 638.2894, found 638.2892.

4.2.8 14-O-[(4-(Piperidine-2-L (-) carboxamide)-6-hydroxypyrimidine-2-yl) thioacetyl] mutilin) (6e)

White solid, 0.62 g (yield 50%); IR (KBr): 3448, 2918, 2849, 1729, 1654, 1617, 1458, 1438, 1279, 1362, 1265, 1151, 1240, 1115, 1042, 1081, 984, 957, 747 cm^{-1}. ^{1}H NMR (400 MHz, CDCl3) δ 6.49 (dd, J=17.2, 11.3 Hz, 1H), 5.96 (s, 1H), 5.73 (d, J=8.5 Hz, 1H), 5.32 (d, J=11.0 Hz, 1H), 5.30-5.01 (m, 3H), 4.29-3.99 (m, 1H), 3.79 (dd, J=41.0, 16.3 Hz, 2H), 3.35 (s, 1H), 2.27 (s, 6H), 2.06

(d, J = 17.5 Hz, 5H), 1.63 (d, J = 10.3 Hz, 5H), 1.45 - 1.17 (m, 9H), 1.14 (s, 4H), 0.86 (d, J=6.8 Hz, 3H), 0.74 (d, J=6.7 Hz, 3H); ^{13}C NMR (101 MHz, $CDCl_3$) δ 216.99, 170.41, 167.89, 167.61, 164.80, 139.28, 129.94, 128.13, 117.06, 90.73, 74.66, 69.73, 60.39, 58.20, 45.48, 44.54, 44.01, 41.92, 36.81, 36.04, 34.49, 34.09, 30.47, 26.93, 26.40, 24.86, 21.41, 16.84, 16.55, 14.91, 14.21, 11.43; HRMS (ES) calcd [M+H]$^+$ for $C_{32}H_{46}N_4O_6S$ 615.3210, found 615.3213.

4.2.9 14 - O - [(4 - (Piperidine - 2 - (D) - carboxamide) - 6 - hydroxypyrimidine - 2- yl) thioacetyl] mutilin (6f)

White solid, 0.60 g (yield 48%); IR (KBr): 3447, 1733, 1637, 1577, 1560, 1541, 1508, 1457, 1245, 1151, 1116, 969, 547 cm^{-1}; ^{1}H NMR (400 MHz, $CDCl_3$) δ 6.48 (ddd, J=15.2, 11.0, 4.2 Hz, 1H), 5.96 (s, 1H), 5.74 (t, J=8.6 Hz, 1H), 5.45-5.27 (m, 1H), 5.23 - 5.10 (m, 3H), 4.11 (q, J = 7.1 Hz, 1H), 3.91 - 3.64 (m, 2H), 3.34 (d, J = 6.3 Hz, 1H), 2.41-2.14 (m, 6H), 2.11-1.95 (m, 5H), 1.82-1.44 (m, 5H), 1.47 - 1.16 (m, 9H), 1.17 - 1.09 (m, 4H), 0.85 (d, J = 7.0 Hz, 3H), 0.73 (t, J=6.4 Hz, 3H); ^{13}CNMR (101 MHz, $CDCl_3$) δ 216.90, 170.32, 167.87, 166.77, 164.80, 139.22, 117.21, 117.08, 90.78, 83.91, 74.62, 69.70, 60.40, 58.17, 45.46, 44.47, 43.97, 41.88, 36.78, 36.67, 34.48, 34.06, 30.43, 26.90, 26.36, 24.83, 21.42, 21.05, 16.83, 14.89, 14.20, 11.45; HRMS (ES) calcd [M+H]$^+$ for $C_{32}H_{46}N_4O_6S$ 615.3210, found 615.3202.

4.2.10 14-O-[(4-(3-Methyl-2-(L)-amino-butyrylamide)-6-hydroxypyrimidine-2-yl) thioacetyl] mutilin (6g)

White solid, 0.72 g (yield 60%); IR (KBr): 3462, 2931, 1733, 1605, 1544, 1432, 1275, 1152, 1117, 1043, 785, 547 cm^{-1}; ^{1}H NMR (400 MHz, $CDCl_3$) δ 6.48 (ddd, J= 15.2, 11.0, 4.2 Hz, 1H), 5.96 (s, 1H), 5.73 (d, J=8.6 Hz, 1H), 5.40-5.24 (m, 1H), 5.17 (d, J = 17.1 Hz, 3H), 4.11 (q, J = 7.1 Hz, 1H), 3.92-3.63 (m, 2H), 3.34 (d, J=6.3 Hz, 1H), 2.40-2.13 (m, 5H), 2.05 (dd, J = 17.3, 9.5 Hz, 4H), 1.69 (ddd, J=11.8, 11.2, 3.3 Hz, 6H), 1.55-1.20 (m, 12H), 1.13 (s, 4H), 0.85 (d, J=7.0 Hz, 3H), 0.74 (d, J=6.8 Hz, 3H); ^{13}C NMR (101 MHz, $CDCl_3$) δ 216.92, 167.05, 163.88, 159.63, 157.52, 139.20, 130.90, 128.84, 117.24, 94.45, 75.38, 74.59, 70.06, 67.78, 58.10, 45.44, 44.37, 43.95, 41.87, 36.72, 36.01, 34.45, 33.24, 30.41, 26.89, 26.27, 25.61, 24.83, 16.86, 14.88, 11.47; HRMS (ES) calcd [M+H]$^+$ for $C_{32}H_{48}N_4O_6S$ 617.3367, found 617.3360.

4.2.11 14-O-[(4-(3-Methyl-2-(D)-amino-butyrylamide)-6-hydroxypyrimidine-2-yl) thioacetyl] mutilin (6h)

White solid, 0.69 g (yield 58%); IR (KBr): 3463, 3363, 2922, 1731, 1604, 1544, 1433, 1278, 1152, 1117, 1042, 785.40 cm^{-1}; ^{1}H NMR (400 MHz, $CDCl_3$) δ 6.48 (dd, J= 17.3, 11.0 Hz, 1H), 5.74 (d, J=7.9 Hz, 1H), 5.46 (s, 1H), 5.33 (d, J=10.9 Hz, 1H), 5.23-4.95 (m, 3H), 4.01 (dd, J = 14.5, 7.4 Hz, 1H), 3.93-3.62 (m, 2H), 3.34 (d, J=6.0 Hz, 1H), 2.42-2.09 (m, 5H), 2.00 (dd, J = 15.7, 9.6 Hz, 4H), 1.69 (ddd, J=23.1, 21.8, 14.5 Hz, 6H), 1.57-1.19 (m, 12H), 1.14 (s, 4H), 0.86

(d, J=6.8 Hz, 3H), 0.73 (d, J=6.8 Hz, 3H); ^{13}C NMR (101 MHz, $CDCl_3$) δ 216.93, 167.03, 163.87, 159.62, 157.51, 139.20, 130.91, 128.84, 117.25, 94.41, 75.39, 74.59, 70.05, 67.78, 58.10, 45.44, 44.36, 43.95, 41.86, 36.72, 36.01, 34.45, 33.24, 30.41, 26.89, 26.26, 25.61, 24.83, 16.86, 14.87, 11.47; HRMS (ES) calcd $[M+H]^+$ for $C_{32}H_{48}N_4O_6S$ 617.3367, found 617.3361.

4.3 MIC determination

The MIC studies were performed on MRSA, MRSE, *B.subtilis* and *E.coli* using the agar dilution method according to the National Committee for Clinical Laboratory Standards (NCCLS). Compounds were dissolved in 15%-30% DMSO in water to prepare a solution that had a concentration of 1.28mg/mL. Tiamulin fumarate used as reference drug was directly dissolved in 10mL distilled water. All the solutions were then diluted two-fold with distilled water to provide 11 dilutions (final concentration is 0.625mg/mL). The dilutions (2mL) of each test compound/drug were incorporated into 18mL hot Mueller-Hinton agar medium, which resulted in the final concentration of each dilutions decreasing tenfold.

Inoculums of MRSA, MRSE, B.subtilis and E.coli were prepared from blood slants and adjusted to approximately 10^5-10^6 CFU/mL with sterile saline (0.90% NaCl). A 10mL amount of bacterial suspension was spotted onto Mueller-Hinton agar plates containing serial dilutions of the compounds/drugs. The plates were incubated at 36.5 C for 24-48 h. The MIC was defined as the minimum concentration of the compound needed to completely inhibit bacterial growth. The same procedure was repeated in triplicate.

4.4 In vivo efficacy in mouse model

Male and female mice were rendered neutropenic upon treatment with 150mg/kg cyclophosphamide intraperitoneally for four days and with 100mg/kg for one day prior to inoculation, respectively. The neutropenic mice (10 per group) received a 0.5mL MRSA inoculum of 10^9 CFU/mL via intraperitoneal (ip) injection. About 1 h after infection, the mice were then intravenously (iv) administered compound 6a dissolved in 0.5mL vehicle (soybean lecithin: sterile water=1 : 30) at doses of 2, 3, 4.46, 6.64 and 10mg/kg body weight. Tiamulin fumaratewas used as a control in the same manner at the same doses as 6a. The survival of the mice at 4 d after infectionwas used as the end-point, and the ED_{50} was calculated by the method described by Reed Muench[21] using the Hill equation.

ACKNOWLEDGEMENTS

This work was financed by Basic Scientific Research Funds in Central Agricultural Scientific Research Institutions (No.1610322014003), The Agricultural Science and Technology Innovation Program (ASTIP) (CAAS-ASTIP-2014-LIHPS-04) and the Science and Technology Development Plan Project of Lanzhou (2013-4-90).

SUPPLEMENTARY DATA

Supplementary data associated with this article can be found in the online version, at http: //dx.doi.org/10.1016/j.ejmech.2015.06.034. These data include MOL files and InChiKeys of the most important compounds described in this article.

References

[1] H.A.Kirst, Expert Opin.Drug Discov.8 (2013) 479-493.

[2] F.Kavanagh, A.Hervey, W.J.Robbins, J.Proc.Natl.Acad.Sci.U.S.A.37 (1951) 570-574.

[3] S.M.Poulsen, M.Karlsson, L.B.Johansson, B.Vester, Mol.Microbiol.41 (2001) 1091-1099.

[4] F.Schluenzen, E.Pyetan, P.Fucini, A.Yonath, J.M.Harms, Mol.Microbiol.54 (2004) 1287-1294.

[5] C.Davidovich, A.Bashan, T.Auerbach-Nevo, R.D.Yaggie, R.R.Gontarek, A.Yonath, Proc.Natl. Acad.Sci.U.S.A.104 (2007) 4291-4296.

[6] R.F.Shang, Y.Liu, Z.J.Xin, W.Z.Guo, Z.T.Guo, B.C.Hao, J.P.Liang, Eur.J.Med.Chem.63 (2013) 231-238.

[7] M.A.Fischbach, C.T.Walsh, Science 325 (2009) 1089-1093.

[8] R.Novak, D.M.Shlaes, Curr.Opin.Investig.Drugs 11 (2010) 182-191.

[9] Y.Z.Tang, Y.H.Liu, J.X.Chen, Mini-Rev.Med.Chem.12 (2012) 53-61.

[10] R.Novak, Ann.N.Y.Acad.Sci.1241 (2011) 71-81.

[11] M.N.Moody, L.K.Morrison, S.K.Tyring, Skin Ther.Lett.15 (2010) 1-4.

[12] R.F.Shang, J.T.Wang, W.Z.Guo, J.P.Liang, Curr.Top.Med.Chem.13 (2013) 3013-3025.

[13] I.Dreier, L.H.Hansen, P.Nielsen, B.Vester, Bioorg.Med.Chem.Lett.24 (2014) 1043-1046.

[14] X.Y.Wang, Y.Ling, H.Wang, J.H.Yu, J.M.Tang, H.Zheng, X.Zhao, D.G.Wang, G.T.Chen, W.Q.Qiu, J.H.Tao, Bioorg.Med.Chem.Lett.22 (2012) 6166-6172.

[15] Y.Z.Tang, Y.H.Liu, J.X.Chen, Mini Rev.Med.Chem.12 (2012) 53-61.

[16] C.Y.Ling, L.Q.Fu, S.Gao, W.J.Chu, H.Wang, Y.Q.Huang, X.Y.Chen, Y.S.Yang, J.Med. Chem.57 (2014) 4772-4795.

[17] R.F.Shang, X.Y.Pu, X.M.Xu, Z.J.Xin, C.Zhang, W.Z.Guo, Y.Liu, J.P.Liang, J.Med.Chem.57 (2014) 5664-5678.

[18] R.F.Shang, G.H.Wang, X.M.Xu, S.J.Liu, C.Zhang, Y.P.Yi, J.P.Liang, Y.Liu, Molecules 19 (2014) 19050-19065.

[19] I.M.Lagoja, Chem.Biodivers.2 (2005) 1-50.

[20] B.Cosimelli, G.Greco, M.Ehlardo, E.Novellino, F.Da Settimo, S.Taliani, C.La Motta, M. Bellandi, T.Tuccinardi, A.Martinelli, O.Ciampi, M.L.Trincavelli, C.Martini, J.Med.Chem.51 (2008) 1764-1770.

[21] L.J.Reed, H.Muench, Am.J.Hyg.27 (1938) 493-497.

（发表于《European Journal of Medicinal Chemistry》，院选 SCI，IF：3.447）

Evaluation of Arecoline Hydrobromide Toxicity after a 14-Day Repeated Oral Administration in Wistar Rats

WEI Xiao-juan *☆, ZHANG Ji-yu☆, NIU Jian-rong*,
ZHOU Xu-zheng*, LI Jian-yong*, LI Bing*

(Key Laboratory of New Animal Drug Project of Gansu Province/Key Laboratory of Veterinary Pharmaceutical Development of Ministry of Agriculture/ Lanzhou Institute of Husbandryand Pharmaceutical Sciences of CAAS, Lanzhou 730050, China)

Abstract: A subchronic toxicity test was conducted in rats on the basis of a previous acute toxicity test to evaluate the safety of arecoline hydrobromide (Ah), to systematically study its pharmacological effects and to provide experimental support for a safe clinical dose. Eighty rats were randomly divided into four groups: a high-dose group (1000mg/kg), medium-dose group (200mg/kg), low-dose group (100mg/kg) and blank control group. The doses were administered daily via gastric lavage for 14 consecutive days. There were no significant differences in the low-dose Ah group compared to the control group ($P>0.05$) with regard to body weight, organ coefficients, hematological parameters and histopathological changes. The high-dose of Ah influenced some of these parameters, which requires further study. The results of this study indicated that a long-term, continuous high dose of Ah was toxic. However, it is safe to use Ah according to the clinically recommended dosing parameters. The level of Ah at which no adverse effects were observed was 100mg/kg/day under the present study conditions.

INTRODUCTION

Arecoline is an alkaloid extracted from areca nuts (*Areca catechu L.*)[1]. Arecoline and its derivatives have been used to treat Alzheimer's disease[2], diarrhea, intestinal worms and other gastrointestinal disorders, gonorrhea, visual ailments, fever, dysentery, hysteria, schizophrenia, headaches, dental disorders, octopus bites and elephant diseases and have also been used as abortifacients[3]. One study demonstrated that the alkaloids and polyphenols from arecanuts could be used to enhance the healing of burn wounds, leg ulcers and skin graft surgery[4]. In addition, arecoline

---

* These authors contributed equally to this work.

☆ These authors are co-first authors to this work.

has demonstrated various pharmacological activities, such as hypoglycemic activity[5], vascular relaxation[6] and molluscicidal activity[7,8].

Parasites, especially parasites found in the digestive system, can cause significant harm to livestock, resulting in low productivity, low fertility and, in some cases, death. To date, only a few types of deinsectization drugs are available in China and have been repeatedly used, which has led to drug resistance. Therefore, it is critical to develop a deinsectization drug with good efficiency and low drug resistance. Arecoline is highly effective against cysticercus in vitro[9] and has an inhibitive effect on *Fasciola hepatica*[10]. However, the arecoline content in *Areca catechu L.* extracts ranges from only 0. 3% to 0. 7%, which, coupled with the low efficiency of traditional extraction methods, greatly limits its clinical application. Additionally, while arecoline as a liquid is not convenient to administer, its salt is appropriate for clinical use. The proposed synthetic process can achieve a yield of more than 78%. The benefits of this new technology include a high yield, the easy purification of the product, a simple operation process, and easy scale-up for industrial production[11].

Because Ah is being developed for the treatment and prevention of parasitic diseases, it is important to characterize its subchronic toxicity in animals following 14 consecutive days of oral administration. The present study aimed to assess the subchronic toxicity of Ah, specifically after daily oral administration for 14 consecutive days in rats. Blood and tissue samples were collected at various time points for hematology, clinical chemistry and pathology analyses. As an evaluation of preclinical safety, this study will provide guidance for the design of further preclinical toxicity studies and clinical trials of Ah.

MATERIALS AND METHODS

Chemicals and reagents

Arecoline hydrobromide (methyl 1,2,5,6-tetrahydro-1-methyl-3-pyridinecarboxylate hydrobromide, Ah) transparent crystals (purity: 99. 5% by RE-HPLC) were prepared at the Lanzhou Institute of Husbandry and Pharmaceutical Sciences of CAAS (Lanzhou, China).

Animals

The study was approved by the Ethics Committee of Animal Experiments at the Institute of Husbandry and Pharmaceutical Sciences of CAAS in Lanzhou, China. EightyWistar rats of both sexes, with a clean grade [Certificate No.: SCXK (Gan) 2008-0075] and an initial body weight of 150-160g, were purchased from the animal breeding facilities at Lanzhou University (Lanzhou, China). The animals were individually housed to allow recording of individual food consumption and to avoid bias from hierarchical stress. The animals were kept in plastic Macrolon cages (Suzhou Fengshi Laboratory Animal Equipment Co., Ltd, Huangqiao town city Suzhou, China) of an appropriate size with a stainless steel wire cover and chopped bedding. The light/dark cycle was 12/12 h, the temperature was (22±2)℃ and the relative humidity was 55±10%. Standard compressed rat feed from the animal breeding facilities at Lanzhou University and drinking water were supplied *ad libitum*. This study was conducted in strict accordance with the recommendations in the Guide for the

Care and Use of Laboratory Animals of the National Institutes of Health. All surgeries were performed under sodium pentobarbital anesthesia, and all efforts were made to minimize suffering. The animals were allowed a 2-week quarantine and acclimation period prior to the start of the study.

Dosing

The selected high, medium and low doses were 1000, 200 and 100mg/kg bw, respectively, based on the results of acute toxicity testing and preliminary studies in rats. Ah was dissolved in distilled water in blackpaper-covered vials and aseptically administered to each rat intragastrically once per day based on the individual daily body weight for 14 consecutive days. The rats were randomly assigned to four groups: three test groups (n=20 rats per group, male: female=1 : 1) and a vehicle group that functioned as the control group (n=20 rats, male: female=1 : 1).

Study design

Once exposure was initiated, the animals were inspected daily regarding their general condition and any clinical abnormalities. The bedding was changed daily, at which time the rats were submitted to in-hand observations, including their reactions to being handled, and their body weight and food consumption were recorded.

At the end of the drug administration period, the rats in each group were euthanized via exsanguination of the femoral artery and were then necropsied. During necropsy, blood was collected, and the following organs were dissected: liver, spleen, thymus, heart, lungs, stomach, duodenum, ileum, jejunum, colon, cecum, ovaries, uterus, kidney, adrenals, brain and testes. All organs were visually inspected and weighed directly either after dissection or after fixation to reduce mechanical damage. All organs were fixed in standard formalin for further histological processing based on previously described methods[12].

Hematology, clinical biochemistry, the visceral index and histopathology were analyzed at the end of the drug administration period. A portion of each blood sample was treated with EDTA-Na_2 and analyzed for hematological indices, including red blood cellcount (RBC), white blood cellcount (WBC), lymphocyte count (LYM), granulocyte count (GRA), hemoglobin (HGB), hematocrit (HCT), mean corpuscular volume (MCV), mean corpuscular hemoglobin (MCH), mean corpuscular hemoglobin concentration (MCHC), platelet count (PLT), platelet distribution width (PDW), mean platelet volume (MPV), thrombocytocrit (PCT) and platelet larger cell ratio (P-LCR), using a hematology analyzer (Cell Dyn1200, Abbott, USA). In addition, an automatic biochemistry analyzer (XL-640, ERBA, Germany) was used to examine the sera obtained from the remaining blood samples for alanine aminotransferase (ALT), aspartate aminotransferase (AST), alkaline phosphatase (AKP), total protein (TP), albumin (ALB), globulin (GLO), total bile acid (TBA), blood urea nitrogen (BUN), creatinine (CREA), glucose (GLU), triglyceride (TG), cholesterol (CHOL) and lactate dehydrogenase (LDH).

Histopathology

All tissues collected during necropsy were fixed in 10% neutral buffered formalin. Following fixation (and subsequent weighing, vide supra), the organs sampled for histological examination were dehydrated, paraffinized and embedded according to standard sampling and trimming procedures. Four-micron sections were stained with hematoxylin and eosin using an automated method. Microscopic observations were performed by an initial unblinded comparison of the control and high-dose samples. Blind and/or semi-quantitative scoring was used when changes were detected in the initial inspection.

Statistics

Continuous data, including body weight, food consumption, hematology, blood biochemistry, and organ weights, are expressed as the mean ± standard deviation. The differences in the ratios of the organ weight to the body weight were analyzed using either ANOVA with LSD or Dunnett's test (SPSS 21.0 software, Chicago, IL, USA). Other data were analyzed using a repeated measures ANOVA built into ageneral linear model (SPSS 21.0). Inter-group comparisons were conducted using a multivariate general linear model. P-values<0.05 were considered statistically significant. For non-continuous data and comparisons of histopathological changes, statistical tests were ranked and examined using the Kruskal—Wallis test. If significant, the Wilcoxon rank sum test was applied for comparison with the control group. The male and female rats were separately evaluated.

Observations of the clinical curative effect

To observe the clinical curative effect, 252 infected dogs were treated orally with the recommended doses of arecoline hydrobromide 15-24 hours after fasting. Within 72 h of administration, stool samples were selected and examined using the saturated salt solution float method to check polypides.

RESULTS

In vivo observations

No deaths were observed in any group during the administration period. Compared to the control group, rats treated with Ah showed a decrease in some responses, including water and food consumption, and the male rats consumed more food than the female rats in all groups.

All of the rats in the test groups exhibited a significant reduction in body weight gain compared to the control animals ($P<0.01$; Table 1). The test group rats had significantly lower body weights during Ah administration, which indicated that long-term and continuous dosing had a toxic effect on the rats (Table 1). Interestingly, the effects in the female rats were more obvious than those in the male rats. There were no further clinical anomalies, and the dosing was well tolerated.

Hematology

According to the hematological analysis (Tables 2 and 3), the hematocrit and HGB levels and the leukocyte, lymphocyte and erythrocyte countsin female rats at all doses were significantly decreased following 14 days of Ah administration compared to the control rats ($P<0.01$). However, in the male rats, no significant changes in these indices were observed, with the exception of the HGB level in the high-dose group ($P>0.05$).

Table 1 Body weight changes (mean ± standard deviation) in male and female rats before and after Ah administration

Group	Sex	Changes in body weight (g)
Control	Male	56. 60±5. 03
	Female	45. 20±13. 72
100mg/mL	Male	43. 70±4. 00**
	Female	22. 20±4. 71**
200mg/mL	Male	42. 78±5. 54**
	Female	20. 90±10. 66**
1 000mg/mL	Male	41. 80±5. 88**
	Female	19. 98±13. 72**

Note: Significant difference compared with male controls, ** $P<0.01$.

doi: 10. 1371/journal.pone.0120165. t001

Table 2 Hematological parameters (mean ± standard deviation) in male rats intragastrically administered Ah daily for 14 consecutive days

Dose	Control	100 (g/kg)	200 (mg/kg)	1 000 (g/kg)
No.of animals examined	10	10	10	10
RBC (10^{12}/L)	5. 92±0. 51	6. 19±0. 84	6. 28±0. 53	6. 36±0. 59
WBC (10^{9}/L)	3. 25±0. 83	3. 29±1. 20	3. 56±0. 86	2. 93±0. 84
GRA (10^{9}/L)	0. 01±0. 01	0. 07±0. 05*	0. 07±0. 05*	0. 03±0. 03
LYM (10^{9}/L)	2. 99±0. 84	4. 48±1. 21*	3. 74±0. 58	2. 38±0. 70
MID (10^{9}/L)	0. 25±0. 03	0. 74±0. 20*	0. 75±0. 27*	0. 52±0. 31
MCV (fL)	73. 25±1. 65	70. 13±0. 85**	72. 34±1. 01	71. 11±1. 34*
HCT (%)	43. 43±4. 78	43. 46±6. 38	45. 44±3. 91	38. 17±4. 52
HGB (g/L)	187. 00±4. 58	181. 30±16. 57	189. 89±9. 55	156. 00±17. 08**
MCH (pg)	433. 16±35. 32	420. 21±24. 56	419. 27±20. 19	409. 47±11. 60
MCHC (g/L)	31. 69±1. 90	29. 45±1. 44**	30. 32±1. 32	29. 11±0. 71**
PLT (109/L)	1190. 00±76. 18	972±180. 60*	1159. 89±123. 12	1017. 90±66. 70
MPV (fL)	7. 63±0. 14	7. 93±0. 50	7. 83±0. 29	7. 66±0. 31
PDW (%)	16. 17±0. 49	15. 32±1. 40	15. 56±0. 89	16. 15±1. 03
PCT (%)	0. 91±0. 53	0. 77±0. 15	0. 90±0. 09	0. 78±0. 07
P-LCR (%)	17. 73±1. 45	19. 70±3. 42	19. 31±2. 02	17. 91±2. 23

Note: Compared with control, ** $P<0.01$; * $P<0.05$. The same as below.

doi: 10. 1371/journal.pone.0120165. t002

Table 3 Hematological parameters (mean ± standard deviation) in female rats intragastrically administered Ah daily for 14 consecutive days

Dose	Control	100 (g/kg)	200 (mg/kg)	1 000 (g/kg)
No.of animals examined	10	10	10	10
RBC (10^{12}/L)	6. 53±0. 51	5. 64±0. 34*	5. 22±0. 83**	5. 20±0. 42**
WBC (10^9/L)	8. 35±1. 22	4. 19±1. 16**	2. 97±1. 36**	2. 92±0. 70**
GRA (10^9/L)	0. 07±0. 04	0. 05±0. 05	0. 04±0. 06	0. 03±0. 04
LYM (10^9/L)	7. 77±1. 04	2. 43±1. 00**	3. 64±1. 19**	2. 48±0. 61**
MID (10^9/L)	0. 51±0. 14	0. 49±0. 18	0. 51±0. 26	0. 41±0. 14
MCV (fL)	71. 18±0. 52	70. 22±1. 10	70. 25±0. 83	70. 95±0. 63
HCT (%)	46. 50±3. 56	36. 66±5. 86**	39. 62±2. 38*	36. 90±3. 15**
HGB (g/L)	200. 67±9. 29	172. 40±7. 75**	159. 30±21. 06**	148. 90±10. 58**
MCH (pg)	432. 24±14. 80	436. 27±18. 14	435. 50±9. 62	401. 07±0. 25*
MCHC (g/L)	30. 77±1. 16	30. 63±1. 30	30. 59±0. 76	28. 66±0. 70**
PLT (109/L)	974. 00±93. 55	949. 20±238. 16	1124. 20±67. 75	786. 70±175. 63
MPV (fL)	7. 80±0. 32	7. 75±0. 48	7. 91±0. 15	7. 95±0. 21
PDW (%)	15. 63±1. 04	15. 88±1. 47	15. 29±0. 48	15. 16±0. 62
PCT (%)	0. 76±0. 51	0. 73±0. 19	0. 89±0. 53	0. 63±0. 13
P-LCR (%)	20. 06±2. 35	18. 81±3. 54	19. 64±1. 06	20. 23±1. 56

doi: 10. 1371/journal.pone.0120165. t003

Blood biochemistry

The blood chemistry results are shown in Tables 4 and 5. Both female and male rats exhibited markedly decreased CHOL and AST in the high-dose group compared to the control animals ($P<0.05$). AKP was increased in the medium-and high-dose groups compared to the control animals, and the levels of AKP in the medium-and high-dose groups were significantly increased in both sexes after 14 consecutive days of dosing compared to the pre-dosing levels ($P<0.05$ and $P<0.01$, respectively).

Table 4 Biochemical parameters (mean ± standard deviation) in male rats intragastrically administered Ah daily for 14 consecutive days

Dose	Control	100 (g/kg)	200 (mg/kg)	1 000 (g/kg)
No.of animals examined	10	10	10	10
ALT (U/L)	68. 56±1. 53	67. 89±6. 72	70. 78±5. 21	70. 78±12. 20
AST (U/L)	303. 70±31. 75	307. 50±22. 07	298. 70±33. 64	261. 60±53. 98*
AKP (U/L)	118. 50±10. 10	118. 40±10. 94	138. 70±9. 83**	141. 80±12. 67**
TP (g/L)	71. 03±0. 76	70. 58±2. 59	71. 41±3. 22	80. 53±7. 28**

(continued)

Dose	Control	100 (g/kg)	200 (mg/kg)	1 000 (g/kg)
ALB (g/L)	37. 80±1. 39	36. 38±2. 66	36. 09±1. 78	35. 67±2. 32*
GLO (g/L)	32. 39±1. 39	33. 15±1. 65	33. 24±2. 70	39. 34±6. 34*
TBA (μmol/L)	26. 47±0. 32	27. 84±5. 06	32. 75±7. 17	48. 74±15. 90**
BUN (mmol/L)	19. 67±14. 77	15. 03±5. 21	13. 23±3. 16	10. 41±2. 15*
CREA (μmol/L)	81. 33±102. 17	35. 44±34. 06	30. 59±27. 31	14. 37±2. 84
GLU (mmol/L)	0. 77±0. 12	1. 17±0. 35	1. 30±0. 55**	1. 48±0. 44**
TG (mmol/L)	0. 92±0. 66	0. 96±0. 11	0. 97±0. 25	1. 14±0. 18**
CHOL (mmol/L)	2. 42±0. 18	2. 33±0. 22	2. 09±0. 28**	1. 64±0. 18**
LDH (U/L)	1 285. 67±167. 50	1 364. 30±303. 92	1 193. 00±151. 99*	1 151. 10±123. 84*

doi: 10. 1371/journal.pone.0120165. t004

Table 5 Biochemical parameters (mean ± standard deviation) in female rats intragastrically administered Ah daily for 14 consecutive days

Dose	Control	100 (g/kg)	200 (mg/kg)	1 000 (g/kg)
No.of animals examined	10	10	10	10
ALT (U/L)	74. 30±4. 62	76. 60±11. 76	73. 80±6. 09	71. 10±17. 99
AST (U/L)	361. 70±50. 10	345. 30±49. 72	340. 50±50. 10	315. 20±28. 53*
AKP (U/L)	104. 40±10. 94	109. 60±17. 65	119. 40±12. 67*	123. 00±19. 37*
TP (g/L)	77. 09±0. 70	75. 51±4. 43	71. 16±3. 74*	70. 58±2. 30*
ALB (g/L)	37. 50±0. 61	36. 17±2. 39	37. 92±3. 40	38. 19±2. 68
GLO (g/L)	40. 72±1. 11	39. 34±3. 62	33. 24±2. 52**	32. 39±1. 95**
TBA (μmol/L)	31. 21±10. 02	24. 59±2. 23	29. 98±7. 49	31. 44±16. 17
BUN (mmol/L)	14. 30±2. 89	13. 99±1. 27	14. 17±1. 34	12. 59±3. 28
CREA (μmol/L)	24. 06±7. 42	17. 29±8. 34	18. 78±3. 57	22. 07±12. 94
GLU (mmol/L)	2. 30±0. 50	1. 11±0. 35**	1. 15±0. 25**	1. 17±0. 30**
TG (mmol/L)	1. 40±0. 32	1. 09±0. 31	1. 29±0. 39	0. 93±0. 27*
CHOL (mmol/L)	2. 09±0. 14	2. 09±0. 25	2. 10±.021	1. 52±0. 26**
LDH (U/L)	1238. 00±163. 89	1281. 70±218. 85	1175. 90±203. 26	1156. 50±115. 26

doi: 10. 1371/journal.pone.0120165. t005

TP, GLO, GLU, and TG levels were all altered in the high-dose groups in both sexes; however, in the males, these indicators were increased, whereas they were decreased in the females compared to the controls ($P<0.05$). ALB, BUN, and LDH levels were all decreased in the highdose male group compared to the control group ($P<0.05$). However, in the females, no differences were observed between the three test groups and the control group ($P>0.05$). Only ALT and CREA showed no change in either sex in any test group compared to the control groups.

In addition, the levels of blood GLU in the male rats were significantly increased in the medium-and high-dose groups compared to the pre-dosing levels ($P<0.01$). In contrast, the levels of blood GLU in the females were significantly decreased in all of the test groups compared to the control group ($P<0.01$). The levels of CHOL in the high-dose group were significantly decreased in both sexes ($P<0.01$) compared to the control group. Thus, the changing trends of AKP and CHOL in both sexes were the same. The differences in the blood chemistry between males and females are difficult to explain.

Organ weight

In the male rats, the weight of the liver and spleen increased following Ah administration in a dose-dependent manner, and a significant difference was observed between the high-dose group and the control group ($P<0.05$; Table 6). In the female rats, the weight of the liver, kidney and brain significantly increased following Ah administration ($P<0.05$; Table 7). The other organs did not exhibit any dose-dependent changes. Only the liver weight in the highdose group was significantly increased in both sexes after 14 consecutive days of dosing compared to the control ($P<0.01$).

Table 6 Relative organ weight (mean±standard deviation) in males intragastrically administered Ah daily for 14 consecutive days

Dose	Control	100 (g/kg)	200 (mg/kg)	1 000 (g/kg)
No.of animals examined	10	10	10	10
Liver (mg/g)	0.034±0.000	0.035±0.0035	0.042±0.00	0.069±0.097 **
Kidney (mg/g)	0.008±0.000	0.008±0.001	0.008±0.001	0.009±0.001 **
Adrenal gland (mg/g)	0.0002±0.0000	0.0002±0.0000	0.0002±0.0000	0.0002±0.0000
Spleen (mg/g)	0.002±0.000	0.003±0.000 *	0.003±0.000 *	0.003±0.000 *
Testis (mg/g)	0.012±0.002	0.012±0.002	0.012±0.003	0.012±0.001
Brain (mg/g)	0.013±0.001	0.013±0.001	0.013±0.001	0.013±0.002
Heart (mg/g)	0.005±0.000	0.005±0.001	0.005±0.001	0.005±0.001
Thymus (mg/g)	0.002±0.000	0.002±0.001	0.002±0.000	0.002±0.001
Lung (mg/g)	0.006±0.001	0.006±0.000	0.006±0.000	0.007±0.001 *
Testis (mg/g)	0.012±0.002	0.012±0.002	0.012±0.003	0.012±0.001

doi: 10.1371/journal.pone.0120165.t006

Table 7 Relative organ weights (mean ± standard deviation) in female rats intragastrically administered Ah daily for 14 consecutive days

Dose	Control	100 (g/kg)	200 (mg/kg)	1 000 (g/kg)
No.of animals examined	10	10	10	10
Liver (mg/g)	0.033±0.001	0.035±0.005*	0.036±0.018*	0.038±0.020**
Kidney (mg/g)	0.007±0.001	0.008±0.001*	0.008±0.001*	0.008±0.001*
Adrenal gland (mg/g)	0.0003±0.0000	0.0003±0.0000	0.0003±0.0000	0.0003±0.0000
Spleen (mg/g)	0.003±0.000	0.003±0.000	0.003±0.000	0.003±0.000
Ovary (mg/g)	0.000±0.000	0.001±0.001*	0.000±0.000	0.000±0.000
Uterus (mg/g)	0.002±0.000	0.002±0.000	0.002±0.001	0.002±0.002
Brain (mg/g)	0.012±0.001	0.014±0.001*	0.014±0.002*	0.014±0.001*
Heart (mg/g)	0.004±0.000	0.005±0.001	0.005±0.000	0.005±0.001
Thymus (mg/g)	0.003±0.000	0.003±0.001	0.003±0.000	0.002±0.001
Lung (mg/g)	0.006±0.000	0.006±0.001	0.007±0.001*	0.006±0.000

doi: 10.1371/journal.pone.0120165.t007

Table 8 Histopathological findings for male rats treated with Ah for 14 days

Organ	Findings	Dose			
		Control	Low	Medium	High
Male	No.of animals	10	10	10	10
Liver	Granular degeneration, slight	1	2	5	8
	Vacuolar degeneration, slight	0	1	4	8
	Hepatic congestion, slight	0	3	5	9
Kidney	Granular degeneration	1	1	5	10
	Necrosis, focal	0	1	6	10
	Congestion, moderate	0	0	2	10
Heart	Congestion, focal	0	0	2	10
	Myocardial degeneration	0	1	3	10
	edema	1	1	1	10
spleen	Edema, slight	1	1	3	10
	Necrosis, focal	0	1	4	8
Brain	Degeneration, nerve cells	0	0	2	8
	Edema	0	1	2	9
	Congestion	0	0	0	10
Duodenum	Inflammation with eosinophilic infiltration	0	0	3	10
Cecum	Inflammation with eosinophilic infiltration	0	0	5	10
Colon	Inflammation with eosinophilic infiltration	0	0	5	10

doi: 10.1371/journal.pone.0120165.t008

Table 9 Histopathological findings for female rats treated with Ah for 14 days

Organ	Findings	Dose			
		Control	Low	Medium	High
Female	No.of animals	10	10	10	10
Liver	Granular degeneration, slight	1	3	6	8
	Vacuolar degeneration, slight	1	1	4	8
	Hepatic congestion, slight	0	2	5	9
Kidney	Granular degeneration	1	2	3	10
	Necrosis, focal	0	1	5	10
	Congestion, moderate	0	0	2	10
Heart	Congestion, focal	0	0	3	10
	Myocardial degeneration	1	1	4	10
	edema	0	1	1	10
spleen	Edema, slight	1	0	4	10
	Necrosis, focal	0	1	5	8
Brain	Degeneration, nerve cell	0	0	3	8
	Edema	0	1	2	9
	Congestion	1	2	4	10
Duodenum	Inflammation with eosinophilic infiltration	0	0	2	10
Cecum	Inflammation with eosinophilic infiltration	0	1	4	10
Colon	Inflammation with eosinophilic infiltration	0	0	3	10

doi: 10. 1371/journal.pone.0120165. t009

Histopathology

Microscopic examination of the organs was performed on animals from the four dose groups.The histopathological findings are summarized in Tables 8 and 9 and are shown in Fig. 1 – 24. No abnormal changes in either the organs or tissues were observed in the low-dose groups compared to the control groups.These results demonstrated that Ah did not induce remarkable histopathological alterations in the low-dose animals.

In the medium- and high-dose groups, treatment-related pathological findings were noted in the colon, spleen, heart, liver, kidney and brain. Acute catarrhal enteritis with eosinophilic infiltration was evident in 17 of the 20 rats in the high-dose group (Fig. 1–Fig. 4).Degeneration, necrosis and exfoliation of epithelial cells in the intestinal epithelial cells was observed, along with an increase in the number of goblet cells.In the spleens, the histopathological changes were characterized by slight degeneration, necrosis, and dropsy (Fig. 5 – Fig. 8). The kidney presented moderate to severe granular degeneration in therenal tubular epithelial cells, focal necrosis and congestion in both genders (Fig. 9–Fig. 12).In the liver, hepatocellular damage was evident, includingminor granule and vacuolar denaturation, necrosis and interstitial connective tissue proliferation (Fig. 13 – Fig. 16). Myofibrillar tissue exhibited degeneration, necrosis and fibrinoid necrosis (Fig. 17–Fig. 20).Cellular necrocytosis was serious, the nuclei showed signs of karyolysis and were

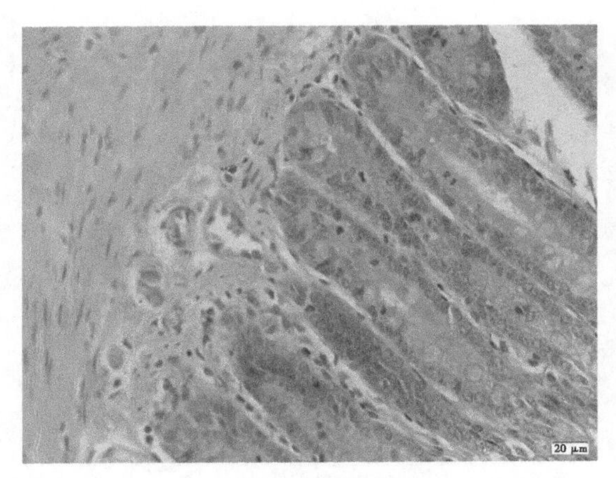

Fig. 1 Photomicrograph of duodenum tissue of control group stained with HE

There is no abnormal

doi: 10.1371/journal.pone.0120165.g001

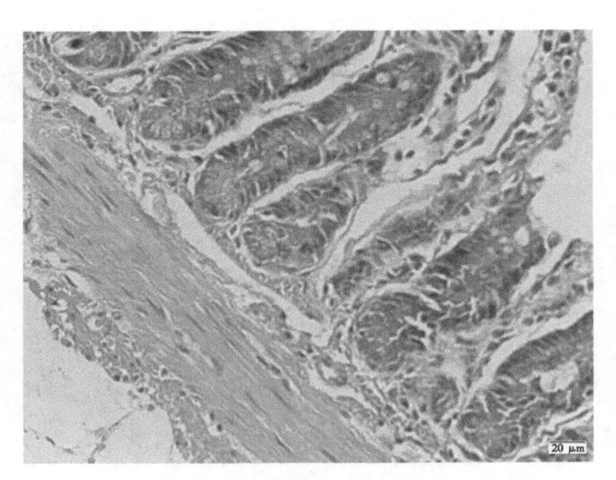

Fig. 2 Photomicrograph of duodenum tissue of low dose group stained with HE

There is no abnormal

doi: 10.1371/journal.pone.0120165.g002

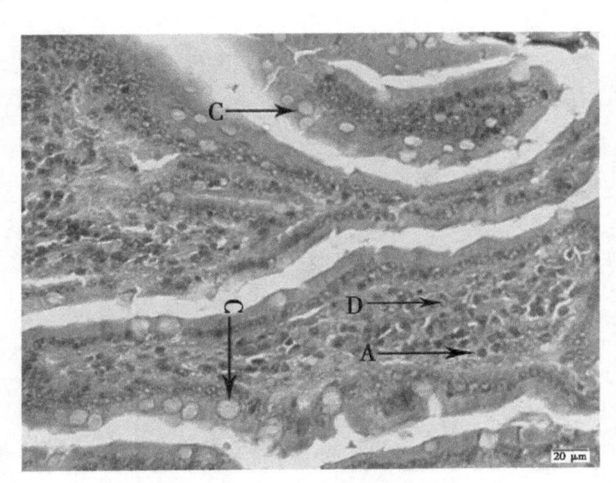

Fig. 3 Photomicrograph of duodenum tissue of medium dose group stained with HE

A: Acute catarrhal enteritis with eosinophilic infiltration; C: Goblet cell hyperplasia; D: Congestion in the lamina propria.

doi: 10.1371/journal.pone.0120165.g003

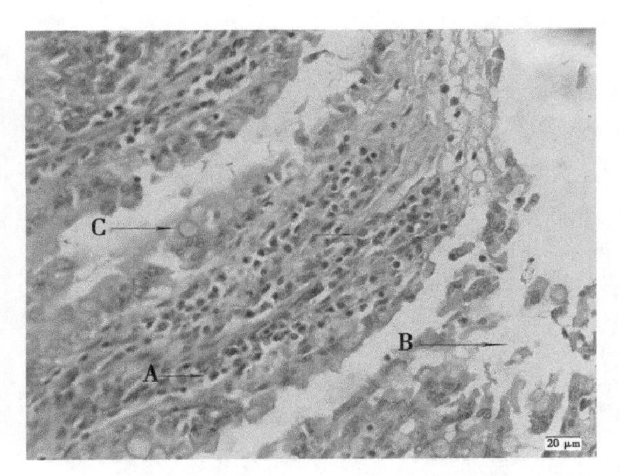

Fig. 4 Photomicrograph of duodenum tissue of high dose group stained with HE

A: Acute catarrhal enteritis with eosinophilic infiltration; B: Degeneration, necrosis and exfoliation of intestinal epithelial cells; C: Goblet cell hyperplasia.

doi: 10.1371/journal.pone.0120165.g004

disappearing, and the cytoplasms were bursting. In the brain, mild congestion, edema and nerve cell degeneration were identified (Fig. 21–Fig. 24). There were statistically significant differences between the high-dose groups and control groups in terms of lesions.

Observations of the clinical curative effect

Only worms with a scolex were regarded as a complete polypide. Among 252 infected dogs, 246 dogs were collected complete polypide. The species was identified according to Beveridge et al [13]. A total of six tapeworm species were identified: *Dipylidium caninum*, *Taenia hydatigena*, *Taenia*

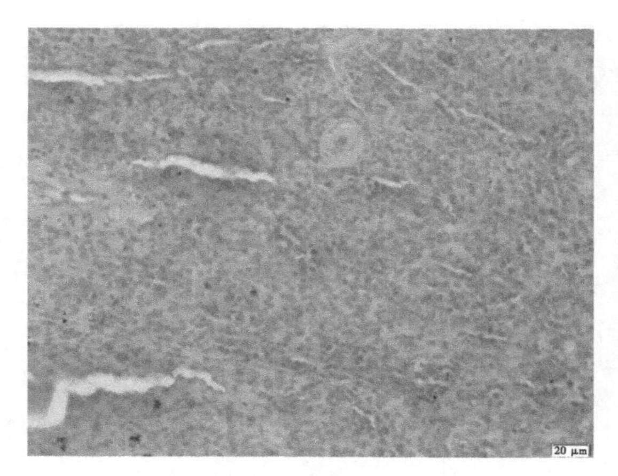

Fig. 5 Photomicrograph of spleen tissue of control group stained with HE

There is no abnormal.

doi: 10. 1371/journal.pone.0120165. g005

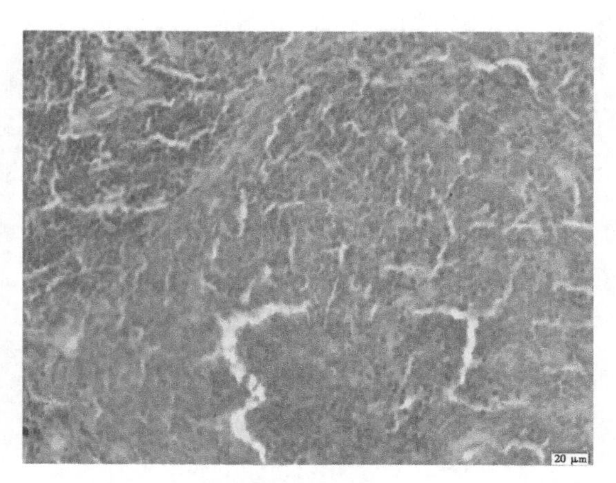

Fig. 6 Photomicrograph of spleen tissue of low dose group stained with HE

There is no abnormal.

doi: 10. 1371/journal.pone.0120165. g006

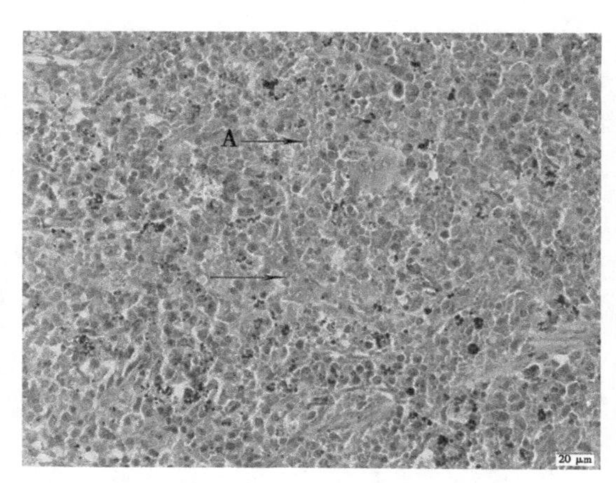

Fig. 7 Photomicrograph of spleen tissue of medium dose group stained with HE

A: Passive congestion.

doi: 10. 1371/journal.pone.0120165. g007

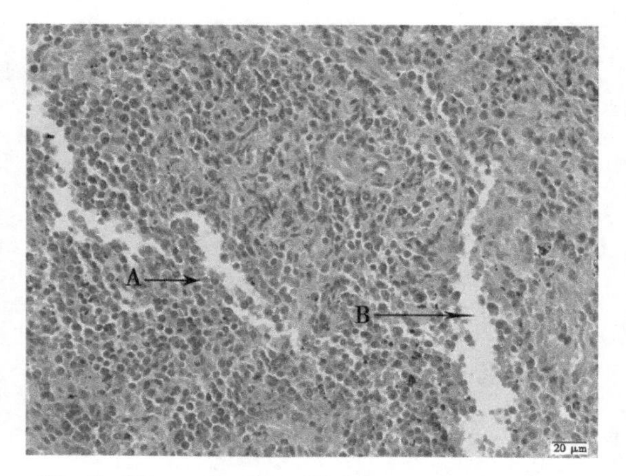

Fig. 8 Photomicrograph of spleen tissue of high dose group stained with HE

A: Passive congestion; B: Edema.

doi: 10. 1371/journal.pone.0120165. g008

pisiformis, *Taenia multiceps*, *Echinococcus granulosus and Mesocestoides lineatus*.

DISCUSSION AND CONCLUSION

Ah requires a deeper evaluation of its efficacy and safety because of its growing demand for reported medicinal use. We conducted a 14-day toxicity study to evaluate the safety of Ah at different doses. The selections of Ah dosage levels were primarily based on our preliminary test. Changes in body weight have been used as an indicator of adverse effects of drugs and chemicals[14]. Decreased body weight and food consumption were identified in all treated rats after 14 days of Ah administration. These findings suggest that oral Ah administration had some effect on the growth and functions of rats at the concentrations studied.

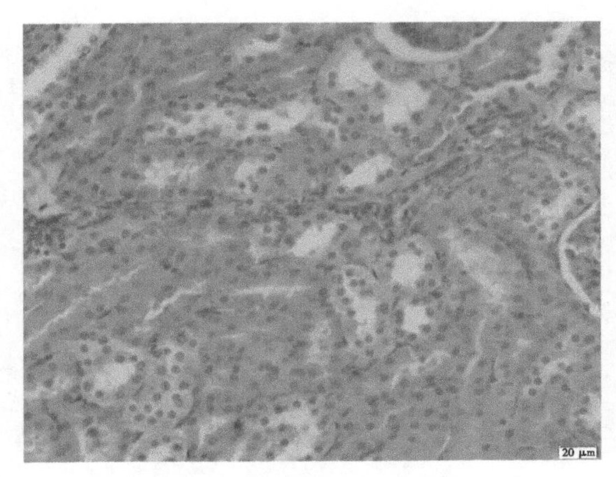

Fig. 9 Photomicrograph of kidney tissue of control group stained with HE

There is no abnormal.

doi: 10.1371/journal.pone.0120165.g009

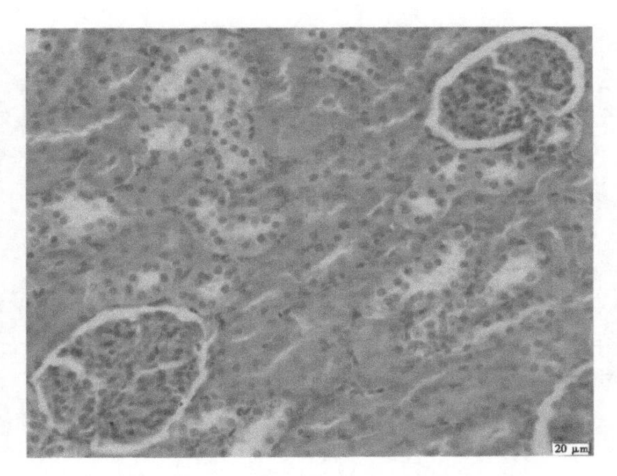

Fig. 10 Photomicrograph of kidney tissue of low dose group stained with HE

There is no abnormal.

doi: 10.1371/journal.pone.0120165.g010

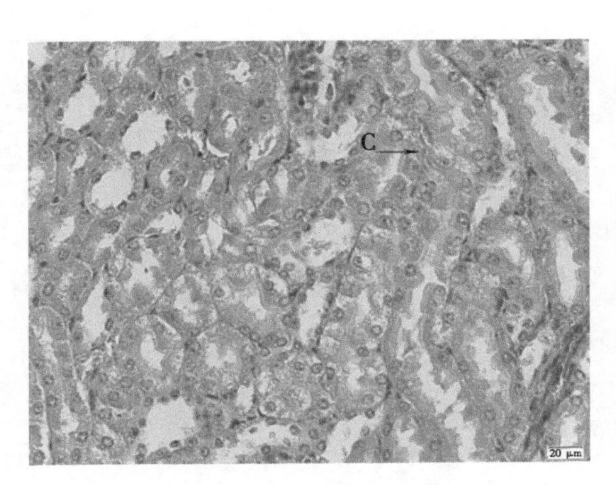

Fig. 11 Photomicrograph of kidney tissue of medium dose group stained with HE

C: Congestion.

doi: 10.1371/journal.pone.0120165.g011

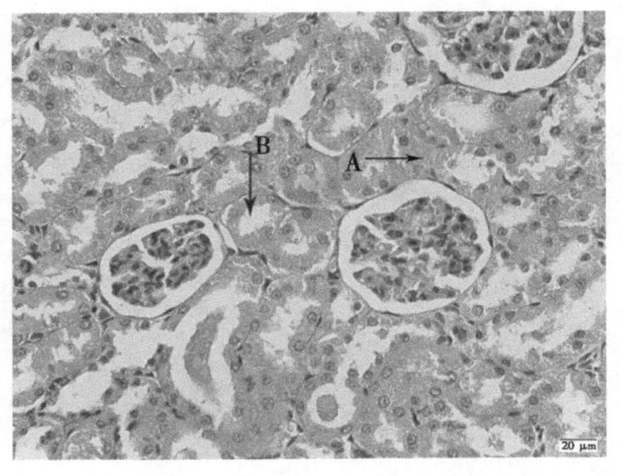

Fig. 12 Photomicrograph of kidney tissue of high dose group stained with HE

A: Granulardegeneration in renal tubular epithelial cells; B: Focalnecrosis.

doi: 10.1371/journal.pone.0120165.g012

The hematopoietic system is one of the most sensitive systems used to assess drug toxicity in humans and animals[15].The present study indicated that there were significant differences in hemoglobin, RBC, platelet, and total and differential leukocyte countsin the high-dose group, which indicates that Ah had specific effects on either the circulating blood cells or their production.

Liver damage results in the elevation of both ALT and AST levels in the blood.In addition, the identification of ALT in the serum is considered to be the first sign of cell and liver damage[16].Creatinine is a good indicator of renal function, i.e., an increase in creatinine indicates there is obvious damage to functional nephrons.There were significant differences in ALP, ALT, AST, creatinine, blood urea, total bilirubin and total proteinlevels, as well as the A/G ratio, in the high-

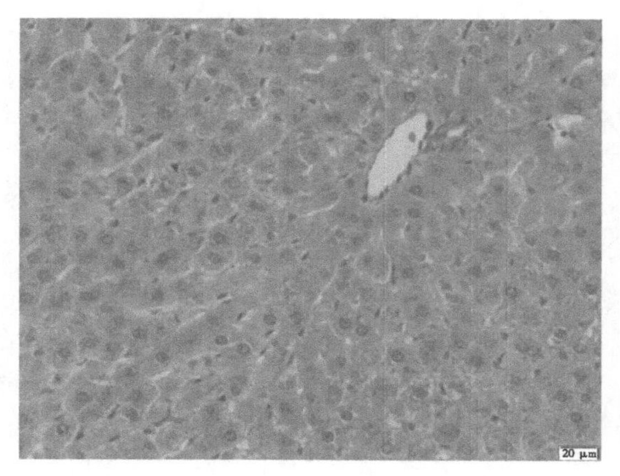

Fig. 13 Photomicrograph of liver tissue of control group stained with HE

There is no abnormal

doi: 10. 1371/journal.pone.0120165. g013

Fig. 14 Photomicrograph of liver tissue of low dose group stained with HE

There is no abnormal

doi: 10. 1371/journal.pone.0120165. g014

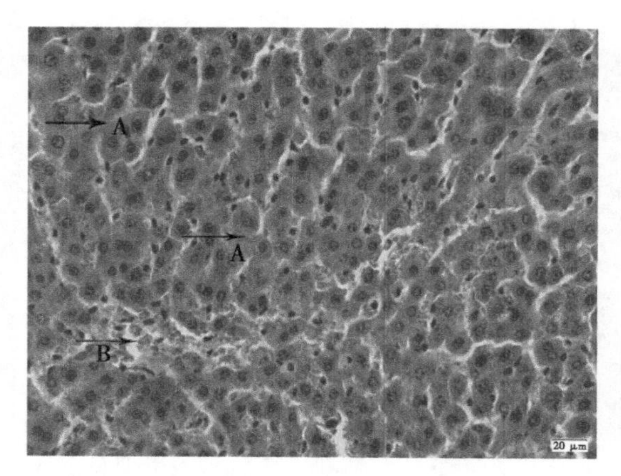

Fig. 15 Photomicrograph of liver tissue of medium dose group stained with HE

A: Granular degeneration of liver cells;

B: Congestion.

doi: 10. 1371/journal.pone.0120165. g015

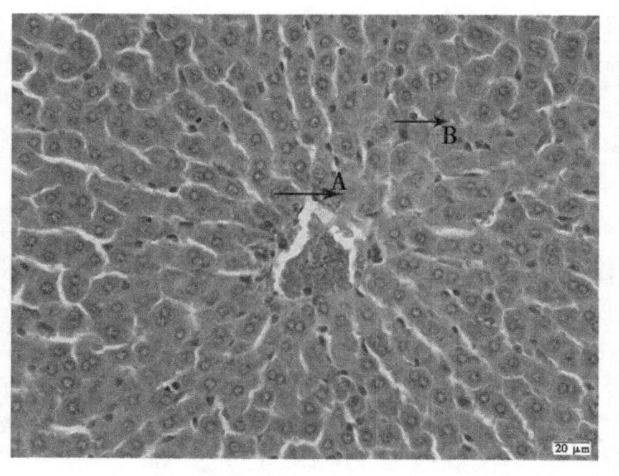

Fig. 16 Photomicrograph of liver tissue of high dose group stained with HE

A: Granular degeneration of liver cells;

B: Congestion.

doi: 10. 1371/journal.pone.0120165. g016

dose animals compared to the controls.

The liver is the site of cholesterol degradation and glucose synthesis, and it generates free glucose, which is secreted into the blood, from hepatic glycogen stores[17]. The subchronic toxicity test is one of the most important aspects of drug safety evaluation and the primary basis for the approval of a drug for clinical application. In general, continuous and repeated dosing is conducted to observe toxicity reactions as well as hematological, blood biochemical and pathological changes in experimental animals. It is also important to analyze the dosetoxicity relationship, the nature and degree of the toxicity reaction in the primary target organs, the reversibility of the toxicity reactions and the accumulation of toxicity. Compared to the control group, the rats in the high-dose group in this study

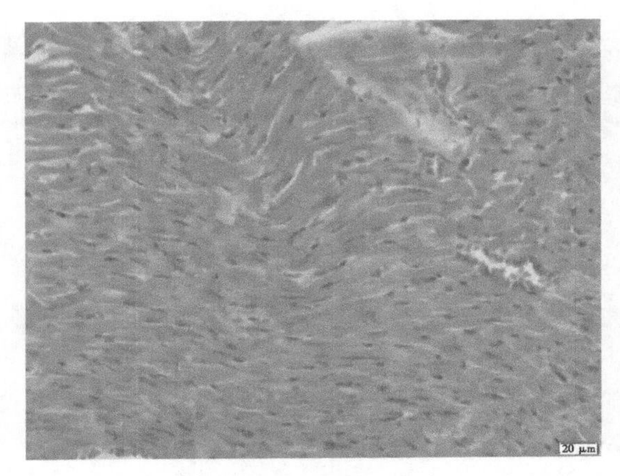

Fig. 17 Photomicrograph of heart tissue of control group stained with HE

There is no abnormal.

doi: 10.1371/journal.pone.0120165. g017

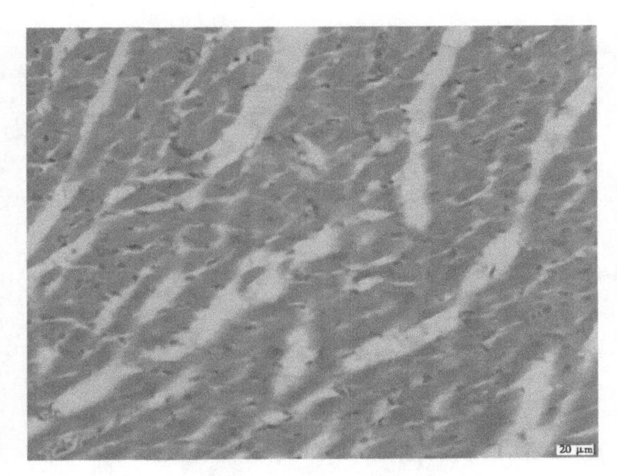

Fig. 18 Photomicrograph of heart tissue of low dose group stained with HE

There is no abnormal.

doi: 10.1371/journal.pone.0120165. g018

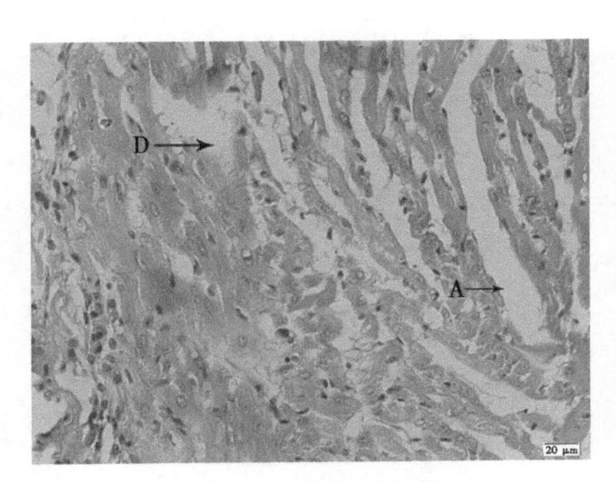

Fig. 19 Photomicrograph of heart tissue of medium dose group stained with HE

A: Interstitial hydrops; D: Myocardial necrosis.

doi: 10.1371/journal.pone.0120165. g019

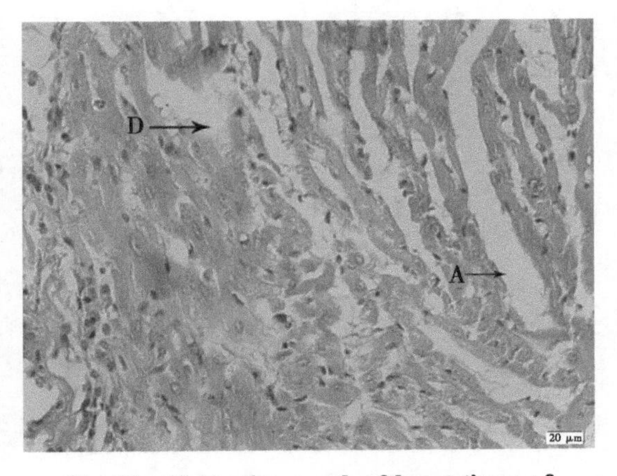

Fig. 20 Photomicrograph of heart tissue of high dose group stained with HE

A: Interstitial hydrops; B: Congestion;

C: Myocardial degeneration

D: Myocardial necrosis.

doi: 10.1371/journal.pone.0120165. g020

exhibited weight loss and significantly increased liver and kidney organ coefficients.

The rats in the high-dose group were depressed and presented with a loss of appetite. Body weight change in experimental animals is a basic parameter that reflects apoisoning effect and the toxicity of the poison.The organ coefficient is an important index of the target organ affected by the test compound.In this study, the results indicated that the toxicity was lowest in the low-dose group; however, the toxicity increased with increasing doses of Ah.

Pharmacological studies can ascertain the toxicity of the test drug more precisely by determining whether the related physiological and biochemical parameters are altered under the effects of the test-

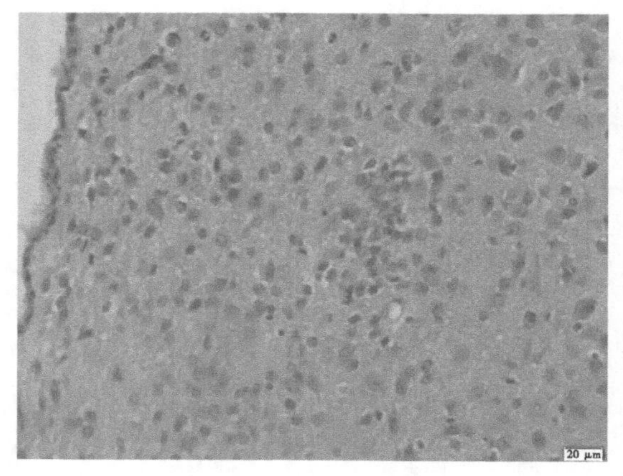

Fig. 21 Photomicrograph of brain tissue of control group stained with HE

There is no abnormal.

doi: 10.1371/journal.pone.0120165. g021

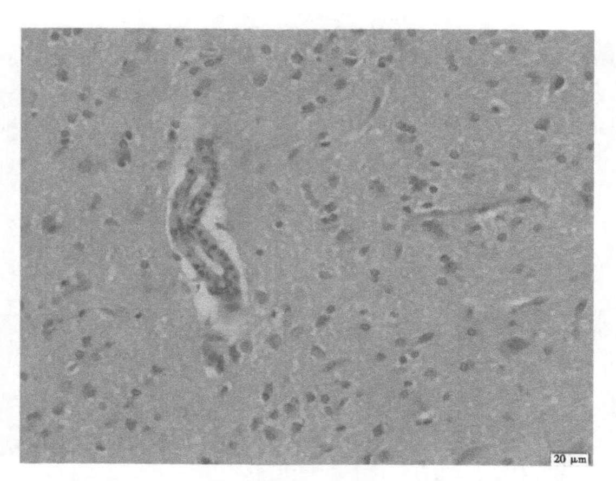

Fig. 22 Photomicrograph of brain tissue of low dose group stained with HE

There is no abnormal.

doi: 10.1371/journal.pone.0120165. g022

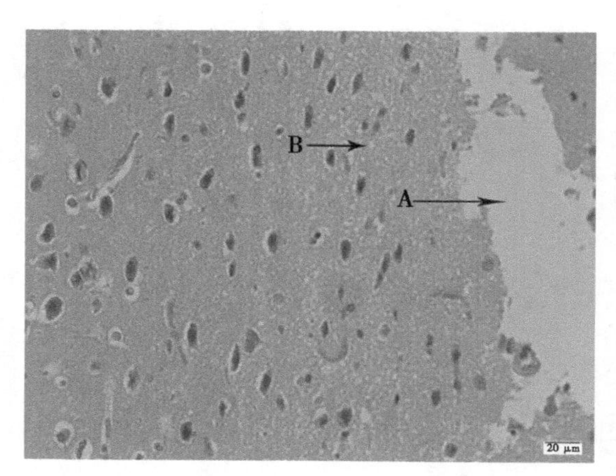

Fig. 23 Photomicrograph of brain tissue of medium dose group stained with HE

A: Edema; B: Coagulative necrosis.

doi: 10.1371/journal.pone.0120165. g023

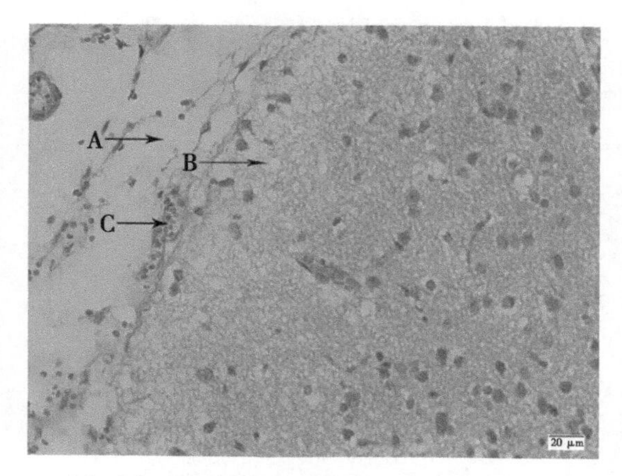

Fig. 24 Photomicrograph of brain tissue of high dose group stained with HE

A: Edema; B: Coagulative necrosis; C: Congestion.

doi: 10.1371/journal.pone.0120165. g024

ed drug.In this study, the rats in the high-dose group exhibited clear inhibitory effects on blood biochemical parameters, including ALT, TP and BUN levels.It was demonstrated that the toxicity reaction was intensified at the higher dosage, and clinical dosing should occur within a safe dosage range.A histopathological examination directly reflects drug toxicity.Our test results demonstrated that with a continuous increase in the Ah dose, organ damage was aggravated, with different degrees of congestion, degeneration and necrosis observed.The results accurately reflected the liver and kidney changes, whereas the results for the changes in the lung were poor.During the test, severe hemorrhage and congestion were observed in the livers of some rats, which could have been caused by damage due to an inappropriate lavage.

Based on the results of our analysis and the earlier acute toxicity test, long-term administration of high doses of Ah is likely to be toxic.The dosage should be set according to the clinically recommended dose to ensure safe dosing.This study provides a theoretical foundation for clinically safe dosing of Ah.

AUTHOR CONTRIBUTIONS

Conceived and designed the experiments: JZ XW.Performed the experiments: XW JN.Analyzed the data: BL JL.Contributed reagents/materials/analysis tools: XZ.Wrote the paper: XW.

References

[1] Arjungi KN.Areca nut: a review.Arzneimittel-forschung.1976, 26 (5): 951-956. PMID: 786304.

[2] Christie JE, Shering A, Ferguson J, Glen AI.Physostigmine and arecoline: effects of intravenous infusions in Alzheimer presenile dementia.Br J Psychiatry.1981, 138: 46-50. PMID: 7023592.

[3] Burkhill IH.In dictionary of the economicproducts of the Malay Peninsula.In: Burkhill IH, editor.London: Crown Agents for the Colonies.1935, Vol.1, 222.

[4] Azeez S, Amudhan S, Adiga S, Rao N, Udupa LA.Wound healing profile of Areca catechu extracts on different wound models in Wistar rats.Kuwait Medical J.2007, (40) 1: 48-52.

[5] Chempakam B.Hypoglycemic activity of arecoline in betel nut Areca catechu L.Indian J Exp Biol.1993, 31 (5): 474-475. PMID: 8359856.

[6] Kuo FC, Wu DC, Yuan SS, Hsiao KM, Wang YY, Yang YC, et al.Effects of arecoline in relaxing human umbilical vessels and inhibiting endothelial cell growth.J Perinat Med.2005, 33 (5): 399-405. PMID: 16238534.

[7] Feng Q, Li Gl, Yang Y, Gao J.Studies on the increasing-effect components for molluscicides in nut of *Areca catechu*.Zhong Yao Cai.1999, 22: 572-574. PMID: 12571896.

[8] Jaiswal P, Singh V, Singh D.Enzyme inhibition by molluscicidal component of *Areca catechu and Carica papaya* in the nervous tissue of vector snail *Lymnaea acuminata*.Pestic Biochem Physiol.2008, 92: 164-168.

[9] Zhao WA, Li ZM, Wang BX.Morphological observation of semen areca and white pepper against cysticercus cellulosae in vitro.Modern Journal of Integrated Traditional Chinese and Western Medicine.2003, 12 (3): 237-238.

[10] Zha Chuanlong, Chen Guangyu, Wu Meijuan.The role of *magnolia officinalis Rehd etWils* and Betel nut on *fasciola hepatica* in vitro.J Nanjing University Traditional Chinese Medicine.1990, 6 (4): 266-269.

[11] Li Bing, Zhou Xu-Zheng, Li Jian-Yong, Yang Ya-Jun, Niu Jian-Rong, Wei Xiao-Juan, et al.Determination and pharmacokinetic studies of arecoline in dogplasma by liquid chromatography—tandem mass spectrometry, Journal of Chromatography B. 2014, 969: 12-18 . doi: 10. 1016/j. jchromb. 2014. 07. 043 PMID: 25140901.

[12] Van der Ven LT, Verhoef A, van de Kuil T, Slob W, Leonards PE, Visser TJ, et al.A 28-day oral dose toxicity study enhanced to detect endocrine effects of hexabromocyclododecane in wistar rats. Toxicol Sci.2006, 94 (2): 281-292. PMID: 16984958.

[13] Beveridge I, Gregory GG.The Identification of *Taenia* Species from Australian Carnivores.J Aust Veter. 1976, 52 (8): 369-373.

[14] Tofovic SP, Jackson EK.Effects of long-term caffeine consumption on renal function in spontaneously hypertensive heart failure prone rats.J Cardiovasc Pharmacol.1999, 33 (3): 360-366. PMID: 10069669.

[15] Rahman MF, Siddiqui MK, Jamil K. Effects of vepacide (Azadirachta indica) on aspartate and alanine aminotransferase profiles in a subchronic study with rats. Hum Exp Toxicol. 2001, 20 (5): 243-249. PMID: 11476156.
[16] Hilaly JE, Israili ZH, Lyouss B. Acute and chronic toxicological studies of *Ajuva iva* in experimental animals. J. Ethnopharmacol. 2004, 91 (1): 43-50. PMID: 15036466.
[17] Kaplan A, Jack R, Opheim KE, Toivola B, Lyon AW. Clinical Chemistry: Interpretation and Techniques. 4th Edition. Williams & Wilkins, USA. 1995.

（发表于《PLOS ONE》，院选 SCI，IF：3.234）

Exploring Differentially Expressed Genes and Natural Antisense Transcripts in Sheep (*Ovis aries*) Skin with Different Wool Fiber Diameters by Digital Gene Expression Profiling

YUE Yao-jing*, GUO Ting-ting*, LIU Jian-bin, GUO Jian, YUAN Chao, FENG Rui-lin, NIU Chun-e, SUN Xiao-ping, YANG Bo-hui**

(Lanzhou Institute of Husbandry and Pharmaceutical Sciences, Chinese Academy of Agricultural Sciences, Lanzhou 730050, China)

Abstract: Wool fiber diameter (WFD) is the most important economic trait of wool. However, the genes specifically controlling WFD remain elusive. In this study, the expression profiles of skin from two groups of Gansu Alpine merino sheep with different WFD (a super-fine wool group [FD = (18.0 ± 0.5) μm, *n* = 3] and a fine wool group [FD = (23.0± 0.5) μm, n = 3]) were analyzed using next-generation sequencing-based digital gene expression profiling. A total of 40 significant differentially expressed genes (DEGs) were detected, including 9 up-regulated genes and 31 down-regulated genes. Further expression profile analysis of natural antisense transcripts (NATs) showed that more than 30% of the genes presented in sheep skin expression profiles had NATs. A total of 7 NATs with significant differential expression were detected, and all were down-regulated. Among of 40 DEGs, 3 DEGs (AQP8, Bos d2, and SPRR) had significant NATs which were all significantly down-regulated in the super-fine wool group. In total of DEGs and NATs were summarized as 3 main GO categories and 38 subcategories. Among the molecular functions, cellular components and biological processes categories, binding, cell part and metabolic process were the most dominant subcategories, respectively. However, no significant enrichment of GO terms was found (corrected P-value > 0.05). The pathways that were significantly enriched with significant DEGs and NATs were mainly the lipoic acid metabolism, bile secretion, salivary secretion and ribosome and phenylalanine metabolism pathways ($P<0.05$). The results indicated that expression of NATs and gene transcripts were correlated, suggesting a role in gene regulation. The discovery of these DEGs and NATs could facilitate enhanced selection for super-fine wool sheep through gene-assisted selection or targeted gene manipulation in the future.

---

* These authors contributed equally to this work.

** lzyangbohui@163.com.

INTRODUCTION

Fine wool sheep, also called Merino, is a world famous sheep breed that is known to produce high-quality fine wool. Fine wool sheep are distributed primarily in Australia, China, New Zealand, South Africa, Uruguay, Argentina and other countries[1]. Wool fiber diameter (WFD) is the most important economic trait of merino sheep and determines 75% of the value of wool fibers. The WFD variation-induced profit accounted for 61% of the total profits of wool[2,3]. Wool is formed from keratinocytes derived from a progenitor population at the base of the hair follicle (HF)[4]. The morphogenesis and growth of HF in sheep has been extensively studied since the 1950's and the developmental processes at the cellular level are reasonably well understood[5-7]. Dermal papilla (DP) cells are a population of mesenchymal cells at the base of the HF[8], and provide signals that contribute to specifying the size, shape and pigmentation of the wool[9]. It is well known that WFD is significantly associated with size of DP and matrix in mammals[4,6,10. 17], and is largely specified post-initiation, during the period of HFs growth and morphogenesis[18]. To illustrate the molecular mechanisms of controlling WFD, the expression profiles of different stage of fetal and adult sheep skin have also been generated by sequencing of expressed sequence tags (ESTs) and cDNA microarray[19,20]. However, the size of DP and matrix in mammals is also markedly influenced by genetic[10,11,21], physiological[13], nutrition[22], hormones[12] during the anagen phase of the hair cycle. Up to now, there are no studies on the molecular mechanisms of controlling WFD during the anagen phase, and the genes specifically controlling WFD remain elusive[23].

Knowledge of the genes controlling development of DP and matrix come from studies of the morphogenesis and cycle of HF of mice and human[24-28], It involves a series of signaling between the matrix and the dermal papill, such as Wnt/beta-catenin, EDA/EDAR/NF-κB, Noggin/Lef-1, Ctgf/Ccn2, Shh, BMP-2/4/7, Dkk1/Dkk4 and EGF[4,29]. The mutation, epigenetic modification and post-translational modification of any ligand or receptors in these pathways maybe affect WFD[30,31]. Therefore, in addition to the current efforts on proteinencoding genes, the attention should also be paid to a novel regulatory factor, non-coding RNA (ncRNA), such as micro RNA, long non-coding RNA and natural antisense transcripts (NATs)[32]. Among these ncRNAs, NATs are not only large in quantity but also play important roles in gene expression regulation in organisms[33]. NATs refer to a class of non-coding RNAs that are produced inside organisms under natural conditions and are expressed in many species[34. 36]. NATs play an important role in transcript regulation at the mRNA and/or protein levels and in regulating various physiological and pathological processes, such as organ formation, cell differentiation and disease[37,38]. However, the roles of NATs in controlling WFD have not been described.

The next-generation sequencing (NGS)-based digital gene expression (DGE) profiling technologies developed in recent years constitute a revolutionary change in traditional transcriptome technology. Compared with ESTs and cDNA microarray, the strength of DGE profiling is that it is an "open system" and has better capability to discover and search for new information, providing a new way to identify novel genes and NATs that specifically control WFD[39]. However, the success or effectiveness of the search is largely dependent on the completeness and stitching quality of the genome sequences of the species studied[40].It is exciting that the International Sheep Genomics Consortium (ISGC) has achieved initial success in assembling the reference sheep genome. The total length of the assembled genome, Sheep Genome v3. 1, has reached 2. 64 Gb, with only 6. 9% gaps[41].Needless to say, the release of a high-quality sheep genome reference sequence provides important resources for using NGS to study the skin transcriptome of sheep with different WFD and find genes and NATs that specifically control WFD[42,43].

In this study, NGS-based DGE profiling was conducted to analyze the differentially expressed genes (DEGs) and NATs in the skin of Gansu Alpine fine wool sheep with different WFD. A total of 47 significant DEGs and NATs were detected, including 9 up-regulated genes, 31 down-regulated genes and 7 down-regulated NATs. These DEGs and NATs may be useful in further study on molecular markers of controlling WFD in fine wool sheep.

RESULTS

Sequencing and assembly

NGS was performed on 6 individuals of the super fine wool group (sample nos.65505, 65530, 65540; $n=3$) and the fine wool group (sample nos.5Y127, 5Y212, 5Y339; $n=3$), and raw data greater than 5Mb were obtained for all individuals (Table 1), submitted to the NCBI BioProject database, with the accession number PRJNA274817. There was a 3′ adaptor sequence in the raw data, along with small amounts of low-quality sequences and various impurities. Impurity data were removed from the raw data, and clean tags greater than 4. 7 Mb were obtained. The distinct tag number of each individual was greater than 0. 11 Mb (Table 1).

Table 1 Summary of tag numbers based on the DGE data from Gansu Alpine fine wool sheep skin with different WFD

Sample ID	Raw Data		Clean Tag	
	Total number	Distinct Tag number	Total number	Distinct Tag number
5Y127	5 259 561	355 014	4 917 435	119 886
5Y212	5 364 832	350 472	5 033 144	117 880
5Y339	5 145 452	336 695	4 857 460	120 525
65505	5 244 848	314 710	4 930 422	110 879
65530	5 485 701	343 795	5 184 375	117 642
65540	5 282 704	373 209	4 941 669	143 800

doi: 10. 1371/journal.pone.0129249. t001

The sheep genome reference sequence database includes 19, 346 gene sequences, including 18, 940 genes with "CATG" loci and accounting for 97. 90% of the total number of genes. The total number of reference tags in the reference tag database is 173, 207, including 169, 843 unambiguous reference tags, accounting for 98. 06% of the total number of tags. All clean tags were compared with reference genes and reference genomes, and the results indicated that 87. 45%,

87.32%, 88.63%, 88.97%, 88.72% and 87.38% of the total numbers of clean tags of the 6 individuals from two groups (sample nos.5Y127, 5Y212, 5Y339, 65505, 65530, and 65540, the same hereinafter), respectively, could be matched to the reference tags; of these, 49.05%, 50.26%, 51.86%, 52.41%, 51.48% and 50.37% of the clean tags could be uniquely located in the reference sequences (sense and antisense), and the ratios of distinct tag were 36.00%, 36.63%, 37.64%, 36.61%, 37.33% and 33.22% (Table 2), respectively; the numbers of unambiguous tagmappedgenes were 10, 391, 10, 209, 10, 666, 10, 434, 10, 137, 10, 298, 10763 and 9, 965, respectively. (Table 2) These uniquely located sequences indicated that the key genes that regulate the WFD may exist in genes expressed in the individuals with different WFD.

Table 2 Summary of unambiguous tag mapping to gene and unambiguous tag-mapped genes (sense & anti-sense)

Sample ID	Unambiguous Tag Mapping to Gene				Unambiguous Tag-mapped Genes	
	Total number	Total of clean tag (%)	Distinct Tag number	Distinct Tag of clean tag (%)	number	% of ref genes
5Y127	2 412 224	49.05	43 158	36.00	10 391	53.71
5Y212	2 529 818	50.26	43 175	36.63	10 209	52.77
5Y339	2 519 260	51.86	45 369	37.64	10 666	55.13
65505	2 583 990	52.41	40 595	36.61	10 434	53.93
65530	2 668 978	51.48	43 910	37.33	10 137	52.40
65540	2 489 004	50.37	47 775	33.22	10 298	53.23

doi: 10.1371/journal.pone.0129249.t002

There were a large number of possible unknown tags that are in the gene expression profile libraries of these six individuals. Accounting for 12.55% (617, 198), 12.68% (638, 292), 11.37% (552, 279), 11.03% (543, 900), 11.28% (585, 017) and 12.62% (623, 705) of the total number of clean tags were unannotated, respectively; the proportions of distinct tag number of unknown tag were 18.77% (22, 503), 18.79% (22, 152), 16.17% (19, 493), 18.02% (19, 975), 17.25% (20, 292) and 20.43% (29, 382), respectively. These results suggested that there were many unknown genes in sheep skin tissue that may also play important roles in the regulation of HF development and wool growth.

Analysis of the expression profiling of NATs

Sense-antisense regulation is an important method for controlling gene expression. If the clean tags can be matched to the antisense strand of the gene, then it suggests that there are also transcripts for the antisense strand of this gene and that this gene may be subjected to sense-antisense regulation[44]. In the present study, we found that in the gene expression profiling libraries of the 6 individuals of the two groups, the percentages of genes with antisense transcripts were, respectively, 5.83%, 6.01%, 5.91%, 5.84%, 6.00% and 6.20% of the total numbers of the genes in the library, including 247, 816 (5.04%), 262, 353 (5.21%), 249, 595 (5.14%), 247, 940 (5.03%), 266, 926 (5.15%) and 265, 599 (5.37%) tags that could be exactly matched, respectively (Table 3); the number of tag species accounted for, respectively, 13.31%, 13.31%, 13.46%, 13.20%, 13.65% and 11.94% of the total numbers of tag species in the expression profile libraries of the 6 individuals from the 2 groups, including 15, 192 (12.67%), 14, 966 (12.70%), 15, 483 (12.85%), 13, 957 (12.59%), 15, 289 (13.00%) and 16, 346 (11.37%) tag species that could be exactly matched, respectively (Table 3); there were 6472 (33.45%), 6398 (33.07%), 6585 (34.04%), 6227 (32.19%), 6346 (32.80%) and 6758 (34.93%) genes with NATs in the gene expression profiles of the 6 individuals of the 2 groups (Table 3).The sense-antisense transcript ratios of all 6 expression profile libraries exhibited similar trends.The ratios between the tag number, distinct tag number and tag-mapped genes of sense and antisense transcripts were 10 : 1, 2 : 1 and 5 : 3, respectively; the spearman r between the sense and antisense transcripts expression were 0.5330.557.

Table 3 Summary of unambiguous tag mapping to anti-sense genes and unambiguous tag-mapped anti-sense genes

Sample ID	Unambiguous tag mapping to anti-sense genes				Unambiguous tag-mapped anti-sense genes	
	Total number	Total of clean tag (%)	Distinct Tag number	Distinct Tag of clean tag (%)	number	% of ref genes
5Y127	247 816	5.04	15 192	12.67	6 472	33.45
5Y212	262 353	5.21	14 966	12.70	6 398	33.07
5Y339	249 595	5.14	15 483	12.85	6 585	34.04
65505	247 940	5.03	13 957	12.59	6 227	32.19
65530	266 926	5.15	15 289	13.00	6 346	32.80
65540	265 599	5.37	16 346	11.37	6 758	34.93

doi: 10.1371/journal.pone.0129249.t003

Detection of DEGs, NATs and validation

The clean tags of all six individuals were aligned with the reference tag database, allowing a maximum of one base mismatch.Unambiguous tags were annotated, and the number of raw clean tags that corresponded to the same gene was counted and then standardized to obtain the standardized expression level of each gene in the skin transcriptome of the 6 individuals.Noiseq software was used to select DEGs, NATs that exhibited different expression levels between S and F group.However, there were only 47 significant DEGs and NATs [log2Ratio (S/F≥1, q-value≥0.8)].9 genes were up-regulated, 31 genes and all 7 NATs were down-regulated (Table 4, Fig 1).Among the down-regulated genes, *NT5C3L* (Gene ID 101109197) showed the greatest expression difference [log2Ratio (S/F) = -10.59, q value=0.89]; among the up-regulated genes, *CCNA*2 (Gene ID 100144758) showed the greatest expression difference (log2Ratio (S/F) = 6.04, q value= 0.83).38 significant expression genes had NATs, and only 2 had no found antisense transcripts: *CCNA*2 (Gene ID: 100144758) and prolactin-inducible protein homolog (Gene ID: LOC101114011). 3 significant DEGs (*AQP*8, Gene ID: 101108013; *Bos d*2, Gene ID:

101116281, and *SPRR*, Gene ID: 443313) had significant NATs. All 3 of these genes and their NATs were significantly down-regulated in the super-fine wool group (Table 4). 10 DEGs, NATs were used to validate selected differentially expressed transcripts identified from DGE profiling by Real-time PCR. The results from the real-time PCR confirmed the expression pattern of DGEs and NATs at two different groups in Gansu Alpine fine wool sheep.

Table 4 Summary of DEGs and NATs between two groups

Kinds of Transcriptes	GeneID	log2Ratio (S/F)	Probability	Symbol
DEGs	101109197	−10. 5912106	0. 886583566	NT5C3L
	101116281	−7. 40151954	0. 969520529	LOC101116281
	101120858	−4. 57602606	0. 868500902	LOC101120858
	443313	−4. 45466762	0. 848327046	Small proline-rich
	101109718	−4. 16402747	0. 835055492	CA4
	101108013	−4. 13509115	0. 838991854	AQP8
	101118004	−3. 99924015	0. 86634137	LOC101118004
	101115395	−3. 90149163	0. 829820677	PGLYRP1
	101114011	−3. 8813167	0. 855789733	LOC101114011
	101107368	−3. 8266707	0. 827278443	SLC25A35
	101104026	−3. 72227757	0. 855311355	S100A8
	101115563	−3. 4495271	0. 829328632	ABCC11
	101109387	−3. 38125687	0. 87068777	GLYATL2
	101106121	−3. 3527916	0. 870715106	LOC101106121
	101115336	−3. 29504885	0. 839237877	DNASE1L2
	101116409	−3. 16407151	0. 86568531	LOC101116409
	101113168	−3. 1562348	0. 826690722	LOC101113168
	100137068	−2. 97890634	0. 857115521	LOC100137068
	101116799	−2. 80442661	0. 859794434	LOC101116799
	NM_ 001009395. 1	−2. 67922661	0. 847588978	—
	101102714	−2. 61084761	0. 822795364	LOC101102714
	101102540	−2. 3882578	0. 861489257	RPL39
	101102697	−2. 14152458	0. 81083593	DNAJC12
	101108654	−2. 10679611	0. 847930676	LOC101108654
	101116537	−1. 89220493	0. 803933629	LOC101116537
	101121216	−1. 86897593	0. 842094473	LOC101121216
	101112716	−1. 85095925	0. 834194413	KRT7
	101121307	−1. 5666126	0. 800489312	ACSM3
	101111121	−1. 46170294	0. 809141108	CD82
	101114256	−1. 40678444	0. 802238806	ACTG2
	101109430	−1. 32738037	0. 80115904	KRT1

(continued)

Kinds of Transcriptes	GeneID	log2Ratio (S/F)	Probability	Symbol
	101105583	1. 377957974	0. 802293478	GSDMA
	101110063	1. 491887466	0. 816426111	HSPA2
	100135694	1. 587840025	0. 830968782	RPS27A
	443218	1. 679295263	0. 826485703	FOS
	101113964	1. 819875496	0. 838185446	PDCD6IP
	101119862	2. 058276167	0. 826212345	LIPK
	101120443	2. 31282643	0. 827415122	LOC101120443
	101105188	4. 991770667	0. 839825597	DAP
	100144758	6. 03994716	0. 834235416	CCNA2
NATs	101116281	-9. 15903036	0. 941608145	Bos d2
	443313	-7. 587465008	0. 819808271	SPRR
	101108013	-7. 807354922	0. 844108795	AQP8
	101120353	-5. 685396543	0. 88975588	major allergen I polypeptide chain 2-like
	101120550	-6. 896227669	0. 975067811	SMC1A
	101113086	-9. 556506055	0. 957344034	primary amine oxidase
	101113693	-8. 661778098	0. 914297923	ABP

doi: 10. 1371/journal.pone.0129249. t004

Gene ontology (GO), Kyoto Encyclopedia of Genes and Genomes (KEGG) and other databases were used for the functional analysis and signaling pathway annotations of these DGEs and NATs. Among all of the significant DEGs and NATs, 44 genes were annotated, and 3 genes located in the genome could not be annotated effectively. In total of 47 DEGs and NATs were summarized as three main GO categories and 38 subcategories (Fig 2). Among the molecular functions category, the top three were involved in binding and catalytic activity. Regarding cellular components, cell part, cell, organelle, membrane, organelle parts were the dominant groups. Within biological processes category, metabolic process and cellular process were the most dominant group. However, no significant enrichment of GO terms was found (corrected P-value >0. 05). The pathways that were significantly enriched with significant DEGs and NATs were mainly the lipoic acid metabolism, bile secretion, salivary secretion and ribosome and phenylalanine metabolism pathways ($P<0.05$).

DISCUSSION

Wool is produced via synthetic processes by HFs, which are embedded in the skin of sheep[45]. There are two types of HFs, named primary HF and secondary HF, which are different in appearance and function. Secondary HFs of the fine wool sheep are the main hair follicle and are critical determinants of mean fiber diameter and other wool characteristics. WFD is highly correlated with the size of DP of HF[17], whose origin can be traced to the dermal condensate, one of the earliest features of the developing HF[19]. The charactering of the molecular controls of HF initiation, morpho-

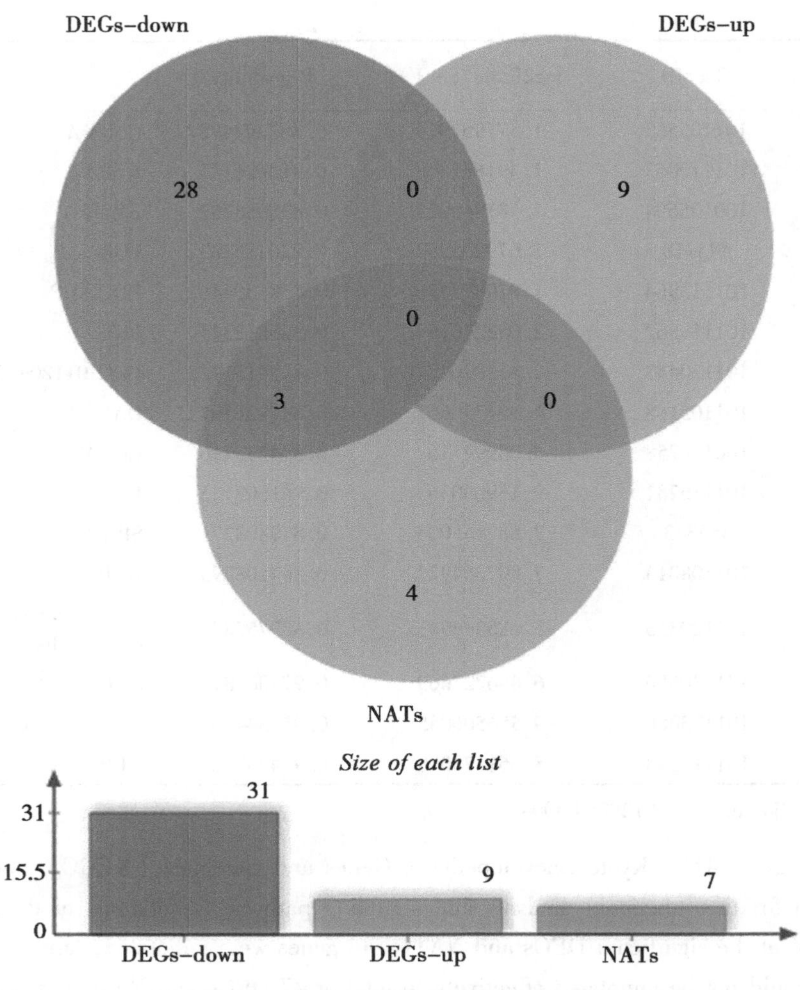

Fig. 1 The numbers of DEGs between two groups

Between two groups, there were 9 upregulated genes and 31 downregulated genes and 7 NATs.doi: 10. 1371/journal.pone.0129249. g001

genesis, branching and growth can facilitate enhanced selection for new sheep breeds with lower WFD[20].In order to understand the molecular mechanisms of controlling WFD, Adelson et al. (2004) constructed three cDNA libraries from fetal and adult sheep skin, obtained 2, 345 noredundant EST sequences and identified 61 ESTs expressed in the adult HF, which constituted a high priority candidate gene subset for further work aimed at identifying genes useful as selection markers or as targets for genetic engineering[19].Norris et al. (2005) found 50 up- and 82 down-regulated genes with increased fetal and inguinal expression relative to adult by compared skin gene expression profiles between fetal day 82, day 105, day 120 and adult HF anagen stage using a combined ovine.bovine skin cDNA microarray[20].However, comparing to sequencing of expressed sequence tags (ESTs) and cDNA microarray, DGE profiling has many unique advantages[46,47].It is highly accurate and has a very low detection limit, giving it a very wide range of applications[48], such as detection of new transcripts[49], functional research of non-coding RNA[50,51].In this study, six

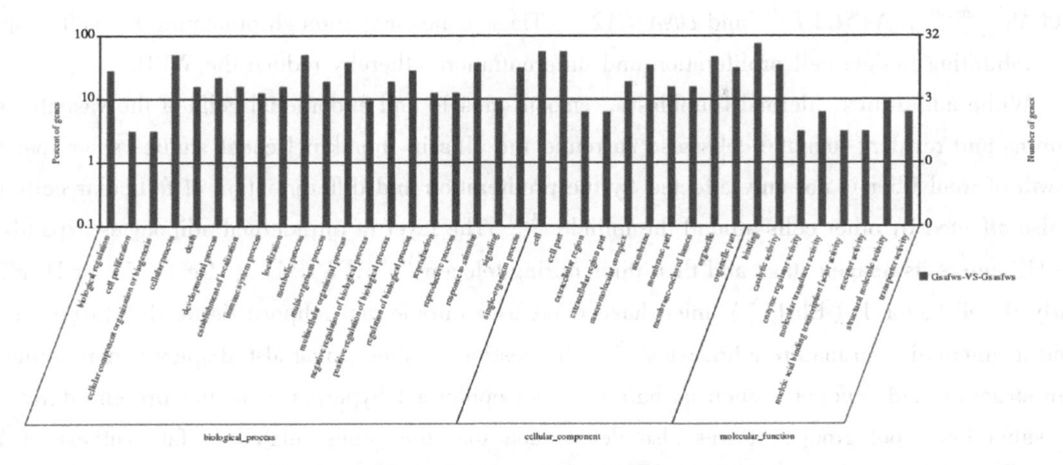

Fig. 2 GO functional analysis of DEGs and NATs

The results were summarized in three main categories: biological process, cellular component and molecular function. Among these groups, the terms binding, cell part and metabolic process were dominant in each of the three main categories, respectively.

doi: 10. 1371/journal.pone.0129249. g002

DGE profiling were conducted to analyze the DEGs and NATs in the skin of Gansu Alpine fine wool sheep with different WFD at the same age, gender, and nutrition level during the anagen. More than 4. 7 Mb of clean tags were obtained for each sheep, with more than 0. 11 Mb of distinct tags. A total of 47 significant DEGs and NATs were detected, including 9 up-regulated genes, 31 down-regulated genes and 7 down-regulated NATs.

The HF comprises several concentric epithelial structures[29]. Wool fiber is enveloped by two epithelial sheaths, known as the inner root sheath (IRS) and the outer root sheath (ORS). At the distal end of the HF, IRS cells undergo apoptosis and liberate the wool fiber[29,52]. ORS cells contain proliferating cells derived from stem cells in the bulge that feed into the matrix compartment of the bulb[53]. The matrix is a proliferative zone located at the proximal end of the HF surrounding the DP[52]. During the anagen phase, the DP cells are thought to direct the matrix cells to proliferate and differentiate into the multiple cell types that form the wool fiber and its channel, the IRS[9]. During postnatal life, hair follicles show patterns of cyclic activity with periods of active growth and hair production (anagen), apoptosis-driven involution (catagen), and relative resting (telogen)[53-55]. However, Merino HFs are predominantly in anagen throughout growth, different to many animals such as the mouse, rabbit, and guinea-pig[30]. It is also well known that WFD is significantly associated with sizes of DP and matrix in mammals[4,6,10-17], which are markedly influenced by genetic[10,11,21], physiological[13], nutrition[22], hormones[12] during the anagen phase of the hair cycle. However, the genes specifically controlling the size of DP and matrix remain elusive. In the present study, the up-regulated genes in the super fine wool group included 5 genes associated with the cell cycle and apoptosis: *GSDMA3*[56], *HSPA2*[57], *RPS27A*[58], *PDCD6IP*[59], *DAP*[60], *CCNA2*[61] and *FOS*[62]; the down-regulated genes included 7 genes associated with promotingfollicle cell proliferation and differentiation: *KRT1*[63,64], *KRT7*[65,66], *HSD11B1*[67,68],

S100A8[41,69,70], *NT5C3 L*[71] and *DNAJC12*[72].These genes may through promoting HF cell apoptosis, inhibiting follicle cell proliferation and differentiation, thereby reduce the WFD.

White adiposities, dermal fibroblasts, smooth muscle and endothelial cells of the vasculature, neurons and resident immune cells also surround the HFs in the skin.Recent studies show that the growth of wool fiber is not only affected by the proliferation and differentiation of follicular cells but is also affected by other cells around the follicle[73,74].The layer of intradermal adipocytes expands as the HF enters its anagen stage and then thins during telogen[75,76].FATP4$^{-/-}$, Dgat1$^{-/-}$, or Dgat2$^{-/-}$ Early B cell factor 1 (Ebf1$^{-/-}$) mice have decreased intradermal adipose tissue due to defects in lipid accumulation in mature adipocytes[77-79].Interestingly, these mice also display abnormalities in skin structure and function, such as hair loss and epidermal hyperplasia.In the present study, in the super fine wool group, it was also determined that the genes related to fat synthesis (*ABCC11*[80], *GLYATL2*[81], and *ACSM3*[82]) were significantly down-regulated and that *LIPK*, which is involved in lipolysis, was significantly up-regulated[83].

Genes with NATs are very common in animal and plant genomes, accounting for 7% to 9% of transcripts in plants and 5% to 30% of transcripts in animals, except for nematodes[84].The present study indicated that more than 30% of the genes had NATs, which is consistent with the findings of the above mentioned studies.In the present study, correlation analysis was conducted on the sense and antisense transcripts in the DGEs of skin, and the results showed that the correlation coefficients between the sense and antisense transcripts were between 0. 533 and 0. 557, indicating that the WFD maybe were regulated by both sense and antisense transcripts.NATs regulate sense transcripts through positive or negative feedback by forming a sense-antisense bidirectional structure[85].In this study, 3 out of 40 significantly differentially expressed genes had significantly different NATs: *AQP8*, Bos d2, and *SPRR* (type II small proline-rich protein).*AQP8* belongs to the aquaporin subfamily of the *AQP* family[86]; it is located mainly in a variety of tissues and on the surfaces of acinar cells in various glands and plays important roles in gland secretion and the transportation of water, urea, ammonia and hydrogen peroxide[87-90].*Bos d* 2 belongs to the lipocalin family of proteins[91].In the skin sections, *Bos d2* was found in the secretory cells of apocrine sweat glands and the basement membranes of the epithelium and HF[92].It is assumed that *Bos d2* is a pheromone carrier.[92].The *SPRR* proteins comprise a subclass of specific cornified envelope precursors encoded by a multigene family clustered within the epidermal differentiation complex region[93].Two *SPRR1*, seven *SPRR2*, one *SPRR3* and one *SPRR4* genes are located within a 300-kb area of the EDC[94,95] and are expressed in the epidermis, HFs and capillaries[96,97].Several studies have suggested that the *SPRRs* are related to increased epithelial proliferation and to malignant processes[98].Other than functioning as structural proteins, *SPRRs* also regulate gene expression levels, inhibit cell proliferation and promote differentiation[99].The present study also found that *AQP8*, *Bos d2*, *SPRR* and their antisense transcripts were significantly down-regulated in the super fine wool group, suggesting that *AQP8*, *Bos d2* and *SPRR* and their antisense transcripts were regulated by positive feedback. However, the mechanism by which *AQP8*, *Bos d2* and *SPRR* regulate theWFD needs to be further studied.

MATERIALS AND METHODS

Sheep skin sampling

Gansu Alpine fine wool sheep were bred in the Huang Cheng District of Gansu Province, China, by cross breeding Mongolian or Tibetan sheep with Xinjiang FineWool sheep and then with some fine wool sheep breeds from the Union of Soviet Socialist Republics, such as Caucasian sheep and Salsk sheep. The breed was approved by the Gansu provincial government in 1980. Gansu Alpine fine wool sheep were obtained from a sheep stud farm located in Zhangye city, Gansu Province. All experimental and surgical procedures were approved by the Institutional Animal Care and Use Committee, Lanzhou Institute of Husbandry and Pharmaceutical Sciences, Peoples Republic of China. Six unrelated 3 years old ewes at different WFD, and also as different DP size were selected and divided into super fine wool group (S) (WFD = (18.0±0.5) μm; Diameter of secondary DP size = (3.2±0.2) μm) and fine wool group (F) (WFD = (23.0±0.5) μm; Diameter of secondary DP size = (4.1±0.2) μm) (Fig 3). A piece of midside skin (2 mm in diameter) was collected via punch skin biopsy under local anesthesia using 1% procaine hydrochloride immediately placed in liquid nitrogen and stored at −80℃ for subsequent analysis.

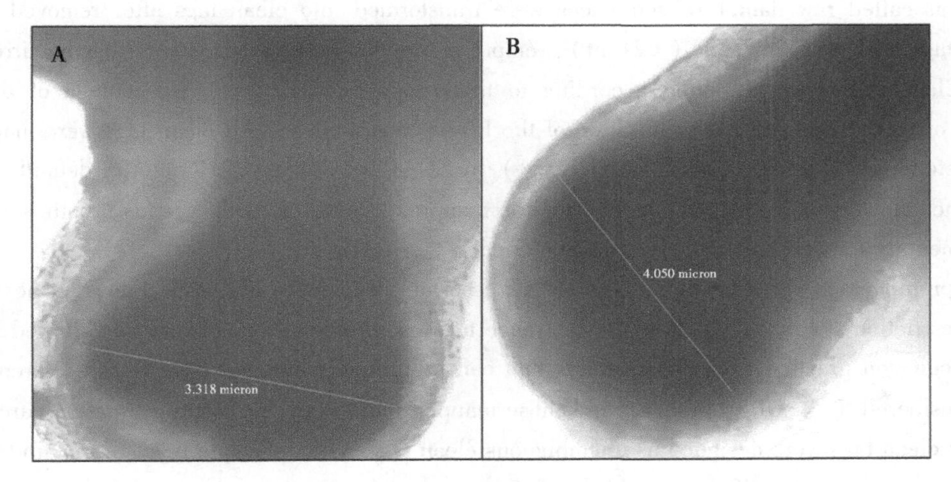

Fig. 3 The size of secondary DP cells between two groups

The diameter of super fine wool sheep secondary DP cells = (3.2±0.2) μm (A), the diameter of fine wool sheep secondary DP cells = (4.1±0.2) μm (B).

doi: 10.1371/journal.pone.0129249.g003

Total RNA extraction, library construction and deep sequencing

Total RNA was isolated from the tissues using the RNeasy Maxi Kit (Qiagen, Hilden, GER) according to the manufacturer instructions. RNA quality was verified using a 2100 Bioanalyzer RNA Nanochip (Agilent, Santa Clara, CA, USA), and the RNA Integrity Number (RIN) value was >8.5. Then, the RNA was quantified using the Nano Drop ND−2000 Spectrophotometer (NanoDrop, Wilmington, DE, USA). Sequence tags were prepared using the Illumina Digital Gene Expression Tag Profiling Kit, according to the manufacturer's protocol. Sequencing libraries were prepared from 1μg of total RNA using reagents from the NlaIII Digital Gene Expression Tag Profiling kit (Illumina Inc., San Diego, CA, USA). mRNA was captured on magnetic oligo (dT) beads and reverse transcribed into double-stranded cDNA (SuperScript II, Invitrogen, Carlsbad, CA, USA). The cDNA was cleaved using the restriction enzyme *NlaIII*. An adapter sequence containing the recognition sequence for the restriction enzyme *MmeI* was ligated to the *NlaIII* cleavage sites. The adapter-ligated cDNA was digested with *MmeI* to release the cDNA from the magnetic bead, while leaving 17 bp of sequence in the fragment. The fragments were dephosphorylated and purified by phenol.chloroform. A second adapter was ligated at the *MmeI* cleavage sites. Adapter-ligated cDNA fragments were amplified by PCR, and after 15 cycles of linear PCR amplification, 95 bp fragments were purified by 6% TBE PAGE gel electrophoresis, denatured and the single-chain molecules were fixed onto the Illumina Sequencing Chip. A single-molecule cluster sequencing template was created through in situ amplification. Nucleotides labeled with different colors were used to perform sequencing by the sequencing by synthesis method; each tunnel can generate millions of raw reads with a sequencing length of 35 bp.

Determination of gene expression levels and detection of DEGs and NATs

Sequencing-received raw image data were transformed by base culling into sequence data, which was called raw data. Raw sequences were transformed into clean tags after removed all low quality tags such as short tags (<21 nt), empty reads, and singletons (tags that occurred only once). Clean tags were classified according to their copy number (as a percentage of the total number of clean tags) and the saturation of the library was analyzed. All clean tags were mapped to sheep reference sequences (version: Oarv3. 1) by SOAP2 (version: 2. 21) with default parameters, and allowing one mismatches. To monitor mapping events on both strands, both sense and complementary antisense sequences were included[100].

A preprocessed database of all possible 17 base-long sequences of Oarv3. 1 located next to the *NlaIII* restriction site was created as the reference tags, and only one mismatch was allowed. Following the common practice, only clean reads that can be uniquely mapped back to the reference tags were considered ("−r 0" option), and those mapped more than one location were discarded. Remainder clean tags were designed as unambiguous clean tags. When multiple types of remainder clean tags were aligned to the different positions of the same gene, the gene expression levels were represented by the summation of all. The number of unambiguous clean tags for each gene was calculated and then normalized to TPM (number of transcripts per millionclean tags) = clean tag number corresponding to each gene number of total clean labels in the sample×1 000 000[101,102].

All TPM of 3 samples each group were integrated, and NOISeq (version: 2. 8. 0)[103] with default parameters, which is a novel nonparametric approach for the identification of DEGs and NATs, was used to detect the differentially expressed transcripts between super fine wool group (S) and fine wool group (F). To measure expression level changes between two groups, NOISeq takes into consideration two statistics: M (the $\log_2$−ratio of the two conditions) and D (the absolute value of the difference between conditions). The probability thresholds were $P \geq 0.8$ and the TMM value in the lower expressed sample was ≥ 1. The higher the probability, the greater the change in expression between two groups. Using a probability threshold of 0. 8 means that the gene is 4 times more likely to be differentially expressed than non-differentially expressed[104].

Strand-specific real-time quantitative RT-PCR

To confirm the differentially expressed sense and antisense transcripts between super fine wool group and fine wool group, ten genes were randomly selected to verify their expression levels of gene and NATs transcripts in skin by strand-specific qRT-PCR according to the protocol described in Haddad et al. (2007)[105]. Primers for real-time PCR were designed with Primer Express 3.0 (Applied Biosystems) (Table 5). GAPDH was used as a reference control. Real time PCR was performed using SYBR Green master mix (TianGen) on the CFX96 Real-Time System (BioRAD, USA). The reaction was performed using the following conditions: denaturation at 95℃ for 3 min, followed by 40 cycles of amplification (95℃ for 30s, 60℃ for 30s, and 72℃ for 30s). Relative expression was calculated using the delta-delta-Ct method.

Table 5 Relevant information of gene and primer sequences for strand-specific RT-PCR

Genes name	Primer sequences (5′→3′)	GenBank accession No	Produce size (bp)
Bos d2	ACAGTAGTGGCACAGAGAGC	XM_ 004021893. 1	136
	TTGTGGTCTTCTCGCCATCA		
SPRR	CAAAAATGCCCTCCTGTGCC	NM_ 001009773. 1	91
	CCTGACCAGATAAAAGCTGATGC		
AQP8	TCATCCTGACGACACTGCTG	XM_ 004020851. 1	146
	ATTCATGCACGCTCCAGACA		
KRT1	CCCTGGATGTGGAGATTGCC	XM_ 004006320. 1	119
	ACTGATGCTGGTGTGACTGG		
KRT7	ACATCAAGCTGGCTCTGGAC	XM_ 004006333. 1	108
	CCACAGAGATGTTCACGGCT		
S100A8	GCTGACGGATCTGGAGAGTG	XM_ 004002523. 1	163
	GAACCAAGTGTCCGCATCCT		
GAPDH	GCTGAGTACGTGGTGGAGTC	NM_ 001190390. 1	136
	GGTTCACGCCCATCACAAAC		
AQP8-NAT	CCTTGGGAAATCCTTGCAGC		137
	GAAAAGCCCGTTGCAGACTC		
Bos d2-NAT	TTGGCGGAAGTATGTCGCAG		112
	CAGCTCAGCCAGAATCGTCA		
SPRR-NAT	CCATGACCCGCTTTGAAGA		132
	CAATGCCAGCAGAAGTGCC		

doi: 10. 1371/journal.pone.0129249. t005

GO and KEGG enrichment analysis of differentially expressed transcripts

Gene ontology (GO), an international standardized gene functional classification system, offers a dynamic-updated controlled vocabulary and strictly defined concept to comprehensively describe the properties of genes and their products in any organism[106]. Kyoto Encyclopedia of Genes and Genomes (KEGG) database is a database resource for understanding functions and utilities of the biological system, such as the cell, the organism and the ecosystem from molecular information, especially for large-scale molecular datasets generated by genome sequencing and other high throughput experimental technologies (http: //www.genome.jp/kegg/)[107]. all DEGs and NATs were mapped to GO-terms in GO database, looking for significantly enriched GO terms in DEGs comparing to the genome background GOseq R package (version: 1.18.0)[108], while significantly enriched metabolic pathways or signal transduction pathways in DEGs and NATs were identified via pathway enrichment analysis using KEGG, public pathway-related database, and comparing with the whole genome background by KOBAS (version: 2.0)[109].

In all tests, P-values were calculated using Benjamini-corrected modified Fisher's exact test and ≤0.05 was taken as a threshold of significance.The calculating formula is:

$$P = 1 - \sum_{i=0}^{m-1} \frac{\binom{M}{i}\binom{N-M}{n-i}}{\binom{N}{n}}$$

Where N is the number of all genes with GO or KEGG annotation; n is the number of DEGs or NATs in N; Mis the number of all genes that are annotated to the certain GO terms or specific pathways; m is the number of DEGs or NATs in M.

SUPPORTING INFORMATION

S1 Table.Summary of tags mapping to gene & genome and unknown tag. (DOCX)

S2 Table. Summary of tags mapping to anti-sense genes and tag-mapped anti-sense genes. (DOCX)

ACKNOWLEDGMENTS

This work was supported by the Central Level, Scientific Research Institutes for Basic R & D Special Fund Business (Grant no.1610322015014); the Earmarked Fund for Modern China Wool & Cashmere Technology Research System (Grant no.nycytx-40-3); and the National Natural Science Foundation for Young Scholars of China (Grant no.31402057).

AUTHOR CONTRIBUTIONS

Conceived and designed the experiments: YJY TTG BHY. Performed the experiments: YJY TTG JBL XPS. Analyzed the data: YJY TTG CY. Contributed reagents/materials/analysis tools: RLF CEN JG.Wrote the paper: YJY BHY.

References

[1] Galbraith H. Animal fibre: connecting science and production Foreword. Animal.2010, 4 (9): 1447-1450. doi: 10.1017/S1751731110000741 WOS: 000281263500001.

[2] Liu H, Zhou Z-Y, Malcolm B. China's Wool Import Demand: Implications for Australia. Australasian Agribusiness Review.2011, 2011 (1442-6951): 19.

[3] Iman NY, Johnson CL, Russell WC, Stobart RH.Estimation of genetic parameters for wool fiber diameter measures.Journal of animal science.1992, 70 (4): 1110-5. PMID: 1582941.

[4] Chi W, Wu E, Morgan BA.Dermal papilla cell number specifies hair size, shape and cycling and its reduction causes follicular decline.Development.2013, 140 (8): 1676-83. doi: 10. 1242/dev.090662 PMID: 23487317; PubMed Central PMCID: PMC3621486.

[5] Hardy MH, Lyne AG.The pre-natal development of wool follicles in Merino sheep.Australian Journal of Biological Sciences.1956, 9 (9): 423-41.

[6] Moore GP, Jackson N, Isaacs K, Brown G.Pattern and morphogenesis in skin.Journal of theoretical biology.1998, 191 (1): 87-94. doi: 10. 1006/jtbi.1997. 0567 PMID: 9593659.

[7] Carter HB, Hardy MH.Studies in the biology of the skin and fleece of sheep.Commonwealth of Australia Council for Scientific and Industrial Research.1947, 215 (215).

[8] Driskell RR, Clavel C, Rendl M, Watt FM.Hair follicle dermal papilla cells at a glance.Journal of cell science.2011, 124 (Pt 8): 1179-82. doi: 10. 1242/jcs.082446 PMID: 21444748; PubMed Central PMCID: PMC3115771.

[9] Enshell-Seijffers D, Lindon C, Kashiwagi M, Morgan BA.beta-catenin activity in the dermal papilla regulates morphogenesis and regeneration of hair. Developmental cell. 2010, 18 (4): 633-42. doi: 10. 1016/j.devcel.2010. 01. 016 PMID: 20412777; PubMed Central PMCID: PMC2893731.

[10] Moore GP, Jackson N, Lax J.Evidence of a unique developmental mechanism specifying both wool follicle density and fibre size in sheep selected for single skin and fleece characters. Genetical research. 1989, 53 (1): 57-62. PMID: 2714646.

[11] Yano K, Brown LF, Detmar M.Control of hair growth and follicle size by VEGF-mediated angiogenesis. The Journal of clinical investigation. 2001, 107 (4): 409-17. doi: 10. 1172/JCI11317 PMID: 11181640; PubMed Central PMCID: PMC199257.

[12] Elliott K, Stephenson TJ, Messenger AG.Differences in hair follicle dermal papilla volume are due to extracellular matrix volume and cell number: implications for the control of hair follicle size and androgen responses. The Journal of investigative dermatology. 1999, 113 (6): 873-7. doi: 10. 1046/j. 1523-1747. 1999. 00797. x PMID: 10594724.

[13] Otberg N, Richter H, Schaefer H, Blume-Peytavi U, SterryW, Lademann J. Variations of hair follicle size and distribution in different body sites.The Journal of investigative dermatology.2004, 122 (1): 14-9. doi: 10. 1046/j.0022-202X.2003. 22110. x PMID: 14962084.

[14] Scobie. DR, Young. SR, editors. The relationship between wool follicle density and fibre diameter is curvilinear.Proceedings of the New Zealand Society of Animal Production; 2000: New Zealand Society of Animal Production.

[15] Adelson DL, Hollis DE, Brown GH.Wool fibre diameter and follicle density are not specified simultaneously during wool follicle initiation. Australian Journal of Agricultural Research. 2002, 53 (9): 1003-1009.

[16] Alcaraz MV, Villena A, Perez de Vargas I. Quantitative study of the human hair follicle in normal scalp and androgenetic alopecia. Journal of cutaneous pathology. 1993, 20 (4): 344-9. PMID: 8227610.

[17] Ibrahim L, Wright EA.A quantitative study of hair growth using mouse and rat vibrissal follicles.I.Dermal papilla volume determines hair volume.Journal of embryology and experimental morphology.1982, 72: 209-24. PMID: 7183740.

[18] Adelson DL, Hollis DE, Brown GH.Wool fibre diameter and follicle density are not specified simultaneously during wool follicle initiation. Aust J Agr Res. 2002, 53 (9): 1003. 9. doi: 10. 1071/

Ar01200 WOS: 000178095800002.

[19] Adelson DL, Cam GR, DeSilva U, Franklin IR.Gene expression in sheep skin and wool (hair).Genomics.2004, 83 (1): 95-105. PMID: 14667813.

[20] Norris BJ, Bower NI, Smith WJM, Cam GR, Reverter A.Gene expression profiling of ovine skin and wool follicle development using a combined ovine-bovine skin cDNA microarray.Aust J Exp Agr.2005, 45 (7-8): 867-77. doi: 10. 1071/Ea05050 WOS: 000231472300017.

[21] Lei M, Guo H, Qiu W, Lai X, Yang T, Widelitz RB, et al. Modulating hair follicle size with Wnt10b/DKK1 during hair regeneration. Experimental dermatology. 2014, 23 (6): 407 - 13. doi: 10. 1111/exd.12416 PMID: 24750467.

[22] Seiberg M, Liu JC, Babiarz L, Sharlow E, Shapiro S.Soymilk reduces hair growth and hair follicle dimensions.Experimental dermatology.2001, 10 (6): 405-13. PMID: 11737259.

[23] Jager M, Ott CE, Grunhagen J, Hecht J, Schell H, Mundlos S, et al.Composite transcriptome assembly of RNA-seq data in a sheep model for delayed bone healing. Bmc Genomics. 2011, 12: 158. doi: 10. 1186/1471 - 2164 - 12 - 158 PMID: 21435219; PubMed Central PMCID: PMC3074554.

[24] Botchkarev VA, Paus R.Molecular biology of hair morphogenesis: development and cycling.Journal of experimental zoology Part B, Molecular and developmental evolution.2003, 298 (1): 164-80. doi: 10. 1002/jez.b.33 PMID: 12949776.

[25] Schmidt-Ullrich R, Paus R. Molecular principles of hair follicle induction and morphogenesis. BioEssays: news and reviews in molecular, cellular and developmental biology.2005, 27 (3): 247-61. doi: 10. 1002/bies.20184 PMID: 15714560.

[26] Bostjancic E, Glavac D.Importance of microRNAs in skin morphogenesis and diseases. Acta dermatovenerologica Alpina, Panonica, et Adriatica.2008, 17 (3): 95-102. PMID: 18853072.

[27] Wang X, Tredget EE, Wu Y.Dynamic signals for hair follicle development and regeneration.Stem cells and development.2012, 21 (1): 7-18. doi: 10. 1089/scd.2011. 0230 PMID: 21787229.

[28] Cadau S, Rosignoli C, Rhetore S, Voegel J, Parenteau-Bareil R, Berthod F.Early stages of hair follicle development: a step by step microarray identity.European journal of dermatology: EJD.2013. doi: 10. 1684/ejd.2013. 1972 PMID: 23567059.

[29] Millar SE. Molecular mechanisms regulating hair follicle development. J Invest Dermatol. 2002, 118 (2): 216-25. doi: 10. 1046/j.0022-202x.2001. 01670. x PMID: 11841536.

[30] Rogers GE.Biology of the wool follicle: an excursion into a unique tissue interaction system waiting to be re-discovered. Experimental dermatology. 2006, 15 (12): 931 - 49. WOS: 000241766500001. PMID: 17083360.

[31] Purvis IW, Franklin IR.Major genes and QTL influencing wool production and quality: a review.Genetics, selection, evolution: GSE. 2005, 37 Suppl 1: S97 - 107. doi: 10. 1051/gse: 2004028 PMID: 15601598; PubMed Central PMCID: PMC3226268.

[32] Carthew RW, Sontheimer EJ.Origins and Mechanisms of miRNAs and siRNAs.Cell.2009, 136 (4): 642-55. doi: 10. 1016/j.cell.2009. 01. 035 WOS: 000263688200016. PMID: 19239886.

[33] Carmichael GG. Antisense starts making more sense. Nat Biotechnol. 2003, 21 (4): 371 - 2. doi: 10. 1038/Nbt0403-371 WOS: 000182082400015. PMID: 12665819.

[34] Katayama S, Tomaru Y, Kasukawa T, Waki K, Nakanishi M, Nakamura M, et al. Antisense transcription in the mammalian transcriptome. Science. 2005, 309 (5740): 1564 - 6. doi: 10. 1126/science.1112009 WOS: 000231715000050. PMID: 16141073.

[35] Carninci P, Kasukawa T, Katayama S, Gough J, Frith MC, Maeda N, et al. The transcriptional

landscape of the mammalian genome.Science.2005, 309 (5740): 1559-63. doi: 10. 1126/science. 1112014 WOS: 000231715000049. PMID: 16141072.

[36] Clark MB, Amaral PP, Schlesinger FJ, Dinger ME, Taft RJ, Rinn JL, et al.The Reality of Pervasive Transcription. Plos Biol. 2011, 9 (7). ARTN e1000625 doi: 10. 1371/journal. pbio. 1000625 WOS: 000293219800001.

[37] Faghihi MA, Modarresi F, Khalil AM, Wood DE, Sahagan BG, Morgan TE, et al.Expression of a noncoding RNA is elevated in Alzheimer's disease and drives rapid feed-forward regulation of beta-secretase. Nat Med. 2008, 14 (7): 723 - 30. doi: 10. 1038/Nm1784 WOS: 000257452700019. PMID: 18587408.

[38] Gupta RA, Shah N, Wang KC, Kim J, Horlings HM, Wong DJ, et al.Long non-coding RNA HOTAIR reprograms chromatin state to promote cancer metastasis. Nature. 2010, 464 (7291): 1071-U148. doi: 10. 1038/Nature08975 WOS: 000276635000045. PMID: 20393566.

[39] Pariset L, Chillemi G, Bongiorni S, Spica VR, Valentini A. Microarrays and high-throughput transcriptomic analysis in species with incomplete availability of genomic sequences.New Biotechnol.2009, 25 (5): 272-9. WOS: 000267621800003. doi: 10. 1016/j.nbt.2009. 03. 013 PMID: 19446516.

[40] Shendure J.The beginning of the end for microarrays? Nat Methods.2008, 5 (7): 585-7. WOS: 000257166700006. doi: 10. 1038/nmeth0708-585 PMID: 18587314.

[41] Jiang Y, Xie M, Chen W, Talbot R, Maddox JF, Faraut T, et al.The sheep genome illuminates biology of the rumen and lipid metabolism.Science.2014, 344 (6188): 1168-73. doi: 10. 1126/science.1252806 PMID: 24904168; PubMed Central PMCID: PMC4157056.

[42] Martin JA, Wang Z.Next-generation transcriptome assembly.Nat Rev Genet.2011, 12 (10): 671-82. WOS: 000294944500008. doi: 10. 1038/nrg3068 PMID: 21897427.

[43] Pritchard CC, Cheng HH, Tewari M.MicroRNA profiling: approaches and considerations.Nat Rev Genet.2012, 13 (5): 358-69. doi: 10. 1038/nrg3198 PMID: 22510765.

[44] Katayama S, Tomaru Y, Kasukawa T, Waki K, Nakanishi M, Nakamura M, et al.Antisense transcription in the mammalian transcriptome. Science. 2005, 309 (5740): 1564 - 6. doi: 10. 1126/science.1112009 PMID: 16141073.

[45] Galbraith H.Fundamental hair follicle biology and fine fibre production in animals.Animal: an international journal of animal bioscience. 2010, 4 (9): 1490-509. doi: 10. 1017/S175173111000025X PMID: 22444696.

[46] Nagalakshmi U, Wang Z, Waern K, Shou C, Raha D, Gerstein M, et al.The transcriptional landscape of the yeast genome defined by RNA sequencing.Science.2008, 320 (5881): 1344-9. WOS: 000256441100046. doi: 10. 1126/science.1158441 PMID: 18451266.

[47] Cloonan N, Forrest ARR, Kolle G, Gardiner BBA, Faulkner GJ, Brown MK, et al.Stem cell transcriptome profiling via massive-scale mRNA sequencing. Nat Methods. 2008, 5 (7): 613-9. doi: 10. 1038/Nmeth.1223 WOS: 000257166700014. PMID: 18516046.

[48] Wang Z, Gerstein M, Snyder M.RNA-Seq: a revolutionary tool for transcriptomics.Nat Rev Genet. 2009, 10 (1): 57-63. WOS: 000261866500012. doi: 10. 1038/nrg2484 PMID: 19015660.

[49] Denoeud F, Aury JM, Da Silva C, Noel B, Rogier O, Delledonne M, et al. Annotating genomes with massive-scale RNA sequencing.Genome biology.2008, 9 (12).WOS: 000263074100014.

[50] Wang ET, Sandberg R, Luo SJ, Khrebtukova I, Zhang L, Mayr C, et al.Alternative isoform regulation in human tissue transcriptomes.Nature.2008, 456 (7221): 470-6. WOS: 000261170500031. doi: 10. 1038/nature07509 PMID: 18978772.

[51] Maher CA, Kumar-Sinha C, Cao XH, Kalyana-Sundaram S, Han B, Jing XJ, et al.Transcriptome

sequencing to detect gene fusions in cancer. Nature. 2009, 458 (7234): 97 - U9. WOS: 000263836000040.doi: 10. 1038/nature07638 PMID: 19136943.

[52] Legue E, Nicolas JF.Hair follicle renewal: organization of stem cells in the matrix and the role of stereotyped lineages and behaviors.Development.2005, 132 (18): 4143-54. doi: 10. 1242/Dev.01975 WOS: 000232579600012. PMID: 16107474.

[53] Sennett R, Rendl M.Mesenchymal-epithelial interactions during hair follicle morphogenesis and cycling. Seminars in cell & developmental biology.2012, 23 (8): 917-27.doi: 10.1016/j.semcdb.2012.08.011 PMID: 22960356; PubMed Central PMCID: PMC3496047.

[54] Purvis IW, Jeffery N.Genetics of fibre production in sheep and goats.Small Ruminant Res.2007, 70 (1): 42-7. WOS: 000246550200005.

[55] Botchkarev VA, Paus R.Molecular biology of hair morphogenesis: Development and cycling.J Exp Zool Part B.2003, 298B (1): 164-80. WOS: 000185162700011. PMID: 12949776.

[56] Lei M, Bai X, Yang T, Lai X, Qiu W, Yang L, et al.Gsdma3 is a new factor needed for TNF-alpha-mediated apoptosis signal pathway in mouse skin keratinocytes. Histochemistry and cell biology. 2012, 138 (3): 385-96. doi: 10. 1007/s00418-012-0960-1 PMID: 22585037.

[57] Filipczak PT, PiglowskiW, Glowala-Kosinska M, Krawczyk Z, Scieglinska D.HSPA2 overexpression protects V79 fibroblasts against bortezomib-induced apoptosis.Biochemistry and cell biology = Biochimie et biologie cellulaire.2012, 90 (2): 224-31. doi: 10. 1139/o11-083 PMID: 22397456.

[58] Wang H, Yu J, Zhang L, Xiong Y, Chen S, Xing H, et al.RPS27a promotes proliferation, regulates cell cycle progression and inhibits apoptosis of leukemia cells. Biochem Biophys Res Commun. 2014, 446 (4): 1204-10. doi: 10. 1016/j.bbrc.2014. 03. 086 PMID: 24680683.

[59] Strappazzon F, Torch S, Chatellard-Causse C, Petiot A, Thibert C, Blot B, et al.Alix is involved in caspase 9 activation during calcium-induced apoptosis.Biochem Bioph Res Co.2010, 397 (1): 64-69. doi: 10. 1016/j.bbrc.2010. 05. 062 WOS: 000279292800012.

[60] Levy-Strumpf N, Kimchi A. Death associated proteins (DAPs): from gene identification to the analysis of their apoptotic and tumor suppressive functions. Oncogene. 1998, 17 (25): 3331 - 40. WOS: 000078048200014. PMID: 9916995.

[61] Das E, Jana NR, Bhattacharyya NP.MicroRNA-124 targets CCNA2 and regulates cell cycle in STHdh (Q111) /Hdh (Q111) cells.Biochem Bioph Res Co.2013, 437 (2): 217-24. doi: 10. 1016/j. bbrc.2013. 06. 041 WOS: 000323018100006. PMID: 23796713.

[62] Reiner G, Heinricy L, Brenig B, Geldermann H, Dzapo V.Cloning, structural organization, and chromosomal assignment of the porcine c-fos proto-oncogene, FOS. Cytogenetics and cell genetics. 2000, 89 (1-2): 59-61. 15565. PMID: 10894939.

[63] Fonseca DJ, Rojas RF, Vergara JI, Rios X, Uribe C, Chavez L, et al.A severe familial phenotype of Ichthyosis Curth-Macklin caused by a novel mutation in the KRT1 gene.Brit J Dermatol.2013, 168 (2): 456-8. doi: 10. 1111/j.1365-2133. 2012. 11181. x WOS: 000314470600047.

[64] Glotzer DJ, Zelzer E, Olsen BR.Impaired skin and hair follicle development in Runx2 deficient mice. Developmental biology. 2008, 315 (2): 459 - 73. doi: 10. 1016/j. ydbio. 2008. 01. 005 PMID: 18262513; PubMed Central PMCID: PMC2280036.

[65] Sandilands A, Smith FJ, Lunny DP, Campbell LE, Davidson KM, MacCallum SF, et al. Generation and characterisation of keratin 7 (K7) knockout mice. PloS one. 2013, 8 (5): e64404. doi: 10. 1371/journal. pone. 0064404 PMID: 23741325; PubMed Central PMCID: PMC3669307.

[66] Smith FJ, Porter RM, Corden LD, Lunny DP, Lane EB, McLean WH.Cloning of human, murine,

and marsupial keratin 7 and a survey of K7 expression in the mouse. Biochem Biophys Res Commun. 2002, 297 (4): 818-27. PMID: 12359226.

[67] Tiganescu A, Walker EA, Hardy RS, Mayes AE, Stewart PM. Localization, Age- and Site-Dependent Expression, and Regulation of 11 beta-Hydroxysteroid Dehydrogenase Type 1 in Skin. J Invest Dermatol. 2011, 131 (1): 30 - 6. doi: 10.1038/Jid.2010.257 WOS: 000285290300010. PMID: 20739946.

[68] Yang K, Smith CL, Dales D, Hammond GL, Challis JR. Cloning of an ovine 11 beta-hydroxysteroid dehydrogenase complementary deoxyribonucleic acid: tissue and temporal distribution of its messenger ribonucleic acid during fetal and neonatal development. Endocrinology. 1992, 131 (5): 2120-6. doi: 10.1210/endo.131.5.1425412 PMID: 1425412.

[69] Korndorfer IP, Brueckner F, Skerra A. The crystal structure of the human (S100A8/S100A9) (2) heterotetramer, calprotectin, illustrates how conformational changes of interacting alpha-helices can determine specific association of two EF-hand proteins. J Mol Biol. 2007, 370 (5): 887-98. doi: 10.1016/j.jmb.2007.04.065 WOS: 000247904500008. PMID: 17553524.

[70] Nacken W, Roth J, Sorg C, Kerkhoff C. S100A9/S100A8: Myeloid representatives of the S100 protein family as prominent players in innate immunity. Microsc Res Techniq. 2003, 60 (6): 569-80. doi: 10.1002/Jemt.10299 WOS: 000181885200005. PMID: 12645005.

[71] Shen Z, Fahey JV, Bodwell JE, Rodriguez-Garcia M, Rossoll RM, Crist SG, et al. Estradiol Regulation of Nucleotidases in Female Reproductive Tract Epithelial Cells and Fibroblasts. PloS one. 2013, 8 (7). ARTN e69854 doi: 10.1371/journal.pone.0069854 WOS: 000322433300071.

[72] De Bessa SA, Salaorni S, Patrao DFC, Neto MM, Brentani MM, Nagai MA. JDP1 (DNAJC12/Hsp40) expression in breast cancer and its association with estrogen receptor status. Int J Mol Med. 2006, 17 (2): 363-7. WOS: 000234760900025. PMID: 16391838.

[73] Plikus MV, Baker RE, Chen CC, Fare C, de la Cruz D, Andl T, et al. Self-Organizing and Stochastic Behaviors During the Regeneration of Hair Stem Cells. Science. 2011, 332 (6029): 586-9. doi: 10.1126/science.1201647 WOS: 000289991100048. PMID: 21527712.

[74] Botchkarev VA, Yaar M, Peters EMJ, Raychaudhuri SP, Botchkareva NV, Marconi A, et al. Neurotrophins in skin biology and pathology. J Invest Dermatol. 2006, 126 (8): 1719-27. doi: 10.1038/sj.jid.5700270 WOS: 000241359100008. PMID: 16845411.

[75] Chase HB, Montagna W, Malone JD. Changes in the skin in relation to the hair growth cycle. The Anatomical record. 1953, 116 (1): 75-81. PMID: 13050993.

[76] Hausman GJ, Martin RJ. The development of adipocytes located around hair follicles in the fetal pig. Journal of animal science. 1982, 54 (6): 1286-96. PMID: 7107536.

[77] Herrmann T, van der Hoeven F, Grone HJ, Stewart AF, Langbein L, Kaiser I, et al. Mice with targeted disruption of the fatty acid transport protein 4 (Fatp 4, Slc27a4) gene show features of lethal restrictive dermopathy. The Journal of cell biology. 2003, 161 (6): 1105-15. doi: 10.1083/jcb.200207080 PMID: 12821645; PubMed Central PMCID: PMC2173002.

[78] Chen HC, Smith SJ, Tow B, Elias PM, Farese RV Jr. Leptin modulates the effects of acyl CoA: diacylglycerol acyltransferase deficiency on murine fur and sebaceous glands. The Journal of clinical investigation. 2002, 109 (2): 175-81. doi: 10.1172/JCI13880 PMID: 11805129; PubMed Central PMCID: PMC150839.

[79] Stone SJ, Myers HM, Watkins SM, Brown BE, Feingold KR, Elias PM, et al. Lipopenia and skin barrier abnormalities in DGAT2-deficient mice. J Biol Chem. 2004, 279 (12): 11767-76. doi: 10.1074/jbc.M311000200 PMID: 14668353.

[80] Yoshiura K, Kinoshita A, Ishida T, Ninokata A, Ishikawa T, Kaname T, et al.A SNP in the ABCC11 gene is the determinant of human earwax type. Nat Genet. 2006, 38 (3): 324-30. doi: 10.1038/Ng1733 WOS: 000235589600014. PMID: 16444273.

[81] Waluk DP, Schultz N, Hunt MC.Identification of glycine N-acyltransferase-like 2 (GLYATL2) as a transferase that produces N-acyl glycines in humans. Faseb J. 2010, 24 (8): 2795-803. doi: 10.1096/Fj.09-148551 WOS: 000285005400020. PMID: 20305126.

[82] Hagedorn C, Telgmann R, Brand E, Nicaud V, Schoenfelder J, Hasenkamp S, et al.SAH gene variants are associated with obesity-related hypertension in caucasians-the PEGASE study-.Eur Heart J. 2007, 28: 249-50. WOS: 000208702201231.

[83] Holmes RS, Cox LA, VandeBerg JL.Comparative studies of mammalian acid lipases: Evidence for a new gene family in mouse and rat (Lipo). Comp Biochem Phys D. 2010, 5 (3): 217-26. doi: 10.1016/j.cbd.2010.05.004 WOS: 000281314400005.

[84] Lapidot M, Pilpel Y. Genome-wide natural antisense transcription: coupling its regulation to its different regulatory mechanisms. Embo Rep. 2006, 7 (12): 1216-22. doi: 10.1038/sj.embor.7400857 WOS: 000242795500009. PMID: 17139297.

[85] Wahlestedt C.Natural antisense and noncoding RNA transcripts as potential drug targets.Drug Discov Today.2006, 11 (11.12): 503-8. doi: 10.1016/j.drudis.2006.04.013 WOS: 000238523500006.

[86] Ishibashi K, Kuwahara M, Kageyama Y, Tohsaka A, Marumo F, Sasaki S.Cloning and functional expression of a second new aquaporin abundantly expressed in testis.Biochem Bioph Res Co.1997, 237 (3): 714-8. doi: 10.1006/bbrc.1997.7219 WOS: A1997XV79600045.

[87] Ma TH, Yang BX, Verkman AS.Cloning of a novel water and urea-permeable aquaporin from mouse expressed strongly in colon, placenta, liver, and heart. Biochem Bioph Res Co. 1997, 240 (2): 324-8. doi: 10.1006/bbrc.1997.7664 WOS: A1997YG93500013.

[88] Holm LM, Jahn TP, Moller ALB, Schjoerring JK, Ferri D, Klaerke DA, et al.NH3 and NH4+ permeability in aquaporin-expressing Xenopus oocytes. Pflug Arch Eur J Phy. 2005, 450 (6): 415-28. doi: 10.1007/s00424-005-1399-1 WOS: 000232339900007.

[89] Calamita G, Moreno M, Ferri D, Silvestri E, Roberti P, Schiavo L, et al.Triiodothyronine modulates the expression of aquaporin-8 in rat liver mitochondria.J Endocrinol.2007, 192 (1): 111-20. doi: 10.1677/Joe-06-0058 WOS: 000244958100012. PMID: 17210748.

[90] Bienert GP, Moller ALB, Kristiansen KA, Schulz A, Moller IM, Schjoerring JK, et al. Specific aquaporins facilitate the diffusion of hydrogen peroxide across membranes. J Biol Chem. 2007, 282 (2): 1183-92. doi: 10.1074/jbc.M603761200 WOS: 000243295200044. PMID: 17105724.

[91] Mantyjarvi R, Parkkinen S, Rytkonen M, Pentikainen J, Pelkonen J, Rautiainen J, et al.Complementary DNA cloning of the predominant allergen of bovine dander: A new member in the lipocalin family.J Allergy Clin Immun.1996, 97 (6): 1297-303. doi: 10.1016/S0091-6749 (96) 70198-7 WOS: A1996VD60700018. PMID: 8648026.

[92] Rautiainen J, Rytkonen M, Syrjanen K, Pentikainen J, Zeiler T, Virtanen T, et al.Tissue localization of bovine dander allergen Bos d 2. J Allergy Clin Immun. 1998, 101 (3): 349-53. WOS: 000072593900012. PMID: 9525451.

[93] Gibbs S, Fijneman R, Wiegant J, van Kessel AG, van De Putte P, Backendorf C.Molecular characterization and evolution of the SPRR family of keratinocyte differentiation markers encoding small proline-rich proteins. Genomics. 1993, 16 (3): 630-7. doi: 10.1006/geno.1993.1240 PMID: 8325635.

[94] Ishida-Yamamoto A, Kartasova T, Matsuo S, Kuroki T, Iizuka H.Involucrin and SPRR are synthe-

sized sequentially in differentiating cultured epidermal cells.J Invest Dermatol.1997, 108 (1): 12-6. PMID: 8980279.

[95] Candi E, Melino G, Mei G, Tarcsa E, Chung SI, Marekov LN, et al. Biochemical, structural, and transglutaminase substrate properties of human loricrin, the major epidermal cornified cell envelope protein.J Biol Chem.1995, 270 (44): 26382-90. PMID: 7592852.

[96] Wang L, Baldwin RL 6th, Jesse BW. Identification of two cDNA clones encoding small proline-rich proteins expressed in sheep ruminal epithelium. The Biochemical journal. 1996, 317 (Pt 1): 225-33. PMID: 8694768; PubMed Central PMCID: PMC1217467.

[97] Akiyama M, Smith LT, Yoneda K, Holbrook KA, Shimizu H. Transglutaminase and major cornified cell envelope precursor proteins, loricrin, small proline-rich proteins 1 and 2, and involucrin are coordinately expressed in the sites defined to form hair canal in developing human hair follicle. Exp Dermatol.1999, 8 (4): 313-4. PMID: 10439243.

[98] Carregaro F, Stefanini AC, Henrique T, Tajara EH. Study of small proline-rich proteins (SPRRs) in health and disease: a review of the literature. Archives of dermatological research. 2013, 305 (10): 857-66. doi: 10.1007/s00403-013-1415-9 PMID: 24085571.

[99] Zimmermann N, Doepker MP, Witte DP, Stringer KF, Fulkerson PC, Pope SM, et al. Expression and regulation of small proline-rich protein 2 in allergic inflammation. American journal of respiratory cell and molecular biology. 2005, 32 (5): 428 - 35. doi: 10.1165/rcmb. 2004 - 0269OC PMID: 15731505.

[100] Wang QQ, Liu F, Chen XS, Ma XJ, Zeng HQ, Yang ZM. Transcriptome profiling of early developing cotton fiber by deep-sequencing reveals significantly differential expression of genes in a fuzzless/lintless mutant. Genomics. 2010, 96 (6): 369 - 76. doi: 10.1016/j. ygeno. 2010.08.009 PMID: 20828606.

[101] t Hoen PAC, Ariyurek Y, Thygesen HH, Vreugdenhil E, Vossen RHAM, de Menezes RX, et al. Deep sequencing-based expression analysis shows major advances in robustness, resolution and inter-lab portability over five microarray platforms. Nucleic acids research. 2008, 36 (21). WOS: 000261299700030.

[102] Morrissy AS, Morin RD, Delaney A, Zeng T, McDonald H, Jones S, et al. Next-generation tag sequencing for cancer gene expression profiling. Genome research. 2009, 19 (10): 1825-35. WOS: 000270389700015. doi: 10.1101/gr.094482.109 PMID: 19541910.

[103] Tarazona S, Garcia-Alcalde F, Dopazo J, Ferrer A, Conesa A. Differential expression in RNA-seq: A matter of depth. Genome research. 2011, 21 (12): 2213 - 23. WOS: 000297918600020. doi: 10.1101/gr.124321.111 PMID: 21903743.

[104] Xing K, Zhu F, Zhai LW, Liu HJ, Wang ZJ, Hou ZC, et al. The liver transcriptome of two full-sibling Songliao black pigs with extreme differences in backfat thickness. Journal of animal science and biotechnology.2014, 5. Artn 32 doi: 10.1186/2049-1891-5-32 WOS: 000339281900001.

[105] Haddad F, Qin AQX, Giger JM, Guo HY, Baldwin KM. Potential pitfalls in the accuracy of analysis of natural sense-antisense RNA pairs by reverse transcription-PCR. Bmc Biotechnol. 2007, 7. Artn 21 doi: 10.1186/1472-6750-7-21 WOS: 000246709200001.

[106] Ashburner M, Ball CA, Blake JA, Botstein D, Butler H, Cherry JM, et al. Gene ontology: tool for the unification of biology. The Gene Ontology Consortium. Nat Genet.2000, 25 (1): 25-9. doi: 10.1038/75556 PMID: 10802651; PubMed Central PMCID: PMC3037419.

[107] Kanehisa M, Araki M, Goto S, Hattori M, Hirakawa M, Itoh M, et al. KEGG for linking genomes to life and the environment. Nucleic acids research. 2008, 36: D480 - D4. doi: 10.1093/Nar/

Gkm882 WOS：000252545400086. PMID：18077471.

[108] Young MD，Wakefield MJ，Smyth GK，Oshlack A.Gene ontology analysis for RNA-seq：accounting for selection bias.Genome biology.2010，11（2）：R14. doi：10. 1186/gb-2010-11-2-r14 PMID：20132535；PubMed Central PMCID：PMC2872874.

[109] Mao XZ，Cai T，Olyarchuk JG，Wei LP.Automated genome annotation and pathway identification using the KEGG Orthology（KO） as a controlled vocabulary.Bioinformatics.2005，21（19）：3787-3793. doi：10. 1093/bioinformatics/bti430 WOS：000232596100013. PMID：15817693.

（发表于《PLOS ONE》，院选 SCI，IF：3. 234）

Preventive Effect of Aspirin Eugenol Ester on Thrombosis in κ-Carrageenan-Induced Rat Tail Thrombosis Model

MA Ning, LIU Xi-wang, YANG Ya-jun, LI Jian-yong*, MOHAMED Isam, LIU Guang-rong, ZHANG Ji-yu

(Key Lab of New Animal Drug Project of Gansu Province, Key Lab of Veterinary Pharmaceutical Development, Ministry of Agriculture, Lanzhou Institute of Husbandry and Pharmaceutical Science of CAAS, Lanzhou, 730050, China)

Abstract: Based on the prodrug principle, aspirin eugenol ester (AEE) was synthesized, which can reduce the side effects of aspirin and eugenol. As a good candidate for new antithrombotic and anti-inflammatory medicine, it is essential to evaluate its preventive effect on thrombosis. Preventive effect of AEE was investigated in κ-carrageenan-induced rat tail thrombosis model. AEE suspension liquids were prepared in 0.5% sodium carboxymethyl cellulose (CMC-Na). AEE was administrated at the dosage of 18, 36 and 72mg/kg. Aspirin (20mg/kg), eugenol (18mg/kg) and 0.5% CMC-Na (30mg/kg) were used as control drug. In order to compare the effects between AEE and its precursor, integration of aspirin and eugenol group (molar ratio 1 : 1) was also designed in the experiment. After drugs were administrated intragastrically for seven days, each rat was injected intraperitoneally with 20mg/kg BW κ-carrageen dissolved in physiological saline to induce thrombosis. The length of tail-thrombosis was measured at 24 and 48 hours. The blank group just was given physiological saline for seven days without κ-carrageenan administrated. The results indicated that AEE significantly not only reduced the average length of thrombus, PT values and FIB concentration, but also reduced the red blood cell (RBC), hemoglobin (HGB), hematocrit (HCT) and platelet (PLT). The effects of AEE on platelet aggregation and anticoagulant *in vitro* showed that AEE could inhibit adenosine diphosphate (ADP)-induced platelet aggregation as dosedependence but no notable effect on blood clotting. From these results, it was concluded that AEE possessed positive effect on thrombosis prevention *in vivo* through the reduction of FIB, PLT, inhibition of platelet aggregation and the change of TT and PT values.

---

* E-mail: lijy1971@163.com.

INTRODUCTION

Since aspirin was invented, its use has become wide spread in the prevention and treatment of inflammation, headache, fever and cardiovascular disease[1-6]. Eugenol, the principle chemical component of clove oil extracted from dry alabastrum of Eugenia *Caryophyllata Thumb*, has been recognized as safe essential oil by Food and Chemical Administration. Eugenol has been known for its various therapeutic effects, including anticoagulation, antivirus, analgesia, antipyretic, anti-inflammation, antibacterial, anti-platelet aggregation, antioxidation, anti-diarrhea, anti-hypoxia, antiulcer, and inhibition of intestinal movement and arachidonic acid metabolism[7-11].

However, the adverse reactions of aspirin such as gastrointestinal damage, have limited the long-term use of this classical drug[12]. Eugenol, which containing phenolic hydroxyl group, is irritant and vulnerable to oxidation. In order to reduce their side effect and improve therapeutic effect and stabilization, aspirin eugenol ester (AEE) was synthesized[13]. The acute toxicity, teratogenicity, metabolism, pharmacodynamics, stability and mutagenicity of AEE have been investigated. The acute toxicity of AEE indicated that the toxicity of AEE is less than that of aspirin and eugenol, only 0.02 times the toxicity of aspirin and 0.27 times the toxicity of eugenol in mice[13]. A 15-day oral dose toxicity study showed that no-observed-adverse-effect level (NOAEL) value of AEE was considered to be 50mg/kg/day under the study conditions[14]. In the Ames test and the mouse bone-marrowmicronucleus assay, the AEE did not show any mutagenesis[15]. Metabolites of AEE in vivo and in vitro also have been confirmed by HPLC – MS/MS, which indicated that AEE was decomposed into salicylic acid and eugenol[16]. Moreover, the evaluation for its anti-inflammatory, analgesic and antipyretic effects through animal model showed that AEE possessed similarity effects with aspirin, but lasted for a longer time[17-18].

Cardiovascular diseases including hypertension, coronary heart disease, atherosclerosis and acute myocardial infarction are the major cause of morbidity and mortality. Intravascular thrombosis is one of the main causes in multiple cardiovascular diseases[19-20]. The physiological process of thrombosis is very complicated, which influenced by many factors such as blood vessel injury, blood flow, platelet adhesion and aggregation. Thrombosis starts with the injury of blood vessel, and then platelets and fibrin are used to form a blood clot. Even when blood vessel is not injured, blood clots may form in the body under certain conditions[21-22]. Thrombus formation will lead ischemic necrosis, which is a key factor for infarction. Inflammation and thrombosis were also essential pathogenetic factors in the initiation of atherosclerosis. Because of the key role of thrombosis in cardiovascular diseases, thrombosis should be responsible for much morbidity and mortality[23]. However, owing to the side effects and inconvenience of currently available agents, new antithrombotic and thrombolytic agents are still needed.

The purpose of this study was to investigate the preventive effect of AEE on κ-carrageenaninduced rat tail thrombosis model. The κ-carrageenan induced tail thrombosis model in rat is useful for evaluating whether compounds have antithrombotic effects in drug discovery stage[24-26]. The researchers could use this model to observe the progression of thrombosis visually and directly in a time-dependent manner. Moreover, this model is simple and no-invasive on lab animals. Based on these advantages, this model was chosen in the experiment. Hematological analysis and blood coagulation parameters examination including thrombin time (TT), prothrombin time (PT) and fibrinogen (FIB), were also carried out to evaluate the influence of AEE on animal thrombosis model.

MATERIALS AND METHODS

Chemicals and reagents

AEE, transparent crystal (purity: 99.5% with RE-HPLC), was prepared in Key Lab of New Animal Drug Project of Gansu Province, Key Lab of Veterinary Pharmaceutical Development of Agricultural Ministry, Lanzhou Institute of Husbandry and Pharmaceutical Sciences of CAAS. CMC-Na was supplied by Tianjin Chemical Reagent Company (Tianjin, China). Aspirin and Tween-80 were obtained from Aladdin Industrial Corporation (Shanghai China). Eugenol was supplied by Sinopharm Chemical Reagent Co., Ltd. (Shanghai China). ADP and κ-carrageenan was purchased from Sigma (St.Louis, USA). Prothrombin time (PT) reagent, thrombin time (TT) reagent and fibrinogen (FIB) reagent were obtained from Sysmex Corporation (Fukuoka, Japan). All chemical reagents were of analytical reagent grade.

Animals

Ninety Wistar male rats with clean grade, aged 7 weeks and weighing 150160 g, were purchased from animal breeding facilities of Lanzhou Army General Hospital (Lanzhou, China). Animals were housed in stainless steel cages in a ventilated room. Light/dark regime was 12/12 h and living temperature is (22 ± 2)℃ with relative humidity of (55 ± 10)%. Standard compressed rat feed from Beijing Keao Xieli Co., Ltd (Beijing China) and drinking water were supplied ad libitum. The study was performed in compliance with the Guidelines for the care and use of laboratory animals as described in the US National Institutes of Health and approved by Institutional Animal Care and Use Committee of Lanzhou Institute of Husbandry and Pharmaceutical Science of CAAS. Animals were allowed a 2-week quarantine and acclimation period prior to start of the study.

Drug preparation

AEE and aspirin suspension liquids were prepared in 0.5% of CMC-Na. Eugenol and Tween-80 at the mass ratio of 1 : 2 mixed with the distilled water. κ-carrageenan was dissolved in physiological saline.

Dose

Total of ninety rats were randomly divided into following groups (n = 10). Group 1 was served as blank group received physiological saline. Group 2 was served as animal disease model. Group 3 was given 5% CMC-Na 30mg/kg as negative control. Group 4 was received 20mg/kg aspirin as positive group. The high-, medium-, and low-doses of AEE were selected as 72mg/kg, 36mg/kg and 18mg/kg, organized into groups 5 to 7 respectively. Group 8 was given eugenol at the dosage of 18mg/kg. Group 9 was received both aspirin and eugenol (aspirin: 20mg/kg, eugenol: 18 mg/kg). After different drugs were administrated intragastrically for seven days, the rats in groups 2 to 9 were intraperitoneally injected with κ-carrageenan to induce thrombosis.

Carrageenan-induced rat tail thrombosis model

Based on animal welfare, rationality and simplity of thrombosis model, κ-carrageenan was used to induce rat tail thrombosis for investigating preventive effect of AEE. The apparent formation character of rat tail thrombosis was swelling and redness which were observed from the experiment. For optimal model, κ-carrageenan dose and environment temperature, infarction time were tested and selected in the preliminary study. Environment temperature of 20℃ was selected for the establishment of thrombosis model. The dose of κ-carrageenan at 20mg/kg BW can guarantee all the carrageenan treated group to appear thrombus and have a suitable thrombus length in the tail. There was a little infarction in the tip end of rat tail after carrageenan injection at 4 hours, and then redness and swell were observed in the tail. The length of infraction was increased with the time elapsed and be stable at 24 hours after κ-carrageenan injection. Therefore, 24 hours was time point which thrombosis model was established successfully.

In the experiment, at half an hour from the last treatment with different drugs, each rat was intraperitoneally injected with 20mg/kg BW κ-carrageenan dissolved in physiological saline. Rats were observed for the formation of thrombosis, swelling and redness at 20°C environment temperature. Thrombus lengths were measured and photographed at 24 and 48 h.

Hematological analysis and coagulation parameters examination

In order to compare the effects between AEE and its precursor, aspirin group, eugenol group and integration of aspirin and eugenol (molar ratio 1 : 1) group were designed as control groups in the experiment, same molar quantity with medium-dose of AEE.

After drugs were administrated intragastrically for seven days, thrombosis disease model was induced by carrageenan, and then the length of tail thrombosis was measured and photographed at 24 and 48 h. After the last measurement of thrombosis length, rats were anesthetized with 10% chloral hydrate and blood samples were taken from heart for hematological analysis and blood coagulation parameters examination.

2 ml blood samples were collected into EDTA-K_2 vacuum tubes for hematological analysis and measured on a Sysmex XE-2100 analyzer. Same amount of blood samples were collected and diluted by 3.2% sodium citrate on the proportion of 9 : 1 in vacuum tubes, then serum be got after centrifuging for 15 minutes in 3000 rpm. The serum was measured for blood coagulation parameters examination on Sysmex CS-5100.

Assay of anticoagulant effect of AEE *in vitro*

The method applied to anticoagulant assay was according to the related reference and made little modification[27]. 1 ml fresh rat blood samples were collected, and different doses of AEE were added into these samples. In order to solve the problem of indissolvableness of AEE, dimethyl sulfoxide (DMSO) was used as vehicle. The control was added with 200μL DMSO in 1 ml blood. The final concentration of AEE in blood samples were 1μg/ml, 10μg/ml, 100μg/ml and 200μg/ml, respectively. After mixed gently, the mixture was incubated at 37℃ for 30 min. Then the anticoagulant effects of AEE on rat fresh blood were observed.

Assay of platelet aggregation *in vitro*

The *in vitro* activity studies on anti-platelet aggregation of AEE have been done by using Born test[28]. Rat blood samples gathering from the rat heart were collected into vacuum tubes which contained 3. 8% sodium citrate (1 : 9, v/v). Platelet aggregation was assessed in plateletrich plasma (PRP) obtained by centrifugation of citrated whole blood at room temperature for 10 min at 1000 rpm. The aggregation rate was measured by platelet aggregation analyzer (Chrono-log Model: 700) after stimulation with ADP (5 μM) using platelet-poor plasma (PPP) to set zero. The PPP was obtained by centrifugation of PRP at room temperature for 10 min at 3000 rpm.

The solution of AEE and aspirin dissolved in DMSO (5μL) was added into PRP (250μL), and the same volume of DMSO with no test compound was added as a reference sample. After incubation of rat platelets with AEE at concentrations ranging from 50μg/ml to 200μg/ml for 10 minutes, the platelet aggregation activities was assessed using ADP (5 μM). In the experiment, aspirin was used as positive control drug. Results were recorded at maximal aggregation after the addition of ADP. Data are expressed as percentage of maximal aggregation.

Statistical analysis

The statistical analyses were carried out using SAS 9. 2 (Statistics Analysis System USA). Data obtained from experiment was expressed as mean ± standard deviation (SD). Statistical differences between the treatments and the control were evaluated by one-way ANOVA. Pvalue less than 0. 05 were considered to indicate statistical significance.

RESULTS

Effects of AEE on rat thrombosis model

The full-length of rat tail was measured in the experiment, and then the data was analyzed by Student's t-test. The results showed that there was no difference in the tail full-length. After 4 hours of κ-carrageenan treatment, swelling and redness were observed in model group, while swelling and redness did not occur in the group which was given AEE at the dosage of 72mg/kg. The lengths of thrombosis at 24 and 48 hours were shown in Table 1. There were not significant differences in aspirin group when compared with CMC-Na group at 24 and 48 hours. However, there were significant differences in eugenol and AEE groups at the time interval (24-48h) ($P<0.01$ in eugenol and AEE groups), which showed that eugenol and AEE possessed obvious antithrombotic effect in this thrombosis model. From 24 to 48 hours, the thrombosis lengths in model group were increased but other groups remained unchanged. There was no difference between model group and CMC-Na group in thrombosis length at 48 hours, which indicated that CMC-Na had no influence on thrombosis formation at 48 hours. The average thrombus lengths at 48 hours in low-dose and medium-dose AEE group were 9. 63 cm and 8. 96 cm, respectively, which showed the stronger antithrombotic effect of AEE with the amount increased. However, the average lengths between medium-dose and high-dose AEE groups were not significant different.

Table 1 Preventive effects of AEE in κ-carrageenan-induced rat tail thrombosis model (n=10)

Groups	Dose (mg/kg)	Tail full-length (cm)	Thrombosis length (cm)	
			24 Hours	48 Hours
Model	—	15.88±0.38	11.04±1.63[aa]	13.63±1.21
CMC-Na	30	15.79±0.78	12.88±1.26	12.83±1.25
Aspirin	20	15.88±0.31	11.75±1.75	11.75±1.54
Low-Dose AEE	18	15.42±0.56	9.46±1.54[aa]	9.63±1.79[bb]
Medium-Dose AEE	36	15.25±0.63	8.83±1.16[aa]	8.96±0.93[bb]
High-Dose AEE	72	15.50±0.71	9.00±1.76[aa]	9.04±1.70[bb]
Integration	20+18	15.25±0.50	13.46±0.72	13.50±0.64
Eugenol	18	15.41±0.47	11.25±1.22[aa]	11.25±1.52[bb]

Note: After different drugs were administrated intragastrically for seven days, rat tail thrombosis model was induced by κ-carrageenan at a dose of 20mg/kg BW, and then the length of tail thrombosis was measured at 24 and 48 h. Values are presented as mean ± SD (n=10).

[aa]$P<0.01$ significant difference from CMC-Na group at 24 hours.

[bb]$P<0.01$ significant difference from CMC-Na at 48 hours. Integration: integration of aspirin and eugenol (molar ratio 1 : 1).

doi: 10.1371/journal.pone.0133125.t001

The results showed that AEE could inhibit significantly thrombus formation in κ-carrageenan-induced thrombosis model in rats (seen in Fig. 1). The representative thrombus figures of each group at 24 hours were shown in Fig 2. The average thrombus length in model group was 13.63 cm at 48 hours, whereas the average lengths of thrombus in three AEE groups were 9.63, 8.96 and 9.04 cm at 48 hours. From the results, when compared with the aspirin and eugenol groups, AEE could significantly reduce the thrombus length, which demonstrated that there was a significant difference between AEE and its precursor. Interestingly, there was no significant difference between integration of aspirin and eugenol (molar ratio 1 : 1) group and model group. Therefore, integration of aspirin and eugenol had no antithrombotic effect, even aspirin and eugenol as two precursors of AEE. The results in vivo suggested that AEE can prevent tail thrombosis formation induced by κ-carrageenan.

Effects of AEE on Coagulation parameters

Anticoagulant activity of AEE was evaluated by the classical coagulation assays prothrombin time (PT), thrombin time (TT) and fibrinogen (FIB). When compared κ-carrageenan-induced tail thrombosis model group with blank group, the results showed that PT and FIB values increased and TT values reduced significantly. Compared with CMC-Na group, the PT values in aspirin, medium- and high-dose AEE groups were decreased significantly ($P<0.01$) (seen in Table 2). For TT values, the results in aspirin, eugenol, low- and medium-dose AEE groups were similar to the CMC-Na group. However, the TT values in high-dose AEE group and integration of aspirin and eugenol (ratio 1 : 1) group were decreased in comparison with CMC-Na group. In addition, the PT and TT values in AEE groups were dose-independent. The concentration values of FIB in eugenol, AEE and integra-

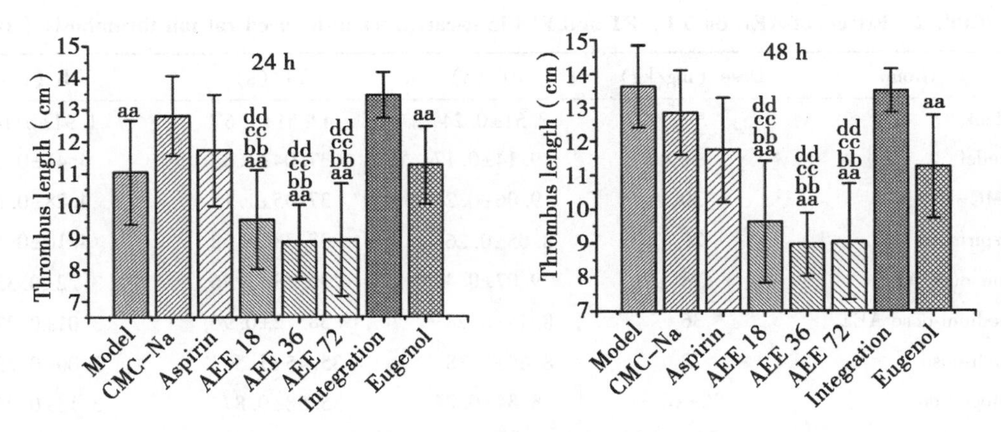

Fig. 1 Preventive effects of aspirin, eugenol and AEE on κ-carrageenan-induced rat tail thrombosis length at 24 and 48 h

Note: Each rat was intraperitoneally injected with 20mg/kg κ-carrageenan dissolved in physiological saline after seven days drug administration, and then the lengths of thrombosis were measured at 24 and 48 hours. Integration: integration of aspirin and eugenol (molar ratio 1 : 1). All groups were compared with CMC-Na group and three AEE groups were compared with its precursor, aspirin, eugenol and integration groups. $^{aa}P<0.01$, $^{a}P<0.05$, significant difference from CMC-Na group. $^{bb}P<0.01$, $^{b}P<0.05$, significant difference from aspirin group. $^{cc}P<0.01$, $^{c}P<0.05$, significant difference from eugenol group. $^{dd}P<0.01$, $^{d}P<0.05$, significant difference from integration group.

doi: 10.1371/journal.pone.0133125.g001

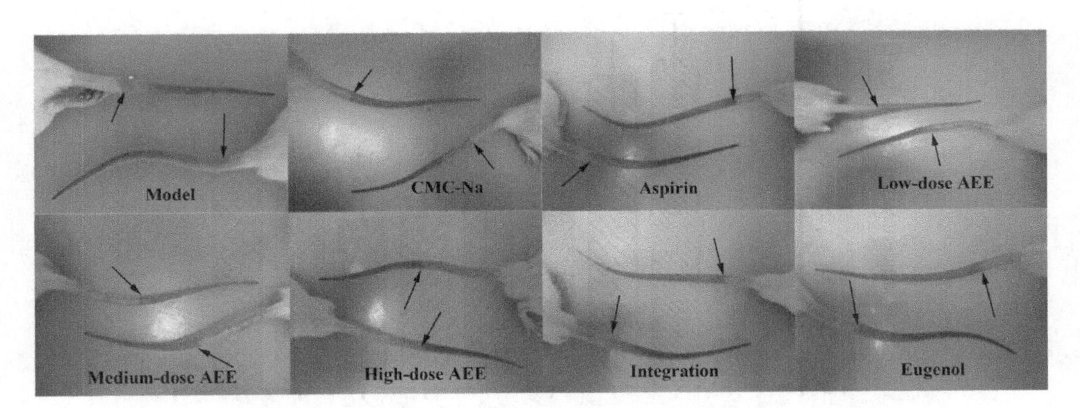

Fig. 2 Preventive effects of aspirin, eugenol and AEE on rat tail thrombus induced by κ-carrageenan at 24 hours

Note: Date represents two of ten animals for each group. Swelling and redness was pointed out by arrow. Integration: integration of aspirin and eugenol (molar ratio 1 : 1).

doi: 10.1371/journal.pone.0133125.g002

tion groups were significantly decreased in comparison with model group (seen in Fig 3).

Table 2 Effects of AEE on TT, PT and FIB in κ-carrageenan-induced rat tail thrombosis (n=10)

Group	Dose (mg/kg)	PT (s)	TT (s)	FIB (g/L)
Blank	—	8.51±0.24##	42.51±1.6##	1.91±0.14##
Model	—	9.14±0.17	37.04±1.33	4.41±0.41
CMC-Na	30	9.06±0.24	37.95±1.11	4.28±0.38
Aspirin	20	8.65±0.26**	38.38±1.32	4.10±0.20
Low-dose AEE	18	9.07±0.44	36.78±1.46	3.22±0.35**
Medium-dose AEE	36	8.45±0.28**	38.72±0.96	3.01±0.22**
High-dose	72	8.65±0.38**	35.45±1.50**	3.00±0.23**
Integration	20+18	8.84±0.24	35.58±0.87**	3.18±0.25**
Eugenol	18	9.28±0.32	37.07±0.83	3.61±0.18**

Note: 2 ml blood samples in vacuum tube were diluted by 3.2% sodium citrate on the proportion of 9 : 1, then serum be got after centrifuging for 15 min in 3000 rpm for blood coagulation parameters examination.

Values are presented as mean ± SD (n=10).

##$P<0.01$ significant difference from model group.

** $P<0.01$, significant difference from CMC-Na group. PT: prothrombin time, TT: thrombin time, FIB: fibrinogen. Integration: integration of aspirin and eugenol (molar ratio 1 : 1).

doi: 10.1371/journal.pone.0133125.t002

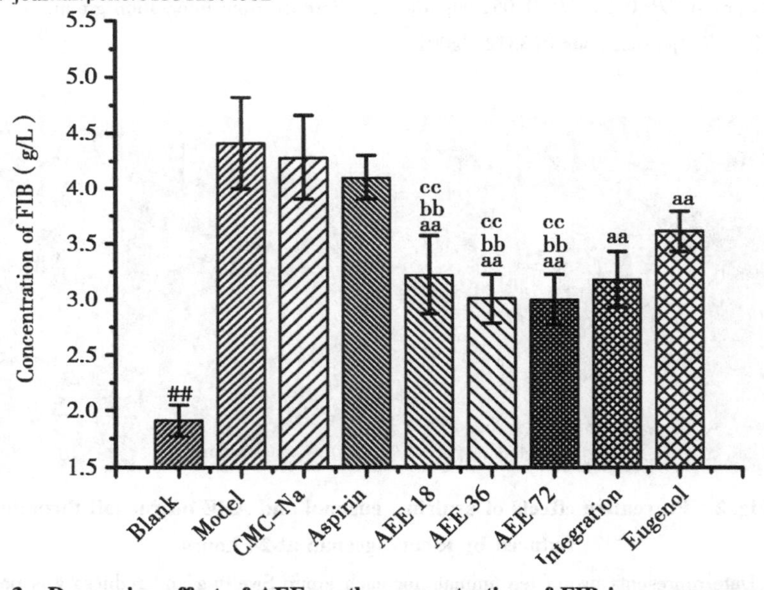

Fig. 3 Decreasing effect of AEE on the concentration of FIB in κ-carrageenan-induced rat tail thrombosis model (n=10)

Note: After the last measurement of thrombosis length, serum samples were got for coagulation parameters examination. FIB: fibrinogen. Integration: integration of aspirin and eugenol (molar ratio 1 : 1). ##$P<0.01$, #$P<0.05$ significant difference from model group. aa$P<0.01$, a$P<0.05$, significant difference from CMC-Na group. bb$P<0.01$, b$P<0.05$, significant difference from aspirin group. cc$P<0.01$, c$P<0.05$, significant difference from eugenol group.

doi: 10.1371/journal.pone.0133125.g003

Prolongation of TT suggests the inhibition of thrombin activity or fibrin polymerization. From the experimental results, there was no statistical difference in drug administration group except for high-dose AEE and integration group. TT values in high-dose AEE and integration group were shorter than those in CMC-Na group. When the body sustained thrombotic disease, TT value should become short. The thrombosis formation in rat tail and the change of FIB concentration may be the reasons for the decrease of TT value.

Prolongation of PT demonstrates the inhibition of the extrinsic pathway of coagulation. The PT values in model group were prolonged in comparison with blank group, which is very interesting. This result may be caused by the thrombosis formation which consumed the coagulation factors. Coagulation factors play a key role in activating prothrombin and the lack of coagulation factors can lead the prolongation of PT value. AEE decreased the PT values through the inhibitation of thrombosis formation. The decrease of FIB concentration in AEE group is helpful to inhibit the conversion of fibrinogen to fibrin. Therefore, AEE possessed better effect on thrombosis prevention from the concentration reduction of FIB.

Anticoagulant effect of AEE *in vitro*

The anticoagulant effects of AEE on fresh rat blood were tested. However, the results showed that AEE had no notable effect on clotting of rat blood, even the concentration of AEE at 200μg/ml. The bloods were coagulated and formed dark red blood clots at the bottom of the tubes (seen in Fig 4). There was no difference between the control and drug treatment groups. Unlike heparin sodium or EDTA - K_2, these results indicated that AEE might have different mechanism for inhibiting thrombosis formation.

Effects of AEE on platelet aggregation *in vitro*

To assess the effect of AEE on platelet aggregation, classical light transmission aggregometry was performed after incubation of rat platelets with AEE at different concentrations ranging from 50μg/mL to 200μg/mL. As shown in Fig 5, compared with normal group, DMSO appears no significant effect on the platelet aggregation. The results of DMSO on platelet aggregation in this experiment were similar as the previous study[29]. After incubation for 10 minutes, AEE significantly inhibited platelet aggregation in comparison with normal control group. Platelet aggregation rates were approximately 24.5% at 50μg/mL, 20.6% at 100μg/mL and 18.2% at 200μg/mL, respectively, which showed that AEE dose-dependently inhibited ADPinduced platelet aggregation. In the experiment, aspirin was used as positive control drug. The results showed that the inhibitory effect of AEE was less potent at concentrations of 100μL/mL in comparison to equimolar doses of aspirin.

Hematology

In the hematological analysis (seen in Table 3), leukocyte count (WBC) and monocyte count (MONON) in model group were significantly increased compared with the blank group, but platelets (PLT) was sharply reduced. No other changes were observed between blank and model groups. The increase of WBC and MONON may be caused by inflammation which was induced by κ-carrageenan. The formation of tail thrombosis could deplete circulating PLT to result in the sharp reduction of PLT.

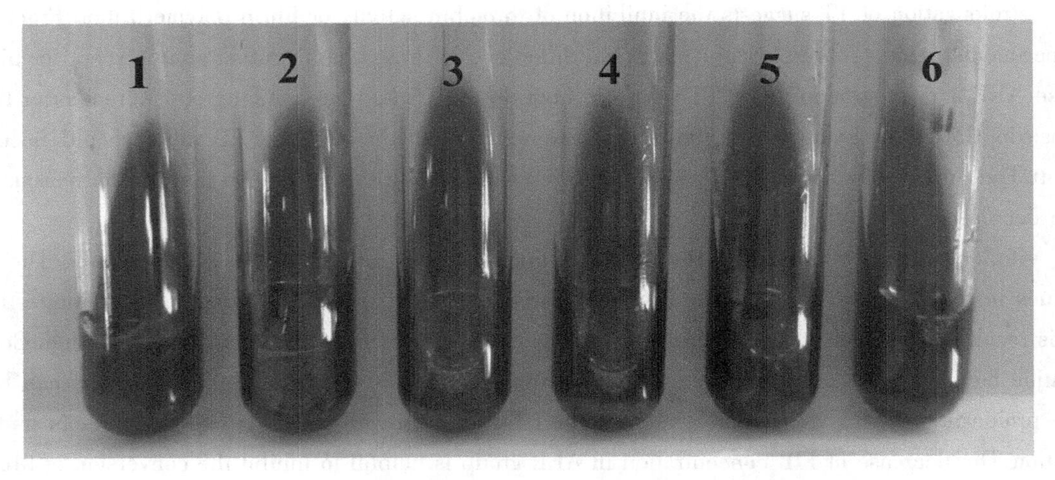

Fig. 4 Drug effect of AEE with different concentrations on fresh rat blood *in vitro*

Note: The bloods were coagulated and formed dark red blood clots at the bottom of the tubes, which indicated that AEE had no notable effect on clot of rat blood. 1: normal fresh rat blood; 2: rat blood with 200μL DMSO; 3: rat blood with 1μg AEE in 200μL DMSO; 4: rat blood with 10μg AEE in 200μL DMSO; 5: rat blood with 100μg AEE in 200μL DMSO; 6: rat blood with 200 μg AEE in 200μL DMSO.

doi: 10. 1371/journal.pone.0133125. g004

When compared with the CMC-Na group, the value of red blood cell (RBC), hemoglobin (HGB) and hematocrit (HCT) were significantly reduced in AEE groups. In regard to RBC, aspirin and eugenol had no effect on the reduction of RBC. However, the results in AEE and integration groups showed significant difference in comparison with aspirin and eugenol. The results suggested that AEE and integration of aspirin and eugenol can reduce RBC value in κ-carrageenan induced thrombosis model. The results of HGB and HCT were similar to RBC owing to the close relationship among HGB, RBC and HCT. With the RBC reduced, the values of HCT, and HGB were reduced correspondingly. Therefore, AEE may have a positive effect on blood viscosity through the reduction of HCT value and FIB concentration. HCT is associated with blood viscosity and blood flow as one of crucial factor. In the experiment, the reduction of HCT values in AEE groups may be caused by the reduction of RBC. Based on the experimental results, it is essential to investigate the effect of AEE on blood viscosity parameters in the further study.

PLT count was decreased in AEE group when compared with CMC-Na group. There was no statistical difference in eugenol group, which may suggest eugenol made no influence on PLT. The values in AEE groups showed significant difference in comparison with eugenol group. Notably, PLT value in integration group was the lowest in all groups. The mean PLT values in AEE groups were also lower than aspirin and eugenol groups. The influence of AEE on PLT in κ-carrageenan induces thrombosis model was similar with the results in 15-day oral dose toxicity study[14]. The reduction of PLT values in AEE group may also make contribution on the inhibition of thrombosis formation and reduction of blood viscosity.

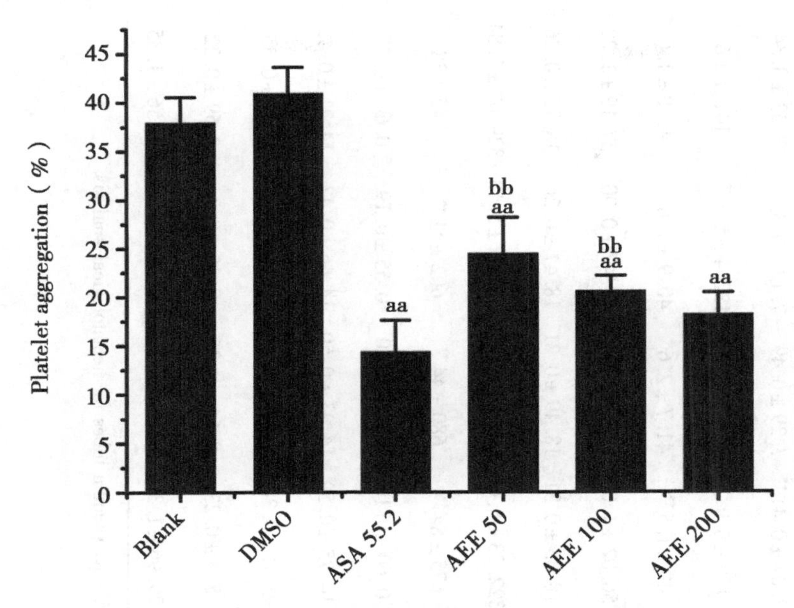

Fig. 5 Effect of AEE on platelet aggregation *in vitro*

Note: After incubation of rat platelets with AEE at concentrations ranging from 50μg/ml to 200μg/ml for 10 minutes, the platelet aggregation activities was assessed using ADP (5 μM). Data are expressed as mean ± SD (n = 8). $^{aa}P<0.01$, $^{a}P<0.05$, significant difference from DMSO group. $^{bb}P<0.01$, $^{b}P<0.05$, significant difference from aspirin group. DMSO: dimethyl sulphoxide; ASA: acetyl salicylic acid (aspirin).

doi: 10. 1371/journal.pone.0133125. g005

DISCUSSION

Based on the design of prodrug principal, AEE is confirmed to be effective against the symptoms of inflammation, fever and soreness[17-18]. The previous study showed that this compound could reduce significantly the value of TG, TC and platelet counts at the dosage of 50mg/kg[14], which is very beneficial to the therapy of atherosclerosis and thrombosis. Meanwhile, there were no histopathological changes in any organ from the rats given AEE at the dose of 50mg/kg. This result suggested that AEE was a safe drug with relatively less gastrointestinal (GI) toxicity. Based on this fact, it is essential to evaluate the preventive effect of AEE in animal thrombosis model.

Carrageenan is straight-chained sulfur-containing macromolecular polysaccharides that are composed of repeating units of D-galactose and 3, 6−anhydro-D-galactose. Kappa-carrageenan is widely used to induce tissue inflammation and tail thrombosis in different animals such as mouse and rat. According to the pathological studies, κ-carrageenan can cause local blood vessel inflammation and endothelial cell injury through releasing inflammation factors, which may lead to the formation of thrombus[30]. Inflammation, especially in blood vessel, is considered to make contribution to thrombosis. On the contrary, blood clots can cause inflammation[31]. In our experiment, the increase in leukocyte count and monocyte count could prove that κ-carrageenan cause acute and systemic inflammation in model group. However, there was no significant difference for two indexes between drug

Table 3 Hematological findings in rats administrated intragastrically with AEE

Variable	Unit	Blank	Model	CMC - Na	Aspirin	Low AEE	Mid AEE	High AEE	Integration	Eugenol
WBC	10^9/L	5.38 ±0.98##	7.68 ±1.64	7.17 ±1.78	8.32 ±1.6	7.38 ±2.09	7.95 ±1.92	7.89 ±1.36	7.04 ±1.40	6.97 ±1.32
RBC	10^12/L	8.14 ±0.45	7.80 ±0.45	7.69 ±0.40	7.86 ±0.32	7.20 ±0.50**	7.04 ±0.41**	7.29 ±0.49*	7.12 ±0.33**	7.53 ±0.34
HGB	g/L	153.7 ±7.6	148.8 ±9	142.4 ±7.8	149 ±6.2*	136 ±8.7*	129 ±6.9**	134.8 ±8.8*	132.4 ±5.1**	141.1 ±6
HCT	%	46.1 ±1.6	44.0 ±2.5	43.6 ±2.4	43.4 ±1.5	41.1 ±1.6**	40 ±1.5**	41.7 ±2.6*	40.9 ±1.6**	43.0 ±1.6
MCV	fL	56.64 ±1.46	56.47 ±0.5	56.54 ±0.84	55.23 ±0.98	57.02 ±1.24	56.02 ±1.01	57.14 ±1.30	57.47 ±0.90	57.19 ±1.22
MCH	pg	18.88 ±0.26	19.08 ±0.24	18.48 ±0.26	18.95 ±0.24	18.92 ±0.24	18.35 ±0.31	18.49 ±0.21	18.47 ±0.26	18.72 ±0.26
MCHC	g/L	333.44 ±7.09	337.91 ±2.3	326.75 ±3.42	343.15 ±5.3**	332.77 ±7.95	322.23 ±6.64	323.64 ±6.54	321.79 ±5.13	329.08 ±2.84
PLT	10^9/L	851 ±68##	703 ±49	782 ±49	681 ±55**	641 ±80**	675 ±84**	680 ±44**	602 ±41**	745 ±38
MONON	10^9/L	0.11 ±0.08##	0.53 ±0.16	0.59 ±0.24	0.59 ±0.14	0.58 ±0.19	0.60 ±0.16	0.59 ±0.10	0.55 ±0.19	0.64 ±0.17
RDW - CV	%	14.91 ±0.54	14.34 ±0.43	14.11 ±0.29	14.26 ±0.67	14.25 ±0.66	14.75 ±0.59	14.23 ±0.50	14.66 ±0.47	14.21 ±0.48
PDW	%	9.48 ±0.29	9.31 ±0.27	9.46 ±0.44	9.01 ±0.35	9.54 ±0.54	9.51 ±0.38	9.65 ±0.41	9.27 ±0.42	9.69 ±0.35
MPV	fL	8.43 ±0.14	8.29 ±0.12	8.37 ±0.18	8.2 ±0.16	8.52 ±0.22	8.4 ±0.23	8.64 ±0.26	8.35 ±0.3	8.59 ±0.22
RDW - SD	%	29.09 ±1.81	28.53 ±0.86	28.05 ±0.68	27.57 ±1.6	28.45 ±1.12	29.96 ±1.38	28.39 ±0.93	29.59 ±0.85	28.56 ±1.35

Note: After the last measurement of thrombosis length, two ml blood samples were collected into EDTA - K_2 vacuum tubes for hematological analysis.

##$P < 0.01$.

#$P < 0.05$ significant difference from model group.

** $P < 0.01$.

* $P < 0.05$Significant difference compared with CMC - Na group. Integration: integration of aspirin and eugenol (molar ratio 1 : 1).

doi: 10.1371/journal.pone.0133125.t003

treatment groups and model group. These results may be from the thrombosis which kept continuous appearance of inflammation and led two indexes to be in higher status.

Thrombosis starts with adhesion of PLT to the vessel wall, followed by aggregation of additional PLT[32-33]. PLT reactions are also strong stimulus to the necessary clotting factors[34]. Therefore, PLT play a key role in the formation of thrombosis. As cyclooxygenase inhibitors, aspirin is used therapeutically in the prevention of cardiovascular disease[35]. Aspirin inhibit PLT aggregation through blocking TXA_2 synthesis which is a potent inducer of PLT aggregation and causes vasoconstriction[36-37]. In our previous study, the metabolites of AEE in vivo and *in vitro* have been confirmed[16]. In total, five metabolites were detected, including salicylic acid, salicylic acid glucuronide, salicylic acid glycine, eugenol glucuronide and eugenol sulfate. Salicylic acid, which is also the same metabolite of aspirin, is found as one of the major metabolites. It is deduced that AEE has similar effects like aspirin in preventing cardiovascular disease. So the effects of AEE ensued not from AEE itself but from salicylic acid and eugenol decomposed by the enzymes after absorption, which showed their original activities again and acted synergistically. 15-day oral dose toxicity study showed that AEE at the dose of 50mg/kg is able to reduce significantly the number of PLT[14]. In hematological analysis under the experiment, the index of PLT count was significant different between administration groups and CMC-Na group. The results showed that AEE had positive effects on PLT reduction, and the action of AEE on PLT reduction was stronger than aspirin and eugenol. Before κ-carrageenan was administrated, PLT may partly reduced by AEE. The different length thromboses deplete residual available PLT after κ-carrageenan administrated. This may be one of the reasons to explain partly why shorter thrombotic length appeared in AEE groups and longer thrombotic length appeared in model and other groups.

Thrombin is at the center of the process of blood clotting[38-39]. In order to control thrombin and regulate blood clotting carefully, thrombin is built as an inactive precursor, prothrombin[40]. When the tissue is cut, the blood flows out of the blood vessels and encounters tissue factor. After that, tissue factor activates a few molecules of Factor VII. These then activate a lot of Factor X. Finally, these activate even more thrombin. Thrombin catalyzes conversion of fibrinogen to fibrin to assemble into large stringy networks[41]. These networks then trap lots of blood cells to form the dark red scab. Under the present experiment, PT was prolonged in model group and shortened in AEE groups. AEE inhibited the thrombosis formation, and then improved the level of coagulation factors and reduced PT values. For FIB, AEE reduced significantly its concentration in rat bloods. The reduction of FIB inhibited the network of fibers which is essential for thrombosis formation. TT in model group was shorter than in blank group, which may be caused by FIB increase. TT values were shorter in high-dose AEE group in comparison with CMC-Na group, which suggested that high-dose AEE may had a negative impact on inhibition of thrombin activity or fibrin polymerization. The reasons for the difference of TT value in high-dose AEE and integration group were needed to be investigated in the further study.

In this study, the antiplatelet and anticoagulant effects of AEE *in vitro* were examined. The effects of AEE on platelet function was determined by measuring ADP-induced platelet aggregation. The results showed that AEE produced obviously antiplatelet effect on ADP-induced platelet aggregation.The inhibitory effect of AEE was less potent in comparison with aspirin at equimolar dose, which indicated that AEE might have different mechanism for inhibiting thrombosis formation.Meanwhile, AEE showed no anticoagulant effect on fresh rat blood.This result may suggest that metabolic processes in the body is necessary for the drug effect of AEE.

After strict biological and toxicological research and test, pure CMC had been approved as food additive by WHO and FAO[42].In order to eliminate the effect of CMC-Na, CMC-Na was also administrated as control group.Therefore, the antithrombotic effect of AEE is not related to CMC-Na. To be notice, the average lengths in CMC-Na group at 24 hours was longer than model group and platelet count was higher than other groups in hematological analysis.These results may indicate that CMC-Na has negative influence on antithrombosis at 24 hours but no influence at 48 hours.Tween 80 is widely used as emulsifier in food industry and drug production.In general, the body has a great tolerance to Tween 80[43]. Because the amount of Tween 80 was fewer, the biological preventive effect of Tween 80 on thrombosis in this experiment is not investigated.

From the total results of the experiment, 36mg/kg as medium-dose AEE may be approved for antithrombotic use in rat tail thrombosis model induced by κ-carrageenan.The average thrombosis length in medium-dose AEE group was shorter than that in low-dose AEE group.Meanwhile, there was no statistical difference in the thrombosis length between medium and high dose AEE groups. Moreover, there was no difference between medium-dose and highdose AEE groups on FIB and PLT reduction.

All in all, AEE displayed strong preventive effect on thrombosis in κ-carrageenan-induced rat tail thrombosis model and showed better antithrombotic effect than its precursor, which indicated that AEE could be a good candidate for antithrombotic agent.From the experiment results, the preventive effect of AEE on thrombosis may come from the reduction of FIB concentration, PLT, inhibition of platelet aggregation and the change of PT and TT values.Certainly, more studies such as evaluation for inflammation biomarker, platelet aggregation, blood stream changes and coagulation factors are needed to characterize AEE and to investigate its action mechanism.

ACKNOWLEDGMENTS

The work was supported by science-technology innovation engineering of CAAS (CAASASTIP-2014-LIHPS) and the National Natural Science Foundation of China (No.31402254).

AUTHOR CONTRIBUTIONS

Conceived and designed the experiments: NM JYL.Performed the experiments: NM XWL IM. Analyzed the data: NM JYL XWL. Contributed reagents/materials/analysis tools: NM XWL GRL YJY JYZ.Wrote the paper: NM JYL.

References

[1] Vane J, Botting R.Inflammation and the mechanism of action of anti-inflammatory drugs.FASEB J.1987, 1: 89-96. PMID: 3111928.

[2] Schoemaker RG, Saxena PR, Kalkman EA.Low-dose aspirin improves in vivo hemodynamics in con-

scious, chronically infarcted rats.Cardiovasc Res.1998, 37: 108-114. PMID: 9539864.

[3] Flower R, Gryglewski R, Herbaczynska-Cedro K, Vane JR.Effects of anti-inflammatory drugs on prostaglandin biosynthesis.Nat New Biol.1972, 238: 104-106. PMID: 4505422.

[4] Hussain M, Javeed A, Ashraf M, Zhao Y, Mukhtar MM, Rehman MU.Aspirin and immune system. Int Immunopharmacol.2012, 12: 10-20. doi: 10.1016/j.intimp.2011.11.021 PMID: 22172645.

[5] Lim K, Han C, Xu L, Isse K, Demetris AJ, Wu T.Cyclooxygenase-2-derived prostaglandin E2 activates beta-catenin in human cholangiocarcinoma cells: evidence for inhibition of these signaling pathways by omega 3 polyunsaturated fatty acids. Cancer Res.2008, 68: 553-560. doi: 10.1158/0008-5472.CAN-07-2295 PMID: 18199552.

[6] Wan QL, Zheng SQ, Wu GS, Luo HR. Aspirin extends the lifespan of Caenorhabditis elegans via AMPK and DAF-16/FOXO in dietary restriction pathway. Exp Gerontol. 2013, 48: 499-506. doi: 10.1016/j.exger.2013.02.020 PMID: 23485446.

[7] Tragoolpua Y, Jatisatienr A. Anti-herpes simplex virus activities of Eugenia caryophyllus (Spreng.) Bullock & S. G. Harrison and essential oil, eugenol. Phytother Res. 2007, 21: 1153-1158. PMID: 17628885.

[8] Chami N, Bennis S, Chami F, Aboussekhra A, Remmal A.Study of anticandidal activity of carvacrol and eugenol in vitro and in vivo.Oral Microbiol Immunol.2005, 20: 106-111. PMID: 15720571.

[9] Gill AO, Holley RA.Mechanisms of bactericidal action of cinnamaldehyde against Listeria monocytogenes and of eugenol against L.monocytogenes and Lactobacillus sakei.Appl Environ Microbiol.2004, 70: 5750-5755. PMID: 15466510.

[10] Raghavendra RH, Naidu KA.Spice active principles as the inhibitors of human platelet aggregation and thromboxane biosynthesis.Prostaglandins Leukot Essent Fatty Acids.2009, 81: 73-78. doi: 10.1016/j.plefa.2009.04.009 PMID: 19501497.

[11] Yogalakshmi B, Viswanathan P, Anuradha CV. Investigation of antioxidant, anti-inflammatory and DNA-protective properties of eugenol in thioacetamide-induced liver injury in rats. Toxicology. 2010, 268: 204-212. doi: 10.1016/j.tox.2009.12.018 PMID: 20036707.

[12] Fernandez AG, Salcedo C, Palacios JM.Aspirin, salicylate and gastrointestinal injury.Nat Med.1995, 1: 602-603. PMID: 7585130.

[13] Li J, Yu Y, Wang Q, Zhang J, Yang Y, Li B, et al.Synthesis of aspirin eugenol ester and its biological activity.Medicinal Chemistry Research.2012, 21: 995-999.

[14] Li J, Yu Y, Yang Y, Liu X, Zhang J, Li B, et al.A 15-day oral dose toxicity study of aspirin eugenol ester in Wistar rats. Food Chem Toxicol. 2012, 50: 1980-1985. doi: 10.1016/j.fct.2012.03.080 PMID: 22516304.

[15] Li J, Kong X, Li X, Yang Y, Zhang J.Genotoxic evaluation of aspirin eugenol ester using the Ames test and the mouse bone marrow micronucleus assay. Food Chem Toxicol. 2013, 62: 805-809. doi: 10.1016/j.fct.2013.10.010 PMID: 24140966.

[16] Shen Y, Liu X, Yang Y, Li J, Ma N, Li B.In vivo and in vitro metabolism of aspirin eugenol ester in dog by liquid chromatography tandem mass spectrometry. Biomed Chromatogr. 2015, 29: 129-137. doi: 10.1002/bmc.3249 PMID: 24935248.

[17] Li J, Yu Y, Wang Q, Yang Y, Wei X, Zhou X, et al.Analgesic roles of aspirin eugenol ester and its mechanisms.Anim.Husband.Vet.Med.2010, 42: 20-24.

[18] Ye D, Yu Y, Li J, Yang Y, Zhang J, Zhou X, et al.Antipyretic effects and its mechanisms of aspirin eugenol ester.Chinese Journal of Pharmacology and Toxicology.2011, 2: 151-155.

[19] Hsiao G, Yen MH, Lee YM, Sheu JR.Antithrombotic effect of PMC, a potent alpha-tocopherol ana-

logue on platelet plug formation in vivo.Br J Haematol.2002, 117: 699-704. PMID: 12028044.

[20] Jorgensen L.The role of platelets in the initial stages of atherosclerosis.J Thromb Haemost.2006, 4: 1443-1449. PMID: 16839335.

[21] Schaff M, Tang C, Maurer E, Bourdon C, Receveur N, Eckly A, et al.Integrin alpha6beta1 is the main receptor for vascular laminins and plays a role in platelet adhesion, activation, and arterial thrombosis.Circulation.2013, 128: 541-552. PMID: 23797810.

[22] Wu Z, Xu Z, Kim O, Alber M.Three-dimensional multi-scale model of deformable platelets adhesion to vessel wall in blood flow.Philos Trans A Math Phys Eng Sci.2014, 372.

[23] Schussheim AE, Fuster V.Thrombosis, antithrombotic agents, and the antithrombotic approach in cardiac disease.Prog Cardiovasc Dis.1997, 40: 205-238. PMID: 9406677.

[24] Arslan R, Bor Z, Bektas N, Mericli AH, Ozturk Y.Antithrombotic effects of ethanol extract of Crataegus orientalis in the carrageenan-induced mice tail thrombosis model.Thromb Res.2011, 127: 210-213. doi: 10.1016/j.thromres.2010.11.028 PMID: 21183208.

[25] Hagimori M, Kamiya S, Yamaguchi Y, Arakawa M.Improving frequency of thrombosis by altering blood flow in the carrageenan-induced rat tail thrombosis model.Pharmacol Res.2009, 60: 320-323. doi: 10.1016/j.phrs.2009.04.010 PMID: 19394423.

[26] Yan F, Yan J, Sun W, Yao L, Wang J, Qi Y, et al.Thrombolytic effect of subtilisin QK on carrageenan induced thrombosis model in mice.J Thromb Thrombolysis.2009, 28: 444-448. doi: 10.1007/s11239-009-0333-3 PMID: 19377880.

[27] Yuan J, Yang J, Zhuang Z, Yang Y, Lin L, Wang S.Thrombolytic effects of Douchi fibrinolytic enzyme from Bacillus subtilis LD-8547 in vitro and in vivo.BMC Biotechnol.2012, 12: 36. doi: 10.1186/1472-6750-12-36 PMID: 22748219.

[28] BORN GV.Aggregation of blood platelets by adenosine diphosphate and its reversal.Nature.1962, 194: 927-929. PMID: 13871375.

[29] Liu XJ, Shi XX, Zhong YL, Liu N, Liu K.Design, synthesis and in vitro activities on anti-platelet aggregation of 4-methoxybenzene-1, 3-isophthalamides.Bioorg Med Chem Lett.2012, 22: 6591-6595.doi: 10.1016/j.bmcl.2012.09.001 PMID: 23010272.

[30] Begum R, Sharma M, Pillai KK, Aeri V, Sheliya MA.Inhibitory effect of Careya arborea on inflammatory biomarkers in carrageenan-induced inflammation.Pharm Biol.2014, 1-9.

[31] Sadowski M, Zabczyk M, Undas A.Coronary thrombus composition: Links with inflammation, platelet and endothelial markers. Atherosclerosis. 2014, 237: 555-561. doi: 10.1016/j.atherosclerosis.2014.10.020 PMID: 25463088.

[32] Gong Y, Lin M, Piao L, Li X, Yang F, Zhang J, et al.Aspirin Enhances Protective Effect of Fish Oil against Thrombosis and Injury-induced Vascular Remodeling.Br J Pharmacol.2014.

[33] Sugiyama K, Iyori M, Sawaguchi A, Akashi S, Tame JR, Park SY, et al.The crystal structure of the active domain of Anopheles anti-platelet protein, a powerful anti-coagulant, in complex with an antibody. J Biol Chem. 2014, 289: 16303-16312. doi: 10.1074/jbc.M114.564526 PMID: 24764297.

[34] Chesnutt JK, Han HC.Simulation of the microscopic process during initiation of stent thrombosis.Comput Biol Med.2014, 56C: 182-191.

[35] Sutcliffe P, Connock M, Gurung T, Freeman K, Johnson S, Ngianga-Bakwin K, et al.Aspirin in primary prevention of cardiovascular disease and cancer: a systematic review of the balance of evidence from reviews of randomized trials.PLoS One.2013, 8: e81970. doi: 10.1371/journal.pone.0081970 PMID: 24339983.

[36] Burdan F, Chalas A, Szumilo J. Cyclooxygenase and prostanoids-biological implications. Postepy Hig Med Dosw (Online). 2006, 60: 129–141.

[37] Gambero A, Marostica M, Becker TL, Pedrazzoli JJ. Effect of different cyclooxygenase inhibitors on gastric adaptive cytoprotection induced by 20% ethanol. Dig Dis Sci. 2007, 52: 425–433. PMID: 17226071.

[38] Davie EW, Fujikawa K, Kisiel W. The coagulation cascade: initiation, maintenance, and regulation. Biochemistry. 1991, 30: 10363–10370. PMID: 1931959.

[39] Kern A, Varnai K, Vasarhelyi B. Thrombin generation assays and their clinical application. Orv Hetil. 2014, 155: 851–857. doi: 10.1556/OH.2014.29899 PMID: 24860049.

[40] Stubbs MT, Bode W. A player of many parts: the spotlight falls on thrombin's structure. Thromb Res. 1993, 69: 1–58. PMID: 8465268.

[41] Philippou H. Thrombin activatable fibrinolysis inhibitor (TAFI): more complex when it meets the clot. Thromb Res. 2014, 133: 1–2. doi: 10.1016/j.thromres.2013.10.034 PMID: 24225483.

[42] Jin Z, Li W, Cao H, Zhang X, Chen G, Wu H, et al. Antimicrobial activity and cytotoxicity of N-2-HACC and characterization of nanoparticles with N-2-HACC and CMC as a vaccine carrier. Chemical Engineering Journal. 2013, 221: 331–341.

[43] Ema M, Hara H, Matsumoto M, Hirata-Koizumi M, Hirose A, Kamata E. Evaluation of developmental neurotoxicity of polysorbate 80 in rats. Reprod Toxicol. 2008, 25: 89–99. PMID: 17961976.

（发表于《PLOS ONE》，院选 SCI，IF：3.234）

Differentially Expressed Genes of LPS Febrile Symptom in Rabbits and that Treated with Bai-Hu-Tang, a Classical Cnti-febrile Chinese Herb Formula

ZHANG Shi-dong [1,2]*, WANG Dong-sheng[1,2], DONG Shu-wei[1,3], YANG Feng[2,3], YAN Zuo-ting[1,3]

(1.Lanzhou Institute of Husbandry and Pharmaceutical, Science of Chinese Academy of Agricultural Sciences, Lanzhou 730050, China; 2.Key Laboratory of Veterinary Pharmaceutics Discovery, Ministry of Agriculture, Lanzhou 730050, China; 3.Key Laboratory of New Animal Drug Project of Gansu Province, Lanzhou 730050, China)

Abstract: *Ethnopharmacological relevance*: Bai-Hu-Tang (BHT) has beentraditionally used to clear heat and engender fluids.

Aim of the study: To reveal the alteration of differentially expressed genes (DEGs) between lipopolysaccharide (LPS) febrile syndrome in rabbits and treatment with BHT which is a classical anti-febrile formula in traditional Chinese medicine.

Materials and methods: Febrile model was induced by LPS injection (*i.v.*) in rabbits, and BHT was gavaged to another group of febrile rabbits.Aftersacrifice of animals, total RNA of liver tissue was isolated, processed, and hybridized to rabbit cDNA microarrays obtained from Agilent Co. The data of DEGs were obtained by lazer scanning and analyzed with Cluster program 3. 0. Then bioinformatic analysis of DEGs was conducted through gene ontology (GO) annotation and Kyoto Encyclopedia of Genes and Genomes (KEGG) pathways. In addition, expression levels of four relative genes were detected by quantitative real time ployenzyme chain reaction (qRT-PCR) to validate the accuracy of microarrays.

Result: The results demonstrated that genes expression pattern could be clustered into three groups significantly, and there were 606 up-regulated genes and 859 down-regulated genes in the model group, and 106 up-regulated genes and 429 down-regulated genes in BHT treated group.There were 286 DEGs existed as the common in two experimental groups. Enrichment analysis of GO annotations indicated that DEGs in model and BHT treated animals mainly referred catalytic activity and oxidoreductase activity for metabolic processes located in the membrane system at intracellular part, and binding activities increased significantly in treatment with BHT. Enrichment of KEGG analysis showed that the pathways of phagosome

---

* Corresponding author.E-mail: sdzhang2006@ 126.com.

and protein processing in endoplasmic reticulum contained the most altered genes in the LPS group, but the percentage of phagosome pathway almost doubled in BHT group. Most DEGs involved in the LPS signal recognition system was up-regulated in LPS group, but partly decreased in BHT group. RT-PCR results of eight relative genes were consistent with the results of microarrays.

Conclusion: DEGs of LPS febrile syndrome mainly involved oxidoreductase and catalytic activity of the *metabolic* processes, and pathways of processing protein for pyrotoxin recognition; BHT mostly regulated the DEGs in the phagosome pathway to clear LPS in the liver, and partly interfered with gene expression in LPS recognition system. The study provided an important pioneering result on gene expression profiling research, and will facilitate the clinical care or further studies of the formula.

Key words: Bai-Hu-Tang; LPS febrile; Microarrays; Differentially expressed gene; Bioinformatics

1 INTRODUCTION

Traditional Chinese medicine (TCM) is a highly valuable and rich medical tradition employing syndrome differentiation as a basis for the treatment of disease, and well established and time-honored practice in China for more than a thousand years. Syndrome, also known as *Zhenghou* in Chinese, is a dynamic reflection of microcosmic changes of organism in certain phases of disease development. It reflects the nature of a disease, and is the basis of diagnosis and treatment based on an overall analysis of symptom and sign in TCM (Zhang and Wang, 2006). With the development of life science, it is an inevitable trend for syndrome research to reveal the essence of the syndrome at gene level (Liu et al., 2008), and it can provide a molecular scientific basis for diagnosis and treatment based on TCM to study alteration of gene expression (Chai and Huang, 2010).

Qifen Syndrome is a representative of febrile disease in TCM. As discussed in our last paper (Zhang et al., 2013), Chinese researchers reported that the injection of LPS can induce a strong fever symptom that was analogous to Qifen Syndrome in rabbit (Yu et al., 2010), and the animal model was usually used to study antipyretic herb medicine and ethnopharmacology (Yang, 2002; Yu et al., 2010; Ai et al., 2011; Ni and Wei, 2012). Actually, LPS can lead to inflammatory cascade reaction in visceral organ and systemic inflammatory response by intravenous or intraperitoneal injection, and eventually results in multiple organ dysfunction syndrome (Aldridge et al., 2009). Multiple organ dysfunction syndrome (MODS) is characterized by the development of progressive and potentially reversible physiological dysfunction in 2 or more or gans or organ systems induced by a variety of acute physiologic insults (Wang and Ma, 2008). This pathological process of reversible physiological dysfunction in organ is verys imilar to the Qifen syndrome (Liu and Wang, 1999; Ni and Wei, 2012). Therefore, LPS injection has been the most common method to establish and research Qifen Syndrome in animals in China (Wang et al., 2010; Ni and Wei, 2012).

Bai-Hu-Tang (or White Tiger Decoction) is a classical Chinese herbal formula with a long history of use. It is a quintessential prescription for treatment of warmdisease in TCM, especially to Qifen Syndrome. The formula originated from the treatise on febrile disease of *Shang Han Lun* (or Treatise on Cold-Attack) compiled by Zhong-Jing Zhang who lived in Eastern Han Dynasty in China. It is formulated with four herbs of liquorice, anemarrhena rhizome, gypsum and rice, and is classically used to clear excess heat and promote the generation of body fluids (Charles, 2000; Zhu, 2007). In recent years, its new usage refers to many clinical diseases, including diabetes mellitus (Chen et al., 2008), eczema, pruritus, some anxiety and emotional disorders (Fred and Bob, 2004). The formula is also widespread use concern and practiced in European countries (European Herbal and Traditional Medicine Practitioners Association, 2007). It was also reported that BHT was a valuable antifebrile and anti-inflammatory natural drug in LPS fever symptom (Jia et al., 2013). For this reason, BHT would be worthy of further investigation for discovering the mechanism in modern life sciences.

DNA microarrays are a high-throughput method used to survey the relative amount of annotated transcripts (gene expression) in a genome (Zhu et al., 2006; Campbell et al., 2006), has opened a new era of whole-genome transcriptional analysis in molecular biology (Draghici et al., 2006). The temporal profile of gene transcription alterations provides markers of biological processes taking place in a particular cell or tissue underlying disease or toxicity development. However, there is no report employing a cDNA microarray approach to evaluate in detail the LPS febrile symptom and BHT treatment at transcriptomic level. Against the above background, the study herein attempted to determine hepatic transcriptomic profiles using cDNA microarray in LPS fever rabbit and that treated with BHT. Differently expressed genes (DEGs) were analyzed using bioinformatics methods in order to explore the biological meaning of DEGs. The research achievement of this study may put forward some new views of febrile syndrome induced by LPS, and support BHT's traditional use at molecular levels.

2 MATERIALS AND METHODS

2.1 Herbal materials and BHT extraction

The herbs of Liquorice (sliced root of *Glycyrrhiza glabra* L. in *Leguminosae*), Anemarrhena Rhizome (sliced root of *Anemarrhena asphodeloides* Bunge in *Liliaceae*), Gypsum (crystal of calcium sulfate) and Rice (nonglutinous rice, polished seed of *Oryza sativa* L. in *Gramineae*) were purchased from Antaitang Pharmaceutical Co., Ltd., China, identified by Dr. Zuoting Yan from Lanzhou Institute of Animal & Veterinary Pharmaceutics of Chinese Academy of Agricultural Sciences. As shown in Fig. 1, all of the herb samples were deposited in a TCM Specimen Room with vouchers number No.100720 for Liquorice, No.101208 for Anemarrhena Rhizome, No.100906 for Gypsum, and No.101202for Rice.

Glycyrrhizae Radix 7.2 g, Anemarrhena Rhizome 21.8 g, Gypsum 60.2 g and Rice 10.8 g were extracted together by refluxing with boiling water twice. They were boiling water (1 : 10, w/v) for 2 h, and water (1 : 5, w/v) for 1 h, respectively. The two-time juice was blended together, filtered with three-layer gauzes, and concentrated to 100 ml experimental decoction. The experi-

Fig.1 Four herb voucher specimens are deposited in the institute herbarium

Note: The specimen bottle was (a) Rice, (b) Liquorice, (c) Anemarrhena Rhizome, and (d) Gypsum.The pie charts were macro-photography of herb specimens from the specimen bottles of a, b, c, andd.

mental juice was autoclaved at 121℃ for 15 min, and kept in airtight containers at 4℃ until used.

2.2 Animals and treatment

Adult New Zealand white rabbits weighted between 2.0kg and 2.5kg were used in the experiments, and housed individually in rabbit stocks under controlled room humidity of 50%±5% and a temperature of (25±1)℃ witha 12 h light and 12 h darkness cycle.Animals were fed commercial stock diet and water *ad libitum*, and allowed to stabilize for at least 3 d in new surroundings before any experiments.All experimental animals were obtained from the animal center of Lanzhou Institute of Biological Products (Lanzhou, China).The rabbits were randomly divided into three groups (6 animals/group).The group I was served as a control, and received only physiological saline; the group II was fever model, which received only LPS (15μg/kg body weight) by intravenous injection (*i.v.*) into the ear vein; the group III was gavaged with BHT (a representative dose of 7 ml/kg body weight) at the same time of LPS intravenous injection.Rabbits then received standard diet and water *ad libitum*.After LPS injection and BHT gavage for 6 h, the rabbits were anesthetized with an intraperitoneal injection of sodium pentobarbital (60mg/kg) for 5 min, and sacrificed immediately.The animal protocol described and the studies here were approved by the Animal Care Committee of Lanzhou Institute of Husbandry and Pharmaceutical Science of Chinese Academy of Agricultural Sciences.

3 cDNA MICROARRAY PROCESSING

After collection of rabbit livers, some liver tissue was taken out, flushed with saline (0.9% NaCl), and finely cut into pieces in RNA later (Takara Biotechnology, China) at 4℃, frozen at -20℃. The hepatic RNA for each rabbit in one group was separately extracted, and then equal amounts of RNA extraction from two animal livers were mixed as one sample. The RNA was quantified by a spectrophotometer (Nanodrop2000, Thermo Scientific, USA), and integrity of total RNA (0.3mg) was determined by formaldehyde-denatured gel electrophoresis (1.2%). A total of nine Agilent 6053122 Rabbit Genome Arrays for each group were applied. Processes for microarray analysis, including RNA extraction, synthesis of double-strandedc DNA, labeling of transcript and fragment, hybridization, washing, staining and scanning of hybridization products, were carried out by CapitalBio Corporation (Beijing, China) according to the Agilent protocol. Briefly, total RNA was extracted and cleaned using TRIzol reagent (Invitrogen, Gaithersburg, MD, USA) and NucleoSpin RNA clean-up kit (Macherey-nagel, Germany), respectively. After validation of RNA samples, double stranded cDNA was synthesized using Superscript II reverse transcriptase, *Escherichia coli* DNA polymerase and *E. coli* DNA ligase. Then, cDNA was used to synthesize biotin-labeled cRNA using a cRNA amplification and labeling kit (CaptialBio, Beijing, China). After cleaning with the genechip samplecle anup liquid, approximately 15mg of cRNA was fragmented and suspended in hybridization buffer. The cRNA fragment was hybridized to the microarray at 42℃ and 50 rpm in hybridization oven 640 (Affymetrix, USA) overnight. After hybridization, the slide was washed with washing buffer (0.03mol/L sodium chloride-sodium citrate, 0.2% sodium dodecyl sulfonate), and then scanned in LuxScan 10KA (CapitalBio Corporation, China). Primary data collection and analysis were carried out using LuxScan 3.0 (CapitalBio Corporation, China). The intensity of each spot represented a unique genes relative expression level, and the date of fluorescence light was acquired. Normalization and comparison analyses of probe values were performed according to the Agilent recommended protocol. Fold changes in gene expression of the treated group were calculated relative to those of the control group. DEGs were identified using the statistical Pvalue of one-way analysis of variance (ANOVA) analysis and fold change, and *P value*<0.05 was considered statistically significant.

3.1 Bioinformatics Analysis

Biological effects of DEGs were identified by gene ontology (GO), pathway, cluster and network analyses. GO analysis could assign genes to various functional categories in terms of GO types: biological process, molecular function, and cellular component. In this study, GO analysis of DEGs was performed using the molecule annotation system 3.0 (MAS 3.0, http://bioinfo.capitalbio.com/mas/) and GO database (http://www.geneontology.org/). Then DEGs were mapped to different biological pathways according to functional categories of the Kyoto Encyclopedia of Genes and Genomes (KEGG) pathway database (http://www.genome.ad.jp/kegg/pathway.html). The significantly altered GO terms and KEGG pathways were identified based on the criteria of having three DEGs and a hypergeometric test. DEGs identified by GO and KEGG analysis were applied to obtain more biological information by cluster analysis. Cluster analysis was carried out by the Cluster program 3. 0 (http: //rana. lbl. gov/EisenSoftware. htm), and the results were visualized and browsed using TreeView program. Venn diagram data were calculated using the Venny web application (http: //bioinfogp.cnb.csic.es/tools/venny/index.html). *P* value and *Q* value used in statistical analysis were calculated by a hyper geometric distribution probability formula, and they reflected whether the statistical result was true or false. In this analysis, both of *P* and *Q* were <0.05 was considered statistically significant.

3.2 Quantitative real-time PCR analysis

Eight genes, related to the inflammation response, defense response, TLR signaling pathway, and acute phase response, were selected for RT-PCR assays to verify the chip data. They were toll-like receptor4 (TLR4), LPS banding protein (LBP), toll-like receptor3 (TLR3), C-reactive protein (CRP), complement 3α (C3α), haptoglobin (HAP), Serum amyloid protein (SAA), and acute phase protein (APP). The gene-specific primers were as follows: forward primer 5′-CAATGGCTCCGGCATGTGC-3′ and reverse primer5′-CGCTTGCTCTGGGCCTCG-3′ for β-actin; forward primer 5′-GCAAGATTTGCAGGCAGATT-3′ and reverse primer 5′-CATCACATCCAA-CATCCCAGC-3′ for LBP; forward primer5′-CATGGAGAAGCTGCTGTGGTG-3′ and reverse primer 5′-GTGAAGGTTTGAGTGGCTTC-3′ for CRP; forward primer5′-GTGGAGACACACCT-GACCTC-3′ and reverse primer 5′-GAAAGGTCCAGGTGCTCAAGG-3′ for TLR4; forward primer5′-GAGCCAGAGTGTGCCTTATC-3′ and reverse primer 5′-GTTGGTGGGCAGGTCATCAG G-3′ for TLR3; forward primer 5′-AACCCTGGACCCAGAAACC-3′ and reverse primer 5′-TGATCTGGGTCATCTTCTCGCG-3′ for C3α; forward primer 5′-GCCGCGGACTTTGGCAAC-3′ and reverse primer 5′-GCTTCTCACTGTTCAAGGC-3′ for HAP; forward primer 5′-CTATCAC-CAATGCCACCCTG-3′ and reverse primer 5′-AGCAGCAAGGTGTCCTCCTC-3′ for APP; forward primer 5′-TCTCTTTGCAGGGCTGCACCTGTC-3′ and reverse primer 5′-TGAGTGGCCTCAC-CAATGAAGGAG-3′ for SAA. Briefly, 2 μg total RNA was reversely transcribed into cDNA by use of an RT-PCR kit (Takara, Dalian, China) following the manufacturer's directions. The information of PCR primers was shown in Table 1. PCR was performed for 39 cycles using the following conditions: preincubation at 95℃ for 30 s, denaturation at 95℃ for 5 s, annealing at 60-62℃ for 25 s, and elongation at 72℃ for 25 s. The PCR products were also verified by ethidium bromide-stained agarose gel electrophoresis (data not shown). Each real-time PCR sample was performed for target gene and the housekeeping gene (β-actin). Real-time PCR was performed for quantification of relative expression of four target genes, and using the quantitative PCR Super Mix in a final reaction volume of 25 μl in 2×SYBR green (Takara, Dalian, China) according to the manufacturer's protocol. The data were expressed as the number of threshold cycle (Ct). The relative quantification of target genes was determined by calculating the ratio between the concentration of the target gene and that of housekeeping gene with the method of $2^{-\Delta\Delta Ct}$. Each PCR result was a statistical analysis of the experiment in triplicate.

Table1 Biological function and classification of genes involved in the LPS recognition system inliver tissue

Gene title	ene symbol	UniGene ID	Functional annotation
Family of LPS binding protein			
LPS binding protein	LBP	M35534	Inflammatory response
Surfactant protein B	SP-B	U17106	Metabolic process
Surfactant protein A	SP-A	J03542	Metabolic process
Serum amyloid protein	SAP	NM_ 001082302	Response to inflammation
Apolipoprotein B-48	APOB	M17780	Metabolic process
Apolipoprotein A-I	APOA	X06658	Lipid transport, localization, and homeostasis
Cationic antimicrobial peptides18	CAP18	M73998	Defense response
C reactive protein	CRP	M13497	Inflammatory response
Family of LPS receptor			
Cluster of differentiation antigen 14	CD14	D16545	
Cluster of differentiation antigen 36	CD36	AF412572	Response to lipid, bacterium, and drug
Vascular cell adhesion molecule-1	VCAM-1	AY212510	Response to stress, stimulus, and wounding
Selectin L	SELL	U26535	Biological adhesion
Integrin-α6	ITGA6	AB231856	Integrin-mediated signaling pathway
Toll-like receptor4	TLR4	AY101394	TLR4 signaling pathway
Activated cluster of LPS receptor			
Membrane metallo-endopeptidase	MME	X05338	Metabolic process
Prostaglandin Ereceptor2	PTGER2	AY166779	Cellular response to prostaglandin stimulus
Nuclear receptorsubfamily3, group C1	NR3C1	AY161275	Intracellular receptor signaling pathway
cAMP-dependent protein kinase	PKA	AF367429	cAMP-mediated signaling
Alanine: glyoxylate aminotransferase	AGT	M84647	Alpha-amino acid catabolic process

4 RESULTS

4.1 Integrity of total RNA

Every sample RNA was detected in formaldehyde-denatured gel electrophoresis, and result indicated that there were three notable bands in every electrophoresis lane.They were 28S rRNA, 18S rRNA, and 5S rRNA, respectively (Fig.2).For every RNA sample, the value of 28S: 18S rRNA was approximate to 1, and total quantity of RNA over 8 μg.Moreover, the ration of A260/A280 from every sample was between 1.8 and 2.0 (data not shown).Theses howed that the quality, purity, and quantity of RNA sample were up to the standards of microarray, and there were no protein and DNA residual in total RNA sample.Therefore, the total RNA sample was complied with microarray result.

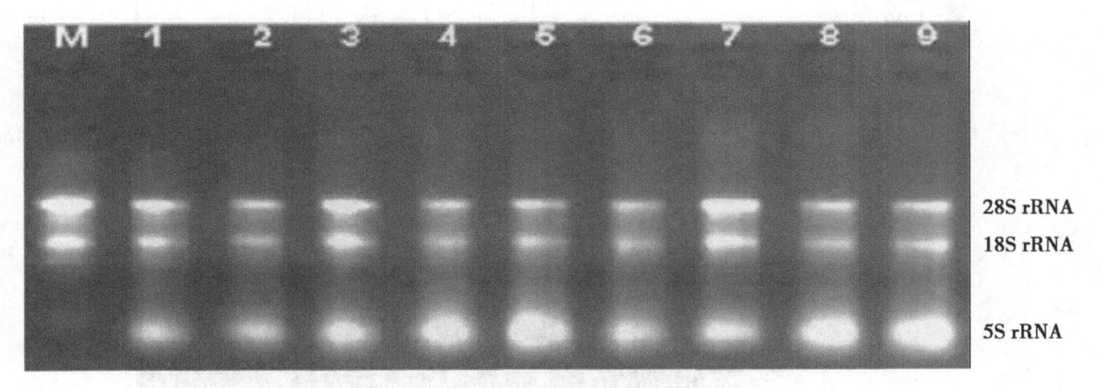

Fig.2 Integrity detection of total RNA

Note: M was detection of rRNA from human Hela cells; total RNA of lane 1-3 were control groups, 4-6 were model groups, and 7-9 were BHT prevented groups.

4.2 Clustering analysis

All differentially expressed genes were clustered through the method of two-way clustering.Individual liver tissue gene expression profiles may predict the LPS febrile symptom and the effect of BHT treatment.Expression levels were indicated by acolor scale for each of three group rabbits.Visual inspection of liver gene expression profiles (Fig.3) displayed that every triplicate could be readily distinguished as controls, febrile group, and BHT treatment, respectively.

4.3 Numbers of DEGs

In microarray studies, a fold-change cutoff of 1.5 or 2.0 was commonly chosen to identify DEGs (Rouse et al., 2007; Wei et al., 2008).It was highlighted in Fig.4, 1467 DEGs were identified from 5919 positive genes in the LPS group compared with the control group.Among these genes, 606 genes were up-regulated, and 859 genes were down-regulated.However, there were just 535 DEGs in BHT group compared to the control group, and 106 genes were up-regulated, 429 genes were down-regulated.From the data in Venn diagram, two experimental groups had 286 common DEGs. Thereinto, 38 up-regulated genes in the LPS group were altered invertedly in BHT groups, and BHT failed to influence another 238 DEGs in expression.

4.4 Functional classifications of DEGs

Functional annotations of the DEGs that were significantly altered after LPS injection and BHT treatment were initially performed using molecular annotation system 3.0. Three main types of annotations were obtained from the gene ontology consortium website: cellular components, metabolic functions, and biological process. Enrichment analysis of biological processes showed that both of LPS and BHT primarily affected metabolic process (Fig.5A and D).Enrichment analysis of cellular component displayed that DEGs mainly located in intracellular cytoplasm and membrane-bounded organelle (Fig. 5B and E). Furthermore, from the perspective of molecular functions enrichment analysis, products of DEGs mainly take part in oxidoreductase activity and catalytic activity in LPS-

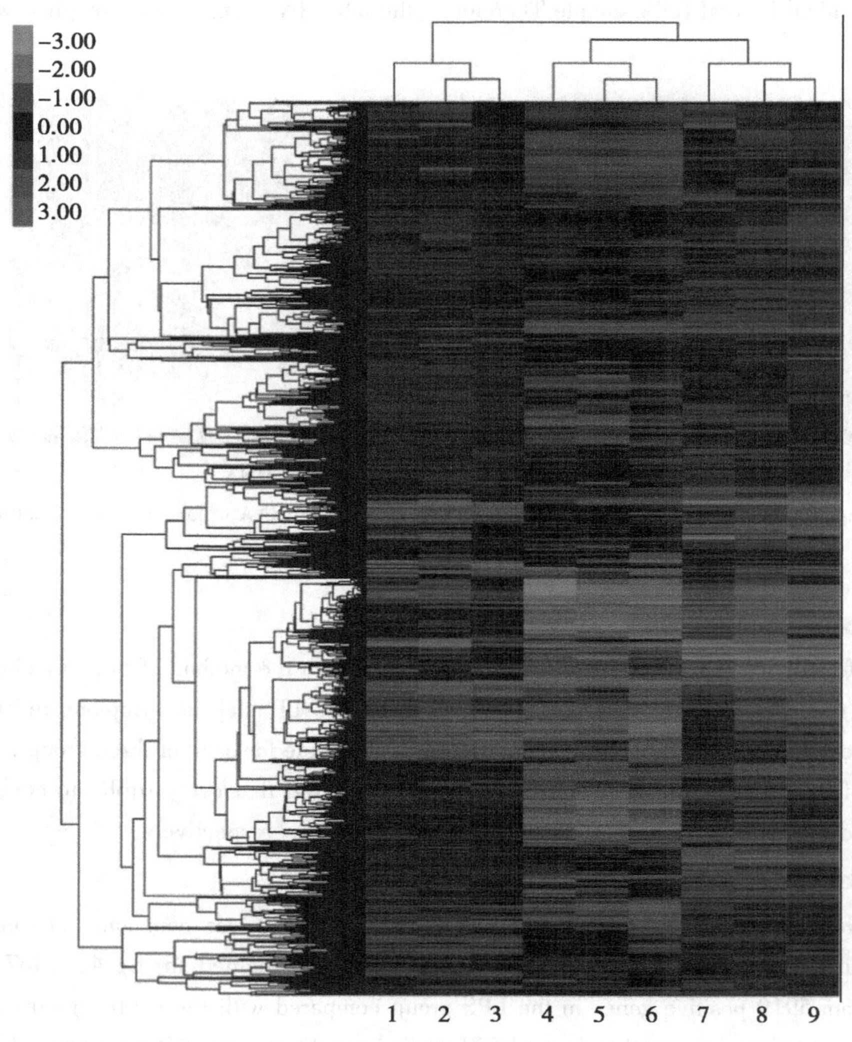

Fig.3 Hierarchical clustering of differentially expressed genes in rabbit

Note: The log2-transformed ratios are shown in a red-green color scale, where red represents a higher expression and green represents a lower expression. The numbers represented different experimental groups. 1-3 were control in triplicate; 4-6 were febrile model in triplicate; 7-9 were BHT-treated group in triplicate. (For interpretation of the references to color in this figure legend, the reader is referred to the web version of this article.)

injected animals (Fig.5C), but changed into oxidoreductase activity, nucleotide binding, organic small molecule binding, and catalytic activity was the most dominant item in BHT-treated animals.

4.5 Kyoto encyclopedia of genes and genomes pathways

The top 11 pathways of DEGs were highlighted from the LPS febrile group and BHT treated group as shown in Fig.6. Pathway analysis of DEGs was conducted based on KEGG. From the data in Fig.6A, it was apparent that transcript data were focused on protein processing in the endoplasmic reticulum and phagosome when animals were injected with LPS. And in Fig.6B, the pathway of phagosome significantly increased by almost half in BHT-treated animal.

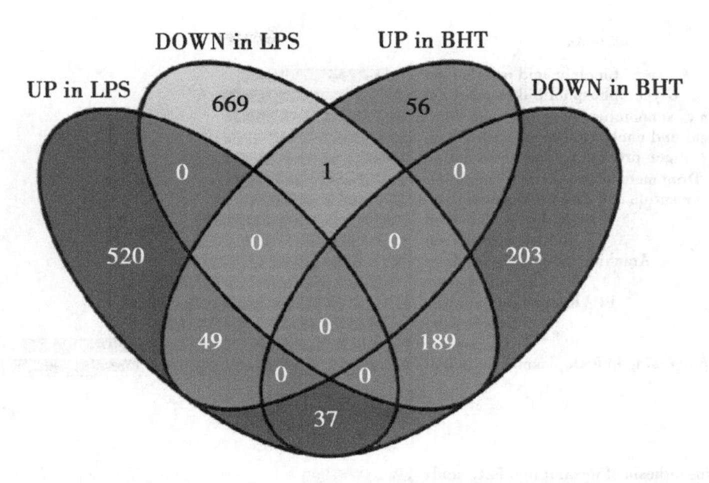

Fig.4 A Venn diagram showing the distribution of shared DEGs compared with control in LPS group and BHT treatment group, respectively

Note: Blue part was upregulated genes in the LPS group (UP in LPS), yellow part was down-regulated genes in the LPS group (DOWN in LPS); while green part was up-regulated genes in BHT-treated group (UP in BHT), red part was down-regulated genes in BHT-treated group (DOWN in BHT). The overlap sections were the numbers of common genes in different groups. (For interpretation of the references to color in this figure legend, the reader is referred to the web version of this article.)

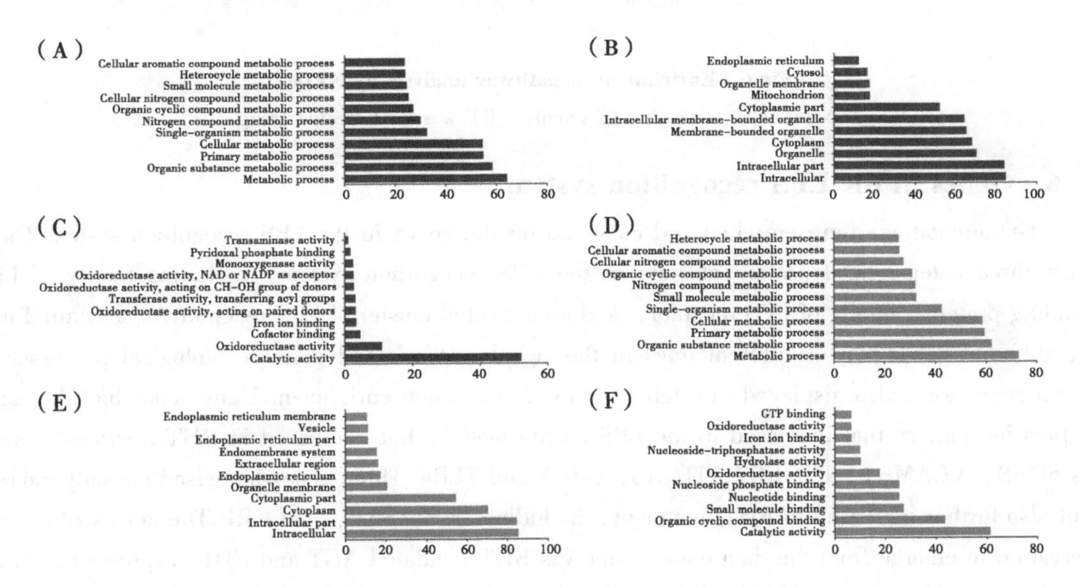

Fig.5 Classification of the differentially expressed genes based on their functional annotations using the Gene Ontology

Note: (A and D) GO biological processes; (B and E) GO cellular components; (C and F) GO molecular functions. (A) -(C) were LPS groups, and (D) -(F) were BHT treated groups. Vertical coordinates were the items of different functional annotations, and abscissas were the percentage of genes referred to the corresponding annotation in all DEGs.

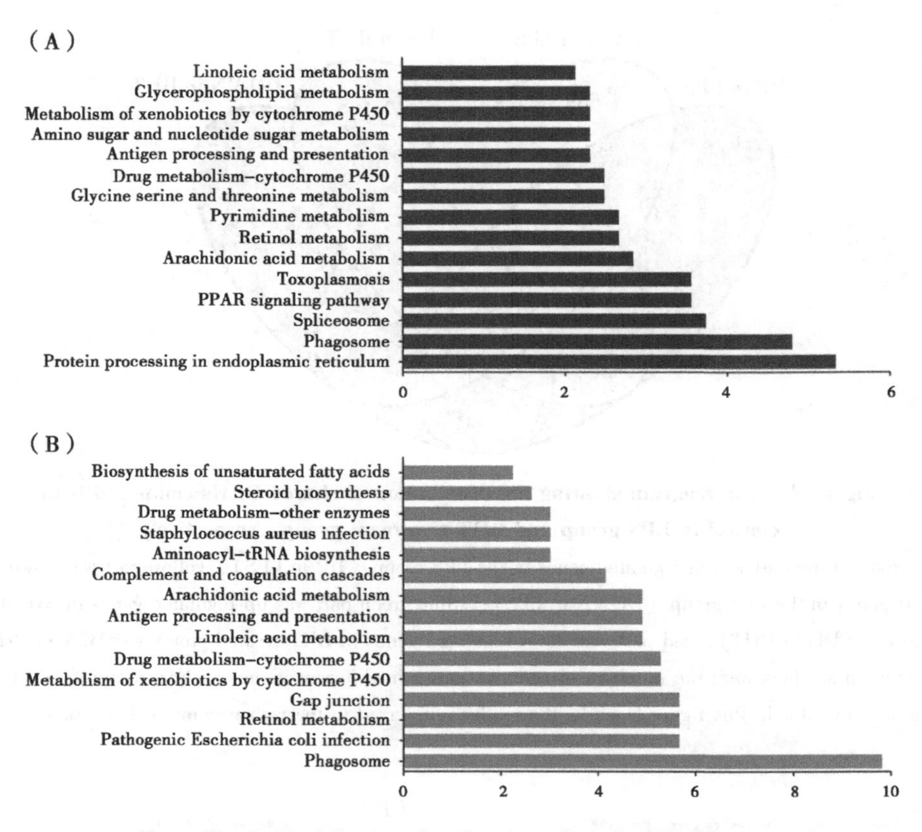

Fig.6 Enrichment of pathway analysis by KEGG

Note: (A) was the LPS group; (B) was BHT treated group.

4.6 Genes in the LPS recognition system

GO annotations were enriched and collected for the genes in the LPS recognition system. There were three categories of protein clusters in the LPS recognition system, including family of LPS binding protein, family of LPS receptor, and an activated cluster of LPS receptor. As shownin Table 1, different genes played different roles in the signal system, and possible biological processes of these genes were also displayed in a table from GO annotation enrichment. Many genes had the same expression pattern that increased in the LPS febrile model, but decreased in BHT treatment, such as SP-B, VCAM-1, PTGER2, ITGA6, SP-A and TLR4. Three genes increased not only model, but also further increased in BHT treatment, including LBP, SAP, and CRP. The most striking observation to emerge from the data comparison was BHT regulated AGT and CD14 expressed at lower levels than control, especially for significantly increased AGT in LPS group (Fig.7).

4.7 Quantitative real-time PCR for genes expression

To verify the results of microarray analysis, mRNA expression of eight genes was examined independently by qRT-PCR. The results showed that differential expression changed as the same trend as gene microarray analysis, even though the magnitude and statistical significance of changes differed between the two methods. Fig.8 highlighted the relative expression level of eight genes, respec-

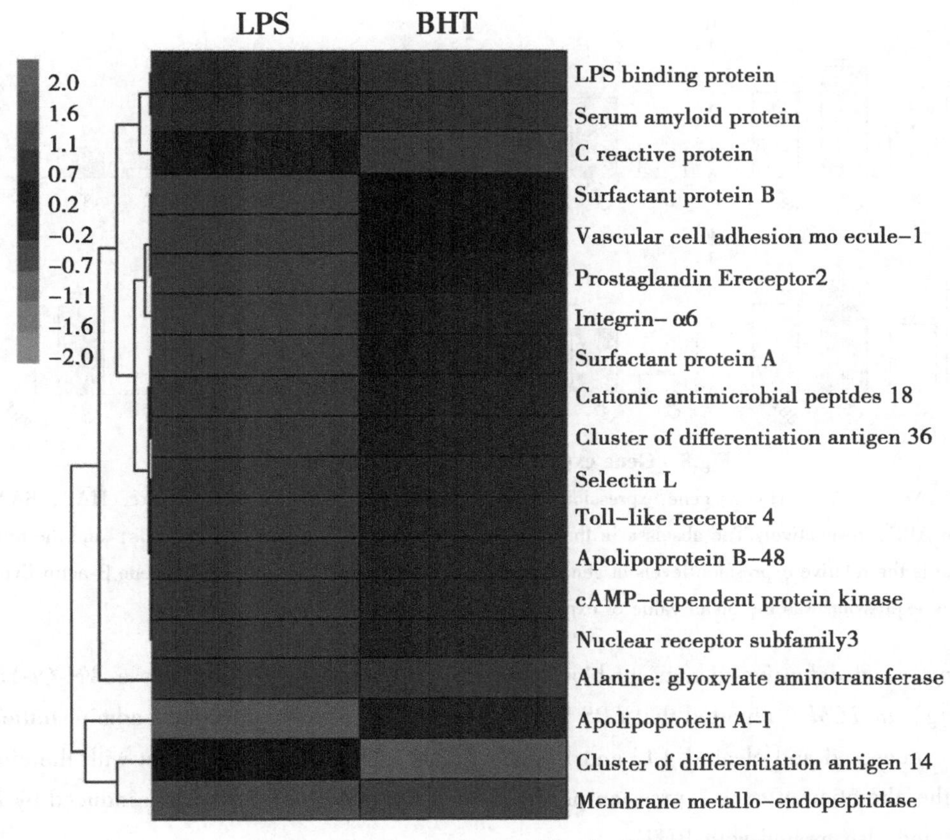

Fig.7 Expression profile of genes involved in the LPS recognition system

tively.RT-PCR results of eight DEGs were consistent with the results of microarrays.

5 DISCUSSION

For ethnopharmacologist, traditional medicine is now entering a new era when gene science is benefiting human health.Omics in the study of TCM syndrome may be certainly a forefront subject for ethnopharmacologist to reveal the mystery.Microarray based determination of altered gene expression might be an alternative approach, and variation in gene expression was increasingly used for exploration of the impact of natural products (Van Poecke et al., 2007).Although the BHT has caused wide public concern and practiced in many countries, its mechanism in modern life science is still not be discovered.So study on gene expression alteration may allow to searching for candidate genes and molecular factors involved in the mechanism of TCM pattern and drug action.The liver is the largest internal organ in mammals, and is important for the maintenance of normal physiological functions of other tissues and organs (Chen and Zeng, 2011).However, modern medicine indicates that liver is primarily responsible for biotransformation and detoxification of xenobiotics (Wu et al., 2013).In TCM theory, the liver is at middle energizer (in Chinese *Zhong Jiao*), governing blood store and free flow of Qi, has a particularly important role in response to pathogenesis transmission according to the principle of triple-energizer syndrome differentiation (in Chinese *Sanjiao*

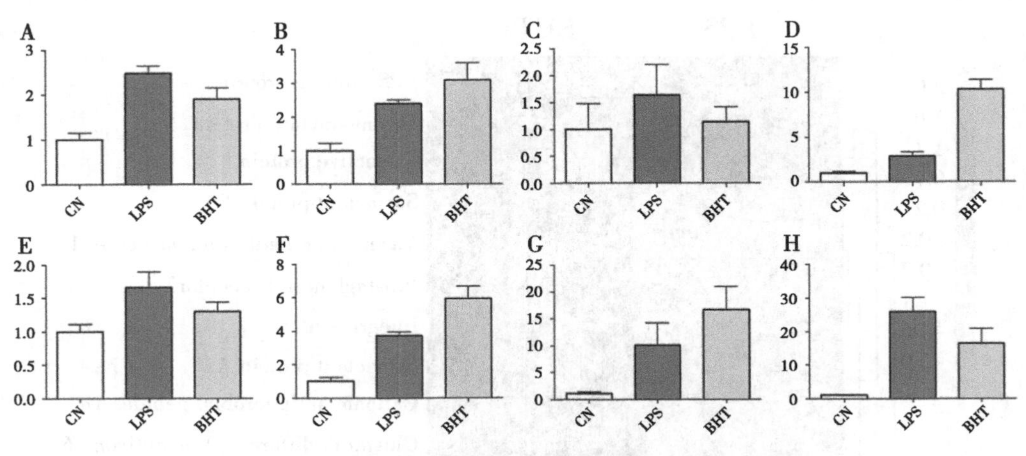

Fig.8 Gene expression verification by qRT-PCR

Note: (A)-(H) was gene expression levels of TLR4, LBP, TLR3, CRP, C3α, HAP, SAA, and APP, respectively.The abscissa in the pictures is the group of experimental animals, and the ordinate is the relative expression levels of genes, which were compared to housekeeping gene β-actin.Every gene expression was the mean value of experiments in triplicate.

Bianzheng) and defensive-qi-nutrient-blood syndrome differentiation (in Chinese *Wei-Qi-Ying-Xue Bianzheng*) *in TCM.It* reported that *LPS* firstly arrived at liver after intravenous administration, and then was processed and cleared (Li and Blatteis, 2004).This study was set out with the aim of assessing the alteration of gene expression profile in liver tissue of the fever rabbits induced by LPS injection, and also treated with BHT.

In this study, comparison LPS group and the BHT group with control showed that there were 1465 and 1401 genes altered expression levels, respectively. It was interesting to note that in all changed genes there were just 276 common genes of two groups.Among these common genes, 38 upregulated genes were reversely expressed by BHT, and other 238 altered genes were not re-changed significantly (Fig.4).Based on the data from microarray analysis, DEGs and associated biological processes were analyzed to evaluate the toxice ffect of LPS and cure effect of BHT.The top eleven altered biological processes and cellular components obtained by GO analysis showed that metabolic processes, which located in the membrane system at intracellular part, were primary body reaction to LPS, and they were also increased by the treatment of BHT (Fig. 5D and E). This may be directly associated with the function of biotransformation and detoxification of xenobiotics in the liver. Except for the above analysis, molecular function analysis showed that altered genes mainly referred to catalytic activity and oxidoreductase activity. It was interesting to note that binding activities increased significantly in treatment with BHT, including organic cyclic compound binding, small molecule binding, nucleotide binding, and nucleoside phosphate binding (Fig.5F).Another important finding was the alteration of KEGG pathway comparing LPS group with BHT group. Similarly, top eleven pathways of altered genes showed that phagosome and protein processing in endoplasmic reticulum contained the most altered genes in the LPS group (Fig.6A), but the percentage of phagosome pathway almost doubled in BHT group, and the pathway of protein processing in endoplasmic reticulum failed to rise in the top 11 (Fig. 6B). Phagosomal accumulation often increases nonspecific uptake and retention in the liver (Choi, 2014), and numerous phagosomes may increase quantities of rough endoplasmic reticulum and prominent Golgi zones related to the synthesis and secretion of different kinds of plasma proteins for LPS recognition and drug metabolism. These results manifested that the organism responded to LPS by processing and activating the proteins for catalytic activity and metabolic process, BHT mainly enhanced the nonspecific phagocytosis for LPS clearance.

Actually, liver was a main LPS-processing and clearance organs *in vivo* (Li et al., 2012). When entered into the liver, LPS can be recognized, distinguished, and transported by organism, then the signal transduction system was mediated, inactivated, and evoked corresponding physiological response (Dauphinee and Karsan, 2006). TLRs play a central role in the activation of the innate system as a result of the recognition of bacterial pathogenassociated molecular patterns, especially LPS (Li et al., 2012). TLR4 recognizes LPS in association with cofactors such as CD14 and LBP (Lee et al., 2012; Naser et al., 2011), then triggers the activation of different intracellular signaling cascades, such as activation of NF-κB (Haller et al., 2002). The downstream effects of LPS recognition elicit effector mechanisms aimed at pathogen eradication. During bacterial infection, the innate immune system detects a restricted number of bacterial structures, such as LPS, and activates signaling pathways conveying an inflammatory reaction aimed at eradication of the pathogen. It was important for LPS-inducing disease to be prevented and treated at this crucial stage (Ding and Jiang, 2007). In order to better understand the gene transcription alterations of this signal process taking place in liver tissue, DEGs involved in the LPS recognition signal system were highlighted. We compared the gene expression between the LPS group and BHT group, and noticed that TLR4 and LBP were expressed much higher in LPS group, but CD14 almost no change. However, LBP expression further increased in BHT group, but TLR4 barely risen, and CD14 was down-regulated significantly (Fig. 7). Most other signal recognition-associated genes showed a similar pattern that higher expression in LPS group and partly decreased in BHT group. These results provided evidence that LPS recognition significantly took place at the level of gene expre ssion, and BHT interfered with the signal recognition system and downstream events by inhibiting or weakening the gene expression. This may be amechanism of BHT for treatment for febrile disease. Of particular interest is the notion that acute-phase proteins of SAP and CRP increased sharply in both of two groups. Though they were hallmarks of systemic inflammation (Menschikowski et al., 2013), as LPS binding proteins they may contribute to both of inflammatory response and drug metabolism.

6 CONCLUSION

This study identified 1465 DEGs in the liver of febrile rabbit, and 535 DEGs after treatment with BHT. Bioinformatics analysis of the DEGs demonstrated that BHT could enhance the catalytic activity and molecular binding activities, promote phagosome pathway increase significantly. This may be one important mechanism of anti-febrile function of BHT. The study provided an important pioneering result on gene expression profiling research. It will facilitate the clinical care or further studies of BHT at level of gene expression regulation. Moreover, research on differentially expressed proteomics (DEPs) of BHT about antifebrile effect is under study.

ACKNOWLEDGMENTS

This study was supported by the National Natural Science Foundation of China (31402244) and the Central Scientific Research Institutes for Basic Research Fund of China (1610322015012). Gratitude is also sent to the anonymous and editor whose suggestions greatly improved the manuscript.

References

[1] Ai, B.S., Zhao, G.R., He, Y.S., Xiao, B.Y., 2011. Experimental study on feasibility of Wei-qi-ying-xue syndrome in rabbit with endotoxemia. Inf. Tradit. Chin. Med. 28, 23-25.

[2] Aldridge, J.R., Moseley, C.E., Boltz, D.A., Negovetich, N.J., Reynolds, C., Franks, J., Brown, S.A., Doherty, P.C., Webster, R.G., Thomas, P.G., 2009. TNF/iNOS-produ-cing dendritic cells are the necessary evil of lethal influenza virus infection. Proc. Natl. Acad. Sci. U. S. A. 106, 5306-5311.

[3] Campbell, A.M., Zanta, C.A., Heyer, L.J., Kittinger, B., Gabric, K.M., Adler, L., Schulz, B., 2006. DNA microarray wet lab simulation brings genomics into the high school curriculum. CBE Life Sci. Educ. 5, 332-339.

[4] Chai, K.F., Huang, X.L., 2010. Study on gene expression of Yin-deficiency and excess of heat syndrome in type 2 diabetes mellitus. Chin. J. Tradit. Med. Sci. Technol. 17, 185-189.

[5] Charles, C., 2000. Bitter realities: applying Wenbing principles in acute respiratory tract infections. J. Chin. Med. 63, 12-17.

[6] Chen, C.C., Hsiang, C.Y., Chiang, A.N., Lo, H.Y., Li, C.I., 2008. Peroxisome proliferatoractivated receptor gamma transactivation-mediated potentiation of glucose uptake by Bai-Hu-Tang. J. Ethnopharmacol. 118, 46-50.

[7] Chen, X., Zeng, F., 2011. Directed hepatic differentiation from embryonic stem cells. Protein Cell 2, 180-188.

[8] Choi, H.S., 2014. Nanoparticle assembly: building blocks for tumour delivery. Nat. Nanotechnol. 9, 93-94.

[9] Dauphinee, S.M., Karsan, A., 2006. Lipopolysaccharide signaling in endothelial cells. Lab. Investig. 86, 9-22.

[10] Ding, N., Jiang, Y., 2007. Bacterial lipopolysaccharide recognition system. Chin. J. Biochem. Mol. Biol. 23, 711-717.

[11] Draghici, S., Khatri, P., Eklund, A.C., Szallasi, Z., 2006. Reliability and reproducibility issues in DNA microarray measurements. Trends Genet. 22, 101-109.

[12] European Herbal and Traditional Medicine Practitioners Association, 2007. The Core Curriculum for Herbal and Traditional Medicine (Second Edition), Gloucestershire, pp.69.

[13] Fred, J., Bob, F., 2004. Herb Toxicities and Drug Interactions: A Formula Approach. Blue Poppy Press, Colombia, pp.44-45.

[14] Haller, D., Russo, M.P., Sartor, R.B., Jobin, C., 2002. IKKβ and phosphatidylinositol 3-Kinase/Akt participate in non-pathogenic gram-negative enteric bacteriainduced RelA phosphorylation and NF-κB activation in both primary and intestinal epithelial cell. J. Biol. Chem. 277, 38168-38178.

[15] Jia, L.L., Li, R., Ma, J., Fan, Y., Li, H.B., 2013. Effects of Bai-Hu decoction on fever induced by lipopolysaccharide.KaohsiungJ.Med.Sci.29, 128-132.

[16] Li, Q., Liu, Y., Che, Z., Zhu, H., Meng, G., Hou, Y., Ding, B., Yin, Y., Chen, F., 2012. Dietary L-arginine supplementation alleviates liver injury caused by *Escherichia coli* LPS in weaned pigs.Innate Immun.18, 804-814.

[17] Li, Z., Blatteis, C.M., 2004. Fever onset is linked to the appearance of lipopolysaccharide in the liver.J.Endotoxin Res.10, 39-53.

[18] Liu, Q.H., Lin, L.Z., Zhou, D.H., 2008. Progression of traditional chineseme dicine Zheng-hou genomics.Chin.Arch.Tradit.Chin.Med.26, 1293-1295.

[19] Liu, J.Y., Wang, Q.X., 1999. Diagnosis and treatment of warm disease-Qifen syndrome.Chin.J.Clin. 27, 2-5.

[20] Lee,H.S., Hattori,T., Park, E.Y., Stevenson,W., Chauhan,S.K., Dana, R., 2012.Expression of toll-like receptor 4 contributes to corneal inflammation in experimental dry eye disease.Investig.Ophthalmol.Vis.Sci.53, 5632-5640.

[21] Menschikowski, M., Hagelgans, A., Fuessel, S., Mareninova, O.A., Asatryan, L., Wirth, M. P., Siegert, G., 2013. Serum amyloid A, phospholipase A2-IIA and C-reactive protein as inflammatory biomarkers for prostate diseases.Inflamm.Res.62, 1063-1072.

[22] Naser, S.A., Romero, C., Urbina, P., Naser, N., Valentine, J., 2011. Cellular infiltration and cytokine expression correlate with fistulizing state in Crohn's disease. Clin. Vaccine Immunol. 18, 1416-1419.

[23] Ni, Q.Q., Wei, K.A., 2012. Study on feasibility of the Qi-aspect pattern and nutrientblood-aspect pattern model of rabbits made by injecting LPS.Guid.J.Tradit.Chin.Med.Pharm.18, 9-12.

[24] Rouse, R.L., Boudreaux, M.J., Penn, A.L., 2007. In utero environmental tobacco smoke exposure alters gene expression in lungs of adult BALB/c mice.Environ.Health Perspect.115, 1757-1766.

[25] Van Poecke, R.M., Sato, M., Lenarz-Wyatt, L., Weisberg, S., Katagiri, F., 2007. Natural variation in RPS2-mediated resistance among Arabidopsis accessions: correlation between gene expression profiles and phenotypic responses.Plant Cell19, 4046-4060.

[26] Wang, H., Ma, S., 2008. The cytokine storm and factors determining the sequence and severity of organ dysfunction in multiple organ dysfunction syndrome.Am.J.Emerg.Med.26, 711-715.

[27] Wang, Z.L., Zhao, S.J., Chen, Z.Q., Tu, J.G., Wu, L.J., Wei, X.L., Li, S.M., Tian, M. C., 2010. Effect of a traditional Chinese formula on temperature and serum antioxidant indicates in rabbit with Qifen Syndrome.Chin.J.Vet.Med.46, 50-52.

[28] Wei, L.L., Sun, B.J., Song, L.R., Nie, P., 2008. Gene expression profiles in liver of zebrafish treated with microcystin-LR.Environ.Toxicol.Pharmacol.26, 6-12.

[29] Wu,B., Liu,S., Cheng,S.P., Zhang,Y., Zhang,X.X., 2013.Hepatic gene expression analysis of mice exposed to raw water from Meiliang Bay,Lake Taihu,China.J.Appl.Toxicol.33,1416-1423.

[30] Yang, J., 2002. Development of animal model with epidemic febrile disease.Chin.J.Lab.Anim.Sci.12, 61-64.

[31] Yu, Z.M., Wu, J.J., Zhao, G.F., Zhang, X.W., Li, X.P., Zhong, J.X., 2010. Endue rabbit experimental model of epidemic febrile disease by lipopolysaccharide infusion.Chin.Arch.Tradit.Chin. Med.28, 286-288.

[32] Zhang, B.L., Wang, X.H., 2006. Current research on syndromes.Re-educ.Med.China 20, 1-4.

[33] Zhang, S.D., Wang, D.S., Wang, X.R., Li, S.H., Li, J.Y., Li, H.S., Yan, Z.T., 2013. Aqueous extract of Bai-Hu-Tang, a classical Chinese herb formula, prevents excessive immune

response and liver injury induced by LPS in rabbits.J.Ethnopharmacol.149, 321-327.

[34] Zhu, B.M., Xu, F., Yoshinobu, B.B., 2006. An evaluation of linear RNA amplication in cDNA microarray gene expression analysis.Mol.Genet.Metab.87, 71-79.

[35] Zhu, L.C., 2007. *Shang Han Lun* is the footstone of dialectical cure in TCM.Tradit.Chin.Med.J.6, 14-19.

（发表于《 Journal of Ethnopharmacology》，院选 SCI，IF：2.998）

Determination and Pharmacokinetic Studies of Aretsunate and its Metabolite in Sheep Plasma by Liquid Chromatography-tandem Mass Spectrometry

LI Bing[1], ZHANG Jie[2], ZHOU Xu-zheng [1], LI Jian-yong[1], YANG Ya-jun [1], WEI Xiao-juan[1], NIU Jian-rong[1], LIU Xi-wang[1], LI Jin-shan[1], ZHANG Ji-yu [1 *]

(1.Key Laboratory of Veterinary Pharmaceutical Development, Ministry of Agriculture/ Key Laboratory of New Animal Drug Project of Gansu Province/Lanzhou Institute of Husbandry and Pharmaceutical Sciences of CAAS, Lanzhou 730050, China; 2.Fulin Animal Science and Veterinary Medicine Officer, Chongqing 408000, China)

Abstract: A rapid and sensitive high-performance liquid chromatography–tandem mass spectrometry (LC–MS/MS) method was developed and validated to simultaneous quantify artesunate and its metabolite in sheep plasma.The plasma samples were prepared by liquid–liquid extraction.Chromatographic separation was achieved on a C18 column (250mm×4. 6mm, 5μm) using methanol: water (60 : 40, v/v) (the water included 1mM ammonium acetate, 0. 1% formic acid, and 0. 02% acetic acid) as the mobile phase. Mass detection was carried out using positive electrospray ionization in multiple reaction monitoring mode.The calibration curve was linear from 1ng/mL to 400ng/mL (r^2 = 0. 9992 for artesunate, r^2 = 0. 9993 for its metabolite).The intra- and inter-day accuracy and precision were within the acceptable limits of ±10% at all concentrations for both artesunate and its metabolite.The recoveries ranged from 92% to 98% at the three concentrations for both.In summary, the LC–MS/MS metho described herein was fully successfully applied to pharmacokinetic studies of artesunate nanoemulsion after intramuscular delivery to sheep.

Key words: Antiparasitic; Artesunate nanoemulsion; LC/MS/MS; Pharmacokinetics; Sheep plasma

1 INTRODUCTION

Theileriosis, caused by various intraerythrocytic protozoan parasites of the genus *Theileria*, is a tick-borne disease of domestic and wild animals.Theileriosis, also known as blood cryptosporidiosis, is a blood protozoosis caused by parasites to infect leukocytes and erythrocytes[1], is a tick-borne disease of domestic and wild animals[2].This epidemic disease is causing serious damage to the cattle

* Corresponding author. E-mail: infzjy@ sina.com.

and sheep industries.It is strongly seasonal and local, and present in many countries, research carried out over many years has shown these to be distributed mainly in Northern China but also sometimes in Southern China[3].*Theileria lestoquardi* is a highly pathogenic parasite of sheep and goats., which occurred in south-eastern Europe, northern Africa, western and central Asia[4], India[5] and in West, Central and North China such as Qinghai, Gansu, Hubei, Liaoning and Inner Mongolia[6-8].Non-pathogenic or mildly pathogenic *Theileria* spp.of small ruminants include *Theileria separata*, *Theileria ovis* and *Theileria recondite*[9].Clinical symptoms in sick sheep include high fever, depression, anemia, conjunctival yellowish discoloration, superficial swollen lymph nodes, limb rigidity, difficulty walking, rapid weight loss, weakness, and finally failure and death[10]. Treatment of the disease is problematic, as there are few effective medicines.Despite the large number of studies on this disease, efficient new drugs are still lacking.The drugs presently in use are mainly berenil, chloroquine phosphate, artesunate, uranidin, and imidocarb. Berenil is the most commonly used, but its popularity has lessened due to the major side effects that can accompany its injection.

Artesunate is a semi-synthetic derivative of artemisinin, extracted from the plant species of *Artemisia*.Artesunate is a highly efficacious anti-malarial[11,12] and also has anti-tumor activity[13].It is also effective for treating some non-malarial parasites[14,15], for example, it is an effective therapeutic for *Theileria annulata* infection of cattle and sheep.However, artesunate still presents a few problems, including limited bioavailability, a short half-life, and also poor water solubility[16], which makes it difficult to deliver effectively to the lesion site and intracellular space.Further, at present there is only an oral tablet and bulk drugs for clinical use.Also, the poor stability of the drug's sodium salt causes its low bioavailability and necessitates frequent patient medication, leading to poor tolerance and efficacy, and greatly limiting its clinical application.New preparations of derivatives with better pharmacokinetic profiles are needed to overcome these issues.

In our previous study, we have successfully prepared artesunate nanoemulsion for the first time in China as an intramuscular preparation that increases the solubility of artesunate and its sodium salt.It has a clear, transparent appearance, an average emulsion droplet particle size of 60 nm, a good degree of dispersion, thermodynamic stability, storage stability, and long-placing non-hierarchical[17].Acute toxicity studies of oral artesunate nanoemulsion in mice indicate that it is nontoxic[17]. Artesunate nanoemulsion has significant preventive and treatment effects in Taylor piroplasmosis of goats[18].

In order to define the pharmacokinetic profile of artesunate nanoemulsion, an high performance liquid chromatography.tandem mass spectrometry (LC.MS/MS) method for the determination of artesunate and its metabolite in sheep plasma needs to be established.Because the artemisinin and its analogs contains no conjugated groups in the structure (Fig.1), they have no appropriate chromophores for use in characterization techniques their quantitation sometimes requires the use of lengthy derivatization techniques, and these derivatizing conditions are not necessarily stable for all artemisinin analogs and can cause the parent compound to decompose[19].Hence, we chose LC.MS/MS as an analytical method that is accurate and sensitive, using a simple liquid.liquid extraction as the organic phase is evaporated and reconstructed before analysis approach for artesunate nanoemulsion.

This method has been successfully applied to characterize the pharmacokinetics of artesunate nanoemulsion in sheep following intramuscular administration at 5mg/kg.

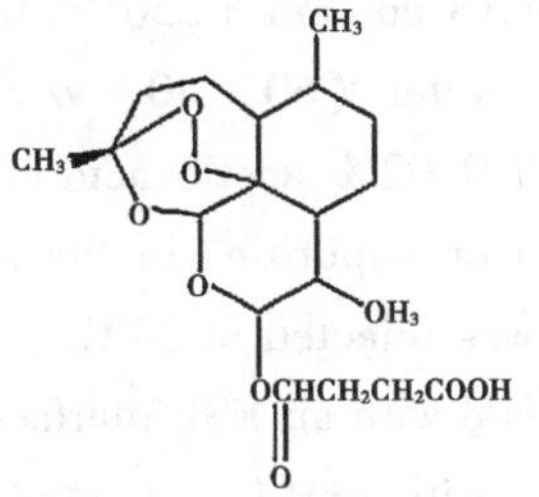

Artesunate (AR, molecular weight: 384 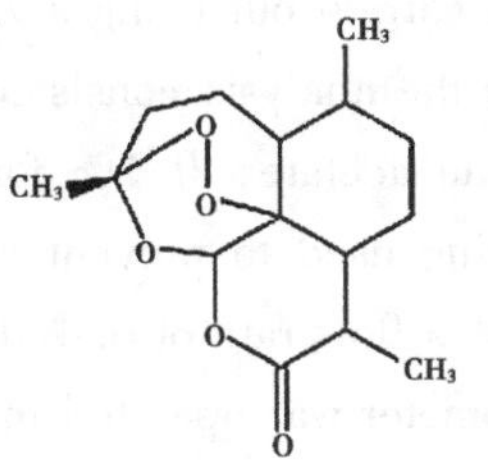)　　Artemisinin (AS, molecular weight: 282)

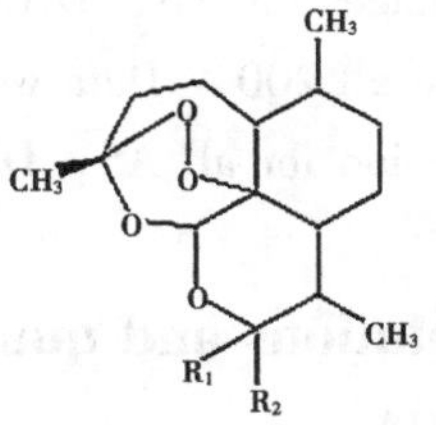

Dihydroartemisinin (DHA, molecular weight: 284)

α -R_1=HR_2=OH, β -R_1=OHR_2=H

Fig.1 Chemical structures of artesunate, artemisinin and dihydroartemisinin

2 EXPERIMENTAL

2.1 Reagents and chemicals

Artesunate (AR), dihydroartemisinin (DHA), and internal standard (IS) artemisinin (AS) were provided by the National Institute for the Control of Pharmaceutical and Biological Products (Beijing, China) with the batch numbers of 100200 - 200202, 100184 - 200401, and 100202 - 200603, respectively, the purity of AR, DHA and IS are > 99%. Acetonitrile and methanol (HPLC grade) were purchased from Fisher Chemical (Waltham, USA). Acetic acid, ethyl acetate, ammoniumacetate, and formic acid were analytically pure and were purchased from Sinopharm Group Chemical Reagent Co., Ltd. (Ningbo, China). Water was purified through a Milli-Q Plus water system (Millipore Corporation, Bedford, MA, USA) before use. Artesunate nanoemulsion (containing artesunate 5%) was supplied by Lanzhou Institute of Animal Science and Veterinary Pharmaceutics.

2.2 Equipment

The LC-MS/MS equipment (1200-6410A) consisted of a LC system with a binary pump-SL and a triple quadrupole mass spectrometer with electrospray ionization (ESI) (Agilent Technologies, Inc., Santa Clara, CA, USA). Data were recorded, and the system was controlled using MassHunter software (version B.01.04, Agilent Technologies).

2.3 Chromatographic and mass spectrometer conditions

The analysis was carried out using a Agilent Zorbox C18 column (250×4.6mm, 5μm); the mobile phase used for the analysis consisted of methanol: water (60 : 40, v/v) (the water included 1mM ammonium acetate, 0.1% formic acid, and 0.02% acetic acid). The mobile phase was filtered before being used to prevent entry of bubbles or impurities in the system. The mobile phase was delivered at a flow rate of 0.4mL/min and 5L was injected at 30℃.

The mass spectrometer was operated in positive ion mode with an ESI interface. Quantitation was performed by multiple reaction monitoring (MRM). In the positive mode, theMS/MSsetting parameters were as follows: capillary voltage 4 kV, cone voltage 40 V, source temperature 100℃, and desolvation temperature 350℃ with a desolvation nitrogen gasflowof 11L/min and a cone gasflow of 9L/min. The optimized fragmentation voltages for AR, DHA and IS were all 100 V, and the Delta electron multiplier voltage (EMV) were all 200 V. Data were collected in multiple reaction monitoring (MRM) mode using [M+ Na] + ion for all AS, DHA and IS with a collision energy of 25 eV, 18 eV and 20 eV, respectively.

2.4 Preparation of standard solutions and quality control samples

2.4.1 Standard solution of AR and DHA

Precisely 2mg of AR and of DHA were, respectively, placed in separate 10mL brown volumetric flasks, then methyl alcohol was added to form stock solutions of 200μg/mL. A series of mixture of ARandDHAworking standard solutions were prepared by dilutions of the stock solution with methyl alcohol to obtain the following concentrations: 0.1ng/mL, 0.2ng/mL, 1ng/mL, 4ng/mL, 20ng/mL, 100ng/mL, 200ng/mL, 400ng/mL.

2.4.2 Internal standard solution

Precisely 4mg of AR was weighed and placed in a 50mL brown volumetric flask, to which methyl alcohol was then added, forming a solution of 80μg/mL. Then, 2.5mL of that solution was placed in a 50mL volumetric flask, followed by methyl alcohol to make a 4μg/mL internal standard solution.

All of the solutions were stored at 4℃ and brought to room temperature before use. Plasma calibration standards of 1−400ng/mL were prepared by spiking 100μL aliquots of blank plasma with 10μL of each of the standard solutions. Quality control (QC) samples were prepared in the same way, with four levels of 4 (QC-low), 100 (QC-med), 400 (QC-high), and 1ng/mL (QC-LLOQ). Both the calibration standard and the QC samples were applied in the method validation and the pharmacokinetic study.

2.5 Sample preparation

Plasma aliquots (100μL) were spiked with 10μL of methanol and artemisinin (10μL of 4μg/mL solution) as an internal standard in centrifuge tubes (when preparing calibration and quality control (QC) samples, the standard solution was added instead of methanol), and mixed. The centrifuge tubes were initially primed with 200μL of ethyl acetate, followed by vortex concussion for 1min and centrifugation for 10 min at the speed of 12, 000 rpm. The organic phase was evaporated by use of a stream of 25℃ nitrogen. The residue was reconstituted with 100μL ofmethanol and immediate vortex concussion for 20s, placed for 10min, and injected into the LC.MS/MS system after filtration through a 0.45μm Millipore filter.

2.6 Validation

This method was validated in terms of linearity, specificity, LLOQ, recovery, intra- and inter-day variation, accuracy and precision, and stability of the analyte during sample storage and processing procedures.

2.6.1 Selectivity

Selectivity was evaluated by comparing the chromatograms of six different batches of the blank plasma with the corresponding standard plasma samples spiked with AR and DHA and the internal standard[19].

2.6.2 Linearity and LLOQ

A calibration curve was constructed from plasma standards at six concentrations of AR and DHA, ranging from 1ng/mL to 400ng/mL. A calibration curve was constructed by plotting the peak area ratio of AR/IS versus the nominal concentration of AR and the peak area ratio of DHA/IS versus the nominal concentration of DHA. The correlation coefficient and linear regression equation were used for the determination of the analyte concentration in the samples. A weighted ($1/X^2$) linear least-squares regression was used as the mathematical model. The LLOQ was determined as the lowest concentration that produced an S/N of 5[20]. The limit of detection (LOD) was determined as the lowest concentration that produced an S/N of 3[20].

2.6.3 Accuracy and precision

Intra-day accuracy and precision were evaluated by analyzing theQCconcentrations at four levels (4ng/mL, 100ng/mL, 400ng/mL and 1ng/mL; Table 1) with six determinations per concentration on thesameday. The inter-day accuracy and precision were evaluated by the analysis of the QC concentrations at four levels (4ng/mL, 100ng/mL, 400ng/mL, and 1ng/mL; Table 1) with six determinations per each concentration over 3 days. Precision and accuracy were based on the criterion that the relative standard deviation (RSD) for each concentration should be not more than 15%, except for LLOQ (not to be more than 20%)[21].

Table 1 Intra and inter day precision and accuracy of AR and DHA ($n=6$) in sheep plasma

Concentration (ng/mL)		Intra-day precision and accuracy ($n=6$)		Inter-day precision and accuracy ($n=6$)	
		Accuracy (%) ±SD	RSD (%)	Accuracy (%) ±SD	RSD (%)
AR	1	91.6±2.4	2.6	92.1±2.9	3.1
	4	93.5±2.5	2.7	92.5±2.5	2.7
	100	102.3±2.2	2.3	96.9±2.2	2.3
	400	98.3±1.3	1.3	98.9±2.3	2.3

(continued)

Concentration (ng/mL)		Intra-day precision and accuracy ($n=6$)		Inter-day precision and accuracy ($n=6$)	
		Accuracy (%) ±SD	RSD (%)	Accuracy (%) ±SD	RSD (%)
DHA	1	91. 2±2. 1	2. 3	93. 5±2. 9	3. 1
	4	95. 6±1. 7	1. 8	94. 7±2. 3	2. 4
	100	96. 4±2. 7	2. 8	101. 6±2. 6	2. 6
	400	99. 1±1. 5	1. 5	98. 5±1. 3	1. 3

2. 6. 4 Recovery and matrix effect

The recovery was determined in quadruplicate by comparing processed QC samples at three levels (low, med, high) with reference solutions in blank plasma extract at the same levels.

The matrix effect was determined by comparing the peak areas of the post-extracted spiked sample with those of the standards containing equivalent amounts of the AS and DHA prepared in the mobile phase, respectively.The experiments were performed at the three levels in six different batches.

2. 6. 5 Stability[22]

The stabilities of AR and DHA in sheep plasma were assessed by analyzing replicates ($n=6$) of QC samples at concentrations of 4ng/mL, 100ng/mL and 400ng/mL during the sample storage and processing procedures. Freshly prepared stability-QC samples were analyzed by using a freshly prepared standard curve for the measurement for all stability studies.The stabilities of stock solutions of AR and DHA were analyzed at room temperature for 24h and at 4℃ after 1month.The short-term stability was assessed after exposure of the plasma samples to ambient temperature for 24 h.The long-term stability was assessed after storage of the plasma samples at −20℃ for 60 days.The freeze/thaw stability was determined after three freeze/thaw cycles (room temperature to −20℃).The sample stability in the autosampler tray was evaluated at 4℃ for 24h; this sample stability evaluation imitates the residence time of the samples in the autosampler for each analytical run.

2. 6. 6 Formulation

The injection of artesunate nanoemulsion was prepared as follows:

The component substances for 6% ethyl oleate, 24% Tween−80, 11. 5% *n*-butanol, 53. 5% ultra-pure water, 5% artesunate were weighed.The ethyl oleate, Tween−80, *n*-butanol, and artesunate were combined and stirred under ambient conditions until the drug was dissolved. Ultrapure water was slowly added, dropwise, to the mixture with stirring.At a certain concentration of water, the system becomes a clarified, translucent, pale yellowO/W type intravenous-formulation artesunate nanoemulsion.The intravenous formulation was administered to sheep ($n=12$) at a dose of 5mg/kg.

2. 7 Animal studies

The assay method described above was used to study the pharmacokinetics of artesunate nanoemulsion in sheep plasma after intramuscular administration.All the experimental procedures were approved and performed in accordance with the Guidelines for the Care and Use of Laboratory Animals of the Lanzhou Institute of Animal Science and Veterinary Pharmaceutics. Healthy Small Tail Han sheep (35±1.27) kg were obtained from the Small Tail Han sheep-breeding base (Kangle, Gansu, China). These sheep were housed in a standard, environmentally controlled animal room (temperature: 25℃±2℃, humidity: 50%±20%) with a natural light.dark cycle for 1 week before the experiment. The sheepwere fasted for 12 h before dosing but allowed free movement and access to water during the whole experiment. All sheep ($n = 12$) were dosed at a dosage of 5mg/kg. After a single dose was administered by intramuscular administration, blood samples (3mL) were collected in heparinized tubes via the jugular vein at 0.083h, 0.167h, 0.25h, 0.333h, 0.5h, 1.5h, 1h, 2h, 4h, 6h, 8h, 10h, 12h, 16h, 20h and 24h. After all blood samples were centrifuged at 12, 000rpm for 10 min, the plasma samples were collected, and then immediately stored in a −20℃ freezer until analysis by LC.MS/MS.

The pharmacokinetic parameters were calculated by use of WinNonlin professional software version 5.2 (Pharsight, Mountain View, CA, USA). A compartmental model was utilized for data fitting and parameter estimation. The essential pharmacokinetic model was confirmed by Akaike Information Criterion (AIC)[23] for the best characterization. Plasma AUC, plasma clearance ($C_{L/F}$), peak plasma concentration (C_{max}), elimination rate constant (K) and apparent volume of distribution (V_{ss}) were all obtained from observed data. Half-life ($t_{1/2}$) was calculated directly according to the pharmacokinetic parameters.

3 RESULTS AND DISCUSSION

3.1 Mass spectrometric detection

In order to optimize positive ESI mode conditions, AR, DHA, and IS were dissolved in methanol, and then infused into the mass spectrometer for scans in positive ion mode. WhenAR, DHA, and IS were injected directly into the mass spectrometer along with the mobile phase, the analytes yielded predominantly $[M+Na]^+$ ions at m/z 407.2 for AR, m/z 307.2 for DHA, and at m/z 305.2 for IS (Fig.2). Each of the precursor ions was subjected to collision-induced dissociation to determine the resulting product ions from the product ion mass spectra. The most abundant and stable fragment ions were generated at m/z 261.1 for AR, m/z 163.0 for DHA, and m/z 151.1 for IS. Thus, the mass transitions chosen for quantitation were m/z 407.2 to 261.10 for AR, m/z 307.2 to 163.0 for DHA, and m/z 305.2 to 151.1 for IS.

3.2 Chromatographic separation

High-performance liquid chromatography with MS/MS separations was run using column packed with a small amount of mobile phase and a shorter analysis time. A 250mm column subjected to an flow rate of 0.4mL/min isocratic elution of the mobile phase for 8.5 min was used for the chromatographic separation. A mobile phase consisting of a mixture of methanol: water (the water included 1mM ammonium acetate, 0.1% formic acid, and 0.02% acetic acid) was found to be suitable for the separation and ionization of AR, DHA, and the internal standard. Formic acid was found to increase the ionization of all three compounds. Under optimized LC and MS conditions, AR, DHA, and IS were separated with retention times of 5. 45min, 6. 21min and 8. 12min, respectively, and the endogenous substances in plasma does not interfere with target detection material (Fig.3).

3. 3 Sample preparation

Recoveries of AR and DHA using methanol (14%), acetonitrile (12%), and chloroform (46%) were found to be less than that from ethyl acetate (95%).Therefore, ethyl acetate was selected as the extraction solvent for plasma due to its high recovery and sensitivity.

3. 4 Optimization of the intramuscular preparations

The traditional tablets of artesunate cannot overcome the firstpass effect of the liver.Further, as artesunate has poor solubility in water, it is difficult to deliver it effectively to the lesion and intracellular space.In addition, the poor stability of its sodium salt, which causes low bioavailability, necessitates frequent patient medication, leading to poor tolerance and efficacy and greatly limits its clinical use.Our aim was to find a new preparation of artesunate that would increase the solubility of artesunate and its sodium salt.A composition of 6% ethyl oleate, 24% Tween-80, 11. 5% n-butanol, 53. 5% ultra-pure water, and 5% artesunate was chosen to formulate an artesunate nanoemulsion for intramuscular administration, which was proved to be safe.

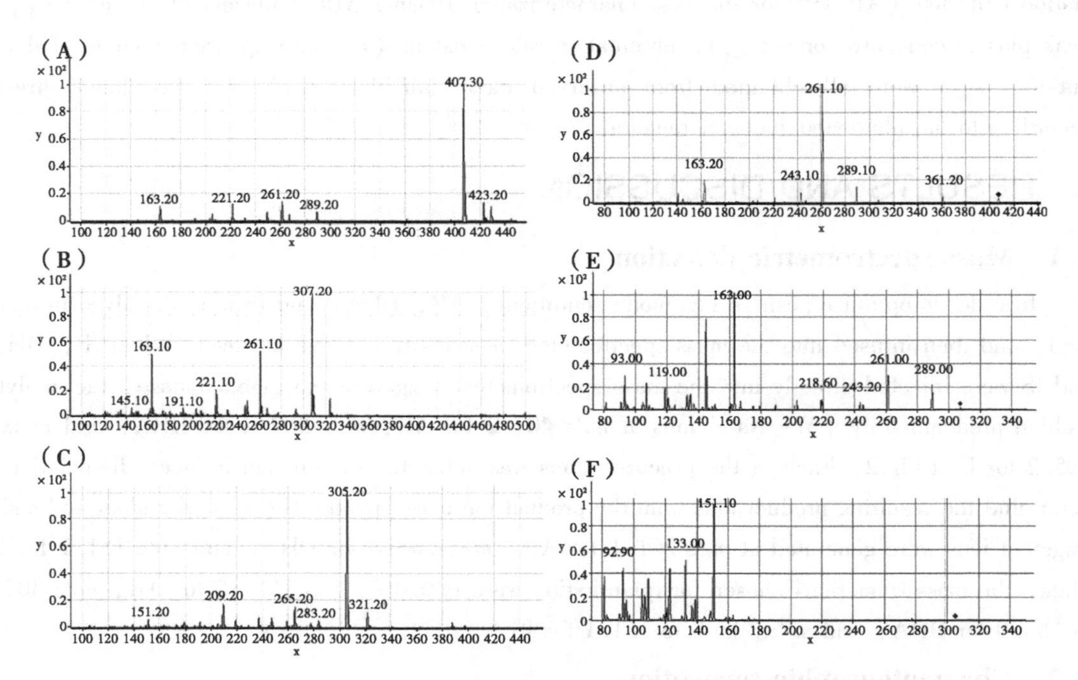

Fig.2 Full mass spectra scan for artesunate (A), dihydrortemisinin (B), and internal standard artemisinin (C); product ions for artesunate (D), dihydrortemisinin (E), and internal standard artemisinin (F)

3. 5 Method validation

3. 5. 1 Selectivity

The specificity of the method was evaluated by analyzing individual blank plasma samples from six different sources.All samples were found to have no interference from endogenous substances affecting the retention times of AR, DHA, and IS. There was good base-line separation of AR, DHA, and the IS extracted from the sheep plasma. Representative chromatograms of blank plasma, blank plasma spiked with AR, DHA, and the IS are shown in Fig.3.

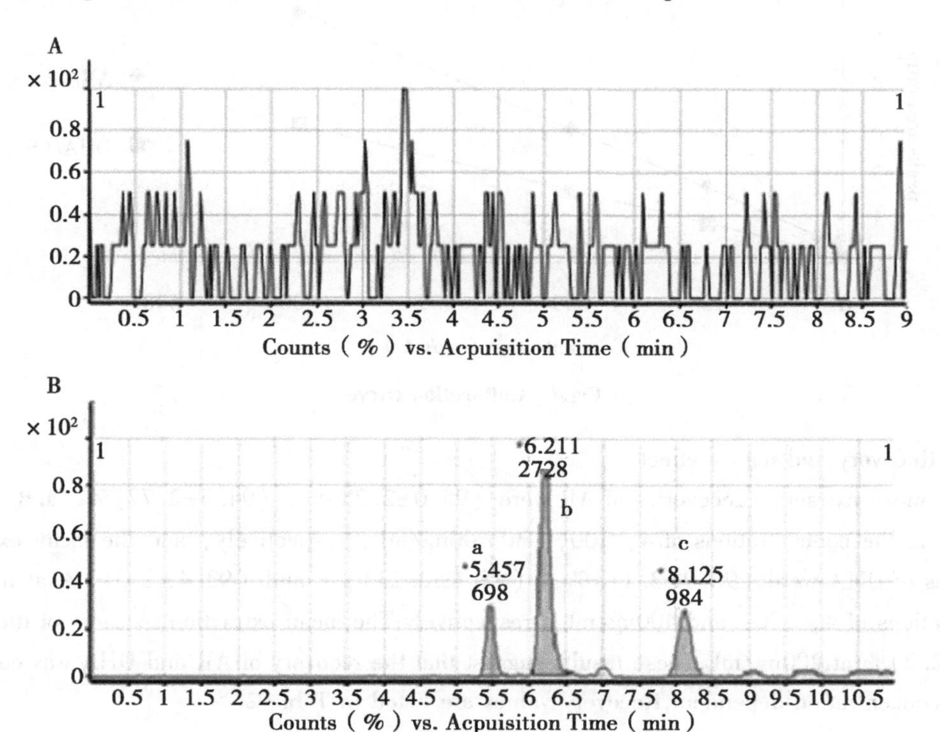

Fig.3 Chromatograms

Note: (A) blank plasma, (B) chromatograms of artesunate, dihydroartemisinin, and internal standard artemisinin in plasma).

3.5.2 Linearity and LLOQ

A calibration curve was constructed from plasma standards at six concentrations of AR and DHA, ranging from 1ng/mL to 400ng/mL. The ratio of peak areas of AR and DHA to that of the IS was used for quantification. The calibration model was selected based on the analysis of the data by linear regression with intercepts and a $1/X^2$ weighting factor. A typical equation of the calibration curve for AR was: $y=0.0130x.0.3212$ ($r^2=0.9992$), wherey is the peak-area ratio of AR to IS and x is the plasma concentration of AR. A typical equation of the calibration curve for DHA was: $y=0.0073x+0.1246$ ($r^2=0.9993$), where y is the peak-area ratio of DHA to IS and x is the plasma concentration of DHA. The calibration curve is shown in Fig.4. The lower limits of quantitation (LLOQ) for AR and DHA were both established at 1ng/mL, with an accuracy of 98.5%. The LOD was found to be 0.1ng/mL of AR and 0.2ng/mL of DHA.

3.5.3 Accuracy and precision

The intra-day and inter-day precision and accuracy of QC samples (4, 100, 400ng/mL and 1ng/mL) is summarized in Table 1. These data demonstrate that the current method has satisfactory accuracy, precision, and reproducibility for the quantification of AR and DHA in sheep plasma.

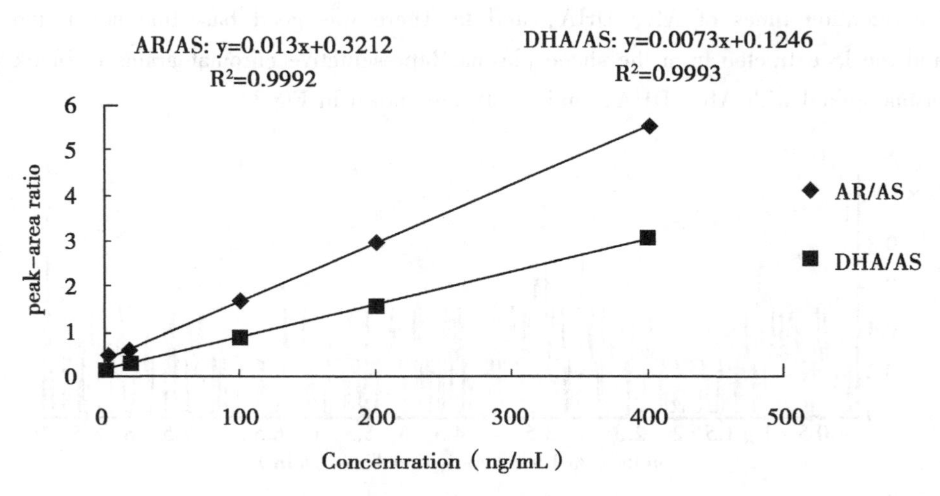

Fig.4 Calibration curve

3. 5. 4 Recovery and matrix effect

The mean extraction recoveries of AR were (95. 0±2. 28)%, (94. 5±2. 77)%, and (97. 7±2. 69)% at the concentrations of 4, 100, and 400ng/mL, respectively, and the mean extraction recoveries of DHA were (92. 9±3. 11)%, (95. 1±2. 23)%, and (93. 4±2. 31)% at the three concentrations of 4, 100, and 400ng/mL, respectively.The mean extraction recovery of the IS was (96. 3±2. 2)% at 200ng/mL.These results suggest that the recovery of AR and DHA was consistent and not concentration-dependent.Recovery values are listed in Table 2.

Table 2 Recovery of AR and DHA (n=6) from sheep plasma

	Spiked concentration (ng/mL)	Recovery (%, n=6)[a]	
		Mean±SD	RSD (%)[b]
AR	4	95. 0±2. 2	2. 3
	100	94. 5±2. 7	2. 9
	400	97. 7±2. 6	2. 7
DHA	4	92. 9±3. 1	3. 3
	100	95. 1±2. 2	2. 3
	400	93. 4±2. 3	2. 5

Note: [a] Recovery = ratio of response of spiked standard before extraction to that after extraction.

[b] RSD, relative standard deviation

The matrix effects ranged from (93. 2±2. 5)% to (96. 4±1. 9)% for AR and (92. 7±2. 2)% to (95. 6±1. 3)% for DHA at the three concentrations of 4, 100, and 400ng/mL, respectively, while the matrix effect of the IS was (93. 8±2. 7)%.This means that limited matrix effects were observed.

3. 5. 5 Stability

The stability tests on AR and DHA include stability data from freeze/thaw, short-term, autosampler, and long-term stability tests.These data are shown in Tables 3 and 4. The results demonstrate that no stability issues were observed in any of the experiments. AR and DHA were stable after being placed in sheep plasma for three cycles when stored at −20℃ and thawed to room temperature, and were stable to repeated exposure to room temperature for 24 h, in the autosampler tray at 4℃ over 24 h, and when stored at −20℃ for 60 days. AR and DHA were also stable in stock solutions at room temperature for 24 h and at 4℃ for 1 month. Taking all these points into consideration, we conclude that AR and DHA can be stored and processed under routine laboratory conditions without special attention.

Table 3 Stability of AR in sheep plasma samples under various conditions ($n=6$)

Storage conditions	Concentration (ng/mL)	Accuracy (%) ±SD	RSD (%)
Three freeze.thaw cycles	4	90. 1±1. 1	1. 2
	100	94. 3±2. 3	2. 4
	400	95. 4±2. 6	2. 7
At room temperature for 24 h	4	93. 5±1. 2	1. 3
	100	104. 3±2. 5	2. 4
	400	97. 4±1. 9	2. 0
At 20℃ for 60 days	4	92. 1±1. 2	1. 3
	100	93. 3±3. 5	3. 8
	400	94. 8±2. 6	2. 7
At 4℃ in the autosampler for 24 h	4	93. 5±0. 9	1. 0
	100	93. 6±2. 5	2. 7
	400	94. 8±3. 1	3. 3
At 4℃ for 1 month	4	97. 6±1. 4	1. 4
	100	103. 4±2. 1	2. 0
	400	96. 4±1. 9	2. 0

Table 4 Stability of DHA in sheep plasma samples under various conditions ($n=6$)

Storage conditions	Concentration (ng/mL)	Accuracy (%) ±SD	RSD (%)
Three freeze.thaw cycles	4	89. 6±1. 4	1. 6
	100	93. 2±2. 5	2. 7
	400	94. 4±2. 3	2. 4
At room temperature for 24 h	4	94. 9±0. 9	0. 9
	100	96. 1±2. 2	2. 3
	400	97. 2±3. 3	3. 4
At 20℃ for 60 days	4	92. 8±1. 6	1. 7
	100	93. 1±2. 2	2. 4
	400	93. 6±3. 0	3. 2
At 4℃ in the autosampler for 24 h	4	93. 6±1. 5	1. 6
	100	93. 2±3. 2	3. 4
	400	96. 7±2. 4	2. 5

(continued)

Storage conditions	Concentration (ng/mL)	Accuracy (%) ±SD	RSD (%)
At 4℃ for 1 month	4	102. 3±1. 1	1. 1
	100	96. 2±3. 3	3. 3
	400	97. 0±2. 7	2. 8

3. 6 Application to pharmacokinetic studies

The present method was successfully validated and applied to quantitate AR and DHA in plasma samples after intramuscular administration of artesunate nanoemulsion to Small Tail Han sheep at doses of 5mg/kg. The pharmacokinetics of AR were investigated in cattle, humans, and dogs, as plasma samples from sheep pharmacokinetic studies were not yet available. Fig.5 shows the mean plasma concentration vs. time profile for AR and DHA in sheep after intramuscular artesunate nanoemulsion. Based on AIC (the fitted values is 53), the plasma concentration-time curves for AR and DHA were adequately fitted with a one compartment model, and the major pharmacokinetic parameters were calculated by use of this model and are listed in Table 5. The PK data analysis showed that AR and DHA have a faster clearance rate and smaller volume of distribution. After intramuscular administration of artesunate nanoemulsion to sheep, AR can be quickly converted to its active metabolite DHA, but the peak concentration (C^{max}) ofAS was higher than that of DHA (the same as the result reported by[24]), and AR was not all converted to DHA. The short half-life ($t_{1/2}$) of AR and DHA indicates that the compound is removed rapidly from the blood. No significant differences were found in comparing males and females (data not shown). This study gives us some useful information to serve as a basis for further research on artesunate nanoemulsion. Method development and evaluation of the pharmacokinetic properties of this formulation will aid the preparation of new formulations of similar drugs with improved pharmacokinetic profiles.

Table 5 Pharmacokinetic parameters of AR and DHA after intramuscular injection (n=12).

Parameter	Mean±SD	
	AR	DHA
^{a}AUC (ng h/mL)	195. 8±27. 8	168. 2±20. 0
^{b}K ($1\ h^{-1}$)	0. 69±0. 08	0. 63±0. 04
$^{c}C_{max}$ (ng/mL)	158. 9±18. 3	137. 5±27. 4
$^{d}C_{L/F}$ (L/h/kg)	30±1. 0	30±2. 0
$^{e}t_{1/2}$ (h)	1. 0±0. 1	1. 1±0. 1
$^{f}V_{ss}$ (mL/kg)	0. 10±0	0. 1±0

Note: ^{a} AUC, area under the concentration.time curve. ^{b} K, elimination rate constant. ^{c} C_{max}, peak plasma concentration. ^{d} $C_{L/F}$, plasma clearance. ^{e} $t_{1/2}$, elimination half life. ^{f} V_{ss}, apparent volume of distrib

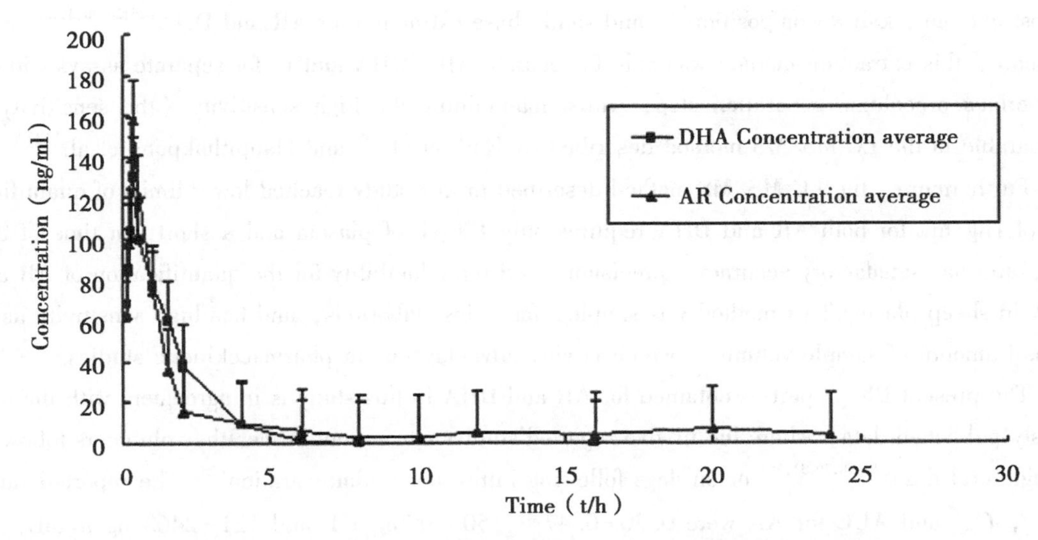

Fig.5 Mean plasma concentration.time profile after intramuscular of artesunate nanoemulsion to sheep ($n = 12$) at 5mg/kg

4 DISCUSSION

A few methods for the quantification of AR and DHA have been published. As these compounds are thermally labile and do not contain anultraviolet (UV) visible or fluorescent chromophore[25], analysis by gas chromatography (GC) and high-performance liquid chromatography (HPLC) have proven difficult. Lai et al.[26] described a HPLC method with UV, which get a lower limit of quantifications for AS and DHA at 20 ng/1mL with a long cycle time of 14 min.

Bangchang et al.[27], Lai et al.[28] and Navaratnam et al.[29] described a HPLC method with electrochemical detection (ECD). The drawbacks of the ECD technique are that it requires rigorously controlled anaerobic conditions and deoxygenation of biologic samples as well as the mobile phase which can be very difficult to establish and maintain. ECD methods are also labor-intensive. The HPLC method with ECD described by Bangchang et al.[27] can detect AR and DHA at concentrations as low as 5ng/mL and 3ng/mL, respectively, but requires a large sample volume (1mL of plasma). Mohamed et al.[30] developed a GC-MS-SIM method. The total run timewas 20. 5 minwith a solvent delay time of 7. 5 min, and the limits of quantification is 5ng/mL using 1mL of plasma for both AR and DHA.

A LC.MS method for quantification of AR and DHA in human plasma was able to reach limits of quantification of 1ng/mL using 0. 5mL of plasma with a total run time of 21 min, which was a long retention time for assay[31]. Recently a comparison between an ECD and a LC.MS/MSmethod indicated good correlation but superior sensitivity for the LC.MS/MS method reaching lower limits of quantification of 2 and 4ng/mL forDHAand ARS, respectively, using 100L plasma[25].

The advantages of the method developed in the present study over that previously reported are the sample extraction procedure using liquid. liquid extraction (using only a single extraction with ethyl acetate) was simple and less laborious when compared with the previously described methods of post column alkali decomposition[32] and solid phase extraction for AR and DHA[26,28,30,31,33]. Furthermore, this extraction method was able to separate AR, DHA and IS for separate assays without a laborious precolumn separation step, whilst maintaining the high sensitivity (the sensitivity is comparable to the LC.MS/MS method described by Naik et al.[31] and Hanpithakpong et al.[33]).

Furthermore, the LC.MS/MS method described in this study reached lower limits of quantification of 1ng/mL for both AR and DHA requires only 100μL of plasma and a short run time of 8.5 min, and has satisfactory accuracy, precision, and reproducibility for the quantification of AR and DHA in sheep plasma. This method was simple, fast, less laborious, and has high sensitivity using a small amount of sample volumes, which is very advantageous in pharmacokinetic studies.

The present PK properties obtained for AR and DHA in this study is in agreement with the previously published data, where the in vivo reported studies carried out on healthy volunteers following a single oral dose[26-28,30,31,34] or on dogs following intravenous administration[25]. The reported range of $t_{1/2}$, C_{max} and AUC for AR were 0.30-0.47 h, 50-387ng/mL and 121-2463 ng h/mL, respectively, and for DHA were 0.75-1.69 h, 35-1003ng/mL and 573-3262 ng h/mL, respectively, following a single oral dose of 20-300mg AR. However, the $t_{1/2}$ for AR (1 h) is more than 2 times compare to the published data, which shows that the nanoemulsion could slow the elimination rate and prolong the action time of AR in sheep. It is clear from the published data that there is wide inter-individual variation in the pharmacokinetic data. This has been attributed to the biotransformation of AR to its active metabolites, which is greatly affected by the differences in metabolic rates between individuals and between different species[35].

5 CONCLUSION

The pharmacokinetic analysis of artesunate nanoemulsion relies on a highly sensitive assay, which can determine AR and DHA in plasma after intramuscular injection. The limited volumes of plasma and interference from the biological matrix all add to the complexity of the trace analysis of artesunate. In this study, a rapid and sensitive LC-MS/MS method was developed, validated, and successfully applied to evaluate the pharmacokinetic parameters after intramuscular of artesunate nanoemulsion to sheep. The assay uses AS as the internal standard. The sample preparation is simple and relatively quick. The analysis requires only 100μL of plasma and a short run time of 8.5 min, which is very advantageous in a pharmacokinetic study. The method has excellent sensitivity, linearity, precision, and accuracy. Currently, tablets are the only available preparation of artesunate for clinical use. This LC-MS/MS assay is an excellent technique with which to further evaluate the pharmacokinetic properties and the therapeutic potential of the new nanoemulsion preparation of the antiparasitic agent artesunate.

ACKNOWLEDGMENTS

This work was supported by the earmarked fund for the China Agriculture Research System (cars-38), and the Special Fund for Agro-scientific Research in the Public Interest (No. 201303038-4).

References

[1] D.Duh, V.Punda-Polic, T.Trilar, T.Avsic-Zupanc, Vet.Parasitol.151 (2008) 327-331.

[2] H.Yin, G.Guan, M.Ma, J.Luo, B.Lu, G.Yuan, Q.Bai, C.Lu, Z.Yuan, P.Preston, Vet.Parasitol. 107 (2002) 29-35.

[3] J.Luo, W.Lu, Trop.Anim.Health Prod.29 (1997) 4S-7S.

[4] G.Uilenberg, Curr.Top.Vet.Med.Anim.Sci.14 (1981) 4-37.

[5] R.S.Sisodia, Livestock Advisor 6 (1981) 15-19.

[6] Q.Liu, Y.Q.Zhou, G.S.He, M.C.Oosthuizen, D.N.Zhou, J.L.Zhao, Trop.Anim.Health Prod.42 (2010) 109-114.

[7] H.Yin, L.Schnittger, J.Luo, U.Seitzer, J.S.Ahmed, Parasitol.Res.101 (Suppl.2) (2007) S191-S195.

[8] L.Yu, S.Zhang, W.Liang, C.Jin, L.Jia, Y.Luo, Y.Li, S.Cao, J.Yamagishi, Y.Nishikawa, S. Kawano, K.Fujisaki, X.Xuan, J.Vet.Med.Sci.73 (2011) 1509-1512.

[9] G.R.Razmi, H.Eshrati, M.Iran, Vet.Parasitol.140 (2006) 239-243.

[10] T.Dehe, Chin.J.Tradit.Vet.Sci.4 (2010).

[11] E.J.Reddy, P.S.Rao, M.L.Narasu, J.Biotechnol.1501 (2010) S91.

[12] V.Dhingra, R.K.Vishweshwar, N.M.Lakshmi, Life Sci.66 (2000) 279-300.

[13] N.P.Singh, H.Lai, Life Sci.70 (2001) 49-56.

[14] D.Boulanger, Y.Dieng, B.Cisse, F.Remoue, F.Capuano, J.L.Dieme, T.Ndiaye, C.Sokhna, J.F. Trape, B.Greenwood, F.Simondon, Trans.R.Soc.Trop.Med.Hyg.101 (2007) 113-116.

[15] Y.K.Goo, M.A.Terkawi, H.Jia, G.O.Aboge, H.Ooka, B.Nelson, S.Kim, F.Sunaga, K.Namikawa, I. Igarashi, Y.Nishikawa, X.Xuan, Parasitol.Int.59 (2010) 481-486.

[16] M.Gabriels, J.Plaizier-Vercammen, J.Pharm.Biomed.Anal.31 (2003) 655-667.

[17] H.W.Hu, X.Z.Zhou, Y.Li, J.Y.Zhang, J.Agric.Sci.36 (2008) 13648-13649 (13652).

[18] X.Z.Zhou, J.Y.Zhang, J.Y.Li, J.S.Li, X.J.Wei, J.R.Niu, B.Li, H.W.Hu, Chin.J.Vet.Med.46 (2010) 54-56.

[19] D.Pabbisetty, A.Illendula, K.M.Muraleedharan, A.G.Chittiboyina, J.S.Williamson, M.A.Avery, B.A.Avery, J.Chromatogr.B: Anal.Technol.Biomed.Life Sci.889. 890 (2012) 123-129.

[20] D.Pabbisetty, A.Illendula, K.M.Muraleedharan, A.G.Chittiboyina, J.S.Williamsonb, M.A.Avery, B.A.Avery, J.Chromatogr.B: Anal.Technol.Biomed.Life Sci.125 (2012) 889-890.

[21] H.Fan, R.Li, Y.Gu, D.Si, C.Liu, J.Chromatogr.B: Anal.Technol.Biomed.Life Sci.105 (2012) 889-890.

[22] V.P.Shah, K.K.Midha, J.W.Findlay, H.M.Hill, J.D.Hulse, I.J.McGilveray, G.McKay, K.J. Miller, R. N. Patnaik, M. L. Powell, A. Tonelli, C. T. Viswanathan, A. Yacobi, Pharm. Res. 17 (2000) 1551-1557.

[23] S.I.Chien, J.C.Yen, R.Kakadiya, C.H.Chen, T.C.Lee, T.L.Su, T.H.Tsai, J.Chromatogr.B: Anal.Technol.Biomed.Life Sci.917-918 (2013) 62-70.

[24] R.S.Miller, Q.Li, L.R.Cantilena, K.J.Leary, G.A.Saviolakis, V.Melendez, B.Smith, P.J.Weina, Malar.J.11 (2012) 255.

[25] Y.Gu, Q.Li, V.Melendez, P.Weina, J.Chromatogr.B: Anal.Technol.Biomed.Life Sci.867 (2008) 213-218.

[26] C.S.Lai, N.K.Nair, S.M.Mansor, P.L.Olliaro, V.Navaratnam, J.Chromatogr.B: Anal.Technol.Biomed.Life Sci.857 (2007) 308.

[27] K.N.Bangchang, K.Congpuong, L.N.Hung, P.Molunto, J.Karbwang, J.Chromatogr.B: Biomed.Sci. Appl.708 (1998) 201-207.

[28] C.S.Lai, N.K.Nair, S.M.Mansor, P.L.Olliaro, V.Navaratnam, J.Chromatogr.B: Biomed.Sci.Appl. 877 (2009) 558-562.

[29] V. Navaratnam, M. N. Mordi, S. M. Mansor, J. Chromatogr. B: Biomed. Sci. Appl. 692 (1) (1997) 157.

[30] S.S.Mohamed, S.A.Khalid, S.A.Ward, T.S.M.Wan, H.P.O.Tang, M.Zheng, R.K.Haynes, G.Edwards, J.Karbwang, J.Chromatogr.B: Biomed.Sci.Appl.731 (1999) 251-260.

[31] H.Naik, D.J.Murry, L.E.Kirsch, L.Fleckenstein, J.Chromatogr.B: Anal.Technol.Biomed.Life Sci. 816 (2005) 233-242.

[32] K.T.Batty, T.M.Davis, L.T.Thu, T.Q.Binh, T.K.Anh, K.F.Ilett, J.Chromatogr.B: Biomed.Appl. 677 (2) (1996) 345.

[33] W.Hanpithakpong, B.Kamanikom, A.M.Dondorp, P.Singhasivanon, N.J.White, N.P.J.Day, N. Lindegardh, J.Chromatogr.B: Anal.Technol.Biomed.Life Sci.876 (2008) 61-68.

[34] Y.Liu, X.Zeng, Y.H.Deng, L.Wang, Y.Feng, L.Yang, D.Zhou, J.Chromatogr.B 877 (2009) 465-470.

[35] S.Parikh, J.-B.Ouedraogo, J.A.Goldstein, P.J.Rosenthal, D.L.Kroetz, Clin.Pharmacol.Ther.82 (2007) 203.

（发表于《Journal of ChromatographyB》，院选 SCI，IF：2. 729）

Simple and Sensitive Monitoring of β_2-agonist Residues in Meat by Liquid Chromatography-tandem Mass Spectrometry Using a QuEChERS with Preconcentration as the Sample Treatment

XIONG Lin*, GAO Ya-qin, LI Wei-hong, YANG Xiao-lin, SHIMO Shimo-peter

(Lanzhou Institute of Husbandry and Pharmaceutical Sciences, Chinese Academy of Agricultural Sciences/Laboratory of Quality & Safety Risk Assessment for Livestock Product (Lanzhou), Ministry of Agriculture, Lanzhou 730050, China)

Abstract: A liquid chromatography with tandem mass spectrometric detection (LC-MS/MS) method was established for the simultaneous determination of the levels of 10 β_2-agonists in meat. The samples were extracted using an aqueous acidic solution and cleaned up using a Quick, Easy, Cheap, Effective, Rugged and Safe (QuEChERS) technique utilising a DVB-NVP-SO_3Na sorbent synthesised in-house. First, the β_2-agonist residueswere extracted in an aqueous acidic solution, followed by matrix solid-phase dispersion for clean-up. The linearities of the method were $R^2 = 0.9925-0.9998$, with RSDs of 2.7%-15.3% and 73.7%-103.5% recoveries. Very low limits of detection (LOD) and quantitation (LOQ) of 0.2-0.9μg/kg and 0.8-3.2μg/kg, respectively, were achieved for spiked meat. The values obtainedwere lower than the maximumresidue limits (MRLs) established by the EU and China. These results clearly demonstrate the feasibility of the proposed approach. The evaluated method provided reliable screening, quantification and identification of 10 β2_agonists in meat.

Key words: QuEChERS; β_2-Agonists; LC-MS/MS; Feed additive; Meat; DVB-NVP-SO_3Na

1 INTRODUCTION

The security of meat is a global concern (Andrée, Jira, Schwind, Wagner, & Schwägele, 2010; Jenson & Sumner, 2012; Sofos, 2014), andmethods to detect drug residues have becomea domestic and international research focus. The presence of feed additive residues is a significant factor affecting the safety of meat. Effective determination methods are needed to guarantee the safety of meat (Baert et al., 2012; Castro-Puyana & Herrero, 2013). One culprit in recent meatsafety

* Corresponding author. E-mail: xionglin807@sina.com.

scandals has been feed additives, which are type of veterinary drugs used in food animals to production and leanness of meat (Cronly et al., 2010; Liu, Zhang, Hu, & Cheng, 2013). β_2-Agonists which can induce the production of leaner meat are widely used as feed supplements to increase feed efficiency and the rate of weight gain in food-producing animals (Garcia, Paris, Gil, Popot, & Bonnaire, 2011; Murat, Usama, Ibrahim, Nilgun, & Nusret, 2013; Sato et al., 2010). However, the misapplication or illegal use of β_2-agonists can result in meat residues that are harmful to animal and human health by inducing heart palpitations, diarrhoea, muscle tremors and even malignant tumours (Du, Zhang, Fu, Zhao, & Chang, 2013; Du, Fu, et al., 2013; Zheng et al., 2010). Scandals involving β_2-agonist residues in meat have caused concern and disappointment among consumers throughout China. Therefore, to monitor food safety, sensitive and specific analytical methods are needed to determine the level of β_2-agonists inmeat (Wang, Li, Zhou, & Yang, 2010). Analytical methods are essential to assess human exposure to β_2-agonists and supportthe enforcement of laws and regulations. The methods for monitoring β_2-agonist residues include HPLC (Morales-Trejo, León, Escobar-Medina, & Gutiérrez-Tolentino, 2013; Sirhan, Tan, & Wong, 2011; Xu, Hu, Hu, & Li, 2010), GC-MS (Corcia, Morra, Pazzi, & Vincenti, 2010; Gallo et al., 2007) and LC-MS or LC-MS/MS (Du, Wu, Yang, & Yang, 2012; González-Antuña et al., 2013; Shao et al., 2009;). Of these methods, LC-MS/MS is the primacy technique (Wang, Wang, & Cai, 2013; Wang, Zeng, et al., 2013; Wang, Zhao, et al., 2013). Recently, β_2-agonist detection has been performed using several different techniques. Immuno-techniques based on enzyme immunoassay (EIA) (Boyd, Heskamp, Bovee, Nielen, & Elliott, 2009) via enzymelinked immunosorbent assay kits (ELISA) (Weilin, Jennifer, Carolyn, & David, 2010) have been performed for the detection of the illegal use of β_2-agonists. However, adverse matrix effects are common in such immuno-techniques (Jean-Philippe, Philippe, Bruno, & Franc, 2002), and all positive samples must be subjected to the confirmation by credible tests, such as GC-MS or LC-MS (Eshaq (Isaac), Sin, Sami, Gene, & Michael, 2003). In addition, immuno-techniques are not suitable for all β_2-agonists according to the Council Directive 96/23/EC (Commission of the European Communities, 1996). Immunotechniques can only be applied to detect a single β_2-agonist compound or a group of β_2-agonists with similar chemical structures (Shao et al., 2009). There are fewstudies on themulti-residue determination of β_2-agonists by screening methods. Some authors have presented screening methods dedicated to serum and tissue samples using HPLC with electrochemical detection at earlier stages. However, according to Commission Decision 2002/657/EC, the confirmation of the presence of an analyte by LC-DAD or LC-ED alone is not sufficient. Confirmation methods mainly relying on GC-MS. GC-MS can achieve excellent selectivity. Thus, for the unambiguous confirmation of β_2-agonists in foodstuffs, GC-MS is the preferred technique, with detection limits higher than those obtained using HPLC/DAD (Wang et al., 2010). However, GC-MS methods require derivation because of the high polarity and low volatility of β_2-agonists (Wang et al., 2010), and this step is time-consuming, tedious, laborious and expensive (Shao et al., 2009). Because of the relatively complicated operation and poor stability of derivatives, the application of GC/MS for the detection of β_2-agonists has decreased every year (Fan, Miao, Zhao, Chen, & Wu, 2012). By contrast, LC-MS or LC-MS/MS methods, which avoid the derivatisation step required in GC-MS, possess an excellent qualitative capacity and high sensitivity in quantitative determination and have been developed as the main analytical method for β_2-agonists. However, these methods are expensive and require a time-consuming and complicated sample-pretreatment procedure (Moragues & Igualada, 2009). Consequently, analytical laboratories are increasingly interested in developing newanalyticalmethods that aremore rapid and enable highersample throughput.

In 2003, Anastassiades described the QuEChERS (Quick, Easy, Cheap, Effective, Rugged and Safe) method for the multi-class and multiresidue analysis of pesticides in various food matrices with high water content (Anastassiades, Lehotay, Stajnbaher, & Schenck, 2003). This method minimises the time required for extraction and clean-up processes and is inexpensive; in addition, the QuEChERS procedure reduces the sample size and required quantities of laboratory glassware and requires low solvent consumption (Xu et al., 2010). Consequently, the QuEChERS procedure has become a popular method that is widely used in the pesticide analysis due to its simplicity, low cost and relatively high efficiency results (Kinsella et al., 2011; Léon et al., 2012; Pérez-Burgos et al., 2012). This procedure has been applied elsewhere for the extraction of residues of veterinary drugs, such as quinolones, macrolides and sulphonamides (Aguilera-Luiz, Vidal, Romero-González, & Frenich, 2012; Domínguez-Álvarez, Mateos-Vivas, García-Gómez, Rodríguez-Gonzalo, & Carabias-Martínez, 2013; Lopes et al., 2012; Park et al., 2012; Wilkowsk & Biziuk, 2011). The positive ion-exchange resin DVB-NVP-SO_3Na has been broadly used in solid-phase extraction (SPE) clean-up and is highly effective in cleaning up drug residues (Hernández-Mesa, García-Campaña, & Cruces-Blanco, 2014; Ruiz et al., 2013). The clean-up agents generally employed in the QuEChERS procedure are PSA and C18 (Wozniak, Zuchowska, & Zmudzki, 2013; Yan et al., 2013), and there has been no report on the use of positive ion-exchange resins as clean-up agents for the determination of the level of drug residues using a QuEChERS procedure with preconcentration as the sample treatment. β_2-Agonists are alkaline compounds and thus can be dissolved in an aqueous acidic solution, followed by clean-up using a cleaning agent (such as DVB-NVP-SO_3Na). In this paper, we synthesised DVB-NVP-SO_3Na for use in the purification of β_2-agonists from meat sample using QuEChERS procedure.

The purpose of this study was to develop a rapid multi-residue method for the determination of the levels of 10 β_2-agonists (cimaterol, terbutaline, salbutamol, fenoterol, ractopamine, clorprenaline, clenbuterol, tulobuterol, phenylethanolamine A and penbuterol) in meat samples. Their molecular structures of these compounds are shown in Fig. 1. Extraction was performed using a QuEChERS-like technique, followed by direct analysis by LC-MS/MS.

2 MATERIALS AND METHODS

2.1 Chemicals, reagents and solution

Analytical drug standards of high purity, including fenoterol hydrobromide (purity>99.0), penbutolol (purity>96.0%), clenbuterol hydrochloride (purity>98.5%), ractopamine hydrochloride (purity > 97.0%), salbutamol free base (purity > 99.0%) and terbutalin sulphate (purity>98.5%) were obtained fromDr. Ehrenstorfer GmbH (Augsburg, Germany). Tulobuterol

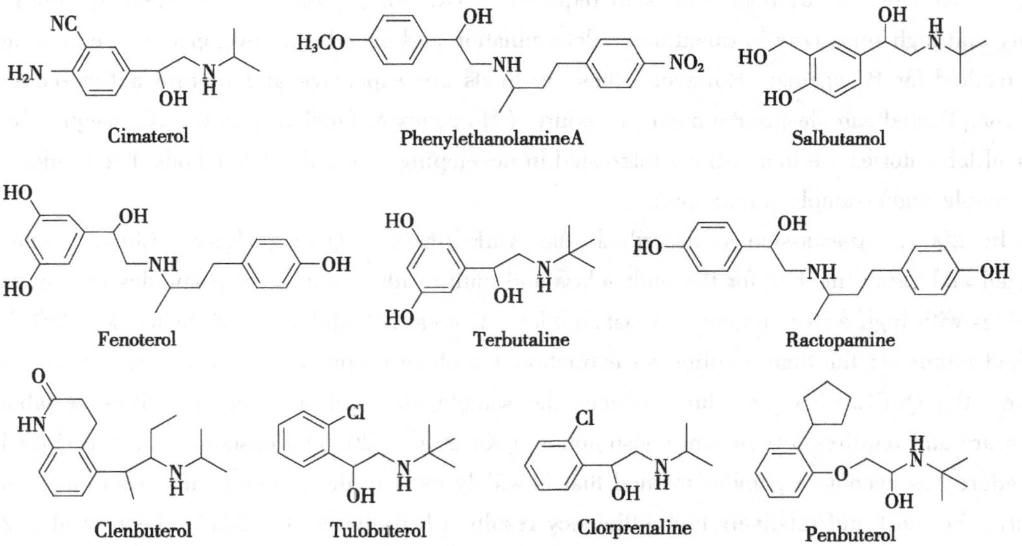

Fig.1 Molecular structures of the 10 β_2-agonists evaluated in this study

hydrochoride (purity>99.3%), phenylethanolamine A (purity>99.7%), clorprenaline (purity>99.7%) and cimaterol (purity>99.7%) were obtained from WITEGA Laboratorien Berlin-Adlershof GmbH (Berlin, Germany). Stock solutions of the individual compounds were prepared in methanol (HPLC grade) at 100 μg/L and were stored at −20℃ in the dark place. The stock solutions of the standards were stored for 6 months. A mixedstandard solution of the 10 β_2-agonists (1mg/L) was prepared using the individual stock solutions, diluted in a mobile phase solution [methanol.water (10/90, v/v) containing 0.1% formic acid] and stored at 4℃ for six months. This mixed-standard solution was used for spiking in the recovery experiments and preparing the calibration standards.

HPLC-grade methanol (MeOH), acetonitrile (ACN) and formic acid (HCOOH) were obtained from the Tedia Company Inc. (Fairfield, OH, USA). A Milli-Q ultrapure water system (Molsheim, France) was used to obtain the HPLC-grade water. Divinylbenzene, N-vinylpyrrolidone, sodium carboxymethyl cellulose, n-dodecane, azodiisobutyronitrile (AIBN) and sulphuric acid were of analytical grade and were obtained from Sinopharm (Shanghai, China). All other chemicals and solvents used were of analytical grade, unless otherwise stated, were purchased from Sigma-Aldrich (Santa Clara, CA, USA), and were used without further purification.

2.2 Instrumentation and apparatus

An Agilent 1200-6410 LC-MS/MS apparatus (Agilent, USA), equipped with an ESI ion source was used in this study. LC separation was performed using an Agilent Eclipse Plus C_{18} column (100mm×3.0mm, 1.8μm). Detection was performed on a TQ triple quadrupole massspectrometric detector with an electrospray ionisation (ESI) interface. A JSM-6510A analytical scanning electron microscope (JEOL, Japan) was utilised. An electronic analytical balance (accurate to 0.1mg and 0.01g, model BSA224S-CW, Sartorius, Germany), a Labinco L24 vortex mixer (Breda, Netherlands), an Omnifuge 2.ORS centrifuge (Osterode, Germany) and N100DR nitrogen evaporators

(Peak Scientific, UK) were used in the sample-pretreatment process.

2.3 Synthesis of the clean-up agent, the positive ion-exchange resin (DVB.NVP.SO_3Na)

DVB.NVP.SO_3Na (5) was prepared as presented in Scheme 1 (Brousmiche et al., 2008; Maciejewska & Gawdzik, 2008; Maciejewska & Osypiuk-Tomasik, 2013). First, the intermediates DVB-NVP (3) was prepared by the polymerising divinylbenzene (2) using N-vinyl pyrrolidone (1). Sodium carboxymethyl cellulose (CMC-Na) (0.3750 g) was dissolved in water (75mL) in a three-necked flask. Thirteen millilitres of divinylbenzene (2) and 8mL of N-vinyl pyrrolidone (1) were completely dissolved in toluene (18mL) and n-dodecane (3.1mL) in a beaker, and azodiisobutyronitrile (AIBN, 0.1388 g) was added as catalyst. The liquid in the beaker was poured into a three-necked flask, and the mixture was then stirred and refluxed at 70℃ for 12 h. Theproduct was filtered, washed with water and methanol, and dried in a vacuum oven at 80℃. Next, DVB-NVP-SO_3H (4) was obtained by treating DVB-NVP (3) with concentrated sulphuric acid in refluxing dichloromethane for 9 h at 80℃. The product was filtered, washed with water, and dried in a vacuum oven at 80℃. Finally, DVB-NVP-SO_3H was mixed with the appropriate sodium hydroxide solution, and the product was filtered and washed with water, to yield the target compound DVB-NVP-SO_3Na (5). The microstructure (Fig.2) of the resultingmaterialwas investigated using scanning electronmicroscopy (SEM) to enable a correlation of structure with performance. The particle-size distribution of DVB-NVP-SO_3Na was primarily 30.80 μm with an acceptable degree of sphericity and dispersibility.

2.4 Sample collection and preparation

Residue-free frozen samples of beef muscle, beef liver, goat muscle and goat liver were obtained from local supermarkets (Lanzhou, China) and kept frozen (at-20℃) until analysis. The samples were prepared using a modified QuEChERS-based approach that was very simple and straightforward (Fig.3). Some of the agonists present in animal tissue, such as ractopamine and clenbuterol, are present as conjugates aftermetabolism; therefore, for detection, theymust be restored to the form of β_2-agonists through enzymatic hydrolysis, which is generally performed using β-glucuronidase-arylsulphatase. The meat samples were cut into pieces and homogenised in a blender. A 5 g comminuted sample was transferred to a 50mL centrifuge tube and mixed with 4mL of 0.05 M acetate buffer (pH 5.2). After the addition of 50μL of β-glucuronidase-arylsulphatase, the sample was incubated at 42℃ for 6 h before extraction. The crude extract was centrifuged at 10 000 r/min for 10min at room temperature. The supernatant was transferred to another tube and mixed with 5mL 0.1mol/L perchloric acid. The mixture was vortexed for 10min, followed by centrifugation at 8 000r/min for 10min. The supernatant was then transferred to another tube, and a specific quantity of DVB-NVP-SO_3Na was added, after which the tube was immediately capped and briefly shaken by hand. The tube was then vortexed for 2min and subsequently centrifuged at 5 000 r/min for 10min, after which the supernatant was discarded. Six millilitres of a mixture of ammonia and methanol (3/97, v/v) was then added, and the solution was vortexed for 10min, followed by centrifugation for 10min at 5 000r/min. The supernatant was collected and evaporated under a stream of nitrogen at

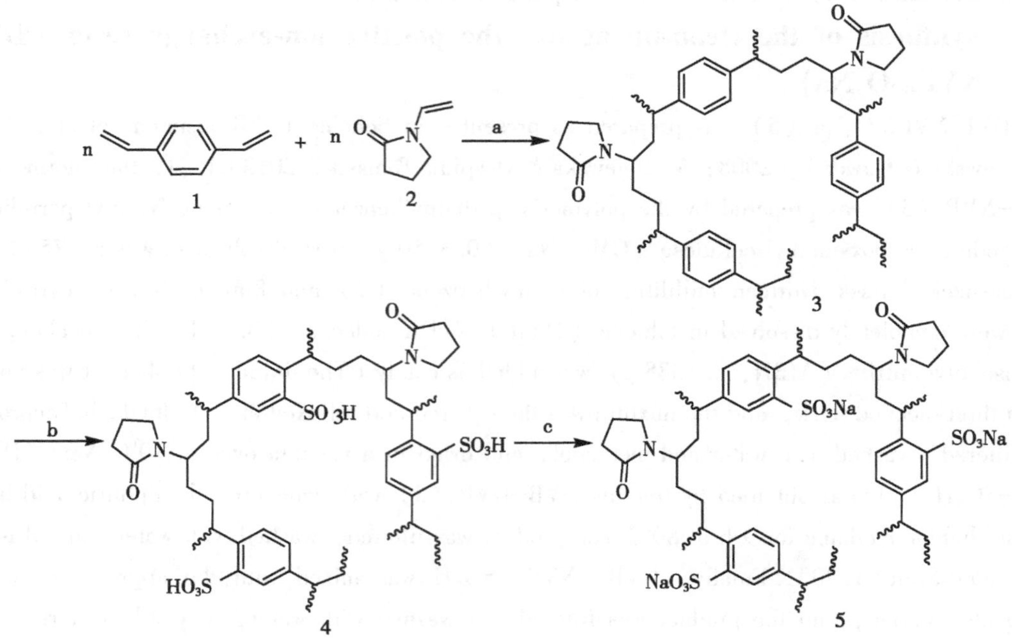

Scheme 1 Steps in the synthesis of a clean-up agent composed of positive ion-exchange resin

Reagents and conditions: (a) CMC-Na, toluene, n-dodecane, AIBN, reflux at 70℃ for 12 h; (b) concentrated sulphuric acid, dichloromethane, reflux at 80℃ for 9 h; (c) neutralisation with sodium hydroxide solution.

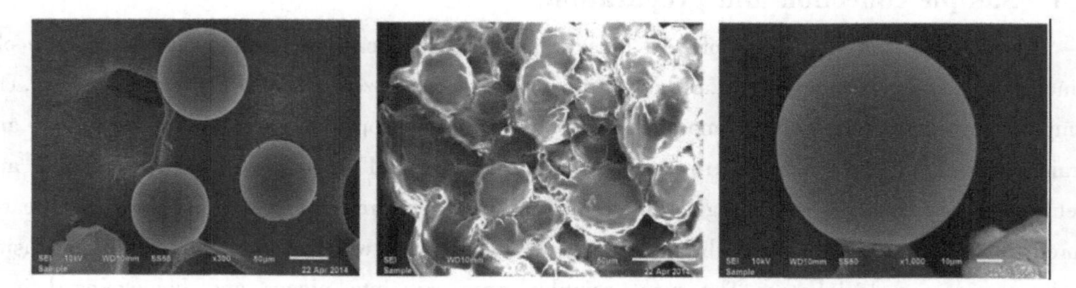

Fig.2 Scanning electron micrographs of DVB-NVP-SO_3Na

50℃, after which both the samples and the standards were dissolved in 0.4mL of the mobile phase (methanol-water (10/90, v/v) containing 0.1% formic acid). The clean-up step was completed by filtering the samples through 0.22μm Millipore Millex-GV filters and collecting the filtrates in LC vials.

2.5 LC operating conditions

Amodel 1200 Agilent LC system (Agilent, USA) was used. LC separation was performed using an Agilent Eclipse Plus C_{18} column (100mm × 3.0mm) with a 1.8 μm particle size. The injection volume was 10μL, and the eluent flow rate was 0.30mL · min^{-1}. The mobile phase consisted of (A) methanol containing 0.1% formic acid and (B) water containing 0.1% formic acid. The

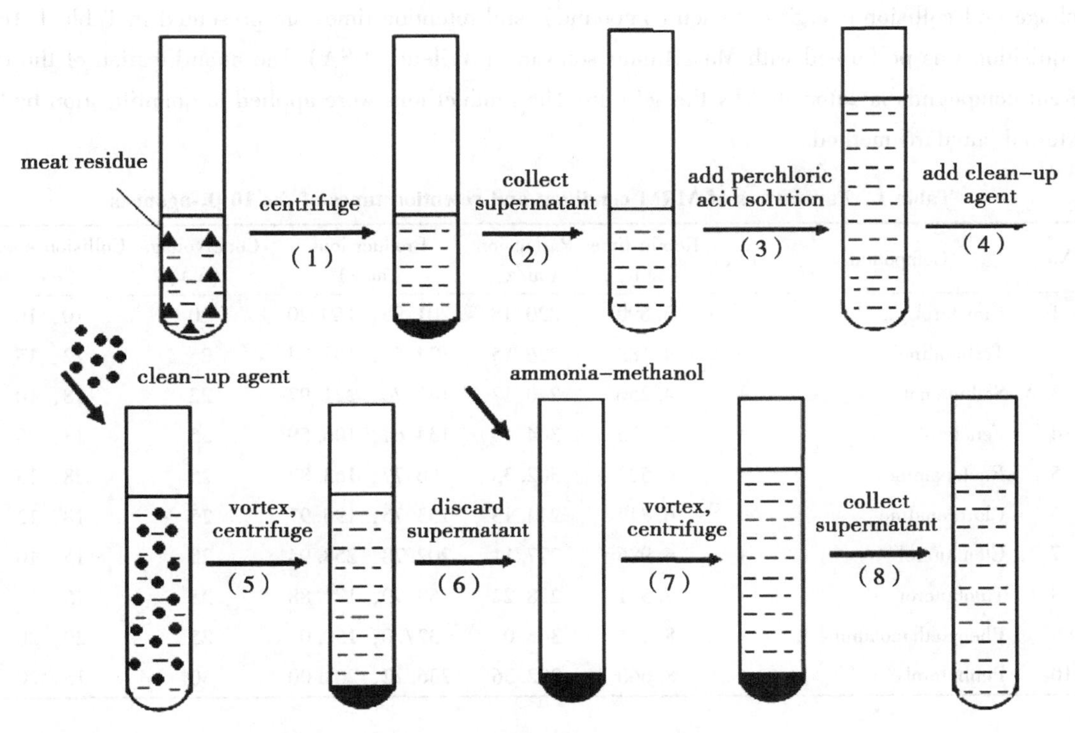

Fig.3 Scheme of sample preparation

Note: (1) The meat residues were extracted by vortexing the mixture for 1 min, followed by centrifugation for 5 min. (2) The supernatant was collected. (3) Five millilitres of a 0.1mol/L perchloric acid solution was added. (4) DVB-NVP-SO_3Na was added. (5) The tubes were vortexed for 2 min and then centrifuged for 10 min. (6) The supernatant was discarded, and 6mL of an ammonia-methanol mixture was added. (7) The mixture was vortexed for 10 min and then centrifuged for 10 min. (8) The supernatant was collected.

following gradient elution protocol was used: initial, 90% B, decreased to 50% B in 5min, decreased to 10% B in 6 min and maintained till 10 min, increased to 90% B in 13 min.The column-was equilibrated for 2min between each analytical run.The total run time was 15 min.The column and the ambient temperature were maintained at 30±1℃.A 10μL of the sample extract was injected into the chromatographic system.

2.6 Mass spectrometry

The MS system used was a 6410 Triple Quad mass spectrometer (Agilent, USA).High-purity nitrogen was used as the drying and nebulising gas and argon was used as the collision gas.Tuning experiments were performed in ESI+mode.The ionisation parameters for the heated ESI source were optimised by increasing the possible intensities of all analytes.The capillary voltage was 3.2 kV and the source block and desolvation temperature were 110 and 350℃, respectively.Nitrogen was used as the cone gas and the desolvation gas at flow rates of 50 and 660 L/h, respectively.Themonitoring-modewas themultiple reaction monitor (MRM).Direct infusion of the standard solution was performed to optimise the multiple reaction monitoring (MRM) transitions and associated acquisition parameters.The parameters (the ionisation mode, precursor ion, product ions, cone voltage and collision energies of each compound) and retention times are presented in Table 1. Data acquisition was performed with MassHunter software (Agilent, USA). The quantification of the different compounds is automated by the software. The product ions were applied to quantification by the external standard method.

Table 1 Parameters of MRM condition and retention times of the 10 β_2-agonists

No.	Compound	ESI	Retain time (min)	Parent ion (m/z)	Product ions (m/z)	Cone voltage (V)	Collision energy (eV)
1	Cimaterol	+	3. 890	220. 18	201. 95, 160. 20	20	10, 16
2	Terbutaline	+	4. 183	226. 15	124. 67, 151. 74	25	22, 15
3	Salbutamol	+	4. 256	240. 17	147. 70, 221. 97	22	18, 10
4	Fenoterol	+	5. 225	304. 15	134. 61, 106. 59	35	18, 30
5	Ractopamine	+	6. 523	302. 33	106. 77, 163. 87	25	28, 15
6	Clorprenaline	+	6. 847	214. 13	153. 75, 195. 97	25	18, 12
7	Clenbuterol	+	6. 986	277. 11	202. 78, 258. 94	25	15, 10
8	Tulobuterol	+	7. 531	228. 22	153. 90, 171. 88	25	15, 12
9	Phenylethanolamine A	+	8. 236	345. 0	327. 0, 150. 0	25	20, 20
10	Penbuterol	+	8. 660	292. 36	236. 22, 201. 00	30	15, 20

3 RESULTS AND DISCUSSION

3. 1 Optimization of analytical parameters

To optimise the chromatographic analysis, several gradient profiles were assessed to improve the peak separation and minimise the run time. Unless otherwise stated, the sample-pretreatment method used was that presented in Section 2. 4, and the determination conditions for LC andMSwere those presented in Sections 2. 5 and 2. 6, respectively. Different mobile phases (CH_3OH/H_2O, CH_3CN/H_2O) were evaluated under fixed conditions to obtain optimal separation. Eight of the 10 β_2-agonists were separated from the baseline in 9 min using CH_3OH/H_2O as the mobile phase, but only 4 β_2-agonists were separated using CH_3CN/H_2O as the mobile phase. Methanol provided increased sensitivity and resolution, particularly when acidified using formic acid. The gradient described above enabled the separation of all compounds within 10 min. Because the chromatographic column has a substantial influence on the analytical results, the separation efficiencies of various columns (Thermo Hypersil GOLD column (100mm × 2. 1mm, 5μm), Agilent XDB C_{18} column (50mm× 2. 1mm, 1. 8μm) and Agilent Eclipse Plus C_{18} column (100mm × 3. 0mm, 1. 8μm) were studied. Due to its short length, sample separation could not be achieved in 20 min using the Agilent XDB C_{18} column. Five chromatographic peakswere separated and 2 compounds were identified using the Thermo Hypersil GOLD column, representing 20% of the studied compounds, whereas 9 chromatographic peaks were separated and 8 compounds were identified using the Agilent Eclipse Plus C_{18} column, representing 80% of the studied compounds. Therefore, the Agilent Eclipse Plus C_{18} column was selected to separate the samples. This column exhibited high column efficiency, a high degree of separation, and very high selectivity for analysing β_2-agonists; only the retention times of Terbutaline and Salbutamol were very similar under the optimal conditions, and the other compounds were separated from the baseline and exhibited acceptable peak shapes. A total ion current chromatogram of a mixture of 10 β_2-agonists obtained from a blank beef-muscle sample spiked at approximately 0. 8μg/kg is shown in Fig.4.

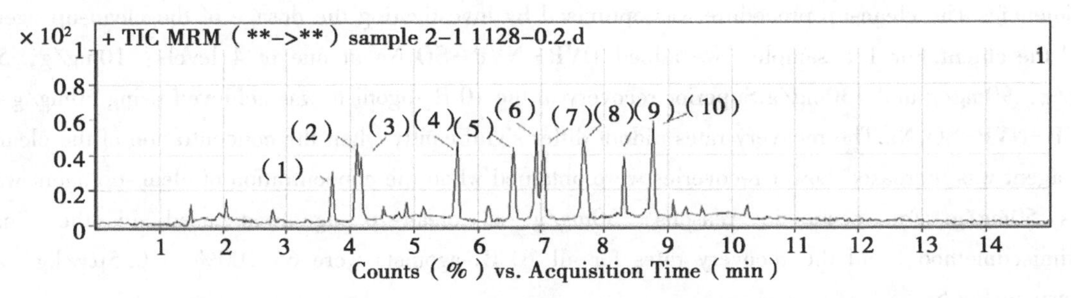

Fig.4 Total ion current chromatogram for a mixture of 10 β_2-agonists obtained from a blank beef-muscle sample spiked at 0. 8μg/kg.The β_2-agonists were

Note: (1) cimaterol; (2) terbutaline; (3) salbutamol; (4) fenoterol; (5) ractopamine; (6) clorprenaline; (7) clenbuterol; (8) tulobuterol; (9) phenylethanolamine A and; (10) penbuterol.

3. 2 Optimization of sample preparation

Sample preparation plays an important role in an analyticalmethod. To date, various pretreatment methods have been proposed for monitoring the illegal use of β_2-agonists.For extraction and purification, such techniques as immunoaffinity chromatography (Shen et al., 2007), liquid. liquid extraction (LLE) (Keskin, Özer, & Temizer, 1998), solid phase extraction (SPE) (Wang, Wang, et al., 2013; Wang, Zeng, et al., 2013) and matrix solid-phase extraction (MSPD) (Qiao & Du, 2013) are used.Due to the complexity of the biological matrices and the trace levels in real samples, these methods cannot fully remove salts and endogenous compounds, such as fat, phospholipids and aliphatic acid, from the sample, leading to possible matrix effects may.In addition, these techniques were the most tedious and time-consuming steps and constitute themost likely source of imprecision and inaccuracy in the analysis as a whole (Du et al., 2014). Because most clean-up methods for β_2-agonists in food are based on solid phase extraction (SPE) using different sorbents, we opted to compare the QuEChERS and SPE techniques in this work. Mixed mode solid-phase extraction cartridges, such as C_8 or C_{18}, combined with a strong cation exchange cartridge are commonly used in the cleanup procedure of β_2-agonists in animal tissues, improving the extraction and purification performance (Fan et al., 2012).However, the general SPE cleanup techniques lack specificity and selectivity and can retain analytes and interferents together, leading to interference and matrix effects and reducing the determination reliability.In addition, the large quantities of organic solvents, including acetonitrile and methanol [Eshaq (Isaac) et al., 2003; Jean-Philippe et al., 2002], may cause environmental pollution.Generally speaking, over 20 h is required for the determination of one sample (Shao et al., 2009) in SPE.In contrast, the established QuEChERS pretreatment procedurewithout covering was very simple and economic and required only small amounts of organic solvents.Only 8 h was needed to prepare and test ameat sample. Thus, thismethod can reduce the sample volume, solvent consumption and analytical time needed. QuEChERS is therefore a green technique for sample preparation.

To develop an excellent detection method, sample preparation procedure was required to successfully extract and clean-up β_2-agonists in the sample. Beef muscle was selected as the initial matrix for method validation. The samplewas spikedwith the 10 β_2-agonists to determine the extraction efficiencies. The clean-up procedure was optimised by investigating the dosage of the clean-up agent and the eluent. For 1 g sample, we added DVB-NVP-SO_3Na at one of 4 levels: 10mg/g, 50 mg/g, 90mg/g and 130mg/g. Superior recovery of the 10 β_2-agonists was achieved using 50mg/g of DVB-NVP-SO_3Na. The recovery rates didnot differ significantly when the concentration of the clean-up agent was increased. Lower recoveries were obtained when the concentration of clean-up agent was less 50mg/g. For economic reasons, 50mg/g of clean-up agentwas used in the final optimisedmethod, and the recovery rates for all 10 β_2-agonists were 65. 100% at 0. 5μg/kg, as shown in Fig.5.

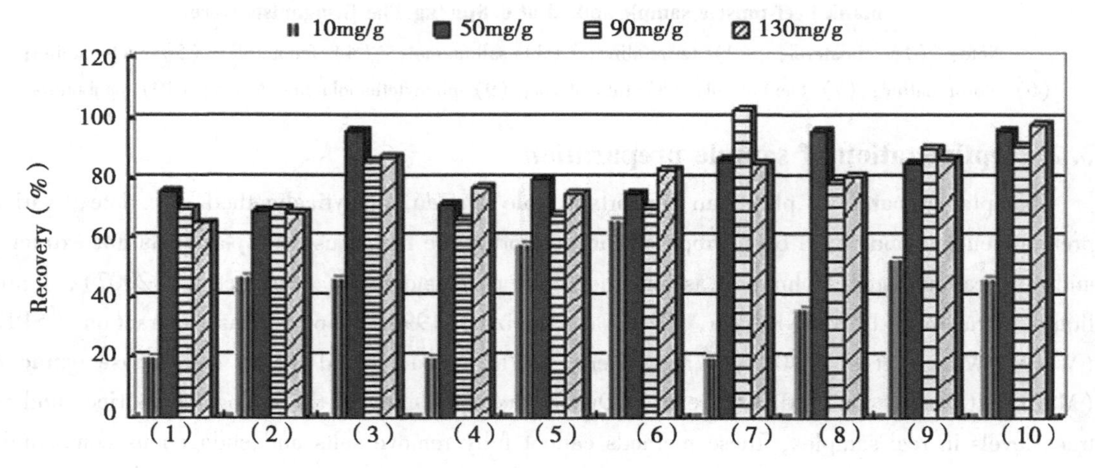

Fig.5 Effect of the dosage of DVB-NVP-SO_3Na on the recovery rates of the target compounds. Veterinary drug code

Note: (1) cimaterol; (2) terbutaline; (3) salbutamol; (4) fenoterol; (5) ractopamine; (6) clorprenaline; (7) clenbuterol; (8) tulobuterol; (9) phenylethanolamine A; (10) penbuterol.

Under similar conditions, the volume of eluent also affected the clean-up of the target compounds. We evaluated ammonia.methanol (3/97, v/v) as the eluent at 3 volumes: 2mL/g, 4mL/g and 6mL/g. The relationship between the volume of the eluent and the recovery rates of the target compounds is shown in Fig. 6. Superior recovery rates of the 10 β_2-agonists were obtained using 6mL/g of eluent. The recovery rates did not differ significantly when the volume of eluent was increased, and a lower volume of eluent led to lower recovery rates. For economic reasons, an eluent volume of 6mL/g was used in the final optimised method, and the recovery rates for all 10 β_2-agonists were 70. 105%.

3.3 Method validation

The blank meat employed in the validation experiments was livestock (beef and goat) muscle

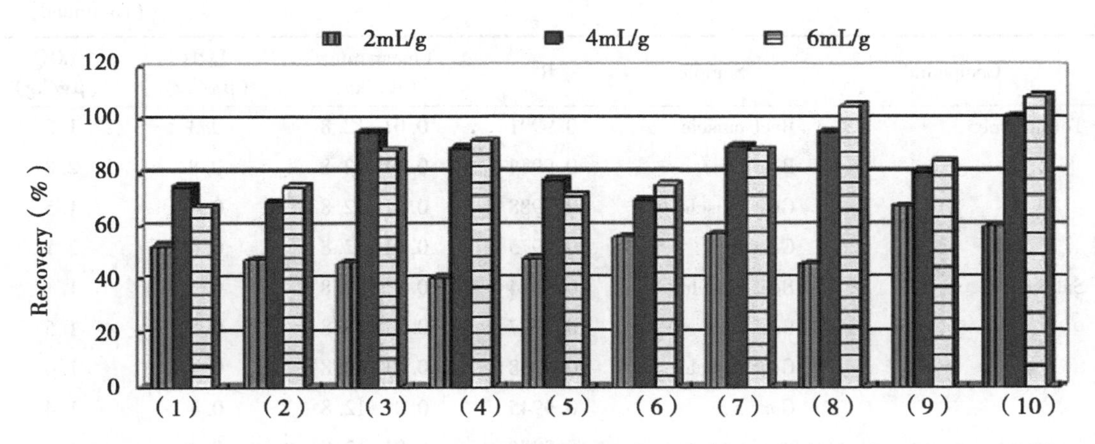

Fig.6 Effect of the volume of eluent ammonia.methanol (3/97, v/v) on the recovery rates of the target compounds.Veterinary drug code

(1) cimaterol; (2) terbutaline; (3) salbutamol; (4) fenoterol; (5) ractopamine; (6) clorprenaline; (7) clenbuterol; (8) tulobuterol; (9) phenylethanolamine A; (10) penbuterol.

and liver purchased at different supermarkets. The method validation experiments were performed using samples that had been previously analysed by the Chinese standard method (China's Ministry of Agriculture Bulletin No. 1025 – 18 – 2008, 2008) to verify that they were blank samples. Fig. 7 shows the chromatograms of 10 β_2-agonists that were spiked at 0. 8μg/kg in goat muscle prepared using the optimised pretreatment conditions. The results indicated that good chromatographic resolution was achieved.

3. 3. 1 Linearity

Analysis of the experimental data validated the good linearity of this method for the detection of β_2-agonists in meat over a range that depended on the target compound. The linearity of the chromatographic responsewas evaluated usingmatrix-matched calibration curves for a blank meat sample using 6 calibration points: 0. 01μg/kg、0. 05μg/kg、0. 2μg/kg、0. 80μg/kg、3. 2μg/kg and 12. 8μg/kg. Calibration curves were constructed, and squared regression coefficients (R^2) were calculated for each compound. The results are shown in Table 2. Statistical analysis of the analytical results for the standards demonstrated a satisfactory relationship between the concentration and the response, and the correlation coefficients (R^2) obtained using this method ranged from 0. 9925 to 0. 9998.

Table 2 Linear ranges, regression coefficients, LODs and LOQs for the β_2-agonists

Compound	Sample	R^2	Linear range (μg/kg)	LOD (μg/kg)	LOQ (μg/kg)
Cimaterol	Beef muscle	0. 9978	0. 01–12. 8	0. 3	1. 1
	Beef liver	0. 9986	0. 01–12. 8	0. 3	1. 2
	Goat muscle	0. 9985	0. 01–12. 8	0. 3	0. 9
	Goat liver	0. 9970	0. 01–12. 8	1. 0	3. 2

(continued)

Compound	Sample	R^2	Linear range (μg/kg)	LOD (μg/kg)	LOQ (μg/kg)
Terbutaline	Beef muscle	0. 9991	0. 01-12. 8	0. 3	1. 2
	Beef liver	0. 9954	0. 01-12. 8	0. 8	2. 8
	Goat muscle	0. 9988	0. 01-12. 8	0. 4	1. 5
	Goat liver	0. 9975	0. 01-12. 8	0. 6	2. 1
Salbutamol	Beef muscle	0. 9974	0. 01-12. 8	0. 3	1. 1
	Beef liver	0. 9987	0. 01-12. 8	0. 5	1. 5
	Goat muscle	0. 9968	0. 01-12. 8	0. 4	1. 4
	Goat liver	0. 9945	0. 01-12. 8	0. 4	1. 4
Fenoterol	Beef muscle	0. 9956	0. 01-12. 8	0. 4	1. 3
	Beef liver	0. 9925	0. 01-12. 8	0. 6	1. 9
	Goat muscle	0. 9943	0. 01-12. 8	0. 2	0. 8
	Goat liver	0. 9956	0. 01-12. 8	0. 3	0. 9
Ractopamine	Beef muscle	0. 9936	0. 01-12. 8	0. 3	1. 2
	Beef liver	0. 9956	0. 01-12. 8	0. 4	1. 3
	Goat muscle	0. 9968	0. 01-12. 8	0. 3	1. 0
	Goat liver	0. 9978	0. 01-12. 8	0. 4	1. 4
Clorprenaline	Beef muscle	0. 9958	0. 01-12. 8	0. 3	0. 8
	Beef liver	0. 9986	0. 01-12. 8	0. 3	1. 1
	Goat muscle	0. 9980	0. 01-12. 8	0. 2	0. 9
	Goat liver	0. 9978	0. 01-12. 8	0. 4	1. 4
Clenbuterol	Beef muscle	0. 9981	0. 01-12. 8	0. 3	1. 2
	Beef liver	0. 9991	0. 01-12. 8	0. 3	1. 1
	Goat muscle	0. 9968	0. 01-12. 8	0. 4	1. 3
	Goat liver	0. 9964	0. 01-12. 8	0. 4	1. 5
Tulobuterol	Beef muscle	0. 9998	0. 01-12. 8	0. 3	0. 9
	Beef liver	0. 9949	0. 01-12. 8	0. 6	2. 1
	Goat muscle	0. 9958	0. 01-12. 8	0. 5	1. 8
	Goat liver	0. 9945	0. 01-12. 8	0. 6	2. 0
Phenylethanolamine A	Beef muscle	0. 9968	0. 01-12. 8	0. 3	1. 0
	Beef liver	0. 9963	0. 01-12. 8	0. 5	1. 6
	Goat muscle	0. 9987	0. 01-12. 8	0. 4	1. 4
	Goat liver	0. 9948	0. 01-12. 8	0. 5	1. 6
Penbuterol	Beef muscle	0. 9939	0. 01-12. 8	0. 5	1. 5
	Beef liver	0. 9930	0. 01-12. 8	0. 5	1. 6
	Goat muscle	0. 9974	0. 01-12. 8	0. 3	1. 1
	Goat liver	0. 9957	0. 01-12. 8	0. 4	1. 1

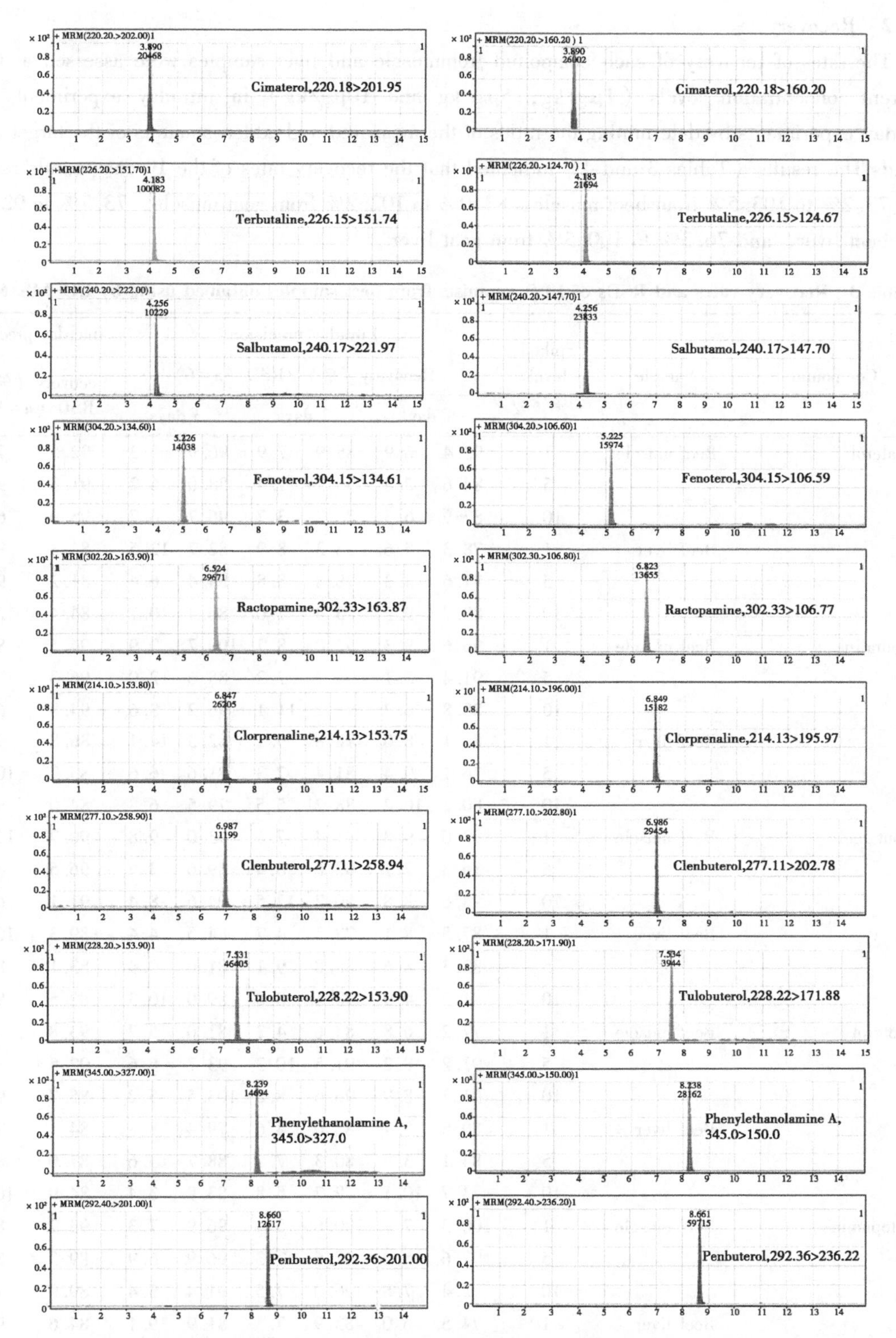

Fig.7 Extract chromatograms of the 10 β_2-agonists spiked at 0. 8μg/kg in goat meat

3.3.2 Recovery

The rates of recovery of each compound frommuscle and liver samples were assessed at three different concentration levels (1μg/kg, 5μg/kg and 10μg/kg) in intraday experiments and interday experiments by determining the ratios of the measured and added amounts of the target compounds.The results (Tables 3 and 4) indicated that the recovery rates of the 10 compounds ranged from 77.2% to 103.5% from beef muscle, 83.0% to 102.2% from goat muscle, 73.7% to 92.2% from beef liver, and 76.3% to 100.5% from goat liver.

Table 3 Recovery rates and RSDs of 10 β_2-agonists from beef samples obtained using by LC-MS/MS

Compound	Sample	Spiked level (μg/kg)	Intraday precision Recovery (%) /RSD (n=6)						Interday precision	
			1 day		2 days		3 days		ecovery (%)/ RSD (n=18)	
Cimaterol	Beef muscle	1	93.4	6.9	85.9	7.9	96.9	6.3	92.5	7.0
		5	84.6	2.8	95.7	7.4	88.0	6.2	91.3	9.3
		10	87.9	6.3	83.2	3.7	96.7	6.7	95.9	6.8
	Beef liver	1	78.3	5.6	78.3	8.9	82.7	12.5	83.3	8.9
		5	81.6	4.3	78.8	8.8	84.4	6.4	81.2	9.4
		10	88.7	9.2	75.4	7.0	86.4	10.7	84.4	12.9
Terbutaline	Beef muscle	1	96.6	9.3	92.0	8.7	100.7	9.9	94.5	8.5
		5	91.4	8.1	94.7	7.2	88.6	12.0	96.3	7.6
		10	88.8	6.7	93.2	11.1	96.2	8.6	94.8	6.9
	Beef liver	1	82.1	8.0	78.8	9.2	82.3	14.1	86.8	9.1
		5	76.1	9.4	81.4	7.3	81.6	6.6	85.7	10.8
		10	80.1	10.2	88.2	5.5	79.5	6.8	84.0	9.8
Salbutamol	Beef muscle	1	93.0	9.8	91.4	7.6	93.0	9.8	90.7	11.9
		5	86.8	7.1	92.6	6.4	89.6	3.5	96.5	6.7
		10	94.6	8.8	90.7	12.5	94.6	8.4	91.1	6.3
	Beef liver	1	87.5	8.1	79.3	4.7	84.5	4.4	89.3	10.0
		5	83.3	8.6	88.8	9.1	81.6	9.4	83.1	8.6
		10	81.5	4.9	78.1	8.0	89.0	10.3	82.5	9.1
Fenoterol	Beef muscle	1	77.2	6.8	83.6	4.3	81.0	5.7	93.8	7.9
		5	91.2	8.2	91.5	10.3	93.7	6.6	92.5	8.7
		10	91.3	8.9	96.6	8.8	103.5	5.2	85.7	9.8
	Beef liver	1	80.5	5.7	83.6	4.6	79.4	8.8	84.1	7.6
		5	84.1	3.9	80.3	7.2	88.7	3.6	87.0	8.1
		10	73.7	10.1	79.7	5.8	83.9	3.4	84.9	10.2
Ractopamine	Beef muscle	1	100.3	7.8	100.6	5.4	96.9	7.3	93.2	8.2
		5	97.6	5.4	97.6	7.3	96.9	6.9	89.3	8.3
		10	92.4	7.8	96.0	7.5	91.4	5.4	89.9	7.9
	Beef liver	1	74.5	6.0	83.9	7.5	81.9	9.1	84.6	9.0
		5	82.6	5.8	83.4	5.0	82.0	4.6	87.3	7.8
		10	84.2	4.5	81.4	4.9	86.2	4.9	83.9	14.0

(continued)

Compound	Sample	Spiked level (μg/kg)	Intraday precision Recovery (%) /RSD (n=6)						Interday precision ecovery (%)/ RSD (n=18)	
			1 day		2 days		3 days			
Clorprenaline	Beef muscle	1	91. 7	9. 5	94. 7	8. 1	93. 5	12. 8	92. 5	7. 9
		5	100. 8	6. 9	87. 2	10. 2	92. 3	7. 7	92. 1	8. 3
		10	95. 1	8. 2	88. 1	7. 2	100. 0	8. 3	89. 7	8. 8
	Beef liver	1	79. 3	6. 1	81. 6	6. 0	85. 5	8. 5	87. 8	8. 3
		5	80. 2	7. 3	80. 8	8. 5	82. 5	9. 5	82. 6	12. 6
		10	84. 6	10. 0	79. 8	6. 8	82. 3	6. 3	83. 8	9. 7
Clenbuterol	Beef muscle	1	92. 5	10. 3	90. 5	6. 9	88. 9	8. 7	89. 0	9. 0
		5	91. 2	6. 9	95. 4	8. 7	87. 3	11. 2	92. 8	9. 9
		10	93. 4	7. 1	93. 9	6. 5	93. 2	11. 4	90. 7	7. 3
	Beef liver	1	83. 4	7. 1	81. 9	7. 0	83. 3	9. 9	86. 7	9. 0
		5	88. 0	4. 7	86. 8	9. 3	92. 4	7. 1	92. 7	9. 9
		10	87. 8	10. 9	85. 6	6. 3	88. 6	11. 7	88. 5	10. 3
Tulobuterol	Beef muscle	1	86. 6	10. 4	86. 0	7. 1	91. 5	9. 3	92. 5	8. 5
		5	90. 3	6. 8	89. 5	11. 3	94. 4	12. 3	86. 4	7. 5
		10	92. 1	7. 9	92. 0	9. 2	98. 4	7. 4	84. 6	10. 5
	Beef liver	1	81. 4	6. 9	87. 9	2. 1	84. 1	7. 8	92. 2	5. 4
		5	80. 9	6. 5	84. 0	7. 6	77. 7	8. 5	91. 7	7. 9
		10	78. 8	4. 8	85. 9	7. 4	88. 4	9. 7	85. 7	8. 9
Phenylethanolamine A	Beef muscle	1	83. 6	13. 0	98. 0	7. 8	89. 2	9. 1	90. 1	9. 8
		5	91. 5	12. 1	85. 0	7. 1	98. 2	12. 0	92. 0	6. 3
		10	98. 2	12. 1	93. 3	7. 2	102. 5	8. 8	89. 2	8. 8
	Beef liver	1	84. 9	10. 4	79. 7	8. 2	87. 2	13. 4	81. 9	11. 0
		5	79. 6	5. 6	77. 9	9. 3	84. 6	8. 1	86. 2	10. 9
		10	85. 0	5. 0	86. 2	6. 8	82. 2	7. 3	86. 4	11. 9
Penbuterol	Beef muscle	1	91. 7	4. 5	96. 8	11. 0	90. 9	7. 5	92. 7	8. 0
		5	101. 9	9. 3	90. 0	8. 2	92. 1	9. 6	89. 7	8. 1
		10	103. 4	7. 1	90. 5	5. 5	93. 8	6. 7	90. 2	9. 9
	Beef liver	1	81. 1	5. 8	90. 8	5. 8	81. 1	6. 1	84. 3	11. 0
		5	91. 3	9. 6	83. 6	7. 4	87. 8	5. 9	84. 3	9. 3
		10	81. 4	3. 4	80. 5	7. 6	80. 5	9. 5	82. 9	14. 9

Table 4 Recovery rates and RSDs of 10 β_2-agonists from goat samples obtained using LC.MS/MS

Compound	Sample	Spiked level (μg/kg)	Intraday precision Recovery (%) /RSD (n=6)						Interday precision ecovery (%)/ RSD (n=18)	
			1 day		2 days		3 days			
Cimaterol	Goat muscle	1	91.8	8.8	84.9	5.6	83.1	5.1	88.0	11.3
		5	89.6	8.4	94.2	9.5	97.3	7.8	87.2	11.0
		10	93.6	7.3	85.4	10.2	91.8	10.8	88.3	12.8
	Goat liver	1	91.5	8.5	85.5	10.0	92.5	7.6	87.5	11.2
		5	91.8	8.5	85.5	9.9	92.5	7.6	81.4	9.4
		10	100.5	11.4	93.4	7.0	100.5	6.9	87.2	11.9
Terbutaline	Goat muscle	1	91.2	6.8	86.3	6.6	87.3	6.1	91.5	7.6
		5	85.1	9.3	90.4	5.0	91.7	8.0	91.2	7.5
		10	86.9	4.8	92.6	8.4	91.0	7.5	86.2	9.3
	Goat liver	1	82.3	9.6	76.5	6.9	82.3	14.0	89.8	9.8
		5	82.5	6.4	81.4	3.3	81.4	7.3	87.0	7.5
		10	77.2	9.3	78.8	5.1	80.3	5.0	89.2	8.9
Salbutamol	Goat muscle	1	93.4	6.7	92.6	3.9	98.9	9.9	91.7	10.3
		5	102.2	10.0	101.2	9.3	90.3	4.4	94.6	6.4
		10	95.9	8.8	92.1	5.1	94.5	8.5	91.7	7.2
	Goat liver	1	78.7	4.1	83.4	9.7	80.5	7.8	84.9	7.3
		5	83.9	3.0	86.4	6.9	80.1	8.5	89.5	7.8
		10	81.0	10.0	81.8	3.3	85.4	7.3	86.3	8.6
Fenoterol	Goat muscle	1	93.8	11.3	94.3	11.2	98.3	10.4	91.8	7.8
		5	90.1	8.4	87.4	9.7	84.2	6.4	90.6	6.6
		10	89.5	10.5	89.6	4.9	99.9	7.5	89.2	10.0
	Goat liver	1	83.5	9.4	88.4	7.9	82.8	5.3	84.2	6.9
		5	88.7	6.4	89.7	4.9	90.6	4.5	89.4	6.4
		10	87.0	11.0	81.4	7.5	81.0	3.7	87.6	7.7
Ractopamine	Goat muscle	1	96.8	8.4	85.5	8.0	90.6	4.1	93.6	9.3
		5	86.5	3.4	98.5	9.0	92.9	8.5	92.7	8.7
		10	89.2	3.7	89.2	4.6	93.1	7.1	88.5	49.0
	Goat liver	1	79.6	6.1	85.8	3.5	83.6	2.2	86.5	8.9
		5	89.1	10.5	85.3	4.2	83.6	7.8	85.9	8.9
		10	82.4	3.3	82.1	7.5	80.3	6.1	86.6	6.8
Clorprenaline	Goat muscle	1	84.1	11.0	94.7	8.1	95.7	6.0	92.7	8.6
		5	93.0	7.1	88.0	5.3	100.3	9.3	88.2	11.3
		10	95.9	7.2	97.5	6.7	101.2	7.8	92.2	8.7
	Goat liver	1	78.3	9.3	86.0	3.9	78.7	7.3	80.9	12.1
		5	85.1	2.4	83.2	3.9	78.8	6.8	85.7	8.9
		10	82.5	15.3	79.5	7.5	86.8	12.7	86.8	7.9

(continued)

Compound	Sample	Spiked level (μg/kg)	Intraday precision						Interday precision	
			Recovery (%) /RSD (n=6)						ecovery (%)/ RSD (n=18)	
			1 day		2 days		3 days			
Clenbuterol	Goat muscle	1	97.9	7.1	86.6	8.6	88.9	6.2	90.3	9.7
		5	86.1	5.0	88.4	9.2	97.9	8.8	90.6	10.1
		10	94.5	9.3	96.6	7.3	89.7	7.6	91.2	7.5
	Goat liver	1	83.0	3.7	76.3	12.1	88.7	4.3	84.8	9.6
		5	90.5	7.4	86.6	6.1	78.8	4.7	89.2	8.3
		10	83.9	3.1	83.2	7.9	83.2	4.0	84.2	8.2
Tulobuterol	Goat muscle	1	95.8	5.8	102.1	6.1	96.5	6.3	92.0	8.2
		5	94.7	10.9	99.8	4.2	95.6	6.2	89.0	8.8
		10	98.9	5.6	99.8	8.5	96.5	7.4	92.3	8.6
	Goat liver	1	83.3	13.8	83.0	8.5	81.1	3.4	85.3	9.6
		5	78.6	6.1	85.4	4.7	79.7	7.6	84.3	6.4
		10	85.4	8.8	81.6	4.4	86.7	83.7	90.0	8.5
Phenylethanolamine A	Goat muscle	1	98.0	3.3	89.0	5.4	93.8	7.0	90.5	8.8
		5	95.3	9.7	93.3	9.4	99.6	5.9	91.4	7.2
		10	94.6	6.9	95.4	9.3	90.0	5.0	93.7	6.6
	Goat liver	1	89.7	9.4	92.6	8.3	78.3	9.2	89.1	9.1
		5	86.6	3.7	82.0	5.1	84.3	5.7	90.3	9.6
		10	83.8	2.6	78.7	7.0	82.7	3.8	86.4	9.7
Penbuterol	Goat muscle	1	92.2	10.3	83.0	6.0	87.8	6.0	93.0	8.7
		5	91.9	8.6	87.3	5.4	99.7	7.7	91.2	8.5
		10	91.0	6.8	90.7	8.6	89.6	9.8	91.9	8.4
	Goat liver	1	81.8	7.3	82.3	9.3	78.6	10.9	85.2	7.3
		5	82.9	14.4	78.5	10.3	84.7	7.7	91.4	10.5
		10	85.4	6.7	84.3	10.2	76.5	6.7	93.5	9.2

3.3.3 Precision

The precision of the method was determined using repeated spiked samples and expressed as the relative standard deviation (RSD). We determined themethod precision and accuracy by evaluating the intraday and interday measurements by analysing the spiked samples on the same (n=18) and three different days (n=6; the samples were evaluated in sets of six replicates evaluated on three consecutive days). The results are presented in Tables 3 and 4. The intraday and interday RSDs of samples for beefmuscle ranged from6.3% to 11.9% and 2.8% to 12.5%, respectively, and the intraday and interday RSD values for beef liver ranged from 5.4% to 12.6% and 3.4% to 12.5%. The intraday and interday RSDs for goat muscle ranged from 6.5% to 12.8% and 3.3% to 11.3%, respectively, and the corresponding values for goat liver ranged from 6.4% to 12.1% and 2.4% to 15.3%, respectively. Thus, all RSD values were less than 15% at three different concentration levels (1μg/kg, 5μg/kg and 10μg/kg), except for clorprenaline in goat liver, which had an RSD value of greater than 15% at 10μg/kg.

3. 3. 4 LOD and LOQ

The sensitivity of the method was evaluated based on the values of LOD and LOQ.The LOQ was determined using spiked samples, with the minimum detectable amount of analyte set at a signal-to-noise ratio (S/N) of 10. The LODs and LOQs (Table 2) of the target compounds achieved using the proposed method were within the range of 0. 2–0. 4μg/kg and 0. 8–1. 5μg/kg for spiked beef muscle, 0. 3–0. 8μg/kg and 1. 1–2. 8μg/kg for spiked beef liver, 0. 2–0. 5μg/kg and 0. 8–1. 8μg/kg for spiked goat muscle, and 0. 3–0. 9μg/kg and 0. 9–3. 2μg/kg for spiked goat liver.

3. 3. 5 Comparison with the existing method

Thesemethod validation valueswere comparablewith those reported in previous papers.Eshaq (Isaac) et al. (2003) determined ractopamine in animal tissues by LC. MS/MS. The recoveries of ractopamine ranged between 80% and 117% in porcine tissues and 85% and 114% in bovine tissues.Li et al. (2014) reported a method for the multi-residue analysis of 14 β–agonists in weight-reducing dietary supplements using LC–MS/MS. Fourteen β–agonists were separatedwithin 15min with an LOD of 1. 8–23. 1μg/kg.The average recoveries in spiked dietary supplements varied from 67. 1% to 107. 3%.The within-day repeatabilities ranged from 1. 0% to 16. 6%, and the between-day repeatabilities ranged from 2. 6% to 18. 9%.Shao et al. (2009) developed a method for the simultaneous analysis of 16 illegal residual β–agonists in pork tissues, including liver, kidney and meat.The analytes were quantified by UPLC–MS/MS.The average recoveries of each compound ranged from 75. 2% to 118%. The RSDs were < 18%. The within-day reproducibility ranged from 3. 4% to 12. 0%, and the between-day reproducibility ranged from 8. 2% to 14. 8%. CCalpha and CCbeta ranged from 0. 02 to 0. 79μg/kg and from 0. 04 to 1. 62μg/kg, respectively. Fan et al. (2012) developed a method for the simultaneous determination of 25 β_2-agonists in animal foods using LC–MS.The recoveries of each compound were in the range of 46. 6%–118. 9%, and the RSDs were in the range of 1. 9%–28. 2%.The decision limits of 48 analytes in muscles, liver, and kidney samples ranged from 0. 05 to 0. 49μg/kg, and the detection capability ranged from 0. 13 to 1. 64μg/kg. In summary, in the existing literature, the average recoveries ranged between 46. 0% and 118. 0%, with RSDs of 1. 0%–28. 0%.The LOD and LOQ were 0. 02–0. 79μg/kg and 0. 04–23. 1μg/kg, respectively.In contrast, the RSDs and recoveries of our method were 2. 7%–15. 3% and 73. 7%–103. 5% respectively, with LOD and LOQ value of 0. 2–0. 8μg/kg and 0. 8–1. 6μg/kg (except for cimaterol and tulobuterol), respectively.The precision of our method is better than those of the currently existing methods, and its sensitivity is equivalent to those of the existingmethods.

3. 3. 6 Matrix effects

Optimisation of sample preparation methods can reduce the level of matrix-induced interference but cannot eliminate it.Using the same pretreatment method and instruments, samples of goat liver, beef liver, goatmuscle and beefmusclewere cleaned-up and tested.Thematrix effects in these samples were evaluated by the LC–MS/MS analysis of blank samples and spiked samples at 0. 8μg/kg (Fig. 8).The liver sample exhibited more impurity-containing peaks.By contrast, the muscle sample had fewer impurity-containing peaks.These results demonstrate that the pretreatment method was more effective for purifying muscle samples than liver samples. The composition of liver is more complicated than that of muscle and includes large amounts of glycogen, fats, proteins, carbohydrates and vitamins, whereas themain component of muscle is protein. The peak areas of the impurities were very small, and the chromatographic peaks of the target compounds in the spiked sample were very obvious; no interfering peaks were observed at the retention times of the 10 β_2-agonists.

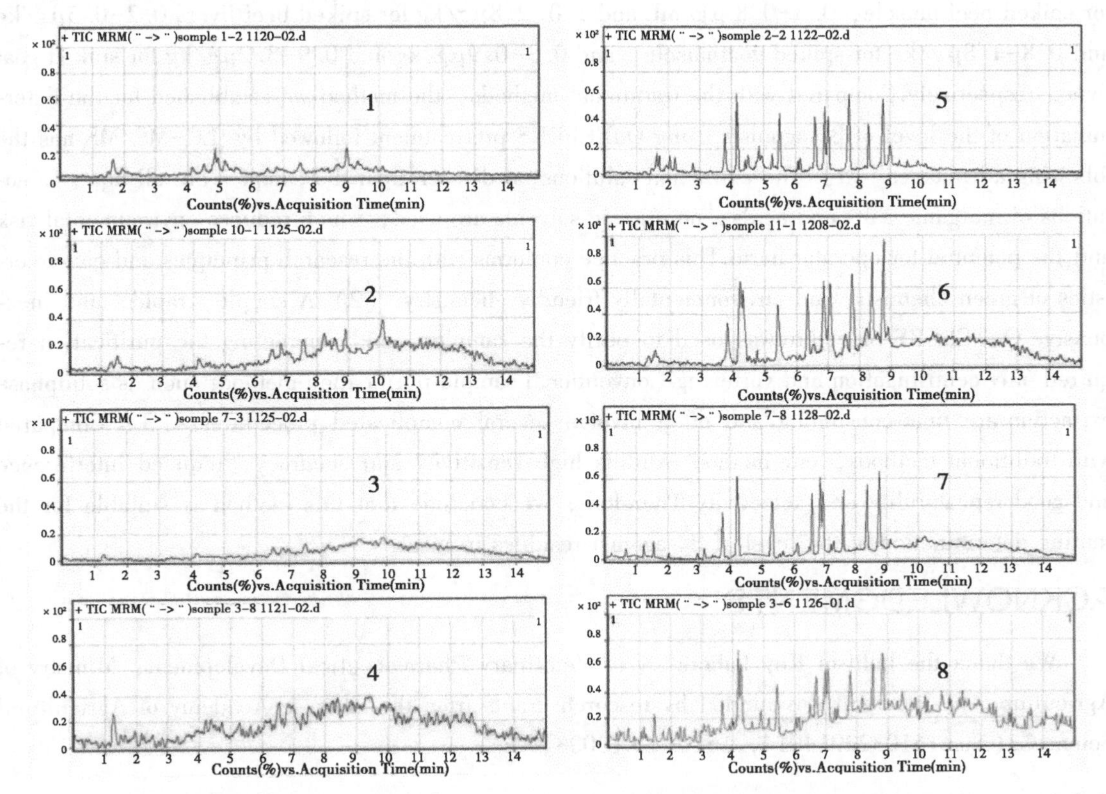

Fig.8 Comparison of the LC.MS/MS chromatograms of the black and spiked samples

Note: 1. beefmuscle; 2. beef liver; 3. goatmuscle; 4. goat liver; 5. spiked beefmuscle; 6. spiked beef liver; 7. spiked goat muscle; 8. spiked goat liver.

3.4 Application of the method to real samples

The effectiveness of the proposed method was verified by analysing more than 50 samples of cattle muscle obtained from local dairy farmers. All samples were processed according to the methods described, and no residues of the target analytes were detected in any of the 50 samples.

4 CONCLUSIONS

A LC-MS/MS method using a QuEChERS procedure with preconcentration as the sample treatment was established for the analysis of trace residues of 10 β_2-agonists in meat. The extraction and purification conditions for the β_2-agonist residue analysis in meat were discussed in depth. We synthesised DVB-NVP-SO_3Na for application as the clean-up agent for the β_2-agonists in the QuEChERS protocol.

Statistical analysis of the analytical results obtained using the standards demonstrated a satisfactory relationship between concentration and response, and the correlation coefficients (R^2) of this method ranged from0. 9925 to 0. 9998. The recovery rates for the 10 compounds ranged from 77. 2% to 103. 5% in beef muscle, 83. 0% to 102. 2% in goat muscle, 73. 7% to 92. 2% in beef liver, and 76. 3% to 100. 5% in goat liver.The RSD values were less 15%.The LOD and LOQ of the target compounds using the proposed method were within the range of 0. 2–0. 4μg/kg and 0. 8–1. 5μg/kg for spiked beef muscle, 0. 3–0. 8 μg/mL and 1. 0–2. 8μg/kg for spiked beef liver, 0. 2–0. 5μg/kg and 0. 8–1. 8μg/kg for spiked goatmuscle, and 0. 3–0. 9μg/kg and 0. 9–3. 2μg/kg for spiked goat liver, respectively.Compared with the traditional methods, the method we established for the determination of the level of β_2-agonists using QuEChERS pretreatment followed by LC–MS/MS has the following advantages. (1) The extraction solutions used to pretreat the samples are all aqueous solutions of inorganic salts and acids; no organic solvents are used, which reduces environmental risk and the potential for operator harm.This practice conforms with the research principles and characteristics of green chemistry and environmentally friendly chemistry. (2) A simple, rapid, and inexpensive QuEChERS procedurewas used to purify the samples, and completing the purification required only centrifugation and vortexing.Conventional sample-purification methods such as solidphase extraction are time-consuming and often involve several complicated procedures. (3) Compared with traditional methods, our method exhibits high sensitivity and accuracy, reduced interference and good repeatability and selectivity.Therefore, we conclude that this method is suitable for the routine determination of the level of β_2-agonist residues in meat.

ACKNOWLEDGMENTS

We thank the help of Key Laboratory of Veterinary Pharmaceutical Development, Ministry of Agriculture.This research was funded by research grants from the Chinese Academy of Agricultural Sciences (Nos.1610322014014, 1610322013008).

References

[1] Aguilera-Luiz, M.M., Vidal, J.L.M., Romero-Gonzalez, R., & Frenich, A.G. (2012).Multiclass method for fast determination of veterinary drug residues in baby food by ultra-high-performance liquid chromatography.tandem mass spectrometry.*Food Chemistry*, 132: 2171–2180.

[2] Anastassiades, M., Lehotay, S.J., Stajnbaher, D., & Schenck, F.J. (2003).Fast and easy multiresidue method employing acetonitrile extraction/partitioning and "dispersive solid-phase extraction" for the determination of pesticide residues in produce.*Journal of AOAC International*, 86: 412–431.

[3] Andree, S., Jira, W., Schwind, K.-H., Wagner, H., & Schwagele, F. (2010).Chemical safety of meat and meat products.*Meat Science*, 86: 38–48.

[4] Baert, K., Huffel, X.V., Jacxsens, L., Berkvens, D., Diricks, H., Huyghebaert, A., & Uyttendaele, M. (2012).Measuring the perceived pressure and stakeholders' response that may impact the status of the safety of the food chain in Belgium.*Food Research International*, 48: 257–264.

[5] Boyd, S., Heskamp, H.H., Bovee, T.F.H., Nielen, M.W.F., &Elliott, C.T. (2009).Development, validation and implementation of a receptor based bioassay capable of detecting a broad range of β–agonist drugs in animal feeding stuffs.*Analytica Chimica Acta*, 637: 24–32.

[6] Brousmiche, D.W., O'Gara, J.E., Walsh, D.P., Lee, P.J., Iraneta, P.C., Trammell, B.C., &

Mallet, C.R. (2008).Functionalization of divinylbenzene/N-vinylpyrrolidone copolymer particles: Ion exchangers for solid phase extraction.*Journal of Chromatography A*, 1191 (1.2): 108-117 (2008).

[7] Castro-Puyana, M., & Herrero, M. (2013).Metabolomics approaches based on mass spectrometry for food safety, quality and traceability.*TrAC Trends in Analytical Chemistry*, 52: 74-87.

[8] China's Ministry of Agriculture Bulletin No. 1025 - 18 - 2008 (2008). *Determination of β_2-agonists residues in animal derived food by liquid chromatography.tandem mass spectrometry.*

[9] Commission of the European Communities (1996).Council directive 96/23/EC.*Official Journal of the European Communities*, *No.L*125, 10.

[10] Corcia, D.D., Morra, V., Pazzi, M., & Vincenti, M. (2010).Simultaneous determination of β_2-agonists in human urine by fast-gas chromatography/mass spectrometry: Method validation and clinical application.*Biomedical Chromatography*, 24: 358-366.

[11] Cronly, M., Behan, P., Foley, B., Malone, E., Earley, S., Gallagher, M., Shearan, P., & Regan, L. (2010).Development and validation of a rapid multi-classmethod for the confirmation of fourteen prohibited medicinal additives in pig and poultry compound feed by liquid chromatography.tandem mass spectrometry.*Journal of Pharmaceutical and Biomedical Analysis*, 53 (4): 929-938.

[12] Domínguez-Álvarez, J., Mateos-Vivas, M., García-Gómez, D., Rodríguez-Gonzalo, E., & Carabias-Martinez, R. (2013).Capillary electrophoresis coupled tomass spectrometry for the determination of anthelmintic benzimidazoles in eggs using a QuEChERSwith preconcentration as sample treatment. *Journal of Chromatography A*, 1278: 166-174.

[13] Du, W., Fu, Q., Zhao, G., Huang, P., Jiao, Y.Y., Wu, H., Luo, Z.M., & Chang, C. (2013a).Dummy-template molecularly imprinted solid phase extraction for selective analysis of ractopamine in pork.*Food Chemistry*, 139: 24-30.

[14] Du, X.D., Wu, Y.L., Yang, H.J., & Yang, T. (2012).Simultaneous determination of 10 β_2-agonists in swine urine using liquid chromatography.tandem mass spectrometry and multi-walled carbon nanotubes as a reversed dispersive solid phase extraction sorbent.*Journal of Chromatography A*, 1260: 25-32.

[15] Du, W., Zhang, S.R., Fu, Q., Zhao, G., & Chang, C. (2013b).Combined solid-phase microextraction and high-performance liquid chromatography with ultraviolet detection for simultaneous analysis of clenbuterol, salbutamol and ractopamine in pig samples. *Biomedical Chromatography*, 27: 1775-1781.

[16] Du, W., Zhao, G., Fu, Q., Sun, M., Zhou, H.Y., & Chang, C. (2014).Combined microextraction by packed sorbent and high-performance liquid chromatography.ultraviolet detection for rapid analysis of ractopamine in porcine muscle and urine samples.*Food Chemistry*, 145: 789-795.

[17] Eshaq (Isaac), S., Sin, C.C., Sami, J., Gene, A., & Michael, K.H. (2003).Determination of ractopamine in animal tissues by liquid chromatography-fluorescence and liquid chromatography/tandem mass spectrometry.*Analytica Chimica Acta*, 483: 137-145.

[18] Fan, S., Miao, H., Zhao, Y.F., Chen, H.J., & Wu, Y.N. (2012).Simultaneous detection of residues of 25 β_2-agonists and 23 β-blockers in animal foods by high-performance liquid chromatography coupled with linear ion trap mass spectrometry.*Journal of Agricultural and Food Chemistry*, 60: 1898-1905.

[19] Gallo, P., Brambilla, G.Gianfranco, Neri, B., Fiori, M., Testa, C., & Serpe, L. (2007). Purification of clenbuterol-like β_2-agonist drugs of new generation from bovine urine and hair by α_1-acid glycoprotein affinity chromatography and determination by gas chromatography.mass spectrometry. *Analytica Chimica Acta*, 587: 67-74.

[20] Garcia, P., Paris, A.-C., Gil, J., Popot, M.-A., & Bonnaire, Y. (2011). Analysis of b-agonists by HPLC/ESI-MSn in horse doping control.*Biomedical Chromatography*, 25: 147-154.

[21] González-Antuña, A., Domínguez-Romero, J. C., García-Reyes, J. F., Rodriguez-González, P., Centineoc, G., Alonso, J.I.G., & Molina-Díazb, A. (2013).Overcoming matrix effects in electrospray: Quantitation of β_2-agonists in complex matrices by isotope dilution liquid chromatography.mass spectrometry using singly ^{13}C-labeled analogues.*Journal of Chromatography A*, 1217: 3612-3618.

[22] Hernández-Mesa, M., García-Campaña, A.M., & Cruces-Blanco, C. (2014). Novel solid phase extraction method for the analysis of 5-nitroimidazoles and metabolites in milk samples by capillary electrophoresis.*Food Chemistry*, 145: 161-167.

[23] Jean-Philippe, A., Philippe, M., Bruno, Le B., & Franc, O.A. (2002).Identification of ractopamine residues in tissue and urine samples at ultra-trace level using liquid chromatography.positive electrospray tandem mass spectrometry.*Journal of Chromatography B*, 774: 59-66.

[24] Jenson, I., & Sumner, J. (2012). Performance standards and meat safety. Developments and direction.*Meat Science*, 92 (3): 260-266.

[25] Keskin, S., Ozer, D., & Temizer, A. (1998).Gas chromatography.mass spectrometric analysis of clenbuterol from urine.*Journal of Pharmaceutical and Biomedical Analysis*, 18: 639-644.

[26] Kinsella, B., Byrne, P., Cantwell, H., Cormack, M.M., Furey, A., & Danaher, M. (2011). Determination of the new anthelmintic monepantel and its sulfone metabolite in milk and muscle using a UHPLC.MS/MS and QuEChERS method.*Journal of Chromatography B*, 879: 3707-3713.

[27] León, N., Roca, M., Igualada, C., Martins, C.P.B., Pastor, A., & Yusá, V. (2012).Wide-range screening of banned veterinary drugs in urine by ultra high liquid chromatography coupled to high-resolution mass spectrometry.*Journal of Chromatography A*, 1258: 55-65.

[28] Li, J.H., Zhang, Z.H., Liu, X., Yan, H., Han, S., Zhang, H.Y., Zhang, S., & Cheng, J. (2014).Analysis of fourteen agonists in weight reducing dietary supplements using QuEChERS-base extraction followed by high resolution UPHPLC-MS.*Food Analytical Methods*, 7: 977-985.

[29] Liu, X.P., Zhang, W.F., Hu, Y.N., & Cheng, H.F. (2013).Extraction and detection of organoarsenic feed additives and common arsenic species in environmental matrices by HPLC.ICP-MS.*Microchemical Journal*, 108: 38-45.

[30] Lopes, R.P., Passos, É.E.D.F., Filho, J.F.D.A., Vargas, E.A., Augusti, D.V., &Augusti, R. (2012).Development and validation of a method for the determination of sulfonamides in animal feed by modified QuEChERS and LC.MS/MS analysis.*Food Control*, 28: 192-198.

[31] Maciejewska, M., & Gawdzik, J. (2008).Preparation and characterization of sorption properties of porous microspheres of 1-vinyl-2-pyrrolidone-divinylbenzene. *Journal of Liquid Chromatography & Related Technologies*, 31 (7): 950-961.

[32] Maciejewska, M., & Osypiuk-Tomasik, J. (2013).Sorption on porous copolymers of 1-vinyl-2-pyrrolidone-divinylbenzene.*Journal of Thermal Analysis and Calorimetry*, 114 (2): 749-755.

[33] Moragues, F., & Igualada, C. (2009).How to decrease ion suppression in a multiresidue determination of β-agonists in animal liver and urine by liquid chromatographymass spectrometry with ion-trap detector.*Analytica Chimica Acta*, 637: 193-195.

[34] Morales-Trejo, F., León, S.V., Escobar-Medina, A., & Gutiérrez-Tolentino, R. (2013).Application of high-performance liquid chromatography.UV detection to quantification of clenbuterol in bovine liver samples.*Journal of Food and Drug Analysis*, 21: 414-420.

[35] Murat, F.U., Usama, A., Ibrahim, L., Nilgun, G.G., & Nusret, E. (2013).Dispersive liquid. liquid microextraction based on solidification of floating organic drop combined with field-amplified sam-

ple injection in capillary electrophoresis for the determination of beta (2) -agonists in bovine urine.*Electrophoresis*, 34: 854-861.

[36] Park, K.H., Choi, J.-H., El-Aty, A.M.A., Cho, S.-K., Park, J.-Y., Kwon, K.S., Park, H. R., Kim, H.S., Shin, H.-C., Kim, M.R., & Shim, J.-H. (2012).Development of QuEChERS-based extraction and liquid chromatography.tandem mass spectrometry method for quantifying flumethasone residues in beef muscle.*Meat Science*, 92: 749-753.

[37] Pérez-Burgos, R., Grzelak, E. M., Gokce, G., Saurina, J., Barbosa, J., & Barrón, D. (2012). Quechers methodologies as an alternative to solid phase extraction (SPE) for the determination and characterization of residues of cephalosporins in beef muscle using LC-MS/MS.*Journal of Chromatography B*, 899: 57-65.

[38] Qiao, F.X., & Du, J.J. (2013).Rapid screening of clenbuterol hydrochloride in chicken samples by molecularly imprinted matrix solid-phase dispersion coupled with liquid chromatography. *Journal of Chromatography B*, 923-924: 136-140.

[39] Ruiz, A., Mardones, C., Vergara, C., Hermosín-Gutiérrez, I., Baer, D.V., Hinrichsen, P., Rodriguez, R., Arribillaga, D., & Dominguez, E. (2013).Analysis of hydroxycinnamic acids derivatives in calafate (*Berberis microphylla* G.Forst) berries by liquid chromatography with photodiode array and mass spectrometry detection.*Journal of Chromatography A*, 1281: 38-45.

[40] Sato, S., Nomura, S., Kawano, F., Tanihata, J., Tachiyashiki, K., & Imaizumi, K. (2010). Adaptive effects of the β_2-agonist clenbuterol on expression of β_2-adrenoceptor mRNA in rat fast-twitch fiber-rich muscles.*Journal of Physiological Sciences*, 60: 119-127.

[41] Shao, B., Jia, X.F., Zhang, J., Meng, J., Wu, Y.N., Duan, H.J., & Tu, X.M. (2009). Multi-residual analysis of 16 b-agonists in pig liver, kidney and muscle by ultra performance liquid chromatography tandem mass spectrometry.*Food Chemistry*, 114: 1115-1121.

[42] Shen, J.Z., Zhang, Z., Yao, Y., Shi, W.M., Liu, Y.B., & Zhang, S.X. (2007).Time-resolved fluoroimmunoassay for ractopamine in swine tissue.*Analytical and Bioanalytical Chemistry*, 387 (4): 1561-1564.

[43] Sirhan, A.Y., Tan, G.H., &Wong, R.C.S. (2011).Method validation in the determination of aflatoxins in noodle samples using the QuEChERS method (Quick, Easy, Cheap, Effective, Rugged and Safe) and high performance liquid chromatography coupled to a fluorescence detector (HPLC-FLD).*Food Control*, 22: 1807-1813.

[44] Sofos, J.N. (2014).Safety of food and beverages: Meat and meat products.*Encyclopedia of Food Safety*, 3: 268-279.

[45] Wang, L., Li, Y.Q., Zhou, Y.K., & Yang, Y. (2010).Determination of four β_2-agonists in meat, liver and kidney by GC-MS with dual internal standards.*Chromatographia*, 71: 737-739.

[46] Wang, X., Wang, S.J., & Cai, Z.W. (2013a).The latest developments and applications of mass spectrometry in food-safety and quality analysis.*TrAC Trends in Analytical Chemistry*, 52: 170-185.

[47] Wang, L.Q., Zeng, Z.L., Wang, X.F., Yang, J.W., Chen, Z.H., & He, L.M. (2013b).Multiresidue analysis of nine β-agonists in animal muscles by LC-MS/MS based on a new polymer cartridge for sample cleanup.*Journal of Separation Science*, 36: 1843-1852.

[48] Wang,G.M., Zhao, J., Peng, T., Chen, D.D., Xi, C.X., Wang, X., & Zhang, J.Z. (2013c). Matrix effects in the determination of β-receptor agonists in animal-derived foodstuffs by ultra-performance liquid chromatography tandem mass spectrometry with immunoaffinity solid-phase extraction. *Journal of Separation Science*, 36: 796-802.

[49] Weilin, L.S., Jennifer, F.T., Carolyn, J.H., & David, J.S. (2010).Depletion of urinary

zilpaterol residues in horses as measured by ELISA and UPLC－MS/MS. *Journal of Agricultural and Food Chemistry*, 58: 4077-4083.

[50] Wilkowsk, A., & Biziuk, M. (2011). Determination of pesticide residues in food matrices using the QuEChERS methodology. *Food Chemistry*, 125: 803-812.

[51] Wozniak, B., Zuchowska, I.M., & Zmudzki, J. (2013). Determination of stilbenes and resorcylic acid lactones in bovine, porcine and poultry muscle tissue by liquid chromatography. negative ion electrospray mass spectrometry and QuEChERS for sample preparation. *Journal of Chromatography B*, 940: 15-23.

[52] Xu, Z.G., Hu, Y.F., Hu, Y.L., & Li, G.K. (2010). Investigation of ractopamine molecularly imprinted stir bar sorptive extraction and its application for trace analysis of β_2-agonists in complex samples. *Journal of Chromatography A*, 1217: 3612-3618.

[53] Yan, H., Liu, X., Cui, F.Y., Yun, H., Li, J.H., Ding, S.Y., Yang, D.J., & Zhang, Z.H. (2013). Determination of amantadine and rimantadine in chicken muscle by QuEChERS pretreatment method and UHPLC coupled with LTQ Orbitrap mass spectrometry. *Journal of Chromatography B*, 938: 8-13.

[54] Zheng, H., Deng, L. G., Lu, X., Zhao, S. C., Guo, C. Y., Mao, J. S., Wang, Y. T., & Yang, G. S. (2010). UPLC - ESI - MS - MS determination of three β_2-agonists in pork. *Chromatographia*, 72: 79-84.

（发表于《Meat science》，SCI 院选，IF：2.615）

Review of Platensimycin and Platencin: Inhibitors of β-Ketoacyl-acyl Carrier Protein (ACP) Synthase III (FabH)

SHANG Ruo-feng[1], LIANG Jian-ping[1], YI Yun-peng[1], LIU Yu[1], WANG Jia-tu[2]*

(1. Key Laboratory of New Animal Drug Project of Gansu Province/Key Laboratory of Veterinary Pharmaceutical Development, Ministry of Agriculture/Lanzhou Institute of Husbandry and Pharmaceutical Sciences of CAAS, Lanzhou 730050, China;
2. Affiliated Hospital of Gansu University of Chinese Medicine, Lanzhou 730000, China)

Abstract: Platensimycin and platencin were successively discovered from the strain *Streptomyces platensis* through systematic screening. These natural products have been defined as promising agents for fighting multidrug resistance in bacteria by targeting type II fatty acid synthesis with slightly different mechanisms. Bioactivity studies have shown that platensimycin and platencin offer great potential to inhibit many resistant bacteria with no cross-resistance or toxicity observed *in vivo*. This review summarizes the general information on platensimycin and platencin, including antibacterial and self-resistant mechanisms. Furthermore, the total synthesis pathways of platensimycin and platencin and their analogues from recent studies are presented.

Key words: Platensimycin; platencin; Drug resistance; Antibacterial activities; Synthesis; Analogues

1 INTRODUCTION

Many available drugs have reduced or lost their curative effect, leading to increased morbidity and mortality, because of the emergence and spread of multidrug resistant bacteria[1]. The drastic increase in pathogenic bacterial resistance, especially in multiresistant bacteria, is one of the most serious problems endangering human health and urgently needs for effective solutions[2,3]. Chemical modifications of existing scaffolds have afforded antibiotics with improved activity and have served well the development of new and effective antibiotics in past decades. However, such modifications are becoming increasingly challenging[4]. Therefore, the discovery of novel antibiotic chemical scaf-

---

* Authors to whom correspondence should be addressed; E-mails: Wang_ jiatu@ sina.com;

Academic Editor: Peter J. Rutledge

Received: 9 July 2015/Accepted: 28 August 2015 / Published: 3 September 2015

folds with new modes of action to overcome drug resistance is crucial[5].Natural products are an important tool for this approach, particularly platensimycin (1) and platencin (2) (Fig. 1).

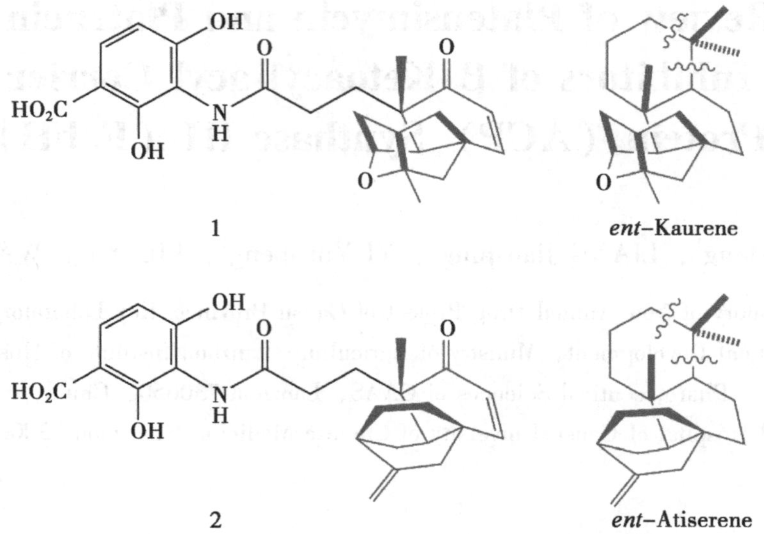

Fig. 1 Structures of platensimycin 1 and platencin 2 and their biosynthetic relationship to *ent*-kaurene and *ent*-atiserene

The natural products platensimycin and platencin, discovered by employing a novel antisense differential sensitivity screening strategy, were reported recently from soil bacterial strains of *Streptomyces platensis*[6-8].Platensimycin inhibits fatty acid acyl carrier protein synthase II (FabF) selectively[6], whereas platencin is a balanced dual inhibitor of both FabF and fatty acid acyl carrier protein synthase III (FabH)[9]. Because of their unique antibacterial mechanism, these natural products show potent, broad-spectrum Gram-positive *in vitro* activity, including against antibiotic-resistant bacteria, such as methicillin-resistant *Staphylococcus aureus* (MRSA), vancomycin-intermediate *S. aureus* (VISA), and vancomycin-resistant *Enterococci* (VRE)[4]. Importantly, both platensimycin and platencin exhibit no cross-resistance to clinically relevant pathogens and show *in vivo* efficacy without toxicity[10].

This review will provide an overview of the isolation, antibacterial activities, biosynthetic machineries, and antibacterial and self-resistant mechanisms of platensimycin and platencin.Moreover, total synthesis and their analogues in recent years are also described.

2 ISOLATION, ANTIBACTERIAL ACTIVITIES, AND BIOSYNTHESIS

Platensimycin was first discovered by the Merck research group through a systematic screening of approximately 250 000 natural product extracts as part of their target-based whole-cell screening strategy using an antisense differential sensitivity assay[11].This product was isolated (2-4mg/L) from the fermentation broth of *Streptomyces platensis* (MA7327 and MA7331) by a two-step process, a capture step followed by reversed-phase high performance liquid chromatography (HPLC), and the structure was elucidated by 2D nuclear magnetic resonance (NMR) methods and confirmed by X-ray crystallographic analysis[11].Platensimycin was ultimately proven to be a new class of antibiotics with no cross-resistance in other classes of antibiotic-resistant bacteria[12,13].It is a selective inhibitor for *S.aureus* FabF and *Escherichia coli* (*E.coli*) FabF/B enzymes with IC_{50} values of 48nm and 160nm, respectively[14].*In vitro* antibacterial studies have shown that platensimycin is effective against Gram-positive bacteria with minimal inhibitory concentration (MIC) values of 0. 1-0. 32mg/mL[15].

Another extract of a new strain of *Streptomyces platensis* (MA7339) was isolated from a soil sample and identified by the same research group after the discovery of platensimycin.Platencin was discovered and isolated by bioassay-guided fractionation of this extract[4,16]. Unlike platensimycin, platencin possesses a polycyclic enone skeleton bearing an exocyclic double bond instead of an ether linkage, and is a balanced dual inhibitor of both FabH (IC_{50} = 9. 2 mm) and FabF (IC_{50} = 4. 6 mm). Platencin shows broad-spectrum antibacterial activity against various bacterial strains, including drug resistant bacteria such as MRSA, linezolid-resistant *S.aureus*, vancomycin intermediate *S.aureus* and VRE with MIC values ≥0. 06-4mg/mL[14].

Feeding experiments using ^{13}C precursors suggest that the aminobenzoic acid moieties of platensimycin and platencin were biosynthesized from pyruvate and acetate via the tricarboxylic acid cycle and phosphoenolpyruvate in the strains of *Streptomyces platensis*.The C - 17 polycyclic enone acid moiety of platensimycin was biosynthesized via a non-mevalonate terpenoid pathway by the oxidative excision of three carbons from *ent*-kaurene (Fig. 1)[17,18]. Recent studies showed that the biosynthesis of platencin in *Streptomyces platensis* MA7327 and MA7339 is controlled by *ent*-atiserene (Fig. 1) synthases, a new pathway for diterpenoid biosynthesis that is different from *ent*-kaurene synthases[19].

3 ANTIBACTERIAL AND SELF-RESISTANT MECHANISM

Fatty acids form the building blocks of many cellular structures and are required for energy storage in bacteria[9].The synthesis of fatty acids in bacteria is conducted by type II bacterial fatty acid synthesis (FASII) which is different from the synthesis found in eukaryotes (mammals and fungi; type I).Type II FAS consists of many discrete enzymes- involved in condensation, reduction and dehydration[20].Platensimycin and platencin interfere with the synthesis of fatty acids by selectively inhibiting FabF, an elongation-condensing enzyme whose main function is to add acetate units to the growing fatty acid chain[9].A FabH/FabF PAGE elongation assay with the crude *S.aureus* cytosolic proteins, which closely mimics the events inside living cells, suggested that platensimycin is a selective FabF inhibitor, whereas platencin is a dual inhibitor with similar inhibition efficiency for both FabF and FabH (Fig. 2) and thus inhibits fatty acid synthesis in *S. aureus* through a synergistic effect[4].

Pathogens have generally gained their resistance genes by horizontal gene transfer from non-pathogenic bacteria, with one potential source being antibiotic producing bacteria that developed highly effective mechanisms to avoid suicide[21].Therefore, it is important to understand the self-resistance mechanisms within platensimycin and platencin producing organisms.A standard disk diffusion assay identified the *ptmP*3 or *ptnP*3 gene within the platensimycin or platencin biosynthetic cluster in the S.*platensis* MA7327 and MA7339 strains as the major resistance conferring element. The FabF gene within the housekeeping fatty acid synthase locus was identified as the second resistance conferring element. PtmP3/PtnP3 and FabF, therefore, confer platensimycin and platencin resistance by target replacement (*i.e.*, FabF and FabH by PtmP3) and target modification (*i.e.*, a platensimycin-insensitive variant of FabF), respectively (Fig. 2). PtmP3/PtnP3 also represents an unprecedented mechanism for fatty acid biosynthesis in which FabH and FabF are functionally replaced by a single condensing enzyme (*i.e.*, a platensimycin-insensitive variant of FabF)[10].

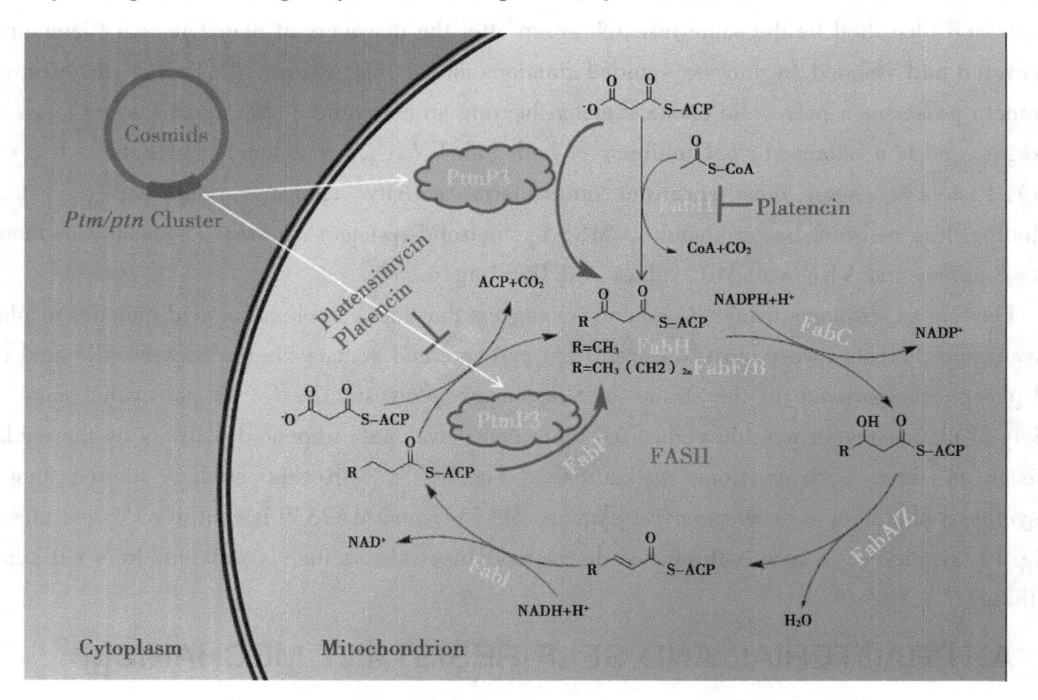

Fig. 2 A diagram of the bacterial fatty acid synthesis cycle (FASII)

Note: Highlighted in red are (i) platensimycin inhibiting FabF/B and platencin dually inhibiting FabF and FabH, and (ii) the two complementary mechanisms of platensimycin and platencin resistance inS.*platensis* by target replacement and target modification. Adapted from Peterson et al.[10]

4 RECENT TOTAL SYNTHESIS

Structural studies revealed that platensimycin and platencin consist of a compact aliphatic tetracyclic ketolide core and polycyclic enone core, respectively, that were connected through a propionate tether to a highly oxygenated aromatic ring[9,22]. Therefore, the syntheses of platensimycin or platencin are generally divided into two sub-targets: the aromatic amine 3 and the carboxylic acid 4 or 6 constructed by tetracyclic cage 5 or Nicolaou's intermediate 7 are the key intermediates for synthesizing platensimycin or platencin, respectively (Scheme 1)[23]. Since the discovery of platensimycin and platencin, many concise synthetic routes targeting their complex molecular framework, especially the aliphatic tetracyclic ketolide core or polycyclic enone core, have been developed.

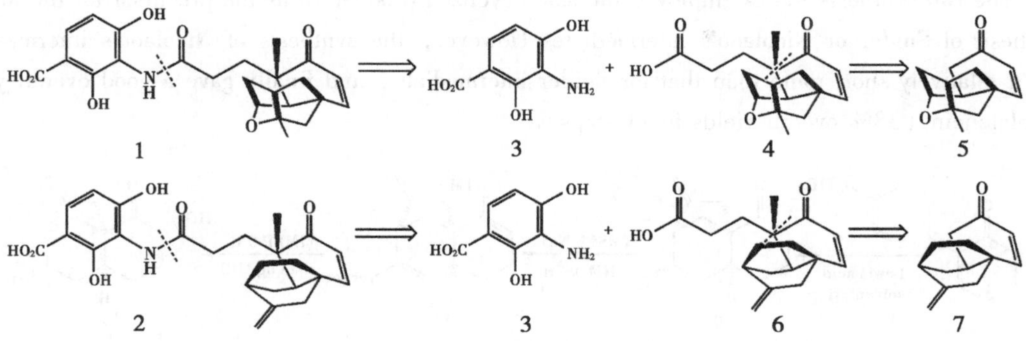

Scheme 1 The retrosynthetic analysis of platensimycin 1 and platencin 2

4.1 Zhu's total syntheses of (±)-platensimycin and platencinusing a cascade cyclization approach

Bifunctional Lewis acids are highly useful for developing cascade cyclization reactions because they can induce cyclization reactions via forming σ- and/or π-complexes with the substrates, as well as the intermediate (s) generated *in situ*[24]. In 2012, a Chinese research group offered a new route for the synthesis of platensimycin (Scheme 2) using a mild and efficient bifunctional Lewis acid-induced cascade cyclization reaction for rapid construction of the tricyclic core of *ent*-kaurenoids[25]. With ZnBr2 as the bifunctional Lewis acid, enone 8 and diene 9 underwent cascade cyclization smoothly at room temperature and provided the tricyclic diketone 10 in one pot with good yields (86%) and high diastereoselectivity. The resulting diketone 10 was converted to the di-tetramethylsilane (TMS) enol ether 11. Subsequent epoxidation using magnesium monoperoxyphthalate provided R-hydroxy ketone 12 selectively in 65% yield. The reduction of the ketone moiety of 12 followed by elimination of the resulting diol afforded 13 in good yields. Silyl enol formation followed by magnesium monoperoxyphthalate (MMPP) epoxidation of 13 provided R-hydroxyl ketone 14 as a single diastereomer, which can be equilibrated to the more stable R-hydroxyl ketone 15. Alcohol 15 was then acetylated and deacetoxylated using SmI2 to afford ketone 16. Finally, the reduction of ketone 16 with K-selectride followed by trifluoroacetic acid treatment afforded the Snider intermediate 17 (32% overall yields in 11 steps), which could be converted to 1 according to the literature report[26,27].

Again using the Lewis acid induced cascade cyclization approach to platencin, with minor modifications (Scheme 3), Zhu et al.[28] synthesized the tricyclic diketone 10, which was further converted to enone 18 by oxidation. Then, enone 18 was reduced using Luche reduction conditions to diol 19 in a single diastereomer. The allylic alcohol of 19 was thus selectively oxidized and protected with *t*-butyldimethylsilyl (TBS) to afford enone 21, followed by stereoselective conversion to 23 via the Wharton transposition protocol. After protecting and functional group manipulations, compound 23 was converted to the bicycle octane 25 using Yoshimitsu's procedures. Finally, oxidative removal of the *p*-methoxybenzyl (PMB) ether followed by oxidation of the resulting alcohol completed the synthesis of Nicolaou's intermediate 7, which can also be converted to platencin according to the literature procedures[29].

The two synthesis routes employed the same cyclized product 10 as the precursor for the formal synthesis of Snider or Nicolaou's intermediates. However, the synthesis of Nicolaou's intermediate was a relatively short route than that for Snider intermediate, and finally gave a good overall yield for platencin (38% overall yields in 11 steps).

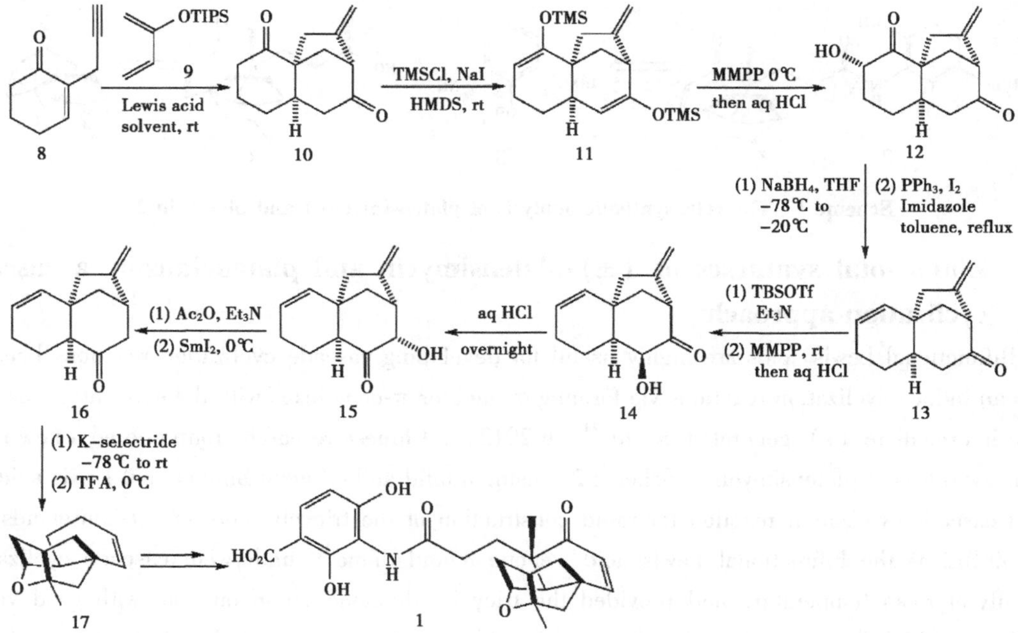

Scheme 2 A bifunctional Lewis acid induced cascade cyclization to the synthesis of platensimycin

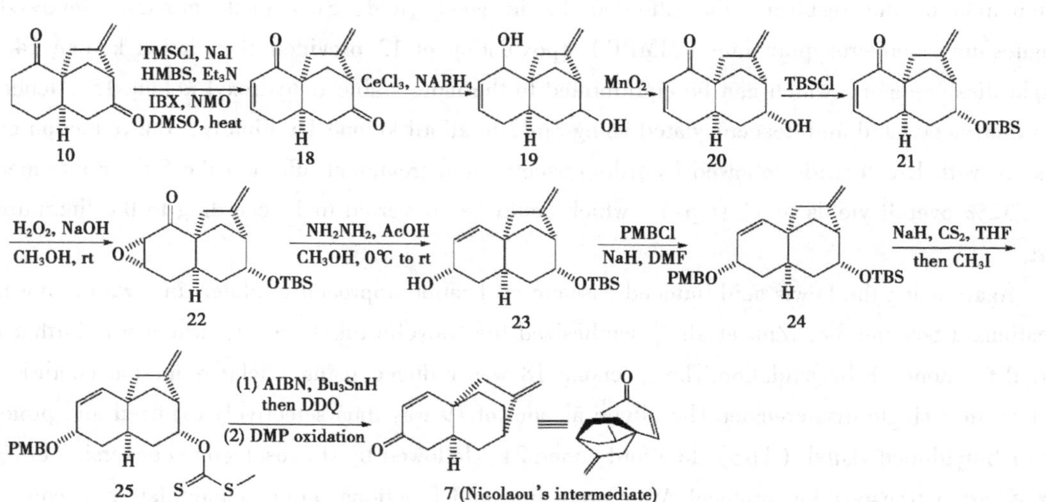

Scheme 3 Lewis acid induced cascade cyclization to the synthesis of platencin

4.2 Horii's synthesis of tetracyclic cage 5

To date, researchers have placed particular emphasis on the diverse approaches for chemical synthesis of the tetracyclic cage 5 because it is an important intermediate for approaching 1[23]. As shown in Scheme 4[30], epoxide 27 was readily obtained from 26 by a known three-step sequence, followed by regioselective trans-diaxial ring opening and dehydration of 27 to 28. The Diels-Alder reaction of 28 with Rawal's diene B gave 29 as a single isomer. The lactone 30, which was obtained by protection of the ketone functionality of 29, was treated with MeMgCl in tetrahydrofuran (THF) / hexamethylphosphoramide to give a diol intermediate, the secondary hydroxy group of which was then protected as its *t*-butyldimethylsilyl ether 31. The nitrile 31 was reduced with diisobutyl aluminum hydride to give aldehyde 32, and the subsequent addition of MeLi to 32 afforded 33 as an inconsequential 1 : 1 mixture of diastereomers. The ring closing metathesis (RCM) precursor 34 obtained by subjecting the diol 23 to an excess amount of Martin's sulfurane could be achieved efficiently by slowly adding the second-generation Grubbs catalyst in the presence of 1, 4-benzoquinone at an elevated temperature to afford the tricyclic product 35. Finally, treatment of 35 with hydrochloric acid to remove the silyl and acetal protecting groups induced concurrent intramolecular etherification, providing tetracyclic cage 5. Although this was a new method for construction tricyclic diketone 29 as a single isomer, the preparation of the cyano lactone 28 inevitably prolonged the synthesis and gave a poor (6.6%) overall yield for the tetracyclic cage 5.

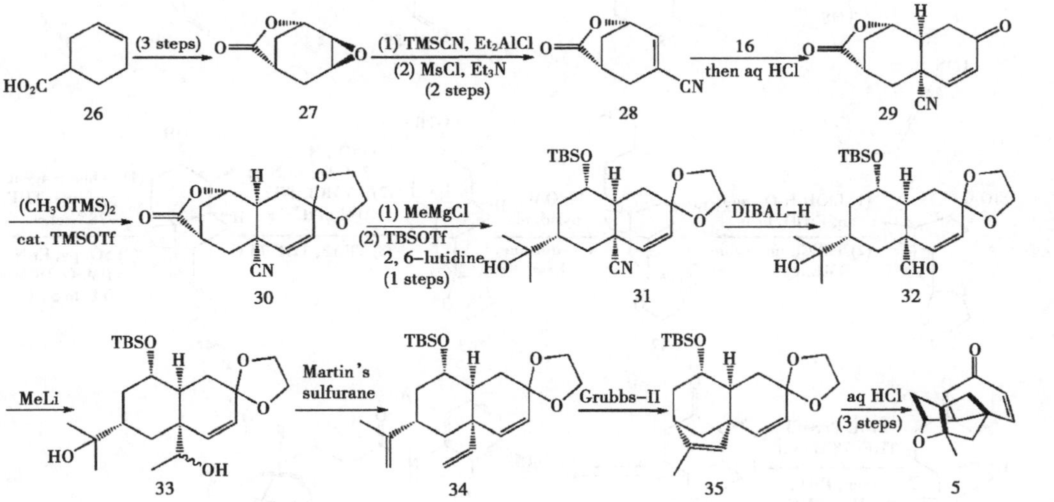

Scheme 4 Stereoselective synthesis of the tetracyclic cage 5

4.3 Gamal's synthesis of platencin

Gamal et al. established a new route to platencin via decarboxylative radical cyclization, shown in Scheme 5[31]. Enone 38 was prepared by adding Grignard reagent 37 to enone 36, followed by simple dehydration of the resultant tertiary alcohol. Then, enone 38 was converted to 39 under conventional sequential Michael reaction conditions. The introduction of oxygen functionality was conducted by SeO_2-mediated propargylic oxidation leading to alcohol 40. Ketalization of the carbonyl group of 40under azeotropic dehydration conditions furnished ketal 41, which was hydrolyzed with LiOH and further silylated with *t*-butyldi-methylsilyl chloride/imidazole to give the disilylated material 42. The decarboxylative radical cyclization of 42 then produced the desired compound 43. After hydrogenation with PtO2 in the presence of triethylamine, compound 43 was converted to a saturated product, and its ketal and the silyl ether were sequentially cleaved using 1 N HCl followed by tetrabutylammonium fluoride to provide hydroxyketone 44. Hydroxyketone 44 was then subjected to Tebbe olefination followed by Parikh-Doering oxidation to produce ketone 45. The conventional phenylselenylation of ketone 45 with lithium diisopropylamide/phenylselenyl chloride, followed by oxidative elimination with H_2O_2, successfully gave enone 46 which could be converted to platencin[29].

The decarboxylative radical cyclization of alkynyl silyl ester 42 was the key step for constructing the tricyclic core of platencin. With Pb (OAc)$_4$ in the presence of pyridine in refluxing 1, 4-dioxane, the key decarboxylation allowed the rapid construction of the twisted polycyclic compound 43, but with a poor (30%) yeild.

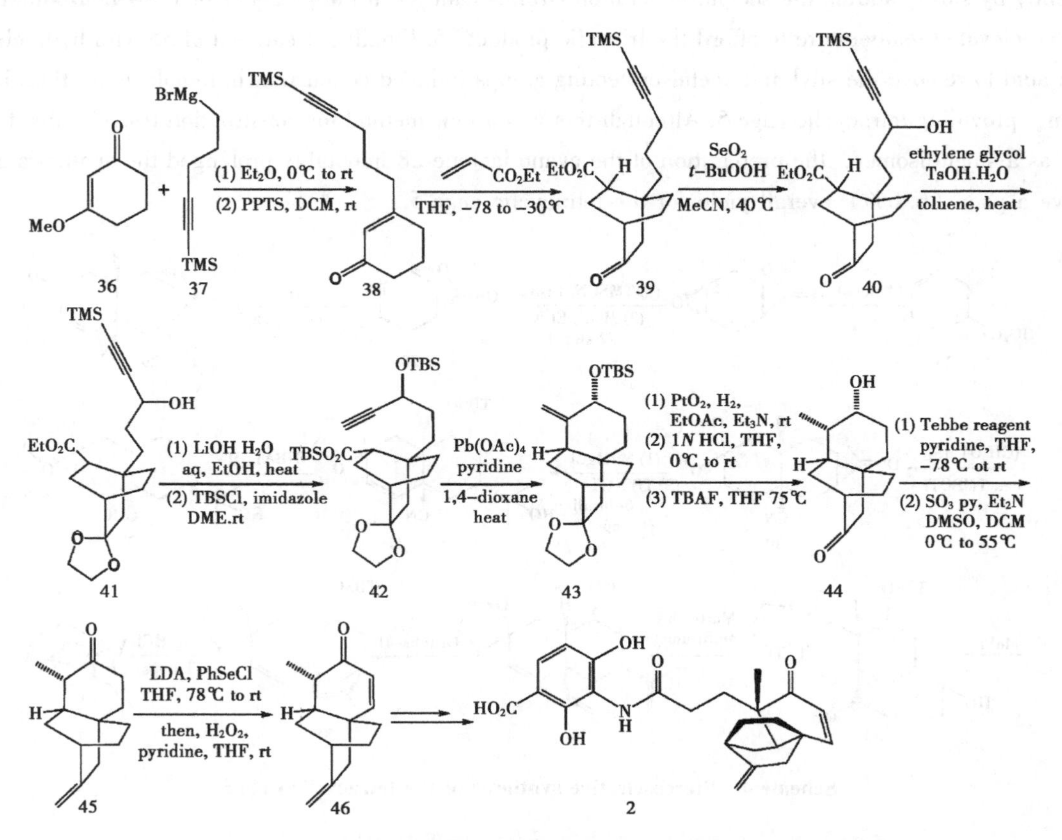

Scheme 5 Synthesis of platencin via decarboxylative radical cyclization

4.4 Chang's synthesis of platencin

Chang et al.used an intramolecular Diels-Alder (IMDA) approach for the direct assembly of the tricarboxylic framework associated with platencin, outlined in Scheme 6[32].

The commercially available amino-diol 47 was converted into 1,3-dioxane 48 under the conditions defined by Nordin and Thomas[33].The condensation of 48 with aldehyde 49 in the presence of trifluoroacetic anhydride gave the imine 50, which was deprotonated using lithium diisopropylamide. The resulting anion was alkylated with the benzyl ether 51, followed by an acidic workup that gave the desired aldehyde 52. Treating compound 52 with ethynylmagnesium bromide in THF gave the anticipated propargyl alcohol 53. The sequential treatment of compound 53 with the readily prepared Lewis acidic hydrostannane Bu2Sn (OTf) H followed by *n*BuLi gave the required Z-configured alkene 54. The tetraene 56 was obtained as an approximately 1 : 1 mixture of diastereoisomers by the treatment of an approximately 1 : 1 mixture of 54 and acetonide 55 with [Pd $(PPh_3)^4$] in the presence of CuI and CsF.Compound 56 was oxidized to the corresponding ketone 57 using the Dess-Martin periodinane in the presence of pyridine.Then, the anticipated intramolecular Diels-Alder adduct 58 was obtained from 57 under-reflux in toluene.Compound 58 was exposed to dihydrogen in the presence of 5% Pd to give the crystalline ketal 59, which was oxidized with pyridinium chlorochromate to give carboxylic acid 60. Cleavage of 60 gave the diol 61, which could be regioselectively monooxidized using the sterically demanding oxammonium salt to afford acyloin 62. After esterification, the acyloin 62 was converted to benzoate 63 and subsequently generated the diketo ester 64 through a samarium iodide-promoted reduction reaction.Subjecting compound 64 to a Wittig olefination reaction using Ph3P $=CH_2$ gave the desired alkene 65, which could finally be converted to 6 by the saponification of ester 65 using aqueous sodium hydroxide in THF.

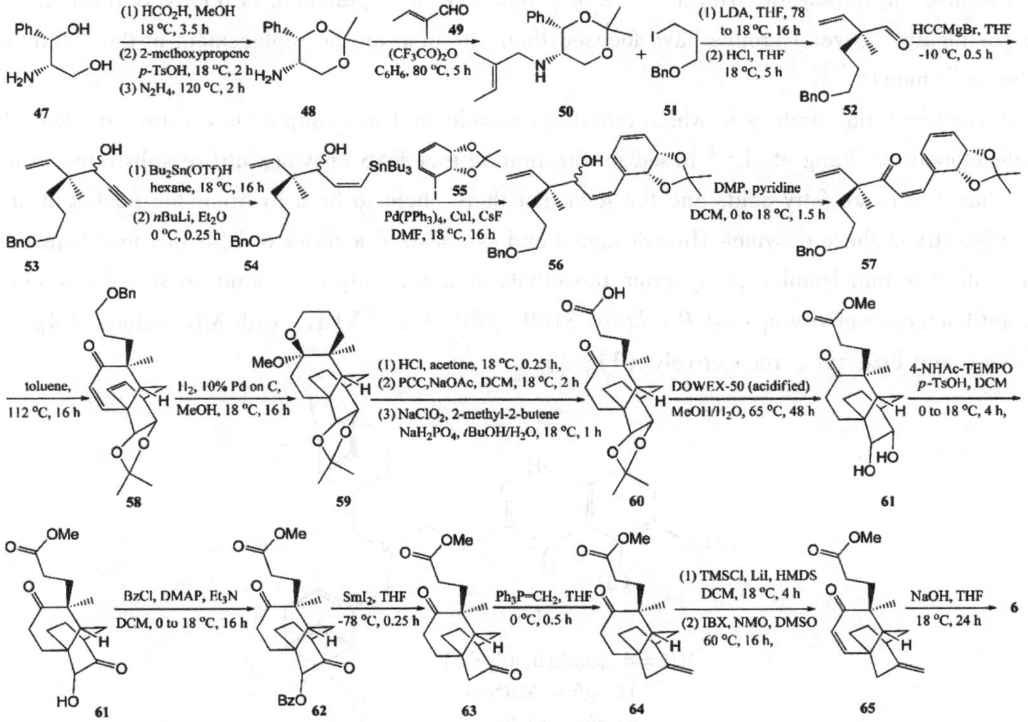

Scheme 6 Synthesis the tricarbocyclic framework of platencin using an intramolecular Diels-Alder approach

Although the original total synthesis of platencin reported by Nicolaou and co-workers in 2008[29] used the same IMDA approach to establish the framework 7 in an enantioselective manner, the pivotal aspect of this work was the use of the enantiomerically pure *cis*-1, 2-dihydrocatechol (which is obtained by the whole-cell biotransformation of iodobenzene using a genetically engineered form of *E.coli*) as the starting material[32]. Furthermore, derivatives of enone 7 that already incorporated a suitably constituted quaternary carbon center adjacent to the carbonyl residue and thus were more readily elaborated to (—) -platencin, were generated by this modified route in as highly controlled a manner as possible.

5 RECENT ANALOGUES AND THEIR ANTIBACTERIAL ACTIVITIES

Since the discovery of platensimycin and platencin, significant effort has been undertaken to prepare platensimycin and platencin analogues with high antibacterial activities by organic synthesis. Meanwhile, structure activity relationship (SAR) studies were performed to find more effective analogues with high efficiency. The SAR studies have revealed that modifications on the 2, 4-dihydroxybenzoic acid moiety of platensimycin or platencin usually diminish most of the biological activity of this antibiotic[34-36]. In sharp contrast, decorations or modifications on the caged ring often retain good biological activity[34,37].

Because the tetracyclic/tricyclic core of platensimycin or platencin is relatively difficult to prepare chemically, several groups have focused their attention on the replacement of the "difficult-to-synthesize" moiety[38].

A combinatorial strategy in which dihydroxybenzoic acid was coupled to a variety of alkyl chains was developed by Wang et al.[38] based on the finding that FAS enzymes utilize substrates with long alkyl chains to make fatty acids and the idea that there ought to be a hydrophobic pocket at or near the active site of these enzymes. They designed and synthesized a series of 2,4-dihydroxybenzoic acid esters with a terminal amino group. From bioactivity studies, only compound 66 showed excellent *in vitro* antibacterial activity against *B.subtilis* (3160), MRSA and VRE, with MIC values of 4μg/mL, 16μg/mL and 8μg/mL, respectively (Fig. 3).

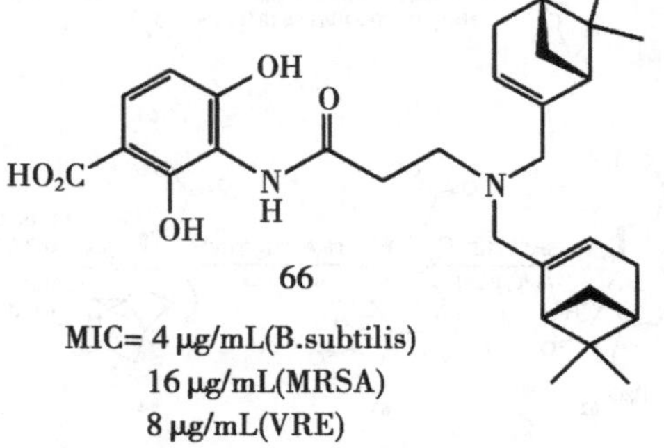

OH
O
HO_2C
N
H
OH
N
66
MIC= 4 μg/mL(B.subtilis)
16 μg/mL(MRSA)
8 μg/mL(VRE)

Fig. 3 Structure and antibacterial activity (MIC, μg/mL) of compound 66

Plesch's group also focused on the development of novel simplified platensimycin analogues and thereby synthesized a series of compounds containing the anilide group and the ketone, both identified as essential for binding of the active components to the Fab F complex[39]. After a standardized agar diffusion assay, ketones 67 and 68 showed a broad spectrum of antibacterial activity, comparable to the antibiotics tetracycline and clotrimazol (Table 1).

Table 1 Agar diffusion assay for compounds 67, 68, tetracycline and clotrimazol (50mg/disc)

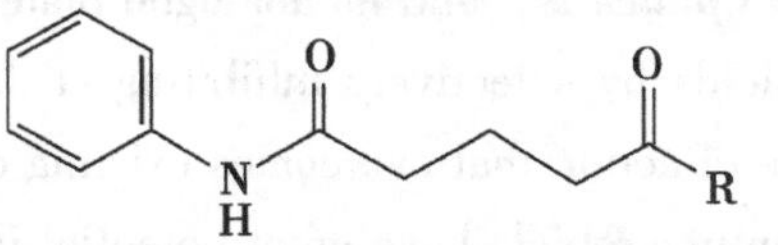

Strains	67: R=Ph	68: R=*Trans* H_3C-CH=CH-	Tetracycline	Clotrimazol
Escherichia coli	11	10	25	0
Pseudomonas antimicrobia	14	10	23	0
Staphylococcus equorum	0	11	23	9
Streptococcus entericus	16	14	12	11
Candida glabrata	20	17	nt	15
Aspergillus niger	0	6	nt	15
Yarrowia lipolytica	0	7	nt	20
Hyphopichia burtonii	10	15	nt	17

Note: Zones of inhibition (in mm); nt, not tested; 0, no measurable zone of inhibition

Fisher et al. described an approach to platensimycin analogue design using structure-based ligand design (SBLD), a powerful approach that can facilitate the discovery of bioactive small molecules when high quality structural information is available[40]. They designed and synthesized a series of platensimycin analogous and determined the affinities of these compounds for the C163Q mutant of FabF using a WaterLOGSY[41] competition binding assay. The resulting designed compounds were shown to bind in the platensimycin binding site of the C163Q mutant of FabF, and, in several cases, compounds 69, 70, 71 and 72 had higher affinity (lower dissociation constants) than the reference compound 73 (Table 2).

Table 2 Dissociation constants of compounds 69, 70, 71, 72 and reference compound 73 for the C163Q mutant of FabF

Structure	R	Compounds	Kd/μM
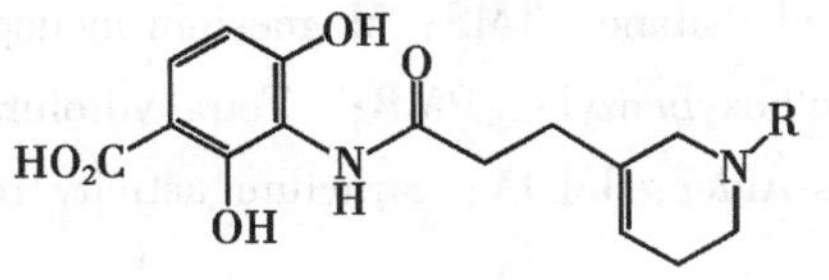	$-SO_2NHBOC$	69	50±10
	$-SO_2NH_2$	70	100±20
	-CONH*i*Pr	71	110±30
	3-pyridyl	72	120±20
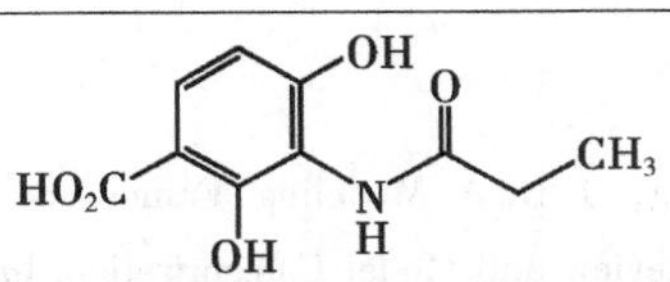		73	650±90

6 CONCLUSIONS

Multidrug resistance has become a serious global concern. Research on new antibacterial drugs that are effective against resistant pathogenic bacteria is highly crucial. This article describes the efforts of chemists to overcome multidrug resistance in the case of the new antibiotic platensimycin isolated from extracts of *Streptomyces platensis*, and its analogue platencin. These natural products interfere with the synthesis of fatty acids by selectively inhibiting -FabF, and are of particular interest because they exhibit a new mode of action that overcomes existing drug resistance. Antibacterial studies have shown that these new antimicrobials have great potential in inhibiting MRSA, VRE, VISA and penicillin-resistant *Streptococcus pneumoniae*. The fascinating molecular architecture, novel mode of action and high antibacterial activities have inspired many different approaches to the synthesis and chemical modifications of platensimycin and platencin in a relatively short time.

ACKNOWLEDGMENTS

This work was supported by grants from Basic Scientific Research Funds in Central Agricultural Scientific Research Institutions (No. 1610322014003) and The Agricultural Science and Technology Innovation Program (CAAS-ASTIP-2014-LIHPS-04).

AUTHOR CONTRIBUTIONS

The manuscript was conceived by all authors. Jianping Liang, Yunpeng Yi, Yu Liu collected the documents and wrote the draft manuscript which was subsequently edited by Ruofeng Shang and Jiatu Wang.

CONFLICTS OF INTEREST

The authors declare no conflict of interest.

ABBREVIATION

Fatty acid acyl carrier protein synthases II: FabF; Fatty acid acyl carrier protein synthases III: FabH; Methicillin-resistant *Staphylococcus aureus*: MRSA; Vancomycin-intermediate *S. aureus*: VISA; Vancomycin-resistant *Enterococci*: VRE; High performance liquid chromatography: HPLC; Nuclear magnetic resonance: NMR; *Escherichia coli*: *E. coli*; Minimal inhibitory concentration: MIC; Type II bacterial fatty acid synthesis: FASII; Tetramethylsilane: TMS; Magnesium monoperoxyphthalate: MMPP; *t*-Butyldimethylsilyl: TBS; *p*-Methoxybenzyl: PMB; Tetrahydrofuran: THF; Ring closing metathesis: RCM; Intramolecular Diels-Alder: IMDA; Structure activity relationship: SAR.

References

[1] Spicknall, I. H.; Foxman, B.; Marrs, C. F.; Eisenberg, J. N. A Modeling Framework for the Evolution and Spread of Antibiotic Resistance: Literature Review and Model Categorization. *Am. J. Epi-*

*demiol.*2013, 178: 508-520.

[2] Brown, D.G.; May-Dracka, T.L.; Gagnon, M.M.; Tommasi, R.Trends and exceptions of physical properties on antibacterial activity for gram-positive and gram-negative pathogens. *J. Med. Chem.* 2014, 57: 10144-10161.

[3] Shang, R.F.; Wang, J.T.; Guo, W.Z.; Liang, J.P.Efficient antibacterial agents: A review of the synthesis, biological evaluation and mechanism of pleuromutilin derivatives.*Curr.Top.Med.Chem.*2013, 13: 3013-3025.

[4] Wang, J.; Kodali, S.; Lee, S.H.; Galgoci, A.; Painter, R.; Dorso, K.; Racine, F.; Motyl, M.; Hernandez, L.; Tinney, E.; et al.Discovery of platencin, a dual FabF and FabH inhibitor with *in vivo* antibiotic properties.*Proc.Natl.Acad.Sci.USA.*2007, 104: 7612-7616.

[5] Kirst, H.A.Developing new antibacterials through natural product research.*Expert Opin.Drug Discv.* 2013, 8: 479-493.

[6] Wang, J.; Soisson, S.M.; Young, K.; Shoop, W.; Kodali, S.; Galgoci, A.; Painter, R.; Parthasarathy, G.; Tang, Y.; Cummings, R.; et al.Platensimycin is a selective FabF inhibitor with potent antibiotic properties.*Nature* 2006, 441: 358-361.

[7] Young, K.; Jayasuriya, H.; Ondeyka, J.G.; Herath, K.; Zhang, C.; Kodali, S.; Galgoci, A.; Painter, R.; Brown-Driver, V.; Yamamoto, R.; et al. Discovery of FabH/FabF Inhibitors from Natural Products.*Antimicrob.Agents Chemother.*2006, 50: 519-526.

[8] Zhang, C, W.; Ondeyka, J.; Herath, K.; Jayasuriya, H.; Guan, Z.Q.; Zink, D.L.; Dietrich, L.; Burgess, B.; Ha, S.N.; Wang, J.; et al.Platensimycin and platencin congeners from Streptomyces platensis.*J.Nat.Prod.*2011, 74: 329-340.

[9] Martens, E.; Demain, A.L.Platensimycin and platencin: Promising antibiotics for future application in human medicine.*J.Antibiot.*2011, 64: 705-710.

[10] Peterson, R.M.; Huang, T.; Rudolf, J.D.; Smanski, M.J.; Shen, B.Mechanisms of self-resistance in the platensimycin- and platencin- producing *Streptomyces platensis* MA7327 and MA7339 strains.*Chem. Bio.*2014, 21: 389-397.

[11] Singh, S.B.; Jayasuriya, H.; Ondeyka, J.G.; Herath, K.B.; Zhang, C.; Zink, D.L.; Tsou, N. N.; Ball, R.G.; Basilio, A.; Genilloud, O.; et al.Isolation, Structure, and Absolute Stereochemistry of Platensimycin, A Broad Spectrum Antibiotic Discovered Using an Antisense Differential Sensitivity Strategy.*J.Am.Chem.Soc.*2006, 128: 11916-11920.

[12] Habich, D.; von Nussbaum, F.Platensimycin, a new antibiotic and "superbug challenger" from nature.*Chem.Med.Chem.*2006, 1: 951-954.

[13] Barton, S.New antibiotic on the horizon? *Nat.Rev.Microbiol.*2006, 4, doi: 10. 1038/nrd2098.

[14] Palanichamy, K.; Kaliappan, K.P.Discovery and syntheses of "superbug challengers" —platensimycin and platencin.*Chem.Asian J.*2010, 5: 668-703.

[15] Singh, S.B.; Young, K.New antibiotic structures from fermentations.*Expert Opin.Ther.Pat.*2010, 20: 1359-1371.

[16] Jayasuriya, H.; Herath, K.B.; Zhang, C.; Zink, D.L.; Basilio, A.; Genilloud, O.; Diez, M. T.; Vicente, F.; Gonzalez, I.; Salazar, O.; et al.Isolation and Structure of Platencin: A FabH and FabF Dual Inhibitor with Potent Broad-Spectrum Antibiotic Activity.*Angew. Chem.* 2007, 46: 4768-4772.

[17] Herath, K.B.; Attygalle, A.B.; Singh, S.B.Biosynthetic studies of platensimycin.*J.Am.Chem.Soc.* 2007, 129: 15422-15423.

[18] Jayasuriya, H.; Herath, K.B.; Ondeyka, J.G.; Zink, D.L.; Burgess, B.; Wang, J.; Singh,

S. B. Structure of homoplatensimide A: A potential key biosynthetic intermediate of platensimycin isolated from Streptomyces platensis. *Tetrahedron Lett.* 2008, 49: 3648–3651.

[19] Smanski, M.J.; Yu, Z.; Casper, J.; Lin, S.; Peterson, R.M.; Chen, Y.; Wendt-Pienkowski, E.; Rajski, S.R.; Shen, B. Dedicated *ent*-kaurene and *ent*-atiserene synthases for platensimycin and platencin biosynthesis. *Proc. Natl. Acad. Sci. USA.* 2011, 108: 13498–13503.

[20] Payne, D.J.; Warren, P.V.; Holmes, D.J.; Ji, Y.; Lonsdale, J.T. Bacterial fatty-acid biosynthesis: A genomics-driven target for antibacterial drug discovery. *Drug Disc. Today* 2001, 6: 537–544.

[21] Hopwood, D.A. How do antibiotic-producing bacteria ensure their self-resistance before antibiotic biosynthesis incapacitates them? *Mol. Microbiol.* 2007, 63: 937–940.

[22] McGrath, N.A.; Bartlett, E.S.; Sittihan, S.; Njardarson, J.T. A concise ring-expansion route to the compact core of platensimycin. *Angew. Chem. Int. Ed. Engl.* 2009, 48: 8543–8546.

[23] Saleem, M.; Hussain, H.; Ahmed, I.; van Ree, T.; Krohn, K. Platensimycin and its relatives: A recent story in the struggle to develop new naturally derived antibiotics. *Nat. Prod. Rep.* 2011, 28: 1534–1579.

[24] Yamamoto, Y. From sigma- to pi- electrophilic Lewis acids. Application to selective organic transformations. *J. Org. Chem.* 2007, 72: 7817–7831.

[25] Zhu, L. Z.; Han, Y. J.; Du, G. Y.; Lee, C. S. A bifunctional Lewis Acid induced cascade cyclization to the tricyclic core of ent-kaurenoids and its application to the formal synthesis of (±) -platensimycin. *Org. Lett.* 2013, 15: 524–527.

[26] Nicolaou, K.C.; Li, A.; Edmonds, D.J. Total synthesis of platensimycin. *Angew. Chem. Int. Ed. Engl.* 2006, 45: 7086–7090.

[27] Zou, Y.; Chen, C.H.; Taylor, C.D.; Foxman, B.M.; Snider, B.B. Formal synthesis of (+/−) -platensimycin. *Org. Lett.* 2007, 9: 1825–1828.

[28] Zhu, L.Z.; Zhou, C.S.; Yang, W.; He, S.Z.; Cheng, G.J.; Zhang, X.H.; Lee, C.S. Formal syntheses of (±) -platensimycin and (±) -platencin via a dual-mode Lewis acid induced cascade cyclization approach. *J. Org. Chem.* 2013, 78: 7912–7929.

[29] Nicolaou, K.C.; Tria, G.S.; Edmonds, D.J. Total synthesis of platencin. *Angew. Chem. Int. Ed. Engl.* 2008, 47: 1780–1783.

[30] Horii, S.; Torihata, M.; Nagasawa, T.; Kuwahara, S. Stereoselective approach to the racemic oxatetracyclic core of platensimycin. *J. Org. Chem.* 2013, 78: 2798–2801.

[31] Moustafa, G.A.; Saku, Y.; Aoyama, H.; Yoshimitsu, T. A new route to platencin via decarboxylative radical cyclization. *Chem. Commun.* 2014, 50: 15706–15709.

[32] Chang, E.L.; Schwartz, B.D.; Draffan, A.G.; Banwell, M.G.; Willis, A.C. A chemoenzymatic and fully stereo-controlled total synthesis of the antibacterial natural product (±) -platencin. *Chem. Asian. J.* 2015, 10: 427–439.

[33] Nordin, I. C.; Thomas, J. A. An improved synthesis of (4*S*, 5*S*) -2, 2-dimethyl-4-phenyl-1, 3-dioxan-5-amine *Tetrahedron Lett.* 1984, 25: 5723–5724.

[34] Nicolaou, K.C.; Stepan, A.F.; Lister, T.; Li, A.; Montero, A.; Tria, G.S.; Turner, C.I.; Tang, Y.; Wang, J.; Denton, R.M.; et al. Design, synthesis, and biological evaluation of platensimycin analogues with varying degrees of molecular complexity. *J. Am. Chem. Soc.* 2008, 130: 13110–13119.

[35] Wang, J.; Lee, V.; Sintim, H.O. Efforts towards the Identification of Simpler Platensimycin Analogues—The Total Synthesis of Oxazinidinyl Platensimycin. *Chem. Eur. J.* 2009, 15: 2747–2750.

[36] Tiefenbacher, K.; Gollner, A.; Mulzer, J. Syntheses and antibacterial properties of iso-platencin,

Cl-iso-platencin and Cl-platencin: Identification of a new lead structure. *Chem. Eur. J.* 2010, 16: 9616-9622.

[37] Shen, H.C.; Ding, F.X.; Singh, S.B.; Parthasarathy, G.; Soisson, S.M.; Ha, S.N.; Chen, X.; Kodali, S.; Wang, J.; Dorso, K.; et al. Synthesis and biological evaluation of platensimycin analogs. *Bioorg. Med. Chem. Lett.* 2009, 19: 1623-1627.

[38] Wang, J.X.; Sintim, H.O. Dialkylamino-2, 4-dihydroxybenzoic acids as easily synthesized analogues of platensimycin and platencin with comparable antibacterial properties. *Chem. Eur. J.* 2011, 17: 3352-3357.

[39] Plesch, E.; Bracher, F.; Krauss, J. Synthesis and Antimicrobial Evaluation of Novel Platensimycin Analogues. *Arch. Pharm.* 2012, 345: 657-662.

[40] Fisher, M.; Basak, R.; Kalverda, A. P.; Fishwick, C. W.; Turnbull, W. B.; Nelson, A. Discovery of novel FabF ligands inspired by platensimycin by integrating structure-based design with diversity-oriented synthetic accessibility. *Org. Biomol. Chem.* 2014, 12: 486-494.

[41] Dalvit, C.; Fasolini, M.; Flocco, M.; Knapp, S.; Pevarello, P.; Veronesi, M. NMR-Based screening with competition water-ligand observed via gradient spectroscopy experiments: Detection of high-affinity ligands. *J. Med. Chem.* 2002, 45: 2610-2614.

(发表于《Molecules》，院选 SCI，IF: 2.416)

In Vivo Efficacy and Toxicity Studies of a Novel Antibacterial Agent: 14-*O*-[(2-Amino-1,3,4-thiadiazol-5-yl) Thioacetyl] Mutilin

ZHANG Chao, YI Yun-peng, CHEN Jiong-ran, XIN Ren-sheng, YANG Zhen, GUO Zhi-ting, LIANG Jian-ping*, SHANG Ruo-feng*

(Key Laboratory of New Animal Drug Project of Gansu Province/Key Laboratory of Veterinary Pharmaceutical Development, Ministry of Agriculture/Lanzhou Institute of Husbandry and Pharmaceutical Sciences of CAAS, Lanzhou 730050, China)

Abstract: A new pleuromutilin derivative with excellent antibacterial activity, 14-*O*-[(2-amino-1,3,4-thiadiazol-5-yl) thioacetyl] mutilin (ATTM), may serve as a possible lead compound for the development of antibacterial drugs. However, *in vivo* efficacy and toxicity evaluations of this compound have not been performed. In this study, we evaluated the efficacy of ATTM by measuring the survival of mice after a lethal challenge with methicillin-resistant *Staphylococcus aureus* (MRSA), and the 50% effective dose (ED_{50}) was 5.74mg/kg by the intravenous route. In an oral single-dose toxicity study, ATTM was orally administered to mice at different doses and the 50% lethal dose (LD_{50}) was calculated to be 2304.4mg/kg by the Bliss method. The results of the subchronic oral toxicity study in rats showed no mortality, exterior signs of toxicity, or differences in the total weight gain or relative organ weights between the treated groups and control group after administration. The hematological and serum biochemical data showed no differences between the treated and control groups, except for the levels of alkaline phosphatase (ALP), creatinine (CR) and blood glucose (GLU), which were significantly different in the high-dose group. The differences in the histopathological findings between the treated groups and the control group were not considered to be treatment-related. Our results indicated that the no observed adverse effect level (NOAEL) for ATTM was 5mg/kg in this study.

Key words: ATTM; ED50; Acute toxicity; Subchronic toxicity

---

* Authors to whom correspondence should be addressed; E-mails: liangjp1963@163.com (J. L.); shangrf1974@163.com (R.S.).

Academic Editor: Peter J. Rutledge

Received: 26 *January* 2015 / *Accepted*: 18 *March* 2015 / *Published*: 24 *March* 2015

1 INTRODUCTION

The increasing resistance of bacteria to the major classes of antibacterial drugs is becoming a serious threat to public health. Drug-resistant bacteria, especially methicillin-resistant *Staphylococcus aureus* (MRSA), currently cause infections in both hospitals and the worldwide community[1]. Therefore, new classes of antibacterial agents that work via novel mechanisms are urgently needed to effectively eradicate drug-resistant bacteria.

The natural compound pleuromutilin (Fig. 1) was first discovered and isolated from cultures of two species of basidiomycetes, *Pleurotusmutilus* and *P.passeckerianus*, in 1951[2]. Modifications of pleuromutilin have led to three drugs: tiamulin, valnemulin, and retapamulin (Fig. 1)[3,4]. Extensive efforts were made to synthesize three other compounds, BC-3781, BC-3205 and BC-7013 (Fig. 1), for human use after the success of retapamulin[5,6]. Chemical footprinting studies showed that tiamulin and valnemulin bound to the bacterial ribosome at the peptidyl transferase center (PTC), thereby inhibiting the synthesis of the peptide bond by hindering the correct location of the amino acid on the tRNA[7,8]. Further studies demonstrated that the interactions of the tricyclic core of tiamulin are mediated through hydrophobic interactions and hydrogen bonds, which are formed mainly by the nucleotides of domain V[9,10].

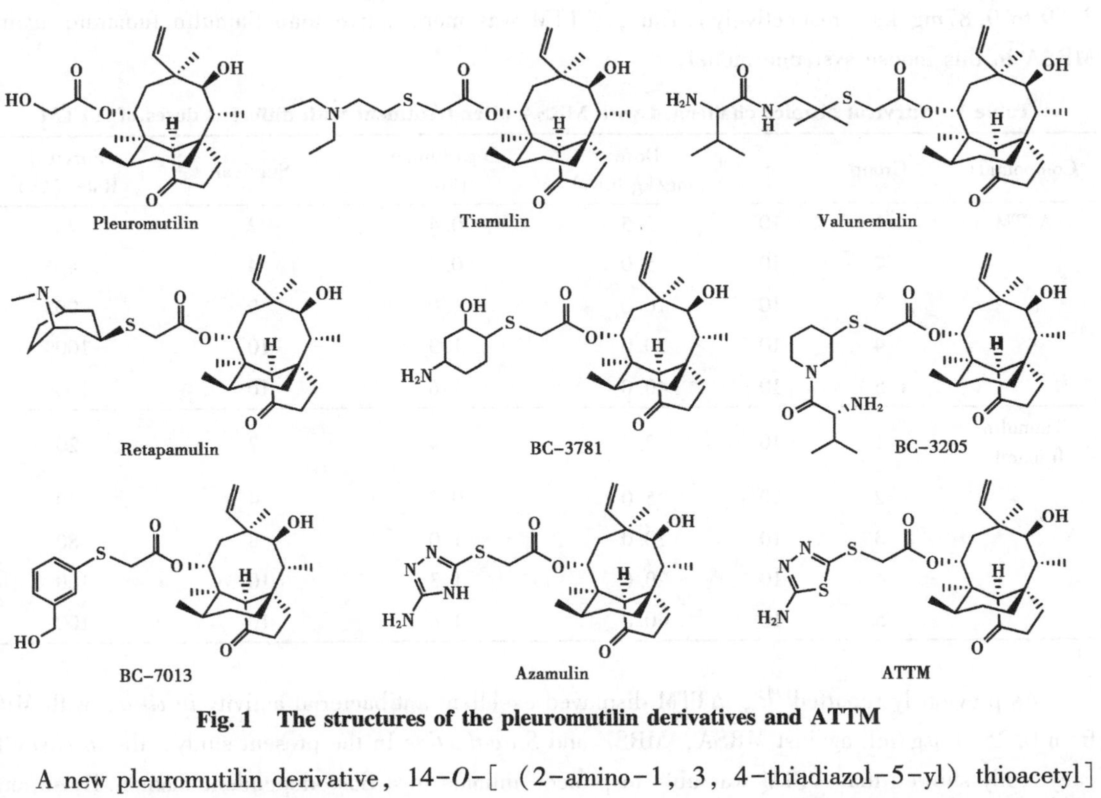

Fig. 1 The structures of the pleuromutilin derivatives and ATTM

A new pleuromutilin derivative, 14-*O*-[(2-amino-1,3,4-thiadiazol-5-yl) thioacetyl] mutilin (ATTM, Fig. 1) is composed of a rather rigid 5-6-8 tricyclic carbon skeleton and a thiadiazole moiety[11]. This compound was structurally similar to azamulin (Fig. 1), which was designed for human use during the early 1980s but did not undergo further clinical trials because of its poor solubility in water and its strong inhibition of human cytochrome P450s[12,13].ATTM was first synthesized and evaluated in our lab.The *in vitro* antibacterial studies showed that it displayed excellent antibacterial activity against MRSA, methicillin-resistant *Staphylococcus epidermidis* (MRSE) and *Streptococcus agalactiae* (*S. agalactiae*). Preliminary pharmacokinetics studies in rats showed that ATTM may be able to serve as a possible lead compound for the development of antibacterial drugs[14].This present report describes a subsequent pharmacological investigation of ATTM, including studies of the *in vivo* efficacy and acute toxicity in mice, and subchronic toxicity in rats.

2 RESULTS AND DISCUSSION

2.1 *In Vivo* Efficacy of attm

A *staphylococcal* systemic infection model was used to evaluate the efficacy of ATTM by determining the survival of mice after a lethal challenge with MRSA.Tiamulin fumarate was chosen as the reference drug.After being infected with MRSA, the mice were intravenously treated with different doses of ATTM or tiamulin fumarate dissolved in vehicle.The survival rates of all groups are summarized in Table 1. Treatment with ATTM and tiamulin fumarate led to dose-dependent protection and the survival of the mice infected with MRSA, with a 50% effective dose (ED_{50}) of 5. 74mg/kg and 5. 95mg/kg body weight (b.w.); the confidence level of 95% was 3. 75mg/kg to 8. 78mg/kg and 3. 59 to 9. 87mg/kg, respectively).Thus, ATTM was more active than tiamulin fumarate against MRSA in this mouse systemic model.

Table 1 Survival of mice challenged with MRSA after treatment with different doses of ATTM

Compounds	Group	n	Dose (mg/kg b.w.)	Logarithmic Dose	Survival	Survival Rate (%)
ATTM	1	10	2. 5	0. 4	2	20
	2	10	5. 0	0. 7	4	40
	3	10	10. 0	1. 0	9	90
	4	10	20. 0	1. 3	10	100
	5	10	40. 0	1. 6	10	100
Tiamulin fumarate	1	10	2. 5	0. 4	2	20
	2	10	5. 0	0. 7	4	40
	3	10	10. 0	1. 0	8	80
	4	10	20. 0	1. 3	10	100
	5	10	40. 0	1. 6	10	100

As previously reported[14], ATTM displayed excellent antibacterial activity *in vitro*, with MICs from 0. 25-1 μg/mL against MRSA, MRSE and *S.agalactiae*.In the present study, the *in vivo* efficacy study showed that ATTM was able to protect animals in a dose-dependent fashion.These antibacterial studies have demonstrated that ATTM might act as a potent antibacterial drug.

2.2 Acute oral toxicity study

The results of the oral single-dose toxicity study are summarized in Table 2. No animals died after receiving an oral dose of up to 948. 15mg/kg b.w.of ATTM. Conversely, all animals died when given the oral dose of 4800mg/kg b.w.of ATTM. The approximate 50% lethal dose (LD_{50}) in mice was determined to be 2304. 4mg/kg by the Bliss method, and the confidence level of 95% was 1861. 4mg/kg to 2870. 5mg/kg. No adverse effects or clinical signs of toxicity were observed for the surviving animals during the study. A necropsy of the surviving mice revealed no gross pathological finding and no significant differences in the liver, lungs, kidneys, heart, stomach, intestine, spleen or adrenal glands. However, drug precipitation was found in the stomachs of the mice that died during the early days after administration. Some areas of hemorrhage on the liver, lungs and small intestine were also found in some of mice that died during the treatment.

Table 2 Oral single-dose toxicity of ATTM in mice

Group	n	Dose (mg/kg b.w.)	Logarithmic Dose	Mortality	Mortality Rate (%)
1	10	948. 15	3. 68	0	0
2	10	1 422. 22	3. 51	2	20
3	10	2 133. 33	3. 33	4	40
4	10	3 200. 00	3. 15	7	70
5	10	4 800. 00	2. 98	10	100
vehicle control group	10	20 (mL/kg b.w.)	-	0	0

The acute toxicity study of ATTM was conducted in mice to establish the potential for acute toxicity and to provide information pertaining to the upper dose limit that could be used in longer-term feeding studies. On the basis of the results, the estimated LD_{50} of ATTM to mice is 2304. 4mg/kg/b. w. Under the conditions of this study, a unique composition of ATTM and dimethyl sulfoxide (DMSO) did not produce any acute oral toxicological effects. This indicates that ATTM has relatively low toxicity and high potential for development as a new drug.

2. 3 Subchronic oral toxicity

2. 3. 1 Clinical signs, body weights, and food consumption

After administration, no mortality or exterior signs of toxicity were observed in any of the dosing groups or the vehicle control group during the 28-day treatment period. The animals exhibited normal behavior and a normal physical condition, with no significant abnormalities in the clinical signs observed throughout the study.

The body weights of animals of both sexes are shown in Fig. 2. No significant differences were observed in any of the groups at any time point. The body weight and food consumption are shown in Table 3. There was no difference in the total weight gain and the average daily food intake between thetreated and control groups.

Table 3 Body weight gain and food consumption of rats orally administered ATTM for 28 days

Item	Control Group	ATTM-Treated Groups (mg/kg b.w./d)			Saline Group
		Low-Dose (5)	Middle-Dose (25)	High-Dose (125)	
Females					
Total body weight gain (g)	74.4±7.5	66.06±5.8	63.07±6.3	87.4±4.9	65.7±8.5
Daily food consumption (g)	118.6±7.2	107.1±13.6	106.9±10.8	106.8±15.7	109.7±9.9
Males					
Total body weight gain (g)	124.8±10.4	123.8±9.6	121.1±8.2	114.5±9.9	121.0±5.6
Daily food consumption (g)	148.4±10.2	134.9±7.5	147.7±12.6	147.7±12.6	157.1±8.6

Note: The values expressed as mean ± SD (n=10/sex/dose).

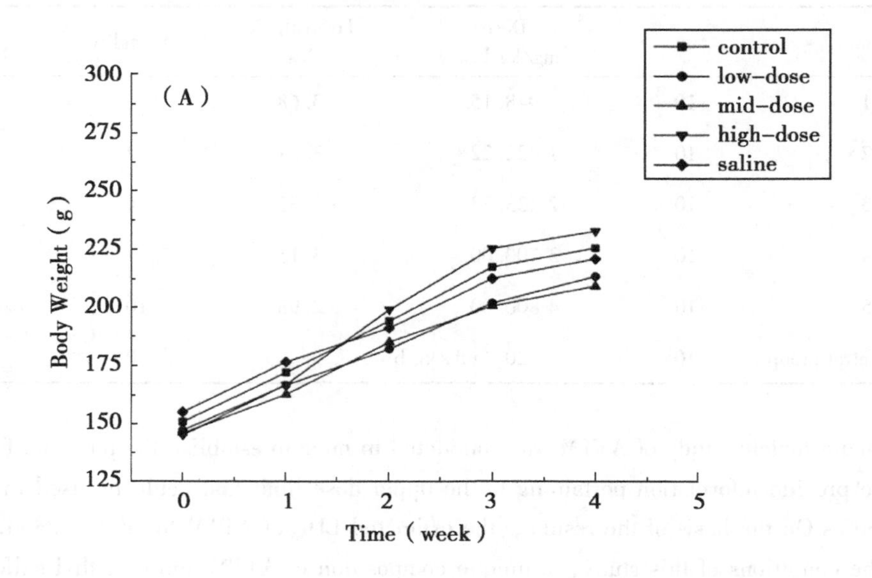

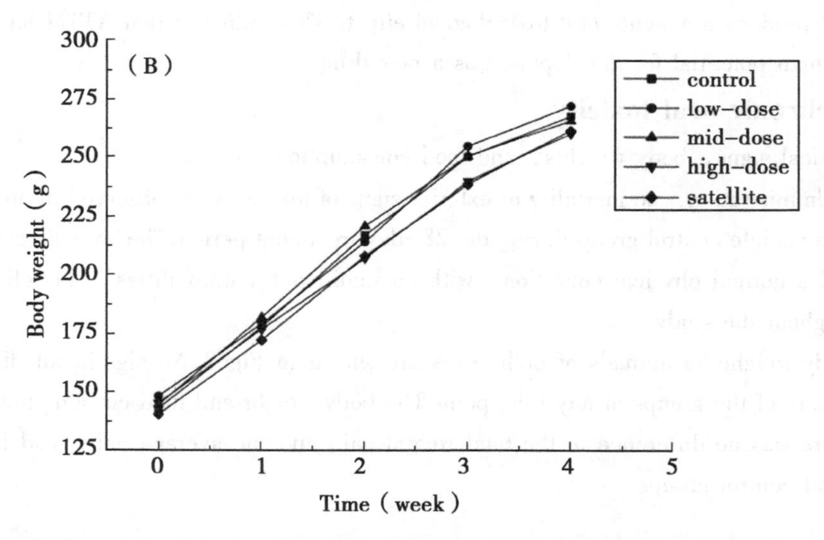

Fig. 2 Body weight of female (A) and male (B) rats after a 28-day treatment with ATTM at 5, 25 and 125mg/kg b.w.per day (n=10)

2.3.2 Hematological and serum biochemical data

The hematological and serum biochemical data on the examination on day 29 are summarized in Tables 4 and 5. There were no significant differences of the hematological data between the treated groups and control group. The level of alkaline phosphatase (ALP) was significantly decreased in the female high-dose group ($p<0.01$). The creatinine (CR) and blood glucose (GLU) levels were significantly increased in both the female and male high-dose groups ($p<0.05$).

Blood biochemical determinations that investigate major toxic effects on specific tissues, specifically the kidneys and liver, may provide useful information about the mechanism of toxicity of the agent. Some enzymes and proteins can indicate the impact on the renal function (such as CR)[15] or liver (such as ALP and GLU)[16]. Significant differences in the CR, ALP and GLU levels were found in the treated groups compared with the control group. These results suggest that ATTM may alter both the renal and hepatic function, indicating that the rats' kidneys and livers in the high-dose group were injured after the administration of 125mg/kg b.w. of ATTM.

Table 4 Hematological findings in rats orally administered ATTM for 28 days

Hematological Parameters	Control group	ATTM-Treated Groups (mg/kg b.w./d)			Saline Group
		Low-Dose (5)	Middle-Dose (25)	High-Dose (125)	
Females					
WBC (10^9/L)	14.2±4.5	10.9±3.6	9.3±7.5	10.2±2.5	15.2±3.8
RBC (10^{12}/L)	8.33±0.78	7.47±0.65	7.85±0.48	7.7±0.54	8.9±0.74
HGB (g/L)	180±7.4	179±6.3	171±5.2	170±7.8	180±7.1
HCT%	0.504±1.4	0.392±1.7	0.464±1.9	0.414±1.6	0.519±1.4
MCV (fL)	60.5±0.89	60.6±0.92	59.1±0.95	61.8±0.91	58.3±0.93
MCH (pg)	21.6±0.90	21.5±0.88	21.8±0.87	22.4±0.79	21.6±0.92
MCHC (g/L)	357±0.52	355±0.48	369±0.58	362±0.41	347±0.61
PLT (10^9/L)	1 211±127.83	1 154±153.47	1 137±184.26	1 224±192.38	1 237±83.54
Males					
WBC (10^9/L)	10.6±2.8	10.9±2.3	12.5±4.6	12.2±3.5	13.3±3.9
RBC (10^{12}/L)	8.37±0.64	8.21±0.61	8.64±0.69	8.7±0.66	8.18±0.65
HGB (g/L)	182±5.4	172±5.6	189±5.9	178±6.3	179±7.1
HCT%	0.520±1.3	0.461±1.6	0.535±1.9	0.474±1.2	0.485±1.7
MCV (fL)	62.1±0.75	56.2±0.72	61.9±0.79	62.8±0.64	59.3±0.68
MCH (pg)	21.7±0.83	21.0±0.95	21.9±0.78	22.4±0.74	20.7±0.87
MCHC (g/L)	350±0.56	373±0.62	353±0.64	362±0.58	348±0.64
PLT (10^9/L)	1 152±118.54	1 143±178.56	1 229±139.72	1 224±157.56	1 215±137.83

Note: The values expressed as mean ± SD (n=10/sex/dose).

Table 5 Blood biochemical parameters of rats orally administered ATTM for 28 days

Biochemical Parameters	Control Group	ATTM-Treated Groups (mg/kg b.w./d)			Saline Group
		Low-Dose (5)	Middle-Dose (25)	High-Dose (125)	
Females					
ALT (U/L)	52±5.2	48±4.6	45±5.8	44±6.1	64±6.5
ALP (U/L)	266±14.5	233±13.6	251±16.8	153±19.2**	278±18.4
AST (U/L)	110±23.7	113±27.5	112±26.5	106±23.4	112±24.6
T-Bil (μmol/L)	4.70±0.36	4.3±0.54	4.83±0.82	3.71±0.71	3.68±0.59
LDH (U/L)	761±12.8	770±11.6	785±13.2	797±14.1	721±13.7
TC (mmol/L)	2.6±0.21	2.6±0.17	2.3±0.19	2.7±0.20	1.9±0.45
HDL-c (mmol/L)	1.83±0.14	1.86±0.17	1.70±0.12	2.03±0.16	1.80±0.19
LDL-c (mmol/L)	0.47±0.07	0.43±0.05	0.36±0.08	0.35±0.04	0.32±0.09
TG (mmol/L)	0.68±0.31	0.60±0.45	0.74±0.41	0.76±0.62	0.80±0.58
CK (U/L)	2178±33.6	2127±37.6	2160±41.4	2007±32.9	1983±36.4
CR (μmol/L)	66±4.7	66±4.6	65±5.3	73±6.2*	68±4.5
Urea (mmol/L)	5.4±5.6	6.0±7.8	8.8±8.1	6.1±6.5	5.7±6.2
UA (μmol/L)	79.5±0.24	82.6±0.37	80.2±0.19	80.2±0.56	83.9±0.47
TP (g/L)	60.3±4.7	57.9±3.8	63.6±5.3	65.6±4.2	69.1±4.9
ALB (g/L)	30.2±2.4	22.4±2.7	33.6±3.2	31.4±3.8	30.8±3.0
GLU (mmol/L)	4.24±0.35	3.94±0.37	5.09±0.32	6.7±0.39*	5.11±0.31
Ca (mmol/L)	2.70±0.10	2.90±0.08	2.49±0.12	2.41±0.14	2.44±0.13
P (mmol/L)	2.7±0.24	2.7±0.33	3.9±0.45	3.3±0.38	2.4±0.51
Males					
ALT (U/L)	41±5.4	49±6.4	48±6.1	44±4.8	47±3.9
ALP (U/L)	103±14.7	116±15.4	110±17.4	113±18.9	117±16.8
AST (U/L)	117±23.5	112±25.6	110±21.9	106±26.5	116±27.8
T-Bil (μmol/L)	5.12±0.34	3.44±0.37	4.49±0.36	3.71±0.45	4.8±0.52
LDH (U/L)	518±12.3	512±12.7	515±11.8	517±13.6	517±14.2
TC (mmol/L)	2.3±0.25	2.3±0.29	2.2±0.21	2.7±0.19	3.2±0.35
HDL-c (mmol/L)	1.77±0.15	1.73±0.16	1.59±0.13	2.03±0.17	2.24±0.15
LDL-c (mmol/L)	0.29±0.08	0.35±0.06	0.27±0.04	0.35±0.05	0.52±0.10
TG (mmol/L)	0.65±0.36	0.69±0.39	0.56±0.32	0.76±0.47	1.17±0.35
CK (U/L)	2 208±42.5	2 178±32.6	2 116±35.7	2 207±36.2	2 287±37.9
CR (μmol/L)	69±4.7	67±4.5	68±5.2	73±6.7*	68±5.8
Urea (mmol/L)	6.6±6.5	6.1±7.2	6.2±5.8	6.1±4.9	6.3±8.1
UA (μmol/L)	89.4±0.51	64.5±0.27	82.9±±0.46	80.2±0.35	84.0±0.49
TP (g/L)	68.7±4.7	60.5±5.6	66.8±3.9	65.6±6.1	64.1±5.8
ALB (g/L)	32.3±2.5	29.5±3.9	32.5±3.2	31.4±4.5	34.9±3.7
GLU (mmol/L)	5.20±3.2	5.22±3.8	5.29±3.3	5.70±3.5*	5.11±3.1
Ca (mmol/L)	2.64±0.13	2.44±0.17	2.39±0.09	2.41±0.13	2.60±0.11
P (mmol/L)	3.0±0.36	2.8±0.29	2.5±0.37	2.4±0.42	2.6±0.24

Note: The values expressed as mean ± SD (n=10/sex/dose). * p<0.05 as compared with control group; ** p<0.01 as compared with control group.

2.3.3 Organ weights

The relative organs weights of male and female rats treated with ATTM orally for 28 days are summarized in Table 6. The relative liver weight was significantly decreased in the high-dose groups in both females and males. No significant differences were observed for other organs in either gender.

Table 6 The relative organ weight of rats orally administered ATTM for 28 days

Item	Control Group	ATTM-Treated Groups (mg/kg b.w./d)			Saline Group
		Low-Dose (5)	Middle-Dose (25)	High-Dose (125)	
Females					
Heart	0.40±0.12	0.39±0.28	0.39±0.26	0.40±0.18	0.37±0.29
Liver	3.06±0.26	3.49±0.35	3.52±0.38	4.01±0.25*	3.30±0.17
Spleen	0.28±0.05	0.23±0.04	0.24±0.03	0.27±0.02	0.26±0.06
Lung	0.60±0.15	0.75±0.13	0.62±0.17	0.61±0.19	0.66±0.21
Kidney	0.74±0.13	0.73±0.17	0.80±0.14	0.80±0.18	0.66±0.17
Thymus	0.20±0.09	0.20±0.06	0.16±0.04	0.20±0.05	0.21±0.08
Ovaries	0.06±0.01	0.07±0.03	0.06±0.04	0.06±0.07	0.04±0.05
Males					
Heart	0.37±0.14	0.34±0.18	0.35±0.26	0.34±0.23	0.31±0.27
Liver	3.14±0.25	3.16±0.28	2.99±0.24	3.98±0.27*	3.03±0.19
Spleen	0.22±0.06	0.17±0.07	0.22±0.03	0.20±0.02	0.27±0.08
Lung	0.57±0.14	0.63±0.13	0.56±0.18	0.60±0.20	0.66±0.25
Kidney	0.86±0.12	0.78±0.15	0.83±0.18	0.80±0.21	0.76±0.19
Thymus	0.15±0.04	0.10±0.07	0.29±0.11	0.19±0.09	0.17±0.12
Testes	1.16±0.12	1.33±0.18	1.27±0.24	1.30±0.32	1.21±0.25

Note: The values expressed as mean ± SD (n = 10/sex/dose). The relative organ body weight ratio = $\frac{\text{Absolute organ weight (g)}}{\text{Body weight of rats on sacrifice day (g)}} \times 100$. * $p<0.05$ as compared with control group.

2.3.4 Histopathology examination

The histological alterations of the heart, liver, spleen, lungs, kidneys, thymus, ovaries or testes were observed after 28 days. Histopathological examinations of the heart, lungs, thymus, ovaries or testes showed no abnormalities in any of the treatment groups (data not shown). The histopathological examination of the liver, spleen and kidneys (Fig. 3) showed that there was occasionally slight degeneration and necrosis in the liver, which was most obvious in the high-dose group. The spleens of the animals in the middle-dose, low-dose and control groups were normal, but there was splenic corpuscle atrophy and an unclear boundary between the red and white pulp in the high-dose group. The kidneys of the middle-dose group presented mild renal tubular cell necrosis and degeneration, but these were most serious in the high-dose group. The kidneys of the low-dose group and control animals were normal.

A histopathological examination of the liver, spleen and kidneys in the high-dose group showed some toxicological effects. However, no significant changes in the low-dose ATTM group were found in the hematological and blood biochemical analysis compared to the control group, which was in agreement with the histopathological finding. These results indicate that ATTM fed at 5mg/kg for 28 days is generally considered to be safe for rats. We inferred that the target organs are the liver, spleen and kidneys, and the no observed adverse effect level (NOAEL) was 5mg/kg in this study.

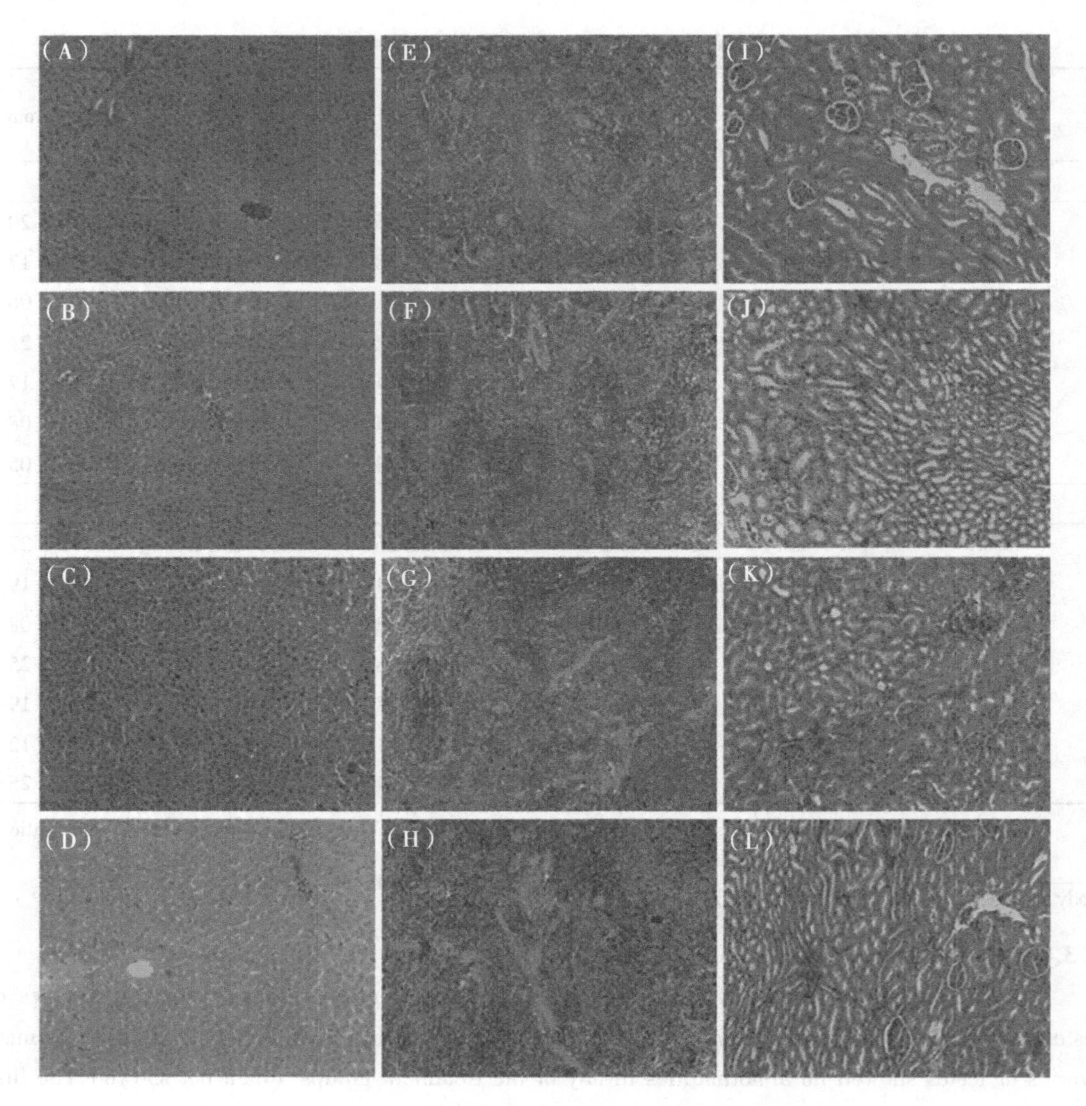

Fig. 3 The results of the histopathological examinations of rats orally administered ATTM for 28 days (H&E × 100)

Note: (A–D) liver; (E–H) spleen; (I–L) kidney. (A, E and I) control group; (B, F and J) low-dose group; (C, G and K) middle-dose group; (D, H and L) high-dose group.

3 EXPERIMENTAL SECTION

3.1 Preparation of ATTM

ATTM was synthesized in our lab, as previously reported [14]. The synthesis of ATTM begins from 22-*O*-tosylpleuromutilin, which was obtained by the reaction of pleuromutilin and p-toluenesulfonyl chloride to activate the 22-hydroxyl of pleuromutilin. The intermediate, 14-*O*-(iodoacetyl)-mutilin was prepared in good yield under reflux for 3 h in acetone. Finally, ATTM was obtained in 82% yield by the nucleophilic attack of 2-amino-5-mercapto-1,3,4-thiadiazole on 14-*O*-(iodoacetyl)-mutilin under alkaline conditions.

3.2 Animals

Adult specific pathogen free (SPF) Kunming mice (weighing 18-22 g) and Sprague-Dawley (SD) rats (aged 3-4 weeks, weighing 140-160 g) were purchased from the Laboratory Animal Center of Lanzhou University (Lanzhou, China) and maintained under controlled temperature conditions (23℃), with a constant 12 h light-dark cycle and free access to food and water. The experimental procedures were performed in accordance with the Ethical Principles in Animal Research and were approved by the Committee for Ethics in the Laboratory Animal Center of Lanzhou University (number: SCXK2013-0002).

3.3 *In Vivo* efficacy in a mouse model

Male and female mice were rendered neutropenic upon treatment for four days with 150mg/kg cyclophosphamide intraperitoneally followed by 100mg/kg for one day, after they had been acclimated for five days. The neutropenic mice (10 per group) received a 0.5mL MRSA inoculum of 10^9 CFU/mL via intraperitoneal (ip) injection. At one hour post infection, the mice were then intravenously (iv) administered ATTM dissolved in 0.5mL vehicle (soybean lecithin: sterile water = 1 : 30) at doses of 2.5mg/kg, 5mg/kg, 10mg/kg and 20mg/kg b.w. Tiamulin fumarate was used as a control in the same manner and at the same doses as ATTM. The survival of mice at 72 h after infection was used as the end point and the ED_{50} was calculated by the method described by Reed Muench[17] using the Hill equation.

3.4 Acute oral toxicity in mice

Thirty male and thirty female SPF Kunming mice were stratified by weight and randomly assigned to six groups: five treatment groups and one vehicle control group, with five male and five female mice for each group. The vehicle control group received DMSO in a volume of 20mL/kg b.w. by oral gavage. The ATTM was dissolved in DMSO and administered to the mice at doses of 948.15mg/kg, 1 422.22mg/kg, 2 133.33mg/kg, 3 200.00mg/kg and 4 800.00mg/kg body weigh, respectively, and all animals were observed twice daily for symptoms and mortality for one week. The vehicle control group was observed at the same time. All of the surviving animals were euthanized at the end of the study, and their vital organs were individually observed for overt pathology by necropsy, and the LD_{50} was calculated by the Bliss method on day 8.

3.5 Subchronic oral toxicity

3.5.1 Experimental design

A total of 100 healthy male and female SD rats (140-160 g) were selected and randomly divided into five groups (20 rats in each group, male : female = 1 : 1) after one week of acclimatization. One group of rats was assigned to serve as control group, and these animals received DMSO in a volume of 4mL/kg b.w., and another group was assigned to a saline group and received physiological saline in the same volume as the DMSO by oral gavage. The other three groups were assigned to ATTM treatment groups, and received ATTM via oral gavage at doses of 5, 25 and 125mg/kg b.w. per day (at the same time each day ± 1 h). The animals in all groups were used for 28-day subchronic toxicity study, and were observed once daily to detect signs of toxicity. The animals' body weight and food consumption were measured every week during the whole observation period, and the gavage volume was adjusted based on the last measured body weight.

3.5.2 Clinical signs, body weights and food consumption

The rats were observed for behavioral changes, mortality, and symptoms and signs of gross toxicity at least once per day during the 28-day subchronic toxicity study. The individual body weights of the rats were measured and recorded at least once a week. The mean weekly body weight gain was calculated for each sex and dose level during the entire test period. Individual food consumption was measured and recorded weekly. The mean daily food consumption was calculated for each sex and dose level for each weekly interval and for the overall test period.

3.5.3 Blood analysis

At the end of the experiment, blood was collected from the femoral artery in ethylenediaminetetraacetic acid (EDTA) coated tubes for hematology studies and in non-oxalate tubes for the separation of serum on the 29th day following a 12-h fast. The hematological analyses included the white blood cell (WBC), red blood cell (RBC) counts, the hemoglobin (HGB) and hematocrit (HCT) levels, mean corpuscular volume (MCV), mean corpuscular hemoglobin (MCH), mean corpuscular haemoglobin concentration (MCHC) and number of platelets (PLT), which were determined using a Poche-100iv Diff instrument (SYSMEX, Kakogawa, Japan).

The other blood collected in non-oxalate tubes was centrifuged at 3500 rpm (15 min at 4℃) and the supernatant (serum) was collected and introduced into new tubes for the subsequent biochemical analyses of the levels of alanine transaminase (ALT), ALP, aspartate transaminase (AST), total bilirubin (T-Bil), lactate dehydrogenase (LDH), total cholesterol (TC), high density lipoprotein (HDL), low density lipoprotein (LDL), triglycerides (TG), creatine kinase (CK), CR, urea nitrogen (Urea), uric acid (UA), total protein (TP), albumin (ALB), GLU, Ca and P. These were measured using reagent kits and a Mindray BS-420 auto hematology analyzer (Mindray Corporation, Shenzhen, China).

3.5.4 Necropsy, organ weight and histopathology

Finally, the rats were anesthetized using an excess of CO_2 anesthesia, followed by exsanguinations. The gross pathological changes were recorded for each rat, and the heart, liver, kidneys, spleen, thymus, lungs, testes and ovaries were collected, weighed and the relative weights (organ weight (g) /100 g b.w.) were recorded. Histopathological examinations were performed on the heart, liver, kidneys, spleen, thymus, lungs, testes or ovaries and small intestines of each rat.

3.6 Statistical analysis

The data were expressed as the mean ± standard deviation (SD) and were analyzed using the SAS statistical software package (Version 9.0; SAS Institute, Cary, NC, USA) to account for the effects of ATTM on the weight, hematological findings and organ effects. A parametric one-way analysis of variance (ANOVA) for repeated measures was used to examine intergroup differences. When a significant difference ($p<0.05$) was found, Tukey's test was used to compare the means.

4 CONCLUSIONS

Historically, the semi-synthesis of new compounds based on natural products, especially complex natural products, has been the predominant avenue to the development of new antibiotics. The chemical modifications of pleuromutilin were made in an attempt to improve the antimicrobial activity and *in vivo* efficacy of the compounds after the identification of the structure of pleuromutilin. ATTM, a new derivative of pleuromutilin bearing a thiadiazole moiety, was synthesized in our lab. The *in vivo* efficacy study showed that ATTM exhibited potent activity, with an ED_{50} of 5.74mg/kg b.w. when used to treat MRSA infected mice. Moreover, a study in mice showed no evidence of acute toxicity after the administration of an oral dose of up to 948.15mg/kg b.w. of ATTM, with an approximate LD_{50} of 2304.4mg/kg determined by the Bliss method. No animals died and no clinical abnormalities were observed that were considered to be associated with the 28-day treatment. Most of the hematological and serum biochemical data showed no differences between the treated groups and control group. Some toxic effects could be found in the liver, spleen and kidneys in the high-dose group by the histopathological examination, but these were not considered treatment-related. Our results show that ATTM fed at 5mg/kg for 28 days is generally considered to be safe for rats.

The above results show that ATTM has a relatively high efficacy *in vivo* and has a low toxicity profile as a new candidate drug. However, other laboratory studies based on protocols by regulatory agencies should be performed to evaluate the drug potential of this compound.

ACKNOWLEDGMENTS

This work was financed by Basic Scientific Research Funds in Central Agricultural Scientific Research Institutions (No. 1610322014003), The Agricultural Science and Technology Innovation Program (ASTIP) and the Science and Technology Development Plan Project of Lanzhou (2013-4-90).

AUTHOR CONTRIBUTIONS

RS, JL, and CZ designed research; YY, JC, RX, ZY, CZ and ZG performed research and analyzed the data; RS and CZ wrote the paper. All authors read and approved the final manuscript.

CONFLICTS OF INTEREST

The authors declare that there are no conflict of interest.

References

[1] Spicknall, I. H.; Foxman, B.; Marrs, C. F.; Eisenberg, J. N. S. A Modeling Framework for the Evolution and Spread of Antibiotic Resistance: Literature Review and Model Categorization. *Am. J. Epidemiol.* 2013, 178: 508-520.

[2] Kavanagh, F.; Hervey, A.; Robbins, W.J. Antibiotic Substances from Basidiomycetes: VIII. Pleurotus Multilus (Fr.) Sacc. and Pleurotus Passeckerianus Pilat. *Proc. Natl. Acad. Sci. USA* 1951, 37: 570-574.

[3] Tang, Y.Z.; Liu, Y.H.; Chen, J.X. Pleuromutilin and its Derivatives-The Lead Compounds for Novel Antibiotics. *Mini Rev. Med. Chem.* 2012, 12: 53-61.

[4] Moody, M.N.; Morrison, L.K.; Tyring, S.K. Retapamulin: What is the role of this topical antimicrobial in the treatment of bacterial infections in atopic dermatitis? *Skin Therapy Lett.* 2010, 15: 1-4.

[5] Sader, H.S.; Paukner, S.; Ivezic-Schoenfeld, Z.; Biedenbach, D.J.; Schmitz, F.J.; Jones, R. N. Antimicrobial activity of the novel pleuromutilin antibiotic BC-3781 against organisms responsible for community-acquired respiratory tract infections (CARTIs). *J. Antimicrob. Chemother.* 2012, 67: 1170-1175.

[6] Shang, R.F.; Wang, J.T.; Guo, W.Z.; Liang, J.P. Efficient antibacterial agents: A review of the synthesis, biological evaluation and mechanism of pleuromutilin derivatives. *Curr. Top. Med. Chem.* 2013, 13: 3013-3025.

[7] Poulsen, S.M.; Karlsson, M.; Johansson, L.B.; Vester, B. The pleuromutilin drugs tianmulin and valnemulin bind to the RNA at the peptidyl transferase centre on the ribosome. *Mol. Microbiol.* 2001, 41: 1091-1099.

[8] Dreier, I.; Hansen, L.H.; Nielsen, P.; Vester, B. A click chemistry approach to pleuromutilin derivatives. Part 3: Extended footprinting analysis and excellent MRSA inhibitionfor a derivative with an adenine phenyl side chain. *Bioorg. Med. Chem. Lett.* 2014, 24: 1043-1046.

[9] Schlunzen, F.; Pyetan, E.; Fucini, P.; Yonath, A.; Harms, J.M. Inhibition of petide bond formation by pleueomutilins: The structure of 50S ribosomal subunit from Deinococcus radiodurans in complex with tiamulin. *Mol. Microbiol.* 2004, 54: 1287-1294.

[10] Davidovich, C.; Bashan, A.; Auerbach-Nevo, T.; Yaggie, R.D.; Gontarek, R.R.; Yonath, A. Induced-fit tightens pleuromutilins, binding to ribosomes and remote interactions enable their selectivity. *Proc. Natl. Acad. Sci. USA* 2007, 104: 4291-4296.

[11] Shang, R.F.; Wang, G.H.; Xu, X.M.; Liu, S.J.; Zhang, C.; Yi, Y.P.; Liang J.P.; Liu, Y. Synthesis and Biological Evaluation of New Pleuromutilin Derivatives as Antibacterial Agents. *Molecules* 2014, 19: 19050-19065.

[12] Hildebrandt, J. F.; Berner, H.; Laber, G.; Turnowsky, F.; Schutze, E. A new semisynthetic pleuromutilin derivative with antibacterial activity: *In vitro* evaluation. *Curr. Chemother. Immunother.* 1982, 4: 346-347.

[13] Ling, C.Y.; Fu, L.Q.; Gao, S.; Chu, W.J.; Wang, H.; Huang, Y.Q.; Chen, X.Y.; Yang, Y.S. Design, Synthesis, and Structure-Activity Relationship Studies of Novel Thioether Pleuromutilin Derivatives as Potent Antibacterial Agents. *J. Med. Chem.* 2014, 57: 4772-4795.

[14] Shang, R.F.; Pu, X.Y.; Xu, X.M.; Xin, Z.J.; Zhang, C.; Guo, W.Z.; Liu, Y.; Liang, J. P. Synthesis and Biological Activities of Novel Pleuromutilin Derivatives with a Substituted Thiadiazole Moiety as Potent Drug-resistant Bacteria Inhibitors. *J. Med. Chem.* 2014, 57: 5664-5678.

[15] Liju, V.B.; Jeena, K.; Kuttan, R. Acute and subchronic toxicity as well as mutagenic evaluation of essential oil from turmeric (*Curcuma longa* L). *Food Chem. Toxicol.* 2013, 53: 52-61.

[16] Brandt, A.P.; Oliveira, L.F.S.; Fernandes, F.B.; Alba, J. Evaluation of prospective hypocholesterolemic effect and preliminary toxicology of crude extract and decoction from *Vitex megapotamica* (Spreng) Moldenke (*V. montevidensis Cham.*) *in vivo*. *Rev. Bras. Farmacogn.* 2009. 19: 388-393.

[17] Reed, I.; Muench, J.H. A simple method of estimating fiftypercent endpoints. *Am. J. Hyg.* 1938, 27:

493-497.

Sample Availability: Sample of the ATTM is available from the authors.

(发表于《molecules》, 院选 SCI, IF: 2. 095)

Antinociceptive and Anti-tussive Activities of the Ethanol Extract of the Flowers of *Meconopsis punicea* Maxim

SHANG Xiao-fei[1], WANG Dong-sheng[1], MIAO Xiao-lou[1], WANG Yu[1], ZHANG Ji-yu[1], WANG Xue-zhi[1], ZHANG Yu [2]* PAN Hu[1]*

(1. Key Laboratory of New Animal Drug Project, Gansu Province/Key Laboratory of Veterinary Pharmaceutical Development of Ministry of Agriculture/Lanzhou Institute of Husbandry and Pharmaceutical Sciences of Chinese Academy of Agricultural Science, Lanzhou 730050, China; 2. Department of Emergency, Lanzhou General Hospital of PLA, Lanzhou 730050, China)

Abstract: Background: As an important traditional Tibetan (veterinary) medicine, the flowers of *Meconopsis punicea* (family *Papaveraceae*) have been used to treat pain, fever, cough, inflammation, liver heat and lung heat of humans and animals by local people for thousands of years. In this paper, we aimed to investigate the antinociceptive and anti-tussive activities of the ethanol extract of *M.punicea* (EEM).

Methods: Firstly, HPLC was used to analyze the main constituents of the ethanol extract of *M.punicea*. In animal experiments, the acetic acid-induced writhing response test, hot plate test, barbiturate-induced sleeping time and formalin tests were used to evaluate the antinociceptive activity. Then, ammonia-induced coughing and sulfur dioxide-induced coughing tests in mice as well as the phenol red secretion in trachea test were used to investigate the anti-tussive activity of the extract. Finally, an acute toxicity study was carried out.

Results: The results showed that alkaloids and flavonoids were the main compounds in the ethanol extract of *M.punicea* flowers. The extract at 125, 250 and 500mg/kg had good antinociceptive and anti-tussive activities in mice with a dose-dependent manner.

Conclusions: These findings suggested that EEM has significant bioactivities, and the active components of *M.punicea* should be studied further.

Key words: *Meconopsis punicea* extract; Antinociceptive activity; Formalin test; Antitussive activity

---

* Correspondence: 262730291@ qq. com; shangxf928@ 126. com

Received: 12 December 2014 Accepted: 13 May 2015

Published online: 22 may 2015

BACKGROUND

Meconopsis punicea Maxim. (Hong Hua Lv Rong Hao) is a perennial herb (family *Papaveraceae*) that grows from 30cm to 70cm tall. In China, as a traditional Tibetan (veterinary) medicine, it grows in alpine scrub and alpine meadows with shady and half-shady slopes at altitudes of 3 000–4 800m and is distributed in the northeastern part of Tibet, southeastern part of Qinghai, western part of Sichuan and southern part of Gansu provinces (Fig.1). The flower of *M.punicea* has been used to treat pain, fever, cough, inflammation, liver heat and lung heat of humans and animals by local people for thousands of years, and about 5 preparations containing *M.punicea* with other medicines have been listed in 'Drug Standard of Ministry of Public Health of the People's Republic of China (Tibetan medicine volume)'. Meanwhile, the beautiful flowers are also used as ornamental plants in the Tibetan region[1].

Fig.1 Picture of *Meconopsis punicea* Maxim

Recently, the chemical composition of *M.punicea* was studied. And as the main components of the aerial part, a serial of alkaloids were identified, such as karachine, valachine, (–) -mecambridine, berberine, protopine and alborine. Meanwhile, flavonoids, including luteolin, apigenin, hydnocarpin and isorhamnetin, were also isolated and reported along with canillic acid, cinnamic acid and other components[2-4]. But up to now, no one has used modern technology to study its pharmacological effects.

In our field investigation of folk veterinary medicine, *M.punicea*, which has used to treat pulmonary disease, pain and inflammation by folk veterinarians for a long period of time in some regions of the Sichuan and Gansu provinces, was identified[5]. In this paper, considering *M.quintuplinervia* and relative species of the genus have been reported to exhibit analgesic and sedative activities[6,7], and the traditional uses of *M.punicea* in alleviating pain and reducing cough and inflammation[5], we systematically evaluated the analgesic, anti-tussive and expectorant activities of the extract of *M.punicea* flowers.

METHODS

Plant material

The herbal sample of M.punicea (34. 02 °N, 102. 74° E, 3531 m altitude) was collected at Rierlang Mountain of Ruoergai County in Sichuan province, China in July of 2011. The raw material was identified by Prof. Zhigang Ma, Pharmacy College of Lanzhou University, China. A voucher specimen with accession number ZSY028 was submitted to the Herbarium of Traditional Chinese Veterinary Medicine (Lanzhou Institute of Husbandry and Pharmaceutical Sciences of CAAS, China).

Extraction

The ethanol extract of *M.punicea* (EEM) was prepared as follows. Briefly, 45 g of *M.punicea* flowers was refluxed in 80℃ with 500 ml of 95% ethanol 3 times, and each reflux time was 2 h. After filtration, the solvent was evaporated in a rotary evaporator. The extract was vacuum dried at 60℃. The yield was 14. 27%.

Drugs

Acetic acid (Tianjing Chemical Reagent Company, Tianjing, China); aspirin (Hefei Jiulian Pharmaceutical Company, Anhui, China); morphine (The First Pharmaceutical Company of Shenyang, Shenyang, China); phenol red (The Third Chemical Reagent Factory of Shanghai, Shanghai, China). The standard compounds of berberine, luteolin and apigenin (purity ≥98%) were obtained from Shanghai R&D Centre for Standardisation of Chinese Medicines (China).

Animals

Male or female Balb/C mice (20 ± 2) g were obtained from the Department of Animal Center, Lanzhou Institute of Biologicals (Lanzhou, China). They were kept in plastic cages at 22 ± 2℃ with free access to pellet food and water. This study was carried out in accordance with the Regulation for the Administration of Affairs Concerning Experimental Animals (State Council of China, 1988), and was approved by the Ethics Committee of Lanzhou Institute of Husbandry and Pharmaceutical Sciences of Chinese Academy of Agricultural Science (Lanzhou, China).

Phytochemical screening

The ethanol extract of *M.punicea* (EEM) was qualitatively tested for the detection of carbohydrates, saponins, flavonoids, tannins, alkaloids, glucosides and steroids following standard procedures[8].

RP-HPLC analysis of EEM was performed on a Waters apparatus (two solvent delivery systems, model 600, and a Photodiode Array detector, model 996), using a gradient solvent system comprised of formic acid in water (pH 3.0) (A) and CH_3CN (B). The gradient profile was as follows: 0-20 min (90%-85% A), 20-35 min (85%-70% A), 35-70 min (70%-40% A) and 70-85 min (40%-10% A), at 0. 5 ml/min. On-line UV spectra were recorded from 200 to 400 nm. Data acquisition and quantification were performed using Millenium 2. 10 version software (Waters). A Sunfire C-18 column (250 mm×4. 6 mm, 5 μm, Waters, Ireland) was maintained at ambient temperature (30.0℃).The mobile phase was filtered through a Millipore 0.45 mm filter and degassed prior to use.The peaks were detected, and berberine, luteolin and apigenin were detected by comparison with chemical standards, which were identified with MS, ^{1}H NMR and ^{13}C NMR.

Evaluation of antinociceptive activity of the ethanol extract.

Acetic acid-induced writhing response

The test was carried out according to a previously described method[9].Mice were randomly divided into five groups with ten animals in each group, namely, the normal control group, reference group, and three groups of EEM.The control group received normal saline (10 ml/kg, i.p.), and the reference group received aspirin (100mg/kg, i.p.).EEM was intraperitoneally injected into each mouse at doses of 125, 250 and 500mg/kg.After 30 min treatment, 0.7% acetic acid (0.1 ml/10 g body weight) was administered intraperitoneally to each mouse.The mice were observed, and the number of abdominal constrictions and stretchings in a period of 0–30 min was counted.

Hot plate test

The test was carried out according to a previously described method[10,11].EEM (125mg/kg, 250mg/kg, 500mg/kg) was administered by intraperitoneal injection.The control group received normal saline (10 ml/kg, i.p.), and the reference group received morphine sulfate (5mg/kg, i.v.).After 30 and 65 min, mice were individually placed on a heated plate at 55 ± 1℃.The latency time of forepaw licking or jumping was determined.

Barbiturate-induced sleeping time

The test was carried out according to a previously described method[12,13].The control group received normal saline (10 ml/kg i.p.), and the reference group received diazepam (12.5mg/kg). EEM (125mg/kg, 250mg/kg, 500mg/kg) was administered to the animals by intraperitoneal injection.After 30 min, sleep was induced by the intraperitoneal administration of 40mg/kg pentobarbital.The latency time to sleep (time to lose the righting reflex) and sleeping time (duration of loss of the righting reflex) were measured.

Formalin test

The test was carried out according to a previously described method[5,14].After treatment for 30 min with normal saline, EEM and positive drugs, 20 μl of 5.0% formalin in normal saline was injected subcutaneously into a hind paw of each mouse.The time of lickings, stampings, and scratchings the injected paw were recorded and separated into the early phase (0–5 min) and late phase (15–40 min) after formalin injection.To elucidate the possible mechanism of action, morphine sulfate (5mg/kg, i.v.) and aspirin (100mg/kg, i.p.) were used as positive controls.

Evaluation of anti-tussive activity of the ethanol extract *Ammonia-induced coughing in mice*

The test was carried out according to a previously described method[15,16].Mice were randomly divided into five groups (n = 10).Thirty minutes after injecting the normal saline, EEM and codeine phosphate, each mouse was placed in a 1 000 ml special glass chamber and exposed to 0.3 ml 25% NH_4OH produced by a nebulizer for 45 s.Then, mice were taken out and put in an open field. The cough frequency and latent period of coughing were recorded for 5 min, and the anti-tussive activity was assessed as the percentage of inhibition of the number of coughs.

Sulfur dioxide-induced coughing in mice

The test was carried out according to a previously described method[15,17], and mice were randomly divided into five groups. After treatment for 30 min with normal saline, EEM and codeine phosphate, a burette containing 2 ml of 50% sulfuric acid solution was fixed to a flask containing 2 g of anhydrous sodium sulfite, and the acid was added to this sulfite to generate sulfur dioxide gas. Meanwhile, each mouse was placed in a 1000mL special glass chamber, and the number of coughs was recorded for 3 min. The anti-tussive activity was assessed as the percentage of inhibition of the number of coughs.

Phenol red secretion in mouse tracheas

The test was carried out according to a previously described method[15,18], and mice were randomly divided into five groups. After treatment for 30 min with normal saline, EEM and NH_4Cl, each mouse was treated with phenol red solution (5% in saline solution, w/v, and 0. 2mL/20 g body weight). Thirty minutes after the application of phenol red solution, mice were sacrificed by cervical dislocation without damaging the tracheas. The trachea was dissected away from adjacent organs, and 2mL of normal saline was used to wash 3 times. After ultrasonication for 15 min, 1mL of a 5% $NaHCO_3$ solution was added to the normal saline, and the optical density was measured at 560 nm using LabTech UV-BlueStar Plus (Beijing LabTech.Inc.China).

Acute toxicity study

The up-and-down or staircase method for acute toxicity testing was carried out as previously described[19]. 500 to 2000mg/kg EEM was administered to mice through the intraperitoneal route with a gradual increase in dose. The behavioral changes and mortality of animals were observed continuously for the first 4 h and 7 days after the drug administration.

Statistical analysis

The data obtained were analyzed using SPSS software program version 18. 0 and expressed as the mean ±S.E.M. Data were analyzed by a one-way ANOVA followed by Student's two-tailed *t*-test for the comparison between test and control and Dunnett's test when the data involved three or more groups. P-values less than 0. 05 ($P<0.05$) were accepted as significant.

RESULTS

Phytochemical screening

After the preliminary phytochemical analysis, we found that alkaloids and flavonoids were the main constituents of EEM. In the HPLC analysis, berberine (1), luteolin (2), and apigenin (3) were found in EEM (Fig.2).

Acetic acid-induced writhing response

The results showed that EEM (500mg/kg, 250mg/kg and 125mg/kg) significantly restrained the writhing response induced by 0. 7% acetic acid with an inhibition rate of 95. 28%, 87. 22% and

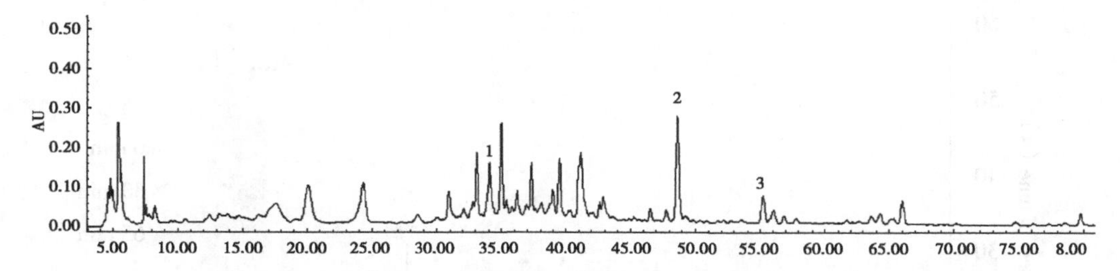

Fig.2 HPLC–DAD chromatogram of the ethanol extract of *M. punicea*

Note: 1. Berberine, 2. Luteolin, 3. Apigenin

75. 56% ($P<0.01$) in a dose-dependent manner. As the positive drug, aspirin (100mg/kg, i.p.) produced a 77. 78% reduction compared to the control. The results are shown in Fig.3.

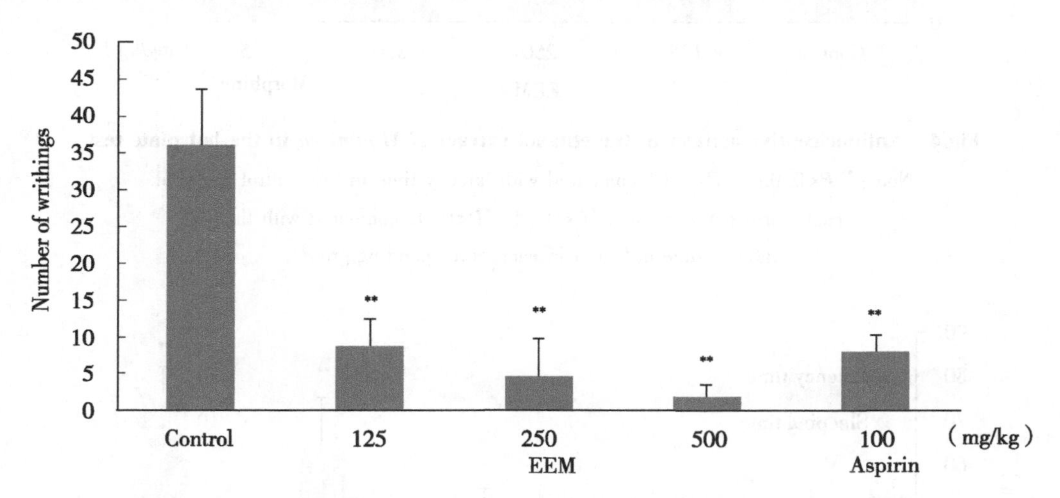

Fig.3 Antinociceptive activity of the ethanol extract of *M.punicea* in the writhing test

Note: ** $P<0.01$ compared with control.

Hot plate test

As shown in Fig.4, EEM had antinociceptive activity in the hot plate test. Compared to the control group at 30 and 65 min, EEM (500mg/kg) could prolong the latency time of mice ($P<0.01$). At the same time, it also prolonged the latency time at 0 min in this group. Morphine sulfate (5mg/kg, i.v.) markedly increased the pain threshold of mice at 30 min or 65 min after treatment ($P<0.01$).

Barbiturate-induced sleeping time

On the basis of the above tests, we carried out the barbiturate-induced sleeping time test. The results showed that at 500mg/kg, EEM significantly decreased the latency time of sleep and prolonged sleeping time of mice compared to the control group, and the latency time and sleeping time of mice were 4. 9 and 62. 1 min ($P<0.01$), respectively. For the diazepam positive control, the latency time and sleeping time were 3. 4 and 75. 3 min ($P<0.01$), respectively (Fig.5).

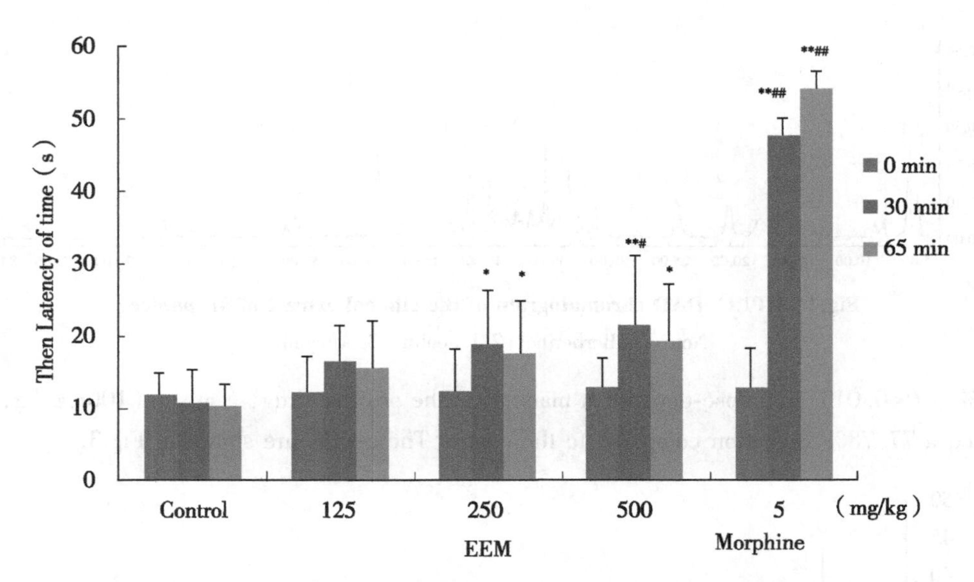

Fig.4 Antinociceptive activity of the ethanol extract of *M.punicea* in the hot plate test

Note: ** $P<0.05$, ** $P<0.01$ compared with latency time in the control group at each corresponding time. ## $P<0.05$, ## $P<0.01$ compared with the latency time at 0 min in each corresponding group.

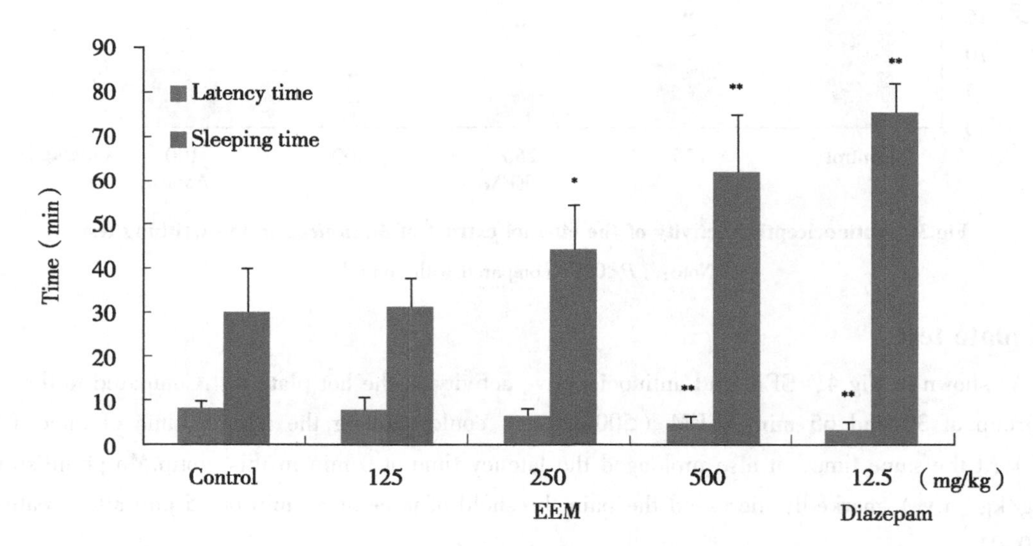

Fig.5 Antinociceptive activity of the ethanol extract of *M.punicea* in the barbiturate-induced sleeping time test

Note: ** $P<0.05$, ** $P<0.01$ compared with latency time in the control group at each corresponding time.

Formalin test

As shown in Fig.6, EEM had antinociceptive activity in the formalin test. In the early (0-5 min) and late phases (15-40 min), EEM could decrease the time of licking, stamping, and scratching induced by formalin in a dosedependent manner, especially at 500mg/kg, with an inhibition of 58.03% and 64.19% ($P<0.01$). Morphine, used as the positive control, produced a marked reduction of 81.91% and 94.94% of the licking time in the early and late phases ($P<0.01$). Aspirin significantly decreased the time of the late phase (70.39%, $P<0.01$) but did not markedly decrease the time of the early phase.

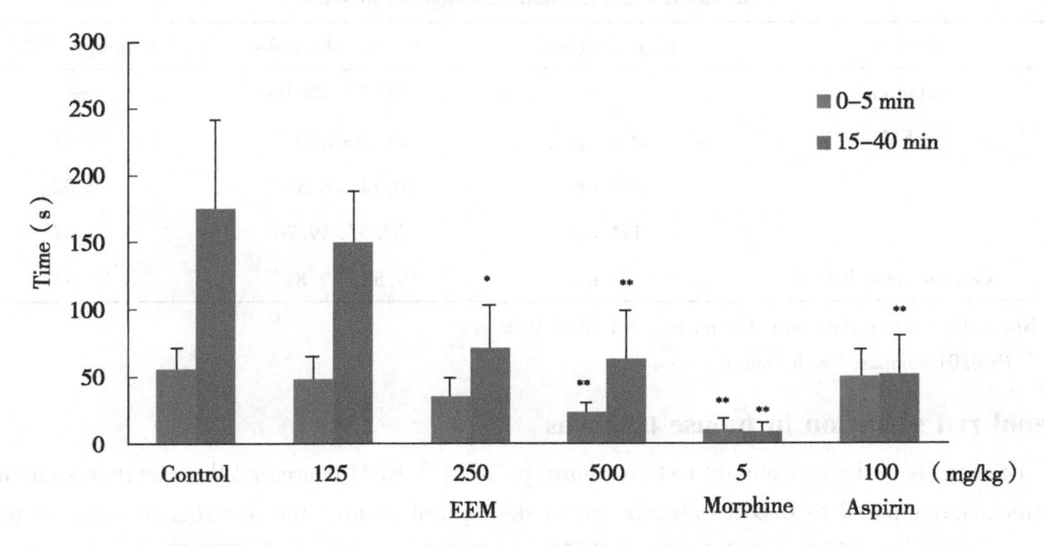

Fig.6 Antinociceptive activity of the ethanol extract of *M.punicea* in the formalin test

Note: ** $P<0.05$, ** $P<0.01$ compared with latency time in the control group at each corresponding time

Ammonia-induced coughing in mice

As shown in Table 1, EEM demonstrated marked antitussive activity. At 500, 250 and 125mg/kg, EEM decreased the number of coughs induced by ammonia, with inhibition rates of 47.88, 34.18 ($P<0.01$) and 14.30% ($P<0.05$), respectively, indicating a dose-dependent effect. Codeine (20mg/kg) had an inhibition of 67.12% ($P<0.01$).

Table 1 Anti-tussive activity of the ethanol extract of *M.punicea* in the ammonia-induced coughs of mice test

Group	Dose (mg/kg)	Number of coughs	Inhibition (%)
Control	—	61.20±6.88	—
EEM	500 i.p.	31.90±7.81**	47.88
	250 i.p.	40.28±8.64**	34.18
	125 i.p.	52.45±7.13*	14.30
Codeine phosphate	20 i.p.	20.12±4.56**	67.12

Note: Each value represents the mean ± S.E.M.of 10 mice.

* $P<0.05$ compared with control. ** $P<0.01$ compared with control.

Sulfur dioxide-induced coughing in mice

As shown in Table 2, EEM demonstrated marked antitussive activity. Compared to the control group (107.71±29.06), 500mg/kg, 250mg/kg and 125mg/kg significantly decreased the number of coughs induced by sulfur dioxide in a dose-dependent manner (15. 71 ± 9. 81, 30. 14 ± 16. 58 and 67. 57±19. 26) ($P<0.01$). Codeine (20mg/kg) inhibited coughing by 81. 82% ($P<0.01$).

Table 2 Anti-tussive activity of the ethanol extract of *M.punicea* in sulfur dioxide -induced coughing of mice

Group	Dose (mg/kg)	Number of coughs	Inhibition (%)
Control	—	107. 71±29. 06	—
EEM	500 i.p.	15. 71±9. 81 **	85. 41
	250 i.p.	30. 14±16. 58 **	72. 02
	125 i.p.	67. 57±19. 26	37. 27
Codeine phosphate	20 i.p.	19. 58 ± 7. 82 **	81. 82

Note: Each value represents the mean ± S.E.M.of 10 mice.

** $P<0.01$ compared with control.

Phenol red secretion in mouse tracheas

The results of the expectorant test are shown in Table 3. EEM decreased the excretive sputum in the trachea. Compared to 0. 0733 absorbance in the control group, the absorbance value of three groups treated with EEM were 0. 1608, 0. 1051 and 0. 0836, respectively. NH_4Cl also showed good expectorant activity ($P<0.01$).

Table 3 Anti-tussive activity of the ethanol extract of M.punicea in the phenol red secretion of mice tracheas

Group	Dose (mg/kg)	Absorbance (A)	Increase (%)
Control	—	0. 0733±0. 019	—
EEM	500 i.p.	0. 1608±0. 0301 **	119. 37
	250 i.p.	0. 1051±0. 0246 *	43. 38
	125 i.p.	0. 0836±0. 0178	13. 32
NH_4Cl	1 500 i.p.	0. 1451±0. 0147 *	97. 95

Note: Each value represents the mean ± S.E.M.of 10 mice.

* $P<0.05$ compared with control. ** $P<0.01$ compared with control.

Acute toxicity

In the acute toxicity test, administration of EEM (500−2000mg/kg) to mice did not cause death or acute behavior changes during the observation periods, and we also did not notice any pathology changes in mice. The LD_{50} was estimated to more than 2000mg/kg, i.p. EEM was safe at the given dose in mice.

DISCUSSION

As an important folk medicine, *Meconopsis* plays an important role in traditional Tibetan (veterinary) medicines. Most plants of this genus are used to clear heat-evil and expel superficial evils, relieve coughing and asthma, eliminate inflammation and relieve pain for a long period of time. *M. quintuplinervia and some species of the genus have been reported to exhibit analgesic and sedative activities*[6]. Until now, the bioactivities of *M.punicea* have not studied. In this paper, based on its traditional use in treating pain, inflammation and coughing, the antinociceptive and antitussive activities of the ethanol extract of *M.punicea* flower were investigated. In the first phase of testing, the results showed that EEM has a significant analgesic activity and decreased the acetic acid-induced writhing response of mice and prolonged the latency time in hot plate test and the sleeping time in the barbiturate-induced sleeping time test. Meanwhile, it markedly decreased the time of licking in both the early and late phase in the formalin test. In the second phase of testing, the results showed that EEM had good antitussive activity. It decreased the number of coughs induced by ammonia and sulfur dioxide and the excretive sputum in the tracheas in a dose-dependent manner.

As a classical non-selective antinociceptive model, acetic acid produces a painful reaction and acute inflammation in the peritoneal area. It indirectly induced the release of endogenous mediators and stimulated nociceptive neurons[20,21]. The results showed that EEM significantly inhibited the acetic acid-induced writhing response in a dose-dependent manner and presented a good peripheral analgesic effect. Then, the hot-plate test was used to evaluate the analgesic activity on the central nervous system[22], and the results indicated that EEM (500mg/kg) markedly prolonged the latency time of mice and showed good central analgesic activity in the hot plate test. In the acetic acid-induced writhing response test, we found that the mice were sedated and slow after administrating EEM. The barbiturate-induced sleeping time test was finally carried out to evaluate the sedative activity of EEM. The results demonstrated that EEM markedly decreased the latency time of sleep and prolonged sleeping time of mice and had strong sedative activity.

Meanwhile, in order to validate the marked analgesic activity of EEM further, the formalin test was carried out. The test could be separated into two different phases in time; the first one was generated in the periphery through the activation of nociceptive neurons by the direct action of formalin, and the second phase occurred through the activation of the ventral horn neurons at the spinal cord level. Thus, it could be discriminated between central and peripheral pain components[16,23]. Centrally acting drugs, such as opioids, inhibited both phases equally, while peripherally acting drugs, such as aspirin and aspirin only, inhibited the late phase[24,25]. The results showed that EEM had a marked antinociceptive effect both the early and the late phase, while aspirin suppressed the later phase only (Fig.6). From the results of the above tests, we suggest that the extract may act both centrally and peripherally to reduce pain.

Anti-tussive animal models were established by a mechanical, electrical and chemical stimulus. In our study, the anti-tussive activity of EEM was evaluated by ammoniainduced and sulfur dioxide-induced coughing in mice. These methods using a chemical stimulus were applied due to the simple procedure that omitted anesthetization; they are frequently used in new drug development[26]. From Table 1 and 2, EEM decreased the number of coughs induced by ammonia and sulfur dioxide. Meanwhile, because most expectorant drugs can increase secretion and dilute the sputum in the respiratory tract so that it could be expectorated easily with ciliary movement[17], the expectorant phenol red secretion in mouse tracheas was tested. As shown in Table 3, EEM markedly enhanced the tracheal phenol red output in a dose-dependent manner.These results were in accordance with the folk clinical uses of *M.punicea* to treat coughs and lung diseases, and the mechanisms should be studied further.

CONCLUSION

The present study demonstrated that the ethanol extract of *M.punicea* flowers had good antinociceptive and antitussive activities *in vivo*.Further studies should be performed to investigate the mechanism of action and the active components of *M.punicea*.

COMPETING INTERESTS

The authors declare that they have no competing interests.

AUTHORS'CONTRIBUTIONS

XS and DW conceived the study, XW, YW and XM determined the index, XS, JZ and YZ wrote the manuscript, HP performed statistical analyses.All the authors read and approved the final version of the manuscript.

ACKNOWLEDGEMENTS

This work was financed by The Special Fund for Agro-scientific Research in the Public Interest (201303040-14), National Science and Technology Infrastructure Program of China (2015BAD11B01).The authors would also like to express their gratitude to the Lanzhou University PhD English writing foreign teacher Allan Grey, who thoroughly corrected the English in the paper.

References

[1] Chinese materia editorial committee.State Chinese Medicine Administration Bureau.Chinese materia, Tibetan volume.Shanghai: Shanghai Scientific and Technical Publishers; 2002: 180.

[2] Liu SY, Wang XK.Chemical Constituents of *Meconopsis punicea*.Trad Chin Med J.1986; 11: 360.

[3] Wu HF, Song ZJ, Zhu HJ, Peng SL, Zhang XF.Chemical Constituents of *Meconopsis punicea*.Nat Prod Res Dev.2011, 23: 202-207.

[4] Xu RS, Ye Y, Zhao WM.Natural Product Chemistry.Beijing: Science Press, 2004.

[5] Shang XF, Tao CX, Miao XL, Wang DS, Tangmuke D, Wang Y, et al.Ethnoveterinary survey of medicinal plants in Ruoergai region, Sichuan province, China. J Ethnopharmacol. 2012, 142: 390-400.

[6] Yang M, Shi XB, Ming K, Sun QL, Chen YG.Advance on the chemical constituents and the pharmacology of the genus *Meconopsis*.Chin Trad Patent Med.2010, 32: 279.

[7] Guo M, Zhao J, Wang ZW, Lu TN, Yu XH.Experimental study of different solvent extracts of *Meconopsis quintupcinervia* Regel on analgesic activity.J Gansu College of TCM.2008, 25: 8.

[8] Ghani A.Medicinal Plants of Bangladesh: Chemical Constituents and Uses, Seconded.Dhaka: The Asiatic Society of Bangladesh, 2003.

[9] Koster R, Anderson M, Beer EJ.Acetic acid for analgesic screening.Fed Proc.1959, 18: 418-420.

[10] Shinde UA, Phadke AS, Nair AM, Mungantiwar AA, Dikshit VJ, Saraf MN.Studies on the anti-in-

flammatory and analgesic activity of Cedrus deodara (Roxb.) Loud.wood oil.J Ethnopharmacol.1999, 65: 21-27.

[11] Turner RA, In: Turner R, Hebborn P, (Eds.).Analgesics: screening methods in pharmacology.Academic Press, New York.1965: 100.

[12] Ferrini R, Miragoli G, Taccardi B.Neuro-pharmacological Studies on SB 5833, a New Psychotherapeutic Agent of the Benzodiazepine Class.Arzneim-Forsch (Drug Res).1974, 24: 2029.

[13] Dos Santos Jr JG, Blanco MM, Do Monte FHM, Russi M, Lanziotti VMNB, Leal LKAM, et al. Sedative and anticonvulsant effects of hydroalcoholic extract of *Equisetum arvense*.Fitoterapia.2005, 76: 508-513.

[14] Kou JP, Sun Y, Lin YY, Cheng ZH, Zheng W, Yu BY, et al.Anti-inflammatory activities of aqueous extract from *Radix Ophiopogon japonicus* and its two constituents. Biol Pharm Bull. 2005, 28: 1234-1238.

[15] Xu SY, Bian RL, Chen X.Pharmacological experiment methodology.Beijing: People's Medical Publishing House, 1991.

[16] Li MX, Shang XF, Zhang RX, Jia ZP, Fan PC, Ying Q, et al.Antinociceptive and anti-inflammatory activities of iridoid glycosides extract of *Lamiophlomis rotata* (Benth.). Kudo Fitoterapia. 2010, 81: 167-172.

[17] Shang JY, Cai XH, Zhao YL, Feng T, Luo XD. Pharmacological evaluation of *Alstonia scholaris*: Anti-tussive, anti-asthmatic and expectorant activities.J Ethnopharmacol.2012, 129: 293-298.

[18] Zhang JL, Wang H, Pi HF, Ruan HL, Zhang P, Wu JZ.Structural analysis and antitussive evaluation of five novel esters of verticinone and bile acids.Steroids.2009, 74: 424-434.

[19] Bruce RD. An up-and-down procedure for acute toxicity testing. Fundament Appl Toxicol. 1985, 5: 151-157.

[20] Sa'nchez-Mateo CC, Bonkanka CX, Hern'andez-P'erez M, Rabanal RM.Evaluation of the analgesic and topical anti-inflammatory effects of *Hypericum reflexum* L.fil.J Ethnopharmacol.2006, 107: 1-6.

[21] Riedel R, Marrassini C, Anesini C, Gorzalczany: Anti-inflammatory and antinociceptive activity of Urera aurantiaca.Phytother Res 2014: doi: 10.1002/ptr.5226

[22] Ferreira MAD, Nunes ODRH, Fontenele JB, Pessoa ODL, Lemos TLG, Viana GSB.Analgesic and anti-inflammatory activities of a fraction rich in ncocalyxone A isolated from *Auxemma oncocalyx*.Phytomed.2004, 11: 315-322.

[23] Tjolsen A, Berge OG, Hunskaar S, Rosland JH, Hole K.The formalin test: an evaluation of the method.Pain.1992, 51: 5-17.

[24] Hunskaar S, Fasmer OB, Hole K.Formalin test in mice: a useful technique for evaluating mild analgesics.J Neurosci Method.1985, 4: 69-76.

[25] Rosland JH, Tjoisen A, Maehle B, Hole K.The formalin test in mice: Effect of formalin concentration.Pain.1990, 42: 235-242.

[26] Han N, Chang CL, Wang YC, Huang T, Liu ZH, Yin J.The *in vivo* expectorant and antitussive activity of extract and fractions from *Reineckia carnea*.J Ethnopharmacol.2010, 131: 220-223.

(发表于《BMC Complementary and Alternative Medicine》，院选 SCI，IF：2.020)

Comparative Proteomics Analysis Provide Novel Insight into Laminitis in Chinese Holstein Cows

DONG Shu-wei[1,2], ZHANG Shi-dong[1], WANG Dong-sheng[1], WANG Hui[1], SHANG Xiao-fei[1], YAN Ping[2], YAN Zuo-ting [1*], YANG Zhi-qiang[1*]

(1. Key Laboratory of Veterinary Pharmaceutical Development of Ministry of Agriculture/ Engineering & Technology Research Center of Traditional Chinese Veterinary Medicine of Gansu Province/Lanzhou Institute of Husbandry and Pharmaceutical Sciences of Chinese Academy of Agricultural Sciences, Lanzhou 730050, China

2. Key Laboratory of Yak Breeding Engineering, Lanzhou 730050, China.)

Abstract: Background: Laminitis is considered as the most important cause of hoof lameness in dairy cows, which causes abundant economic losses in husbandry. Through intense efforts in past decades, the etiology of laminitis is preliminarily considered to be subacute ruminal acidosis; however, the pathogenesis of laminitis needs further research. The differentially expressed proteins (DEP) were detected in plasma of healthy cows and clinical laminitis cows by two-dimensional gel electrophoresis (2-DE) and identified by matrix-assisted laser desorption/ionization time-of-flight mass spectrometry.

Results: Nineteen protein spots were differentially expressed, and 16 kinds of proteins were identified after peptide mass fingerprint search and bioinformatics analysis. of these, 12 proteins were differentially up-regulated and 4 down-regulated. Overall, these differential proteins were involved in carbohydrate metabolism, lipids metabolism, molecular transport, immune regulation, inflammatory response, oxidative stress and so on.

Conclusions: The DEPs were closely related to the occurrence and development of laminitis and the lipid metabolic disturbance may be a new pathway to cause laminitis in dairy cows. The results provide the theory foundation for further revealing the mechanism of laminitis and screening the early diagnostic proteins and therapeutic target.

Key words: Comparative proteomics; 2-DE; Laminitis; Plasma; Dairy cow

BACKGROUND

With China's steady and high-speed economic development, the import amount of dairy cow,

* Correspondence: yanping@caas.cn; yanzuoting@caas.cn

Received: 14 February 2015 Accepted: 8 July 2015

Publishedonline: 23 July 2015

frozen sperm, and embryo are strongly increasing for meeting the dairy industry demands. At the same time, infectious diseases have been effectively controlled, but the animal welfare and some nutrition metabolic diseases are easy to be ignored. Lameness is an increasing disease in dairy cows that are associated with higher production, more intensive feeding, and confined conditions. It has been considered as the third serious disease, following mastitis and reproductive diseases in dairy cows in China. The problem's prevalence is a significant contributor to compromised animal welfare and the economic loss in the dairy industry.

Laminitis - an inflammation of the laminar corium in the hoof wall - is a common underlying cause of lameness, accounting for 41% of the cases of lameness identified in cows[1]. It can result in many kinds of claw horn lesions including sole hemorrhages, white line disease, and sole ulcers in dairy cattle[2], which highly affect a cow's welfare, production, and longevity. Laminitis is divided into two stages depending on clinical presentation: subclinical and clinical stages. The subclinical phase usually does not have any obvious evident syndrome, so the sick cattle is always ignored and misses the best opportunity of the treatment. Thus, laminitis has been an invisible killer of dairy cows. In recent years, many scholars have studied the etiology of laminitis[3-5], laminar morphology[6], metabolism[7], pathophysiology[1], clinical diagnosis[8,9], and comprehensive treatment[10] in laminitis cows. Through intense efforts to understand the root cause of clinical laminitis in past decades, the etiology of laminitis is preliminarily considered to be subacute ruminal acidosis (SARA)[11,12], but the pathogenesis of laminitis is not clearly understood. Laminitis that was been proved refractory to conventional research approaches has been especially frustrating for clinicians because of the dearth of new treatment strategies for age-old diseases[13].

Proteins are the final executants of biological functions in life. Tiny alterations of protein expression happen in all physiological and pathological processes. Thousands of proteins are secreted in plasma by cells or tissues that contain rich information concerning overall pathophysiology of the patient or diseased animal[14]. Therefore, the analysis of the profile of plasma protein alterations is a promising way to find the potential biomarker and shed light on the pathogenesis of disease. At present, comparative proteomics have identified a large number of differentially expressed proteins associated with diseases. Although roles and mechanisms of such proteins in the pathogenesis need to be further proved, at least some of them may be potential biomarkers for early diagnosing diseases. So, the technology based on proteomics has a wide prospect in clinical diagnosis and prevention of diseases.

However, the proteomics application in animal science and veterinary is still in its infancy, and studies have mainly been directed toward production traits[15] and epidemic diseases[16,17]. Proteomics also provides a novel approach to investigate the pathogenesis of laminitis[18]. Hannah analyzed the proteomics of lamellar tissue of equine laminitis and discovered that keratins could serve as serum biomarkers for the developmental phase of endocrinopathic laminitis[19]. So far, only two proteomic studies on claw tissues in dairy cattle have been published[20,21], and very little information are available on the presence of specific proteins in plasma. Laminitis is a local manifestation of systemic metabolism disorder in cows; therefore, some corresponding changes may occur in plasma proteome[22]. To investigate this broad hypothesis, two-dimensional gel electrophoresis (2-DE) coupled with matrix-assisted laser desorption/ionization time-of-flight mass spectrometry (MALDI-TOF MS) was used to detect differentially expressed proteins in plasma, as a first step toward characterizing global alterations in dairy cows with clinical laminitis. The objective of our study was to screen the potential protein biomarkers that monitor the occurrence and the drug target of laminitis and shed light on pathogenesis with the intent of developing novel preventive strategies and therapeutic approaches. The current study provides a novel report of comparing proteomics analysis of plasma in dairy cows with laminitis.

METHODS

Ethics statement

The experiment was conducted in accordance with the good animal practices requirements of the Animal Ethics Procedures and Guidelines of the People's Republic of China. This study was approved by the Institutional Animal Care and Use Committee of Lanzhou Institute of Husbandry and Pharmaceutical Sciences of CAAS (Approval No. LIHPSACUC2011-012).

Case definition of laminitis

A total of 4680 Holstein cows were selected from May to August in 2011in the Reproduction and Breeding Demonstration Center of Chinese Holstein Dairy Cow of Gansu Province, which was the member of the Chinese National technological system of dairy industry. Laminitis was identified with at least one of the following claw horn lesions: double sole, solar ulcer, solar hemorrhage, white line disease, and solar abscess[23]. A case of cow usually has the following clinical signs: redness, heat, pain, or sensitivity to percussion, and swelling of the lamellar hoof, as well as the corresponding systemic symptoms, which are diagnosed by hoof trimmers and veterinarians. A total of 36 adult cows in lactation with acute laminitis were selected as a sick group from the dairy herds during the routine herd trimming, with at least one hoof suffering from laminitis. A total of 15 healthy dairy cows with no evident clinical signs of other diseases were included as a control group. All of the cows enrolled in the study aged 3-5 years, were around 400kg in body weight, and had not received any drug treatment for 1 month before trials. All procedures were followed up by a veterinary assistance and according to the corresponding ethical and animal welfare guidelines.

Collection and selection of blood samples

Blood samples were withdrawn from the jugular vein by cava venepuncture using a 16-gauge needle in 10mL tubes and immediately transferred into sealed vacutainer glass tubes that contained EDTA-K2 as an anticoagulant (for plasma). After collection, samples were placed on ice and transferred to the laboratory in 4h. Plasma was obtained by centrifugation at 2000g for 15min at 4℃. To emphasize proteomic differences between the groups while eliminating potential individual contributions, six cows were randomly selected from the same group and equal volumes of plasma were pooled from them. Two mixed plasma samples were available for the next proteomic experiment, which included comparison between sick and healthy groups.

Preparation of protein samples and depletion of abundant proteins

Plasma samples were analyzed using two commercial kits: Albumin and IgG Removal kit (GE

Healthcare, NJ USA) and 2-D Clean Up kit (GE Healthcare, NJ USA). The most abundant proteins of albumin and immunoglobulin (IgG) in plasma were removed by the immune affinity-based method according to the manufacturer's instructions. After fractionation, samples were desalted using the 2-D Clean Up kit.

The protein precipitate in samples were resuspended in a lysis buffer containing 8 M urea, 2% 3- [(3-cholamidopropyl) dimethylammonio] -1-propanesulfonate (CHAPS) (w/v), 18mM dithiothreitol (DTT), 1% ampholytes (v/v), and bromophenol blue (silver-stained gels) or 30 mM Tris-HCl, and sonicated to completely dissolve the aggregated protein (15s, 0.5 cycles/s, 60% amplitude). Samples were centrifuged at 16000g for 15min at 4℃ in a 5415R centrifugal machine (Eppendorf 5415R, Germany) to get suspensions, which were diluted in the lysis buffer using protease inhibitor (Complete EDTA-free Protease Inhibitor Cocktail Tablets, Roche, Spain) until the final protein concentration was 5-10 μg/μL as determined using the 2-D QUANT KIT (GE Healthcare, NJ, USA). Samples were aliquoted once the experiment was over and storedat . 80℃ until 2-DE analysis.

2-DE procedure

To discover the potential new biomarkers in plasma, 2-DE was performed to screen differentially expressed proteins between healthy and sick cows. Isoelectric focusing (IEF) was run on an Ettan IPG phor II (GE Healthcare, CA, USA) using 24cm nonlinear immobilized pH gradient strips (pH 3-10; GE Healthcare). Protein samples (150 μg) pooled with rehydration solution (8M urea, 2% CHAPS, 20mM DTT, 0.5% (v/v) immobilized pH gradient (IPG) buffer (pH 3-10), and 0.001% bromophenol blue) were placed for 12h at 4℃. The linear ramping mode of the IEF voltage was applied in the focusing program as description in reference[24]. Strips were sequentially incubated for 15min in 10mL equilibration solution with 2.5% (w/v) DTT or iodoacetamide (IAA). Second-dimension electrophoresis was performed on 12.5% sodium dodecyl sulfate gels in an Ettan DALT six apparatus (Amersham Bioscience, Uppsala, Sweden) with constant power at 5W per gel for the first 30min, and then at 12W per gel for 6-7h until the bromophenol blue line reached the bottom of the gels. Gels were treated in triplicate and silver stained according to published procedures[25]. Gels were scanned at 300 dpi resolution (UMAX USB2100XL, Taiwan, China), and the profiles were renamed as the experiment.

Image analysis and protein identification

Differential analysis was performed using Image Master 2D platinum software (Version 5.0, GE Healthcare, CA, USA) for spot detection, quantification, matching, and comparative and statistical analyses. Data were averaged from three independent gels, and the mean and standard deviations were calculated and assessed for statistical significance by normalized intensities of spots. Finally the differentially expressed proteins were defined between the sick and healthy groups with a paired t test if P values were less than 0.05, and the average spot intensity was greater than three-fold.

Protein spots of interest were excised manually from the gel, subjected to destaining and trypsin digestion according to the protocol described by Wu[24], and purified using ZipTip microliter plates (Millipore). MALDI-TOF MS analysis of tryptic peptides was performed on an Ultraflex TOF/TOF instrument (Bruker Daltonics). Proteins were identified by peptide mass fingerprint (PMF) using the Mascot search engine (http://www.matrixscience.com; Matrix Science Ltd., London, UK) and the Swiss-Prot 55.4 database.

Bioinformatics analysis

The identified proteins were searched in the Uniprot database (http://www.uniprot.org/), AgBase (http://www.agbase.msstate.edu/), and published literature for their functions[26]. According to the combined search results, these proteins were divided into different functional groups. Categorical annotation was supplied in the form of gene ontology (GO) biological process (BP), molecular function (MF), cellular component (CC), as well as participation in a Kyoto Encyclopedia of Genes and Genomes (KEGG) pathway and membership in a protein complex as defined by the comprehensive resource of mammalian protein complexes (CORUM)[27].

Validation of differentially expressed protein

To add confidence to the results obtained by 2-DE, 2 of 16 differentially expressed proteins in plasma were measured in healthy and sick groups. The concentrations of haptoglobin were detected by Bovine ELISA kits (Shanghai Institute of Biological enzyme-linked, Shanghai, China). ApoA-I in plasma was detected by immunoturbidimetric method in the automatic biochemical analyzer (Mindary 420, Shenzhen, China) using commercial test kits for human (Mindary, Shenzhen, China). The standard curve was developed with the known ApoA-I concentration (0 g/L, 0.180 g/L, 0.530 g/L, 1.26 g/L and 2.45 g/L). The procedure was performed according to the manufacturer's instructions.

Plasma samples (1 : 2000 dilutions) were added in duplicate to each well in enzyme-linked immunosorbent assay (ELISA) plate precoated with monoclonal antibody (McAb) against bovine haptoglobin, and then 10μL of biotin-labeled McAb and 50μL of streptavidin-HRP conjugates were added to the wells. After incubation at 37℃ for 1h, the ELISA plate was washed for three times using PBST [0.5% (v/v) Tween-20, PBS, pH 7.4]. Coloration was developed by 3', 5, 5'-tetramethylbenzidine solution, and the reaction was stopped with 50μL of 2M H_2SO_4. The absorbance was measured at a wavelength of 450nm. In ELISA test, bovine haptoglobin solutions with known concentrations (800mg/L, 400mg/L, 200mg/L, 100mg/L and 50mg/L) were used to prepare a standard curve according to the ELISA procedure described by the manufacture. Haptoglobin concentration in plasma was calculated according to the sample absorbance and standard curve.

Assay of antioxidant ability of plasma in dairy cows

Plasma samples were analyzed for the total antioxidative capacity (T-AOC), malonaldehyde (MDA), super oxygen dehydrogenases (SOD), and glutathione peroxidase (GSH-Px) using colorimetric assay kits (Nanjing Jiancheng Bioengineering Institute, Jiangsu, China), and detected by microplate reader (Spectra Max M2, Molecular Devices, CA, USA). All samples were tested in duplicate. The T-AOC concentration was determined by the reaction of phenanthroline and Fe^{2+} using spectrophotometer at 520nm. MDA was measured by the thiobarbituric acid method.

SOD activity was determined by inhibiting nitroblue tetrazolium reduction due to superoxide anion generation by a xanthine-xanthine oxidase system. The GSH-Px level was determined using the direct measurement of the remaining GSH after the enzyme-catalyzed reaction. All assay procedures were performed according to the manufacturer's instructions.

Statistical analysis

Statistical analysis was performed using the SPSS statistical package v 17.0. Normality of data was tested using the one-sample Kolmogorov-Smirnov test. Data were analyzed by one-way analysis of variance, and differences between group means were evaluated with the Duncan test. Differences between groups were analyzed with independent t test ($P<0.05$).

RESULTS

Comparative proteomic analysis of plasma samples

Scanning of 2-DE gel maps showed that several spots did not correspond to proteins, which may be due to contamination or gel impurities. Therefore, these spots were eliminated in comparative analysis. Plasma proteomic profiles of healthy and sick groups were analyzed by Image Master. Representative 2-DE profiles of healthy and sick groups are illustrated in Fig. 1a and b. Qualitative analysis revealed approximately 763 and 757 protein spots on each 2-DE images. While the quantitative analysis of the spots' expression showed that a total of 19 stained protein spots (Fig. 1) showed significant changes of at least threefold up- or down-regulated expression in the healthy group (Fig. 2). Among them, 15 spots (M-01 to M-15) were up-regulated while other 4 spots (M-16 to M-19) were down-regulated in response to the occurrence of laminitis.

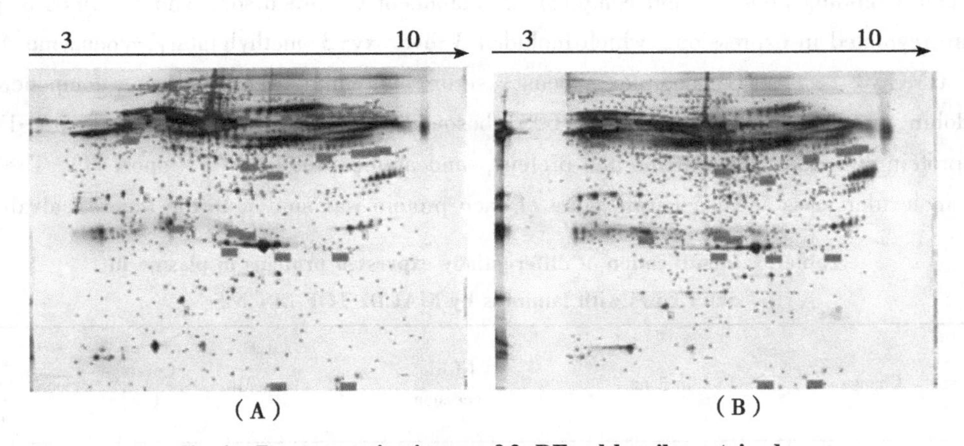

Fig. 1 Representative image of 2-DE gel by silver stained

Note: Identified spots of differentially expression proteins are indicated by green blank with spot number. (A) is from sick cows and (B) is from healthy cow

Protein identification by MS

All differentially expressed protein spots were excised from the 2-DE gels and subjected to tryp-

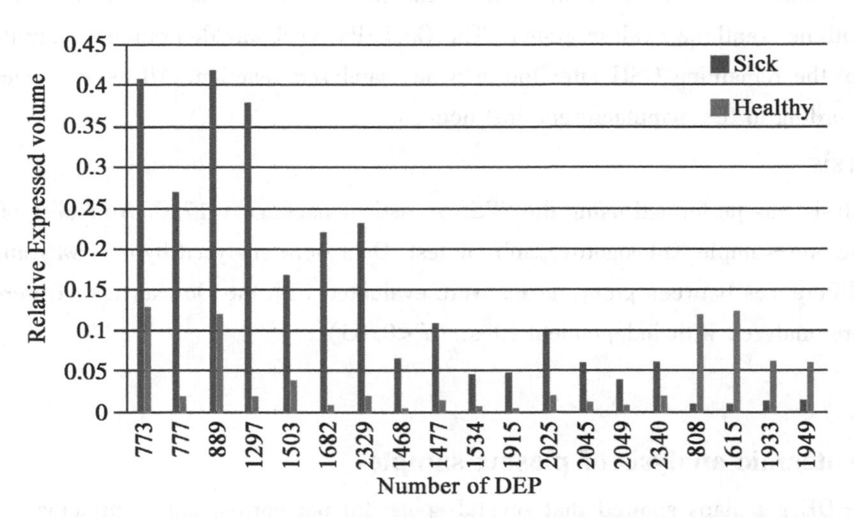

Fig. 2 The relative expressed volume of differentially expression proteins

sin digestion. The peptide mixtures were submitted to MALDI-TOF MSfor identification. Peptide mass fingerprinting was searched by Mascot in NCBInr database (Version 20101030) with *other mammalian* as taxonomy. Nineteen of them were positively identified as 16 kinds of proteins (Table 1). Spot889, spot1468, and spot1477 were identified as the same protein of serum albumin, and spot2025 and spot2329 were the same protein of apolipoprotein A-IV (ApoA-IV) precursor. Finally, 4 kinds of proteins were down-regulated in expression in laminitis cows, which included ectoderm-neural cortex protein 1, glycerol-3-phosphate dehydrogenase 1-like protein, complement component 4 binding protein, and complement component C9 precursor, and 12 kinds of proteins were up-regulated in expression, which included 3-hydroxy-3-methylglutaryl-coenzyme A reductase (HMGCR), SPEG complex locus, serum albumin, complement component C9, haptoglobin, isocitrate dehydrogenase 1, 60S ribosomal protein L5, conglutinin, ApoA-IV, zinc finger protein 300-like, transmembrane protein, and apolipoprotein A-I (apoA-I). The experimental molecular mass and isoelectric point of each protein was similar to the theoreticalvalues.

Table 1 Identification of differentially expressed proteins in plasma in dairy cows with laminitis by MALDI TOF MS/MS

NO.	spot	Change	Protein name	NCBI accession	pI	Mw	Coverage (%)	Score	NO. of peptides identified
1	773	UP	3-hydroxy-3-methylglutaryl-Coenzyme A reductase(HMGCR)	gi|157785597	6.27	99155	33	184	16
2	777	UP	SPEG complex locus	gi|215599780	8.83	357571	6	73	15
3	889	UP	Serum albumin	gi|76445989	6.09	55487	18	78	6
4	1297	UP	Complement component C9	gi|85718632	5.66	63327	20	108	9
5	1468	UP	serum albumin	gi|76445989	6.09	55487	18	73	5
6	1477	UP	serum albumin	gi|76445989	6.09	55487	27	78	8

(continued)

NO.	spot	Change	Protein name	NCBI accession	pI	Mw	Coverage (%)	Score	NO. of peptides identified
7	1503	UP	Haptoglobin, HPT	gi|94966763	7.83	45629	47	136	9
8	1534	UP	Isocitrate Dehydrogenase 1, IDH1	gi|89573995	6.12	41658	49	175	10
9	1682	UP	60S ribosomal protein L5, RPL5	gi|78369655	9.73	34551	28	63	7
10	1915	UP	Conglutinin, CONG_BOVIN	gi|395268	5.82	38370	27	73	4
11	2025	UP	Apolipoprotein A-IV precursor	gi|296480272	5.30	42963	20	92	7
12	2045	UP	Zinc finger protein 300-like, ZNF300	gi|297493133	8.66	27753	59	181	6
13	2049	UP	Transmembrane protein, TMP10	gi59858239	6.85	15875	59	199	3
14	2329	UP	Apolipoprotein A-IV precursor, apoA-IV	gi|296480272	5.30	42963	55	174	5
15	2340	UP	Apolipoprotein A-I, apoA-I, apoA-I	gi|245563	5.57	28415	67	242	7
16	808	Down	Ectoderm-neural cortex protein 1, ENC1	gi|118151210	6.40	67257	46	176	15
17	1615	Down	Glycerol-3-phosphate dehydrogenase 1-like protein, GPD1L	gi|154152135	6.13	38841	42	133	10
18	1933	Down	complement component 4 binding protein, alpha chain-like, C4BP	gi|76677514	6.34	22393	42	123 6	
19	1949	Down	complement component C9 precursor	gi|78369352	5.66	63327	18	101	7

Function enrichment and GO analysis of DEP

To gain a better understanding of the 16 differentially expressed proteins identified in this study, further analysis was performed using bioinformatics. The identified proteins were categorized according to their function based on published literature and Uniprot database. These differential proteins were mainly classified by function into the following categories: carbohydrate metabolism, lipid metabolism, molecular transporter, immune regulation, inflammatory reaction, oxidative stress, and so on. By the GO analysis of AgBase, BPs of DEPs were mainly classified as biological processs (29%), metabolic process (20%), lipid metabolic process (9%), regulation of biological process (8%), transport of biomolecules (7%), response to stress (5%), and so on (Fig. 3a). MFs of DEPs were mainly classified as catalytic activity (27%), binding (25%), nucleotide binding (10%), RNA binding (7%), protein binding (7%), lipid binding (5%), and so on (Fig. 3b). CCs of DEPs were mainly classified as cellular component (26%), cytoplasm (11%), extracellular region (11%), extracellular space (11%), intracellular region (9%), cell (9%), and so on (Fig. 3c).

Validation of 2-DE results

To validate the accuracy of 2-DE results, haptoglobin and ApoA-I in plasma were successfully measured, respectively, by ELISA and immunoturbidimetric method. It indicated that the Mindray human test kits for ApoAI was also suitable for bovine. The results showed that haptoglobin ($P<0.01$) and ApoA-I ($P<0.05$) in the cows with laminitis were significantly higher than those in

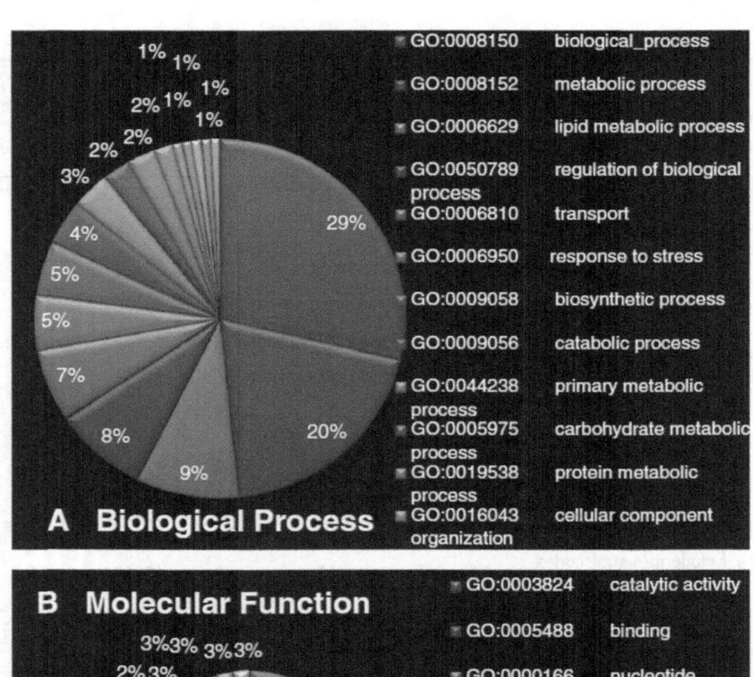

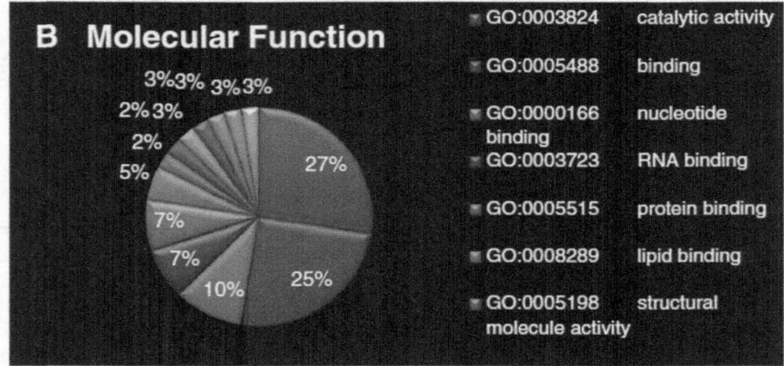

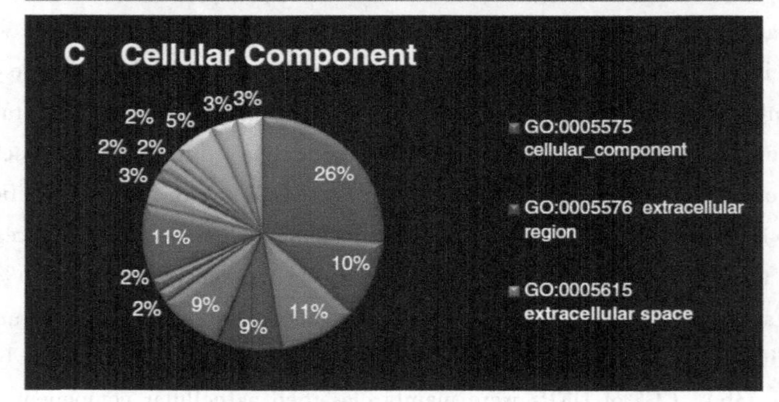

Fig. 3 The GO analysis of differently expressed protein

Note: (A) is biological process; (B) is molecular function and; (C) is cellular component

the healthy cows (Fig. 4). The differences in haptoglobin and ApoA-I concentrations were confirmed again. Although the fold changes were different from the results by 2-DE, the changing trend was consistent with2-DE, enhancing the possibility of the data of 2-DE being reliable.

Analysis of antioxidant ability of plasma in dairy cows

There were significant differences in the concentrations of T-AOC, MDA, and GSH-Px be-

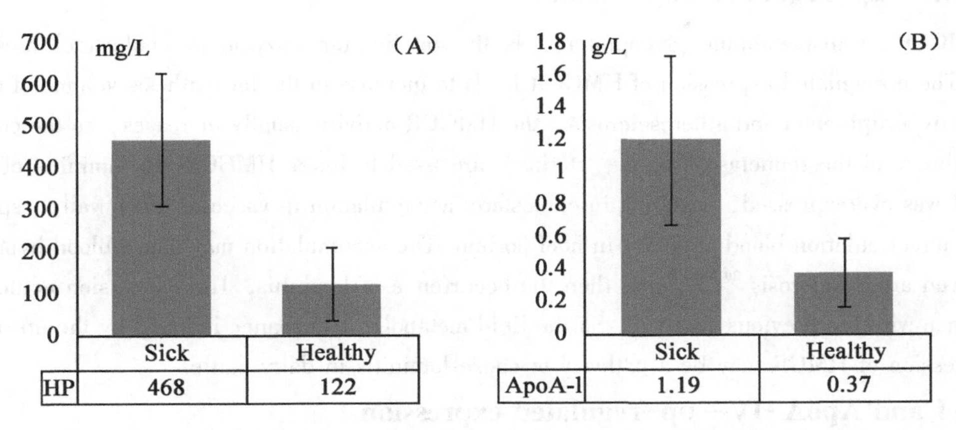

Fig. 4 Concentration of HP (A) and ApoA-I (B) of plasma by ELISA in the sick and healthy cows

Note: All data are expressed as mean±SD, $p<0.05$

tween sick and healthy groups ($P<0.01$), but the concentration of SOD was not different between them (Table 2). The results indicated that the antioxidant ability of plasma decreased and the redox equilibrium was disturbed—caused by oxidative stress—in the sick cows. It was consistent with the proteomic results.

Table 2 Antioxidant ability of plasma in dairy cows

Group	Number	T-AOC umol/L	SOD U/mL	MDA nmol/mL	GSH-Px U/mL
Sick	36	5.82±0.2a	58±7.2a	3.4±0.5a	27±3.4a
Health	15	7.2±1.45b	64±4.6a	2.2±0.1b	41±4.2b

Note: Different letters in the same column mean significant difference ($P<0.01$)

DISCUSSION

Laminitis is a major cause of lameness in dairy cattle, which leads to serious economic loss for producers. But so far, the etiology and pathogenesis of laminitis in cow are not clear. The current situation gives us so many difficulties to prevent the disease. Comparative proteomics provide a novel pathway to resolve the question, Hannah[19] analyzed the proteomics of lamellar tissue of equine laminitis, and the results showed that COMP and keratins could serve as serum biomarkers for the developmental phase of laminitis. This study applied proteomics technology to compare the difference in the expression of plasma proteome between healthy and laminitis cows. A total of 16 DE proteins were discovered of which 4 were down-regulated and 12 were up-regulated in expression. These DE proteins were involved in the pathways such as carbohydrate metabolism, lipids metabolism, molecular transport, immune regulation, inflammatory reaction, oxidative stress, and so on. The result has laid a good foundation to further understand the pathogenesis of laminitis in dairy cow.

HMGCR—up-regulated expression

HMGCR, transmembrane glycoprotein, is the ratelimiting enzyme in cholesterol biosynthesis[28]. The upregulated expression of HMGCR leads to increase in the biosynthesis volume of cholesterol. In hyperlipidemia and atherosclerosis, the HMGCR activity usually increases, so the competitive inhibitors of the reductase (e. g., statins) are used to lower HMGCR. In laminitis of cows, HMGCR was overexpressed, resulting in cholesterol accumulation in vascular inner wall, especially in the microcirculation blood capillary in hoof corium. The accumulation may cause blood hypoperfusion, even arteriosclerosis[29,30], and then the occurrence of laminitis. The conclusion could be in accordance with the previous result[1]. So the lipid metabolic disturbance induced by the up-regulated expression of HMGR may be a pathway to cause laminitis in dairy cattle.

ApoA-I and ApoA-IV—up-regulated expression

Apolipoprotein is constituted of an important component of plasma lipoproteins, which have been found to bind lipopolysaccharide (LPS) in vitro and neutralize its toxic effects[31]. *In vitro*, apoA-IV plays antiinflammatory and antiatherogenic roles after the administration of LPS[32,33]. Therefore, apolipoprotein may play a vital role in perpetuating the inflammatory response. In addition, ApoA-I participates in the reverse transport of cholesterol from tissues to the liver for excretion by promoting cholesterol efflux from tissues and by acting as a cofactor for the lecithin cholesterol acyltransferase. ApoA-IV is a major component of highdensity lipoprotein, which can promote cholesterol efflux to an equal extent from adipose cells[34]. In bovine, subacute ruminal acidosis was considered as the main cause of laminitis with the high level of LPS in blood[3,12]. Therefore, ApoA-I and ApoA-IV were both up-regulated in expression to enhance the ability of the organism to neutralize the LPS toxicity; these proteins can transport cholesterol from tissue to liver for catabolism. In a research on horse with chronic laminitis, ApoA-IV was also found up-regulated in expression to play an anti-inflammatory role[35], which was in accordance with the findings of this study.

Haptoglobin—up-regulated expression

Haptoglobin is an important acute-phase protein, which can assess the innate immune system's systemic response to infection, inflammation, or trauma, including hemoglobin-binding capacity, in maintaining the iron homeostasis in cattle[36]. Haptoglobin also functions as a bacteriostatic agent, which increases rapidly during an infectious disease. At acute clinical mastitis in dairy cows, the haptoglobin concentration increases markedly both in blood and milk[37]. Haptoglobin concentrations increase to sixfold in dairy cows with infectious and metabolic diseases at slaughter compared to animals with minor lesions[38], In cattle, haptoglobin is effective in the diagnosis and prognosis of mastitis, enteritis, peritonitis, pneumonia, endocarditis, and endometritis[39]. Elevations in this protein have also been reported in cows with fatty liver syndrome, at parturition and during periods of starvation and transport stress, but not in laminitis. In this study, haptoglobin expression was found to be up-regulated in the case of laminitis, which prevents loss of iron and plays a vital role in anti-inflammatory reaction.

Conglutinin—up-regulated expression

Conglutinin is a 371-amino acid calcium-dependent serum lectin specific for N-acetylglucosamine and plays an important role in defense mechanisms[40]. Conglutinin of cow is a member of the collectin protein family, which is an effector molecule in innate, non-adaptive immune defense against microorganism pathogens. Conglutinin binds to microbial surface wall and immune complexes through the complement component (C3bi), thereby impeding infectivity or mediating phagocytosis through specific receptors on phagocytes[41]. In the development of laminitis, conglutinin was up-regulated in expression, which indicated that the natural immune ability of cows may be improved.

CONCLUSIONS

The current study provides a novel report on plasma proteomic profiles of laminitis in dairy cows. The results highlight the differentially expressed proteins in blood plasma between healthy dairy cows and clinical laminitis cows, including 12 proteins up-regulated and 4 proteins down-regulated. Overall, these differential proteins were involved in the pathways including carbohydrate metabolism, lipid metabolism, molecular transport, immune regulation, inflammatory reaction, oxidative stress, and so on. The differentially expressed proteins were related to the occurrence and development of laminitis and the lipid metabolic disturbance may be a new pathway to cause laminitis in dairy cows. The results may provide some clues for further revealing the pathogenesis of laminitis and screening the early diagnostic proteins and therapeutic drug target.

ABBREVIATIONS

BP: Biological process; CAAS: Chinese academy of agricultural sciences;

CC: Cellular component; CHAPS: 3- [(3-Cholamidopropyl) dimethylammonio] -1-propanesulfonate; DEP: Differentially expressed proteins; DTT: Dithiothreitol; EDTA-K2: Dipotassium ethylene diamine tetra-acetate; ELISA: Enzyme-linked immunosorbent assay; GO: Gene ontology; GSH-Px: Glutathione peroxidase; IAA: Iodoacetamide; IEF: Isoelectric Focusing; IPG: Immobilized pH gradient; MALDI-TOF MS: Matrix-assisted laser desorption/ionization time-of-flight mass spectrometry;

McAb: Monoclonal antibody; MDA: Malonaldehyde; MF: Molecular function; PBST: Phosphate buffer solution and Tween-20; PMF: Peptide mass fingerprinting; SOD: Super oxygen dehydrogenases; T-AOC: Total antioxidative capacity; 2-D:

Two-dimensional gel; 2-DE: Two-dimensional gel electrophoresis.

COMPETING INTERESTS

The authors declare that they have no competing interest.

AUTHORS' CONTRIBUTIONS

ZTY and ZQY conceived and designed the study, and critically revised the manuscript. SWD, SDZ, and DSW performed the experiments and analyzed the results, and SWD drafted the manuscript. HW and XFS participated in ELISA and manuscript revision. PY helped in the 2-DE experiment. All authors read and approved the final manuscript.

ACKNOWLEDGMENTS

The authors would like to thank Doctor Fuqiang Wang and Shiyang Xu of Nanjing medical University for the MS technical support. We are grateful to Doctor Qinzhe and Doctor Ali for his helpful scientific suggestion. This work was financially supported by the National Science Research Fund of China (No. 31302156), Chinese Central Public Interest Scientific Institution Basal Research Fund (No. 1610322010004), and the 12th five-year plan of Chinese national science and technology support (No. 2012BAD12B03).

References

[1] Boosman R, Nemeth F, Gruys E. Bovine laminitis: clinical aspects, pathology and pathogenesis with reference to acute equine laminitis. Vet Q. 1991, 13 (3): 163-171.

[2] Vermunt JJ. One step closer to unravelling the pathophysiology of claw horn disruption: for the sake of the cows' welfare. Vet J. 2007, 174 (2): 219-220.

[3] Mgasa MN. Bovine pododermatitis aseptica diffusa (laminitis) aetiology, pathogenesis, treatment and control. Vet Res Commun. 1987, 11 (3): 235-241.

[4] Svensson C, Bergsten C. Laminitis in young dairy calves fed a high starch diet and with a history of bovine viral diarrhoea virus infection. Vet Rec. 1997, 140 (22): 574-577.

[5] Sanders AH, Shearer JK, De Vries A. Seasonal incidence of lameness and risk factors associated with thin soles, white line disease, ulcers, and sole punctures in dairy cattle. J Dairy Sci. 2009, 92 (7): 3165-3174.

[6] Danscher AM, Toelboell TH, Wattle O. Biomechanics and histology of bovine claw suspensory tissue in early acute laminitis. J Dairy Sci. 2010, 93 (1): 53-62.

[7] Danscher AM, Enemark HL, Andersen PH, Aalbaek B, Nielsen OL. Polysynovitis after oligofructose overload in dairy cattle. J Comp Pathol. 2010, 142 (2-3): 129-138.

[8] Al-Qudah KM, Ismail ZB. The relationship between serum biotin and oxidant/antioxidant activities in bovine lameness. Res Vet Sci. 2012, 92 (1): 138-141.

[9] Belge F, Bildik A, Belge A, Kilicalp D, Atasoy N. Possible association between chronic laminitis and some biochemical parameters in dairy cattle. Aust Vet J. 2004, 82 (9): 556-557.

[10] Bergsten C. Causes, risk factors, and prevention of laminitis and related claw lesions. Acta Vet Scand Suppl. 2003, 98: 157-166.

[11] Kleen JL, Hooijer GA, Rehage J, Noordhuizen JP. Subacute ruminal acidosis (SARA): a review. J Vet Med A Physiol Pathol Clin Med. 2003, 50 (8): 406-414.

[12] Enemark JM. The monitoring, prevention and treatment of sub-acute ruminal acidosis (SARA): a review. Vet J. 2008, 176 (1): 32-43.

[13] Harris P. Laminitis after 2000 years: adding bricks to our wall of knowledge. Vet J. 2012, 191 (3): 273-274.

[14] Talamo F, D' Ambrosio C, Arena S, Del VP, Ledda L, Zehender G, et al. Proteins from bovine tissues and biological fluids: defining a reference electrophoresis map for liver, kidney, muscle, plasma and red blood cells. Proteomics. 2003, 3 (4): 440-460.

[15] Eckersall PD, de Almeida AM, Miller I. Proteomics, a new tool for farm animal science. J Proteomics. 2012, 75 (14): 4187-4189.

[16] Turk R, Piras C, Kovacic M, Samardzija M, Ahmed H, De Canio M, et al. Proteomics of inflammatory and oxidative stress response in cows with subclinical and clinical mastitis. J Proteomics. 2012,

75 (14): 4412-4428.

[17] Piras C, Soggiu A, Bonizzi L, Greco V, Ricchi M, Arrigoni N, et al. Identification of immunoreactive proteins of Mycobacterium avium subsp. paratuberculosis. Proteomics. 2015, 15 (4): 813-823.

[18] Mankowski JL, Graham DR. Potential proteomic-based strategies for understanding laminitis: predictions and pathogenesis. J Equine Vet Sci. 2008, 28 (8): 484-487.

[19] Galantino-Homer H, Galantino-Homer H, Carter R, Linardi R, Megee S, Engiles J, et al. The big picture: using quantitative proteomics to investigate laminitis pathophysiology. J Equine Vet Sci. 2011, 31: 562-563.

[20] Galbraith H, Flannigan S, Swan L, Cash P. Proteomic evaluation of tissues at functionally important sites in the Bovine Claw. Cattle Practice. 2006, 14: 127-137.

[21] Tolboll TH, Danscher AM, Andersen PH, Codrea MC, Bendixen E. Proteomics: a new tool in bovine claw disease research. Vet J. 2012, 193 (3): 694-700.

[22] Wagner IP, Rees CA, Dunstan RW, Credille KM, Hood DM. Evaluation of systemic immunologic hyperreactivity after intradermal testing in horses with chronic laminitis. Am J Vet Res. 2003, 64 (3): 279-283.

[23] Nordlund KV, Cook NB, Oetzel GR. Investigation strategies for laminitis problem herds. J Dairy Sci. 2004, 87 (Supplement): E27-35.

[24] Wu J, Wang F, Gong Y, Li D, Sha J, Huang X, et al. Proteomic analysis of changes induced by nonylphenol in Sprague-Dawley rat Sertoli cells. Chem Res Toxicol. 2009, 22 (4): 668-675.

[25] Shevchenko A, Wilm M, Vorm O, Mann M. Mass spectrometric sequencing of proteins silver-stained polyacrylamide gels. Anal Chem. 1996, 68 (5): 850-858.

[26] Hajduch M, Casteel JE, Hurrelmeyer KE, Song Z, Agrawal GK, Thelen JJ. Proteomic analysis of seed filling in Brassica napus. Developmental characterization of metabolic isozymes using high-resolution twodimensional gel electrophoresis. Plant Physiol. 2006, 141 (1): 32-46.

[27] Ruepp A, Waegele B, Lechner M, Brauner B, Dunger-Kaltenbach I, Fobo G, et al. CORUM: the comprehensive resource of mammalian protein complexes-2009. Nucleic Acids Res. 2010, 38 (Database issue): D497-501.

[28] Elsik CG, Tellam RL, Worley KC, Gibbs RA, Muzny DM, Weinstock GM, et al. The genome sequence of taurine cattle: a window to ruminant biology and evolution. Science. 2009, 324 (5926): 522-528.

[29] Singh SS, Murray RD, Ward WR. Histopathological and morphometric studies on the hooves of dairy and beef cattle in relation to overgrown sole and laminitis. J Comp Pathol. 1992, 107 (3): 319-328.

[30] Boosman R, Nemeth F, Gruys E, Klarenbeek A. Arteriographical and pathological changes in chronic laminitis in dairy cattle. Vet Q. 1989, 11 (3): 144-155.

[31] Read TE, Harris HW, Grunfeld C, Feingold KR, Kane JP, Rapp JH. The protective effect of serum lipoproteins against bacterial lipopolysaccharide. Eur Heart J. 1993, 14 (Suppl K): 125-129.

[32] Spaulding HL, Saijo F, Turnage RH, Alexander JS, Aw TY, Kalogeris TJ. Apolipoprotein A-IV attenuates oxidant-induced apoptosis in mitotic competent, undifferentiated cells by modulating intracellular glutathione redox balance. Am J Physiol Cell Physiol. 2006, 290 (1): C95-C103.

[33] Recalde D, Ostos MA, Badell E, Garcia-Otin AL, Pidoux J, Castro G, et al. Human apolipoprotein A-IV reduces secretion of proinflammatory cytokines and atherosclerotic effects of a chronic infection mimicked by lipopolysaccharide. Arterioscler Thromb Vasc Biol. 2004, 24 (4): 756-761.

[34] Duverger N, Murry-Brelier A, Latta M, Reboul S, Castro G, Mayaux JF, et al. Functional char-

acterization of human recombinant apolipoprotein AIV produced in Escherichia coli. Eur J Biochem. 1991, 201 (2): 373-383.

[35] Steelman SM, Chowdhary BP. Plasma proteomics shows an elevation of the anti-inflammatory protein APOA-IV in chronic equine laminitis. BMC Vet Res. 2012, 8 (1): 179.

[36] Bowman BH. Haptoglobin. Hepatic plasma proteins. San Diego: Academic; 1993.

[37] Cooray R, Waller KP, Venge P. Haptoglobin comprises about 10% of granule protein extracted from bovine granulocytes isolated from healthy cattle. Vet Immunol Immunopathol. 2007, 119 (3-4): 310-315.

[38] Hirvonen J, Hietakorpi S, Saloniemi H. Acute phase response in emergency slaughtered dairy cows. Meat Sci. 1997, 46 (3): 249-257.

[39] Eckersall PD, Bell R. Acute phase proteins: biomarkers of infection and inflammation in veterinary medicine. Vet J. 2010, 185 (1): 23-27.

[40] Zimin AV, Delcher AL, Florea L, Kelley DR, Schatz MC, Puiu D, et al. A whole-genome assembly of the domestic cow, Bos taurus. Genome Biol. 2009, 10 (4): R42.

[41] Laursen SB, Thiel S, Teisner B, Holmskov U, Wang Y, Sim RB, et al. Bovine conglutinin binds to an oligosaccharide determinant presented by iC3b, but not by C3, C3b or C3c. Immunology. 1994, 81 (4): 648-654.

Regulation Effect of Aspirin Eugenol Ester on Blood Lipids in Wistar Rats with Hyperlipidemia

KARAM Isam, MA Ning, LIU Xi-wang , LI Shi-hong, KONG Xiao-jun, LI Jian-yong, YANG Ya-jun

Abstract: [Background]: Aspirin eugenol ester (AEE) is a promising drug candidate for treatment of inflammation, pain and fever and prevention of cardiovascular diseases with less side effects. The experiment will be conducted to investigate the efficacy of AEE on curing hyperlipidemia in Wistar rats. The rats were fed with high fat diet (HFD) for 8 weeks to induce hyperlipidemia. [Results]: Compared with the model group, the results showed that AEE at 54mg/kg dosage could significantly decrease the hyperlipidemia indexes including triglyceride (TG), low density lipoprotein (LDL) and total cholesterol (TCH) ($p<0.01$), increase high density lipoprotein (HDL) ($p<0.05$) for five weeks drug administration. Meanwhile, simvastatin had same effect on hyperlipidemia indexes such as TG, LDL, TC, but no significant increase in HDL. [Conclusion]: AEE was effective against hyperlipidemia and had better anti-hyperlipidemic effect than its component, acetylsalicylic acid (Aspirin, ASA), eugenol and integration of ASA and eugenol. Under the experimental circumstance, the optimal dose of AEE to cure hyperlipidemia is 54mg/kg for five weeks in Wistar rats.

Key words: Aspirin eugenol ester (AEE); Hyperlipidemia; High fat diet; Rats

BACKGROUND

Hyperlipidemia is a heterogeneous group of disorders characterized by an excess of lipids in the bloodstream. The concentrations of lipids, such as triglycerides (TG), cholesterol (TCH) and low density lipoprotein (LDL) increase, or the level of high density lipoprotein (HDL) decrease in the blood[1].Hyperlipidemia is becoming a major health problem in the world recently even in human and companion animal clinic[2,3].

Acetylsalicylic acid (Aspirin, ASA) has been extended to prevention and treatment of cardiovascular diseases[4-6].Low-dose ASA (5mg/kg) can ameliorate hyperlipidemia induced by high-fat diet (HFD) and hyperinsulinemia in Sprague-Dawley rats[7]. However, the side effects of ASA, such as serious gastrointestinal damage, have limited the long time using of this classic drug[8].Eugenol is the main component of volatile oil extracted from dry alabastrum of *Eugenia Caryophyllata Thumb*. Eugenol has many pharmacological activities, including antibacterial, antiinflammation, analgesia, analgesia, as well as antioxidation[9], antiplatelet aggregation [10], and hyperlipidemia amelioration[11,12].However, the disadvantages of irritation and vulnerablity to oxidation have limited the application of eugenol in practice[13,14].

To overcome the disadvantages of ASA and eugenol, aspirin eugenol ester (AEE) as a novel medicinal compound was designed and synthesized according to the prodrug principles of structure recombination[15].It is a pale yellow and smellless crystal. The stomach, duodenum and ileum showed an increased mucosa height after high-dose AEE exposure ($2000mg \cdot kg^{-1}$) in a 15-day oral dose toxicity study[16], this was mainly due to increased height of the villus. The acute toxicity of AEE is less than that of ASA and eugenol, only 0.02 times the toxicity of ASA and 0.27 times the toxicity of eugenol in mice[17]. AEE could be metabolized into ASA and eugenol *in vitro and in vivo*[18], and had the effects of anti-inflammation, antipyretic, analgesia, and antioxidant[15,19].In addition, the previous research indicated that AEE ($50mg \cdot kg^{-1}$ and $160mg \cdot kg^{-1}$) could reduce the serum TC and TG in rats with standard diet[16].

However, AEE as a promising drug candidate for preventing and treating hyperlipidemia, it is important to characterize its efficacy in rats with HFD. The present study was performed to assess the efficacy of AEE against hyperlipidemia. Meanwhile, this study will provide guidance for the design of further studies and clinical trials of AEE.

RESULTS

Animal disease model

In regard to blood lipid levels, there was no significant difference between Group I and Group II in week 4 and 6 (data not shown). So the administration time of HFD was prolonged. At week 8, there were significant increase of TG, LDL and TC and decrease of HDL of model group ($P<0.01$, seen in Table 1), which confirmed that the hyperlipidemia animal model was successfully established by HFD in the experiment.

Table 1 Blood lipids level in Group I and II at the end of 8th week after using HFD diet

Variables	Unit	TG	TCH	HDL	LDL
Group I	mmol/L	0.77±0.42	1.01±0.25	0.55±0.07	0.12±0.08
Group II	mmol/L	1.51±0.38**	2.18±0.31**	0.44±0.06**	0.52±0.08**

Note: TG: Triglyceride; HDL: High density lipoprotein; LDL: Low density lipoprotein; TCH: Total cholesterol. ** $P<0.01$ significant difference from blank group.

Body weight

From the results of body weight before drug administration, blank group had a lower body weight than other groups ($P<0.01$, seen in Table 2), and this indicated that HFD had a remakable influence on body weight. There was no significant difference among other groups except blank group. After drug administration, the average values of body weight in drug treatment group were less than model group and there was still significant difference between blank and model group ($P<0.01$). The average body weight values in CMC-Na group were similar with model group.

Table 2 Body weights changes of rats before and after drug treatment

Groups	Body weight (g)	
	Before drug treatment	Fifth week after drug treatment
A	276.1 ± 6.4^{aa}	330.1 ± 5.6^{bb}
B	295.4 ± 7.1	358.8 ± 7.2
C	298.4 ± 5.6	347.3 ± 6.8^{b}
D	303.6 ± 7.3	340.8 ± 5.9^{bb}
E	299.5 ± 6.9	340.8 ± 8.1^{bb}
F	301.1 ± 5.9	339.3 ± 6.4^{bb}
G	305.6 ± 6.9	343.4 ± 5.1^{bb}
H	292.6 ± 5.3	321.1 ± 5.9^{bb}
I	306.8 ± 5.9	342.4 ± 7.1^{bb}
J	302.7 ± 6.7	352.3 ± 6.6

Note: A: blank group, B: model group, C: AEE low dose, D: AEE medium dose, E: AEE high dose, F: integration group (acetylsalicylic acid: eugenol, molar ratio 1 : 1, 0.11 mmol), G: acetylsalicylic acid group, H: eugenol group, I: simvastatin group, J: CMC-Na group. $^{aa}P<0.01$ significant difference from model group before administration. $^{b}P<0.05$, $^{bb}P<0.01$ significant difference from model group after administration. The time before drug treatment was the end of 8th week after HFD diet was used and fifth week after drug treatment was the end of 13th week after HFD diet was used.

Anti-hyperlipidemic effect

Blood samples were taken for lipids examination after the drugs were given for two weeks, but there was no significant difference among groups (data not shown). So the time was extended for five weeks. From table 3, the difference effects of drugs on hyperlipidemia appeared after AEE was given for five weeks. Meanwhile, there was no statistical difference between model and CMC-Na group, which indicated that CMC-Na has no effect on hyperlipidemia indexes as a vehicle control.

In the blood lipid analysis (seen in Table 3), following five weeks administration of drugs, TG, TC and LDL were significantly decreased in varying degrees in comparison with model group. These changes meant that there were significant differences between the treated groups and the model group at the end of 13th weeks. In regard to TG index, the results in ASA and simvastatin groups were significantly reduced when compared with model group ($p<0.01$). With the increase of AEE dose, the mean values of TG were decreased which showed that TG values were dosedependent in AEE groups. Meanwhile, TG values in integration group also showed significant difference from model group ($p<0.05$). AEE at three different doses, simvastatin, ASA, eugenol and integration of ASA and eugenol reduced significantly levels of LDL and TC ($p<0.01$) when compared with model group. HDL in model group showed no significant difference from blank group at week 13. Only AEE high dose group significantly increased HDL ($p<0.05$) when compared with model group.

Table 3 The blood lipids levels at the end of 13th week (after drugs administration for five weeks, $n=10$)

Variables	Blank	Model	CMC-Na	Simvastatin	AEE 18	AEE 36	AEE 54	Integration	Aspirin	Eugenol
TG	0.77±0.2 **	1.65±0.22	1.63±0.34	1.12±0.15 **	1.43±0.11	1.36±0.16 *	1.23±0.15 **	1.35±0.17 *	1.27±0.15 **	1.39±0.22
HDL	0.46±0.04	0.52±0.05	0.58±0.02	0.56±0.05	0.55±0.04	0.56±0.03	0.62±0.05 *	0.61±0.02	0.61±0.05	0.56±0.01
LDL	0.11±0.03 **	0.54±0.04	0.51±0.05	0.23±0.04 **	0.31±0.04 **	0.3±0.03 **	0.25±0.04 **	0.31±0.02 **	0.30±0.04 **	0.28±0.05 **
TCH	1.01±0.18 **	2.49±0.14	2.11±0.38	1.61±0.23 **	1.82±0.16 **	1.73±0.21 **	1.55±0.27 **	1.77±0.11 **	1.80±0.10 **	1.68±0.13 **

Note: TG: Triglyceride; HDL: High density lipoprotein; LDL: Low density lipoprotein; TCH: Total cholesterol; Integration: acetylsalicylic acid: eugenol (molar ratio 1 : 1, 0.11 mmol). The unit of TG, HDL, LDL and TC is mmol/L. * $P<0.05$ significant difference from model group. ** $P<0.01$ significant difference from model group.

Optimal dosage of AEE

The results demonstrated that AEE had significant effect on ameliorative hyperlipidemia indexes. Through the evaluation of AEE groups, high dose AEE could decrease more significantly the levels of TG, TC, LDL ($p<0.01$), and at the same time increase HDL level ($p<0.05$). The effects of AEE at high, medium and low dose on LDL and TC were similar (Fig. 1). However, AEE at high dose had more significant effect on TG ($p<0.01$) and HDL ($p<0.05$).

Multiple comparisons

Fig. 2 showed the different effects of drugs on blood lipid levels. When compared with ASA and eugenol group, there was no significant difference between AEE and other groups, except the HDL index in AEE high dose group (Fig. 2). However, when compared with integration group, there was significant difference of LDL value in AEE high dose group ($p<0.05$) and HDL value in AEE medium dose group ($p<0.05$).

Compared with model group, AEE and simvastatin significantly reduced TG, LDL and TC ($p<0.01$). Meanwhile, AEE increased significantly HDL ($p<0.05$), but simvastatin had no significant increase of HDL. The results showed that simvastatin had stronger effect than low and medium dose of AEE on TG and LDL value. From the results, there was no statistical difference between simvastatin and AEE high-dose groups, which suggested that the anti-hyperlipidemic effects of high-dose AEE and simvastatin were similar.

DISCUSSION

Hyperlipidemia, a group of metabolic disorders characterized by the elevated levels of lipids, is a major modifiable risk factor for atherosclerosis and cardiovascular disease[20]. These lipids include cholesterol, cholesterol esters, phospholipids, and triglycerides. Increased levels of LDL are related to the development of atherosclerosis[21,22]. HDL plays an important role in removing cholesterol from tissues and protecting against cardiovascular disease. Hyperlipidemia can be the result of an inherited disease in certain breeds of dogs[23]. In pets, hyperlipidemia most often occurs as a consequence of some disorder, hyperlipidemia even can also occur spontaneously after a meal of high-fat foods, particularly table scraps[24,25]. Hyperlipidemia is seen most commonly in ponies, miniature horses, and donkeys, and less frequently in standardsize adult horses[26,27]. In non-rumi-

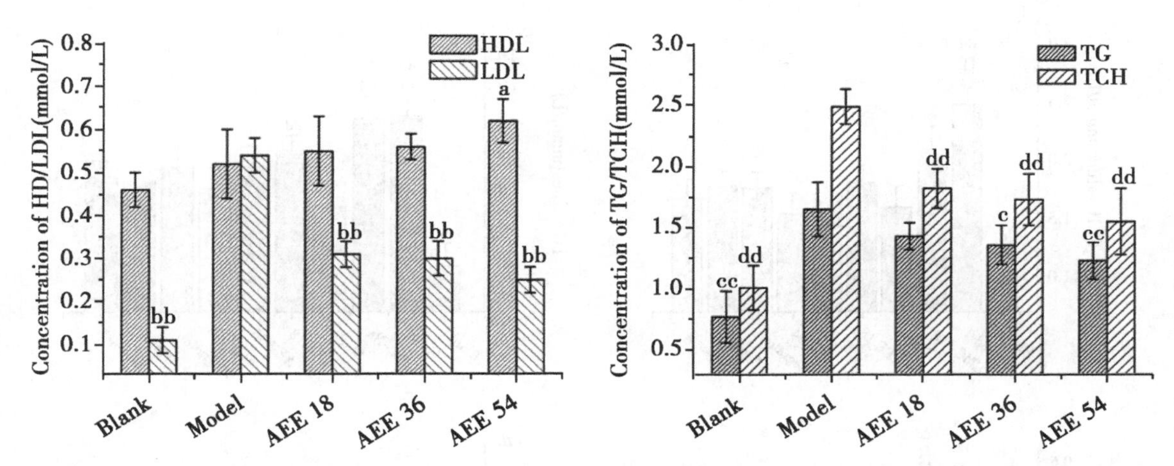

Fig. 1 The influence of AEE different dosage on hyperlipemia indexes after administration for five weeks (*n*=10)

Note: TG: Triglyceride; HDL: High density lipoprotein; LDL: Low density lipoprotein; TCH: Total cholesterol. The blank group and AEE groups were used to compare with the model group. $^{a}P<0.05$, $^{aa}P<0.01$ significant difference of HDL; $^{b}P<0.05$, $^{bb}P<0.01$ significant difference of LDL; $^{c}P<0.05$, $^{cc}P<0.01$ significant difference of TG. $^{d}P<0.05$, $^{dd}P<0.01$ significant difference of TCH. There was no significant difference among three AEE groups in influence on hyperlipidemia indexes

nants, including primates and man, hyperlipidemia may be increased by dietary manipulations such as feeding excessive cholesterol or fats with high saturated fatty acid content[28]. A great number of animal models, such pigeons, chickens, swine, cats, dogs, monkeys, mice, rabbits and rats, have been tested[29]. Commonly used rat strains (i. e. Sprague-Dawley, Wistar) typically have high levels of HDL-C and low levels of LDL-C. HFD is capable of promoting elevations in TC and LDL in Sprague-Dawley or Wistar rats likely by reducing bile acid production[30-32].

There are many chemical drugs that could ameliorate hyperlipidemia such as: statins, fibrates, ezetimibe and nicotinic acid, but most of them are expensive and have undesirable effect[33]. So there are increasing interest in alternative drug for the prevention and treatment of hyperlipidemia. Currently available hyperlipidemic drugs have been associated with a number of side effects. Therefore, now it's important to develop a novel drug that is less toxic, less expensive, which can provide better safety and efficacy on a long term usage.

In this study, the anti-hyperlipidaemic effects of AEE were investigated in rats with induced hyperlipidemia by HFD. The blood lipid indexes were observed during eight weeks. When compared the lipid level between blank and model group at week 8, there was a significant elevation in the levels of serum LDL, TG and TC, and decrease of HDL. Changed levels of these parameters in serum are presumptive markers of hyperlipidemia in serum. LDL is more related to hyperlipidemia and is an important index in these parameters. On other hand, HDL in this experiment decreaseed at week 8 in model group. However, HDL value in model group showed no significant difference from blank group at week 13. This apparently unexpected result could be explained by the light increase of HDL in model group and decrease in blank group, some studies found that the diet content can

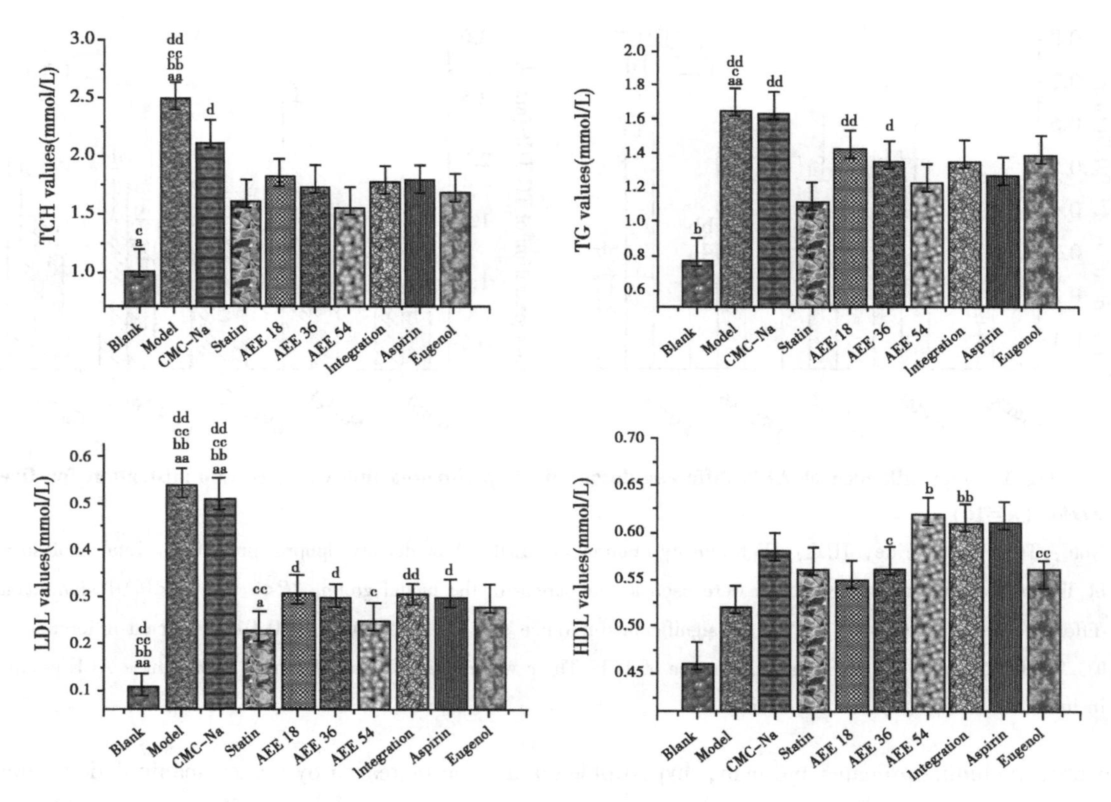

Fig. 2 Effects of different drugs on hyperlipemia indexes after drugs administration for five weeks ($n=10$)

Note: TG: Triglyceride; HDL: High density lipoprotein; LDL: Low density lipoprotein; TCH: Total cholesterol. Integration: aspirin and eugenol at the mole ratio 1 : 1. $^{a}P<0.05$ significant difference from aspirin group. $^{aa}P<0.01$ significant difference from aspirin group. $^{b}P<0.05$ significant difference from eugenol group. $^{bb}P<0.01$ significant difference from eugenol group. $^{c}P<0.05$ significant difference from integration group. $^{cc}P<0.01$ significant difference from integration group. $^{d}P<0.05$ significant difference from simvastatin group. $^{dd}P<0.01$ significant difference from simvastatin group

effect HDL level[34], light decrease of HDL in blank group may due to the feeding long period rich of carbohydrate in standard diet. Some studies found that the increased plasma apoE of apolipoproteins in aged hyperlipidemia rats can increase HDL after a long period[34-38].

From the result in this study, AEE had influence on blood lipid indexes. The changes in TC, TG, LDL and HDL confirmed that AEE had influence on hyperlipidemia. Previous study on AEE showed that it had effect on TG and TC[17]. Compared AEE high dose with simvastatin and CMC-Na, it had positive effect on antihyperlipidemia. AEE high dose significantly decreased TG, LDL, TC and increased HDL index, but simvastatin has no significant increase in HDL, this confirmed that AEE more benefit on curing hyperlipidemia. Based on the results (Table 3), ASA reduced TG significantly meanwhile eugenol had no significant difference on TG, which may speculate that ASA may play a key role on TG reduction from AEE. This may suggest that the effects of simvastatin, AEE and its component on LDL, TC were similar which decreased ($P<0.01$) compared with model group. On the other hand, CMC-Na as a vehicle had no effect on hyperlipidemia, which confirmed that the influence on hyperlipidemia is due to AEE only.

The previous study showed that eugenol had effect on hyperlipidemia since it is probably mediated through inhibition of hepatic cholesterol biosynthesis, reduction of lipid absorption, enhanced catabolism of LDL-cholesterol, and catabolism of TG[11,12]. In our study, the results showed that eugenol could reduce significantly the TG and TCH index (Table 3 $P<0.01$). However, there was no statistical difference between eugenol and model group in the TG index. Noteworthy, ASA could reduce TG, LDL and TCH index (Table 3 $P<0.05$), which showed that ASA possessed positive drug effects on hyperlipidemia. The study on the metabolism of AEE showed that AEE was decomposed into ASA and eugenol. Salicylic acid (SA) from AEE metabolism as a final metabolite excreted in urine was coincided with SA from ASA metabolism. However, the elimination time of SA was longer than ASA, which might explain the efficacy of AEE lasting longer than those of ASA and eugenol. Therefore, the AEE effect on hyperlipidemia can come mainly from eugenol and ASA, and the effect of AEE on hyperlipidemia may result from synergetic action of ASA and eugenol together, which could display stronger drug effects than ASA and eugenol.

CONCLUSIONS

In summary, there were significant differences in blood lipid indexes throughout the experimental period under the present study conditions. For Wistar rats, the optimal dose for curing hyperlipidemia was considered to be 54mg/kg/day administrating for five weeks, further studies should be conducted to investigate its prevention effect and the action mechanism of AEE on antihyperlipidemia.

MATERIALS AND METHODS

Chemicals and reagents

ASA eugenol ester (AEE), transparent crystal (purity: 99.5% with RE-HPLC), was prepared in Key Lab of New Animal Drug Project of Gansu Province, Key Lab of Veterinary Drug Development of Agricultural Ministry, Lanzhou Institute of Husbandry and Pharmaceutical Sciences of CAAS. CMC-Na (carboxyl methyl cellulose sodium) and simvastatin was supplied by Tianjin Chemical Reagent Company (Tianjin, China). ASA and Tween-80 were supplied by Aladdin Industrial Corporation (Shanghai China). Eugenol was supplied by Sinopharm Chemical Reagent Co., Ltd. (Shanghai China). High diet feed (standard rat diet 77.8%, yolk power 10%, lard 10%, cholesterol 2%, bile salts 0.2%) consists of 41.5% lipids, 40.2% carbohydrates, and 18.3% proteins (kcal) and standard rat diet consists of 12.3% lipids, 63.3% carbohydrates, and 24.4% proteins (kcal). The feed was supplied from KeaoXieli Co., Ltd (Beijing, China). The TG, TC, LDL and HDL kits were provided by Ningbo Medical System Biotechnology Co., Ltd (Ningbo, China). Erba XL-640 analyzer (German) was used to measure the blood lipid level.

Animals

Wistar male rats with clean grade, aged 7 weeks and weighing 160-180g were purchased from the animal breeding facilities of Lanzhou Army General Hospital (Lanzhou, China). They were housed in plastic cages of appropriate size (50×35×20cm, ten rats per cage) with stainless steel wire cover and chopped bedding. Light/dark regimen was 12/12 h and living temperature was 22±2℃ with relative humidity of 55%±10%. Rat feed and drinking water were supplied *ad libitum.* The study was performed in compliance with the Guidelines for the care and use of laboratory animals as described in the US National Institutes of Health and approved by Institutional Animal Care and Use Committee of Lanzhou Institute of Husbandry and Pharmaceutical Science of CAAS. Animals were allowed a 2-week quarantine and acclimation period prior to start of the study.

Drug preparation

AEE, simvastatin and ASA suspension liquids were prepared in 0.5% of CMC-Na. Eugenol and Tween-80 at the mass ratio of 1 : 2 mixed with the distilled water.

Group and dosing

After their arrival for two weeks, rats were randomized into two main groups, group I as blank control (n = 10 rats) feed with basal diet and group II (n = 90 rats) feed with HFD for eight weeks. After hyperlipidemia were induced successfully in rats, the HFD group was averagely divided into nine groups and they were model group and eight treatment groups. So it included eight test groups and two control groups as blank and model groups (seen in Table 4).

Table 4 The experimental design in the study

Groups	Number of rats	Food	Drug	Dosage (mg/kg)	Average volume per rat (mL)
A	10	Basal diet	—	—	—
B	10	HFD diet	—	—	—
C	10	HFD diet	AEE	18	0.58
D	10	HFD diet	AEE	36	1.17
E	10	HFD diet	AEE	54	1.72
F	10	HFD diet	ASA+Eugenol	(20+18)	1.19
G	10	HFD diet	ASA	20	0.65
H	10	HFD diet	Eugenol	18	0.54
I	10	HFD diet	Simvastatin	10	0.63
J	10	HFD diet	CMC-Na	20	1.28

Note: HFD: high fat diet. A: blank group, B: model group, C: EE low dose, D: AEE medium dose, E: AEE high dose, F: integration group (ASA: eugenol, molar ratio 1 : 1), G: acetylsalicylic acid (ASA) group, H: eugenol group, I: simvastatin group, J: CMC-Na group. At the end of 4th, 6th, 8th, 10th and 13th week after HFD was used, the rats were fasted for 10-12h and anesthetized with 10% chloral hydrate. Approximately 1.5ml blood samples were taken from tail tip for blood lipids examination.

The dosage for rats was based on individual weekly body weights for five weeks. Drugs were administered intragastrically to each rat. HFD continued during the experiment period.

In order to compare AEE and its precursor, integration of aspirin and eugenol group was designed in the experiment. For the comparability of the results in the experiment, the mole of medium-dose of AEE, aspirin, and eugenol is the same as 0.11mmol. The integration group administrated both aspirin and eugenol, which were also designed at 0.11mmol. The 0.5% of CMC-Na at the dose of 4 ml · kg^{-1} was as the drug vehicle control and the dosage of CMC-Na was close to equal in comparison with AEE and ASA groups. Simvastatin (10mg · kg^{-1}) was choosed as positive control drug to compare with AEE.

Blood sampling

After fasting for 10-12h, rats were anesthetized with 10% chloral hydrate and blood samples were taken from tail tip for blood lipids examination. In order to make sure the volume of the blood sample, the rat tails were immersed in water bath at 45℃ for 5 min to make the blood vessel swelling, then approximately 1.5ml blood sample was collected from the rat tail. The serums were got through centrifuge for 15 min at the speed of 4000 g at 4℃.

Design of the experiment

First hyperlipidemia model disease is needed to be established successfully by HFD. At the beginning of the experiment, the rats were divided into two main groups. Group I feed with standard diet and Group II feed with HFD. The blood samples were taken on 4th, 6th and 8th week to examine blood lipids. After the success of hyperlipidemia animal model, Group II was divided into nine groups, model group and eight treatment groups, and high fat diet was continued during the rest experiment period.

After the administration of drug, blood samples were taken to measure the change in blood lipids on 10th week and 13th week. Blood sample was analyzed with the same method.

STATISTICS

The statistical analyses were carried out using IBM SPSS 19.0 (USA). All data obtained from the experiment are expressed as mean±standard deviation (SD). The difference between Group I and Group II were evaluated by Student's t-Test. Statistical differences between the treatments and the control were evaluated by using one-way ANOVA with Duncan test. P-values less than 0.05 were considered statistically significant.

COMPETING INTERESTS

The authors have declared that no competing interests exist.

AUTHOR CONTRIBUTIONS

Conceived and designed the experiments: IK NM JYL YJY. Performed the experiments: IK NM XWL YJY. Analyzed the data: IK NM JYL XWL. Contributed reagents/materials/analysis tools: NM IK XWL GRL YJY. Wrote the manuscript: IK NM JYL YJY XWL. All authors read and approved the final manuscript.

ACKNOWLEDGEMENT

The work was supported by the National Natural Science Foundation of China (No. 31402254), and special project of Fundamental Scientific Research Professional Fund for Central Public Welfare Scientific Research Institutes (1610322015013).

References

[1] Graham JM, Higgins JA, Taylor T, Gillott T, Wilkinson J, Ford TC, et al. A novel method for the rapid separation of human plasma lipoproteins using self-generating gradients of Iodixanol. Biochem Soc Trans. 1996, 24 (2): 170S.

[2] Xenoulis PG, Steiner JM.Lipid metabolism and hyperlipidemia in dogs.Vet J.2010, 183 (1): 12-21.

[3] Mao J, Xia Z, Chen J, Yu J. Prevalence and risk factors for canine obesity surveyed in veterinary practices in Beijing. China Prev Vet Med. 2013, 112 (3-4): 438-442.

[4] Vane J, Botting R.Inflammation and the mechanism of action of anti-inflammatory drugs.Faseb J.1987, 1 (2): 89-96.

[5] Patrono C.Aspirin and human platelets: from clinical trials to acetylation of cyclooxygenase and back. Trends Pharmacol Sci.1989, 10 (11): 453-458.

[6] Wallenburg HC, Dekker GA, Makovitz JW, Rotmans P. Low-dose aspirin prevents pregnancy-induced hypertension and pre-eclampsia in angiotensin-sensitive primigravidae. Lancet. 1986, 1 (8471): 1-3.

[7] Lin HL, Yen HW, Hsieh SL, An LM, Shen KP. Low-dose aspirin ameliorated hyperlipidemia, adhesion molecule, and chemokine production induced by high-fat diet in Sprague-Dawley rats. Drug Dev Res. 2014, 75 (2): 97-106.

[8] Smith WL, DeWitt DL, Garavito RM. Cyclooxygenases: structural, cellular, and molecular biology. Annu Rev Biochem. 2000, 69: 145-182.

[9] Atsumi T, Fujisawa S, Tonosaki K. A comparative study of the antioxidant/prooxidant activities of eugenol and isoeugenol with various concentrations and oxidation conditions. Toxicol In Vitro. 2005, 19 (8): 1025-1033.

[10] Saeed SA, Simjee RU, Shamim G, Gilani AH. Eugenol: a dual inhibitor of platelet-activating factor and arachidonic acid metabolism. Phytomedicine. 1995, 2 (1): 23-28.

[11] Mnafgui K, Kaanich F, Derbali A, Hamden K, Derbali F, Slama S, et al. Inhibition of key enzymes related to diabetes and hypertension by Eugenol *in vitro* and in alloxan-induced diabetic rats. Arch Physiol Biochem. 2013, 119 (5): 225-233.

[12] Venkadeswaran K, Muralidharan AR, Annadurai T, Ruban VV, Sundararajan M, Anandhi R, et al. Antihypercholesterolemic and Antioxidative Potential of an Extract of the Plant, Piper betle, and Its Active Constituent, Eugenol, in Triton WR-1339-Induced Hypercholesterolemia in Experimental Rats. Evid Based Complement Alternat Med. 2014, 2014: 478973.

[13] Badger DA, Smith RL, Bao J, Kuester RK, Sipes IG. Disposition and metabolism of isoeugenol in the male Fischer 344 rat. Food Chem Toxicol. 2002, 40 (12): 1757-1765.

[14] Thompson DC, Constantin-Teodosiu D, Moldeus P. Metabolism and cytotoxicity of eugenol in isolated rat hepatocytes. Chem Biol Interact. 1991, 77 (2): 137-147.

[15] Li J, Yu Y, Wang Q, Zhang J, Yang Y, Li B, et al. Synthesis of aspirin eugenol ester and its bio-

logical activity. Med Chem Res. 2012, 21 (7): 995–999.

[16] Li J, Yu Y, Yang Y, Liu X, Zhang J, Li B, et al. A 15–day oral dose toxicity study of aspirin eugenol ester in Wistar rats. Food Chem Toxicol. 2012, 50 (6): 1980–1985.

[17] Shen Y, Liu X, Yang Y, Li J, Ma N, Li B. *In vivo* and *in vitro* metabolism of aspirin eugenol ester in dog by liquid chromatography tandem mass spectrometry. Biomed Chromatogr. 2015, 29 (1): 129–137.

[18] Jianyong Li, Yuanguang Yu, Yajun Yang, Xiwang Liu, Jiyu Zhang, Bing Li, et al. Antioxidant activity of eugenol ester for aging model of mice by D–galactose. J Anim Vet Adv. 2012, 11 (23): 4401–4405.

[19] Bauer JE. Evaluation and dietary considerations in idiopathic hyperlipidemia in dogs. J Am Vet Med Assoc. 1995, 206 (11): 1684–1688.

[20] Smith SJ, Jackson R, Pearson TA, Fuster V, Yusuf S, Faergeman O, et al. Principles for national and regional guidelines on cardiovascular disease prevention: a scientific statement from the World Heart and Stroke Forum. Circulation. 2004, 109 (25): 3112–321.

[21] Jacobson MS. Heart healthy diets for all children: no longer controversial. J Pediatr. 1998, 133 (1): 1–2.

[22] Xenoulis PG, Steiner JM. Lipid metabolism and hyperlipidemia in dogs. Vet J. 2010, 183 (1): 12–21.

[23] Whitney MS. Evaluation of hyperlipidemias in dogs and cats. Semin Vet Med Surg (Small Anim). 1992, 7 (4): 292–300.

[24] Bauer JE. Lipoprotein–mediated transport of dietary and synthesized lipids and lipid abnormalities of dogs and cats. J Am Vet Med Assoc. 2004, 224 (5): 668–675.

[25] Durham AE. Clinical application of parenteral nutrition in the treatment of five ponies and one donkey with hyperlipaemia. Vet Rec. 2006, 158 (5): 159–164.

[26] Duhlmeier R, Guck T, Deegen E, Busche R, Sallmann HP. [Effects of excess caloric fat feeding on the lipid metabolism in Shetland ponies].Dtsch Tierarztl Wochenschr. 2003, 110 (4): 170–174.

[27] Nestel PJ, Poyser A, Hood RL, Mills SC, Willis MR, Cook LJ, et al. The effect of dietary fat supplements on cholesterol metabolism in ruminants. J Lipid Res. 1978, 19 (7): 899–909.

[28] Moghadasian MH, Frohlich JJ, McManus BM. Advances in experimental dyslipidemia and atherosclerosis. Lab Invest. 2001, 81 (9): 1173–1183.

[29] Jeong WI, Jeong DH, Do SH, Kim YK, Park HY, Kwon OD, et al. Mild hepatic fibrosis in cholesterol and sodium cholate diet–fed rats. J Vet Med Sci. 2005, 67 (3): 235–242.

[30] Horton JD, Cuthbert JA, Spady DK. Regulation of hepatic 7 alpha – hydroxylase expression and response to dietary cholesterol in the rat and hamster. J Biol Chem. 1995, 270 (10): 5381–5387.

[31] Yokozawa T, Cho EJ, Sasaki S, Satoh A, Okamoto T, Sei Y. The protective role of Chinese prescription Kangen–karyu extract on diet–induced hypercholesterolemia in rats. Biol Pharm Bull. 2006, 29 (4): 760–765.

[32] Thomas HD, Maynard C, Wagner GS, Eisenstein EL. Results from a practice – based lipid clinic model in achieving low density lipoprotein cholesterol goals. N C Med J. 2003, 64 (6): 263–266.

[33] Xu C, Wang X, Wang S. [Effect of soybean protein and high calcium intake on the concentration of serum lipids in hypercholesterolemic rats]. Zhonghua Yu Fang Yi Xue Za Zhi. 2001, 35 (5): 318–321.

[34] Van Lenten BJ, Roheim PS. Changes in the concentrations and distributions of apolipoproteins of the aging rat. J Lipid Res. 1982, 23 (8): 1187–1195.

[35] Brown MS, Goldstein JL. Receptor-mediated control of cholesterol metabolism. Science. 1976, 191 (4223): 150-154.

[36] Hui DY, Innerarity TL, Mahley RW. Lipoprotein binding to canine hepatic membranes. Metabolically distinct apo-E and apo-B, E receptors. J Biol Chem. 1981, 256 (11): 5646-5655.

[37] Sparks CE, Marsh JB. Metabolic heterogeneity of apolipoprotein B in the rat. J Lipid Res. 1981, 22 (3): 519-527.

（发表于《BMC Veterinary Research》，院选 SCI，IF：1.777）

Effects of Long-Term Mineral Block Supplementation on Antioxidants, Immunity, and Health of Tibetan Sheep

WANG Hui[1,3], LIU Zhi-qi[2], HUANG Mei-zhou[1], WANG Sheng-yi[1], CUI Dong-an[1], DONG Shu-wei[1], LI Sheng-kun[1], QI Zhi-ming[1], LIU Yong-ming[1]

(1. Engineering and Technology Research Center of Traditional Chinese Veterinary Medicine of Gansu Province/Key Lab of New Animal Drug Project of Gansu Province/Key Lab of Veterinary Pharmaceutical Development of Ministry of Agriculture/Lanzhou Institute of Husbandry and Pharmaceutical Sciences of Chinese Academy of Agricultural Sciences, Lanzhou 730050, China; 2. Institute of Agro-Products Processing Science and Technology, Chinese Academy of Agricultural Sciences, Beijing 100193, China; 3. College of Veterinary Medicine, Northwest Agriculture and Forestry University, Yangling 712100, China)

Abstract: Tibetan sheep have been observed with mineral deficiencies and marginal deficiencies in Qinghai-Tibetan Plateau. Adequate amounts of essential minerals are critical to maximize the productivity and health of livestock. The objectives of this study were to evaluate the effects of 6 months of mineral block supplementation on the antioxidants, immunity, and health of Tibetan sheep. The study was conducted in Qinghai-Tibetan Plateau. The consumed values of mineral blocks were measured. Blood samples were collected at the end of the experiment to evaluate the trace elements, malondialdehyde (MDA) and glutathione (GSH) activities, and antioxidant enzyme activities. Additionally, levels of IgA, IgG, IgM, IL-2, IL-12, tumor necrosis factor-α (TNF-α), triiodothyronine (T3), tyroxine (T4), and insulin-like growth factor-1 (IGF-1) were determined. The toxic effects of the mineral block were also monitored. For Tibetan sheep, the average consumed value of mineral block was 13.09g per day per sheep. Mineral block supplementation significantly increased the serum levels of Mn, Fe, and Se ($P<0.01$), decreased the level of MDA ($P<0.05$), and increased GSH ac-tivity ($P<0.05$). Additionally, the mineral block-treated sheep blood had greater total antioxidative capacity (T-AOC) and total superoxide dismutase (T-SOD) activities ($P<0.01$ or $P<0.05$) than control sheep. Moreover, the mineral block supplementation improved the levels IgA, IgM, and IGF-1 ($P<0.01$ or $P<0.05$). Additionally, there were no significant histopathological changes in the organs of Tibetan sheep after long-term treatment with the mineral block. The results dem-onstrated that the mineral block was non-toxic and safe; the protective effects of the mineral block might be caused by an increase in the antioxidant defense system, as well as an increase in the benefits from immunity-related parameters.

Key words: Mineralblock; Antioxidant; Immunity; Health; Toxicity; Tibetan sheep

INTRODUCTION

Trace minerals have critical roles in the key interrelated systems of growth, reproduction, immune function, cell replication, skeletal development, oxidative metabolism, and energy metabolism in ruminants[1,2]. Trace mineral deficiency can lead to impaired growth and reproduction and to an increase in disease; serious mineral deficiencies may even cause death[3].

With a population of over 50 million, theTibetan sheep is one of the most important livestock species for providing meat, milk, and income for most of the nomadic and seminomadic people of the Qinghai-Tibetan Plateau[4]. Based on their traditional knowledge, the farmers of the region have developed a low-input sheep production and management system[5]. However, the condition of the livestock is often poor, and the household incomes are at or below poverty levels. Over six million of the poorest people earn< $ 2 per day per person in this area[6]. The natural vegetation of the rangelands and crop residues are often low in energy and digestible proteins and, in most cases, fail to cover livestock maintenance requirements. Additionally, these food sources cannot completely satisfy the mineral requirements of the sheep[7]. The Tibetan sheep in these areas have been observed with mineral deficiencies and marginal deficiencies[4]. Mineral disorders cause greater losses of sheep than in-fectious diseases[8]. It is well known that adequate amounts of essential minerals are critical to maximize the productivity and health of livestock. Therefore, supplementary minerals may play a role in balancing the mineral intake of livestock and lead to an increase in animal production and income on the Qinghai-Tibetan Plateau.

Mineral blocks contain the trace minerals that are essential to the well-being of livestock, and there are trace mineral supplements that are specially created for animals to help rem-edy mineral deficiencies. According to previous studies, the feeding of mineral blocks improved the feed efficiency, milk production, reproductive performance, and body function of animals[9-12]. The application of mineral blocks is easy and efficient, and the sheep can obtain the necessary dose of trace elements to ensure health and maximum productivity. Mineral blocks are inexpensive; therefore, the herders can afford the supplement. Many herders in the Qinghai-Tibetan Plateau region have started using mineral blocks in recent years, and they have observed substantial improvements in the general health of their flocks. Pica animals have disappeared, sheep health has gradually improved, the mortality rate of lambs has significantly decreased, the stocking rates have been reduced, and animals have become more productive. The net incomes of herders have increased. The grazing pressures on the grasslands have been reduced to a certain extent comparedto conditions without using mineral blocks.

The effects of mineral block supplementation on the antioxidant and immune response of Tibetan sheep have not been investigated. Therefore, the objective of this study was to evaluate the oxidative status and immune levels by determining the activities of the antioxidant enzymes and the levels of immune factors in the peripheral blood of Tibetan sheep treated with a mineral block supplement containing copper (Cu), manganese (Mn), iron (Fe), zinc (Zn), selenium (Se), etc., and the bio-safety was also evaluated.

MATERIALS AND METHODS

Experimental animals, treatments, and blood sampling

The study was conducted on two Tibetan sheep farms located in Huangcheng Town, Sunan County, Gansu Province, China (longitude 101° 60′ 19″-101° 95′ 92″, latitude37°62′ 88″-38° 00′ 15″, altitude 2 500-5 254m) from June to November 2013 (Fig. 1a). The management was similar at each farm, and the feeding system was the same with natural grazing in pasture on both farms. The sheep were kept in pens during the experiment.

A total of 100 sheep were randomly allocated to the two treatments (the mineral block supplement treatment or the control; each treatment was 50 sheep) within sex and live weight to ensure that each pen contained sheep that covered the full range of weights available. The sheep in the mineral block supplement treatment group received free-choice commercial mineral blocks (Dingli ®, Chinese Veterinarian Institute Pharmaceutical Factory of Chinese Academy of Agricultural Sciences, Lanzhou, China) to meet their mineral requirements (Fig. 1b, c). The sheep in the control group were left untreated and did not receive mineral blocks. The animals had ad libitum access to pasture grass and water. The mineral block supplement-treated animals were offered commercial mineral blocks throughout the whole experiment. The mineral content of the blocks is shown in Table 1.

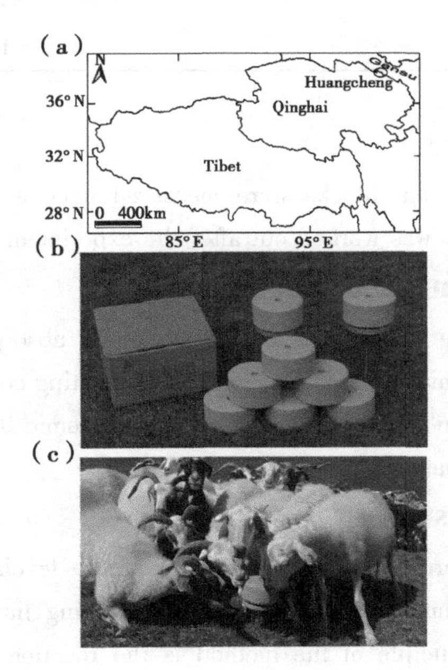

Fig. 1 Location of the sampling site (Huangcheng Town) on the Qinghai-Tibetan Plateau (a); mineral blocks in grassland (b); Tibetan sheep greedily licking mineral block (c)

Blood was sampled after the experiment. A random subset (40sheep, 20 for each treatment) of serum samples from 100 sheep was collected via jugular vein puncture using a sterile needle into

10-mL vacuum tubes without anticoagulant for the serum. The serum was harvested following centrifugation at 3 000rpm for 15 min at 4℃ after blood clotting. The serum samples were stored at −80℃ before analyses. The sheep were 1−5 years of age and represented both sexes-male ($n=14$) and female ($n=26$) —and weighed between 15kg and 40kg. All sheep used in this study were examined and considered clinically healthy. Following the protocols approved by the Institutional Animal Care and Use Committee of Lanzhou Institute of Husbandry and Pharmaceutics Sciences of Chinese Academy of Agricultural Sciences (animal use permit: SCXK20008 − 0003), the animals utilized in these experiments were treated humanely and with respect for the alleviation of suffering.

Table 1 Mineral content of the blocks

Composition	Content
Ca (g/kg)	40
P (g/kg)	15
Mg (g/kg)	10
Na (g/kg)	150
Fe (mg/kg)	3 000
Zn (mg/kg)	800
Mn (mg/kg)	1 000
Cu (mg/kg)	400
Se (mg/kg)	10

Determination of feed intake

The consumed values of mineral blocks were measured once a month. Finally, the average value of feed intake mineral block was worked out after the experiment.

Levels of Cu, Mn, Fe, Zn, and sein serum

Cu, Mn, Fe, and Zn were determined by flame atomic absorption spectrometry (FAAS; ZEEnit 700, Analytik Jena, Germa-ny), and the optimal operating condition was developed as described in our previous publication[13]. The hydride generation atomic fluorescence spectrometry was used for selenium (Se) determination[14].

Determination of oxidative stress

After collecting the blood serum, the determination of the levels of serum malondialdehyde (MDA) was performed using commercial kits (A003−1; Nanjing Jiancheng Bioengineering Institute, Nanjing, China). The principle of the method is the reaction of MDA with thiobarbituric acid (TBA), which produces a red color with a maximum absorption peak of 532nm. The concentration of MDA was calculated by the absorbance coefficient of the MDA-TBA complex and was expressed as millimoles per milligram of serum. The glutathione (GSH) concentration was measured according to the manufacturer's instructions (A006−1; Nanjing Jiancheng Bioengineering Institute, Nanjing, China). The principle of the method is the reaction of GSH with 5, 5′-dithiobis-pnitrobenzoic acid (DTNB), which produces yellow-colored compounds that are detected at 412nm and represent the reduction of GSH. The GSH content was defined as milligram per liter of blood serum.

The assays of activities of the serum total antioxidative capacity (T-AOC), total superoxide dismutase (T-SOD), glutathione peroxidase (GSH-Px), and ceruloplasmin (CP) were performed with commercial kits (T-AOC, A015; T-SOD, A001 - 1; GSH-Px, A005; CP, A029; Nanjing Jiancheng Bioengineering Institute, Nanjing, China) according to the manufacturer's instructions. The oxidative stress-related enzyme activities were evaluated using an ultraviolet spectrophotom-eter (UV-2100, Shimadzu Corporation, Kyoto, Japan). The TAOC was measured by the ferric reducing antioxidant power (FRAP) assay method. The reaction measures reduction of ferric 2, 4, 6-tripyridyl-s-triazine (TPTZ) to a colored product by an antioxidative substance. And, the color that developed was detected at 593nm with a spectrophotometer. T-AOC was expressed in units per milliliter, and one unit (U) was defined as the amount that increased the absorbance by 0.01 at 37℃. T-SOD activity was determined using the system of xanthinexanthine oxidase and nitroblue tetrazolium. One unit (U) of T-SOD activity was defined as the amount that reduced the absorbance at 550nm by 50%. The reaction of GSH-Px kit was initiated by the addition of H_2O_2. A series of enzymatic reactions was activated by GSH-Px which subsequently led to the conversion of GSH to oxidized glutathione (GSSG). The change in absorbance during the conversion of GSH to GSSG was recorded using a spectrophotometer at 412nm[15]. The activity of GSH-Px was expressed as a decrease of 1.0μm GSH per 5min at 37℃, after the non-enzymatic reaction is subtracted. The CP in the serum catalyzes the dianisidine, and resulting violet color was measured at 540nm on spectrophotometer. CP was expressed as millimoles per milliliter.

ELISA assays for IgA, IgG, IgM, IL-2, IL-12, and TNF-α levels in serum

The levels of IgA, IgM, IgG, IL-2, IL-12, and tumor necrosis factor-α (TNF-α) were measured using an ELISA assay kit (BD Biosciences, San Diego, CA, USA) following the manufacturer's instructions. Briefly, 40μL of samples, 10μL of antibodies, and 50μL of streptavidin-HRP were added to 96well plates. The wells were sealed with the sealing membrane, gently shaken, and incubated for 60min at 37℃. The membrane was carefully removed, the solutions were removed, and the plate was washed five times with wash solution. Then, 50μL of the chromogen solutions A and B wasadded and incubated for 10min at 37℃ away from light. Finally, the reactions were stopped with 50μL of stop solution. The absorbance was read at 450nm using a microplate reader (SpectreMax M2, Molecular Devices, USA). The results were expressed as the concentrations of IgA (mg/mL), IgM (mg/mL), IgG (mg/mL), IL-2 (ng/mL), IL-12 (ng/L), and TNF-α (ng/L) in the blood serum. For the IgA assayed, the intra-assay and inter-assay coefficients of variation were 2.5%and 4.6%, respectively, with a sensitivity of 0.4ng/mL. For the IgM assayed, the intra-assay and inter-assay coefficients of varia-tion were 2.8% and 4.8%, respectively, with a sensitivity of 0.3ng/mL. For the IgG assayed, the intra-assay and inter-assay coefficients of variation were 4.6% and 7.2%, respectively, with a sensitivity of 0.5ng/mL. For the IL-2 assayed, the intra-assay and inter-assay coefficients of variation were 3.1% and 5.6%, respectively, with a sensitivity of<10pg/mL. For the IL-12 assayed, the intra-assay and inter-assay coefficients of variation were 4.2% and 7.3%, respectively, with a sensitivity of 1.0pg/

L. Meanwhile, the intra-assay and inter-assay coefficients of variation were 2.4% and 6.2%, respectively, with a sensitivity of 0.5pg/L for the TNF-α assayed.

ELISA assays for T3, T4, and IGF-1 levels in serum

The levels of triiodothyronine (T3), tyroxine (T4), and insulinlike growth factor-1 (IGF-1) were measured with ELISA assay kits accordingto the manufacturer's instructions (BD Biosciences, San Diego, CA, USA). The results were expressed as the concentrations of T3 (nmol/L), T4 (nmol/L), and IGF-1 (ng/mL) in the blood serum. For the T3 assay, the intra-assay and inter-assay coefficients of variation were 2.3% and 6.4%, respectively, with a sensitivity of< 0.3 pmol/L. For the T4 assay, the intra-assay andinter-assay coefficients of variation were 3.2% and 6.6%, respectively, with a sensitivity of 0.76pmol/L. Meanwhile, the intra-assay and inter-assay coefficients of variation were 5.6% and 6.8%, respectively, with a sensitivity of 0.53 pg/mL for the IGF-1 assayed.

Histopathology assessments

Six randomly selected Tibetan sheep from the mineral block treatment were slaughtered after the experiment according to commercial procedures. The intestines, heart, liver, spleen, lung, and kidney were harvested and immediately immersed in 10% neutral buffered formalin for 24h. After fixation, the tissues were dehydrated through a gradient of alcohols, embedded in paraffin using conventional methods[16], sectioned into 5-μm-thick slices, and mounted on glass microscope slides[17]. The slices were stained with hematoxylin-eosin (HE) and were examined by a light microscope (Olympus Bx51 model, Tokyo, Japan) equipped with a camera (Olympus DP20, Tokyo, Japan).

Statistical analyses

The data obtained from the experimental animals were expressed as the means and standard error (±SE). Statistical analyses were performed with one-way analysis of variance (ANOVA) using SPSS software (version 17.0; SPSS, Inc., Chicago, USA), and $P<0.05$ was considered significant.

RESULTS

Mineral block consumed values

The results of consumed mineral block were measured each month, which are shown in Fig. 2. The average mineral block consumed value of experiment sheep was 157.48kg/month, as well as 13.09g per dayper sheep.

Blood mineral levels

The analysis of serum mineral concentrations revealed that Cu and Zn concentrations were not different between the mineral block treatment and the control Tibetan sheep at the end of the experiment ($P>0.05$; Fig. 3). However, the mineral blocktreated sheep had significantly increased concentrations of Mn, Fe, and Se ($P<0.01$) compared with the control sheep (Fig. 3).

Effects on LPO and GSH

The mean concentration of MDA in the serum of control sheep was (3.688±0.413) mmol/mL

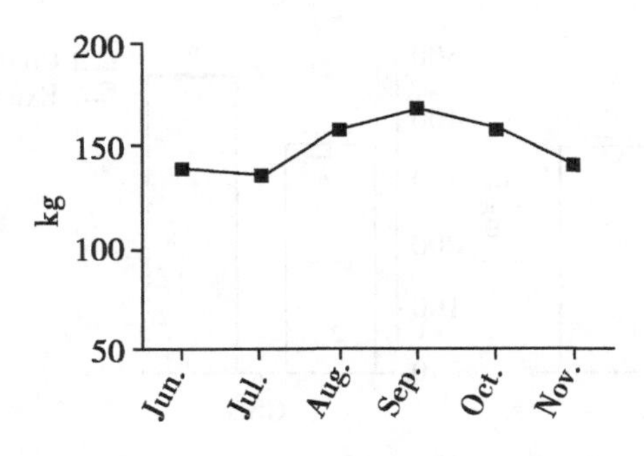

Fig. 2 The values of mineral block consumed by experiment sheep during the experiment period

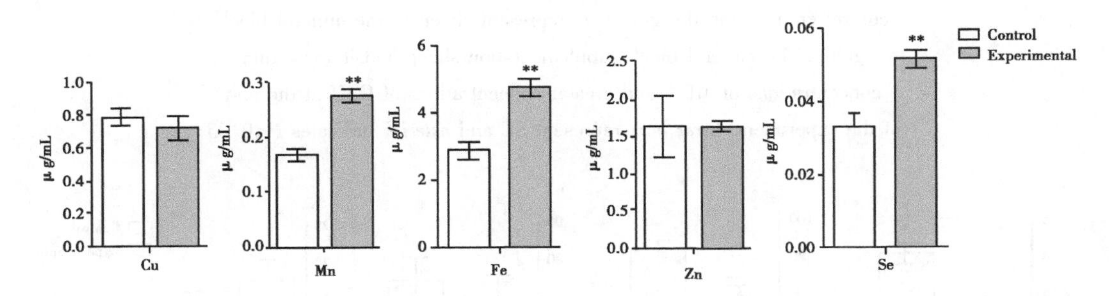

Fig. 3 Effects of the mineral block supplement on the serum trace element concentrations at the end of the experiment

Note: The *gray bars* represent sheep in the control group, and the *black bars* represent sheep in the mineral block group. The mineral block supplementation sheep had greater serum concentrations of Mn, Fe, and Se after the experiment. *Error bars* represent SE and *double asterisk* indicates $P<0.01$

and in the mineral block-treated sheep was (2.987±0.115) mmol/mL at the end of the experiment. The small decline in the concentration of MDA in the mineral block treatment sheep was statistically significant ($P<0.05$; Fig. 4). A significant elevation in the GSH level was observed in the serum ($P<0.05$) of mineral block administered sheep [(0.371±0.060) mg/Lcompared with the serum in control sheep (0.248±0.064) mg/L; Fig. 4].

Effects on antioxidant enzymes

To determine the effects of the mineral block on antioxidant cellular defenses, we evaluated the enzymatic and nonenzymatic antioxidants in serum, including T-AOC, T-SOD, GSH-Px, and CP, which are shown in Fig. 5. In the mineral block sheep, theT-AOC and T-SOD activities were found to be higher in the serum ($P<0.05$) compared with those in the control sheep serum. In contrast, there were no significant differences between the two groups in the levels of GSH-Px activity and CP (Fig. 5, $P>0.05$). These data suggested that the mineral block may protect Tibetan sheep from oxidative injury by enhancing the activity of antioxidant enzymes.

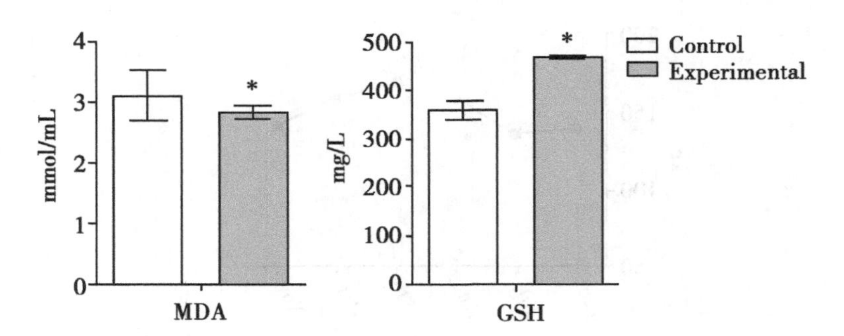

Fig. 4　Effects of the mineral block supplement on the LPO and GSH levels in the serum of Tibetan sheep

Nite: The *white bars* represent sheep in the control group, and the *gray bars* represent sheep in the mineral block group. The mineral block supplementation sheep had lower serum concentrations of MDA and greater concentrations of GSH at the end of the experiment. *Error bars* represent SE and *asterisk* indicates $P<0.05$

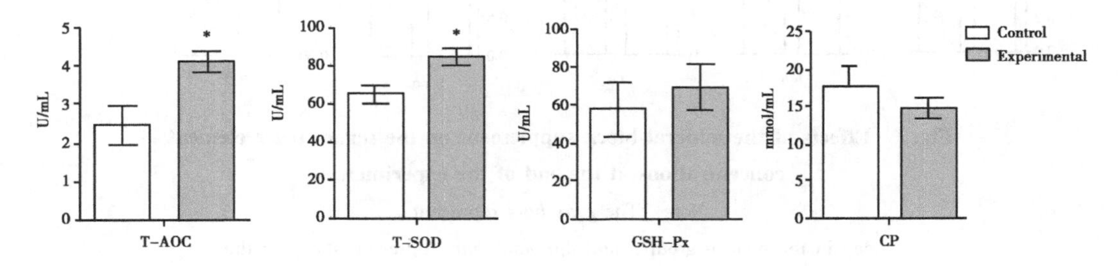

Fig. 5　Effects of the mineral block supplement on the activities of TAOC, T-SOD, GSH-Px, and the levels of CP in the serum of Tibetan sheep

Note: The *white bars* represent sheep in the control group, and the *orange bars* represent sheep in the mineral block group. The mineral block supplementation sheep had greater serum activities of T-AOC and T-SOD at the end of the experiment. *Error bars* represent SE and *asterisk* indicates $P<0.05$ (color figure online)

Determination of the levels of IgA, IgG, IgM, IL-2, IL-12, and TNF-α in serum

In the serum, the immunoglobulins IgA and IgM increased significantly in the sheep treated with the mineral block [(3.562±0.394) mg/mL and (3.290±0.287) mg/mL, respectively] compared with the levels in the control sheep (2.616±0.107mg/mL and 1.714±0.131mg/mL, respectively; $P<0.05$ and $P<0.01$, respectively; Fig. 6). Although there were no significant differences in the levels of IgG, IL-2, IL-12, and TNF-α between the two groups, the levels of these compounds tended to increase in sheep in the mineral block group compared with the control group (Fig. 6).

Determination of the levels of T3, T4, and IGF-1 in serum

The IGF-1 analysis showed that the level of this cytokine in serum increased significantly ($P<$

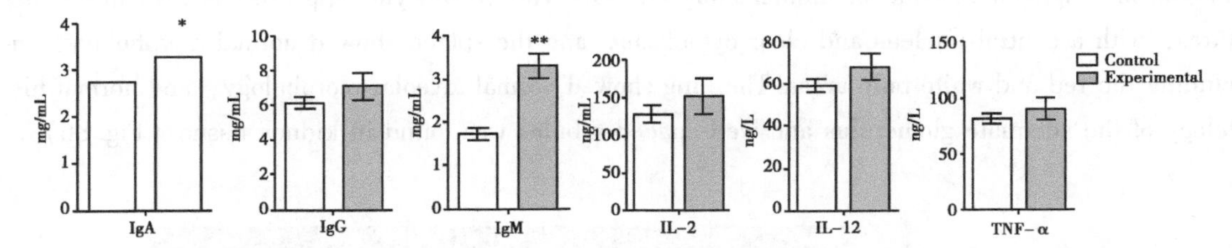

Fig. 6 Effects of the mineral block supplement on the levels of IgA, IgG, IgM, IL-2, IL-12, and TNF-α in the serum of Tibetan sheep

Note: The *white bars* represent sheep in the control group, and the *blue bars* represent sheep in the mineral block group. The mineral block supplementation sheep had greater serum concentrations of IgA and IgM at the end of the study. *Error bars* represent SE and *asterisk* indicates $P<0.05$ and *double asterisk indicates* $P<0.01$ (colorfigure online)

0.05) in the mineral block treatment sheep (118.6±11.50) ng/mL in comparison with that in the control sheep (79.77±4.322) ng/mL; Fig. 7. There were no significant differences in the T3 and T4 levels between the two groups, but the mineral block treatment sheep had increased T4 levels compared with the control sheep (Fig. 7).

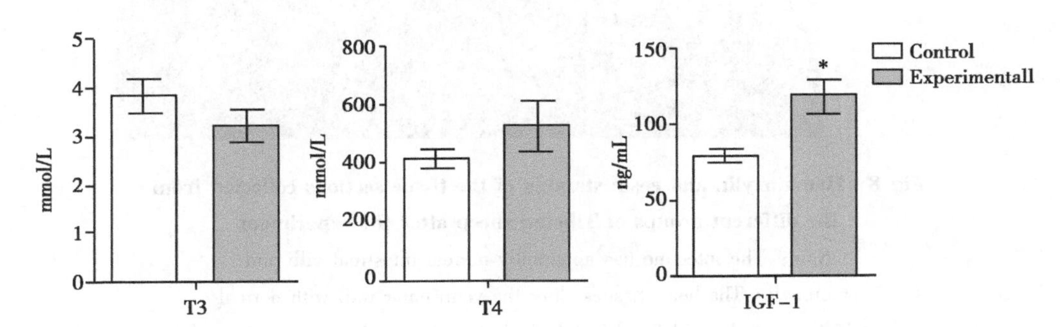

Fig. 7 Effects of the mineral block supplement on the levels of T3, T4, and IGF-1 in the serum of Tibetan sheep

Note: The *white bars* represent sheep in the control group, and the *pink bars* represent sheep in the mineral block group. The mineral block supplementation sheep had a greater serum concentration of IGF-1 at the end of the experiment. *Error bars* represent SE and *asterisk indicates* $P<0.05$ (color figure online)

Histopathological examination

A histopathological study of the major organs harvested from sheep was conducted. In the control group, no significant histopathological changes were observed in the intestine, heart, liver, spleen, lung, and kidney of the sheep (Fig. 8a) compared with the treated sheep. There were no detectable lesions observed for any organ of the sheep after long-term (6 months) treatment with the mineral block. The intestine had apparently normal intestinal villi and enterocytes, and the heart tissues showed the ventricular wall with normal orientation and well-defined histological structures without any signs of vascular or inflammatory changes. The hepatocytes appeared as cord-like structures, with a central nucleus and clear cytoplasm, and the spleen showed normal morphology, including the red and white pulp areas. The lung showed normal alveolar morphology, and normal histology of the adequate glomerulus and well-spaced tubules was found in kidney tissue (Fig. 8b).

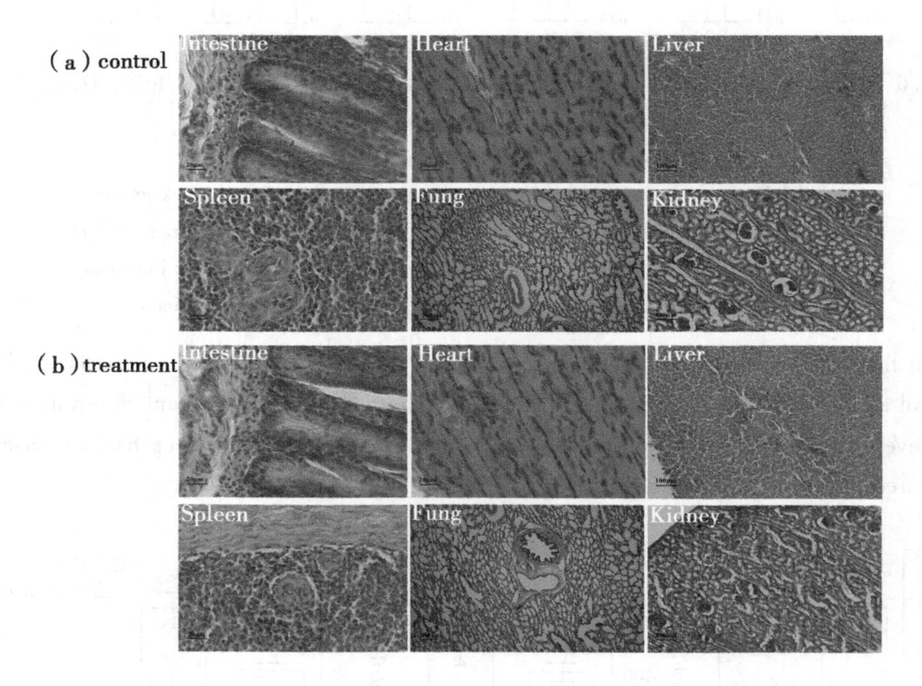

Fig. 8 Hematoxylin and eosin staining of the tissue sections collected from the different groups of Tibetan sheep after the experiment

Note: The intestine has apparently normal intestinal villi and enterocytes. The heart tissues show the ventricular wall with normal orientation and welldefined histological structures without any signs of vascular or inflammatory changes. The hepatocytes appear as cord-like structures, with a central nucleus and clear cytoplasm. The spleen shows normal morphology, including the red and white pulp areas. The lung shows normal alveolar morphology. The kidney shows normal histology with adequate glomerulus and wellspaced tubules (color figure online)

DISCUSSION

Trace minerals have an essential role in ensuring efficient growth, reproduction, and immunocompetence in animals[18]. Marginal and deficient levels of trace minerals can increase the risk of disease in animals[19]. A recent study reported a lower incidence of diarrhea, mastitis, and endometritis for cows with a diet supplemented with an injectable trace mineral solution containing Zn, Cu, Mn, and Se[2,20]. Previous studies indicated that the feeding of urea molasses mineral block improved the feed efficiency, milk production and composition, and reproductive performance of crossbred cows[10,12]; the combination of molasses/mineral feed blocks and medicated blocks had significant effects on blood factors including calcium, creatinine, urea nitrogen, and packed cell volume (PCV) and had no negative effects on body function for grazing goats[11].

Copper, manganese, iron, zinc, and selenium are essential elements required by animals for a number of biochemical functions. Copper deficiency can cause anemia, bone and connective tissue abnormalities, and impaired immune function[21]. Manganese participates in a wide range of redox and adsorptive reactions and plays a significant role in the bioavailability and geochemical cycling of many essential or toxic elements[22]. Iron is needed for embryonic and fetal survival and development, normal growth of animals, and synthesis of Fe-requiring enzymes, whereas Fe-deficient anemia has been associated with long-term adverse effects on cognitive function and motor development[23]. Zinc deficiency causes decreased growth, impaired immune function, and increased susceptibility to infection[24]. Selenoenzymes are ubiquitous in mammalian cells and have major roles as antioxidants that protect cell components and inhibit proinflammatory cell metabolisms by reducing the peroxide tone of intracellular water[25]. The normal values of serum Mn, Fe, and Se in sheep are 0.18–0.25μg/mL, 1–5μg/mL, and 0.08–0.5μg/mL, respectively[26]. Pogge et al. reported that the use of an injectable trace mineral solution increased the plasma concentrations of Mn and Se[27]. Similarly, in this study, serum Mn, Fe, and Se concentrations in the mineral block-treated Tibetan sheep were greater than in the controls over the 6 month period (Fig. 3). But, the level of serum Mn in control was lower than critical value, and the level of serum Se in experiment and control was also lower than critical value. The normal values of serum Cu and Zn in sheep are 0.5–1.5μg/mL and 0.8–1.7μg/mL, respectively[28]. The results of the study indicated that the levels of serum Cu and Zn in experiments were not increased significantly compared with controls, but they were within the normal range, which supports the finding that Se can antagonize with Cu, and Fe and Mn can antagonize with Zn[28]. According to the results of this study, Tibetan sheep in this region had deficiency in trace elements of Mn and Se.

In biological systems, lipid peroxidation generates many aldehyde products, among which MDA is the most recognized marker. The increase in MDA implied that antioxidant defense mechanisms were not sufficient to neutralize oxidative stress, and there was increased oxidative stress and cellular damage[29]. Lipid peroxidation is a cause of cell membrane disruption, cell damage, metabolic disorders, and inflammation[30]. This process is initiated by the hydroxyl and the superoxide radicals and leads to the formation of peroxy radicals (LOO)[26]. Thus, antioxidants capable of scavenging peroxy radicals could prevent lipid peroxidation. GSH can reduce a wide variety of disulfides by transhydrogenation and is an important reactive oxygen species (ROS) scavenger[31]. In this study, the observed decrease in MDA and the concomitant increase of GSH levels in the treated group suggested that the increased peroxidation could be a consequence of depleted GSH stores. The mineral block supplement for Tibetan sheep improved the levels of GSH, which may have resulted from the trace element-mediated reduction of peroxidation activity among cells.

The adaptation of organisms to environmental stress depends on antioxidant defense[32]. The metabolic demands associated with nutritional deficiency can lead to the increased production of reactive oxygen species (ROS)[33]. An imbalance exists between the production of ROS and the antioxidant ability to neutralize these intermediates, which results in cellular damage. ROS can initiate lipid peroxidation, cause cellular damage, and induce apoptotic cell death in various cell types[34]. Immune cells are particularly sensitive to oxidative stress[35]. To maintain the lowest possible levels of ROS in the cell[36], all organisms have evolved antioxidant defenses that can be measured by the T-AOC with both enzymatic and non-enzymatic components such as SOD, GSH-Px, and CP. Trace elements have a specific role in free radical control at the cellular level and influence the antioxidant/free radical balance[18]. Trace minerals are required for the functioning of enzymes involved in the antioxidant defense system and may also affect immune cells through mechanisms distinct from antioxidant properties. Copper acts as a cofactor of several metalloenzymes and other metalloproteins such as CP and Cu-zinc SOD[37]. Zinc induces synthesis of a metalbinding protein that may scavenge hydroxide radicals[38] and may affect immunity via its important role in cell replication and proliferation[35]. The protective role of Se is primarily its antioxidant property because of its role in an antioxidant enzyme GSH-Px[39]. SOD protects cells from the oxygen free radical by catalyzing the removal of O_2^-. Moreover, previous research showed that supplementation with injectable trace minerals was beneficial to calf immunity and oxidative stress status[2,20,40]. Our study found that the long-term administration of the mineral blocks strongly enhanced the TAOC and the activities of T-SOD. Increase in the level of TAOC in the experimental sheep may probably be ascribed to the enhancement of antioxidant enzymes as free radical scavengers during licking mineral blocks. Thus, it is speculated that the increased level of trace minerals in the serum of experimental sheep proved that mineral block can impose a marked effect on the status of mineral nutrients and also represents their coordinated antioxidant role accompanied by antioxidant enzyme activities.

Trace elements are indispensable for life and play an important role in immune function[41]. Immunoglobulin, an important part of the body's immune system, plays an important role in immune regulation and mucosal defense. Serum immunoglobulin levels are determined routinely in clinical practice because they provide key information on the immune status[42]. In our study, the levels of serum IgM and IgA in the mineral block-supplemented Tibetan sheep increased significantly. The increase in immunoglobulins indicated that the mineral block treatment influenced the function of the innate immune system of sheep. Tibetan sheep in this region were deficient in Mn and Se; it is speculated that supplementation with Se and Mn have marked immunostimulating effects. Several studies in agreement with our results of manganese deprivation could impair immunity. The cell-mediated immunity and B cell function can be impaired in selenium deficiency[43], but supplementation with selenium can enhance proliferation of activated cytotoxic T cells and increase NK cell activity[41].

In addition to the immune function, we also detected the levels of some cytokines. Cytokines are secreted by immune cells and are chemical messengers that have important roles in regulating the immune response; some cytokines directly fight pathogens[44]. IL-2 is secreted by activated T cells; IL-2 is widely considered to be a key cytokine in T cell-dependent immune responses[45]. IL-12 is known to be a T cell-stimulating factor, which can stimulate the growth and function of T cells. IL-12 plays an important role in activating natural killer cells and T lymphocytes[46]. TNF-α, which is secreted by activated T cells, has a primary role in the regulation of immune cells[47]. In the present study, we examined the levels of IL-2, IL-12, and TNF-α in the serum and found that these cytokines were increased in the mineral block group compared with control group, but there were no significant differences.

Thyroid hormone circulates in plasma in two primary forms, triiodothyronine (T3) and thyroxine (T4). T3 is a thyroid hormone that affects almost every physiological process in the body, including growth and development, metabolism, body temperature, and heart rate[48,49]. T4, produced by the thyroid gland, helps to control metabolism and growth, and the analysis of T4 is performed as part of an evaluation of thyroid function[48]. IGF-1 is a peptide hormone that is structurally related to insulin and plays an essential role in many physiological functions, including stimulation of growth, reproduction, metabolism, cell proliferation, cell differentiation, apoptosis, ovarian folliculogenesis, and secretory activities [50]. The IGF-1 is a broad-spectrum growth factor, and a positive correlation between plasma IGF-1 level and specific growth rate has been reported for numerous species[51]. In this study, T3 and T4 levels were not significantly affected by the mineral block, but IGF-1 levels were significantly correlated with the mineral block. Our study showed that serum IGF-1 levels may be affected by trace mineral supplementation.

In this study, there were no significant histopathological changes in the intestine, heart, liver, spleen, lung, and kidney of Tibetan sheep throughout the experimental period (6 months). These results collectively suggested that the mineral blocks possess no significant toxicity to sheep. As part of the preclinical safety evaluation of the mineral block, these findings lay the groundwork for additional studies to further investigate preclinical toxicity associated with the mineral block. The experiments above were merely preliminary analyses to prove the bio-safety of the mineral block, and more efforts will be required to systematically test the metabolic clearance rate of the mineral block trace elements in animals.

In conclusion, the mineral block was non-toxic and safe, and the protective effects of the mineral block might be caused by both an increase in the activity of the antioxidant defense system and a decrease in lipid peroxidation activity, as well as the increased benefits from immunity-related parameters. Mineral block supplementation has been widely adopted by farmers in western China and plays an important role in sheep production and improved farmers' incomes. Future research should focus on exploring the effects of mineral block supplements on rumen fermentation and product quality (meat and milk) to maximize the health benefits to Tibetan sheep.

ACKNOWLEDGMENTS

The financial support from the National Key Technology Research and Development Program of the Ministry of Science and Technology of China (No. 2012BAD12B03) and Central Publicinterest Scientific Institution Basal Research Fund (No. 1610322013003) is greatly appreciated.

COMPLIANCE WITH ETHICAL STANDARDS

This study was approved by the Institutional Animal Care and Use Committee of Lanzhou Institute of Husbandry and Pharmaceutics Sciences of Chinese Academy of Agricultural Sciences (animal use permit: SCXK20008-0003).

CONFLICT OF INTEREST

The authors declare that they have no competing interests.

References

[1] Overton TR, Yasui T (2014) Practical applications of trace minerals for dairy cattle. J Anim Sci, 92: 416-426.

[2] Teixeira AG, Lima FS, Bicalho ML et al (2014) Effect of an injectable trace mineral supplement contaioning selenium, copper, zinc, and manganese on immunity, health, and growth of dairy calves. J Dairy Sci, 97: 4216-4226.

[3] Underwood EJ, Suttle NF (1999) The mineral nutrition of livestock. CABI Publishing, London, UK.

[4] Xin GS, Long RJ, Guo XS et al (2011) Blood mineral status of grazing Tibetan sheep in the Northeast of the Qinghai-Tibetan Plateau. Livestock Sci, 136: 102-107.

[5] Wang H, Liu Y, Qi Z et al (2014) The estimation of soil trace elements distribution and soil-plant-animal continuum in relation to trace elements status of sheep in Huangcheng area of Qilian mountain grassland, China. J Integ Agri, 13: 140-147.

[6] Kemp DR, Guodong H, Xiangyang H et al (2013) Innovative grassland management systems for environmental and livelihood benefits. Proc Natl Acad Sci U S A, 110: 8369-8374.

[7] Ben Salem H, Nefzaoui A (2003) Feed blocks as alternative supplements for sheep and goats. Small Rumin Res, 49: 275-288.

[8] Judson GJ, McFarlane JD (1998) Mineral disorders in grazing livestock and the usefulness of soil and plant analysis in the assessment of these disorders. Aust J Exp Agric, 38: 707-723.

[9] Forsberg NE, Al-Maqbaly R, Al-Halhali A et al (2002) Assessment of molasses urea blocks for goat and sheep production in the sultanate of Oman: intake and growth studies. Trop Anim Health Prod, 34: 231-239.

[10] Sahoo B, Bhushan C, Kwatra J et al (2009) Effect of urea molasses mineral block supplementation on milk production of cows (*Bos indicus*) in mid hills of Uttarakhand. Anim Nutr Feed Technol, 9: 171-178.

[11] Kioumarsi H, Yahaya ZS, Rahman AW (2011) The effect of molasses/mineral feed blocks along with the use of medicated blocks on haematological and biochemical blood parameters in Boer goats. Asian J Anim Vet Adv, 6: 1264-1270.

[12] Khadda BS, Lata K, Kumar R et al (2014) Effect of urea molasses minerals block on nutrient utilization, milk production and reproductive performance of crossbred cattle under semi arid ecosystem. Indian J Anim Sci, 84: 302-305.

[13] Wang X, Wang H, Li J et al (2014) Evaluation of bioaccumulation and toxic effects of copper on hepatocellular structure in mice. Biol Trace Elem Res, 159: 312-319.

[14] Lu CY, Yan XP, Zhang ZP et al (2004) Flow injection online sorption preconcentration coupled with hydride generation atomic fluorescence spectrometry using a polytetrafluoroethylene fiber-packed microcolumn for determination of Se (IV) in natural water. J Anal Atom Spectrom, 19: 277-281.

[15] Liu L, Liu Y, Cui J et al (2013) Oxidative stress induces gastric submucosal arteriolar dysfunction in the elderly. World J Gastroenterol, 19: 9439-9446.

[16] Lim AY, Segarra I, Chakravarthi S et al (2010) Histopathology and biochemistry analysis of the interaction between sunitinib and paracetamol in mice. BMC Pharmacol, 10: 14.

[17] Deng H, Zhong Y, Du M et al (2014) Theranostic self-assembly structure of gold nanoparticles for NIR photothermal therapy and X-ray computed tomography imaging. Theranostics, 4: 904–918.

[18] Andrieu S (2008) Is there a role for organic trace element supplements in transition cow health? Vet J, 176: 77–83.

[19] Enjalbert F, Lebreton P, Salat O (2006) Effects of copper, zinc and selenium status on performance and health in commercial dairy and beef herds: Retrospective study. J Anim Physiol Anim Nutr (Berl), 90: 459–466.

[20] Machado VS, Bicalho ML, Pereira RV et al (2013) Effect of an injectable trace mineral supplement containing selenium, copper, zinc, and manganese on the health and production of lactating Holstein cows. Vet J, 197: 451–456.

[21] Olivares M, Araya M, Uauy R (2000) Copper homeostasis in infant nutrition: deficit and excess. J Pediatr Gastroenterol Nutr, 31: 102–111.

[22] Tebo BM, Bargar JR, Clement BG et al (2004) Biogenic manganese oxides: properties and mechanisms of formation. Annu Rev Earth Planet Sci, 32: 287–328.

[23] Hostetler CE, Kincaid RL, Mirando MA (2003) The role of essential trace elements in embryonic and fetal development in livestock. Vet J, 166: 125–139.

[24] Hotz C, Brown KH (2001) Identifying populations at risk of zinc deficiency: the use of supplementation trials. Nutr Rev, 59: 80–84.

[25] Rayman MP (2000) The importance of selenium to human health. Lancet, 356: 233–241.

[26] Gul MZ, Ahmad F, Kondapi AK et al (2013) Antioxidant and antiproliferative activities of *Abrus precatorius* leaf extracts—an in vitro study. BMC Complement Altern Med, 13: 53.

[27] Pogge DJ, Richter EL, Drewnoski ME et al (2012) Mineral concentrations of plasma and liver following injection with a trace mineral complex differ among Angus and Simmental cattle. J Anim Sci, 90: 2692–2698.

[28] Wang JH (2007) Domestic animal medicine. Beijing. China Agricultural Press. pp. 215–234.

[29] Grewal A, Ahuja CS, Singh SP et al (2005) Status of lipidperoxidation, some antioxidant enzymes and erythrocytic fragility ofcrossbred cattle naturally infected with *Theileria annulata*. Vet Res Commun, 29: 387–394.

[30] Barrera G, Pizzimenti S, Dianzani MU (2008) Lipid peroxidation: control of cell proliferation, cell differentiation and cell death. Mol Aspects Med, 29: 1–8.

[31] You H, Chen S, Mao L et al (2014) The adjuvant effect induced by di- (2-ethylhexyl) phthalate (DEHP) is mediated through oxidative stress in a mouse model of asthma. Food Chem Toxicol, 71: 272–281.

[32] Regoli F, Cerrano C, Chierici E (2004) Seasonal variability of prooxidant pressure and antioxidant adaptation to symbiosis in the Mediterranean demosponge *Petrosia ficiformis*. Mar Ecol-Prog Ser, 275: 129–137.

[33] Prasad R, Kowalczyk JC, Meimaridou E et al (2014) Oxidative stress and adrenocortical insufficiency. J Endocrinol, 221: 63–73.

[34] Bayir H, Kagan VE (2008) Bench-to-bedside review: mitochondrial injury, oxidative stress and apoptosis—there is nothing more practical than a good theory. Crit Care, 12: 206.

[35] Spears JW, Weiss WP (2008) Role of antioxidants and trace elements in health and immunity of transition dairy cows. VetJ, 176: 70–76.

[36] Wang H, Liu YM, Qi ZM et al (2013) An overview on natural polysaccharides with antioxidant properties. Curr Med Chem, 20: 2899–2913.

[37] Zhang W , Zhang Y, Zhang SW et al (2012) Effect of different levels of copper and molybdenum supplements on serum lipid profiles and antioxidant status in Cashmere goats. Biol Trace Elem Res, 148: 309-315.
[38] Prasad AS, Bao B, Beck FW et al (2004) Antioxidant effect of zinc in humans. Free Radic Biol Med, 37: 1182-1190.
[39] Tinggi U (2003) Essentiality and toxicity of selenium and its status in Australia: a review. Toxicol Lett, 137: 103-110.
[40] Genther ON, Hansen SL (2014) A multielement trace mineral injection improves liver copper and selenium concentrations and manganese superoxide dismutase activity in beef steers. J Anim Sci, 92: 695-704.
[41] Wintergerst ES, Maggini S, Hornig DH (2007) Contribution of selected vitamins and trace elements to immune function. Ann Nutr Metab, 51: 301-323.
[42] Gonzalez-Quintela A, Alende R, Gude F et al (2008) Serum levels of immunoglobulins (IgG, IgA, IgM) in a general adult population and their relationship with alcohol consumption, smoking and common metabolic abnormalities. Clin Exp Immunol, 151: 42-50.
[43] Suttle NF (2010) Mineral nutrition of livestock, 4th edn. CABI Publishing, Wallingford, Oxfordshire, UK.
[44] Iwasaki A, Medzhitov R (2010) Regulation of adaptive immunity by the innate immune system. Science, 327: 291-295.
[45] Malek TR, Bayer AL (2004) Tolerance, not immunity, crucially depends on IL-2. Nat Rev Immunol, 4: 665-674.
[46] Hsieh CS, Macatonia SE, Tripp CS et al (1993) Development of TH1 $CD4^+$ T cells through IL-12 produced by Listeria-induced macrophages. Science, 260: 547-549.
[47] Locksley RM, Killeen N, Lenardo MJ (2001) The TNF and TNF receptor superfamilies: integrating mammalian biology. Cell, 104: 487-501.
[48] Pittas AG, Lee SL (2003) Evaluation of thyroid function. In Handbook of Diagnostic Endocrinology. Humana Press, pp. 107-129. doi: 10. 1007/978-1-59259-293-7_ 6.
[49] Leung LY, Woo NY (2010) Effects of growth hormone, insulin-like growth factor I, triiodothyronine, thyroxine, and cortisol on gene expression of carbohydrate metabolic enzymes in sea bream hepatocytes. Comp Biochem Physiol A Mol Integr Physiol, 157: 272-282.
[50] Clay LA, Wang SY, Wolters WR et al (2005) Molecular characterization of the insulin-like growth factor-1 (IGF-1) gene in channel catfish (*Ictalurus punctatus*) . Biochim Biophys Acta, 1731: 139-148.
[51] Lee SW, Lee SY, Lee SR et al (2010) Plasma levels of insulin-like growth factor-1 and insulin-like growth factor binding protein-3 in women with cervical neoplasia. J Gynecol Oncol, 21: 174-180.

(发表于《Biological Trace Element Research》，院选 SCI，IF：1.748)

Poly (lactic acid) /Palygorskite Nanocomposites: Enhanced the Physical and Thermal Properties

LIU Yu[1], HAN Shun-yu[2], JUANG Yu-mei[2], LIANG Jian-ping[1], SHANG Ruo-feng[1], HAO Bao-cheng[1], CHENG Fu-sheng [1], ZHANG Sheng-gui[2]

(1. Key Laboratory of New Animal Drug Project, Gansu Province/Key laboratory of Veterinary Pharmaceutical Development, Ministry of Agriculture/ Lanzhou institute of Husbandry and Pharmaceutical Sciences of CAAS, Lanzhou 730050, China; 2. College of Food Science and Engineering, Gansu Agricultural University, Lanzhou 730070, China)

Abstract: Poly (lactic acid) /palygorskite (PA) nanocomposites were fabricated through the melt compounding. The significant finding by differential scanning calorimetry (DSC) showed that PA acted as a nucleating agent and accelerated the crystallization process of neat PLA. Consequently, a PLA crystallization peak appeared in all nanocomposites. With increasing PA content, the crystallization temperature increased from 92.1℃ to 99.6℃. Dynamic Mechanical Analysis (DMA) indicated that the incorporation of PA hindered the motion of the PLA chains in the matrix, thereby increasing the maximum service temperature of the PLA. The storage modulus, G′, of the nanocomposite reached a plateau in the low frequency region which indicated formation of a network structure due to the increased interaction sites. The PLA/PA nanocomposites showed significant improvement in physical and thermal properties. POLYM. COMPOS., 00: 000—000, 2015.© 2015 Society of Plastics Engineers

INTRODUCTION

Thermoplastics demonstrate good properties in packaging and other consumer product application; however, most thermoplastic packaging materials are derived from nonrenewable resources[1]. The production and the use of those plastics have been increasing significantly in the past decades and the White Waste has become worsening. Therefore, there is a considerable need in replacing part or all of the non-biodegradable plastics by biodegradable materials. With the developments of agriculture industry, the use of agricultural products such as corn and starch are no longer in the traditional simple processing mode. In order to add additional value to agricultural products, new applications are found in food packaging materials[2]. Biodegradable materials offer a possible alternative to the traditional non-biodegradable polymers. Among these, poly (lactic acid) is one of the most promising biodegradable, biocompatible, and environmentally friendly thermoplastic for applications in packaging material[3]. PLA exhibits good biodegradability and melt processibility, which can be further modified by blending them with additives in the molten state. These features make PLA an attractive alternative for petroleum-based non-biodegradable materials. Although the tensile strength of PLA is relatively good, the thermal and mechanical properties of PLA need further improvement for agriculture products packaging materials. Compounding with inorganic nanoparticles is an effective method to enhance the properties of the polymer matrix. The main reason for these improved properties, as reported by Ray et al.[4], is the interfacial interaction between matrix and clay surface.

In many applications, increasing the crystallization speed of PLA is desired since in its amorphous form, the range of application of PLA is severely limited by its low glass transition temperature[5]. When temperature is higher than T_g of PLA, only the crystalline phase can provide useful mechanical properties. Thus, the crystalline form is required to increase the temperature resistance of the material [5]. Unfortunately, the normal crystallization rate of PLA is too slow to reach significant crystallinity in the timescale of typical commercial processing. In order to obtain desirable crystallinity, several methods have been carried out. Among these methods, adding a nucleating agent is one of the promising methods to improve the crystallinity. For example, sodium benzoate[6], N, Nethylenebis (12-hydroxystearamide) (EBHSA)[7], Carbon nanotubes (CNT)[8-10], talcum powder[11,12], thermoplastic starch[13], Cellulose nanocrystal (CNC)[14], and PLA stereocomplex[15]. But, the torganic salts of sodium were uesd to accelated the crystallization speed of polyesters such as PET and PC[6]. Based on the above literatures, sodium salts failed to provide significant improvement of the crystallization rate. Both thermoplastic starch and CNC did not significantly affect PLA crystallinity.

Clay has been employed to improve thermal, mechanical and barrier properties of polymers. It is therefore interesting to examine its effect on the crystallization of PLA[16]. In a qualitative study, an increasing crystallization rate was reported in presence of different clay. Palygorskite (PA) is one of the most promoting clay in recent years, it can be found in many applications. PA presented a fibrous morphology and forms bundle. Its diameter is about 20-50nm with a length at about 700-1 200 nm [17]. Li et al.[18] prepared the polyethylene/palygorskite nanocomposites by in situ polymerization, and found that PA had strong effect on crystallization of PE. Wang et al.[19] used the γ-methacryloxypropyl trimethoxysilane (KH-570) as the coupling agent to modify the PA and prepared polypropylene/org- palygorskite nanocomposites by melt blending. The thermal and dynamic mechanical properties were characterized by differential scanning calorimetry and dynamic mechanical analysis. The results indicated that the incorporation of org-ATP also gave rise to an increase of the storage modulus and the changes of the glass transition temperature for PP composites. TEM and XRD results further revealed the addition of palygorskite did not change the crystal structure of PP; however, orgpalygorskite acted as nucleating agents for the crystallization of PP[19].

In a previous work[20], we also found that adding PA nanoparticles could improve the mechanical properties due to the strong hydrogen bonding interactions between the PLA matrix and the PA surface. In this research, PA was introduced into the PLA matrix as a natural nucleating agent, which is a promising natural one-dimensional filler for polymer matrix composites in recent years[21].

The crystallization kinetics of poly (L-lactide) have also been investigated over the past 40 years; therefore, the crystallization mechanisms of neat PLA are fundamentally well understood[22-25]. However, the incorporation of an inorganic particle into the polymer matrix increases the complexity of the crystallization mechanisms. As reported by Krikorian et al.[25], the interactions between nanoparticles and PLA matrix, the content, and the surface properties of nanoparticles are also alter the crystallization kinetics and the degree of crystallinity. There is a further need to reveal the effect of nanoparticles on the crystallization behavior of PLA. The interaction of an inorganic phase with the PLA, the volume percent, shape, and surface properties of the inorganic phase can potentially alter the crystallization kinetics, extent of crystallinity. In addition, the mechanical properties and degradation rates of PLA are strongly dependent on the crystal structure and the degree of crystallinity. The crystalline structure and morphology are also influenced by the thermal history[26]. During processing, the filler in polymeric-based composites may affect the crystallization behavior[27], and there is a further need to investigate the crystallization kinetics to reveal the relationship between the structural features and crystallization conditions.

This study was a continuous work followed by our previous report[20], where the nanocomposites exhibited strong hydrogen bonding interactions and presented several improved properties. Therefore, we not only can master the behavior of PA on PLA crystallization, but also can better understand the results reported in our earlier work. The focus of this study is to reveal the significant nucleating effect of PA on the crystallization behavior of the PLA matrix. In addition, the dynamic mechanical properties, rheology, and thermal stability are also investigated.

EXPERIMENTAL

Materials

Poly (lactic acid) (4032D) was supplied by Nature Works (USA) with a reported average molecular weight of $\overline{M}_w = 13.2 \times 10^4$ g/mol.

Palygorskite (PA) was kindly supplied by Gansu Kaixi environmental engineering, (China). The length-diameter ratio was about 23-30, and the composition as measured by X-ray fluorescence (XRF) was SiO_2 59.34 wt%, Al_2O_3 15.81 wt%, MgO 11.47 wt%, Fe_2O_3 7.36 wt%, CaO 3.12 wt%, K_2O 1.79 wt%, Na_2O 0.53 wt%, TiO_2 0.35 wt%, MnO 0.08 wt%, P_2O_5 0.06 wt%, Cr_2O_3 0.06 wt%, and BaO 0.01 wt%. The PA was used as received without any modifications.

Nanocomposites preparation

The nanocomposites were melt blended at 170℃ and 60 rpm on a RS600 HAAKE mixer. The processing of the PLA/PA nanocomposite with 1 wt% PA was carried out by placing 49.5 g PLA into the mixer at 170℃ for 2 minutes, followed by addition of 0.5 g PA with blending for another 5 minutes. Other nanocomposite samples with 3 wt% PA and 5 wt% PA were processed under the same methods and neat PLA was also processed using the same conditions.

Characterization

The morphologies of PA and PLA/PA nanocomposites were examined by scanning electron microscopy on the JEOL JSM-7401 (Tokyo, Japan) at an accelerating voltage of 1 kV. The PA surface and fracture surfaces were all coated with gold before examination. Transmission electron microscopy analysis was conducted by a JEOL JEM - 2010 (Tokyo, Japan) instrument with an accelerating voltage of 120 kV. PLA/PA nanocomposite was prepared by ultramicrocut on a Leica EMUC 6 microtoming (Germany).

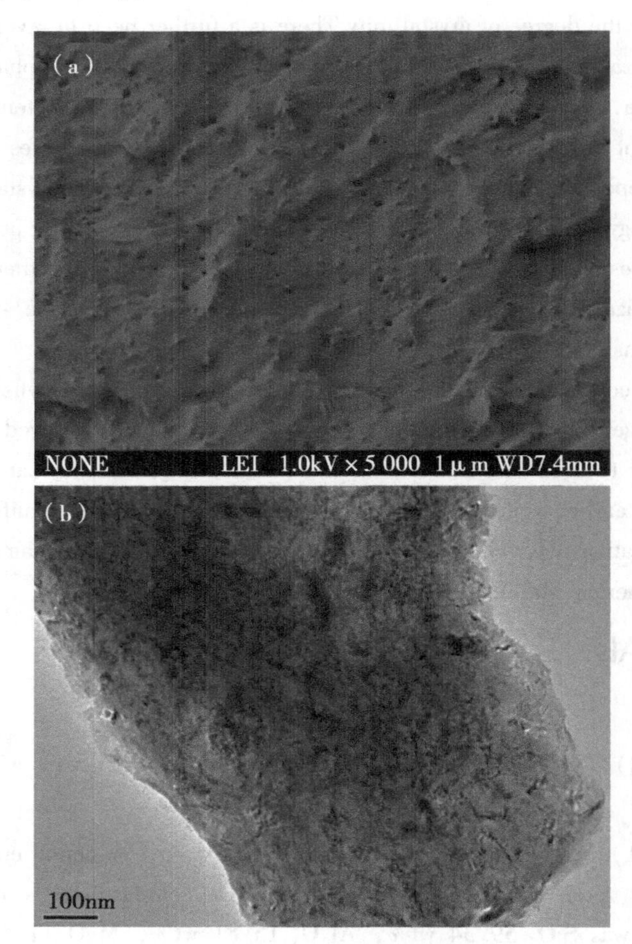

Fig. 1 SEM and TEM images of 3wt% PLA/PA nanocmposite

Dynamic mechanical analysis (DMA) of the nanocomposites was measured in tensile mode by a Rheometric Scientific DMTA V. The measurements were carried out from 30℃ to 80℃ at a constant frequency of 1 Hz, strain amplitude of 0.05%, and a heating rate of 3℃/min.

The dynamic rheological measurements of PLA and PLA/PA nanocomposites were performed in a Malvern gemi 200 stress/strain controlled rheometer. The diameter of the parallel plates was 25 mm with a gap of 1 mm. The operation was in oscillatory shear mode at 170℃ by the frequency sweeps from 10^{-2} to 10^{2} Hz at 1% strain.

Differential scanning calorimetry (DSC) was performed on a Shimadzu DSC - 60 with a nitrogen gas flow rate of 40mL/min. All samples were heated to 210℃ at a heating rate of 10℃/min and maintained for 3 min in order to eliminate any thermal history, and then quickly cooled to the crystallization temperature, and maintained at T_c until the crystallization was finished. After that, the samples were reheated to 210℃ at a rate of 10℃/min. The crystallization temperatures were selected from 110℃ to 130℃ with an interval of 5℃.

RESULTS AND DISCUSION

Physical properties

Scanning Electron Microscopy (SEM) and Transmission Electron Microscopy (TEM) Analyses. The morphology of PA dispersion in PLA matrix were measured by SEM and TEM, respectively. The fracture surface of the PLA/PA nanocomposites was smooth and the boundary between PA and PLA matrix was unconspicuous (Fig.1a). PA was shown as single fiber randomly distributed in PLA matrix. Most PA fibers were dispersed and embedded in the PLA matrix without obvious aggregates. Some unique characteristics shown by uniformly dispersed PA fibers in the PLA matrix (Fig.1b) indicated the good compatibility between PA and PLA when PA content was less than 3 wt%.

Dynamic Mechanical Properties. DMA provides information on mechanical behavior, molecular relaxations as well as interactions taking place in the produced materials as the temperature is varied. Therefore, the dynamic mechanical properties of PLA and PLA/PA nanocomposites were employed to measure the thermomechanical properties of PLA and its nanocomposites. Fig. 2a and b shows the temperature dependence of the storage modulus (E') and damping factor (tanδ) of the neat PLA and PLA/PA nanocomposites at 30-80℃.

Storage modulus corresponds to the elastic response to the deformation. In Fig.2 (a), storage modulus increases with the increasing PA content in the low temperature range. In the test temperature range, nanocomposites exhibit higher storage modulus than that of the neat PLA, indicating that addition of PA particles into PLA matrix results in a remarkable increase of elastic properties for PLA matrix. The SEM and TEM analysis also showed that PA fibers embedded in the PLA matrix, which would have a strong effect on the segmental motion of the PLA chains. On the contrary, all samples reached a rubbery plateau which suggested a rubberlike structure composed of both crystalline and amorphous phases at higher temperatures. The reason, as reported by Krikorian et al.[28], was that the effect of clay on the storage modulus becomes negligible and the nanocomposite stiffness becomes matrix dependent in the high temperature region. Therefore, all samples reached thermal-mechanical stability at high temperatures. The above phenomena, as reported by Di et al.[29], was attributed to the well dispersion of PA in the PLA matrix, and was further confirmed in this research.

An apparent α relaxation appeared as indicated by a sharp decrease of storage modulus in all samples, which attributed to the glass transition of the amorphous phase of PLA. The glass transition is assigned to the energy dissipation possibilities across the free amorphous phase, and the lower T_g values mean an easier mobility of the free amorphous phase in the composites[30]. The glass transition temperature (T_g) obtained from the peak temperatures of the tand curves is presented in Fig. 2 (b). Normally, the T_g tends to increase with the addition of nanoparticles because of the interactions between the polymer chains and the nanoparticles which reduce molecular chain mobility[31,32]. Fig. 2b shows that the intensity of the tanδ peak of the PLA/PA nanocomposites are all lower than that of neat PLA.The possible explanation, as reported by Huda et al.[30], is that there is no restriction to the chain motion in the neat PLA matrix.By contrast, the presence of the PA hinders the chain mobility, resulting in the reduction of sharpness and height of the tanδ peak.In this research, the T_g of all nanocomposites was higher than that of neat PLA, which shifted from 59. 8℃ to 62. 1℃ for 3 wt% PLA/PA nanocomposite.The similar results were once reported by other researchers[30]. The T_g of all the composites shifted to higher temperatures, which is associated with the decreased mobility of the matrix chains.The increase in modulus, together with positive shift in tanδ peak position, can be attributed to physical interaction between the polymer and reinforcements that restrict the segmental mobility of the polymer chains in the vicinity of the nanoreinforcements[33].The DMA results indicate that the incorporation of PA hinders the motions of the PLA chains in the matrix, thereby increasing the maximum service temperature of the PLA.

Rheology.In order to investigate the response of the nanostructure to dynamic shearing, the rheology test was performed on PLA and its nanocomposites because the melt viscosity of polymers is sensitive to the changes in macromolecular chain structure.This research was performed at 170℃ as a function of oscillation frequency. Fig. 3 shows the results of (a) dynamic complex viscosity ($|\eta*|$), (b) storage modulus (G'), and (c) loss modulus (G''), of PLA and PLA/PA nanocomposites.

In Fig.3a, all of the samples present shear thinning behavior with increasing shear frequency.In the low oscillation frequency region, the complex viscosities of all samples present only a small frequency dependence, revealing a Newtonian plateau, except for testing below 0. 05 Hz.Below 0. 05 Hz, the complex viscosities of all samples decreases with decreasing frequency.The reason, as reported by Racha et al.[34] can be attributed to thermal degradation which further confirms that the PLA can be easily degraded, especially at high temperature.This phenomenon is related to the decrease of the average molecular weight.Here it is noted that the zero shear viscosity increases with the addition of PA.All the zeroshear-rate viscosity of the nanocomposites is higher than the neat PLA and this reinforcement effect is more significant at low frequency than at high frequency.

Fig. 3b and c demonstrates the response of frequency on G' and G'' for all the samples.For homogeneous polymers, polymer chains follow to $G' \alpha\ \omega^2$ and $G'' \alpha\ \omega$[35].Therefore, the G' of PLA chains follow this manner.By contrast, the G' of PLA/PA nanocomposites increases monotonically at all frequency region and reaches a plateau at low frequency region, especially when the clay content is more than 3%.The reason for the terminal region in the nanocomposites at high concentration of clay, as reported by Qi et al.[36] is that the intercalation of the clay particles.

In Fig.3b, there is an improvement of G' over the whole frequency range, more pronounces at low frequency region. Therefore, the difference of slope at terminal could be easy to notice. The values of G' of all nanocomposites increases at low frequency which indicates that the addition of clay has effect on the samples'linear dynamic viscoeleastic response, especially in the low frequency region[37].In contrast, the G' of the nanocomposites approaches similar values at high frequency region.This phenomenon indicates that the observed chain relaxation modes are almost unaffected by the presence of the PA particles[38].

The frequency dependence of G'' has a similar tendency of G'.But, the structure of the nano-

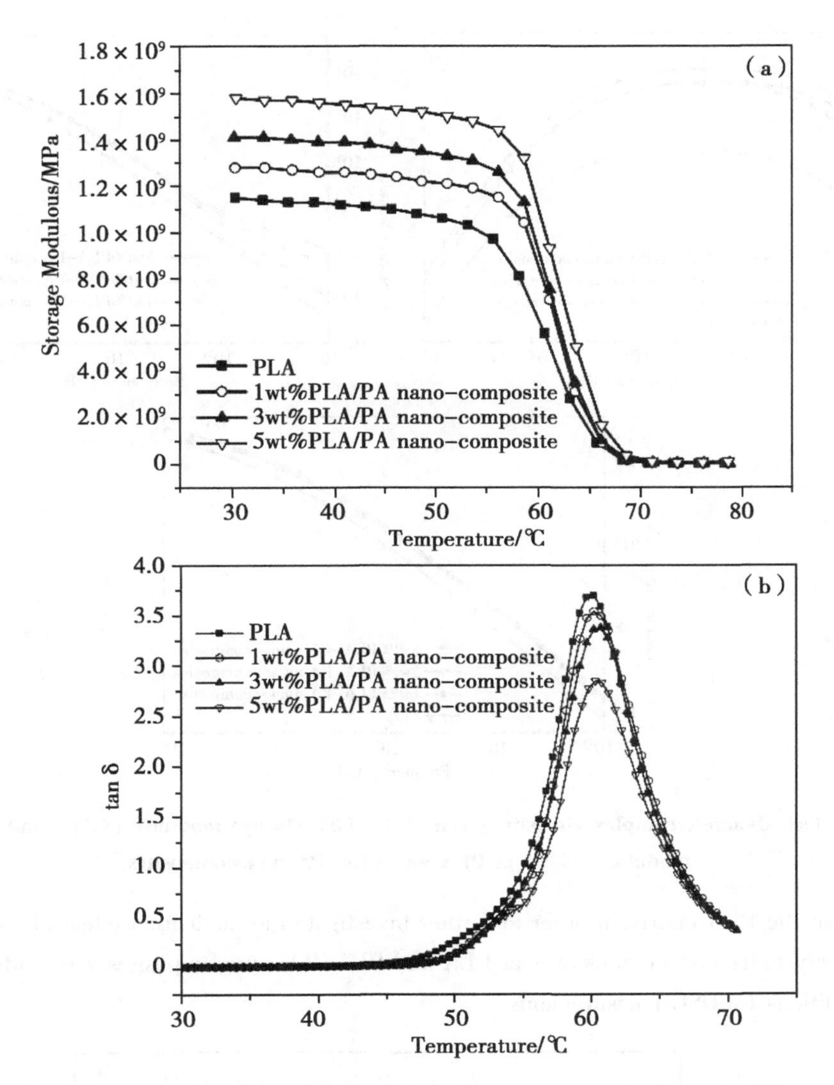

Fig.2 Temperature dependence (a) Storage modulus (E'), and (b) damping factor (tanδ) of PLA and PLA/PA nanocomposites

composites is less sensitively reflected on the loss modulus than on the storage modulus.

Thermal properties

Crystallization Behavior.The crystallization rate of PLA is very slow compared to other polymers, therefore, the enhancement of crystallization rate and thermal stability of PLA is very important to the application of PLA.Increasing the crystallinity by the addition of a nucleating agent to PLA is an effective method for the production PLA-based materials with high thermal stability.In order to investigate the nucleation effect, DSC of the composites was employed in this research.

Fig. 4 presents the first cooling DSC curves of neat PLA and nanocomposites.It can be seen that the neat PLA has no crystallization peak.By contrast, the most encouraging finding was the appearance of a crystallization peak in all nanocomposites.With increasing PA content, the crystallization temperature shifts from 92. 1℃ to 99. 6℃.The above results indicate that the PA has a strong nucle-

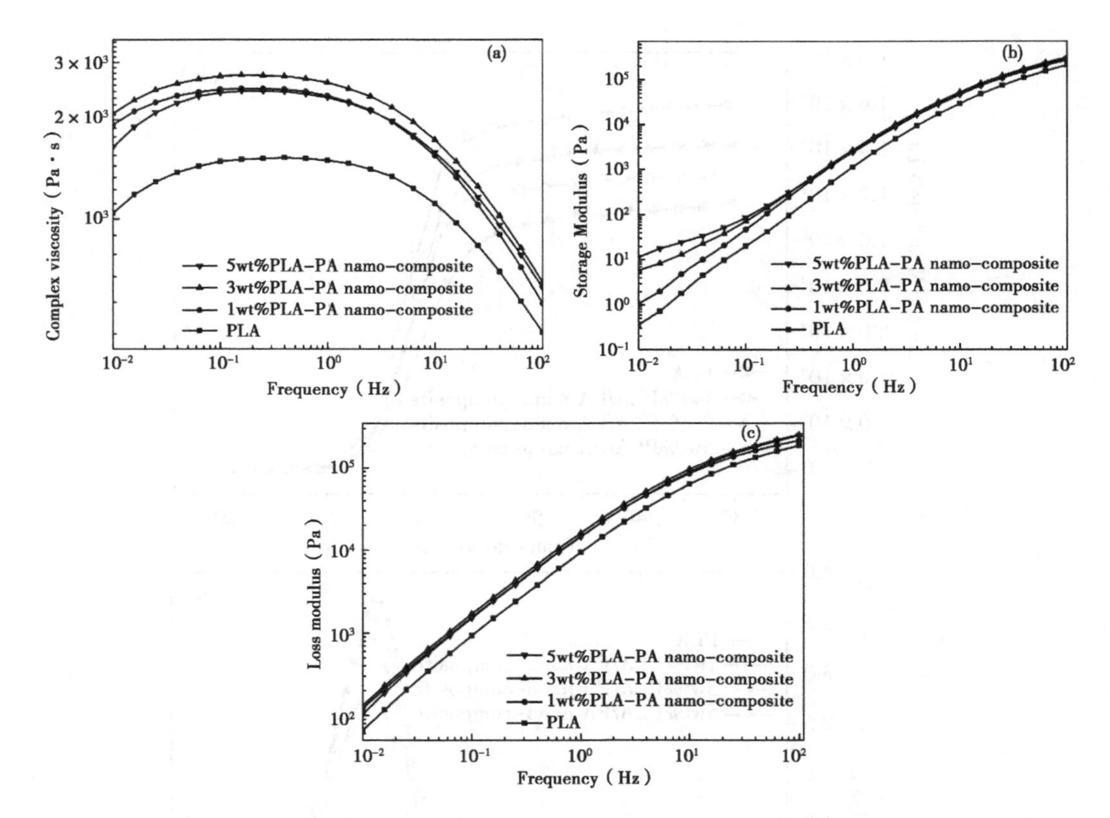

Fig.3 (a) dynamic complex viscosity ($|\eta^*|$), (b) storage modulus (G'), and (c) loss modulus (G''), of PLA and PLA/PA nanocomposites

ating effect on the PLA matrix. In order to further investigate the nucleation effect of PA on the PLA matrix, the crystallization kinetics of neat PLA and PLA/PA nanocomposites was studied under isothermal conditions by DSC measurements.

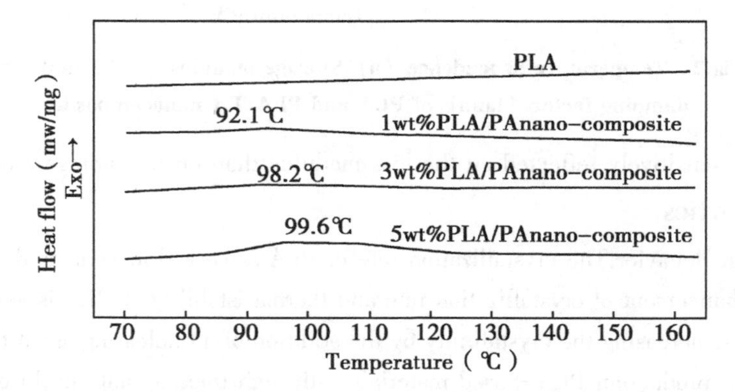

Fig.4 The first cooling DSC curves of neat PLA and nanocomposites

Fig. 5 shows the first heating and second heating DSC curves of neat PLA and nanocomposites. It can be obviously seen that the cold-crystallization peak of nanocomposites shifted to the lower temperature during the heating process, which indicated that the crystallization ability of nanocomposites were higher than the neat PLA.This phenomenon would be a clear indication of the nucleation effect of PA on PLA matrix.The related thermal data are summarized in Table 1, the X_c was also improved by incorporation of PA.The X_c of 5 wt% PLA/PA nanocomposite was lower than the other nanocomposites due to the aggregation of PA, which decreased the crystallization rate.

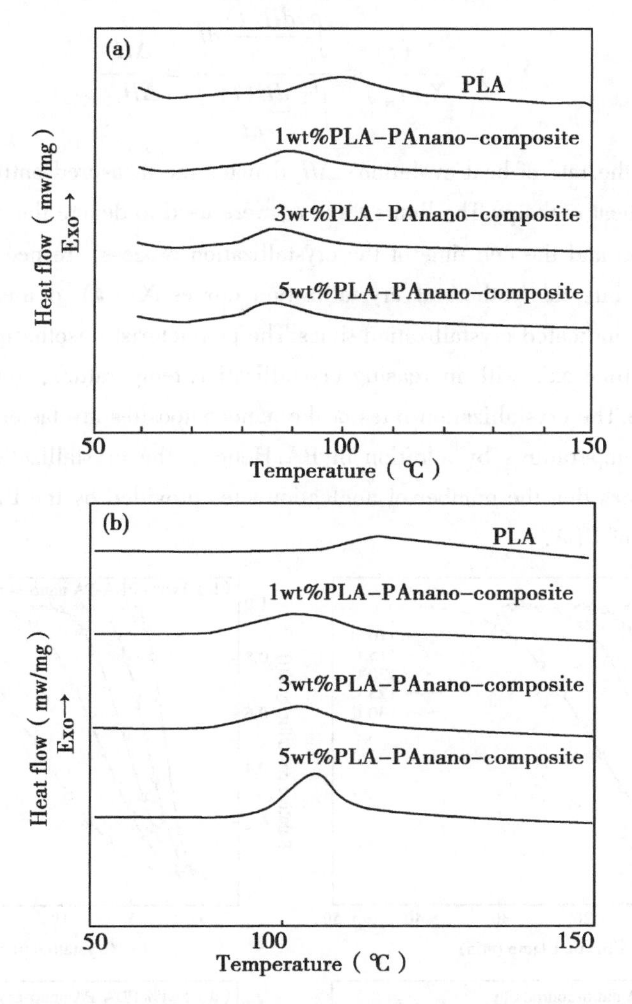

Fig.5 The first heating and second heating DSC curves of neat PLA and nanocomposites

Table 1 Thermal parameters for neat PLA and PLA/PA nanocomposites

Sample	Content				
	T_{cc} (℃)	T_c (℃)	T_m (℃)	ΔH_m (J/g)	X_c (%)
PLA	114.4	—	163.4, 168.2	28.81	31.3
1wt% PLA/PA nanocomposite	103.3	92.1	168.5	35.23	38.3
3wt% PLA/PA nanocomposite	103.5	98.2	166.9	38.14	42.3
5wt% PLA/PA nanocomposite	103.8	99.6	168.3	32.19	36.4

where, $X_c = \frac{\Delta H_m}{(1-\phi)\ \Delta H_m^0} \times 100\%$, ϕ-PA content (%), $\Delta H_m^0 = 93J/g$.

Isothermal Crystallization Kinetics Analysis.The addition of nucleating agents into PLA was investigated in terms of half–time for crystallization under isothermal conditions from 110 to 130℃.The degree of crystallinity X (t) is defined relative to the maximum achievable crystallinity level[5], which was followed as a function of time at fixed temperatures.

$$X(t)=\frac{X_c(t)}{X_c(t_\infty)}=\frac{\int_0^t \frac{dH(t)}{dt}dt}{\int_0^{t_\infty} \frac{dH(t)}{dt}dt}=\frac{\Delta H_t}{\Delta H_{t_\infty}} \tag{1}$$

Where dH/dt is the rate of heat evolution; ΔH_t denotes the measured enthalpy of crystallization; and ΔH_{t_∞} is the total heat enthalpy.The limits t and t_∞ were used to denote the elapsed time during the course of crystallization and the end time of the crystallization process, respectively.

Fig. 6 shows the integral isothermal crystallization curves X (t) of neat PLA and PLA/PA nanocomposites at the indicated crystallization times.The characteristic isotherms are all shifted to the right range along the time axis with increasing crystallization temperature, reflecting a reduction in the crystallization rate.The crystallization rates of the nanocomposites are faster than that of neat PLA at all crystallization temperatures by addition of PA. Hence, the crystallization time is greatly decreased, which deduces that the number of nucleation sites provided by the PA would accelerate the rate of crystallization of PLA.

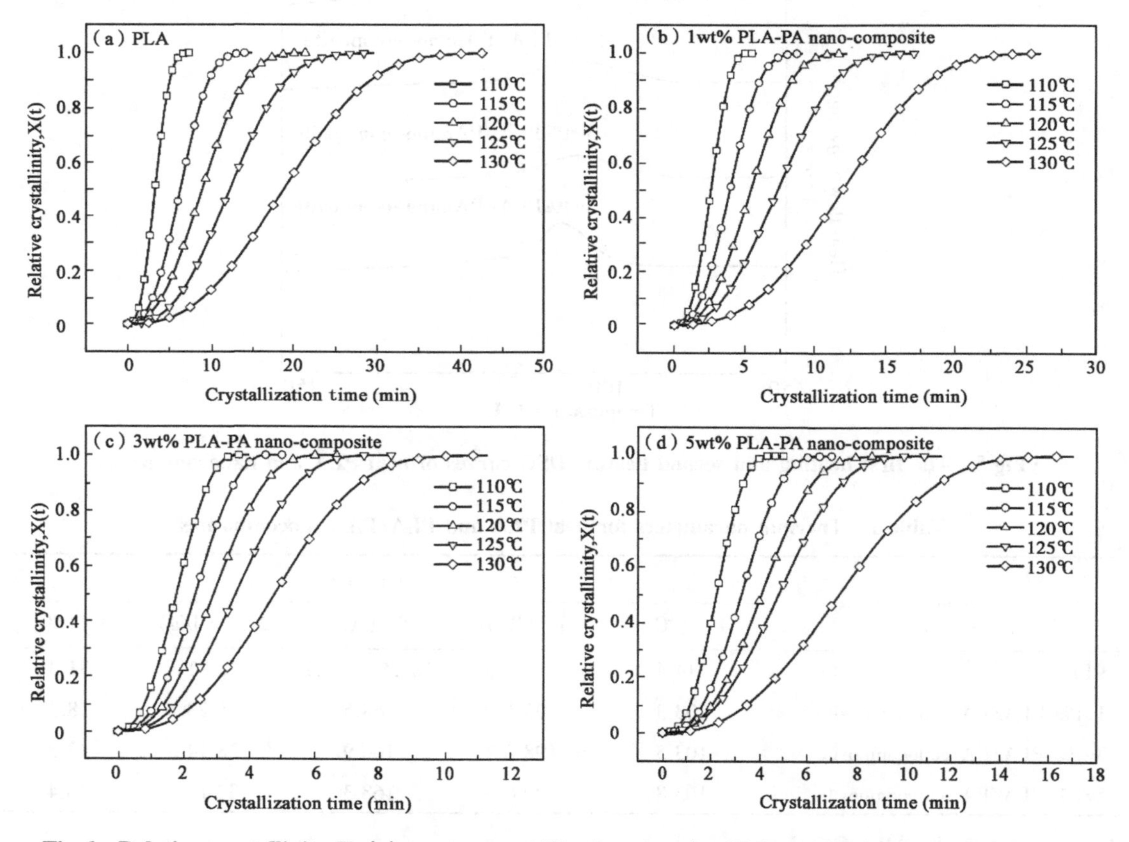

Fig.6 Relative crystallinity X (t) versus crystallization time for PLA and PLA/PA nanocomposites

In this research, the Avrami equation was used to predict the crystallization kinetics of polymers under isothermal conditions for various modes of nucleation and growth[39]. The general form of the equation is

$$X(t) = 1 - \exp(-kt^n) \quad (2)$$

where n is a constant that depends on the type of nucleation and growth of the crystals, k is the crystallization rate constant involving both nucleation and growth rate parameters under isothermal conditions. Equation 2 above can also be rewritten as follows:

$$\lg[-\ln(1 - X(t))] = \lg k + n\lg t \quad (3)$$

Fig. 7 shows the plots of lg [−ln (1−X (t))] versus lg [t] for PLA and its nanocomposites. The parameters of the Avrami equation are got from these plots. Almost all curves show a linear relationship at first, and then the lines subsequently tend to deviate from the straight lines. This is caused by the secondary crystallization, which is due to the spherulite impingement and perfection of internal spherulite crystallization in the later stage of the crystallization process[40]. However, the Avrami method can still be used to roughly characterize the isothermal crystallization behavior of PLA and its blends[41]. Therefore, the n and k should be obtained from the slope and interception of the linear fitting, and the results are listed in Table 2.

Table 2 Isothermal crystallization kinetics parameters for neat PLA and PLA/PA nanocomposites

Samples	T (℃)	n	k (min^{-1})	$t_{1/2}$ (min)
PLA	110	2.69	0.028984	3.24
	115	2.56	0.006484	6.20
	120	2.42	0.003610	8.76
	125	2.64	0.000946	12.15
	130	2.61	0.000341	18.54
1 wt%-PLA/PA nano-composite	110	2.72	0.053605	2.56
	115	2.63	0.019622	3.89
	120	2.63	0.008371	5.35
	125	2.56	0.004403	7.19
	130	2.66	0.001001	11.72
3 wt%-PLA/PA nano-composite	110	2.41	0.186724	1.73
	115	2.61	0.078540	2.30
	120	2.61	0.044104	2.87
	125	2.69	0.023143	3.54
	130	2.65	0.011270	4.73
5 wt%-PLA/PA nanocomposite	110	2.82	0.074459	2.27
	115	2.81	0.025956	3.22
	120	2.74	0.015110	4.04
	125	2.51	0.014293	4.70
	130	2.57	0.002076	9.62

The Avrami exponent n reflects the nucleation and the growth processes[42,43]. For a three-dimensional diffusion-controlled crystal growth, the values of n should be 3. As described in the

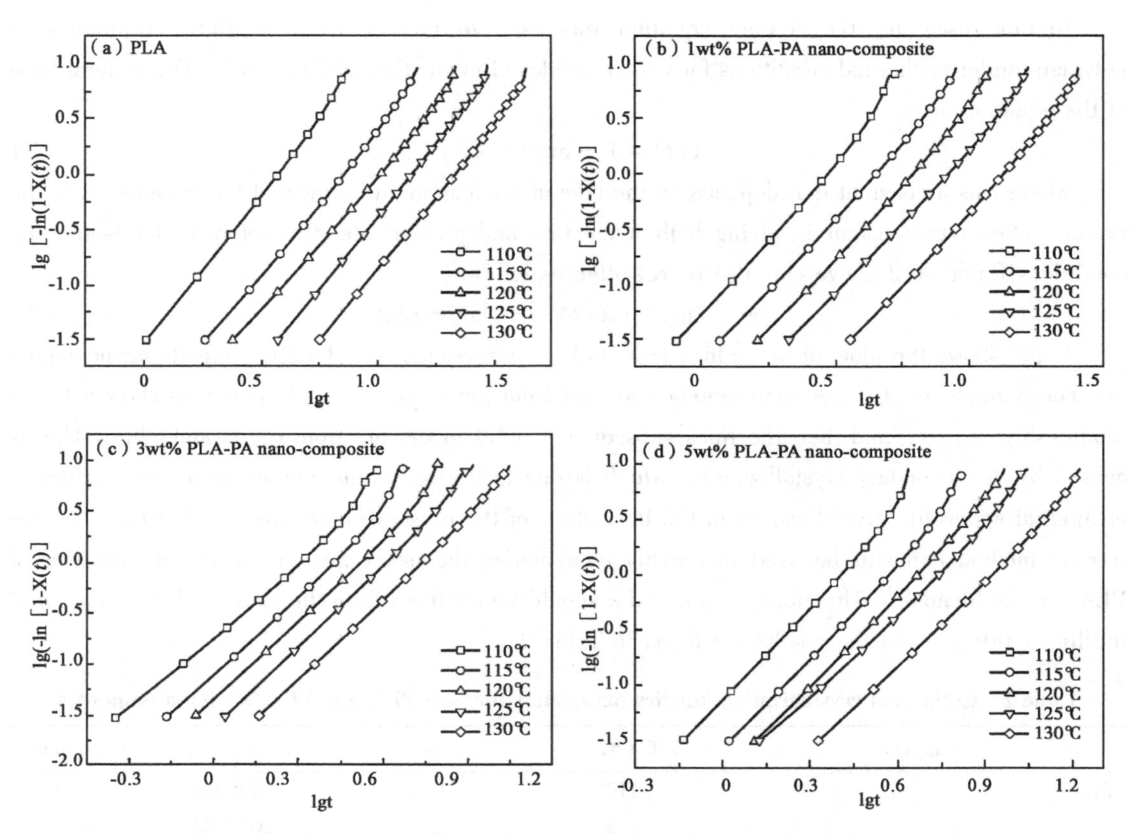

Fig.7 Plots of lg [−ln (1−X (*t*))] versus lg*t* for PLA and PLA/PA nanocomposites

former literature, the n values of neat PLA and nucleated PLA are affected by many factors, such as the nucleation density and restriction of crystalline formation[41,44].Thus, the Avrami exponent n differs from sample to sample, and it also varies with Tc.In this research, the n in the Tc range tested are varied from 2.42 to 2.69 for neat PLA.The reason, as reported by Qi et al.[45], is that the residual catalysts and impurities had provided nucleation sites during the crystallization process. Hence, the crystallization mechanism of neat PLA was correlated to the spherulitic growth form.For nanocomposites, it was slightly increased by the addition of the PA, the same results was once obtained by Qi et al.[45].

Generally, the overall crystallization rate is determined by the rate of nucleation and the crystal growth process, and k and $t_{1/2}$ are evaluated to compare the overall crystallization rate of polymers[46].Both the rate of nucleation and the growth processes reflect on the constant k.In Table 2, the values of k decrease with increasing Tc in all of the cases, which correlates to the nucleation rate and growth processes. The crystallization kinetic constant k for PLA − PA nanocomposites increases by incorporation of the PA nanoparticles, suggesting that PA can act as an effective nucleating agent to accelerate the crystallization rate of PLA.

Normally, the $t_{1/2}$ indicates a minimum crystallization time at a fixed crystallization temperature, and the crystallization half-time $t_{1/2}$ is defined as the time required to reach half of the final crystallinity.The form of the equation is shown as the following manner.

$$t_{1/2} = (\ln 2/k)^{1/n} \quad (4)$$

In Table 2, the neat PLA shows a very slow crystallization rate and the minimum crystallization $t_{1/2}$ is around 18.54 min at 130℃. The $t_{1/2}$ of PLA was greatly reduced by the addition of the PA at all crystallization temperatures, indicating that all of the tested nucleating agents can increase the crystallization rate. The addition of PA effectively increases the crystallization rate, thus the $t_{1/2}$ decreases to 4.73 min for 3wt% PLA/PA nanocomposite. The above results further confirm that PA acts as a heterogeneous nucleation agent and accelerates the crystallization rate of PLA matrix. The PA nucleating ability could be compared to the silica and montmorillonite[47], therefore, PA could be a succedaneum of silica and montmorillonoid as nucleating agent for PLA.

CONCLUSIONS

PLA/PA nanocomposites were prepared through melt compounding in a HAAKE mixer. The increased storage modulus of all PLA/PA nanocomposites in the low temperature region was induced by the incorporation of PA, which had strong effect on their high elastic properties. By contrast, at higher temperatures, all samples reached a rubbery plateau, which suggested a rubberlike structure composed of both crystalline and amorphous phases. The T_g of all nanocomposites was higher than that of neat PLA, which was slightly shifted to the higher temperature. The DMA results indicated that the incorporation of PA hindered the motions of the PLA chains in the matrix and thereby increasing the temperature of use of the PLA.

From the rheology test, all the zero-shear-rate viscosities of the nanocomposites were higher than the neat PLA and this reinforcement effect was more significant at low frequencies than that at high frequencies. The values of G' for all PA reinforced samples increased at low frequencies and became similar in the high frequency region which indicated that adding clay had a significant effect on the samples' linear dynamic viscoeleastic response, especially in the low frequency region.

The significant finding by DSC was that PA acted as nucleating-agent and accelerated the crystallization process of neat PLA. As a result, the crystallization peak appeared in all nanocomposites. With increasing PA content, the crystallization temperature shifted from 92.1℃ to 99.6℃, which has also been further confirmed by isothermal crystallization kinetics analysis. Therefore, PLA/PA nanocomposites revealed the significant improvement in physical and thermal properties.

References

[1] L.T.Lim, R.Auras, and M.Rubino, *Prog.Polym.Sci.*, 33, 820 (2008).

[2] K.Marsh and B.Bugusu, *J.Food Sci.*, 72, R39 (2007).

[3] D.Garlotta, *J.Polym.Environ.*, 9, 63 (2001).

[4] S.S.Ray, *Polymer*, 51, 3966 (2010).

[5] H.Li and M.A.Huneault, *Polymer*, 48, 6855 (2007).

[6] M.Penco, G.Spagnoli, I.Peroni, M.A.Rahman, M.Frediani, W.Oberhauser, and A.Lazzeri, *J.Appl. Polym.Sci.*, 122, 3528 (2011).

[7] J.Y.Nam, M.Okamoto, H.Okamoto, M.Nakano, A.Usuki, and M.Matsuda, *Polymer*, 47, 1340 (2006).

[8] Y.T.Shieh, Y.K.Twu, C.C.Su, R.H.Lin, and G.L.Liu, *J.Polym.Sci.*, *Part B: Polym.Phys.*, 48, 983 (2010).

[9] S.I.Moon, F.Jin, C.J.Lee, S.Tsutsumi, and S.H.Hyon, *Macromol.Symp.*, 224, 287 (2005).
[10] J. - Z. Xu, T. Chen, C. - L. Yang, Z. - M. Li, Y. - M. Mao, B. - Q. Zeng, and B. S. Hsiao, *Macromolecules*, 43, 5000 (2010).
[11] H.Urayama, T.Kanamori, K.Fukushima, and Y.Kimura, *Polymer*, 44, 5635 (2003).
[12] N.Kawamoto, A.Sakai, T.Horikoshi, T.Urushihara, and E.Tobita, *J.Appl.Polym.Sci.*, 103, 198 (2007).
[13] H.Li and M.Huneault, *Int.Polym.Process.*, 23, 412 (2008).
[14] A.Pei, Q.Zhou, and L.A.Berglund, *Compos.Sci.Technol.*, 70, 815 (2010).
[15] S.Brochu, R.E.Prud'Homme, I.Barakat, and R.Jerome, *Macromolecules*, 28, 5230 (1995).
[16] S.Saeidlou, M.A.Huneault, H.Li, and C.B.Park, *Prog.Polym.Sci.*, 37, 1657 (2012).
[17] H.H.Murray, *Appl.Clay Sci.*, 17, 207 (2000).
[18] W.Li, A.Adams, J.Wang, B.Blumich, and Y.Yang, *Polymer*, 51, 4686 (2010).
[19] L.Wang and J.Sheng, *Polymer*, 46, 6243 (2005).
[20] Y.Jiang, S.Han, S.Zhang, J.Li, G.Huang, Y.Bi, and Q.Chai, *Polym.Compos.*, 35, 468 (2014).
[21] B.Pan, Q.Yue, J.Ren, H.Wang, L.Jian, J.Zhang, and S.Yang, *Polym.Test.*, 25, 384 (2006).
[22] T.Miyata and T.Masuko, *Polymer*, 39, 5515 (1998).
[23] P.Pan, W.Kai, B.Zhu, T.Dong, and Y.Inoue, *Macromolecules*, 40, 6898 (2007).
[24] Z.Kulinski and E.Piorkowska, *Polymer*, 46, 10290 (2005).
[25] V.Krikorian and D.J.Pochan, *Macromolecules*, 38, 6520 (2005).
[26] Z.Qiu and W.Yang, *Polymer*, 47, 6429 (2006).
[27] G. Z. Papageorgiou, D. S. Achilias, D. N. Bikiaris, and G. P. Karayannidis, *Thermochim. Acta*, 427, 117 (2005).
[28] V.Krikorian and D.J.Pochan, *Chem.Mater.*, 15, 4317 (2003).
[29] Y.Di, S.Iannace, E.D.Maio, and L.Nicolais, *J.Polym.Sci.*, *Part B*: *Polym.Phys.*, 43, 689 (2005).
[30] M.Huda, L.Drzal, A.Mohanty, and M.Misra, *Compos.Part B*, 38, 367 (2007).
[31] L.Petersson, I.Kvien, and K.Oksman, *Compos.Sci.Technol.*, 67, 2535 (2007).
[32] O.Martin and L.Averous, *Polymer*, 42, 6209 (2001).
[33] M.Jonoobi, J.Harun, A.P.Mathew, and K.Oksman, *Compos.Sci.Technol.*, 70, 1742 (2010).
[34] R.Al-Itry, K.Lamnawar, and A.Maazouz, *Polym.Degrad.Stab.*, 97, 1898 (2012).
[35] J.R.Dorgan, H.Lehermeier, and M.Mang, *J.Polym.Environ.*, 8, 1 (2000).
[36] Z.Qi, H.Ye, J.Xu, J.Peng, J.Chen, and B.Guo, *Colloids Surf.A*, 436, 26 (2013).
[37] L.Shen, Y.Lin, Q.Du, W.Zhong, and Y.Yang, *Polymer*, 46, 5758 (2005).
[38] A.Bhatia, R.K.Gupta, S.N.Bhattacharya, and H.Choi, *J.Appl.Polym.Sci.*, 114, 2837 (2009).
[39] A.T.Lorenzo, M.L.Arnal, J.Albuerne, and A.J.Muller, *Polym.Test.*, 26, 222 (2007).
[40] X.Chen, J.Shi, L.Wang, H.Shi, and Y.Liu, *Polym.Compos.*, 32, 177 (2011).
[41] T.Ke and X.Sun, J.Appl.*Polym.Sci.*, 89, 1203 (2003).
[42] H.Tsuji, H.Takai, and S.K.Saha, *Polymer*, 47, 3826 (2006).
[43] M.L.Di Lorenzo, *Eur.Polym.J.*, 41, 569 (2005).
[44] K.S.Kang, S.I.Lee, T.J.Lee, R.Narayan, and B.Y.Shin, *Korean J.Chem.Eng.*, 25, 599 (2008).
[45] Z.Qi, Y.Tang, J.Xu, J.Chen, and B.Guo, *Polym.Compos.*, 34, 1126 (2013).
[46] R.Vasanthakumari and A.Pennings, *Polymer*, 24, 175 (1983).
[47] G.Papageorgiou, D.Achilias, S.Nanaki, T.Beslikas, and D.Bikiaris, *Thermochim.Acta*, 511, 129 (2010).

(发表于《Polymer Composites》, SCI: 1.632)

Study on Matrix Metalloproteinase 1 and 2 Gene Expression and NO in Dairy Cows with Ovarian Cysts

MUTLAG Ali-M., WANG Xue-zhi, YANG Zhi-qiang, MENG Jia-ren, WANG Xu-rong, ZHANG Jing-yan, QIN Zhe, WANG Gui-bo, LI Jian-xi

(Engineering & Technology Research Center of Traditional Chinese Veterinary Medicine of Gansu Province/Key Laboratory of Veterinary Pharmaceutical Development of Ministry of Agriculture/Key Laboratory of Animal Drug Project of Gansu Province/Lanzhou Institute of Husbandry and Pharmaceutical Sciences of Chinese Academy of Agricultural Sciences, Lanzhou 730050, China)

Abstract: The objective of this study was to investigate how changes in total nitric oxide (NO) and ovarian matrix metalloproteinase (MMP) -1 and MMP-2 correlated with luteinizing hormone (LH) and estradiol (E_2) in infertile dairy cows with ovarian cysts. Holstein cows ($n=21$ infertile cows and 19 fertile) were studied during their estrous phase to minimize hormone fluctuations. Blood LH, E_2 and NO were measured. Expression of the MMP-1 and MMP-2 genes in the ovaries was measured by real-time PCR and immunohistochemistry. LH, E_2 and NO were less in infertile cows with ovarian cysts than in fertile cows ($P<0.05$). The mRNAs of MMP-1 and MMP-2 were less in infertile cows with ovarian cysts than in fertile cows ($P<0.05$). The immunohistochemical results showed that MMP-1 and MMP-2 genes were expressed in different parts of the ovarian tissues, including granulosa and theca cells of preovulatory follicles, epithelial follicular cells of small follicles, stromal cells and endothelial cells of blood vessels. The results showed that a decrease in MMP-1 and MMP-2 gene expression is accompanied with a decrease in NO concentrations in infertile cows affected with ovarian cysts. The expression of these marker genes might be risk factors of infertility in cows and might correlate with the hormonal profile. The present study suggests that the abnormal expression of the MMP-1/2 gene might be an important marker of ovarian follicular cysts in dairy cows.

Key words: Matrix metalloproteinase; Nitric oxide; Luteinizing hormone; Estradiol; Infertility; Ovarian follicular cyst

1 INTRODUCTION

Reproductive performance is one of the most important factors affecting dairy farm profitability (Plaizier et al., 1998). Reproductive efficiency and profitability are maximized by the timely achievement of a subsequent pregnancy (Peng et al., 2011). However, the overall fertility in dairy cows has decreased in recent decades, although milk production per cow has increased. The risk factors associated with infertility are many and can be complex. The risk factors include diseases, poor nutrition, hereditary factors, congenital factors, environmental factors and managerial factors (Walsh et al., 2011). The ovaries have key roles in reproduction, and any impairment in their functions can result in either sterility or infertility (AlDahash and Bensassi, 2009). Ovarian follicular cysts are the most common reproductive disorder in dairy cows (Silvia et al., 2002).

In general, ovaries undergo continuous tissue remodeling during follicular growth, maturation and ovulation. The extracellular matrix (ECM) influences basic cellular processes, such as proliferation, differentiation, migration and adhesion (Portela et al., 2009). According to Whited et al. (2010), matrix metalloproteinases (MMP), a family of zinc-dependent endopeptidases with proteolytic activities against several components of the ECM (Hulboy et al., 1997), have important roles in this process by cleaving ovarian tissue components, releasing growth factors, contributing to the degradation of the ECM, and clearing the way for new growth. Bakke et al. (2004) demonstrated that MMP-1 immunoreactivity is localized to both the granulosa and theca layers of preovulatory follicles in cattle relative to gonadotropin-releasing hormone (GnRH) treatment, with minimal specific gene expression detected in the adjacent ovarian stroma. Imai et al. (2003) reported that MMP-2 increases in the follicular fluid in the preovulatory follicles of fertile cows when the follicles reach maximum size.

Nitric oxide (NO), a highly reactive free radical synthesized from L-arginine, either by a constitutive calciumdependent neuronal NO synthase (nNOS) and endothelial NOS (eNOS) or by a pro-inflammatory cytokine-inducible NOS (iNOS) *in vivo* (Moncada et al., 1991), is an intraand inter-cellular modulator of many biological processes, including vasodilation, maintenance of endothelial cell barrier function and control of apoptosis (Rosselli et al., 1998). The NO is present in the follicular fluid of cattle, and it is produced by granulosa cells in culture; the most active granulosa cells producing NO are those from the small follicles (Basini et al., 1998; Pancarc. et al., 2012). According to Anteby et al. (1996) and Tamanini et al. (2003), NO has a key role in ovarian physiology, including follicular development, ovulation, CL function and ovarian vasodilatation.

To our knowledge, few studies have examined the role of MMP in the ovaries of infertile dairy cows affected by ovarian cysts. The present study aimed to evaluate how the changes in total NO, ovarian MMP-1 and ovarian MMP-2 were correlated with LH and E_2 in infertile dairy cows with ovarian cysts. Particular attention was paid to the quantity and immunolocalization of MMP-1 and MMP-2.

2 MATERIAL AND METHODS

2.1 Animals

This study involved 21 Holstein dairy cows diagnosed with ovarian cysts between 60 and 150 days postpartum and artificially inseminated 4 or 5 times, according to the farm records. For the purpose of the present study, ovarian follicular cysts are defined as anovulatory fluid-filled structures ≥

17mm in diameter that persisted on the ovaries for more than 6 days, according to the protocol described by Silvia et al. (2002). Nineteen cows with no clinically visible pathological conditions and with follicles from which ovulation typically occurred when follicles were 13–17mm in diameter were included in the fertile group (control). The cows were obtained from three commercial dairy farms with similar management practices, as follows: 13 cows from Huazhuang dairy farm, Lanzhou, China (six infertile and seven fertile), 14 cows from Qinwangchuan dairy farm, Lanzhou (eight infertile and six fertile), and 13 cows from Zhenxin dairy farm, Yinchuan (eight infertile and five fertile). The study population was 3–7 years old (two to five lactations), and the voluntary waiting period from calving to first service established for this dairy herd was 60 days after birth. In addition, the enrolled cows had a body condition score between 2.5 and 4.0 according to the criteria of Edmonson et al. (1989).

2.2 Samples and experimental protocol

Blood samples of all cows were collected from the coccygeal vein and transferred to an icebox. The serum was separated in the laboratory by centrifugation at 3 000rpm for 15 min and stored at −20℃ until assayed. For the purpose of the present study, the ovaries with fertile dominant follicles from three different fertile cows (1 317mm in diameter) and a few preovulatory follicles, as well as ovaries with anovulatory fluid-filled structures (≥ 17mm in diameter) and a few preovulatory follicles from three different infertile cows, were collected from the abattoir and immediately cut into small pieces, and then snap-frozen in liquid N_2 and stored at −80℃ until RNA extraction. Other ovarian tissue pieces were fixed in 10% formalin for the immunohistochemical detection of MMP-1 and MMP-2.

2.2.1 Measurement of LH and E_2

The LH and E_2 were measured by ELISA using LH and E_2 kits for cattle (Abnova Company, Taiwan). The assay was performed according to the manufacturer's instructions for each kit. The results were recorded after constructing a standard curve for each hormone.

2.2.2 Measurement of NO

The NO was measured by the Griess reagent method using a kit purchased from Promega (USA) and according to the manufacturer's instructions. Briefly, standard dilutions were prepared to establish a standard curve and allowed the sulfanilamide solution and *N*-(1-naphthyl) ethylenediamine dihydrochloride (NED) solution to equilibrate to room temperature (1 530min). There was 50μL of serum added from each sample to duplicate wells of a 96-well plate. Using a multichannel pipette, 50μL of the sulfanilamide was dispensed to each well, then incubated the plate for 510min at room temperature, protected from the light. There was 50μL of the NED solution dispensed, then incubated again at room temperature for 510min, protected from the light. The solution changed to a purple/magenta color. The absorbance of each well was determined in a plate reader at 535nm and calculated the nitrite concentrations of the samples after constructing a standard reference curve.

Table 1 Oligonucleotide primers used for real-time PCR

Gene	Primer	Sequence	Length (bp)
MMP-1 (NM 174112)	Forward	TGTGTCGTGTTAGTCGCTCAGTC	221
	Reverse	AACCCGCTCCAGTATTCTTGC	
MMP-2 (NM 174745)	Forward	CGAACGCCATCCCTGATAACC	197
	Reverse	TGCTTCCAACTTCACGCTCTT	
GAPDH (NM 001034034)	Forward	GCAAGTTCAACGGCACAGTCA	205
	Reverse	TTGGTTCACGCCCATCACA	

2.2.3 Total RNA extraction

Total cellular RNA was extracted from ovarian tissue with a total RNA extraction kit (Omega Biotech, USA) according to the manufacturer's instructions. Ovarian tissue (100mg) was homogenized briefly in liquid N_2 in a ceramic mortar. When the liquid N_2 had completely evaporated, 1mL of RNA-Solv reagent was added and then transferred the sample to a clean 1.5ml tube. After incubating for 3 min, 0.2mL of chloroform was added, the tube was shaken vigorously for 15 s and then incubated on ice for 10 min. After the tube was centrifuged at 12 000rpm for 10min, no more than 80% of the aqueous part was transferred into a clean tube, to which 1/3 of the volume absolute ethanol was added. After vortexing, 700μL of mixture was transferred into a HiBind RNA column, which was centrifuged at 10 000rpm for 60s at room temperature. Then, the column was moved into a new tube, to which 300μL of RNA washing buffer I was added. This solution was centrifuged at 10 000rpm for 60s, followed by two more wash steps. The flow-through liquid was discarded and 500μL of RNA washing buffer II diluted in ethanol was added, and centrifugation occured as in the final step, the column was transferred to a new tube, 50μL DEPC water was added and then centrifugation occurred at full speed for 1 min. The amount of total RNA was determined using a NanoDrop spectrophotometer (Thermo Scientific, USA) at 260nm. The RNA was reverse-transcribed into cDNA with random primers (PrimeScript RT Reagent Kit, TAKARA Biotechnology Co., China).

2.2.4 Real-time PCR

The quantity of MMP-1 and MMP-2 gene expression was measured by real-time RCR using a SYBR green assay (TAKARA Biotechnology Co., China). Primer pairs for MMP-1 and MMP-2 were designed using Primer Premier 5.0 software based on the sequences of MMP-1 and MMP-2 of cattle (Table 1). The PCR was conducted in a BIO-RAD Multicolor Real-Time PCR machine (USA). The two step-time PCR reaction was conducted at 95℃ for 30s followed by 40 cycles of 5 s at 95℃ and 30s at 60 and 95℃. The melting curve program consisted of temperatures between 60 and 95℃ with a heating rate of 0.05℃/s with a continuous fluorescence measurement. In addition, the excepted size of the amplicon and the absence of nonspecific products were conformed again in agarose. The relative amounts of mRNA of MMP-1 and MMP-2 were normalized to GAPDH, and a melting curve of each gene was plotted.

2.2.5 Immunohistochemistry

The immunohistochemical detection of MMP-1 and MMP-2 was based on the procedure used by Riley et al. (2001) with slight modifications. The MMP-1 and MMP-2 primary antibodies were purchased from Lifespan Biosciences (USA). Streptavidin-biotin-peroxidase compound (SABC), 3, 3′-diaminobenzidine (DAB), biotinylated goat anti rabbit IgG and bovine serum albumin (BSA) were purchased from (Boster Company, Wuhan, China). For staining, ovarian tissues were fixed in 10% formalin, dehydrated in ethanol, embedded in paraffin wax and sectioned at 5μm thick. The staining process began with dewaxing in xylene, rehydration in ethanol and inactivation of endogenous enzymes with 3% H_2O_2 at room temperature for 30min. After washing in distilled water, we added 5% BSA (confining liquid) and incubated the samples at room temperature for 20min, discarding the excess liquid. The sections were incubated in primary antibodies against MMP-1 and MMP-2 for 17h at 4℃ and washed in PBS (pH 7.5) again. Biotinylated goat anti-rabbit IgG (secondary antibody) were then added to the tissues and incubated at 2 037℃ for 20min. The tissue sections were washed in PBS and incubated in SABC at 20℃ to 37℃ for 20min. After a final wash, tissues were stained with DAB for 20min. The tissues were mildly counter-stained with Harris hematoxylin for 1min, followed by differentiation with acid-alcohol, dehydration, clearing and mounting. According to the criterion specified in the protocol of Robinson et al. (2001), immunostaining of tissue sections was assessed semi-quantitatively: +indicated pale staining, ++indicated marked staining, +++signified intense immunostaining, and-meant there was no positive stain.

2.3 Statistical analysis

Real-time PCR was performed in triplicate. Quantitative values are expressed as the mean ±SD. The concentrations of hormones and nitric oxide are presented as the mean ±SEM. An independent *t*-test was used to compare the differences between the fertile and infertile cows. A $P<0.05$ was considered significant.

3 RESULTS

3.1 LH, E_2 and NO results

The concentrations of LH, E_2, and NO in the infertile group compared with the fertile group are presented in Table 2. The concentration of LH was less in the infertile cows than the fertile cows ($P<0.05$). The E_2 was markedly decreased in infertile cows ($P<0.05$). The concentration of NO in infertile cows was less than in fertile cows ($P<0.05$)

Table 2 LH, E_2 and nitric oxide concentrations in infertile and fertile cows

Variables	Infertile cows	Fertile cows
LH (ng/mL) *	12.25±0.32	17.60±0.65
E_2 (pg/mL) *	49.70±8.08	82.53±6.49
Nitric oxide (μmol/L) *	9.76±1.10	15.53±1.42

Note: Data represent the mean ± SEM.

* Differences between infertile and fertile cows for LH, E_2 and NO ($P<0.05$).

3.2 Gene expression of MMP-1 and MMP-2

The quantities of MMP-1 and MMP-2 mRNAs in the ovarian tissue of fertile cows and infertile cows with ovarian cysts were normalized to GAPDH. The relative amount of MMP-1 (0.54±0.07 compared with 0.9±0.08, $P<0.05$) and MMP-2 (0.72±0.076 compared with 1.24±0.071, $P<0.05$) mRNA was less in infertile cows with ovarian cysts than in fertile cows (Fig. 1).

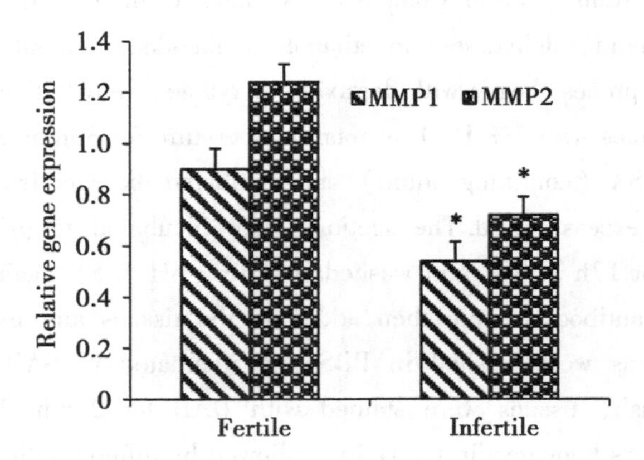

Fig. 1 Expression of MMP-1 and MMP-2 genes in the ovaries of infertile and fertile cows measured by real-time PCR

Note: Gene expression decreased in ovaries of infertile cows compared with fertile cows * $P<0.05$

3.3 MMP tissue distribution

The immunohistochemical results revealed the specific gene expression pattern of MMP-1 and MMP-2 in the ovaries of both fertile and infertile cows. The staining was assigned semi-quantitative grades in the form of+and-symbols and is summarized in Table 3. The MMP-1 immunolocalization in ovaries of fertile cows is shown in Fig. 2 A-D. The MMP-1 was localized in the cytoplasm of the granulosa and theca cells of the large antral follicles from fertile cows (Fig. 2A). Conversely, the MMP-1 was localized but in small amounts in the granulosa and theca cells of large antral follicles from infertile cows with ovarian cysts (Fig. 3A). There was clear localization of MMP-1 in epithelial cells of primordial and primary follicles (Fig. 2B and C), while there was less immunostaining for MMP-1 detected in the infertile cows' primary ovaries (Fig. 3B) and primordial follicles (Fig. 3C). The MMP-1 localization was in the epithelial cells of secondary follicles of ovaries from fertile females, but this immunostaining was clearly less in ovaries of infertile cows (Fig. 2D and Fig. 3D). Ovarian stromal cells and endothelial cells of blood vessels in the ovarian stroma of fertile cows (Fig. 2A and C) had greater MMP-1 immunostainingin the cytoplasm. In contrast, these structures had faint immunostaining for MMP-1 in the cytoplasm of cells from fertile cows (Fig. 3B and D).

Specific positive immunostaining for MMP-2 appeared in ovaries of fertile cows but not infertile cows. The MMP-2 immunostaining showed band-like distribution in the granulosa layer under low magnification (Fig. 4A). However, the large antral follicles in ovaries of infertile cows had less

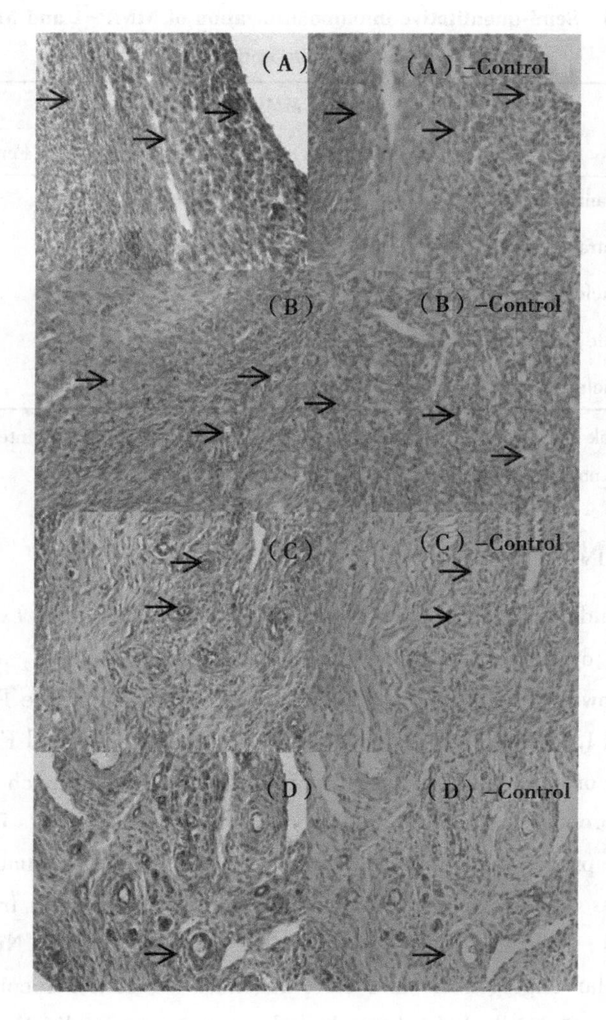

Fig. 2 Immunohistochemical sections showing MMP-1 gene expression in ovaries of fertile cows

Note: (A) Granulosa and theca cells of a large antral follicle and the interstitial cells of the ovarian stroma; (B) Epithelial follicular cells of a primordial follicle. (C) Epithelial follicular cells of a primary follicle. (D) Secondary follicle and endothelial cells of a blood vessel; arrows indicate cells expressing the MMP-1 gene (magnification×400)

dense immunostaining for MMP-2 in theca and granulosa cells under high magnification (Fig. 5A). Cytoplasmic MMP-2 localization appeared in the follicular epithelial cells of primordial follicles and endothelial cells of blood vessels (Fig. 4B). The immunostaining for MMP-2 was strong in the cuboidal follicular epithelial cells of primary follicles, shown in Fig. 4C as a dark brown color, while ovaries of infertile cows had less immunostaining for MMP-2 in the epithelial cells of primordial and primary follicles (Fig. 5B and C). In secondary follicles of fertile cows there was also localization of MMP-2 in the follicular cells, and immunostaining was less in follicular cells of infertile cows (Fig. 4D and Fig. 5D, respectively).

Table 3 Semi-quantitative immunolocalization of MMP-1 and MMP-2 in the ovaries of fertile and infertile cows

Structure	MMP-1		MMP-2	
	Fertile cows	Infertile cows	Fertile cows	Infertile cows
Granulosa cells of large antral follicle	+++	++	+++	+
Theca cells of large antral follicle	++	+	+++	++
Primordial follicle	++	+	++	+
Primary follicle	++	+	+++	++
Secondary follicle	++	+	++	+

Note: + indicates weak staining, ++ indicates marked staining, +++ indicates intense immunostaining, and- indicates that there was no positive stain.

4 DISCUSSION

This is the first study to demonstrate the MMP-1 and MMP-2 gene expression profiles in infertile dairy cows with ovarian cysts. In the present study, the LH, E_2 and NO concentrations were less in infertile cows with ovarian cysts compared to fertile cows. The E_2 is an ovarian marker that regulates FSH and LH secretion through a feedback mechanism, and E_2 can reflect the development and maturation of dominant follicles in many mammals (Danilovich et al., 2002). Silvia et al. (2002) have proposed that the formation of ovarian follicular cysts is a failure of the hypothalamus to induce the preovulatory surge of LH in response to E_2. Adequate LH secretion is needed to support follicular and oocyte maturation and subsequent ovulation from a dominant follicle (Roche et al., 2000). Pancarc et al. (2011) have demonstrated that NO is an important regulatory agent for follicular angiogenesis and steroidogenesis in cattle. According to Rettori et al. (1993), NO induces a pulsatile LH release by releasing prostaglandin E_2 that could activate the exocytosis of LHRH secretory granules into the portal vessels. These findings suggest that little of the feedback signal is represented by E_2 for LH release, and the LH surge from the pituitary gland may have been suppressed by the basal concentrations of NO (or vice versa) in infertile cows with ovarian cysts.

In the present study, the expression of MMP-1 and MMP-2 genes was less in infertile cows with ovarian cysts compared to fertile cows. The MMP are considered the most important enzymes for tissue degradation (Woessner, 1991). According to Smith et al. (2005), MMP have an important role in regulating the theca and granulosa compartments of ovarian follicles in cattle and in preserving a microenvironment that is conducive to follicular function. The MMP-1 can break down the fibrillar collagen forms that confer much of the structural integrity to the ovarian stroma (Chaffin and Stouffer, 1999; Espey, 1994; Riley et al., 2001). Bakke et al. (2004) revealed that the expression of MMP-1 and MMP-13 genes in the thecal layer of preovulatory follicles of cattle and the increase after the GnRH surge are consistent with a potential role for these enzymes in collagenolysis and the ovulation process. This result was also evidenced by the expression of MMP-1 genes in

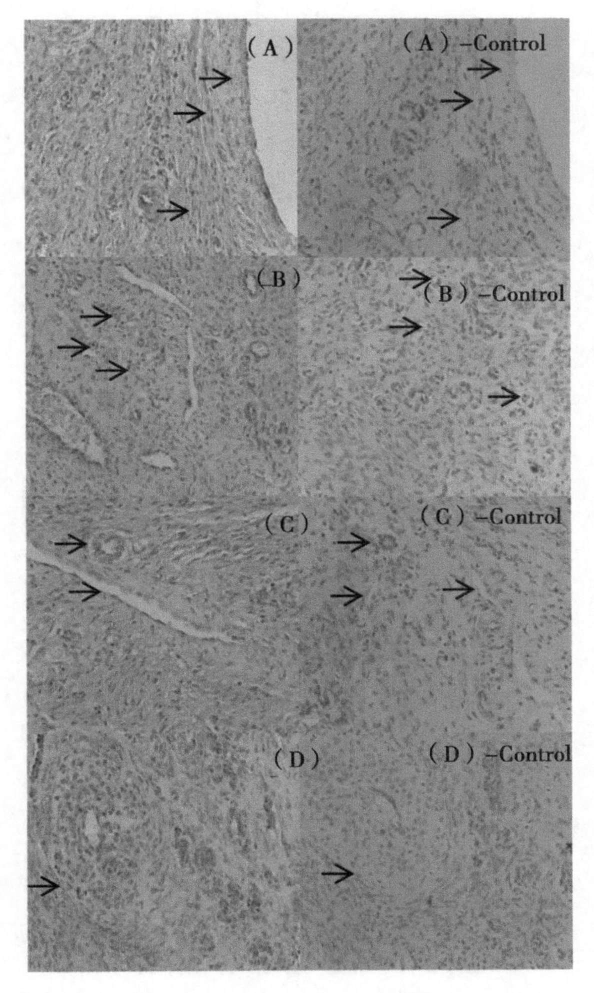

Fig. 3 Immunohistochemical sections showing decreased expression of the MMP-1 gene in the ovary of infertile cows

Note: (A) Granulosa and theca cells of a large antral follicle and the interstitial cells of ovarian stroma; (B) Primary follicle and endothelial cells of the blood vessels; (C) MMP-1 gene expression in the epithelial cells of a primordial follicle; (D) Low expression of MMP-1 genes in the epithelial follicular cells of a secondary follicle; arrows indicate cells expressing the MMP-1 gene (magnification×400)

theca and granulosa cells of the large antral follicles in the present study. The MMP-2, a gelatinase with the ability to degrade and denature basement membrane collagens (Woessner, 1991), may permit the turnover and reconstruction of the follicle wall at times of growth (Cooke et al., 1999) and ovulation (Tadakuma et al., 1993), and it may facilitate tissue remodeling during corpus luteum formation and development (Goldberg et al., 1996). Imai et al. (2003) found that MMP-2 increases as follicular size increases. Häglund et al. (1999) reported that the expression of MMP genes is upregulated in the theca cells of large preovulatory follicles and follicles from which ovulation occurs. Consequently, MMP gene expression should increase in ovarian stroma during the

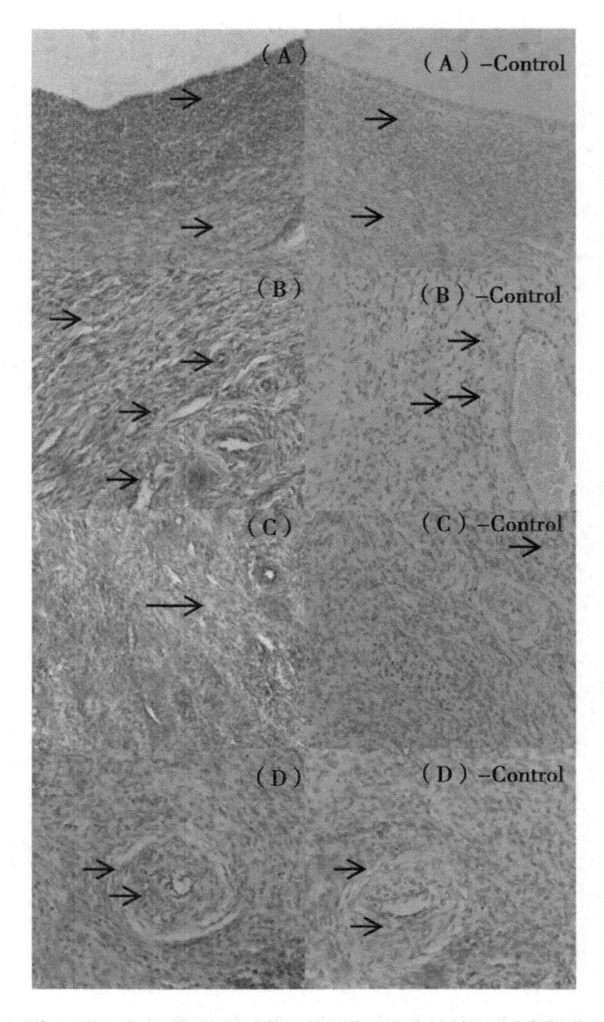

Fig. 4 Immunohistochemical sections showing the expression of MMP-2 genes in the ovaries of fertile cows. The figure illustrates strong gene expression in the different ovarian structures

Note: (A) Theca cells and granulosa cells of a large antral follicle; (B) Epithelial follicular cells of a primordial follicle, endothelial cells of the blood vessels and interstitial stromal cells; (C) Epithelial follicular cells of a primary follicle; (D) Secondary follicle; Arrows indicate cells expressing the MMP-2 gene (magnification of A is×200; B, C and D are ×400)

process of folliculogenesis in fertile cows, and a decrease in MMP gene expression should occur in ovarian tissue and may be an indicator of abnormal follicle growth. Results from the present study suggest that decreased expression of MMP-1 and MMP-2 genes might serve as a marker of ovarian follicular cysts in cows.

5 CONCLUSION

The decrease in MMP-1 and MMP-2 gene expression is accompanied with a decrease in NO

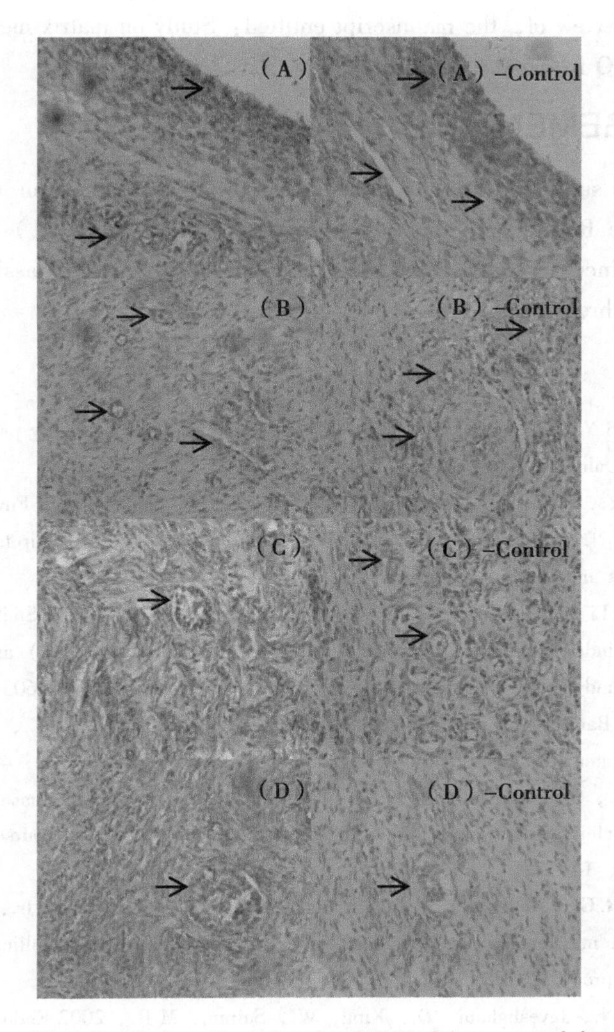

Fig. 5 Immunohistochemical sections showing MMP-2 immunostaining in ovaries of infertile cows

Note: (A) Low MMP-2 gene expression in the granulosa and theca cells of a large antral follicle; (B) MMP-2 gene expression in the ovarian stroma cells, endothelial cells and follicular epithelial cells of a primordial follicle; (C) Epithelial follicular cells of a primary follicle; (D) Less MMP-1 gene expression in a secondary follicle; Arrows indicate cells expressing the MMP-2 gene (magnification ×400).

concentrations in infertile cows with ovarian cysts. These proteins might be useful markers for fertility in cows and the gene expression pattern correlates with hormonal profiles. Given the present results, it is proposed that the improper expression of MMP-1 and MMP-2 genes might be an important marker of ovarian follicular cysts in dairy cows.

CONFLICTS OF INTEREST

We have no proprietary, financial, professional or other personal interest of any nature or kind in any product, service and/or company that could be construed as influencing the position presented in, or the review of, the manuscript entitled: Study on matrix metalloproteinase 1 and 2 gene expression and NO in dairy cows with ovarian cysts.

ACKNOWLEDGEMENTS

This research was supported by the China Dairy Cow Research System (CARS-37), Special Fund for Agroscientific Research in the Public Interest (201303040-01) and the major special projects of Gansu province (1102NKD0202). The authors appreciate the assistance of the managers and employees of the three farms.

References

[1] Al-Dahash, S.Y.A., Bensassi, M.F., 2009.Treatment of some infertility problems in cows using Dalmarelin and Dalmazin.Iraqi.J.Vet.Sci.23, 255-257.

[2] Anteby, E.Y., Hurwitz, A., Korach, O., Revel, A., Simon, A., Finci-Yeheskel, Z., Mayer, M., Laufer, N., 1996.Human follicular nitric oxide pathway: relationship to follicular size, oestradiol concentrations and ovarian blood flow.Hum.Reprod.11, 1947-1951.

[3] Bakke, L.J., Li, Q., Cassar, C.A., Dow, M.P.D., Pursley, J.R., Smith, G.W., 2004.Gonadotropin surge-induced differential upregulation of collagenase-1 (MMP-1) and collagenase-3 (MMP-13) mRNA and protein in bovine preovulatory follicles.Biol.Reprod.71, 605-612.

[4] Basini, G., Baratta, M., Ponderato, N., Bussolati, S., Tamanini, C., 1998.Is nitric oxide an autocrine modulator of bovine granulosa cell function? Reprod.Fertil.Dev.10, 471-478.

[5] Chaffin, C.L., Stouffer, R.L., 1999.Expression of matrix metalloproteinases and their tissue inhibitor messenger nucleic acids in macaque periovulatory granulosa cells: time course and steroid regulation.Biol.Reprod.61, 14-21.

[6] Cooke III, R.G., Nothnick, W.B., Komar, C., Burns, P., Curry Jr., T.E., 1999.Collagenase and gelatinase messenger ribonucleic acid expression and activity during follicular development in the rat ovary.Biol.Reprod.61, 1309-1316.

[7] Danilovich, N., Javeshghani, D., Xing, W., Sairam, M.R., 2002.Endocrine alterations and signaling changes associated with declining ovarian function and advanced biological aging in follicle-stimulating hormone receptor haploinsufficient mice.Biol.Reprod.67, 370-378.

[8] Edmonson, A.J., Lean, I.J., Weaver, L.D., Farver, T., Webster, G., 1989.A body condition scoring chart for Holstein dairy cows.J.Dairy Sci.72, 68-78.

[9] Espey, 1994.Current status of the hypothesis that mammalian ovulation is comparable to an inflammatory reaction.Biol.Reprod.50, 233-238.

[10] Goldberg, M.J., Moses, M.A., Tsang, P.C., 1996.Identification of matrix metalloproteinases and metalloproteinase inhibitors in bovine corpora lutea and their variation during the estrous cycle.J.Anim.Sci.74, 849-857.

[11] Hägglund, A.C., Ny, A., Leonardsson, G., Ny, T., 1999.Regulation and localization of matrix metalloproteinases and tissue inhibitors of metalloproteinases in the mouse ovary during gonadotropin-induced ovulation.Endocrinology 140, 4351-4358.

[12] Hulboy, D.L., Rudolph, L.A., Matrisian, L.M., 1997.Matrix metalloproteinases as mediators of reproductive function.Mol.Hum.Reprod.3, 27-45.

[13] Imai, K., Khandoker, M.A.M.Y., Yonai, M., Takahashi, T., Sato, T., Ito, A., Hasegawa, Y., Hashizume, K., 2003.Matrix metalloproteinases-2 and -9 activities in bovine follicular fluid of

different-sized follicles: relationship to intra-follicular inhibin and steroid concentrations.Domest.Anim. Endocrinol.24, 171-183.

[14] Moncada, S., Palmer, R.M.J., Higgs, E.A., 1991. Nitric oxide: physiology, pathophysiology, and pharmacology.Phamacol.Rev.43, 109-142.

[15] Pancarc., Ş.M., Ar., U.Ç., Atakişi, O., Güngör, Ö., Çiğremiş, Y., Bollwein, H., 2012. Nitric oxide concentrations, estradiol-17β-progesterone ratio in follicular fluid, and COC quality with respect to perifollicular blood flow in cows.Anim.Reprod.Sci.130, 9-15.

[16] Pancarc., Ş.M., Güngör, Ö., Atakisşi, O., Çiiğremisş, Y., Ar., U.Ç., Bollwein, H., 2011. Changes in follicular blood flow and nitric oxide levels in follicular fluid during follicular deviation in cows.Anim.Reprod.Sci.123, 149-156.

[17] Peng, X., Cao, B., Deng, G.Z., Li, C.Y., Ye, L.L., Yu, H., 2011.Echography characteristics of abfertile ovaries in infertile dairy cows.J.Anim.Vet.Adv.10, 1166-1170.

[18] Plaizier, J.C.B., Lissemore, K.D., Kelton, D., King, G.J., 1998.Evaluation of overall reproductive performance of dairy herds.J.Dairy Sci.81, 1848-1854.

[19] Portela, V.M., Veiga, A., Price, C.A., 2009.Regulation of MMP2 and MMP9 metalloproteinases by FSH and growth factors in bovine granulosa cells.Genet.Mol.Biol.32, 516-520.

[20] Rettori, V., Belova, N., Dees, W.L., Nyberg, C.L., Gimeno, M., McCann, S.M., 1993.Role of nitric oxide in the control of luteinizing hormonereleasing hormone release in vivo and in vitro.Proc. Natl.Acad.Sci.90, 10130-10134.

[21] Riley, S.C., Gibson, A.H., Leask, R., Mauchline, D.J.W., Pedersen, H.G., Watson, E.D., 2001.Secretion of matrix metalloproteinases 2 and 9 and tissue inhibitor of metalloproteinases into follicular fluid during follicle development in equine ovaries.Reproduction 121, 553-560.

[22] Robinson, L.L., Sznajder, N.A., Riley, S.C., Anderson, R.A., 2001.Matrix metalloproteinases and tissue inhibitors of metalloproteinases in human fetal testis and ovary.Mol.Hum.Reprod.7, 641-648.

[23] Roche, J.F., Mackey, D., Diskin, M.D., 2000. Reproductive management of postpartum cows. Anim.Reprod.Sci.60, 703-712.

[24] Rosselli, M., Keller, P.J., Dubey, R.K., 1998.Role of nitric oxide in the biology, physiology and pathophysiology of reproduction.Hum.Reprod.Update 4, 3-24.

[25] Silvia, W.J., Hatler, T.B., Nugent, A.M., Laranja da Fonseca, L.F., 2002. Ovarian follicular cysts in dairy cows: an abfertileity in folliculogenesis.Domest.Anim.Endocrinol.23, 167-177.

[26] Smith, M.F., Gutierrez, C.G., Ricke, W.A., Armstrong, D.G., Webb, R., 2005.Production of matrix metalloproteinases by cultured bovine theca and granulosa cells.Reproduction 129, 75-87.

[27] Tadakuma, H., Okamura, H., Kitaoka, M., Iyana, K., Usuku, G., 1993.Association of immunolocalization of matrix metalloproteinase 1 with ovulation in hCG-treated rabbit ovary.J.Reprod.Fertil. 98, 503-508.

[28] Tamanini, C., Basini, G., Grasselli, F., Tirelli, M., 2003. Nitric oxide and the ovary. J. Anim. Sci.81 (Suppl.2), E1-E7.

[29] Walsh, S.W., Williams, E.J., Evans, A.C.O., 2011.A review of the causes of poor fertility in high milk producing dairy cows.Anim.Reprod.Sci.123, 127-138.

[30] Whited, J., Shahed, A., McMichael, C.F., Young, K.A., 2010.Inhibition of matrix metalloproteinase in Siberian hamsters impedes photostimulated recrudescence of ovaries.Reproduction 140, 875-883.

[31] Woessner Jr., J.F., 1991.Matrix metalloproteinases and their inhibitors in connective tissue remodeling. FASEB J.5, 2145-2154.

（发表于《Animal Reproduction Science》，院选 SCI，IF：1.581）

The Complete Mitochondrial Genome of *Hequ horse*

GUO Xian, PEI Jie, CHU Min, WU Xiao-yun, BAO Peng-jia,
DING Gue-zhi , LIANG Chun-nian, YAN Ping
(Lanzhou Institute of Husbandry and Pharmaceutical Sciences,
Chinese Academy of Agricultural Sciences, Lanzhou 730050, China)

Abstract: The complete mitochondrial genome of *Hequ horse* was determined in this study. The mitogenome is 16 656 bp in length and contains 13 protein-coding genes, 22 transfer RNA genes, 2 ribosomal RNA genes, and a D-loop region. The overall base composition of the Hstrand is 32. 20% for A, 28. 55% for C, 13. 38% for G and 25. 86% for T. Tree constructed using MEGA 6 with Maximum-likelihood (ML) methods demonstrated that *Hequ horse* was clustered in subfamily *Equidae*.

Key words: Complete mitochondrial genome; *Hequ horse*; mitochondrial DNA

Hequ horse belonging to *Equidae*, *Perissodactyla*, *mammalia*, and is spread in junction of Gansu, Sichuan and Qinghai province of China (Zhang et al., 1984). In the present study, we sequenced the complete mitochondrial genome of thoroughbred *Hequ horse* with the individual bred in Maqu county, Gansu province, China (34°00′N, 102°04′E). The collection the specimen resides in Lanzhou Institute of Husbandry and Pharmaceutical Sciences, Chinese Academy of Agricultural Sciences. The total genomic DNA was extracted from blood using traditional the phenolchloroform extraction method and sequenced with Illumina Hiseq 2 500 (Johns & Paulus-Thomas, 1989). The protein-coding region was determined by comparing with the sequence of *Naqu horse* (EF597513.1) reported before (Xu et al., 2007). Also, the transfer RNA (tRNA) genes were identified using the tRNAscan-SE1.21 (http: //selab.janelia.org/tRNAscan-SE/). MEGA 6 (MEGA Inc., Englewood, NJ) was used to construct the phylogenetic tree in this study.

The whole mitogenome of *Hequ horse* is 16 656 bp in length (GenBank accession no. KT596764) contains a set of 13 protein-coding genes, 2 ribosomal RNA genes (rRNA), 22 transfer RNA genes (tRNA), and a control region. Its organization and gene order were similar to those of other Equus mitochondrial genomes. The overall base composition of the H-strand is 32. 20% for A, 28. 55% for C, 13. 38% for G and 25. 86% for T. Tree constructed using MEGA 6 (MEGA Inc., Englewood, NJ) phylogenetic analysis is shown in Fig. 1. As showed in maximum-likelihood (ML) tree (Fig. 1), our sequence were clustered *Equidae*, which included *Naqu horse*, *Equus przewalskii*, *Equus asinus*, and *Equus hemionus*.

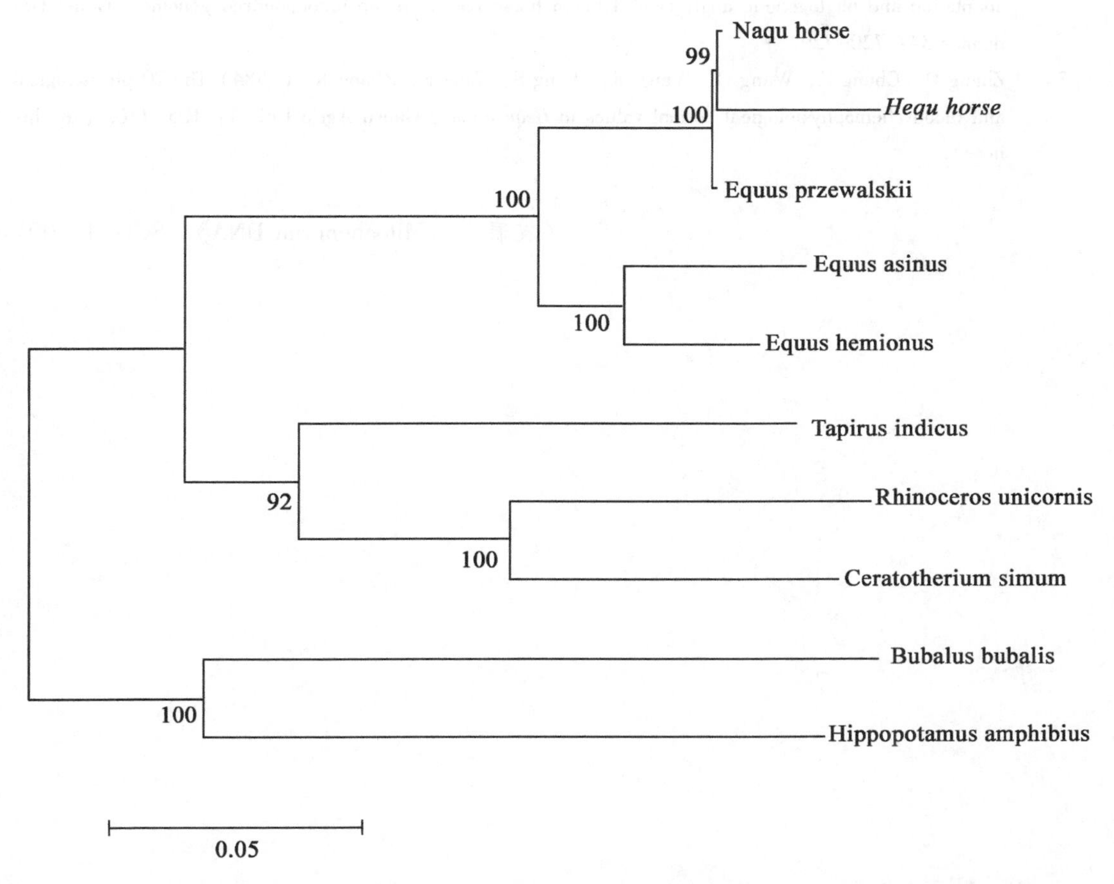

Fig. 1 Genetic analysis of 10 horse strains with MEGA 6 using the coding region of mitochondrial genomedata

Note: Data sources: *Naqu horse* (EF597513.1); *Hequ horse* (KT596764); *Equus przewalskii* (HQ439484.1); *Equus asinus* (NC_ 001788.1); *Equus hemionus* (NC_ 016061.1); *Tapirus indicus* (NC_ 023838.1); *Rhinoceros unicornis* (NC_ 001779.1); *Ceratotherium simum* (NC_ 001808.1); *Bubalus bubalis* (AY488491.1); *Hippopotamus amphibius* (AP003425.1)

DECLARATION OF INTEREST

The authors report no conflicts of interest.The authors alone are responsible for the content and writing of the paper.The work was supported or partly supported by grants from National Natural Science Foundation of China (31301976), Central Public-interest Scientific Institution Basal Research Fund (1610322014010), China Agriculture Research System (CARS-38), and The innovation project of Chinese Academy of Agricultural Sciences (CAAS-ASTIP-2014-LIHPS-01).

References

[1] Johns MB, Paulus-Thomas JE. (1989).Purification of human genomic DNA from whole blood using sodium perchlorate in place of phenol.Anal Biochem 180: 276–278.

[2] Xu S, Luosang J, Hua S, He J, Ciren A, Wang W, Tong X, et al. (2007). High altitude

adaptation and phylogenetic analysis of Tibetan horse based on the mitochondrial genome. J Genet Genomics 34：720–729.

[3] Zhang D, Cheng X, Wang Q, Wang M, Wang S, Tang P, Zhang K. (1984). The 20 physiological and blood-chemophysiological normal values in *Hequ horse*. J Gansu Agric Univ 1：105–110. (In chinese).

（发表于《Mitochondrial DNA》，SCI：1.209）

The Complete Mitochondrial Genome of the Qinghai Plateau Yak *Bos grunniens* (Cetartiodactyla: Bovidae)

GUO Xian, PEI Jie, BAO Peng-jia, CHU Min,
WU Xiao-yun, DING Xue-zhi , YAN Ping

(Key Laboratory of Yak Breeding Engineering of Gansu Province, Lanzhou Institute of Husbandry and Pharmaceutical Sciences, Chinese Academy of Agricultural Sciences, Lanzhou 730050, China)

Abstract: The Qinghai Plateau yak *Bos grunniens* (Cetartiodactyla: Bovidae) is an important primitive local breed in the Qinghai-Tibetan Plateau and adjacent regions. In this study, its complete mitochondrial genome sequence has been assembled and characterized from high-throughput Illumina sequencing data. This genome is 16 322 bp in length, and contains 13 protein-coding genes, 22 tRNA genes, two rRNA genes, and a non-coding D-loop or control region. The nucleotide composition is asymmetric (33.73% A, 25.79% C, 13.19% G, and 27.29% T) with an overall A+T content of 61.02%. The gene arrangement and the composition are similar to most other vertebrates. These data would contribute to our better understanding its population genetics and evolutionary history.

Key words: Bos grunniens; Illumina sequencing; Mitochondrial genome; Next-generation sequencing (NGS); Qinghai Plateau yak

Yaks (*Bos grunniens*) are mainly distributed in the Qinghai-Tibetan Plateau and adjacent regions, and have long been exploited for meat, milk, wool and fuel by Tibetan nomadic pastoralists. The Qinghai Plateau yak is an important primitive local breed in the Qinghai-Tibetan Plateau. In this study, its complete mitochondrial genome has been determined using Illumina sequencing technology.

Total genomic DNA was extracted from the blood sample of an individual using the Genomic DNA Isolation Kit (Tiangen Biotech, Beijing, China), and used for the shotgun library construction according to the manual of the manufacturer for the Illumina Hiseq 2 500 Sequencing System (Illumina, San Diego, CA). A total of 17.20M 125-bp raw reads were retrieved, and quality-trimmed using CLC Genomics Workbench v8.0 (CLC Bio, Aarhus, Denmark). The trimmed reads were then used for the assembly of mitochondrial genome using MITObim v1.8 (University of Oslo, Oslo, Norway; Hahn et al., 2013). The annotation was performed in GENEIOUS R8 (Biomatters Ltd., Auckland, New Zealand) by comparing with the mitochondrial genomes of its congeners. To identify its phylogenetic placement within the family Bovidae, a phylogenetic tree was reconstructed based on the maximum likelihood (ML) analysis of 12 protein-coding genes (PCGs) located on the heavy strand for a panel of 25 species (see Fig. 1 for details) with MEGA6 (MEGA Inc., Englewood, NJ) (Tamura et al., 2013).The resultant tree was rooted with two species within the family Cervidae, i.e., *Cervus elaphus* (GU457435) and *C.nippon* (GU457433).

The circular genome of the Qinghai Plateau yak is 16 322 bp in size, with an asymmetric nucleotide composition (33. 73% A, 25. 79% C, 13. 19% G, and 27. 29% T) and an overall A+T content of 61. 02%.The gene composition and the arrangement are similar to those of most other vertebrates (e.g., Cheng et al., 2011; Douglas et al., 2011; Hassanin et al., 2012; Tu et al., 2014).Most genes are encoded on the heavy strand except for one PCG (*ND6*) and eight tRNAs (*tRNA-Ala*, *-Asn*, *-Cys*, *-Gln*, *-Glu*, *-Pro*, *-Ser*, and *-Tyr*).

All PCGs are initiated with ATR codons.Four distinct stop codons are employed, i.e., AGA for *CYTB*, TAA for *ATP6*, *ATP8*, *COX1*, *COX2*, *ND4L*, *ND5*, and *ND6*, and an incomplete stop codon T for *COX3*, *ND1*, *ND2*, *ND3*, *and ND4*.The two rRNAs are 957bp (12S; 960-1 916) and 1 571 bp (16S; 1 984-3 554) in length, respectively. Twenty-two tRNAs are interspersed throughout the whole genome, and range in length from 60 to 75 bp.The putative control region (D-loop) is located between *tRNA-Phe* and *tRNA-Pro* with a length of 892 bp (1-892).A putative origin of the light stand replication is located between *tRNA-Asn* and *tRNACys* (31bp; 6 051-6 081).

As shown in the phylogenetic tree (Fig. 1), the Qinghai Plateau yak (*Bos grunniens*) and its congeners were grouped into a single clade.Surprisingly, this clade also included the two extant species for the genus *Bison*, i.e.*Bison bonasus and Bison bison*.This may be explained by the fact that recent studies have strongly supported the inclusion of *Bison* in the genus *Bos* (reviewed by Douglas et al., 2011).

DECLARATION OF INTEREST

The work was financially co-supported by the National Natural Science Foundation of China (31301976), the Central Public-Interest Scientific Institution Basal Research Fund (1610322014010), China Agriculture Research System (CARS-38), and the National Science and Technology Support Project (2013BAD16B09).The annotated genome sequence has been deposited into GenBank under the accession number KR011113.The authors report that they have no conflicts of interest, and alone are responsible for the content and writing of the paper.

References

[1] Cheng YZ, Xu TJ, Jin XX, Wang RX. (2011).Complete mitochondrial genome of the yellow drum *Nibea albiflora* (Perciformes, Sciaenidae).Mitochondrial DNA 22: 80-82.

[2] Douglas KC, Halbert ND, Kolenda C, Childers C, Hunter DL, Derr JN. (2011).Complete mitochondrial DNA sequence analysis of *Bison bison* and bison-cattle hybrids: Function and phylogeny.Mitochondrion 11: 166-175.

[3] Hahn C, Bachmann L, Chevreux B. (2013).Reconstructing mitochondrial genomes directly from genomic next-generation sequencing reads-A baiting and iterative mapping approach. Nucleic Acids Res 41: e129.

[4] Hassanin A, Delsuc F, Ropiquet A, Hammer C, Jansen van Vuuren B, Matthee C, Ruiz-Garcia M,

et al. (2012).Pattern and timing of diversification of Cetartiodactyla (Mammalia, Laurasiatheria), as revealed by a comprehensive analysis of mitochondrial genomes.C R Biol 335: 32-50.

[5] Tamura K, Stecher G, Peterson D, Filipski A, Kumar S. (2013).MEGA6: Molecular Evolutionary Genetics Analysis Version 6.0.Mol Biol Evol 30: 2725-2729.

[6] Tu J, Si F, Wu Q, Cong B, Xing X, Yang FH. (2014).The complete mitochondrial genome of the Muscovy duck, *Cairina moschata* (Anseriformes, Anatidae, *Cairina*). Mitochondrial DNA 25: 102-103.

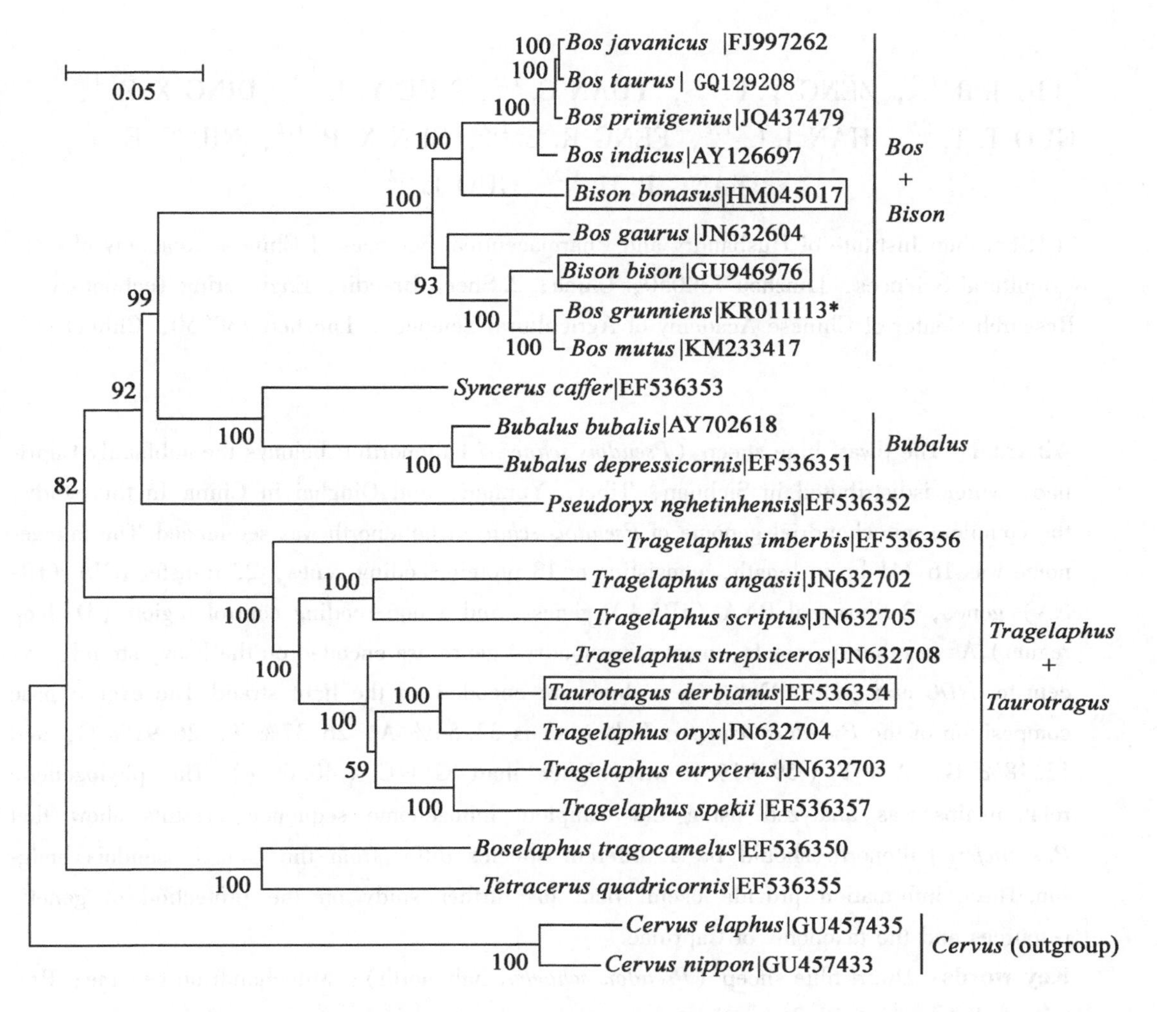

Fig. 1 Phylogeny of 23 species within the family Bovidae based on the maximum likelihood (ML) analysis of 12 protein-coding genes located on the heavy strand (10 830bp)

Note: The GTR+G+I model was employed as suggested by the 'Model Selection (ML)' function of the program MEGA6 (MEGA Inc., Englewood, NJ) (Tamura et al., 2013).The bootstrap values were based on 100 resamplings, and are indicated next to the branches.The tree was rooted with two species within the family Cervidae, i.e., *Cervus elaphus* (GU457435) and *C.nippon* (GU457433).Codon positions included were 1st+2nd+3rd.

（发表于《Mitochondrial DNA》，SCI：1.209）

The Complete Mitochondrial Genome Sequence of the Dwarf Blue Sheep, *Pseudois schaeferi* Haltenorth in China

LIU J. B.[1,2], ZENG Y. F.[1*], YUAN C.[1,2], YUE Y. J.[1,2], DING X. Z.[1], GUO T. T.[1,2], HAN J. L.[1,2], FENG R. L.[1,2], SUN X. P.[1,2], NIU C. E.[1,2], YANG B. H.[1,2], GUO J.[1,2]

(1. Lanzhou Institute of Husbandry and Pharmaceutical Sciences of Chinese Academy of Agricultural Sciences, Lanzhou 730050, China; 2. Sheep Breeding Engineering Technology Research Center of Chinese Academy of Agricultural Sciences, Lanzhou 730050, China)

Abstract: The dwarf blue sheep (*Pseudois schaeferi* haltenorth) belongs the subfamily Caprinae, which isdistributed in Sichuan, Tibet, Yunnan, and Qinghai in China. In this study, the complete mitochondrial genome of *Pseudois schaeferi* haltenorth was sequenced. The mitogenome was 16 741 bp in length, consisting of 13 protein-coding genes, 22 transfer RNA (tRNA) genes, 2 ribosomal RNA (rRNA) genes, and a non-coding control region (D-loop region). As in other mammals, most mitochondrial genes are encoded on the heavy strand, except for *ND*6 and eight tRNA genes which are encoded on the light strand. The overall base composition of the *Pseudois schaeferi haltenorth* is 33.54% A, 26.37% T, 26.91% C, and 13.18% G, A + T (59.91%) was higher than G + C (40.09%). The phylogenetic relationships was analyzed using the complete mitogenome sequence, results show that *P. schaeferi* haltenorth should be a different species differ from the Genus pseudois hodgson. These information provide useful data for further study on the protection of genetic resources and the taxonomy of Caprinae.

Key words: Dwarf blue sheep (*Pseudois schaeferi* haltenorth); Mitochondrion genome; Protein-coding genes; Phylogenetic

The dwarf blue sheep (*Pseudois schaeferi* haltenorth) belongs to the subfamily Caprinae within the Cetartiodactyla, and is the special species in China. Surveys show that it is distributed in deep valleys of Shitang to Shiqu region along the Jinshajiang River in western Sichuan. It is also found in Jiangda, Gongjue, and Mangkang of Tibet. It has been also found Yushu and Nangqian of Qinghai and Deqin Yunnan (Wu et al., 1990; Yi & Liu, 1993). In this work, we first determined the mitochondrial genome of *P. schaeferi* haltenorth (GenBank accession number: KP998469).

The genomic total DNA was exacted from the blood sample of *P. schaeferi* haltenorth, using the genomic DNA isolate kit (Tiangen Biotech, Beijing, China) according to the manufacturer's instructions. Mitochondrial DNA was sequenced by the secondgeneration sequencing technology (Illumina Hiseq 2 500, Illumina Inc., San Diego, CA), and the DNA fragments were assembled by MITObin software (V 1.7) (Hahn et al., 2013). The complete mitogenome is 16 741 bp in length, and consists of 13 proteincoding genes, 22 transfer RNA (tRNA) genes, 2 ribosomal RNA (rRNA) genes and a non-coding control region (D-loop region), the order and composition are similar to most of the other vertebrates (Hu & Gao, 2014). The overall base composition of the mitogenome (A: 33.54%, T: 26.37%, C: 26.91%, and G: 13.18%) has a high A+T content (59.91%).

The total length of the protein-coding gene is 11 403 bp, the percentage of the total is 68.11%. As in other mammals, most mitochondrial genes are encoded on the heavy stand, except for *ND 6* and eight tRNA genes, which are encoded on the light strand. All 13 mitochondrial protein-coding genes share the start codon ATG, except for *ND 2*, *ND 3*, *ND 5*, and *ATPase 8*, which initiate with start codon ATA and GTG, respectively. It is also important to note that the majority of protein-coding genes (five of 13 genes) were inferred to terminate with an incomplete termination codon TA (a) or T (aa) (*ND 1*, *ND 2*, *COX 3*, *ND 3*, and *ND 4*); seven protein-coding genes shared the typical termination codon TAA (*COX1*, *COX2*, *ATPase8*, *ATPase6*, *ND4L*, *ND5*, and *ND6*); Cytb use AGA instead of TAA as the stop codon. In the complete mitochondrial genome, there are totally 72 overlapped nucleotides between neighboring genes in seven locations and the length of overlapped sequence was range from 1 bp to 40 bp. Eighteen pairs of gene were directly adjacent without intergenic or overlapping nucleotides. The control region (D-Loop) locates between $tRNA^{Phe}$ and $tRNA^{Pro}$, the length is 1 087 bp from 15 439 to 16 525 bp. The 2 ribosomal RNA genes, 12S rRNA (958 bp) and 16S *rRNA* (1 577 bp), were located between $tRNA^{Phe}$ and $tRNA^{Leu}$ (UUR) and separated by $tRNA^{Val}$ gene, and the lengths of 22 *tRNA* genes were range from 60 to 75 bp. Among all *tRNA* genes of the *P. schaeferi* haltenorth, except that $tRNA^{Ser}$ (AGN) lack the dihydrouridine stem and loop, other 21 *tRNA* genes could all fold into the typical cloverleaf-shaped secondary structure. There was a small replication non-coding region located between $tRNA^{Asn}$ and $tRNA^{Cys}$ on the light-strand replication length is 32 bp from 5 166 to 5 197.

Because the mitochondrial DNA is a small genome, with the special characteristics of high evolution rate, lack of recombination, no speciality in tissues and maternal inheritance, it has been extensively used for exploring the genealogical history in and among the species. In this study, we constructed a phylogenetic tree (Fig. 1) using the *P. schaeferi* haltenorth mitogenomic (KP998469) and other 21 *Ovis aries* sequences, four *Pseudois* sequences, and six *Capra* sequences, which were download from the GenBank: Ovis aries (KF938337.1, KF938331.1, KP998470, HM236175.1, KF938345.1, KF977847.1, KF938338.1, KF938342.1, HM236174.1, KF938330.1, KF938336.1, KF938332.1, KF938350.1, HM236185.1, KF938348.1, HM236181.1, KF938320.1, KF312238.2, HM236183.1, HM236189.1, and JX101654.1), *Pseudois* (FJ207537.1, NC_ 016689.1, KJ784494.1, and JX101652.1), and *Capra* (FJ207529.1, FJ207527.1, FJ207528.1, FJ207526.1, FJ207525.1, and NC_ 005044.2).

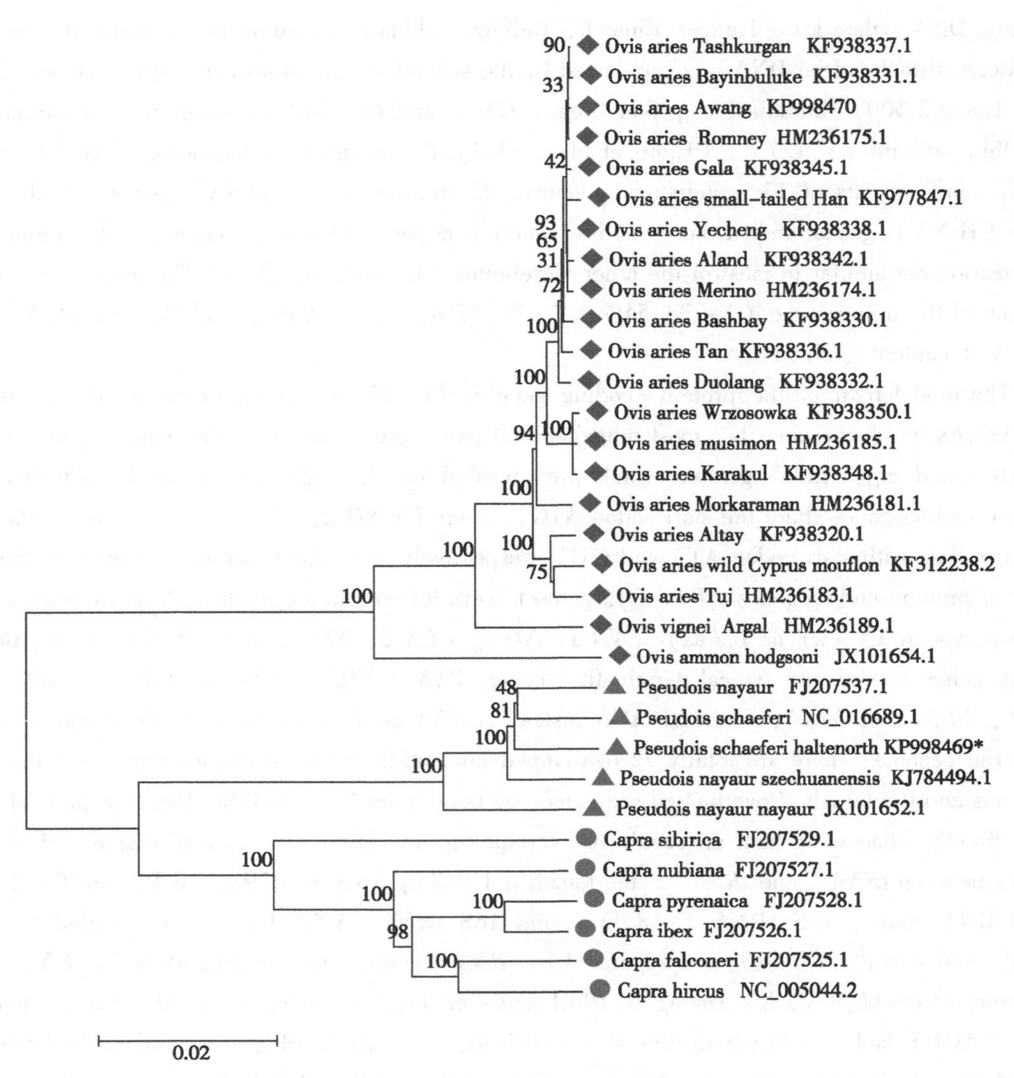

Fig. 1 Neighbor-joining phylogenetic tree of bovindae. The optimal tree with the sum of branch length = 0. 55653866 is shown

Note: The percentage of replicate trees in which the associated origin clustered together in the bootstrap test (1 000 replicates) is shown next to the branches. The tree is above to scale, with branch lengths in the same units as those of the evolutionary distances used to infer the phylogenetic tree. Codon positions included were 1st+2nd+3rd+noncoding. All positions containing gaps and missing data were eliminated. There were a total of 16 695 positions in the final dataset

Evolutionary analyses were conducted in MEGA6 (MEGA Inc., Englewood, NJ), the distances were computed using the Maximum Composite Likelihood method and are in the units of the number of base substitutions per site. Five *Pseudois* breeds clustered with two *Pseudois schaeferi* and three *Pseudois nayaur*, *Capra falconeri* clustered with *Capra hircus*, *Capra pyrenaica* clustered with *Capra ibex*, *Capra nubiana* clustered with *Capra sibirica*, then two group (21 *ovis aries* breeds) clustered with vignei argal and ammon hodgsoni. This result was according to reporting by other scientists (Jiang et al., 2014; Lancioni et al., 2013; Zhong et al., 2010), therefore, we suggest that dwarf blue sheep (*P. schaeferi* haltenorth) should be a different species differ from the *Pseudois*.

DECLARATION OF INTEREST

The authors report that they have no conflicts of interest. The authors alone are responsible for the content and writing of this article. This work was supported by special fund from the Gansu Provincial Funds for Distinguished Young Scientists (1308RJDA015), Gansu Provincial Natural Science Foundation (145RJZA061), and Gansu Provincial Agricultural biotechnology research and application projects (GNSW-2014-21), and the Central Level, Scientific Research Institutes for Basic R & D Special Fund Business (1610322012006 and 1610322015002).

References

[1] Hahn C, Bachmann L, Chevreux B. (2013). Reconstructing mitochondrial genomes directly from genomic next-generation sequencing reads-A baiting and iterative mapping approach. Nucleic Acids Res 41: e129.

[2] Hu XD, Gao LZ. (2014). The complete mitochondrial genome of domestic sheep, *Ovis aries*. Mitochondrial DNA 1-3. DOI: 10.3109/19401736.2014.953076.

[3] Jiang LC, Tan S, Peng R, Zou FD. (2014). The complete mitochondrial genome sequence of the blue sheep, *Pseudois nayaur nayaur* (Cetartiodactyla: Caprinae). Mitochondrial DNA. DOI: 10.3109/19401736.2013.861434.

[4] Lancioni H, Di Lorenzo P, Ceccobelli S, Perego UA, Miglio A, Landi V, Antognoni MT, et al. (2013). Phylogenetic relationships of three Italian merino-derived sheep breeds evaluated through a complete mitogenome analysis. PLoS One 8: e73712.

[5] Wu Y, Yuan CG, Hu JC, Peng JT, Tao PL. (1990). A biological study of dwarf blue sheep. Acta Theriol Sin 10: 185-188.

[6] Yi BG, Liu WL. (1993). Rare value wild animals and their protection. Beijing: China Forestry Publishing House.

[7] Zhong T, Han JL, Guo J, Zhao QJ, Fu BL, He XH, Jeon JT, et al. (2010). Genetic diversity of Chinese indigenous sheep breeds inferred from microsatellite markers. Small Ruminant Res 90: 88-94.

(发表于《Mitochondrial DNA》，院选 SCI，IF：1.209)

The Complete Mitochondrial Genome Sequence of the Wild *Huoba* Tibetan Sheep of the Qinghai-Tibetan Plateau in China

LIU J.B.[1,2], DING X.Z.[1], GUO T.T.[1,2], YUE Y.J.[1,2],
ZENG Y.F.[1], GUO X.[1], CHU M.[1], HAN J.L.[1,2], FENG R.L.[1,2],
SUN X.P.[1,2], NIU C.E.[1,2], YANG B.H.[1,2], GUO J.[1,2], YUAN C.[1,2]

(1. Lanzhou Institute of Husbandry and Pharmaceutical Sciences of Chinese Academy of Agricultural Sciences, Lanzhou 730050, China;
2. Sheep Breeding Engineering Technology Research Center of Chinese Academy of Agricultural Sciences, Lanzhou 730050, China)

Abstract: The wild Huoba Tibetan sheep belongs to the subfamily Caprinae, which distributes in Huoba Town of Tibet Autonomous Region, China. In the present work, we report the complete mitochondrial genome sequence of wild Huoba Tibetan sheep for the first time. The total length of the mitogenome is 16 621 bp, consisting of 13 protein-coding genes, 22 transfer RNA (tRNA) genes, two ribosomal RNA (rRNA) genes, and a non-coding control region (D-loop region). As in other mammals, most mitochondrial genes are encoded on the heavy strand. Its overall base composition is A: 33.64%, T: 27.32%, C: 25.90%, and G: 13.14%, A + T (61.96%) was higher than G + C (39.04%). The phylogenetic relationships was analyzed using the complete mitogenome sequence, results show that wild Huoba Tibetan sheep should be a different species differ from the Ovis aries. These information provide an important data for further study on protection of genetic resources and the taxonomy of Caprinae.

Key words: Genome; mitochondrion; phylogenetic; Wild Huoba Tibetan sheep

The wild Huoba Tibetan sheep belongs to the subfamily Caprinae within the Cetartiodactyla, and is found on the Huoba Town of Tibet Autonomous Region. It is a very important local wild breed of the Qinghai-Tibetan plateau in China, as a year-round grazing animal, it plays a very important role to the economic and native herdsmen. Wild Huoba Tibetan sheep are well suited to the harsh climate and poor pasture conditions characteristic of the hills and plains of the region, the average altitude is 4 614m above sea level, because of the good adaptation and performance, which is attached great importance by the government. The specimen was carried out on natural rangeland typical alpine meadows, located in the Rima Village, Huoba Town, Zhongba County, Rikaze Territory of Tibet Autonomous Region (30°13′822″N, 083°00′249″E), China, and it is situated 5 020m above sea level.In this work, we first determined the mitochondrial genome of wild Huoba Tibetan sheep (GenBank accession no.KP998471).

Ten-milliliter blood sample was collected from one wild Huoba Tibetan sheep (via the jugular vein), and 2mL was rapidly frozen in liquid nitrogen and stored at −80℃ for genomic DNA extraction.The genomic total DNA was exacted from the blood sample, using the genomic DNA isolate kit (Tiangen Biotech, Beijing, China) according to the instructions of the manufacturer, mitochondrial DNA was sequenced by the second generation sequencing technology (Illumina Hiseq 2 500, Illumina Inc., San Diego, CA), and the DNA fragments were assembled by MITObin software (V 1.7) (Hahn et al., 2013).The complete mitogenome is 16 621bp in length, and consists of 13 proteincoding genes, 22 transfer RNA (tRNA) genes, two ribosomal RNA (rRNA) genes, and a non-coding control region (D-loop region), the order and composition are similar to the most of other vertebrates (Hu et al., 2014).The overall base composition of the mitogenome (A: 33.64%, T: 27.32%, C: 25.910%, and G: 13.14%) has a high A+T content (61.96%).The total length of the proteincoding gene is 11 402 bp; the percentage of the total is 68.60%.As in other mammals, most mitochondrial genes are encoded on the heavy stand, except for *ND*6 and eight tRNA genes, which are encoded on the light strand.All 13 mitochondrial protein-coding genes share the start codon ATG, except for *ND*2, *ND*3, *and ND*5, which initiate with start codon ATA.Five protein-coding genes are inferred to terminate with an incomplete termination codon T— (aa) (*ND*1, *ND*2, *COX*3, *ND*3 and *ND*4), while Cytb use AGA instead of TAA as the stop codon.The remaining seven proteincoding genes end with the typical termination codon TAA.The control region (D-Loop) locates between ***tRNA-Phe*** and ***tRNAPro***, the length is 1 183 bp from 15 439 to 16 621, the 12S and 16S rRNA genes were 959 bp (from 69 to 1 027) and 1 576 bp (from 1 095 to 2 670) in length, respectively.The lengths of twenty two tRNA genes were range from 60 to 75 bp.The putative origin of light-strand replication (32 bp in length), as in most vertebrates, was located in a cluster of the $tRNA^{Trp}$-$tRNA^{Ala}$-$tRNA^{Asn}$-$tRNA^{Cys}$-$tRNA^{Tyr}$ region.

Because the mitochondrial DNA is a small genome, with the special characteristics of high evolution rate, lack of recombination, no speciality in tissues, and maternal inheritance, it has been extensively used for exploring the genealogical history in and among the species.In this study, we constructed a phylogenetic tree (Fig. 1) using the wild Huoba Tibetan sheep mitogenomic (KP998471) and other thirteen Ovis aries sequences, *two Pseudois* sequences, and six *Capra* sequences, which were download from the GenBank: *Ovis aries* (KP998470, KF977846.1, KF938341.1, KF938357.1, KF938355.1, KF938356.1, KF938353.1, KF938351.1, EF490456.1, KF938346.1, KF938358.1, KF977845.1, and KF938340.1), *Pseudois* (FJ207537.1 and NC_ 016689.1, and *Capra* (FJ207525.1, FJ207528.1, FJ207526.1, NC_005044.2, FJ207527.1, and FJ207529.1).Evolutionary analyses were conducted in MEGA6 (MEGA Inc., Englewood, NJ), the distances were computed using the Maximum Composite Likelihood method and are in the units of the number of base substitutions per site.Eleven *Ovis aries* breeds clustered with Awang and wild Huoba, Andi clustered with them, *Pseudois nayaur* clustered with *Pseudois schaeferi*, then two groups clustered together, six Capra as another group clustered with them. This result was according to reporting by other scientists (Lancioni et al., 2013; Zhong et al., 2010), therefore, we suggest that wild Huoba Tibetan sheep should be a different species differ from the *Ovis aries*.

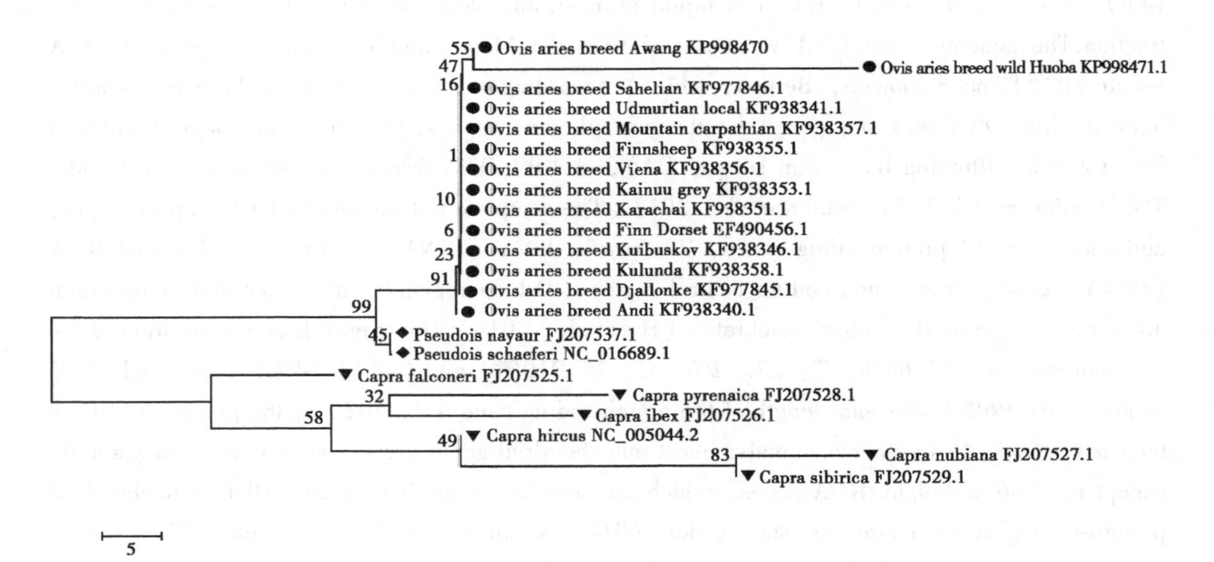

Fig. 1 Neighbor-joining phylogenetic tree of Bovindae. The optimal tree with the sum of branch length = 0.55653866 is shown

Note: The percentage of replicate trees in which the associated origin clustered together in the bootstrap test (1 000 replicates) is shown next to the branches. The tree is drawn to scale, with branch lengths in the same units as those of the evolutionary distances used to infer the phylogenetic tree. Codon positions included were 1st+2nd+3rd+Non-coding. All positions containing gaps and missing data were eliminated. There were a total of 16 575 positions in the final dataset.

DECLARATION OF INTEREST

The authors report no conflicts of interest. The authors alone are responsible for the content and writing of this article. This work was supported by special fund from the Gansu Provincial Funds for Distinguished Young Scientists (1308RJDA015), Gansu Provincial Natural Science Foundation (145RJZA061), and Gansu Provincial Agricultural biotechnology research and application projects (GNSW-2014-21), and the Central Level, Scientific Research Institutes for Basic R & D Special Fund Business (1610322012006 and 1610322015002).

References

[1] Hahn C, Bachmann L, Chevreux B. (2013). Reconstructing mitochondrial genomes directly from genomic next-generation sequencing reads-A baiting and iterative mapping approach. Nucleic Acids Res 41: e129.

[2] Hu XD, Gao LZ. (2014). The complete mitochondrial genome of domestic sheep, *Ovis aries*. Mitochondrial DNA 1-3. DOI: 10.3109/19401736.2014.953076.

[3] Lancioni H, Di Lorenzo P, Ceccobelli S, Perego UA, Miglio A, Landi V, Antognoni MT, et al.

(2013). Phylogenetic relationships of three Italian Merino-Derived Sheep breeds evaluated through a complete mitogenome analysis.PLoS One 8：e73712.

[4] Zhong T, Han JL, Guo J, Zhao QJ, Fu BL, He XH, Jeon JT, et al. (2010).Genetic diversity of Chinese indigenous sheep breeds inferred from microsatellite markers.Small Ruminant Res 90：88-94.

（发表于《Mitochondrial DNA》，SCI：1.209）

Efficacy of an Herbal Granule as Treatment Option for Neonatal Tibetan Lamb Diarrhea under Field Conditions*

LI Sheng-kun, CUI Don-gan , WANG Sheng-yi , WANG Hui, HUANG Mei-zhou, QI Zhi-ming, LIU Yong-ming

(Lanzhou Institute of Husbandry and Pharmaceutical Sciences of Chinese Academy of Agricultural Sciences, Lanzhou 730050, China)

Abstract: Diarrhea is the leading cause of death in neonatal lambs. Herbal remedies are believed to address the conditions. We tested whether an herbal granule had a beneficial efficacy on neonatal lamb diarrhea in this study. The herbal granule was extracted from a combination of *Coptis chinensi*, *Magnolia officinalis*, *Atractylodes lancea*, *Prunus mume and Poria cocos with* a concentration of 1. 0g crude herb/g. Two hundred and sixty-six 3–10day old lambs within 24h from their first onset of diarrhea were randomly divided into one of two treatment groups (A and B), with animals receiving either herbal granule orally in group A ($n = 117$) or oxytetracycline and pepsin orally in group B ($n = 109$) two times daily for amaximum of 5 days at which time lambs were eligible for exit. Thirty lambs with no clinically visible pathological conditions and with no diarrhea diagnosis were availed as the normal controls (group C) to evaluate the growth performance of lamb herds during the 45-days following the treatment. Although statistically non-significant ($P=0.063$), more lambs (103/117) recovered from diarrhea in group A than in group B (85/109). We found that lambs treated with the herbal granule experienced reduced days to recovery from diarrhea (3. 1±0. 8 vs.3. 5±0. 6days, $P<0.01$), reduced the diarrhea-associated mortality (5. 1% vs.13. 8%, $P<0.05$), and reduced recurrence rate (8. 7% vs.24. 4%, $P<0.01$) during the diarrhea episode compared to the controls in group B. The live body weight of lambs were higher in group A than in group B at days 15 ((4. 1±0. 9) kg vs. (4. 1±1. 1) kg, $P<0.05$), 30 ((6. 8±1. 0) kg vs. (6. 3±1. 3) kg, $P<0.01$) and 45 ((10. 7±1. 3) kg vs. (8. 7±1. 7) kg, $P<0.01$) following the treatment. Additionally, the live body weight of lambs at days 45 ((10. 7±1. 3) kg vs. (10. 6±1. 2) kg, $P>0.05$) following the treatment were higher in group A than in group C. Herbal granule used in this study might have a beneficial clinical effect under these study circumstances. Thus, herbal granule could represent a potential effective treatment strategy for neonatal lamb diarrhea.

Key words: Diarrhea; Herbal therapy; Lamb; Alternative remedies

1 INTRODUCTION

Neonatal lamb diarrhea is acommon problem for sheep enterprises worldwide (Scott, 2007).A study at the U.S.Sheep Experiment Station showed that diarrhea accounted for 46 percent of lamb mortality (Schoenian, 2007).Additionally, diarrhea poses a major risk factor for the accumulation of feces on fleece at the breech (perineal region), which in turn results in aggravating the outbreaks of cutaneous myiasis (Morley et al., 1976; French et al., 1994; Hall and Wall, 1995; Bisdorff and Wall, 2008), increasing the risk of carcass contamination with enteric microbes associated with meat spoilage and human food poisoning (Greer et al., 1983; Hadley et al., 1997). Therefore, effective treatments for diarrhea symptom are crucial for animal healthy and economic benefits in sheep enterprises.

Many anti-diarrhea medications, including antibiotics and anthelmintics are commonly used to control of diarrhea in small ruminant animal, however, the problem of the resistance against antibiotics and anthelmintic tend to be increasingly due to overrated use in clinical practices (Hodgson et al., 1999; Dance et al., 1987; Palmer et al., 2001; Besier and Love, 2003; Suter et al., 2005). Antimicrobic treatment for diarrhea in lambs could cause the release of excess endotoxins and delay the clinical evolution of diarrhea lambs (Jiménez et al., 2007). Gómez et al., (2008) demonstrated that administering antibiotics orally may damage the normal flora and actually prolong the diarrhea.Current studies indicates that neonatal diarrhea syndrome has multifactorial in etiology involving the animal, management practices, and infectious agents, e.g.E. *Coli*, *Salmonella* sp., Cryposporidum sp. and Rotavirus (Muñoz et al., 1996; Andrés et al., 2007; Sargison, 2004; Causapé et al., 2002; Wani, et al., 2004). Additionally, neonatal lambs do not have fully functioning immune systems, and which elevates the difficulty for the prevention and control of lamb diarrhea.Thus, diarrhea remains therapeutic challenge in neonatal lambs.Herbal remedies are believed to be effective for many disorders in clinical practices (Jiang, 2005; Grayson, 2011).

According to traditional Chinese veterinary medicine (TCVM) theory, neonatal diarrhea syndrome is always considered a disorder of the entire gastrointestinal tract due to the disharmony between the spleen and the stomach in neonatal small ruminants, and TCVM could provide a practical and effective way to address this condition (Editorial Committee of Encyclopedia of China's agricultural, 1991). The herbal granule used in this study was developed from an herbal formula, which was designed according to the therapeutic principle of "clearing away heat and toxicity materials, relieving diarrhea with astringents" for treatment neonatal diarrhea syndrome in TCVM (Editorial Committee of Encyclopedia of China's agricultural, 1991). The herbal granule consists of *Coptis chinensi*, *Magnolia officinalis*, *Atractylodes lancea*, *Prunus mume and Poria cocos*. Our previous study indicated that administering herbal granule orally does not to be toxic in acute toxicity tests (maximum daily dose.40g/kg bw) in mice according to Hodge and Sterner scale (CCOHS, 2005) (results not published).

The hypothesis of the present study is that neonatal Tibetan lambs affecting with diarrhea treated with the herbal granule will experienc ereduced days to recovery from diarrhea, reduced the diarrhea-associated mortality, and reduced recurrence rate during the diarrhea episode compared to the controls.The aim of this study was to evaluate the efficacy ofthe herbalgranule as an alternative option for lamb diarrhea, and to assess subsequent growth performance in neonatal Tibetan lambs affecting with diarrhea under field conditions.

2 MATERIALS AND METHODS

2.1 Herbal granule preparation

The herbal granule in this study consists of five herbs 80g of *C. chinensi*, 60g of M.*officinalis*, 60g of A.*lancea*, 90g of *P. mume*, and 50g of *P.cocos*. The herb quality criteria were congruent to the Veterinary Pharmacopoeia of the People's Republic of China (Chinese Veterinary Pharmacopoeia Commission, 2010). After pre-processing with washing, drying and chopping, the mixture was decocted with 10 times purified water in 100℃ for 1. 5h for three times, and the extracted liquid were concentrated to 340mL.Finally, the herbal granule was prepared according to a standardized procedure to produce a granule with a final concentration of 1. 0g crude herb/g.

2.2 Study population

The clinical trial was conducted between January and May 2014 in Gansu province, northwest of China, which was conducted in spring as it is the season with the increased incidence of lamb diarrhea syndrome in this area.A total of 5 semi-intensive sheep commercial farms with approximately 600 local Tibetan breed ewes were included in this study, and they were managed under a semi-intensive husbandry system with similar health, nutrition and husbandry practices.All lambs were under veterinary supervision to ensure for colostrum intake directly from their dams in the first 1-6 h after birth, and were reared with the ewes for 4-6 weeks before being weaned.During the experimental period, the enrolled sheep flocks presented natural outbreaks of diarrhea affecting neonatal lambs in the first 1-15days of life, with a prevalence of 15%-40.0%.

2.3 Enrollment criterion

Treatment groups included Tibetan lambs born from January 2014 till May 2014, with lamb diarrhea syndrome diagnosis. Diarrhea was diagnosed based on the parameters fecal consistency score (FCS) and the condition of the Tibetan lambs, and clinical diagnosis was done by a veterinarian (Li) who had no knowledge of the experimental groups. Particularly, the lambs enrolled in this study developed diarrhea for the first time between 3 and 10days of age (Gómez et al., 2008; Joshua et al., 2012). For the purpose of presented study, FCS was between 3 and 5 using a scale of 1 (hard dry fecal pellet) to 5 (liquid/fluid diarrhea) previously described protocol (Greeff and Karlsson, 1997), the degree of a dehydration between 5% and 10% according to a previously described protocol (Hassan, et al., 2014), and lamb attitude score between 1 and 2 using the clinical scoring scale of 1 according to the protocol described by Glover et al. (2013).

2.4 Experimental protocol

This study was designed as a randomized controlled trial. Upon enrollment, all animals were weighed and received a clinical examination for each lamb, including the rectal temperature (℃), fecal consistency, attitude score and degree of dehydration.Particular, the enrolled lambs were clinically examined by a veterinarian prior to enrollment to identify no any other congenital or pathological conditions (e.g.omphalitis, arthritis or pneumonia).During the study period, 266 lambs within 24h from their first onset of diarrhea were randomly divided into one of the two treatment groups (A, $n=117$, B, $n=109$) in blocks that ranged from 3 to 9 lambs depending on the number of lambs eligible at each enrollment day.In addition, 30 lambs with no clinically visible pathological conditions, 3-10days old and average weight of 2.96±0.35kg were availed as controls (group C) to evaluate the effect of the herbal granule on growth performance of the enrolled lamb herds during the 45-days following the treatment.No lambs effecting with diarrhea were used as a control group for ethical reasons in this study.

Therapy began after lamb diarrhea syndrome diagnosis and lambs received with either an oral dose of 1.2g crude herb/kg bw (group A) of the herbal granule dissolved in 250mL of an oral rehydration solution or an oral dose of 100mg/kg bw oxytetracycline and 80mg/kg bw pepsin dissolved in 250mL of an oral rehydration solution (group B) two times daily.All lambs were clinically examined following the treatment and continued receiving their allocated treatment until clinical signs resolved (normal rectal temperature, no diarrhea for 24-48h, and attitude score of 1) or for a maximum of 5 days at which time lambs were eligible for exit.Each group of lambs was kept with their dams in separate boxes, and strict hygiene was maintained to avoid extraneous infections.During this period, the lambs with unfavorable evolution were submitted to a final clinical examination for gross evidence of disease and received etiological treatment.

Lambs were observed three times daily.Detailed symptoms and clinical observations including enrollment time, survival, fecal consistency, general condition, and any other abnormal clinical signs, were recorded in all the animals of different groups.The case was considered as recurrence when another new episode of diarrhea appeared 24-48h following the full remission of a previous episode has been achieved.Prior to treatment (D0), and D15, D30 and D45 following the first treatment, the lambs were weighed to calculate body weight gain using weighing scale with sensitivity of 100g.The average daily weight gains were calculated as the difference between final live body weight at days 45 following the treatment and initial live body weight divided by number of days.Data on survival, recurrence case, days to recovery from diarrhea (from the first sign of diarrhea to the recovery), clinical score (fecal consistency score and attitude score), and body weight gain were recorded to assess the efficacy of the herbal granule for treatment neonatal lambs affecting diarrhea during this trial.

2.5 Efficacy measurements

The primary clinical efficacy of this treatment was the clinical cure in lambs.Clinical cure is defined as normal fecal consistency for at least 24h and lamb attitude score of 1 at exit.Secondary outcome measure was based on live body weight at age 15, 30 and 45 days following the first treatment and the average dail yweight gain at age 45 days for lambs affecting with diarrhea syndrome.

2.6 Exclusion criteria

Lambs were excluded from the study if they had a previous incident of diarrhea since birth or any other congenital or pathological conditions (e.g.an umbilical abscess, pneumonia, omphalitis or arthritis), or previously treated with any antimicrobial drugs.In addition, the lambs were excluded from the study during the days 45 following treatment if their dam died.

2.7 Statistical analysis

Statistical analyses were performed using SPSS for Windows (version 17.0, IBM SPSS Statistics; USA). The baseline characteristics, including enrollment time, fecal consistency score, live body weight, rectal temperature, degree of a dehydration and lamb attitude score between the two groups (A, B), were analyzed by a one-way analysis of variance or a Kruskal-Wallis rank sum test. An ANOVA was used to compare live body weight (g) at day 0, day 15, day 30 and day 45 following the treatment and mean daily weight gain (kg) among the three groups.

The clinical cure rate, mortality rate and recurrence rate were analyzed by logistic regression. The hazard of lambs with diarrhea following the treatment was analyzed by Cox's proportional hazard model. In the model, the time variable availed was the interval in days to recovery from diarrhea and the explanatory variable included diarrhea (yes/no), fecal consistency score, rectal temperature, degree of a dehydration, lamb attitude score, mortality rate, recurrence rate and the interaction between treatment and covariates. All lambs that died or had a fecal consistency score greater than 1 at exit were censored. Interactions between explanatory variables and time, and survival curves were used to evaluate the proportionality of hazard rate. The level of significance was set at $\alpha = 0.05$.

3 RESULTS

3.1 Animals' characteristics

Out of 1 056 lambs during the study period, lamb diarrhea syndrome was diagnosed in 226 according to the enrollment criteria in this study. Of these 226 lambs, 117 lambs were randomly assigned to the herbal granule group (A) and 109 lambs to the control group (B). The baseline characteristics are presented in Table 1. There were no difference in any individual characteristics, including enrollment time, live body weight, lamb attitude score, feces consistency score, rectal temperature, degree of dehydration and twin proportion. Additionally, the average live body weight and enrollment time were 3.38 ± 0.45kg and 4.68 ± 0.3 days in normal control group (C), respectively, and there were no statistically significant differences among the three groups ($P > 0.05$). Six lambs in group A and 15 lambs in group B died at different stages during the trial period. Finally, three groups of animals (A, $n = 111$; B, $n = 94$; C, $n = 30$) were created for further statistical analysis for evaluating the growth performance during the 45 days following the treatment.

3.2 Clinical efficacy

The efficacy of the herbal granule for treatment lamb diarrhea was presented in Table2. The clinical cure rate was higher for lambs in group A compared to group B ($P = 0.063$). Nevertheless, the average duration of diarrhea (days to recovery from diarrhea) was shorter in the group A than in the group B ($P < 0.05$) presented in the clinical cure rate curves (survival curves) (Fig. 1). Moreover, there was a significant reduction in mortality rate ($P < 0.05$) and recurrence rate ($P < 0.01$) for lambs in group A compared to the controls in group B.

Table 1 Characteristics of the enrolled animals

Characteristics	Treatment group (A)	Control group (B)
Enrollment time, days	5. 4±2. 0	5. 4±2. 0
Fecal consistency score	3. 1±0. 7	3. 1±0. 7
Live body weight (kg)	3. 0±0. 7	3. 1±0. 3
Rectal temperature, (℃)	40. 8±0. 7	40. 6±0. 7
Degree of a dehydration (%)	7. 1±1. 8	7. 2±1. 7
Lamb attitude score	1. 8±0. 5	1. 8±0. 5
Twin proportion (n/N)	6 (117)	4 (109)

Table 2 Effect of the herbal granule for treatment diarrhea in lambs

Efficacy parameters	Herbal granule group (A) (n. 117)	Control group (B) (n. 109)
Clinical cure rate[1] (%)	88. 0	78. 9
Days to recovery from diarrhea, days	3. 1±0. 8^{A}	3. 57±0. 6^{B}
Mortality rate (%)	5. 1^{a}	13. 8^{b}
Recurrence rate[2] (%)	8. 7^{A}	24. 4^{B}

Note: Values within the same row marked with different letters in superscript differ significantly: a, b = $P<$ 0. 05; A, B. $P<$0. 01.

[1] Clinical cure defined as normal fecal consistency for at least 24h.

[2] Recurrence rate defined as the number of the recurrence cases divided by the total number of clinical cure cases.

3.3 Body weight gain

Weight data for 205 lambs at enrollment and exit in treatment group (A and B) and 30 lambs in normal control group (C) was available for analysis (Fig. 2). The live body weight were higher in lambs in group A than the controls in group B at days 15 ($P<$0. 05), 30 ($P<$0. 01) and 45 ($P<$0. 01) following the treatment. Furthermore, the live body weight of lambs at days 45 following the treatment were higher in group A than in group C, however there was no significance between the two groups (10. 7±1.3g vs.10. 6±1. 2g, $P>$0. 05).

4 DISCUSSION

The current trial results provided evidence for a beneficial effect of the herbal granule as treatment option for diarrhea in neonatal lambs. As the key component of the herbal granule, *C. chinensi* has demonstrated significant antimicrobial activity against bacteria, yeast, protozoans (Giardia, etc.), viruses and helminths (worms), and are commonly applied to treatment for bacterial diarrhea and intestinal parasites in clinical practices (Lan et al., 2001; Yu et al., 2006). Current study suggests that *P. mume* possesses variety of organic acids (Chen et al., 2006), which was believed to help improve the condition of digest by modulating digestive processes and balancing the intestinal microbiota (Zentek et al., 2013; Goodarzi et al., 2014). Additionally, it can be suggested that *P. cocos* modulates host immunity through anti-inflammatory, immunomodulatory and anti-oxidative (Sun, 2014; Tang et al., 2014; Ríos, 2011; Wang and Yang, 2013). It is possible that

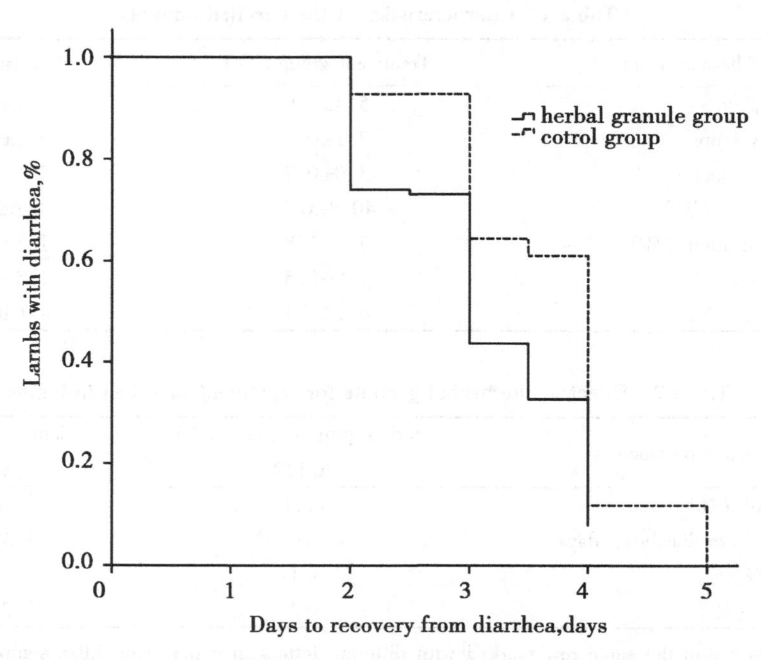

Fig. 1 The proportion of lambs with diarrhea following the herbal granule treatment (survival analyses)

Note: Survival curves provides the clinical cure rate curves of lambs in herbal granule and the control group (B).

herbal granule treatment provided more favorable gut environment for an earlier recovery of diarrhea syndrome through antimicrobial activity, modulating host immunity and improving the condition of digest. In the present study, 103 of 117 lambs were recovery from diarrhea following treatment with the herbal granule, whereas 85 of 109 lambs in group (B) were recovery from diarrhea, although there no differences between two the groups when using a P value of <0.05 ($P=0.063$), herbal granule treatment could significantly reduce the diarrhea-associated mortality and recurrence rate in lambs affecting with diarrhea. Additionally, this efficacy of the herbal granule treatment was also evidenced by shorter days to recovery from diarrhea presented in the clinical cure rate curves (survival curves) (Fig. 1). As previously study reported by Gómez et al. (2008) that antibiotic treatment can delay the clinical evolution of lambs affecting with diarrhea and cause the high mortality rate. Therefore, the herbal granule can be considered as a treatment option for treatment diarrhea in neonatal lambs, and it is interesting for neonatal lamb diarrhea syndrome due to no specific treatment for the conditions in lambs.

In the present study, herbal granule treatment significantly increased the live body weight of lambs at days 15, 30 and 45 following the treatment. Green et al. (1998) demonstrated that an episode of diarrhea can reduce the absolute weight of lambs by multi-level analysis. The presented results suggested that the herbal granule treatment might improve the overall growth performance of lambs affecting with diarrhea. This could be result of the synergistic effect of *C. chinensi's* antimicrobial activity; *M. officinalis* and *A. lancea* having antiinflammatory and analgesia properties (Zhu et al., 1997); *P. mume's* ability to balance the gastrointestinal microbial ecology (Zentek et al., 2013; Goodarzi et al., 2014;); and *P. cocos* having anti-inflammatory, immunomodulatory and anti-oxidative effects (Sun, 2014; Tang et al., 2014; Ríos, 2011; Wang and Yang, 2013). We also found that lambs treated with the herbal granule have similar or even superior growth performance compared to the controls in group C, which indicates the effectiveness of the herbal granule remedy. Additionally, the use of herbal granule to treat neonatal lamb diarrhea syndrome is interesting and warranted to the increased resistance to antimicrobial drug. Therefore, the herbal granule can be considered as an alternative remedy for treatment of diarrhea in lambs.

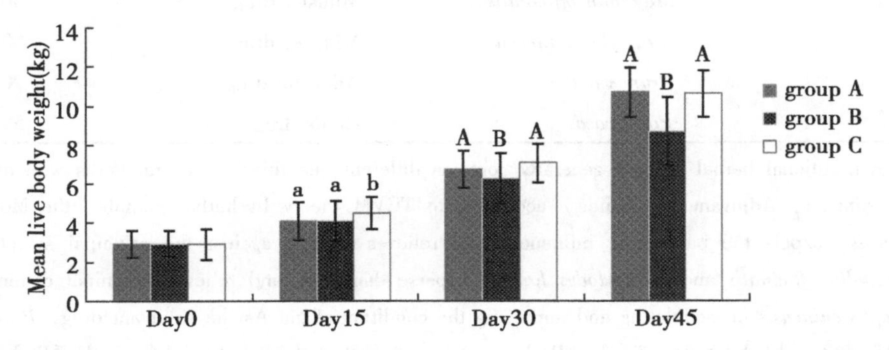

Fig. 2 Live body weight of Tibetan lambs at day 0, day 15, day 30 and day 45 following the treatment

Note: Lambs with diarrhea received with either an oral dose of 1.2g crude herb/kg bw (group A) of the herbal granule dissolved in 250mL of an oral rehydration solution or an oral dose of 100mg/kg bw oxytetracycline and 80mg/kg bw pepsin dissolved in 250mL of an oral rehydration solution (group B) two times daily for a maximum of 5 days at which time calves were eligible for exit, whereas lambs with no clinically visible pathological conditions received no intervention (groupC). Values represent mean±SD. Bars with different letters in same day differ significantly: a, b=P<0.05; A, B=P<0.01.

5 CONCLUSIONS

The present results provided evidence for a beneficial clinical effect of the herbal granule for treatment diarrhea in lambs, including higher clinical cure rate, shorter days to recovery from diarrhea, lower the diarrhea-associated mortality and recurrence rate, and superior growth performance of lamb herds. Thus, this herbal remedy might represent a potential effective treatment strategy for neonatal lamb diarrhea.

ACKNOWLEDGMENTS

This work was funded by the National Key Technology Research and Development of the Ministry of Science and Technology of the People's Republic of China (2012BAD12B03), Special Fund for Agro-scientific Research in the Public Interest (201303040-17) and the Special Fund of the Chinese Central Government for Basic Scientific Research Operations in Commonwealth Research Institutes (1610322015006).

APPENDIX A

See appendix Table 1.

Table 1 Ingredients of the herbal granule used in the present study

Herb name		*Rolein Sheng HuaTang*	Standard dose
Chinese name	Latin name		
Huang lian	*Coptis chinensis*	Monarch drug[a]	60. 0g
Hou po	*Magnolia officinalis*	Minister drug[b]	60. 0g
Cang zhu	*Atractylodes lancea*	Minister drug[b]	80. 0g
Wu mei	*Prunus mume*	Adjuvant drug[c]	90. 0g
Fu ling	*Poria cocos*	Guide drug[d]	50. 0g

Note: A traditional herbal formula generally contains different quantities of several herbs with different roles (Monarch, Minister, Adjuvant and Guide) according to TCVM theory. In herbal granule, the Monarch drug, *Coptis chinensis*, expels the pathogenic influences and relieves toxiticy against the principal symptom of lamb diarrhea.*Magnolia officinalis* and *Atractylodes lancea* disperse dampness and relieve abdominal distention, which assist the *Coptis chinensis* in modulating and improving the condition of gut. As the Adjuvant drug, *Prunus mume* is sour and astringent and helps to stopdiarrhea.*Poria cocos* can strengthen the Spleen and tonify the Middle Burner.The five herbs together expel pathogens and toxic heat, eliminate dampness and active the spleen to help resolve the disorder of the entire gastrointestinal tract of lambs affecting with diarrhea (Xie, et al., 2010).

[a] The Monarch drug is the key component of a herbal formula, and it provides the principal therapeutic effects against a related disease.

[b] The Minister drug is the synergistic secondary element of a herbal formula.It potentiates the effects of the Monarch drug or treats the accompanying symptoms.

[c] The Adjuvant drug is generally intended to enhance the therapeutic effects and mitigate any adverse effects of the Monarch drug and/or Minister drug.

[d] The Guide drug is generally used to harmonize the other herbs; it may also be used to facilitate the uptake of other herbs by a specific organ or target tissue.

References

[1] Andrés, S., Jiménez, A., Sánchez, J., Alonso, J.M., Gómez, L., López, F., Rey, J., 2007. Evaluation of some etiological factors predisposing to diarrhoea in lambs in "La Serena" (Southwest Spain).Small Rumin.Res.70, 272-275.

[2] Besier, R.B., Love, S.C.J., 2003.Anthelmintic resistance in sheep nematodes in Australia: The need for new approaches.Aust.J.Exp.Agric., 1383-1391.

[3] Bisdorff, B., Wall, R., 2008.Sheep blowfly strike risk and management in Great Britain: a survey of current practice.Med.Vet.Entomol.22, 303-308.

[4] Causapé, A.C., Quílez, J., Sánchez-Acedo, C., del Cacho, E., López-Bernad, F., 2002.Prevalence and analysis of potential risk factors for *Cryptosporidium parvum* infection in lambs in Zaragoza (northeastern Spain).Vet. Parasitol.104, 287-298.

[5] CCOHS, 2005.What is an LD50 and LC50? In: Canada's National Occupational Health and Safety Resource.Canadian Centre for Occupational Health and Safety, ON, Canada_ 〈http: //www.ccohs.ca/

oshanswers/chemicals/ld50.html〉 _ .

[6] Chen, Z.G., En, B.T., Zhang, Z.Q., 2006. Simultaneous determination of eight organic acids in Fructus mume by RP-HPLC.ChinaJ.Chin.Mat.Med. 31, 1783-1786.

[7] Chinese Veterinary Pharmacopoeia Commission, 2010.Veterinary Pharmacopoeia of the People's Republic of China.China Agriculture Press, Beijing (in Chinese).

[8] Dance, D.A., Pearson, A.D., Seal, D.V., Lowes, J.A., 1987.A hospital outbreak caused by a chlorhexidine and antibiotic-resistant *Proteus mirabilis*.J. Hosp.Infect.10, 10-16.

[9] Editorial Committee of Encyclopedia of China's agricultural, 1991. China Agricultural Encyclopedia (traditional Chinese veterinary medicine Volume). Agricultural Publishing House, Beijing (in Chinese).

[10] French, N.P., Wall, R., Morgan, K.L., 1994.Ectoparasite control on sheep farms in England and Wales: the method, type and timing of insecticidal treatment.Vet.Rec.135, 35-38.

[11] Glover, A.D., Puschner, B., Rossow, H.A., Lehenbauer, T.W., Champagne, J.D., Blanchard, P.C., Aly, S.S., 2013. A double-blind block randomized clinical trial on the effect of zinc as a treatment for diarrhea in neonatal Holstein calvesunder natural challenge conditions.Prev.Vet.Med. 112, 338-347.

[12] Gómez, L., André, S., Sánchez, J., Alonso, J.M., Rey, J., López, F., Jiménez, A., 2008. Relationship between the treatment and the evolution of the clinical course in scouring Merino lambs from "La Serena" (Southwest Spain). Small Ruminant Res.76, 223-227.

[13] Goodarzi, B.F., Vahjen, W., Mader, A., Knorr, F., Ruhnke, I., Röhe, I., Hafeez, A., Villodre, C., Männer, K., Zentek, J., 2014.The effects of different thermal treatments and organic acid levels in feed on microbial composition and activity in gastrointestinal tract of broilers.Poult.Sci. 93, 1440-1452.

[14] Grayson, M., 2011.Traditional Asian medicine.Nature 480, S81.

[15] Greeff, J.C., Karlsson, L.J.E., 1997. Genetic relationship between fecal worm egg count and scouring in Merino sheep in a Mediterranean environment. In: Proceedings of the Association for the Advancement of Animal Breeding and Genetics, 12, pp.333-337.

[16] Green, L.E., Berriatua, E., Morgan, K.L., 1998.A multi-level model of data with repeated measures of the effect of lamb diarrhoea on weight.Prev.Vet.Med.36, 85-94.

[17] Greer, G., Jeremiah, L., Weiss, G., 1983. Effects of wholesale and retail contamination on the case life of beef.J.Food Prot.46, 842-845.

[18] Gómez, L., André, S., Sánchez, J., Alonso, J.M., Rey, J., López, F., Jiménez, A., 2008. Relationship between the treatment and the evolution of the clinical course in scouring Merino lambs from "La Serena" (Southwest Spain). Small Rumin.Res.76, 223-227.

[19] Hadley, P.J., Holder, J.S., Hinton, M.H., 1997.Effects of fleece soiling and skinning method on the microbiology of sheep carcases.Vet.Rec.140, 570-574.

[20] Hall, M., Wall, R., 1995.Myiasis of humans and domestic animals.Adv.Parasitol.35, 257-334.

[21] Hassan, N., Sheikh, G.N., Hussian, S.A., Nazir, G., 2014.Variation in clinical findings associated with neonatal colibacillosis in lambs before and after treatment.Vet.World 7, 262-265.

[22] Hodgson, J.C., Brebner, J., McKendrick, I.J., 1999.Efficacy of a single dose of oral antibiotic given within two hours of birth in preventing watery mouth disease and illthriff in colostrums-deficient lambs.Vet.Rec.145, 67-71.

[23] Jiang, W.Y., 2005.Therapeutic wisdom in traditional Chinese medicine: a perspective from modern science.Trends Pharm.Sci 26, 558-563.

[24] Jiménez, A., Sánchez, J., Andrés, S., Alonso, J.M., Gómez, L., López, F., Rey, J., 2007.Evaluation of endotoxemia in the prognosis and reatment of scouring Merinol ambs.J.Vet.Med.A54, 103-106.

[25] Joshua, P.A.S., Ryan, U.M., Robertson, I.D., Jacobson, C., 2012.Prevalence and on-farm risk factors for diarrhoea in meat lamb flocks in Western Australia.Vet.J.192, 503-510.

[26] Lan, J., Yang, S.L., Zheng, Y.Q., Shao, J.B., Li, Y., 2001.Advances in studies on *Coptis chinensis*. Chin.Trad.Herb.Drugs32, 1139-1141. (in Chinese).

[27] Morley, F.H., Donald, A.D., Donnelly, J.R., Axelsen, A., Waller, P.J., 1976.Blowfly strike in the breech region of sheep in relation to helminth infection. Aust.Vet.J.52, 325-329.

[28] Muñoz, M., Alvarez, M., Lanza, I., Cármenes, P., 1996.Role of enteric pathogens in the aetiology of neonatal diarrhoea in lambs and goat kids in Spain.Epidemiol.Infect.117, 203-211.

[29] Palmer, D.G., Besier, R.B., Lyon, J., Mitchell, T.J., January 21-25, 2001.Detection of macrocyclic lactone resistance using fecal egg count reduction test-the Western Australian experience.In: Proceedings of the Fifth International Congress for Sheep Veterinarians.Onderstepoort, South Africa.

[30] Ríos, J.L., 2011.Chemical constituents and pharmacological properties of *Poria cocos*. Planta Med.77, 681-691.

[31] Sargison, N.D., 2004.Differential diagnosis of diarrhoea in lambs.In Pract.26, 20-27.

[32] Scott, P.R., 2007.Sheep Medicine, Manson Publishing.London, 99-136.

[33] Sun, Y., 2014.Biological activities and potential health benefits of polysaccharides from *Poria cocos* and their derivatives.Int.J.Biol.Macromol. 68, 131-134.

[34] Schoenian S., 2007. Diarrhea (Scours) in Small Ruminants. (http: //www.thejudgingconnection.com/pdfs/Scours_ in_ Sheep_ and_ Goats.pdf).

[35] Suter, R.J., McKinnon, E.J., Perkins, N.R., Besier, R.B., 2005.The effective life of ivermectin on Western Australian sheep farms-a survival analysis.Prev.Vet.Med.72, 311-322.

[36] Tang, J., Nie, J., Li, D., Zhu, W., Zhang, S., Ma, F., Sun, Q., Song, J., Zheng, Y., Chen, P., 2014. Characterization and antioxidant activities of degraded polysaccharides from *Poria cocos sclerotium*. Carbohyd.Polym.105, 121-126.

[37] Wang, Z.K., Yang, Y.S., 2013. Upper gastrointestinal microbiota and digestive diseases. World J. Gastroenterol.19, 1541-1550.

[38] Wani, S.A., Bhat, M.A., Samanta, I., Ishaq, S.M., Ashrafi, M.A., Buchh, A.S., 2004.Epidemiology of diarrhoea caused by rotavirus and *Escherichia coli* in lambs in Kashmir valley, India.Small Rumin.Res.52, 145-153.

[39] Xie, H.S., Preast, V., Beckford, B.J., 2010. Xie's Chinese Veterinary Herbology Hardcover. A John Wiley and Sons, Ltd., Singapore.

[40] Yu, Y.Y., Wang, B.C., Peng, L., Wang, J.B., Zeng, C., 2006. Advances in Pharmacological Studies of *Coptis Chinesis*. Journal of Chongqing University (Natural Science Edition) 29, 107-111. (in Chinese).

[41] Zentek, J., Ferrara, F., Pieper, R., Tedin, L., Meyer, W., Vahjen, W., 2013.Effects of dietary combinations of organic acids and medium chain fatty acids on the gastrointestinal microbial ecology and bacterial metabolites in the digestive tract of weaning piglets.J Anim.Sci.91, 3200-3210.

[42] Zhu, Z.P., Zhang, M.F., Shen, Y.Q., Wang, H.W., 1997. Pharmacological effect of Cortex *Magnoliae officinalis* on digestion system.China J.Chin. Mater.Med.22, 686-689.

（发表于《Livestock Science》，SCI 院选：1.100）

Prophylactic Strategy with Herbal Remedy to Reduce Puerperal Metritis Risk in Dairy Cows: A Randomized Clinical Trial

CUI Don-gan, WANG Sheng-yi, WANG Lei, WANG Hui, LI Xia, LIU Yong-ming

(Engineering & Technology Research Center of Traditional Chinese Veterinary Medicine of Gansu Province/Lanzhou Institute of Husbandry and Pharmaceutical Sciences of Chinese Academy of Agricultural Sciences, Lanzhou 730050, China)

Abstract: Puerperal metritis is an important disorders usually within 21 days postpartum in dairy cattle that occurs within 21 days postpartum, and herbal remedies are believed to bebeneficial for postpartum female livestock. *Sheng Hua Tang* is a prime example of herbal formula used as a therapeutic aid in prevention or control of postpartum disease for centuries in China. In the present study, we were to evaluate the efficacy of *Sheng Hua Tang* as a prophylactic strategy for lowering puerperal metritis risks and improving reproductive performance in dairy cows under field conditions. A total of 311 clinically healthy cows were randomly allocated to the intervention group or the control group 2-4h after delivery. Treated cows ($n = 58$) received *Sheng Hua Tang* with an oral dose of 0. 36g crude herb/kg bw once daily for three consecutive days, whereas the controls ($n = 53$) received no treatment. The logistic regression and survival analysis were used to analyze the incidence of puerperal metritis and reproduction parameters of cows between the two groups, respectively. The results showed that there was a significant reduction in the incidence of puerperal metritis (12. 1% vs. 33. 3%, $P = 0.01$, odd ratio [OR] 2. 392) between *Sheng Hua Tang* group and the control group. The calving-to-first-service interval ((68. 9±17. 7) days vs. (80. 5±26. 6) days, $P<0.05$) and service per conception (1. 7 vs. 2. 1, $P<0.01$) were lower in cowsin *Sheng Hua Tang* group than the controls. Additionally, *Sheng Hua Tang* treatment effectively elevated the first AI conception proportion (61. 1% vs. 51. 3%, $P<0.05$) and proportion of cows that were pregnant at 305 days in milk (89. 8% vs. 82. 0%, $P < 0.01$) compared with that of controls. The present results would support efforts to the use of *Sheng Hua Tang* immediately after delivery as a prophylactic strategy for lowering puerperal metritis risk and improving the overall reproductive efficiency of dairy herds under these study circumstances. Thus, *Sheng Hua Tang* treatment could represent an effective prophylactic strategy for bovine postpartum care.

Key words: Herbal remedy; Puerperal metritis; Prophylactic strategy; Cow

1 INTRODUCTION

Puerperal metritis is a frequently diagnosed disease conditions within 21 d after parturition in dairy cattle, which is characterized by an abnormally enlarged uterus and a fetid watery red-rown uterine discharge, associated with signs of systemic illness, including fever (≥39.5℃), dullness, inappetance and decreased milk production (Sheldon et al., 2006). It is considered to be one of the most important postpartum diseases in dairy cows due to impairing subsequent fertility and decreasing farm profitability (Markusfeld, 1984; Curtis et al., 1985; Fourichon et al., 2000; Drillich et al., 2003; LeBlanc, 2008; McLaughlin et al., 2013). Up to date, there is a lack of data on the efficacy of treatment of puerperal metritis for improvement of subsequent fertility, even though certain progresses have been made towards the prevention or control of puerperal metritis in cows (LeBlanc, 2008; Haimerl and Heuwieser, 2014). Because of the infectious nature of puerperal metritis, antibiotic treatment is usually believed to be able to be beneficial for this condition (Smith et al., 1998; Drillich et al., 2001; Chenault et al., 2004; Goshen and Shpigel, 2006; Malinowski, et al., 2011). However, Trevisi et al. (2014) have proposed that production diseases of farm animals are complicated by the overuse of antibiotics and the generation of drug-resistant bacteria. Therefore, there is a significant need to encourage prudent use of antibiotics, which stresses the importance of alternative therapies and preventive measures for postpartum uterine disease in dairy cows. Certain herbal remedies are considered beneficial in prevention and control of postpartum disease associated with fertility in livestock in China (Song, 1988).

In traditional Chinese veterinary medicine theory, puerperal metritis and retained placenta both fall within the blood stasis syndrome category (Editorial Committee of Encyclopedia of China's Agricultural, 1991). Blood stasis is believed to be a key aspect of pathogenesis in postpartum diseases according to the traditional Chinese medicine theory (Luo, 1986; Lü et al., 1992) and many herbal agents with blood circulation-activating activities are usually used as a therapeutic aid to treatment postpartum disease in China (Chan et al., 1983; Luo, 1986; Chang et al., 2013). *Sheng Hua Tang*, a classical herbal formula consisting of *Radix Angelicae sinensis*, *Ligustici rhizoma*, *Semen persicae*, *Zingiberis rhizoma* and *Radix glycyrrhizae*, is a good example of a herbal formula that activates the blood circulation and removes blood stasis (Luo, 1986; Chang et al., 2013), and widely used as a therapeutic aid in prevention of postpartum disease in female livestock (Song, 1988). Our previous study (Cui et al., 2014) has demonstrated that the oral *Sheng Hua Tang* administration as a preventive treatment has beneficial effects in lowering the incidence of retained placenta and improving subsequent fertility in cows. It is generally accepted that retained placenta is associated with puerperal metritis (Sandals et al., 1979; Sheldon et al., 2008), and certain therapy for retained placenta can be effective for prevention and control of uterine infection following placental retention in cattle (Drillich et al., 2006; Goshen and Shpigel, 2006). Thus, the hypothesis of the present study is that the use of *Sheng Hua Tang* immediately after delivery as a prophylactic strategy would decrease the odds of puerperal metritis.

In the present study, we are to evaluate the efficacy of *Sheng Hua Tang* as a prophylactic strategy for puerperal metritis, and to assess the reproductive performance of the cows during the subsequent lactation under field conditions.

2 MATERIALS AND METHODS

2.1 *Sheng Hua Tang* preparation

Sheng Hua Tang used in our study was composed of 5 herbs-*Angelica sinensis*, 120.0g; *Ligusticum chuanxiong*, 45.0g; *Prunus persica*, 30.0g; *Zingiber officinale*, 10.0g; and *Glycyrrhiza uralensis*, 10.0g-that was prepared as with a final concentration of 0.5g crude herb/ml according to the procedure established by Cui et al. (2014).

2.2 Animals and enrollment criteria

The study was conducted from January 2011 to January 2013 in a commercial dairy farm of 850 lactating cows in northwestern China, and prior to this study the annual incidence of puerperal metritis in this dairy farm is 35.4%. Cows were housed in free-stall barns with cubicles, rubber mats and slotted floors, and were fed twice daily with a diet of total mixed ration (TMR) consisting of corn silage and grass forage, with corn meal, barley and mineral supplements based on the stage of lactation and milk yield according to NRC (2001). Fresh water was provided *ad libitum*. Cows were milked three times daily.

2.3 Enrollment criteria

The enrolled animals calved from January 2011 to March 2012 and were free of infectious or metabolic diseases and 37 years of age (25 lactations). Additionally, the cows with the risk factors abortion, dystocia, given twins were also enrolled in this study. They had a Body Condition Score (BCS) between 2.5 and 4.0 during the peripartal stage (Edmonson et al., 1989). For the purpose of the present study, dystocia diagnosis was established according to the protocol described by Schuenemann et al. (2011) with assistance score between 1 and 3 using a 14 scale (1 = no assistance; 2 = light assistance by one person without the use of mechanical traction; 3 = assistance with mechanical traction; 4 = surgical procedure). Abortion was defined as the termination of a pregnancy of seven month or more, and the day of abortion considered the same as the day of parturition.

2.4 Exclusion criteria

Animals suffering from diseases such as cesarean section during previous or current calving, displaced abomasum, and laminitis were excluded from the study during the experimental period. In addition, cows with incomplete treatments or other deviations from the treatment protocol were also withdrawn from the present study.

2.5 Study design and clinical examination programs

This study was designed as a randomized controlled trial and our sample size was calculated to attain a power more than 80% in the test of association between puerperal metritis and exposure to *Sheng Hua Tang* 2-4h after delivery. After completing the physical evaluation, all odd-numbered cows were allocated to the intervention group, and even-numbered cows to the control group. According to the treatment protocols described by Cui et al. (2014), the treated cows were received an oral dose of *Sheng Hua Tang* of 430mL/600kg bw (equal to 0.36g crude herb/kg bw) within 2-4 hafter delivery once daily for 3 consecutive days, and the controls received no treatment. Individuals involved in the diagnosis of uterine diseases were blinded to treatment allocation.

During the experimental period, placental expulsion was monitored at 2h intervals by visually examining the perineum and no attempts were made to manually remove the placenta within 12h postpartum. The uterine condition and placental expulsion were confirmed by vaginal and transrectal palpation at 12h postpartum. The diagnosis of puerperal metritis was based on a complete physical examination of the cow including the general condition of the cow, rectal temperature, and palpation of the uterus per rectum to evaluate uterine discharge and uterus status according to the protocol described by Sheldon et al. (2006), cows having fetid, reddish-brown, watery vulvar discharge in combination with a rectal temperature >39.5℃ were diagnosed as puerperal metritis. Evaluation of rectal temperature was performed before palpation per rectum, and vaginal examination was performed manually after thoroughly cleaning and disinfection of the perineal area with antiseptic solution. All cows, with a diagnosis of retained placenta or puerperal metritis conditions, received an oxytetracycline infusion administered to uterus (2 000mg in 250mL of normal saline) every other day for 3 times according to usual management measures at the dairy farm in which we conducted the clinical trial.

All enrolled animals were monitored by clinically visible estrus signs starting from the day 25 following partus. Artificial insemination (AI) was performed within 1 218h following observed estrus. If estrus did not occur at days 65 following birth, it was induced by i. m. injection of 0.4mg cloprostenol sodium (2ml: 0.2mg, Suzhou SUMU animal medical industry Co. Ltd.). Pregnancy diagnosis was performed by rectal palpation of the uterus and its contents between days 40 and 50 following the last AI. Inseminations were performed throughout the year and cows not conceiving to the first AI were re-examined for pregnancy after each subsequent insemination between days 40 and 50 following AI.

The date of parturition, abortion, dystocia, twin, retained placenta, date of estrus, dates of artificial inseminations, number of AI's until pregnancy and health problems were recorded. To evaluate the effect of *Sheng Hua Tang* treatment on the reproduction performance of the cows during the subsequent lactation, thecalving-to-first-service interval, first AI conception proportion, service per conception, calving-to-conception interval and proportion of cows that were pregnant at 305 DIM were calculated.

2.6 Outcomes

The primary measure of outcome was based on the effect of *Sheng Hua Tang* on the risk of occurrence of puerperal metritis in dairy cows based on the above-established diagnosis criteria. The second end point was the effect of *Sheng Hua Tang* treatment on the reproductive performance in cows, included calving-to-firstservice interval, first AI conception proportion, service per conception and proportion of cows that were pregnant at 305DIM.

2.7 Statistical analysis

All analyses were conducted in SPSS (version 17.0, IBM SPSS Statistics; New York,

USA). The baseline characteristics, including BCS at calving, mean parity, mean age, proportion of the twins, proportion of dystocia and proportion of abortion between the two groups, were analyzed by a one-way analysis of variance or a Kruskal-Wallis rank sum test. The incidence of puerperal metritis between the groups was modeled in a logistic regression model with parity, BCS, twinning, dystocia and retained placenta as covariates, and the model produced odds ratios (OR) as estimates of the strength of association between the potential risk factors and puerperal metritis. The calving-to-first-service interval was compared between the cows in the *Sheng Hua Tang* group and the controls witha Cox proportional hazards regression. Values were censored if observations were terminated for reasons beyond the investigator's control (e. g. , data for cattle that were never inseminated and conceived). The survival analysis involved a Cox's proportional hazard model with forward likelihood-ratio testing that included the hazard of cows being inseminated for the first time and cows that were pregnancy during the first 305 days postpartum. In the models, the time variable availed was the interval in days from calving to pregnant and the explanatory variable included twin calvings, dystocia, abortion, BCS and parity. All cows that were not pregnant or never inseminated at 305 DIM were censored. Additionally, specific censured cases were cows that became pregnant, but then aborted and did not become pregnant again within 305 DIM. Interactions between explanatory variables and time, and survival curves were used to evaluate the proportionality of hazard rate. The level of significance was set at $\alpha = 0.05$.

3 RESULTS AND DISCUSSION

3.1 Participant characteristics

In this trial, a total of 311 cows were included in this trial and 158 cows in intervention group and 153 cows in the control group. There were no differences in BCS at calving, parity, age, proportion of the twins, proportion of dystocia and proportion of abortion between the treated and control groups (Table 1). During the study period, one animal with displaced abomasum was excluded in *Sheng Hua Tang* group. Two cows with acute mastitis and one with serious lameness were excluded from the control group. Finally, of 307 cows, data were created for further statistical analysis (157 cows in *Sheng Hua Tang* group and 150 cows in the control group).

3.2 Effect of Sheng Hua Tang on the incidence of puerperal metritis

In the present study, 19 cows treated with *Sheng Hua Tang* with diagnosed puerperal metritis, whereas 50 cows of the control cows, and the difference between the two groups was statistically significant (12.1% vs. 33.3%, $P = 0.01$, OR 2.392). It is possible that *Sheng Hua Tang* treatment had beneficial effect on the uterine recovery process in consequence of a promptly placental detachment during the early postpartum stage (Cui et al., 2014). *Sheng Hua Tang* is usually traditionally used to relieve afterpains and expel lochia following calving in clinical practices (Luo, 1986). Previous studies demonstrated that *Sheng Hua Tang* could induce anti-inflammatory and analgesia responses (Hou et al., 1992), and modulate the immune response (Wang et al., 2004; Li et al., 2011). These effects might improve postpartum overall physical health, and thereby contributing to lower the risk of puerperal metritis. In the present study, there was a reduction of

21. 2% in the incidence of puerperal metritis in cows in *Sheng Hua Tang* group compared to the controls. Thus, this study presented an effective prophylactic approach for puerperal metritis in dairy cows.

Table 1 Baseline demographic characteristics of the enrolled animals

Indexes	Sheng Hua Tang group	Controlgroup
BCS	3. 37±1. 03	3. 52±1. 15
Mean age	5. 34±1. 05	5. 41±1. 14
Mean parity	3. 47±1. 28	3. 57±1. 27
Twin proportion (*n*/N)	5 (158)	4 (153)
Abortion proportion (*n*/N)	4 (158)	3 (153)
Dystocia proportion (*n*/N)	11 (158)	9 (153)

Note: BCS: Bodyconditionscore.

3. 3 Effect of Sheng Hua Tang on subsequent reproductive performance

Table 2 provided a summary of the parameters of reproductive performance in cows. The present study suggested that the oral administration of *Sheng Hua Tang* could lower the number of days to first AI [(68. 9±17. 7) days vs. (80. 5±26. 6) days, $P<0.05$], and enhance the proportion of first AI conception (61. 1% vs. 51. 3%, $P<0.05$). These findings might indicate that more cows resumed the next sequential occurrence of reproductive cycle earlier in the postpartum period. Thatcher and Wilcox (1973) have proposed that a more optimal restoration of the uterine environment could advance the early and frequent occurrences of estrus following parturition, and thus increase reproductive performance in dairy cattle. Hong et al. (2003) have demonstrated that *Sheng Hua Tang* could modulate and restore regular uterine contractions in mice. Oral administration of *Sheng Hua Tang* has beneficial effecton the uterine involution in postpartum women (Ho et al., 2011). From the present results, we speculated that *Sheng Hua Tang* treatment provided more favorable uterine environment for the recovery of reproductive competence, i. e., uterine involution and normal reoccurring estrous cycles, by lowering the risk of the incidence of retained placenta and puerperal metritis during the early postpartum stage. This efficacy of *Sheng Hua Tang* treatment was also evidenced by more cows having conceived during 100 DIM presented in the pregnancy rate curves (survival curves) (Fig. 1). Additionally, *Sheng Hua Tang* treatment effectively reduced the number of services per conception (1. 7% vs. 2. 1%, $P<0.01$) and elevated percentage of cows being pregnant at 305 DIM (89. 8% vs. 82. 0%, $P<0.01$). Thus, *Sheng Hua Tang* treatment as prophylactic strategy would effectively increase potential savings due to fewer inseminations, superior reproductive performance of treated cows and decreased treatment cost.

Table 2 Influence of *Sheng Hua Tang* treatment on the postpartal fertility parameters in cows

Assessment index	*Sheng Hua Tang* group	Control group
No. ofcows	157	150

(continued)

Assessment index	*Sheng Hua Tang* group	Control group
Calving-to-first-service interval[1], days	68. 9±17. 7^{a}	80. 5±26. 6^{b}
First AI conception proportion[2], %	61. 1^{a}	51. 3^{b}
Service per conception[3]	1. 7^{A}	2. 1^{B}
Proportion of cows that were pregnant at 305 DIM[4], %	89. 8^{A}	82. 0^{B}

Note: Different superscript letters mean statistical differences between groups within the same row-a, b = P< 0. 05; A, B=P<0. 01.

DIM: Day in milk.

[1] Calving-to-first-service interval was defined as the number of days from calving to the first service based on the all serviced cows.

[2] Services per conception=Total services/Total number of cows that were pregnant at 305 DIM. Data for any cows culled or dead within 50 days following last service are excluded.

[3] First AI conception proportion was defined as the number of cows that were pregnant in first service divided by the total number of inseminations.

[4] Proportion of cows that were pregnant at 305 DIMexpressed as a proportion of the all inseminated cows in the respective group.

Our study had certain limitations. First, it was not a randomized, placebo-controlled clinical trial. The ear-tag numbers of the enrolled cows were used to allocate to the groups instead of randomization sequence using appropriate methods. Additionally, *Sheng Hua Tang* was given as a decoction, and the controls received no inventions due to lack of appropriate placebos for *Sheng Hua Tang* that had a similar color and taste (brown and bitter). Second, all cows affected with retained placenta or puerperal metritis were treated with intrauterine oxytetracycline infusion in the present study. Intrauterine oxytetracycline treatment is benefit to control local bacterial growth in cows (Haimerl and Heuwieser, 2014); however, oxytetracycline also inhibits matrix metalloproteinases, which play important roles in tissue remodeling during postpartum uterine involution (Curry and Osteen, 2003). Thus, the use of intrauterine oxytetracycline might have a certain effect on the difference in reproduction parameters of cows between the two groups. Even so, the presented results would provide critical information on *Sheng Hua Tang* treatment as a potential prophylactic strategy for the primary health care in postpartum cows.

4 CONCLUSION

In agreement with the working hypothesis of our study, the present results would support efforts to the use of *Sheng Hua Tang* immediately after delivery as a prophylactic strategy for lowering the risk of occurrence of puerperal metritis and improving the overall reproductive efficiency of dairy herds under these study circumstances. Thus, *Sheng Hua Tang* treatment could represent a potential effective prophylactic strategy for bovine postpartum care.

CONFLICT OF INTEREST STATEMENT

None of the authors has any financial or personal relationships that could inappropriately influ-

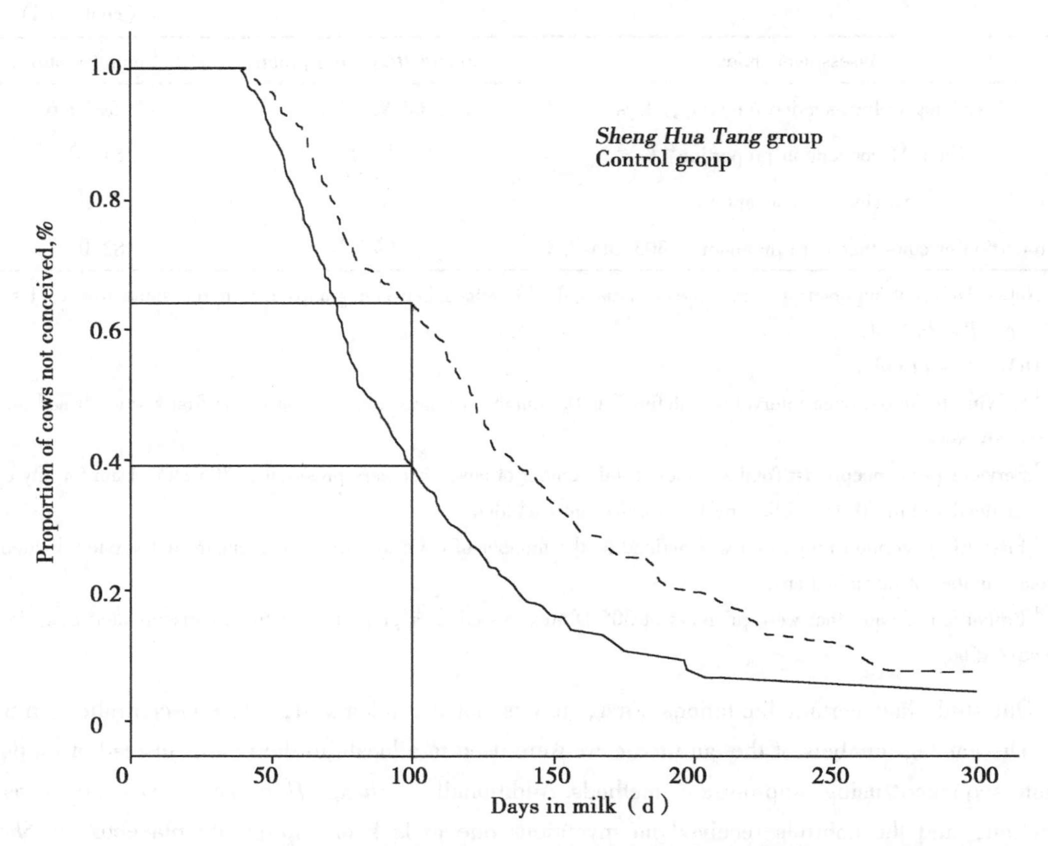

Fig. 1 Survival curves for time to pregnancy up to 305d after parturition in 307 Holstein cows treated ($n=157$) or not ($n=150$) with *sheng Hua Tang* 2–4h after parturition

ence or bias the content of the paper.

ACKNOWLEDGMENT

This research was financially supported by the Special Fund of the Chinese Central Government for Basic Scientific Research Operations in Commonwealth Research Institutes (1610322015006), Special Fund for Agro-scientific Research in the Public Interest (201303040-17) and the National Key Technology Research and Development of the Ministry of Science and Technology of China (2012BAD12B03).

References

[1] Chang, P. J., Tseng, Y. C., Chuang, C. H., Chen, Y. C., Hsieh, W. S., Hurng, B. S., Lin, S. J., Chen, P. C., 2013. Use of herbal dietary supplement Si-Wu-Tang and health-re-lated quality of life in postpartum women: a population-based correlational study. Evid. -Based Complement. Altern., Article ID 790474, 9pages.

[2] Chan, W. C., Wong, Y. C., Kong, Y. C., Chun, Y. T., Chang, H. T., Chan, W. F., 1983. Clinical observation on the uterotonic effect of I-muTs'ao (*Leonurus artemisia*). Am. J. Chin. Med. 11, 77.

[3] Chenault, J. R. , McAllister, J. F. , Chester, T. , Dame, K. J. , Kausche, F. M. , Robb, E. J. , 2004. Efficacy of ceftiofur hydrochloride administered parenterally for the treatment of acute postpartum metritis in dairy cows. J. Am. Vet. Med. Assoc. 224, 1634-1639.

[4] Cui, D.A., Wang, X.Z., Wang, L., Wang, X.R., Zhang, J.Y., Qin, Z., Li, J.X., Yang, Z. Q., 2014.The administration of *Sheng Hua Tang* immediately after delivery to reduce the incidence of retained placenta in Holstein dairy cows.Theriogenology 81, 645-650.

[5] Curry Jr. , T. E. , Osteen, K. G. , 2003. The matrix metalloproteinase system: changes, regulation, and impact throughout the ovarian and uterine reproductive cycle. Endocr. Rev. 24 (4), 428-465.

[6] Curtis, C. R. , Erb, H. N. , Sniffen, C. J. , Smith, R. D. , Kronfeld, D. S. , 1985. Path analysis of dry period nutrition, postpartum metabolic and reproductive disorders, and mastitis in Holstein cows. J. Dairy Sci. 68, 2347-2360.

[7] Drillich, M. , Beetz, O. , Pfüzner, A. , Sabin, M. , Sabin, H. J. , Kutzer, P. , Nattermann, H. , Heuwieser, W. , 2001. Evaluation of a systemic antibiotic treatment of toxic puerperal metritis in dairy cows. J. Dairy Sci. 84, 2010-2017.

[8] Drillich, M. , Pfutzner, A. , Sabin, H. J. , Sabin, M. , Heuwieser, W. , 2003. Comparison of two protocols for the treatment of retained fetal membranes in dairy cattle. Theriogenology 59, 951-960.

[9] Drillich, M. , Mahlstedt, M. , Reichert, U. , Tenhagen, B. A. , Heuieser, W. , 2006. Strategies to improve the therapy of retained fetal membranes in dairy cows. J. Dairy Sci. 89, 627-635.

[10] Editorial Committee of Encyclopedia of China's Agricultural, 1991. China Agricultural Encyclopedia (traditional Chinese veterinary medicine Volume) (in Chinese). Agricultural Publishing House, Beijing.

[11] Edmonson, A.J., Lean, I.J., Weaver, L.D., Farver, T., Webster, G., 1989.A body condition scoring chart for Holstein dairy cows.J.Dairy Sci.72, 68-78.

[12] Fourichon, C. , Seegers, H. , Malher, X. , 2000. Effect of disease on reproduction in the dairy cow: ameta-analysis. Theriogenology 53, 1729-1759.

[13] Goshen, T. , Shpigel, N. Y. , 2006. Evaluation of intrauterine antibiotic treatment of clinical metritis and retained fetal membranes in dairy cows. Theriogenology 66 (9), 2210-2218.

[14] Haimerl, P. , Heuwieser, W. , 2014. Invited review: Antibiotic treatment of metritis in dairy cows: a systematic approach. J. Dairy Sci. 97, 6649-6661.

[15] Ho, M. , Li, T. C. , Su, S. Y. , 2011. The association between traditional Chinese dietary and herbal therapies and uterine involution in postpartum women. Evid. -Based Complement. Altern. , Article ID 918291, 9 pages.

[16] Hong, M., Yu, L., Ma, C., Zhu, Q., 2003.Effect of extract from shenghua decoction on myoelectric activity of rabbit uterine muscle in the latest period of pregnancy (in Chinese). Chin.J.Chin. Mater.Med.28, 1162-1164.

[17] Hou, Z. S. , Shi, J. Z. , Wang, M. Y. , 1992. The study of anti-inflammatory effect of *Sheng Hua Tang* and *Wan Dai Tang* (in Chinese). LiaoningJ. Tradit. Chin. Med. 6, 43-44.

[18] LeBlanc, S. J. , 2008. Postpartum uterine disease and dairy herd reproductive performance: a review. Vet. J. 176, 102-114.

[19] Li, X. , Chi, X. Z. , Wang, L. , Zhou, X. B. , Guo, Y. Q. , Wang, B. , Hao. , Y. , Yao, C. F. , 2011. Shenghua Decoction reversed methotrexate-induced interferon-γ immunosuppression (in Chinese). Chin. J. Chin. Mater. Med. 31, 363-367.

[20] Luo, Y. K. , 1986. Gynaecology of TCM. Shanghai Scientific & Technical Publishers, P135.

[21] Lü X. S. , Ni, B. S. , Sui, B. S. , 1992. Improving the therapeutic efficacy of retained fetal mem-

branes in dairy cows by diagnosis and treatment based on an overall analysis of the illness and the cows' condition (in Chinese). J. Tradit. Chin. Vet. Med. 3, 20-32.

[22] Markusfeld, O., 1984. Factors responsible for post parturient metritis in dairy cattle. Vet. Rec. 114, 539-542.

[23] Malinowski, E., Lassa, H., Markiewicz, H., Kaptur, M., Nadolny, M., Niewitecki, W., Zietara, J., 2011. Sensitivity to antibiotics of *Arcanobacterium pyogenes* and *Escherichia coli* from the uteri of cows with metritis/endometritis. Vet. J. 187, 234-238.

[24] McLaughlin, C. L., Stanisiewski, E. P., Risco, C. A., Santos, J. E., Dahl, G. E., Chebel, R. C., LaGrow, C., Daugherty, C., Bryson, L., Weigel, D., Hallberg, J., Lucas, M. J., 2013. Evaluation of ceftiofur crystalline free acid sterile suspension for control of metritis in high-risk lactating dairy cows. Theriogenology 79, 725-734.

[25] NRC, 2001. Nutrient Requirements of Dairy Cattle (7th). National Academies Press, Washington, DC.

[26] Sandals, W. C. D., Curtis, R. A., Cote, J. F., Martin, S. W., 1979. The effect of retained placenta and metritis complex on reproductive performance in dairy cattle-a case control study. Can. Vet. J. 20, 131-135.

[27] Schuenemann, G. M., Nieto, I., Bas, S., Galvão, K. N., Workman, J., 2011. Assessment of calving progress and reference times for obstetric intervention during dystocia in Holstein dairy cows. J. Dairy Sci. 94, 5494-4501.

[28] Sheldon, I. M., Lewis, G. S., LeBlanc, S., Gilbert, R. O., 2006. Defining postpartum uterine disease in cattle. Theriogenology 65, 1516-1530.

[29] Sheldon, I. M., Williams, E. J., Miller, A. N., Nash, D. M., Herath, S., 2008. Uterine diseases in cattle after parturition. Vet. J. 176, 115-121.

[30] Smith, B. I., Donovan, G. A., Risco, C., Littell, R., Young, C., Stanker, L. H., Elliott, J., 1998. Comparison of various antibiotic treatments for cows diagnosed with toxic puerperal metritis. J. Dairy Sci. 81, 1555-1562.

[31] Song, D. L., 1988. Traditional Chinese veterinary medicine and animal reproduction: a review (in Chinese). J. Tradit. Chin. Vet. Med. 4, 22-34.

[32] Thatcher, W. W., Wilcox, C. J., 1973. Postpartum estrus as an indicator of reproductive status in the dairy cow. J. Dairy Sci. 56, 608-610.

[33] Trevisi, E., Zecconi, A., Cogrossi, S., Razzuoli, E., Grossi, P., Amadori, M., 2014. Strategies for reduced antibiotic usage in dairy cattle farms. Res. Vet. Sci. 96, 229-233.

[34] Wang, J., Xia, X. Y., Peng, R. X., Chen, X., 2004. Activation of the immunologic function of rat Kupffer cells by the polysaccharides of Angelica sinensis (in Chinee). Acta Pharm. Sin. 39, 168-171.

(发表于《Livestock Science》，院选 SCI，IF：1.100)

Analysis of Agouti Signaling Protein (*ASIP*) Gene Polymorphisms and Association with Coat Color in Tibetan sheep (*Ovis aries*)

HAN J. L.[1], YANG M.[2], YUE Y. J.[1], GUO T. T.[1], LIU J. B.[1], NIU C. E.[1], YANG B. H.[1]

(1. Lanzhou Institute of Husbandry and Pharmaceutical Sciences, Chinese Academy of Agricultural Sciences, Lanzhou 730050, China; 2. Institute of Animal Science, Chinese Academy of Agricultural Sciences, Beijing 100094, China)

Abstract: Tibetan sheep, an indigenous breed, have a wide variety of phenotypes and a colorful coat, which make this breed an interesting model for evaluating the effects of coat-color gene mutations on this phenotypic trait. The agouti signaling protein (*ASIP*) gene is a positional candidate gene, as was inferred based on previous study. In our research, *ASIP* gene copy numbers in genomic DNA were detected using a novel approach, and the exon 2 g. 100–104 mutation and copy number variation (CNV) of *ASIP* were associated with coat color in 256 sheep collected from eight populations with different coat colors by highresolution melting curve assay. We found that the relative copy numbers of *ASIP* ranged from one to eight in Tibetan sheep. All of the g. 100–104 genotypes in the populations were in Hardy-Weinberg equilibrium, and there was no relationship between the g. 100–104 genotype and coat color ($P > 0.05$). The single *ASIP* CNV allele was found to be almost entirely associated with solid-black coat color; however, not all solid-black sheep displayed the putative single *ASIP* CNV genotype. From our study, we speculate that the *ASIP* CNV is under great selective pressure and the single *ASIP* CNV allows selection for black coat color in Tibetan sheep, but this does not explain all black phenotypes in Tibetan sheep.

Key words: *ASIP*; Polymorphisms; Copy number variation; Coat color

INTRODUCTION

In mammalian species, coat color is an important breed characteristic and production trait. Coat color is determined by the amount of pigmentation and the ratio of eumelanin (black/brown) and pheomelanin (yellow/red), which are produced by melanocytes (Ito and Wakamatsu, 2003). Tibetan sheep are traditionally reared for production and without any structured genetic selection plan; they often have a white coat, but the head and limbs are covered with variegated wool. Tibetan sheep with variegated wool account for 90% of the total population, while solid-white or solid-black sheep are scarce. Therefore, we considered that Tibetan sheep could be an interesting model for identifying coat-color candidate genes. Although there are over 300 genes identified with known roles in mammalian pigmentation, a few key genes have been identified as major regulators of pigment production in domestic animals (Rieder et al., 2001; Liu et al., 2009; Suzuki, 2013).

The agouti signaling protein gene (*ASIP*) has been shown to be associated with coatcolor production in a variety of animals; previous studies mapped the coat color locus on chromosome 13 for cats (Eizirik et al., 2003), mice (Kuramoto et al., 2001), foxes (Vage et al., 1997), horse (Rieder et al., 2001), cattle (Girardot et al., 2005), pigs (Drogemuller et al., 2006), dogs (Kerns et al., 2004; Schmutz et al., 2007), sheep (Norris and Whan, 2008), and humans (Kwon et al., 1994). The *ASIP* gene encodes a signaling protein (ASIP) that can act as an antagonist of the alpha-melanocyte-stimulating hormone (α-MSH), preventing the binding of α-MSH to melanocortin 1 receptor, which induces a decrease in the levels of cyclic adenosine monophosphate (cAMP). This occurs through a cascade of reactions that inhibit the formation of eumelanin, which results in lighter coat color. In this way, *ASIP* inhibits the production of eumelanin and leads to production of pheomelanin (Aberdam et al., 1998; Virador et al., 2000).

α-MSH combines with ASIP to regulate the formation of eumelanin and pheomelanin in melanoma cells (Hida et al., 2009). Studies have shown that recessive mutations that either impair ASIP protein function or abrogate ASIP expression result in more darkly pigmentedphenotypes. Conversely, dominant mutations arising from upregulated expression of ASIP result in lighter phenotypes in some sheep breeds (Gratten et al., 2010). The dominant white color is caused by deregulated expression of the agouti protein and results from a 190-kbp genomic duplication that places a functional ASIP-coding sequence under the control of a duplicated promoter from the neighboring itchy homolog E3 ubiquitin protein ligase locus (Norris and Whan, 2008). The recessive self-color pattern is caused by the absence of ASIP expression, presumably arising from a *cis*-regulatory mutation that inactivates the promoter (Norris and Whan, 2008; Royo et al., 2008). ASIP expression was not detected in the skin tissue of the recessive black sheep with single copy *ASIP* alleles (Norris and Whan, 2008), and recessive self-color patterns may also result from the expression of functional mutations in exon 2 (Smit et al., 2002; Royo et al., 2008) and exon 4 (Norris and Whan, 2008). The contribution of *ASIP* to coat color patterns of domestic sheep has been confirmed to involve multiple alleles. Herein, we examine the mutation and copy number variation (CNV) of *ASIP* associated with coat color in Tibetan sheep.

MATERIAL AND METHODS

Animal and sample collection

Blood samples were taken from 256 Tibetan sheep belonging to eight local populations in northwest China, most of which were female. The coat color patterns expressed included solid-black, solid-white, and variegated. The eight populations were Guide Black Tibetan sheep and Minxian Black Tibetan sheep, which are two populations of sheep that are covered with black wool (Fig. 1A); breeding Qinghai Tianjun White Tibetan sheep and Qilian White Tibetan sheep, the wool of which is white and sometimes the eye socket has small black spots/regions (Fig. 1B); and Qinghai Ola

Tibetan sheep, Gansu Ola Tibetan sheep, Langkazi Tibetan sheep, and Gansu Ganjia Tibetan sheep, which have white wool and black/brown spots on their body, and black/brown is usually present on the head and legs (Fig. 1C). All eight populations live high in the mountains in harsh climatic conditions.

A solid black B solid white C variegated

Fig. 1 Coat color variation of Tibetan sheep

DNA extraction, amplification, and sequencing of the agouti locus

Genomic DNA was extracted from the collected blood samples using a TIANGEN DNA Mini Kit (TIANGEN Bio, China). A NanoDrop 2000 (Thermo Scientific, USA) was used to quantify the concentration of genomic DNA. Eight pairs of primers, ASIP 1-ASIP 8, were designed to characterize the sheep *ASIP* gene (Table 1), and the reference sequence (GenBank accession No. EU420023.1) was used to design primers for amplification of the 5353-bp coding region. Primers were synthesized by the Beijing Genomics Institution (BGI, China). Some of the PCR primers and conditions were previously reported (Norris and Whan, 2008). The mutations in the coding DNA sequence of *ASIP* were identified by sequencing. Mixed pools of DNA (Taylor et al., 1994) were obtained by mixing the DNA extracted from 10 black sheep and 10 white sheep. PCR products were sequenced by BGI, and single nucleotide polymorphisms were analyzed based on the sequencing peak diagrams.

Table 1 Sequences of PCR primers

Primer name	Primer sequence (5′to 3′)	Products (bp)	Temperature (℃)
ASIP 1	CATTACTGGGGACCTATCAAC AGACAGAAGGGAAATCCAAC	504	55.0
ASIP 2	TGCTTCCTCAGTGCCTACAG CTCTTTTCTTCTTTTCCGCTTC	1 480	57.0
ASIP 3	AGTTCACATTGTCGGGATG GTTTGATGTCAATATATCCTTAG	1 080	57.0
ASIP 4	GTTGCTTGCCACAAGTTCTA CTCCGCCATCAACACTTAAC	2 320	59.0
ASIP 5	GACTCCGAACCTACCGTGAA CGCCCCAACGTCAATAGC	601	61.0
ASIP 6	CATTACTGGGGACCTATCAAC TATCGGCTTGGAGAGTGTTTG	609	56.0

(continued)

Primer name	Primer sequence (5′to 3′)	Products (bp)	Temperature (℃)
ASIP 7	CTGGCCTAAGTCCCAAGATATG CCATTTCCTTCTCCAGTTTAGTTTAG	440	55. 0
ASIP 8	GGAAGGGGAAGACAACAGAGAC ACTAGGCGAAGGGAAAAGAATG	668	56. 0
Agt16	CAGCAATGAGGACGTGAGTTT		
Agt17	GTTTCTGCTGGACCTCTTGTTC	238	61. 0
Agt18	GTGCCTTGTGAGGTAGAGATGGTGTT	242	60. 5

Genotyping polymorphisms

The association between coat pattern and an *ASIP* functional mutation was quantified by genotyping the 5-bp deletion in 256 sheep. *ASIP* coding-region mutations (g. 100−104delAGGA A in exon 2) were genotyped by designing primers near the genome location g. 100−104: D5−F: 5′−TGAGGAAAGCCCAGAGATG−3′ and D5−R: 5′−CAGAAGGGAAATCCAACAGG-3′. The PCR products were distinguished by the peaks of their melting curves. Then, three primers were used for detection of *ASIP* CNV by amplifying both the junction between the duplicated DNA copies and the 5′-breakpoint sequences (Norris and Whan, 2008) (Agt16, Agt17, and Agt18; Table 1). Primers Agt16 and Agt18 spanned a unique 242-bp PCR product, which includes the junction between the duplicated copies, while Agt16 and Agt17 can produce a 238-bp fragment that can cover the 5′-breakpoint sequence (Fig. 2) according to the high-resolution melting curve assay (HRM) melting peaks of products that we can easily be used to detect the CNV.

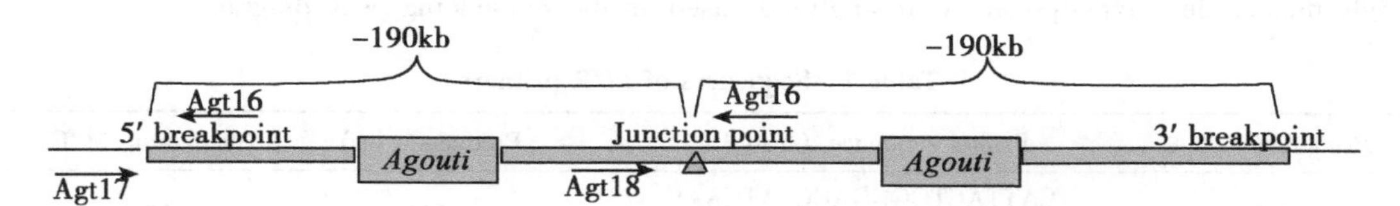

Fig. 2 Structure of the *ASIP* gene duplication in genomic DNA

ASIP CNV numbers assay

To estimate sheep *ASIP* CNV, we made a pMD19-agouti fragment clone plasmid using standard construction techniques. We chose a conserved fragment between intron 3 and exon 4 of *ASIP* using DNA and mRNA sequences (GenBank EU420023. 1 and NM_ 001134303. 1) as references to design a primer to amplify a 193-bp product: *ASIP*-F: 5′-GTTAAGTGTTGATGGCGGAG-3′ and *ASIP*-R: 5′-GTTCTTCATCGGAGCCTTTC-3′. PCR was performed in a 50-μL reaction volume containing 50−100ng DNA, LA Taq (5U/μL) (Takara, China), 0. 5μl 10X LA-PCR buffer, 5μL dNTP mixture, 8μL 10μM *ASIP*-F and *ASIP*-R per 2μL, with ddH_2O added to bring the solution to 50μL. PCR conditions were 4 min at 95℃, 20 s at 95℃, 20 s at 60℃, and 20 s at 72℃ for 40 cycles, and 5 min at 72℃ (Veriti 96 well, Applied Biosystems, USA). The PCR product was subjected to 2% agarose gel electrophoresis, and the target band was recovered and purified using an agarose gel DNA extraction kit (TIANgel, Beijing, China); fragments were connected with a pMD19-T vector (TaKaRa, China). The spliced product was transformed into *E. coli* JM109-competent cells (TaKaRa, China) and coated onto agar plates that contained isopropyl-β-d-thiogalactoside, 5-bromo-4-chloro-3-indolyl β-D-galactoside and ampicillin. The white-positive colonies were screened for culturing, and positive clone plasmids were sequenced for identification.

The plasmid concentration was measured using the NanoDrop 2000, and the copy number of plasmids per microliter was calculated using the following equation: copies/μL = $(6.02\times10^{23})\times(ng/\mu L\times10^{-9})/(\text{DNA length}\times660)$. The plasmid DNA length was 2 885bp, which was composed of the sum of the vector length and the PCR product. The standard plasmid was diluted to a concentration of 10^8 copies/μL, and then it was diluted from 10^7 to 10^2 (copies/μL) by six gradients. The standard curve was made using real-time quantitative PCR with five concentrations (10^2–10^6) as standards (three replicates per concentration). Real-time quantitative PCR was performed in an 18-μL reaction volume containing 9μL 2X SYBR Premix Ex Taq TaKaRa, China), 0.4μL Rox Reference Dye, 0.8μL 10μM upstream and downstream primers, 6μL ddH_2O, and 1μL standard plasmid DNA, and each sample was analyzed three times. Real-time quantitative PCR conditions were 3 min at 95℃, 10s at 95℃, and 30s at 60℃ for 40 cycles (Bio-Rad CFX-96, USA). The mean Ct value of each sample was measured to calculate the copies according to the linear regression equation of the standard curve. Finally, the DNA from thirty sheep randomly selected was checked using this new approach.

Statistical analysis

The association between *ASIP* gene polymorphisms and coat-color phenotype was analyzed using the Fisher exact test (SHEsis; http://analysis.bio-x.cn), and the Hardy-Weinberg equilibrium of genotype frequencies was detected by χ^2 test in SAS version 9.2 (SAS Institute Inc., USA).

RESULTS

Mutations in Tibetan sheep *ASIP*

Thirty-five mutation sites were found in the full-length sequence from exon 2 to the 3′UTR of *ASIP*, including three mutation sites in the exon region (g.100–104 deletion; g.5051G>C; g.5172T>A) and 32 in the intron region (Table S1). We focused our study on a 5-bp deletion in exon 2 (g.100–104delAGGAA, denoted as D5). The D5 deletion would result in a frame shift followed by a premature stop codon that is 63 amino acids downstream of the start site, truncating the agouti protein before the functionally important cysteine signaling domain (amino acids 91–130) (Norris and Whan, 2008), while N5 does not have the AGGAA deletion. HRM was used to genotype the g.100–104 site based on melting peaks, and we only found two genotypes: D5N5 and N5N5 (Fig. 3). The result of the sequencing was consistent with HRM genotype, which just have two genotypes (Fig. 4). We estimated the frequency of the D5N5 and N5N5 alleles in the 256 sheep by applying the Hardy-Weinberg equilibrium. The most frequent allele was N5N5, and D5N5 was less frequent (Table 2). We compared pairwise population measures of the three coat patterns for both types of coat color. χ^2 analysis showed that both g.100–104 genotypes were in Hardy-Weinberg equilibrium. There was no relationship between the g. 100−104 genotype and coat color, as determined by a comparative analysis of correlation (Fisher's exact test, $P > 0.05$).

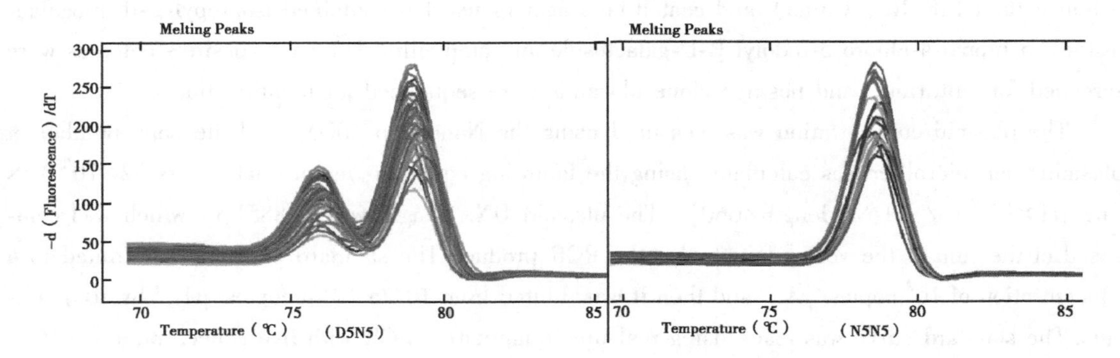

Fig. 3 High-resolution melting curve assay genotyping

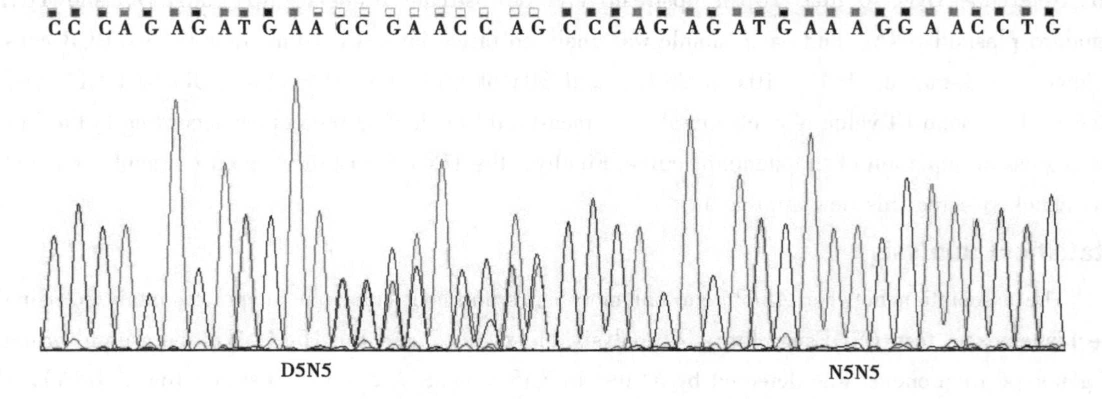

Fig. 4 DNA sequencing of the *ASIP* gene g. 100−104

Table 2 Genotypes of the *ASIP* exon 2 g. 100−104 deletions and the *ASIP* CNV

Coat color	Local strains	Multiple CNVs		Single CNV	
		N5D5	N5N5	N5D5	N5N5
Solid white	QLW (40)	18	21	—	1
	TJW (24)	8	11	2	3
Solid black	GDB (40)	4	18	3	15
	MXB (56)	20	10	10	16
Variegated	GSOL (27)	15	12	—	—
	QHOL (21)	9	12	—	—
	GJ (24)	12	11	—	1
	LKZ (24)	1	23	—	—

Note: QLW = Qilian White Tibetan sheep; TJW = Qinghai Tianjun White Tibetan sheep; GDB = Guide Black Tibetan sheep; MXB = Minxian Black Tibetan sheep; GSOL = Gansu Ola Tibetan sheep; QHOL = Qinghai Ola Tibetan sheep; GJ = Gansu Ganjia Tibetan sheep; LKZ = Langkazi Tibetan sheep.

ASIP CNV analysis

The gradients of 10^2-10^6 copies/μL plasmid DNA were amplified by real-time quantitative PCR according to the values of Ct and log copies. The program drafted a standard curve ($R^2=0.999$, $E=104.6\%$), the linear equation $y=-3.215\log x+35.8$ (Fig. 5). The amplification efficiency of genomic DNA was $E=98\%$, indicating that the standard curve could be used to detect gene copy numbers of *ASIP*.

Only two of the thirty sheep with a single copy allele containing the *ASIP* were detected and were named A1 and A2. We used the *ASIP* copy number of A1 as a control sample to calculate the relative copy number of other samples. The relative copy number of A2 was 0.98; the relative copy numbers of other individuals ranged from two to eight (Table S2).

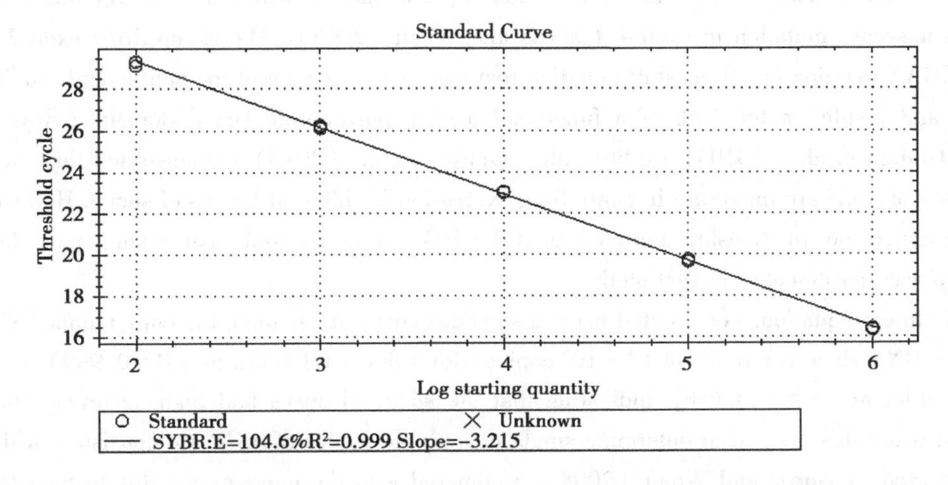

Fig. 5 Standard and dynamic curves by real-time quantitative PCR

Association between coat pattern and *ASIP* CNV

We analyzed the *ASIP* CNV of 256 Tibetan sheep belonging to eight local populations with the HRM assay, which is designed to test the presence of a duplicated copy allele with single or multiple *ASIP* CNVs. According to the different dissolution peaks for genotyping using Lightscanner (Idaho Technology, Inc., USA), there were three types of dissolution peaks, which means they were amplified differently (Fig S1). Only one of the 96 variegated sheep and six of the 64 solid-white sheep had a single *ASIP* CNV allele, while 44 of the 96 solid-black sheep had a single *ASIP* CNV allele (Table 2). The association between the single *ASIP* CNV allele and mostly dark coat color (when all sheep were analyzed together) was highly significant (Fisher's exact test; solid black vs. solid white, $P<0.01$; solid black vs. variegated, $P<0.01$). We found a departure of the CNV allele from Hardy-Weinberg equilibrium in both pairwise-compared populations (χ^2 test; solid black vs. solid white, $P<0.01$; solid black vs. variegated, $P<0.01$), indicating that *ASIP* CNV is under great selective pressure.

DISCUSSION

The coding region of the *ASIP* gene is determined by three exons (2, 3, and 4, according to the nomenclature in mice). The full-length (approximately 5 353bp) sequence from exon 2 to the 3′UTR of *ASIP* was amplified and sequenced in our study. After splicing and assembly, we discovered 35 mutation sites, all of which have been previously reported in domestic sheep (Norris and Whan, 2008). In sheep, the *ASIP* protein consists of 133 amino acids that comprise a signal peptide and functional amino acids. Previous studies showed *ASIP* mutations related to animal coat color but mainly addressed the *ASIP* functional-deletion mutation and missense mutations. *ASIP* alleles causing coat-color variation have also been characterized in foxes, in which exon 2 is completely missing (Vage et al., 1997); rats, which have a 19bp deletion within the coding DNA sequence (Kuramoto et al., 2001); cats, which have a 2bp deletion in exon 2 (Eizirik et al., 2003); horses, which have an 11bp deletion in exon 3 (Rieder et al., 2001); dogs, which have a missense mutation in exon 4 (Schmutz et al., 2007); and sheep, which have a 5bp deletion in exon 2 and a missense mutation in exon 4 (Norris and Whan, 2008). The sheep *ASIP* exon 2 g. 100−104 AGGAA deletion is a frameshift mutation that causes termination of the amino acids coding at locus 64 and results in the lack of a functional mature polypeptide signal domain (Royo et al., 2008; Gratten et al., 2010). Additionally, Gratten et al. (2010) demonstrated that the regulatory exons of *ASIP* are important in controlling expression in different breeds of sheep. However, that study revealed no relationship between g. 100 − 104 genotypes and coat color, and the D5D5 genotype was not detected in that study.

The linear equations constructed using a standard curve to calculate the copy number of *ASIP* in unknown DNA showed a range of 10^2-10^6 copies with a linear relationship ($R=0.999$) and amplification efficiency $E=104.6\%$, indicating that the standard curve had high accuracy. The results obtained using this method to determine single or multiple copies of *ASIP* are consistent with the results reported by Norris and Whan (2008). Compared with the microarray, this method for establishing *ASIP* CNV using SYBR Green real-time quantitative PCR is simple, fast, and economical. Previous data from Norris and Whan (2008) on asymmetric competitive PCR copy-number assays of recessive black and white Merino sheep revealed that recessive black Merino sheep always have a single *ASIP* copy, while white Merino sheep can have two to five *ASIP* copy alleles. Our calculations of relative copy numbers of *ASIP* range from one to eight in Tibetan sheep. Calculation of the *ASIP* copy numbers using this method was consistent with the findings of Norris and Whan (2008); in comparison with the method they designed, our approach can be used to easily detect CNVs in a typical lab like ours.

In sheep, *ASIP* is characterized by CNV, which determines the white and tan (A^{Wt}) agouti allele and white coat color, whereas non-agouti black sheep with a single copy of *ASIP* could carry a regulatory mutation in an unidentified regulatory region, deletions in exon 2, or a missense mutation in exon 4 (Norris and Whan, 2008; Royo et al., 2008; Gratten et al., 2010). Association between coat color and *ASIP* CNVs was highly significant in the populations of solid-black *vs* solid-white or variegated sheep; our study found that the single *ASIP* CNV allele was almost entirely associated with dark coat color. These results confirmed that the *ASIP* locus affects coat color in sheep.

We also found that Tibetan sheep with the three coat-color patterns had both single and multiple

ASIP CNV alleles. Norris and Whan (2008) found that all-white Merinos had at least one duplicated *ASIP* allele, while all of the recessive black Merinos contained only a single allele copy. This was indirectly confirmed by Fontanesi et al. (2011), who showed that a copy of a duplicated allele produces grey coat color in Massese sheep. It was predicted that at least one duplication of the genomic region, including a functional copy of the *ASIP*, is necessary to produce the white phenotype. Our research revealed that one of the 96 variegated sheep and six of the 64 solid-white sheep had a single *ASIP* CNV allele. Interestingly, another study recently identified a copy-number variant in goat *ASIP*, where animals carrying a duplicated *ASIP* allele were not completely associated with white coat color, and black coat color in goats was not completely associated with any other identified missense mutation of this gene (Fontanesi et al., 2010). *ASIP* CNV is under great selective pressure, and it can be speculated that the single copy of the *ASIP* is more associated with Tibetan sheep with black coat color. Thus, we conclude that there are other major pigment genes responsible for the presence of pigmented fibers in Tibetan sheep.

A genome-wide association study (GWAS) of most of the world's sheep breeds revealed that the *KIT*, *ASIP*, and *MITF* genes that determine the sheep coat color have been under strong selection (Kijas et al., 2012). However, *ASIP*, which controls a series of alleles of black and white coat color, was not detected in a new GWAS of sheep selection (Zhang et al., 2013), although they used a different sheep breed. Tibetan sheep live in unique geographic environment that causes natural selection in breed internal, indicating no artificial selection for the phenotypes of coat color. Unlike domestic sheep such as Merino (Norris and Whan, 2008), Dubian, Privorian (Fontanesi et al., 2012), and Massese sheep (Fontanesi et al., 2011), which are likely to have been influenced by strong artificial selection for various wool colors, the wool color of Tibetan sheep is not the main selection trait. *ASIP* is the major regulatory gene for coat color of modern Western sheep, primitive feral-breed Soay sheep (Gratten et al., 2010), and the ancient caprinae species of Barbary sheep (Norris and Whan., 2008). In our study, we revealed that Tibetan sheep may also be affected by *ASIP*, and then presumably there may be other genes in addition to *ASIP* involved in coat-color regulation.

ACKNOWLEDGMENTS

We are grateful for the assistance of the Qinghai Academy of Animal Science and Veterinary Science in the Sample collection. Research supported by the Earmarked Fund for Modern China Wool & Cashmere Technology Research System (#CARS-40-03) Project.

References

[1] Aberdam E, Bertolotto C, Sviderskaya EV, de Thillot V, et al. (1998). Involvement of microphthalmia in the inhibition of melanocyte lineage differentiation and of melanogenesis by agouti signal protein. *J. Biol. Chem.* 273: 19560-19565.

[2] Drogemuller C, Giese A, Martins-Wess F, Wiedemann S, et al. (2006). The mutation causing the black-and-tan pigmentation phenotype of Mangalitza pigs maps to the porcine ASIP locus but does not affect its coding sequence. *Mamm. Genome.* 17: 58-66.

[3] Eizirik E, Yuhki N, Johnson WE, Menotti-Raymond M, et al. (2003). Molecular genetics and evolution of melanism in the cat family. *Curr Biol.* 13: 448-453.

[4] Fontanesi L, Beretti F, Riggio V, Gomez Gonzalez E, et al. (2010). Copy number variation and missense mutations of the Agouti signaling protein (ASIP) gene in goat breeds with different coat colors. *Cytogenet. Genome Res.* 126: 333-347.

[5] Fontanesi L, Dall'Olio S, Beretti F, Portolano B, et al. (2011). Coat colours in the Massese sheep breed are associated with mutations in the agouti signalling protein (ASIP) and melanocortin 1 receptor (MC1R) genes. *Animal.* 5: 8-17.

[6] Fontanesi L, Rustempasic A, Brka M and Russo V (2012). Analysis of polymorphisms in the agouti signalling protein (ASIP) and melanocortin 1 receptor (MC1R) genes and association with coat colours in two Pramenka sheep types. *Small Ruminant Res.* 105: 89-96.

[7] Girardot M, Martin J, Guibert S, Leveziel H, et al. (2005). Widespread expression of the bovine Agouti gene results from at least three alternative promoters. *Pigment Cell Res.* 18: 34-41.

[8] Gratten J, Pilkington JG, Brown EA, Beraldi D, et al. (2010). The genetic basis of recessive self-colour pattern in a wild sheep population. *Heredity* 104: 206-214.

[9] Hida T, Wakamatsu K, Sviderskaya EV, Donkin AJ, et al. (2009). Agouti protein, mahogunin, and attractin in pheomelanogenesis and melanoblast-like alteration of melanocytes: a cAMP-independent pathway. *Pigment Cell Melanoma Res.* 22: 623-634.

[10] Ito S and Wakamatsu K (2003). Quantitative analysis of eumelanin and pheomelanin in humans, mice, and other animals: a comparative review. *Pigment Cell Res.* 16: 523-531.

[11] Kerns JA, Newton J, Berryere TG, Rubin EM, et al. (2004). Characterization of the dog Agouti gene and a nonagouti mutation in German Shepherd Dogs. *Mamm. Genome.* 15: 798-808.

[12] Kijas JW, Lenstra JA, Hayes B, Boitard S, et al. (2012). Genome-wide analysis of the world's sheep breeds reveals high levels of historic mixture and strong recent selection. *PLoS Biol.* 10: e10012582.

[13] Kuramoto T, Nomoto T, Sugimura T and Ushijima T (2001). Cloning of the rat agouti gene and identification of the rat nonagouti mutation. *Mamm. Genome.* 12: 469-471.

[14] Kwon HY, Bultman SJ, Loeffler C, Chen W, et al. (1994). Molecular structure and chromosomal mapping of the human homolog of the agouti gene. *Proc. Natl. Acad. Sci. U. S. A.* 91: 9760-9764.

[15] Liu L, Harris B, Keehan M and Zhang Y. (2009). Genome scan for the degree of white spotting in dairy cattle. *Anim. Genet.* 40: 975-977.

[16] Norris BJ and Whan VA (2008). A gene duplication affecting expression of the ovine ASIP gene is responsible for white and black sheep. *Genome Res.* 18: 1282-1293.

[17] Rieder S, Taourit S, Mariat D, Langlois B, et al. (2001). Mutations in the agouti (ASIP), the extension (MC1R), and the brown (TYRP1) loci and their association to coat color phenotypes in horses (*Equus caballus*). *Mamm. Genome.* 12: 450-455.

[18] Royo LJ, Alvarez I, Arranz JJ, Fernandez I, et al. (2008). Differences in the expression of the ASIP gene are involved in the recessive black coat colour pattern in sheep: evidence from the rare Xalda sheep breed. *Anim. Genet.* 39: 290-293.

[19] Schmutz SM, Berryere TG, Barta JL, Reddick KD, et al. (2007). Agouti sequence polymorphisms in coyotes, wolves and dogs suggest hybridization. *J. Hered.* 98: 351-355.

[20] Smit MA, Shay TL, Beever JE, Notter DR, et al. (2002). Identification of an agouti-like locus in sheep. *Anim. Genet.* 33: 383-385.

[21] Suzuki H (2013). Evolutionary and phylogeographic views on Mc1r and ASIP variation in mam-

mals. *Genes Genet. Syst.* 88：155-164.

[22] Taylor BA，Navin A and Phillips SJ（1994）. PCR-amplification of simple sequence repeat variants from pooled DNA samples for rapidly mapping new mutations of the mouse. *Genomics*. 21：626-632.

[23] Vage DI，Lu DS，Klungland H，Lien S，et al.（1997）. A non-epistatic interaction of agouti and extension in the fox，*Vulpes vulpes*. *Nat. Genet*. 15：311-315.

[24] Virador VM，Santis C，Furumura M，Kalbacher H，et al.（2000）. Bioactive motifs of agouti signal protein. *Exp. Cell Res*. 259：54-63.

[25] Zhang LF，Mousel MR，Wu XL，Michal JJ，et al.（2013）. Genome-wide genetic diversity and differentially selected regions among Suffolk，Rambouillet，Columbia，Polypay，and Targhee sheep. *PLoS One*.8：e659426.

（发表于《Genetics and Molecular Research》，院选 SCI，IF：0.850）

Association between Single-nucleotide Polymorphisms of Fatty acid Synthase Gene and Meat Quality Traits in Datong Yak (*Bos grunniens*)

CHU M.[1,2], WU X. Y.[1,2], GUO X.[1,2], PEI J.[1,2], JIAO F.[3], FANG H. T.[3], LIANG C. N.[1,2], DING X. Z.[1,2], BAO P. J.[1,2], Yan P.[1,2]

(1. Lanzhou Institute of Husbandry and Pharmaceutical Sciences, Chinese Academy of Agricultural Science, Lanzhou 730050, China; 2. Key Laboratory of Yak Breeding Engineering of Gansu Province, Lanzhou 730050, China; 3. Food and Drug Administration of Jinchang, Jinchang, China)

Abstract: Fatty acid synthase (FASN) is a key enzyme in fatty acid anabolism that plays an important role in the fat deposit of eukaryotic cells. Therefore, in this study, we detected 2 novel single-nucleotide polymorphisms (SNPs) in the *FASN* gene in 313 adult individuals of Datong yak using polymerase chain reaction-single strand conformation polymorphism and DNA sequencing techniques. SNP g. 5477C>T is located in intron 3 of *FASN*, and 3 genotypes, HH, HG, and GG, were detected in this mutation site. SNP g. 16930T>A is located in exon 37 of *FASN*, and 2 genotypes, EE and EF, were detected in this site. Association analysis of these 2 SNPs with meat quality traits showed that in SNP g. 5477C>T, yaks with the HH genotype and HG genotype had significantly higher intramuscular fat content than individuals with the GG genotype ($P<0.01$). In SNP g. 16930T>A, yaks with the EE genotype also had significantly higher IMF content than individuals with the EF genotype ($P<0.01$). The results indicate that *FASN* may be used as a candidate gene affecting intramuscular fat content in Datong yaks.

Key words: Fatty acid synthase gene; Meat quality traits; Polymorphism; Yak

INTRODUCTION

Most fatty acids needed for animal body fat deposition originate from the *de novo* fatty acid synthesis, which requires fatty acid synthase (FASN) to catalyze the reaction between acetyl coenzyme A, malonyl-CoA, and nicotinamide adenine dinucleotide phosphate (Stuart et al., 1991; Chakravarty et al., 2004). *FASN* is thought to be a candidate gene in animal breeding for improving fat deposition and meat quality traits and plays a central role in lipogenesis in mammals (Roy et al., 2005; Zhang et al., 2008). The bovine *FASN* gene is located on chromosome 19 (Roy et al., 2001) and is 18 824 base pairs in length, including 42 exons and 41 introns. Reverse transcription-polymerase chain reaction (PCR) and Western blot analysis have shown that FASN expression in the brain, testis, and adipose tissue is higher than that in the liver and heart (Roy et al., 2005). Genetic variants of *FASN* gene have been reported to be associated with fat percentage in bovine milk (Roy et al., 2006). Five single-nucleotide polymorphisms (SNPs) in 3 different cattle populations have been identified, and association mapping results using these SNPs were associated with variation in fatty acid composition in adipose fat and milk fat (Morris et al., 2007).

Yak (*Bos grunniens*) is a domestic animal that lives on the Qinghai-Tibetan Plateau at altitudes between 2 500 and 6 000 m. It is the primary source of meat, milk, and hair in this area of Tibet. Yak meat is high in protein and calories, yet low in fat, and is increasing in popularity because of its good taste. Moreover, yak meat is referred to as "green food" because yaks subsist on the alpine grassland without the pollution of modern industry (Guo et al., 2009). With the development of molecular genetics, genetic marker-assisted selection has been widely used in animal breeding (Yan et al., 2010). Thus, biological genetic marker technology is very important in yak breeding.

These findings suggest that the *FASN* gene plays an important role in fatty acid metabolism. Few studies of the *FASN* gene in yak have been reported. Therefore, the objective of this study was to detect potential SNPs in yak *FASN* gene and explore the associations between SNPs and meat quality traits in yak to select animals with better meat quality.

MATERIAL AND METHODS

Animals

During slaughter season (October-November), a total of 313 blood samples of adult (approximately 4 years old) Datong yaks were collected from domestic slaughter houses before slaughter. Several meat quality traits were measured in the *longissimus dorsi* muscle in a sample taken from the last rib. Intramuscular fat content (IMF) was measured using standard methodology (Honikel, 1998). Muscle pH was measured at 24h *post mortem* (pH24h) using a portable pH meter (Crison Instruments, Barcelona, Spain) (Korkeala et al., 1986). Warner-Bratzler shear force was measured using the Texture Analyser Winopal (Ahnbeck, Germany) with a Warner-Bratzler blade (2.8mm wide). Water holding capacity was estimated by centrifuging 1g muscle placed on tissue paper inside a tube for 4 min at 1 500g. The water remaining after centrifugation was quantified by drying the samples at 70℃ overnight. Water holding capacity was calculated as: dried weight/initial weight× 100. Cooked meat percentage was estimated as the percentage of the roasted sample weight (cooled for 30 min, remove the meat samples stored at room temperature after cooling 15 min). Genomic DNA was extracted using a genomic DNA isolation kit (Tiangene, Beijing, China) according to manufacturer instructions.

Primer design and PCR amplification

Five pairs of PCR primers (Table 1) were designed based on the bovine *FASN* gene (GenBank accession No. NC_000176). A 15-mL reaction volume included 1μL (50ng/μL) template, 7.5mL *Taq* PCR MasterMix, 0.2μL Taq DNA polymerase, 5.7μL ddH$_2$O, and 0.3μL of each primer (10pM/μL), performed in a Tpersonal thermocycler (Biometra, Goettingen, Germany) under the following conditions: 95℃ for 5 min (preliminary denaturation), followed by 35 cycles of 94℃ for 60 s (denaturation), primer annealing at 58.3° and 57℃ (corresponding to 2 primer pairs) for 35s, and 72℃ for 50s, and a final extension at 72℃ for 10 min. PCR products were then electrophoresed on 1.5% agarose gels using 1X TBE buffer (89 mM Tris, 89 mM boric acid, 2 mM EDTA Na2), containing 200ng/mL ethidium bromide to detect the products.

Table 1 Primers used for PCR amplification

Primers	Sequence of primers	Size (bp)	Annealing temperature (℃)
P1	F: GGTTTGACTTCTGCCTCCT	649	61
	R: CCACTGCTCTCACCTGATG		
P2	F: CCAACCGCCTCTCCTTCTT	355	65
	R: ACTCCCCTCTCTGGATGGC		
P3	F: TCGGCAAAGTGGTCATTCAG	767	62.4
	R: CATTGTACTTGGGCTTGTTGA		
P4	F: GGGCACCTTAGGCTTGG	365	59.3
	R: ACTCAGGGGTCTGGTTATCC		
P5	F: CTGAAGGGCACTAAAGACAAA	256	57
	R: GGAGACCAGACTCGGAAGA		

Genotyping

To screen for polymorphisms in the fragments analyzed, PCR amplifications were subjected to single-strand conformation polymorphism analysis. First, 4μL PCR products were mixed with 8μL loading dye (98% formamide, 10 mM EDTA, 0.025% bromophenol blue, and 0.025% xylene-cyanol) after denaturing at 98℃ for 10 min; the mixture was immediately placed on ice for 10min. Next, the mixture was loaded onto polyacrylamide gels and electrophoresed at 250 V for 10 min and then at 200V for 2h in 1X TBE buffer. The polyacrylamide gel was silver-stained with 0.1% $AgNO_3$ and visualized using 2% NaOH (Lan et al., 2007). According to the PCR-single-strand conformation polymorphism band patterns observed using visual light, individual genotypes were defined (Qu et al., 2005). Amplifica tions derived from the 3 allele standards were included on each gel for reference (Byun et al., 2008). Representative PCR products corresponding to different mutation types of homozygous individuals with different genotypes were sequenced. Forward and reverse reactions were both performed to rule out false-positives.

Statistical analysis

Genotypic frequency, allele frequency, Hardy-Weinberg equilibrium, gene homozygosity, gene heterozygosity (HE), effective allele numbers (NE), and polymorphism information content (PIC) were statistically analyzed using the method described by Nei and Roychoudhury (1974) and Nei and Li (1979). Associations between genotypes of SNPs in the *FASN* gene, 5 meat quality traits (water loss, cooked meat percentage, PH_{24}, shear force, and IMF fat content), and genetic effects were analyzed using the general linear model procedure in the SPSS software (Version 18.0; SPSS, Inc., Chicago, IL, USA). The following model was used: $Y_{ijk}=\mu+S_i+N_j+E_{ijk}$, where Y_{ijk} is the trait measured in yak, μ is the population mean, S_i is the fixed effect of gender, N_j is the fixed effect of genotype, and E_{ijk} is random error. Significant differences between least-square means of different genotypes were determined using the Duncan multiple-range test, given that the trait was excluded from the model if its effect was not significant ($P>0.05$). The values are reported as least square means and standard error. P values of 0.05 were considered to be statistically significant.

RESULTS

Genotype distribution and genetic diversity

The PCR products obtained at the most suitable temperature were detected, and the amplification results were satisfactory (Fig. 1A and 1B). The 2 target fragments of the gene were denatured; polymorphisms were identified by polyacrylamide gel electrophoresis (Fig. 2A and 2B). For convenience, the 2 mutations were referred to as sequence variant 1 (SV1) and sequence variant 2 (SV2).

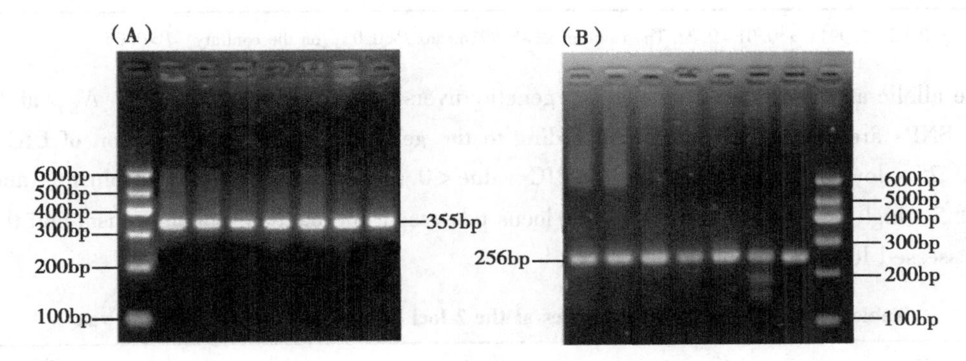

Fig. 1 Detection of PCR products of SV1 (A) and SV2 (B)

Note: A. *Lane M*=marker I (600; 500; 400; 300; 200; 100 bp); *lanes* 1-7=PCR products of SV1 (355 bp). B. *Lane M*=marker I (600; 500; 400; 300; 200; 100 bp); *lanes* 1-7=PCR products of SV2 (256 bp).

At the SV1 locus, 3 different genotypes (HH, HG, and GG) were identified in the Datong yak population. The frequencies of the 2 alleles were near and the heterozygote genotype (HG) showed a high prevalence with a frequency of 0.645. The genotypic frequencies of the SV1 locus in the Datong yak population were not in Hardy-Weinberg disequilibrium ($P<0.05$) (Table 2). At the SV2 locus, only 2 genotypes (EE, EF) were identified. The frequency of allele E was dominant in the Datong yak population and EE genotype was more frequent than the EF genotype. The genotypic frequencies of the SV2 locus in Datong yak population agreed with Hardy-Weinberg disequilibrium ($P>0.05$) (Table 2).

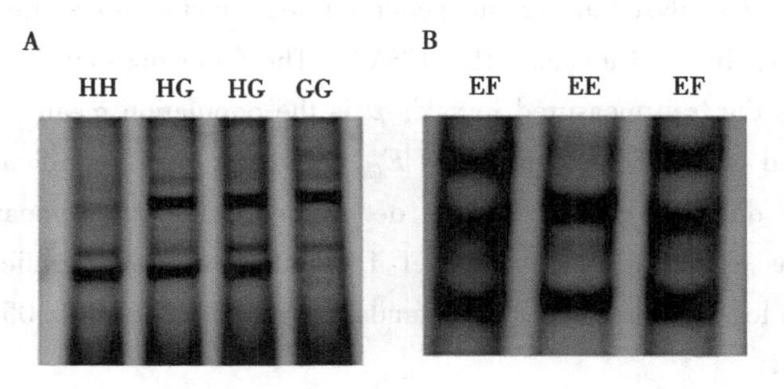

Fig. 2 Electrophoresis pattern of PCR-SSCP of SV1 (A) and SV2 (B)

Note: A. Three genotypes HH, HG, and GG in SV1 locus. B. Three genotypes EE and EF in SV2 locus.

Table 2 Genotypic frequencies at *FASN* gene for the SNPs in Datong yak

Locus	Genotype frequencies			Allele frequencies		x^2
	HH	HG	GG	H	G	
SV1	0.232	0.610	0.158	0.537	0.463	15.88
SV2	EE	EF	FF	E	F	3.160
	0.801	0.199	0	0.900	0.100	

Note: $x^2$0.05=5.991, $x^2$0.01=9.21. The value of x2>5.991means $P<0.05$, on the contrary, $P>0.05$.

The allelic and genotypic frequencies, genetic diversity parameters (H_O, H_E, N_E, and PIC) of the 2 SNPs are shown in Table 3. According to the genetic diversity classification of PIC (PIC value<0.25, low polymorphism; 0.25<PIC value<0.5, intermediate polymorphism; and PIC value>0.5, high polymorphism), the SV1 locus possessed intermediate polymorphism and the SV2 locus possessed low polymorphism.

Table 3 Population genetic indices at the 2 loci ofthe *FASN* gene in Datong yak

Locus	H_O	H_E	N_E	PIC
SV1	0.502	0.498	1.992	0.375
SV2	0.810	0.190	1.234	0.190

Note: H_O=gene homozygosity; H_E=gene heterozygosity; N_E=effective allele numbers; PIC=polymorphism information content. PIC>0.5 means high diversity, 0.25<PIC<0.5 means moderate diversity, PIC<0.25 means low diversity.

Sequence variants identified in the *FASN* gene

Comparison between sequences of the bovine *FASN* gene and the pooled yak DNA samples revealed 2 SNPs, g.5477C>T and g.16930T>A in intron 3 and exon 37, respectively (Fig. 3A and 3B). The SNP g.16930T>A was a synonymous mutation.

Association analysis

The association analysis of the 2 mutations in the *FASN* gene of the Datong yak and meat quality traits are shown in Table 4. For the SV1 locus, the animals with genotype HH and HG showed significantly higher IMF content ($P<0.01$) than those with genotype GG, implying that the H is associated with fat content. For the SV2 locus, significant differences in IMF content ($P<0.01$) were

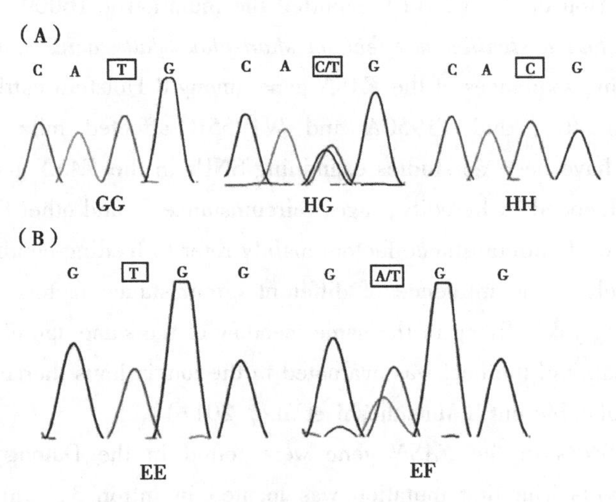

Fig. 3 Sequences of SV1 (A) and SV2 (B)

Note: A. Sequences of 3 genotypes HH, HG, and GG in SV1 locus. B. Sequences of 2 genotypes EE and EF in SV2 locus.

observed between cattle with genotype EE and those with EF. The animals with genotype EE appeared to be superior in fat content compared to those with genotype EF. No significant differences were observed in $PH2_4$, water loss, cooked meat percentage, and shear force for these 2 mutations. The remaining records of meat quality traits showed no significant association ($P>0.05$).

Table 4 Association between genotypes of the 2 SNPs of the *FASN* gene and meat quality traits in Datong yak

Locus	Genotype	Water loss (%)	Cooked meat percentage (%)	pH_{24}	Shear force (kg/cm^2)	IMF content (%)
SV1	HH	28.12±0.38	66.49±0.43	5.37±0.06	4.85±0.62	$1.99±0.64^A$
	HG	27.96±0.55	66.52±0.49	5.37±0.04	4.83±0.86	$1.97±0.54^A$
	GG	27.95±0.48	66.57±0.26	5.34±0.06	4.83±0.67	$1.88±0.52^B$
SV2	EE	28.12±0.73	66.52±0.16	5.37±0.08	4.84±0.67	$1.96±0.62^A$
	EF	28.19±0.52	66.38±0.17	5.36±0.06	4.85±0.91	$1.88±0.63^B$

Note: ^{A,B}Means with different superscript letters are highly significantly different ($P<0.01$). pH_{24} means pH value measured at 24h after slaughtered.

DISCUSSION

As a central multifunctional enzyme responsible for *de novo* fatty acid biosynthesis, FASN is an important candidate gene affecting fat deposit and meat quality traits in animals. Numerous studies have been reported the effect of the *FASN* gene on fat acid in beef and milk. Morris et al. (2007) reported 5 SNPs in the *FASN* gene related to adipose fat and milk fat in 3 bovine breeds. Abe et al. (2009) and Matsuhashi et al. (2011) both reported SNPs in the *FASN* gene in Japanese Black beef; mutations were associated with fatty acid content of C18 : 0, C18 : 1, C14 : 0, C14 : 1,

C16 : 0, and C16 : 1. Hou et al. (2011) reported the mutation g. 16009 A/G in exon 34 of the *FASN gene*; *this SNP had a significant effect on short-chain fatty acid content.* Matsumoto et al. (2012) compared coding sequences of the *FASN* gene among 8 Holstein cattle. Thirteen SNPs were identified in Holstein cattle, and T1950A and W1955R affected milk fat content and C14 index. However, there have been no studies examining SNPs in the *FASN* gene in yak.

Meat quality is influenced by heredity, age, circumstances, and other factors. Heredity factors mainly refer to breed, while circumstance factors mainly refer to feeding conditions such as nutrition and temperature. To exclude the influence of different circumstance factors and age, we selected yaks of 1 breed (Datong yak) living in the same meadow of the same age of 4 years old as experimental animals. The quality of the beef was evaluated in the longissimus thoracis muscle because it is regarded as the most palatable cut (Matsuhashi et al., 2011).

In this study, 2 SNPs in the *FASN* gene were found in the Datong yak population, and affected meat quality traits. The first mutation was located in intron 3, which was not in Hardy-Weinberg equilibrium. The second mutation was located in exon 37, which was in Hardy-Weinberg equilibrium. Allele H in intron 3 and allele E in exon 37 were dominant, and individuals with genotype HH and HG in the first mutation and of genotype EE in the second mutation performed better regarding IMF content than those with genotypes GG and EF. A previous study showed that higher IMF content indicates better tenderness and thus beef flavor. Yak meat has lower IMF content compared with cattle meat. Therefore, increasing the IMF content of yak meat is an effective method for improving meat flavor. Allele H in intron 3 and allele E in exon 37 both had a positive effect on IMF content in yak meat. In this study, no individuals with genotype FF were found in 313 yaks. Datong yak is a cultivated yak breed in China, and after nearly 30 years of artificial selections and cultivation, the yaks with better performance have already been selected for use as breeders. This mating mechanism disrupted the Hardy-Weinberg equilibrium of natural mating mechanism. As a result, alleles associated with high performances were maintained, while alleles associated with low performances were reduced.

The results of this study as well as those of previous studies (Oh et al., 2012; Souza et al., 2012) suggest that *FASN* is an effective candidate gene that can be used as a DNA marker for improving IMF content through marker-assisted selection in yaks.

CONCLUSIONS

This study is the first to investigate the association between polymorphisms in the *FASN* gene and meat traits in Datong yak. Statistical results showed a significant relationship between the *FASN* polymorphisms and IMF content in Datong yak. In conclusion, the study indicated that genotyping of the *FASN* gene may be useful for selecting yaks with higher IMF content and thus improving the beef flavor.

ACKNOWLEDGMENTS

Research supported by the National Beef Cattle Industry Technology & System (#CARS-38), the Special Fund for Agro-Scientific Research in the Public Interest (#201003061), the 'Fivetwelfth' National Science and Technology Support Program (#2012BAD13B05), the Recommend International Advanced Agricutural Science and Technology Plan (#2013-Z11), and the Fundamental Research Funds for Central Non-Profit Research Institutes (#1610322014002).

References

[1] Abe T, Saburi J, Hasebe H, Nakagawa T, et al. (2009). Novel mutations of the FASN gene and their effect on fatty acid composition in Japanese Black beef. *Biochem. Genet.* 47: 397-411.

[2] Byun SO, Fang Q, Zhou H and Hickford JG (2008). Rapid genotyping of the ovine ADRB3 gene by polymerase chain reaction-single-strand conformation polymorphism (PCR-SSCP). *Mol. Cell. Probes* 22: 69-70.

[3] Chakravarty B, Gu Z, Chirala SS, Wakil SJ, et al. (2004). Human fatty acid synthase: structure and substrate selectivity of the thioesterase domain. *Proc. Natl. Acad. Sci. U. S. A.* 101: 15567-15572.

[4] Guo X, Yan P, Liang CN, Pei J, et al. (2009). Developmental situations and countermeasures of Yak Industry in China. *China Cattle Sci.* 35: 55-57.

[5] Honikel KO (1998). Reference methods for the assessment of physical characteristics of meat. *Meat Sci.* 49: 447-457.

[6] Hou YJ, Chang LL and Mao HB (2011). SNPs of FASN gene and their effects on fatty acids composition in Chinese Holstein. *Chin. Cow* 18: 1-4.

[7] Korkeala H, Mäki-Petäys O, Alanko T and Sorvettula O (1986). Determination of pH in meat. *Meat Sci.* 18: 121-125.

[8] Lan XY, Pan CY, Chen H, Zhang CL, et al. (2007). An AluI PCR-RFLP detecting a silent allele at the goat POU1F1 locus and its association with production traits. *Small Ruminant Res.* 73: 8-12.

[9] Matsuhashi T, Maruyama S, Uemoto Y, Kobayashi N, et al. (2011). Effects of bovine fatty acid synthase, stearoyl-coenzyme A desaturase, sterol regulatory element-binding protein 1, and growth hormone gene polymorphisms on fatty acid composition and carcass traits in Japanese Black cattle. *J. Anim. Sci.* 89: 12-22.

[10] Matsumoto H, Inada S, Kobayashi E, Abe T, et al. (2012). Identification of SNPs in the FASN gene and their effect on fatty acid milk composition in Holstein cattle. *Livestock Sci.* 144: 281-284.

[11] Nei M and Li WH (1979). Mathematical model for studying genetic variation in terms of restriction endonucleases. *Proc. Natl. Acad. Sci. U. S. A.* 76: 5269-5273.

[12] Morris CA, Cullen NG, Glass BC, Hyndman DL, et al. (2007). Fatty acid synthase effects on bovine adipose fat and milk fat. *Mamm. Genome* 18: 64-74.

[13] Nei M and Roychoudhury AK (1974). Sampling variances of heterozygosity and genetic distance. *Genetics* 76: 379-390.

[14] Oh D, Lee Y, La B, Yeo J, et al. (2012). Fatty acid composition of beef is associated with exonic nucleotide variants of the gene encoding FASN. *Mol. Biol. Rep.* 39: 4083-4090.

[15] Qu L, Li X, Wu G and Yang N (2005). Efficient and sensitive method of DNA silver staining in polyacrylamide gels. *Electrophoresis* 26: 99-101.

[16] Roy R, Gautier M, Hayes H, Laurent P, et al. (2001). Assignment of the fatty acid synthase (FASN) gene to bovine chromosome 19 ($19q_{22}$) by *in situ* hybridization and confirmation by somatic ce11 hybrid mapping. *Cytogenet. Cell Genet.* 93: 141-142.

[17] Roy R, Taourit S, Zaragoza P, Eggen A, et al. (2005). Genomic structure and alternative transcript of bovine fatty acid synthase gene (FASN): comparative analysis of the FASN gene between

monogastric and ruminant species. *Cytogenet. Genome Res.* 111：65-73.

[18] Roy R，Ordovas L，Zaragoza P，Romero A，et al. （2006）. Association of polymorphisms in the bovine FASN gene with milk-fat content. *Anim. Genet.* 37：215-218.

[19] Souza FR，Chiquitelli MG，da Fonseca LF，Cardoso DF，et al. （2012）. Associations of FASN gene polymorphisms with economical traits in Nellore cattle（*Bos primigenius indicus*）. *Mol. Biol. Rep.* 39：10097-10104.

[20] Stuart S，Andrzej W and Anil KJ（1991）. Structural and functional organization of the animal fatty acid synthase. *Eur. J. Biochem.* 198：571-579.

[21] Yan L，Lin JD，Zhang XJ and Zhang Y（2010）. Application of single nucleotide polymorphism in breeding of pig. *China Anim. Husbandry Vet. Med.* 37：94-98.

[22] Zhang S，Knight TJ，Reecy JM and Beitz DC（2008）. DNA polymorphisms in bovine fatty acid synthase are associated with beef fatty acid composition. *Anim. Genet.* 39：62-70.

（发表于《Genetics and Molecular Research》，院选 SCI，IF：0. 850）

De novo Assembly and Characterization of Skin Transcriptome Using RNAseq in Sheep (*Ovis aries*)

YUE Y.J.[1,2], LIU J.B.[2], YANG M.[2], HAN J.L.[2],
GUO T.T.[2], GUO J.[2], FENG R.L.[2], YANG B.H.[1,2]

(1. College of Animal Science and Technology, Gansu Agricultural University, Lanzhou 730050, China; 2. Lanzhou Institute of Husbandry and Pharmaceutical Sciences, Chinese Academy of Agricultural Sciences, Lanzhou 730050, China)

Abstract: Wool is produced via synthetic processes of wool follicles, which are embedded in the skin of sheep. The development of new-generation sequencing and RNA sequencing provides new approaches that may elucidate the molecular regulation mechanism of wool follicle development and facilitate enhanced selection for wool traits through gene-assisted selection or targeted gene manipulation. We performed *de novo* transcriptome sequencing of skin using the Illumina Hiseq 2000 sequencing system in sheep (*Ovis aries*). Transcriptome *de novo* assembly was carried out via short-read assembly programs, including SOAPdenovo and ESTScan. The protein function, clusters of orthologous group function, gene ontology function, metabolic pathway analysis, and protein coding region prediction of unigenes were annotated by BLASTx, BLAST2GO, and ESTScan. More than 26 266 670 clean reads were collected and assembled into 79 741 unigene sequences, with a final assembly length of 35 447 962 nucleotides. A total of 22 164 unigenes were annotated, accounting for 36.27% of the total number of unigenes, which were divided into 25 classes belonging to 218 signaling pathways. Among them, there were 17 signal paths related to hair follicle development. Based on mass sequencing data of sheepskin obtained by RNA-Seq, many unigenes were identified and annotated, which provides an excellent platform for future sheep genetic and functional genomic research. The data could be used for improving wool quality and as a model for human hair follicle development or disease prevention.

Key words: Sheep (*Ovis aries*); Skin; *De novo* assembly; Transcriptome; RNA sequencing

INTRODUCTION

Wool production is a major agricultural industry throughout the world; the most important wool-producing countries are Australia, China, New Zealand, South Africa, and a number of countries in South America (Galbraith, 2010a). China, for example, as the world's second largest producer of wool, accounting for ~10% of the world's wool production (i.e., 12 million tons) (FAO, 2012). When compared to the quality of Australian wool, the wool from China has a broader micro and a shorter staple; most of the wool produced in China is not suitable for the production of high-end woolen apparel (Liu et al., 2011).The current annual wool processing capacity in China exceeds 400 kt (clean equivalent).However, China is not able to produce enough wool to meet the demand of the processors, resulting in an increase in imports.In 2009, total wool imports into China reached 327 kt, accounting for 33% of the world's output (Liu et al., 2011).Therefore, there is an urgent need to improve the quantity and quality of wool by researching the genetics, endocrinology, and ectoparasitic diseases of sheep (Smith et al., 2010).

Wool is produced via synthetic processes using wool follicles, which are embedded in the skin of sheep (Galbraith, 2010b).The underlying molecular mechanisms regulating wool follicle initiation, development, and cycling can facilitate enhanced selection for wool traits through gene-assisted selection or targeted gene manipulation (Norris et al., 2005).In sheep, wool follicle induction and morphogenesis have been characterized in a number of studies (Mc Cloghry et al., 1992; Bond et al., 1994; Hynd et al., 1999).Gene expression profiles of sheepskin have also been generated by sequencing expressed sequence tages (ESTs) and complementary DNA (cDNA) microarray (Adelson et al., 2004; Norris et al., 2005; Smith et al., 2010; Burgess et al., 2012; Peñagaricano et al., 2012).However, molecular regulation of these developmental processes has been hampered by the paucity of information on the sheep genome and skin transcriptome (Jager et al., 2011).Most of our current and rapidly growing knowledge on the genes controlling skin and fiber development is provided by studies on the hair follicles of mice and humans (Botchkarev and Paus, 2003; Schmidt-Ullrich and Paus, 2005; Bostjancic and Glavac, 2008; Wang et al., 2012). Currently, the development of new-generation sequencing and RNA sequencing (RNA-Seq) provides new approaches that may elucidate the molecular regulation mechanism of hair follicle development (Jager et al., 2011; Okano et al., 2012; Geng et al., 2013).New-generation sequence technologies have opened the door to genome-scale experiments in organisms that lack comprehensive genome or transcriptome information, thus making it possible to assemble novel transcripts and identify differential regulation in a single experiment (Birzele et al., 2010; Sun et al., 2010; Abyzov et al., 2012).

In this study, we performed *de novo* transcriptome sequencing of skin using the Illumina Hiseq 2000 sequencing system in sheep (*Ovis aries*).More than 26 266 670 clean reads were collected and assembled into 79, 741 unigenes. Annotation and gene ontology analyses were then performed on these unigenes, thus providing a valuable resource for future genetic and genomic research on sheep and closely related species

MATERIAL AND METHODS

Animal material

Gansu Alpine fine wool sheep were bred in the Huang Cheng District of Gansu Province, China, by cross breeding Mongolian or Tibetan sheep with Xinjiang Fine Wool sheep and then with some fine wool sheep breeds from the Union of Soviet Socialist Republics, such as Caucasian and Salsk.The breed was approved by the Gansu provincial government in 1980.Since then, to improve wool quality, the breed was hybridized with Australian Merino, New Zealand Merino, and German

Mutton Merino sheep. Today, the modern Gansu Alpine fine wool sheep can be classified into 3 strains according to the fineness of the wool, including superfine [fiber diameter (FD) < 19.0μm], fine (19.0 μm<FD<21.0m) and dual-purpose strains (FD>21.0μm). A body side skin sample (SHEawmTARAAPEI-12) was collected from the superfine strain (FD=16.5μm) of Gansu Alpine fine wool sheep. The sampled tissues were immediately frozen in liquid nitrogen and stored at −80℃ for subsequent analysis.

RNA extraction

Total RNA was isolated from the tissues using the RNeasy Maxi Kit (Qiagen, Hilden, Germany) according to manufacturer instructions. RNA quality was verified using a 2 100 Bioanalyzer RNA Nanochip (Agilent, Santa Clara, CA, USA), and the RNA Integrity Number value was>8.5. Then, the RNA was quantified using the Nano Drop ND-2000 Spectrophotometer (Nano-Drop, Wilmington, DE, USA).

cDNA library construction and sequencing

Illumina sequencing using the GAII platform was performed at the Beijing Genomics Institute (BGI) -Shenzhen, Shenzhen, China (http://www.genomics.cn/index.php) according to manufacturer instructions (Illumina, San Diego, CA, USA). Beads with Oligo (dT) were used to isolate poly (A) messenger RNA (mRNA) after total RNA was collected from eukaryote (prokaryocyte can be treated with a kit to remove ribosomal RNA before the next step). Fragmentation buffer was added for interrupting mRNA to short fragments. Utilizing these short fragments as templates, a random hexamer-primer was used to synthesize the first-strand cDNA. The second-strand cDNA was synthesized using a buffer, dNTPs, RNaseH, and DNA polymerase I. Short fragments were purified with a QiaQuick polymerase chain reaction (PCR) extraction kit and resolved with EB buffer for end reparation and adding the poly (A). Subsequently, the short fragments were connected with sequencing adapters. Then, after agarose gel electrophoresis, the suitable fragments were selected for PCR amplification as templates. Finally, the library was sequenced using Illumina HiSeq™ 2000.

Data filtering and *de novo* assembly

The quality requirement for *de novo* transcriptome sequencing was far higher than that for resequencing because sequencing errors can cause difficulties for the short-read assembly algorithm. Therefore, we carried out a stringent filtering process. First, we removed reads that did not pass the built-in Illumina's software failed-chastity filter according to the relation "failed-chastity ≤ 1" using a chastity threshold of 0.6 on the first 25 cycles. Second, we discarded all reads with adaptor contamination. Third, we ruled out low-quality reads with ambiguous sequences "N". Finally, the reads with more than 10% Q<20 bases were also removed.

De novo assembly

Transcriptome *de novo* assembly was carried out with the short-read assembly program SOAPdenovo (Li et al., 2010). SOAPdenovo first combined reads with a certain length of overlap to form longer fragments without N, which are called contigs. Then, the reads were mapped back to contigs; with paired-end reads, it is able to detect contigs from the same transcript as well as the distances between those contigs. Next, SOAPdenovo connected the contigs using N to represent unknown sequences between 2 contigs and scaffolds were formed. Paired-end reads were used again for gap filling of scaffolds to form sequences with the least number of Ns that could not be extended on either end (i.e., unigenes). When multiple samples from the same species are sequenced, unigenes from the assembly of each sample can be further processed for sequence splicing and redundancy removal by using the sequence clustering software to acquire the longest possible non-redundant (nr) unigenes. In the final step, BLASTx alignment (E value<0.00001) between unigenes and protein databases [e.g., nr database, Swiss-Prot, Kyoto Encyclopedia of Genes and Genomes (KEGG), and clusters of orthologous groups (COG)] was performed, and the best-fit alignments were used to determine the sequence direction of unigenes. If the results obtained from different databases conflicted with each other, a priority order of nr, Swiss-Prot, KEGG, and COG was followed to determine the sequence direction of unigenes. When a unigene did not align to any of the above databases, the ESTScan software (Iseli et al., 1999) was introduced to predict the coding regions and determine the direction of the sequence. For unigenes with sequence directions, we provided their sequences from the 5′- to 3′-end; for those without a direction, we provided their sequences from the assembly software.

Unigene COG function annotation

Unigene annotation provides information on the expression and functional annotation of a unigene. Unigene sequences were first aligned by BLASTx to protein databases (e.g., nr, Swiss-Prot, KEGG, and COG; E value<0.00001), retrieving proteins with the highest sequence similarity with the given unigenes along with their protein functional annotations. COG is a database where orthologous gene products are classified. Every protein in the COG is assumed to have evolved from an ancestral protein, and the whole database is built on coding proteins with complete genomes and evolutionary relationships of bacteria, algae, and eukaryotic creatures. Unigenes were aligned to the COG database to predict and classify possible functions of unigenes.

Unigene Gene Ontology (GO) classification

We can obtain GO functional annotation with nr annotation. GO is an international standardized gene functional classification system that offers a dynamically updated and controlled vocabulary; it is a strictly defined concept to comprehensively describe properties of genes and their products in any organism. GO has 3 ontologies, including molecular function, cellular component, and biological process. The basic unit of GO is the GO-term. Every GO-term belongs to a type of ontology.

With nr annotation, we used the BLAST2GO program (Conesa et al., 2005) to obtain GO annotation of unigenes. BLAST2GO has been cited by other articles (i.e., >150 times) and is a widely recognized GO annotation software. After obtaining the GO annotation for each unigene, we used the WEGO software (Ye et al., 2006) to conduct GO functional classification for all unigenes to investigate the species distributions of gene functions at the macro level.

Unigene metabolic pathway analysis

KEGG is a database that is able to analyze a gene product during metabolic processes and related gene function in cellular processes. KEGG may further research on the genetics of biologically complex behaviors. Using KEGG annotation, we can comment on the pathways for unigenes.

Protein coding DNA sequence (CDS) prediction

Unigenes were first aligned by BLASTx (E value < 0.00001) with protein databases in the priority order of nr, Swiss-Prot, KEGG, and COG. Unigenes aligned with databases with higher priority did not enter the next circle. The alignments ended when all of the circles were completed. Proteins with the highest ranks in the BLAST results were utilized to determine the CDSs of unigenes; then, the coding region sequences were translated into amino sequences with the standard codon table. Thus, both the nucleotide sequences (5'-3') and amino sequences of the unigene coding regions were acquired. Unigenes that could not be aligned with any database were scanned by ESTScan (Iseli et al., 1999) to obtain nucleotide (5'-3') and amino sequences of the coding regions.

RESULTS

Sheepskin transcriptome sequencing and assembly

The amount of sequencing data is an important indicator for completion of transcriptome sequencing. The project extracted tissue mRNA from sheepskin and then conducted the transcriptome sequencing. We obtained 26 266 670 clean reads, and the number of bases obtained by the clean reads was 2 364 000 300 nucleotides (nt) (Table 1). Transcriptome *de novo* assembly was carried out with the short-read assembly program SOAPdenovo using mRNAs, ESTs, and clean reads following a bioinformatic workflow (Li et al., 2010; Jager et al., 2011; NCBI, 2013). A total of 40 300 ovine gene GenBank entries (corresponding to 17 319 unigenes) and 370 194 ovine ESTs were available for assembly. SOAPdenovo output of 853 437 contigs with a final assembly length of 105 879 352 nt. The average size of the contigs was 134 nt. The shortest assembly contig was 50 bp, and the longest contig was 4 881 bp; there were 779 172 contigs that were < 200 bp in length, which accounted for 91.30% of the contigs (Table 2).

Table 1 Output statistics of sequencing

Sample	Total reads	Total nucleotides (nt)	Q20 (%)	N (%)	GC (%)
SHEawmTARAAPEI-12	26 266 670	2 364 000 300	92.92	0.00	48.93

Note: Q20 = quality score of 20; Total nucleotides = total reads 1 x read 1 size + total reads 2 x read 2 size.

Table 2 Contig quality of sheepskin

Length of contig	Number	Percent (%)
75–100 nt	685 968	80.38
100–200 nt	93 204	10.92
200–300 nt	33 428	3.92
300–400 nt	15 112	1.77
400–500 nt	8 197	0.96
≥500 nt	17 528	2.05
N50	90	

(continued)

Length of contig	Number	Percent
Mean	124	
All contig	853 437	
Length of all contig (nt)	105 879 352	

To determine different contigs from the same transcript and the distance between these contigs by paired-end reads, we used the sequence-obtained clean reads for alignment with the contigs. Next, SOAPdenovo connected the contigs using N to represent unknown sequences between 2 contigs; then, scaffolds were constructed. In this study, we obtained 147 155 scaffolds with a final assembly length of 44 758 530 nt. The average size of the scaffolds was 304 nt (Table 3). The length distribution of the assembled scaffolds primarily included small fragments. The number of scaffolds that were<500 bp in length totaled 126 021 (i.e., 85.64%), the proportion of which was significantly reduced. The number of scaffolds that did not contain nucleotide deletions was 122 860 (i.e., 83.49%; Fig. S1).

Table 3 Scaffold quality of sheepskin

Length of scaffold	Number	Percent (%)
100–500 nt	126 021	85.64
500–1 000 nt	16 004	10.88
1 000–1 500 nt	3 775	2.57
1 500–2 000 nt	1 038	0.71
≥2 000 nt	317	0.22
N50	387	
Mean	304	
All scafSeq	147 155	
Length of all scafSeq (nt)	44 758 530	

Paired-end reads were used again for gap filling of scaffolds to obtain sequences with the least Ns and that could not be extended on either end (i.e., unigenes). A total of 79 741 sheepskin unigene sequences with a final assembly length of 35 447 962 nt were obtained, which contained the least Ns and could not be extended on either end. The average sequence length was 445 bp. A total of 74 592 unigenes were<1 000 bp in length (i.e., 93.55%). The N50 length of 508 bp (i.e., half of the assembled bases were incorporated into unigenes with a length at least 508 bp) was obtained; 26.53% (21 151 unigenes) had lengths >500 bp (Table 4). A total of 84.67% of the unigenes had no gaps, and<0.54% of the unigenes had a gap percentage (ratio of number 'N' to unigene length) >20% (Fig. S2). Only one unigene had a base deletion ratio of >0.3. Analysis of the positional distribution of obtained sequences identified several characteristics as follows: 1) sheepskin transcriptome sequencing obtained a relatively large number of reads and a more balanced distribution of the unigene 5′-end; and 2) the number of reads located in the unigene 3′-end was less than that in the 5′-end.For the number of reads with a relative increase in position, there was a linear downward trend in the relative position of 5′ to 3′ that was >0. 6 (Fig. 1).This result is consistent with other studies (Wang et al., 2010; Shi et al., 2011).This indicates that the transcriptome sequencing quality of the sheepskin and that of other non-model organisms was considerable.

Table 4 Unigene quality of sheepskin

Length of unigene	Number	Percent (%)
100–500 nt	58 590	73. 48
500–1 000 nt	16 002	20. 07
1 000–1 500 nt	3 795	4. 76
1 500–2 000 nt	1 036	1. 30
≥2 000 nt	318	0. 40
N50	508	
Mean	445	
All unigenes	79 741	
Length of all unigenes (nt)	35 447 962	

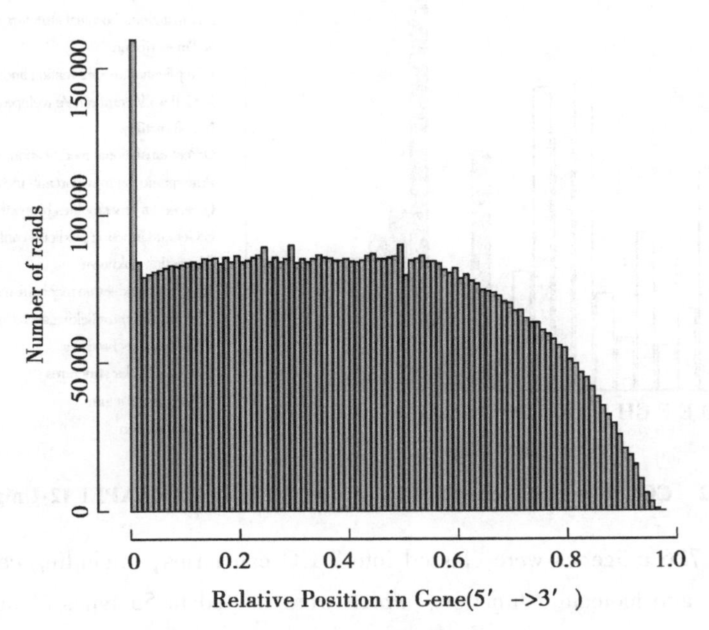

Fig. 1 Randomness of reads from sample

Functional annotation of sheepskin transcriptome

We used the BLASTx comparison to align the unigene sequences with the protein databases nr, Swiss-Prot, KEGG, and COG (E value<0. 00001).We searched for the highest protein sequence similarity for a given unigene and gained functional annotation information on sheepskin unigene proteins.The functional annotation results of sheepskin transcriptome sequencing are shown in Table 1. After alignment with the 4 databases, a total of 22, 164 sheepskin unigenes were annotated, accounting for 36. 27% of the total number of unigenes (i.e., a total of 79 741). Of these, 28 924 were annotated in the nr database; 26 079 in the Swiss-Prot database; 17 113 in the KEGG database; and 6 616 in the COG database. A total of 6 616 unigenes were annotated in all 4 databases.

Through BLAST and the National Center for Biotechnology Information (NCBI) COG databases, we obtained 12 711 sheepskin unigenes via COG functional classification. These were divided into 25 categories. The number of unigenes belonging to the R class (general function prediction only) was 2 083 (i.e., 16. 39%). The J class (chromatin structure and dynamics) contained 1 253 unigenes (i. e., 9. 86%). The L class (energy production and conversion) contained 1 147 unigenes (i. e., 9. 02%). Only one unigene fell into the W class (extracellular structures) (Fig. 2).

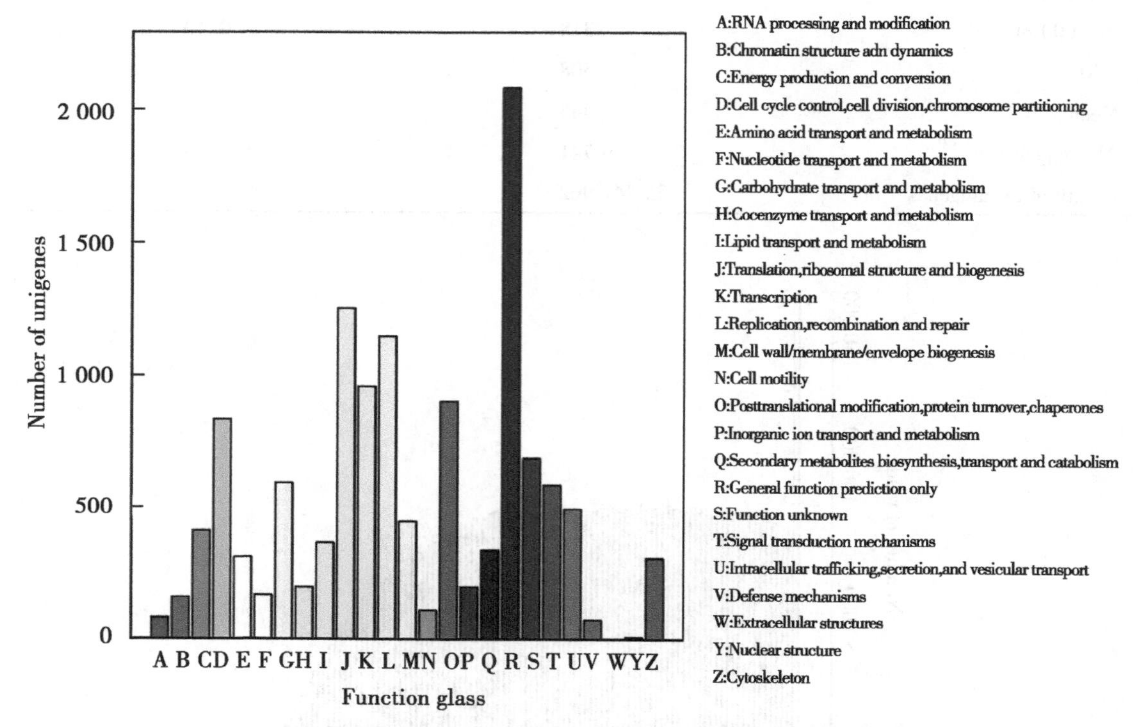

Fig. 2 COG functional classification of the SHEawmTARAAPEI-12-Unigene

More than 11 726 unigenes were classed into 3 GO categories, including cellular component, biological process, and molecular function, which were related to 53 types of biological functions. Under the cellular component category, large numbers of unigenes were categorized as cell (9 669 unigenes), cell part (9 669 unigenes), organelle (6 731 unigenes), and organelle part (3 490 unigenes). For the biological process category, cellular process (8 199 unigenes), biological regulation (4 371 unigenes), and pigmentation (4 094 unigenes) represented the greatest proportion of unigenes. Under the molecular function category, binding (8 799 unigenes) and catalytic activity (4 200 unigenes) were the top 2 most abundant subcategories (Fig. 3).

There were 17 096 unigenes mapped into 218 KEGG pathways. The pathways or maps with the highest representation were metabolic pathways (1 977 unigenes, 11. 56%, ko01100), focal ad-

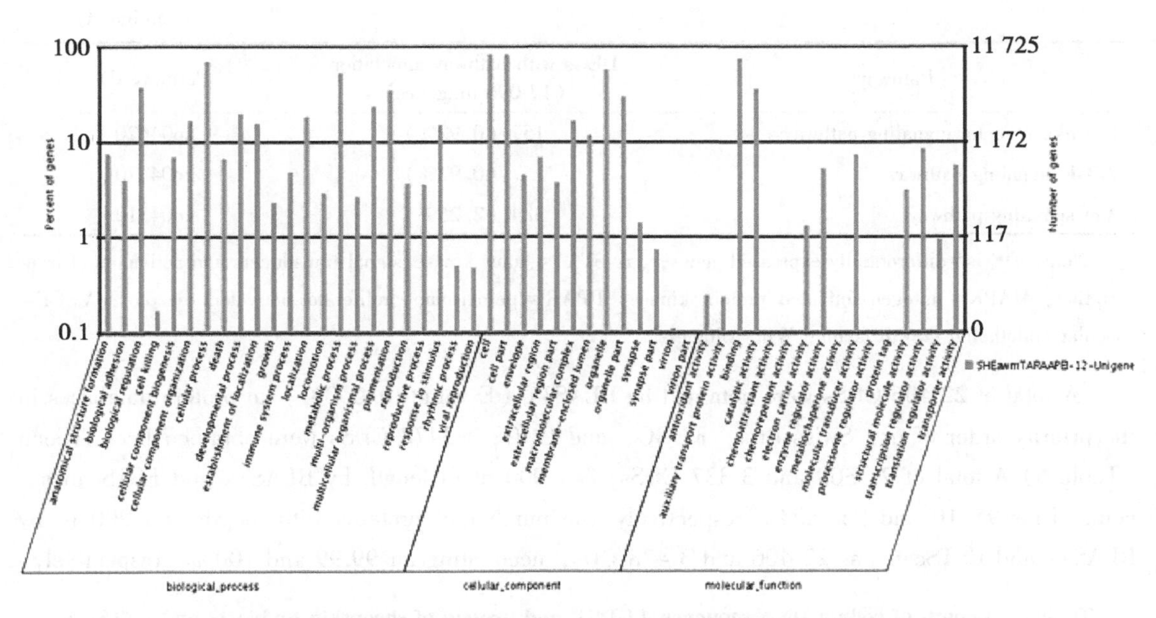

Fig. 3 GO functional classification of the SHEawmTARAAPEI-12-Unigene

hesion (966 unigenes, 5.65%, ko04510), and regulation of the actin cytoskeleton (840 unigenes, 4.91%, ko04810). Signaling pathways related to hair follicle development included, for example, the hedgehog signaling, insulin signaling, Jak-STAT signaling, MAPK signaling, melanogenesis, notch signaling, TGF-beta signaling, Toll-like receptor signaling, VEGF signaling, and Wnt signaling pathways (Botchkarev and Paus, 2003; Adelson et al., 2004; Norris et al., 2005; Schmidt-Ullrich and Paus, 2005; Bostjancic and Glavac, 2008; Smith et al., 2010; Penagaricano et al., 2012; Wang et al., 2012) (Table 5).

Table 5 Signaling pathways related to hair follicle development in sheepskin

Pathway	DEGs with pathway annotation (17 096 unigenes)	Pathway ID
Hedgehog signaling pathway	83 (0.49%)	ko04340
Insulin signaling pathway	317 (1.85%)	ko04910
Jak-STAT signaling pathway	178 (1.04%)	ko04630
Lysine biosynthesis	15 (0.09%)	ko00300
Lysine degradation	259 (1.51%)	ko00310
MAPK signaling pathway	536 (3.14%)	ko04010
Melanogenesis	171 (1%)	ko04916
Melanoma	115 (0.67%)	ko05218
Metabolic pathway	1 977 (11.56%)	ko01100
Notch signaling pathway	164 (0.96%)	ko04330
PPAR signaling pathway	153 (0.89%)	ko03320
TGF-beta signaling pathway	197 (1.15%)	ko04350

(continued)

Pathway	DEGs with pathway annotation (17 096 unigenes)	Pathway ID
Toll-like receptor signaling pathway	157 (0.92%)	ko04620
VEGF signaling pathway	157 (0.92%)	ko04370
Wnt signaling pathway	384 (2.25%)	ko04310

Note: DEGs = differentially expressed genes; Jak-STAT = Janus kinase/signal transducers and activators of transcription; MAPK = mitogen-activated protein kinase; PPAR = peroxisome proliferator-activated receptor; VEGF = vascular endothelial growth factor; Wnt = wingless.

A total of 22 406 CDSs were obtained by BLASTx (E value<0.00001) to protein databases in the priority order of nr, Swiss-Prot, KEGG, and COG; >3478 CDSs were obtained by ESTScan (Table 6).A total of 21 309 and 3 337 CDSs of<1 000 nt in length by BLASTx and ESTScan accounted for 95.10 and 95.95%, respectively.The number of proteins with lengths <1 000 nt by BLASTx and ESTScan was 22 406 and 3 478 CDs, accounting for 99.99 and 100%, respectively.

Table 6 Length of coding DNA sequence (CDS) and protein of sheepskin by blastx and ESTScan

Range of CDS/protein length	Number of CDS sequences in specified range		Number of protein sequences in specified range	
	BLASTx	ESTScan	BLASTx	ESTScan
200	4 306	192	18 594	2 832
300	7 091	1 096	2 330	435
400	3 774	567	885	146
500	1 983	609	353	47
600	1 440	367	142	12
700	1 098	164	58	4
800	695	164	32	1
900	537	108	7	1
1 000	385	70	4	0
1 100	279	47	0	0
1 200	221	29	0	0
1 300	159	16	0	0
1 400	118	17	0	0
1 500	76	14	0	0
1 600	71	8	0	0
1 700	37	1	1	0
1 800	34	3	0	0
1 900	28	2	0	0
2 000	17	0	0	0
2 100	13	2	0	0
2 200	13	1	0	0

(continued)

Range of CDS/protein length	Number of CDS sequences in specified range		Number of protein sequences in specified range	
	BLASTx	ESTScan	BLASTx	ESTScan
2 300	7	0	0	0
2 400	12	0	0	0
2 500	4	0	0	0
2 600	1	1	0	0
2 700	2	0	0	0
2 800	3	0	0	0
2 900	1	0	0	0
3 000	0	0	0	0
>3 000	1	0	0	0

DISCUSSION

Currently, methods used for the acquisition and analysis of transcriptome data are mainly based on serial analysis of gene expression, hybridization chips (gene chip or microarray), and RNA-Seq technology. Gene expression profiles of sheepskin have also been generated by sequencing of ESTs and cDNA microarray (Adelson et al., 2004; Norris et al., 2005; Smith et al., 2010; Burgess et al., 2012). However, hybridization-based microarray technology is restricted to known sequences, which are unable to detect new transcripts. Meanwhile, it is also difficult to detect fusion gene transcripts, polycistronic transcription, and abnormal transcripts, thus limiting its use (Okoniewski and Miller, 2006). Compared to microarray technology, RNA-Seq technology has many unique advantages (Nagalakshmi et al., 2008). It is highly accurate and has a very low detection limitation, making it useful for a wide range of applications (Wang et al., 2009). These applications include the detection of new transcripts (unknown transcripts and rare transcripts) (Denoeud et al., 2008), functional research of non-coding regions (e.g., non-coding RNA research, and small RNA precursor studies), transcript structure (e.g., UTR identification, intron boundaries identification, alternative splicing, and start codon identification), gene transcription, and detection of SNPs of a coding sequence (Cloonan et al., 2008). These applications make RNA-Seq technology a powerful tool for in-depth research on complex transcriptomes. The International Sheep Genome Consortium utilizes the results from the Human Genome Project; moreover, genomic research for the dog and cow is used to derive a virtual sheep genome map (Dalrymple et al., 2007; Archibald et al., 2010). However, these are currently available only for low-coverage genome sequences (Archibald et al., 2010). In order to reveal the molecular mechanism of hair follicle development, we performed *de novo* transcriptome sequencing of skin using the Illumina Hiseq 2000 sequencing system in sheep (*Ovis aries*). We obtained 79 741 unigenes in sheepskin, totaling 35 447 962 bp. The average sequence length was 445 bp. The quality of the sequencing results was equivalent to traditional Sanger sequencing, but the collection of the information was time-consuming when compared to that of traditional Sanger sequencing and microarray (Lobo et al., 2009). Using bioinformatic tools and aligning sequences with the Uniprot, NCBI's nr, COG, Pfam, InterPro, and KEGG 6 databases, 22 164 of the sheepskin unigenes were annotated, which only accounted for 36.27% of the total number of unigenes (total number = 79 741). Unannotated unigenes compared using ESTScan yielded 3 478 CDSs. A total of 35 171 unigenes remained unannotated; therefore, additional research on the assembly and annotation of the sheep transcriptome is needed for the sheep reference genome (Archibald et al., 2010).

The skin is composed of the epidermis, dermis, and subcutaneous tissue that interconnect anatomically (Prost-Squarcioni et al., 2008). Hair follicles develop as a result of epithelial-mesenchymal interactions between epidermal keratinocytes committed to hair-specific differentiation and the clustering of dermal fibroblasts that form follicular papilla. The development of primary, secondary, and derived follicles in the pelage of Merino sheep has been described by Rogers (2006). Primary follicles with arrector pilae muscle, and sebaceous and apocrine glands commence development first (E70) in the form of a central follicle surrounded by 2 lateral primaries. Then, the secondary follicles (So) associated with the primaries appear (E85). Additional follicles develop by branching from the So follicles and are apparent by E105. Branching of the secondary original follicles is critical to the volume and nature of the Merino fleece, as secondary follicles represent the major source of fine fibers (Adelson et al., 2004; Rogers, 2006). During postnatal life, hair follicles show patterns of cyclic activity with periods of active growth and hair production (anagen), apoptosis-driven involution (catagen), and relative resting (telogen) (Botchkarev and Paus, 2003). However, Merino wool follicles are predominant throughout growth, different to many animals such as the mouse, rabbit, and guinea-pig (Rogers, 2006). During the past decade there has been a dramatic increase in our knowledge on the molecular control of fiber initiation and development in mice and humans (Botchkarev and Paus, 2003; Schmidt-Ullrich and Paus, 2005; Bostjancic and Glavac, 2008; Wang et al., 2012; Cadau et al., 2013). Gene expression profiles of sheepskin have also been generated by the sequencing of ESTs and cDNA microarray (Adelson et al., 2004; Norris et al., 2005; Smith et al., 2010; Burgess et al., 2012). However, molecular regulation of these developmental processes has been hampered by the paucity of information on the sheep genome and skin transcriptome (Jager et al., 2011) However, comparative analyses of gene expression patterns in skin suggest that the molecular regulation of the basic mechanisms of follicle initiation and development and fiber production are similar in most mammals. Studies in these mammals have detailed genes with translated products that control hair follicle initiation, development, and regulation of secondary branching events, as described in reviews (Galbraith, 2010b). Such factors defined in the hair follicle include members of the wingless, bone morphogenetic protein, fibroblast growth factor, tumor necrosis factor, transforming growth factor, and the notch and hedgehog signaling pathways. In this study, these regulatory pathways and signaling molecules are also found in the sheep skin. This has only been used for improving wool quality, but can also be used as a model for human hair follicle development or disease prevention (Archibald et al., 2010).

The most important function of the skin is to form an effective barrier between the "inside" and "outside" of an organism. The epidermis comprises the physical, chemical/biochemical (antimicrobial and innate immunity), and adaptive immunological barriers because it is composed of the stratum corneum, nucleated epidermis, lipids, acids, hydrolytic enzymes, antimicrobial peptides and macrophages, and humoral and cellular constituents of the immune system (Proksch et al., 2008). Therefore, in this study, we aligned sequences using the Uniprot, NCBI nr, COG, Pfam, InterPro, and KEGG 6 databases. A total of 22 164 unigenes in the sheepskin were annotated and divided into 25 categories, which included 218 signaling pathways involved in cell composition (23 822 unigenes), biological processes (49 340 unigenes), and molecular function (16 150 unigenes). Among the signaling pathways that are related to diseases, the following number of unigenes and respective signaling pathways were identified: bacteria invasion of epithelial cells: 334; vibrio cholerae infection: 219; epithelial cell signaling in *Helicobacter pylori* infection: 132; pathogenic *Escherichia coli* infection: 379; shigellosis: 362; leishmaniasis: 131; Chagas disease: 173; malaria: 71; and ameobiasis: 661. This study will also provide a foundation for the research and development of livestock breeding and antiectoparasitic drugs (Burgess et al., 2012; Losson, 2012).

CONCLUSION

Based on mass sequencing data of sheepskin obtained by RNA-Seq, many unigenes were identified and annotated, which provides an excellent platform for future genetic and functional genomic research.

ACKNOWLEDGMENTS

Research supported by the Central Level, Scientific Research Institutes for Basic R&D Special Fund Business (#BRF100102), the Earmarked Fund for Modern China Wool & Cashmere Technology Research System (#nycytx-40-3), the Project of the National Natural Science Foundation for Young Scholars of China (#31402057), and the National High Technology Research and Development Program of China ("863" Program; #2008AA101011-2).

References

[1] Abyzov A, Mariani J, Palejev D, Zhang Y, et al. (2012). Somatic copy number mosaicism in human skin revealed by induced pluripotent stem cells. *Nature* 492: 438-442.

[2] Adelson DL, Cam GR, DeSilva U and Franklin IR (2004). Gene expression in sheep skin and wool (hair). *Genomics* 83: 95-105.

[3] Archibald AL, Cockett NE, Dalrymple BP, Faraut T, et al. (2010). The sheep genome reference sequence: a work in progress. *Anim. Genet.* 41: 449-453.

[4] Birzele F, Schaub J, Rust W, Clemens C, et al. (2010). Into the unknown: expression profiling without genome sequence information in CHO by next generation sequencing. *Nucleic Acids Res.* 38: 3999-4010.

[5] Bond JJ, Wynn PC, Brown GN and Moore GP (1994). Growth of wool follicles in culture. *In Vitro Cell Dev. Biol. Anim.* 30A: 90-98.

[6] Bostjancic E and Glavac D (2008). Importance of microRNAs in skin morphogenesis and diseases. *Acta Dermatovenerol. Alp. Pannonica Adriat.* 17: 95-102.

[7] Botchkarev VA and Paus R (2003). Molecular biology of hair morphogenesis: development and cycling.

*J.Exp.Zool.B Mol.Dev.Evol.*298：164–180.

[8] Burgess ST, Greer A, Frew D, Wells B, et al. (2012).Transcriptomic analysis of circulating leukocytes reveals novel aspects of the host systemic inflammatory response to sheep scab mites.*PLoS One* 7：e42778.

[9] Cadau S, Rosignoli C, Rhetore S, Voegel JJ, et al. (2013).Early stages of hair follicle development：a step by step microarray identity.*Eur.J.Dermatol.*23：4–10.

[10] Cloonan N, Forrest ARR, Kolle G, Gardiner BBA, et al. (2008).Stem cell transcriptome profiling via massive-scale mRNA sequencing.*Nat.Methods* 5：613-619.

[11] Conesa A, Götz S, García-Gómez JM, Terol J, et al. (2005).Blast2GO：a universal tool for annotation, visualization and analysis in functional genomics research.*Bioinformatics* 21：3674–3676.

[12] Dalrymple BP, Kirkness EF, Nefedov M, McWilliam S, et al. (2007).Using comparative genomics to reorder the human genome sequence into a virtual sheep genome.*Genome Biol.*8：R152.

[13] Denoeud F, Aury JM, Da Silva C, Noel B, et al. (2008).Annotating genomes with massive-scale RNA sequencing.*Genome Biol.*9：R175.

[14] Galbraith H (2010a).Foreword.Animal fibre：connecting science and production.*Animal* 4：1447–1450.

[15] Galbraith H (2010b).Fundamental hair follicle biology and fine fibre production in animals.*Animal* 4：1490–1509.

[16] Geng R, Yuan C and Chen Y (2013).Exploring differentially expressed genes by RNA-Seq in cashmere goat (*Capra hircus*) skin during hair follicle development and cycling.*PLoS One* 8：e62704.

[17] Hynd PI, Penno NM and Bates EJ (1999). Follicle morphogenesis *in vitro*. *Exp. Dermatol.* 8：350–351.

[18] Iseli C, Jongeneel CV and Bucher P (1999).ESTScan：a program for detecting, evaluating, and reconstructing potential coding regions in EST sequences.*Proc.Int.Conf.Intell.Syst.Mol.Biol.*138–148.

[19] Jager M, Ott CE, Grünhagen J, Hecht J, et al. (2011).Composite transcriptome assembly of RNA-seq data in a sheep model for delayed bone healing.*BMC Genomics* 12：158.

[20] Li R, Zhu H, Ruan J, Qian W, et al. (2010).*De novo* assembly of human genomes with massively parallel short read sequencing.*Genome Res.*20：265–272.

[21] Liu H, Zhou ZY and Malcolm B (2011).China's wool import demand：implications for Australia.*Australasian Agribusiness Rev.*19：16–34.

[22] Lobo AM, Lobo RN, Paiva SR, de Oliveira SM, et al. (2009).Genetic parameters for growth, reproductive and maternal traits in a multibreed meat sheep population.*Genet.Mol.Biol.*32：761–770.

[23] Losson BJ (2012).Sheep psoroptic mange：an update.*Vet.Parasitol.*189：39–43.

[24] Mc Cloghry E, Foldes A, Hollis D, Rintoul A, et al. (1992). Effects of pinealectomy on wool growth and wool follicle density in merino sheep.*J.Pineal Res.*13：139–144.

[25] Nagalakshmi U, Wang Z, Waern K, Shou C, et al. (2008).The transcriptional landscape of the yeast genome defined by RNA sequencing.*Science* 320：1344–1349.

[26] NCBI Resource Coordinators (2013).Database resources of the National Center for Biotechnology Information.*Nucleic Acids Res.*41：D8–D20.

[27] Norris BJ, Bower NI, Smith WJM and Cam GR (2005).Gene expression profiling of ovine skin and wool follicle development using a combined ovine-bovine skin cDNA microarray.*Aust. J. Exp. Agr.*45：867–877.

[28] Okano J, Levy C, Lichti U, Sun HW, et al. (2012).Cutaneous retinoic acid levels determine hair follicle development and downgrowth.*J.Biol.Chem.*287：39304–39315.

[29] Okoniewski MJ and Miller CJ (2006).Hybridization interactions between probesets in short oligo mi-

croarrays lead to spurious correlations.*BMC Bioinformatics* 7: 276.

[30] Peñagaricano F, Zorrilla P, Naya H, Robello C, et al. (2012).Gene expression analysis identifies new candidate genes associated with the development of black skin spots in Corriedale sheep.*J.Appl. Genet.*53: 99-106.

[31] Proksch E, Brandner JM and Jensen JM (2008).The skin: an indispensable barrier.*Exp.Dermatol.* 17: 1063-1072.

[32] Prost-Squarcioni C, Fraitag S, Heller M and Boehm N (2008).Functional histology of dermis.*Ann. Dermatol.Venereol.*135: 1S5-20.

[33] Rogers GE (2006).Biology of the wool follicle: an excursion into a unique tissue interaction system waiting to be rediscovered.*Exp.Dermatol.*15: 931-949.

[34] Schmidt-Ullrich R and Paus R (2005).Molecular principles of hair follicle induction and morphogenesis.*Bioessays* 27: 247-261.

[35] Shi CY, Yang H, Wei CL, Yu O, et al. (2011).Deep sequencing of the *Camellia sinensis* transcriptome revealed candidate genes for major metabolic pathways of tea-specific compounds.*BMC Genomics* 12: 131.

[36] Smith WJ, Li Y, Ingham A, Collis E, et al. (2010).A genomics-informed, SNP association study reveals FBLN1 and FABP4 as contributing to resistance to fleece rot in Australian Merino sheep.*BMC Vet.Res.*6: 27.

[37] Sun C, Li Y, Wu Q, Luo H, et al. (2010).*De novo* sequencing and analysis of the American ginseng root transcriptome using a GS FLX Titanium platform to discover putative genes involved in ginsenoside biosynthesis.*BMC Genomics* 11: 262.

[38] Wang X, Tredget EE and Wu Y (2012).Dynamic signals for hair follicle development and regeneration.*Stem Cells Dev.*21: 7-18.

[39] Wang XW, Luan JB, Li JM, Bao YY, et al. (2010).*De novo* characterization of a whitefly transcriptome and analysis of its gene expression during development.*BMC Genomics* 11: 400.

[40] Wang Z, Gerstein M and Snyder M (2009).RNA-Seq: a revolutionary tool for transcriptomics.*Nat. Rev.Genet.*10: 57-63.

[41] Ye J, Fang L, Zheng H, Zhang Y, et al. (2006).WEGO: a web tool for plotting GO annotations. *Nucleic Acids Res.*34: W293-W297.

(发表于《Genetics and Molecular Research》, SCI: 0.850)

Novel SNP of *EPAS1* Gene Associated with Higher Hemoglobin Concentration Revealed the Hypoxia Adaptation of Yak (*Bos grunniens*)

WU Xiao-yun, DING Xue-zhi, CHU Min, GUO Xian, BAO Peng-jia, LIANG Chun-nian, YAN Ping

(Key Laboratory for Yak Breeding Engineering of Gansu Province, Lanzhou Institute of Husbandry and Pharmaceutical Sciences, Chinese Academy of Agricultural Sciences, Lanzhou 730050, China)

Abstract: Endothelial PAS domain protein 1 gene (*EPAS1*) is a key transcription factor that activates the expression of oxygen-regulated genes. In this study, in order to better understand the effects of *EPAS1* gene on hematologic parameters in yak, we firstly quantified the tissue expression patterns for *EPAS1* mRNA of yak, identified polymorphism in this gene and evaluated its association with hematologic parameters. Expression of *EPAS1* mRNA was detected in all eight tissues (heart, liver, lung, spleen, pancreas, kidney, muscles and ovary). The expressions of *EPAS1* in lung and pancreas were extremely higher than other tissues examined. Three novel single nucleotide polymorphisms (SNPs) (g. 83052 C>T, g. 83065 G>A and g. 83067 C>A) within the *EPAS1* were identified and genotyped in Pali (PL), Gannan (GN) and Tianzhu White (TZW) yak breeds. Significant higher frequencies of the AA and GA genotypes and A allele of the g. 83065 G>A were observed in the PL and GN breeds than that in the TZW breed ($P<0.01$). Association analysis of the PL breed indicated that the g. 83065 G > A polymorphism was significantly associated with hemoglobin (HGB) concentration in yaks ($P<0.05$). Individuals with genotype AA had significantly higher HGB concentration ($P<0.05$) than those with genotype GA and GG. All these results will help our further understanding of biological functional of yak *EPAS1* gene in responding to hypoxia and also indicate *EPAS1* might contribute to the hypoxia adaptation of the yak.

Key words: *EPAS1* gene; MRNA expression; Polymorphism; Hypoxia adaptation; Yak

1 INTRODUCTION

Yak (*Bos grunniens*) is a unique domesticated animal that lives on the Qinghai-Tibetan plateau at altitudes of 2 5006 000 m. Owing to its adaptation to high-altitude environments and its high level of endurance for work, yak is a model animal for understanding the molecular basis of adaptation to high-altitude hypoxia. Many studies have focused on the singular adaptability of the yak to high altitude, and have identified several factors contributing to this trait: Compared with close relatives such as cattle living at lower altitudes, yak has higher red blood cell (RBC) count and hemoglobin (HGB) concentration, and more smooth muscle in the arteries of the lung (Li et al., 2006; Ma et al., 2011). In addition, the endothelial cells of pulmonary arteries in yak are much longer, wider and rounder than the corresponding cells in the domestic cattle (Wang et al., 2006).

Genomic comparison between yak and cattle has identified candidate genes in hypoxia-inducible factor (HIF) signaling pathway which were suffered positive selection in yak, so the HIF pathway might play a key role in the high-altitude adaptation of yak (Qiu et al., 2012). Endothelial PAS domain protein 1 (EPAS1), also known as hypoxia-inducible factor-2α (HIF-2α), belongs to the HIF pathway. It plays a central role in response to hypoxia (Qing and Simon, 2009). *EPAS1* has been implicated in processes such as erythropoiesis, iron homeostasis, pulmonary hypertension and vascular permeability (Maxwell, 2005; van Patot and Gassmann, 2011). Recent genome-wide analyses have identified *EPAS1* as a target of selection and also detected variants of this gene that were associated with differences in HGB concentration in Tibetans (Beall et al., 2010). *EPAS1* was also identified as a selective target gene and association with blood-related phenotypes in Tibetan mastiffs (Li et al., 2014) and indigenous dogs living at high altitudes (Gou et al., 2014). However, there was no report on the polymorphisms in the *EPAS1* gene and their associations with hematologic parameters in yak. Therefore, the aims of this study were to determine expression pattern of *EPAS1* mRNA in various tissues of yak, and investigate polymorphism in this gene and its association with hematologic parameters.

2 RESULTS

2.1 Expression of *EPAS1* among yak tissues

The results showed that *EPAS1* mRNA was detected in all eight tissues tested. The *EPAS1* mRNA levels of yak were extremely high in the lung, followed by the pancreas, kindey, liver, heart, ovary, spleen and muscle (Fig. 1, Table 1).

Table 1 Expression analysis of *EPAS1* mRNA in different tissues of yak performance by real-time PCR

Tissues	Heart	Liver	Lung	Spleen	Pancreas	Kidney	Muscle	Ovary
Relative expression	0.27±0.08C	0.67±0.09C	15.94±0.31A	0.44±0.18C	2.32±0.99B	0.76±0.08C	0.02±0.01C	0.04±0.03C

Note: Values are expressed as means±standard error. Different letters indicate significant difference ($P<0.01$). The same as below.

2.2 SNP identification and genotyping

We successfully amplified the *EPAS1* gene with those primers and the pooled DNA as the template. After sequencing the fragments, six single nucleotide polymorphisms (SNPs) were revealed in the *EPAS1* from yak (Table 2). Three novel SNPs (g. 83052 C>T, g. 83065 G>A and g. 83067 C>A) were located in intron 8 of *EPAS1* and absent in the bovine *EPAS1*, we concluded that these SNPs were specific to yak. Therefore, only these three SNPs were analyzed. Both allelic and geno-

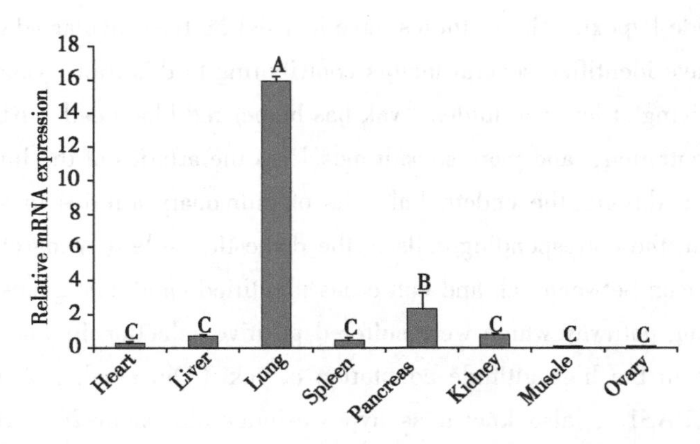

Fig. 1 Expression analysis of *EPAS1* mRNA in different tissues of yak. Values are expressed as means ± standard error. Different letters indicate significant difference (*P*<0. 01)

typic distributions of the g. 83065 G > A were significantly different among the three yak breeds. Significant higher frequencies of the AA and GA genotypes and A allele of the g. 83065 G>A were observed in the Pali (PL) and Gannan (GN) yak breeds than those in the Tianzhu White (TZW) yak breeds (*P*<0. 01) (Table 3).Genotypes of g. 83065 G>A were all in Hardy-Weinberg equilibrium (*P* > 0. 05). No significant differences of allelic and genotypic distributions of the g. 83052 C>T and g. 83067 C>A were observed between the three different breeds. Genotypes of g. 83052 C>T in all yak breed were in agreement with the Hardy-Weinberg equilibrium (*P*>0. 05) except GN breed. Genotypic frequencies of g. 83067 C>A in all yak breeds were in agreement with the Hardy-Weinberg equilibrium (*P*>0. 05) except TZW breed.

Table 2 The SNP information of the *EPAS1* gene in yaks

SNP	Variant type	Variant location
g. 62012	G>A	Intron 4
g. 83052	C>T	Intron 8
g. 83065	G>A	Intron 8
g. 83067	C>A	Intron 8
g. 84949	A>C	Exon 10
g. 84968	C>T	Exon 10

Table 3 Genotypic and allelic frequencies at polymorphic site of *EPAS1* gene among three yak breeds

SNP	Population[1)]	Genotype frequencies (%)			*P* value	Allele frequencies (%)		*P* value	X2 (HWE)
g. 83052 C>T		TT	TC	CC	*P*>0. 05	T	C	*P*>0. 05	
	PL	46. 4	35. 7	17. 9		64. 3	35. 7		*P*>0. 05
	GN	44. 4	37. 4	18. 2		63. 1	36. 9		*P*<0. 05
	TZW	42. 7	42	15. 3		63. 7	36. 3		*P*>0. 05
g. 83065 G>A		GG	GA	AA	*P*<0. 01	G	A	*P*<0. 01	
	PL	48. 2	35. 7A	16. 1A		66. 1	33. 9A		*P*>0. 05

(continued)

SNP	Population[1)]	Genotype frequencies (%)			P value	Allele frequencies (%)		P value	X2 (HWE)
	GN	43.3	40.1A	16.6A		63.4	36.6A		$P>0.05$
	TZW	84	15.3B	0.7B		91.7	8.3B		$P>0.05$
g. 83067 C>A		AA	AC	CC	$P>0.05$	A	C	$P>0.05$	
	PL	44.6	41.1	14.3		65.2	34.8		$P>0.05$
	GN	43.9	40.6	15.5		64.2	35.8		$P>0.05$
	TZW	47.6	38.5	13.9		66.8	33.2		$P<0.05$

Note: [1)] PL, Pali yak breed; GN, Gannan yak breed; TZW, Tianzhu white yak breed. The same as below.

To further investigate the genetic makeup of the three yak breeds, the *EPAS1* SNP data were used for haplotype analysis. This analysis indicated that the three breeds contained six different haplotypes. The frequency of the haplotype CAC was significantly ($P<0.01$) higher in the PL and GN yaks than in the TZW yaks (Table 4). Haplotypes CAA and TGC were absent in TZW yaks. Our results suggested that the haplotypes CAC, CAA and TGC may have a better adaptation ability to the hypoxic environment in plateau.

Table 4 Haplotype distribution of three SNPs in the three yak breeds

Breeds	Haplotye					
	CAA	CAC	CGA	CGC	TGA	TGC
PL	0.036A	0.293A	0.029	0.027A	0.576	0.039A
GN	0.038A	0.290A	0.015	0.029A	0.588	0.040A
TZW	0.000	0.083B	0.031	0.249B	0.637	0.000

2.3 Association analysis

To follow on from the above described results, we conducted a study with PL yaks to investigate whether the different *EPAS1* variant genotypes identified in this study were associated with differences in hematologic parameters. Association analysis revealed that the g. 83065 G>A polymorphism was significantly associated with HGB concentration in yaks (Table 5). Individuals with genotype AA had significantly higher HGB concentration ($P<0.05$) than those with GG and GA genotypes; the RBC count of yak with the AA genotype was also greater than those of yak with the GG and GA genotypes, but the differences were not significant.

Table 5 Associations between single SNP and hematologic parameters in PL yaks

SNPs	Genotypes	Red blood cell count ($10^{12} \cdot L^{-1}$)	Hemoglobin concentration ($g \cdot L^{-1}$)
g. 83052 C>T	TT	4.11±0.90	118.87±7.51
	TC	4.15±0.73	122.22±8.11
	CC	4.63±0.77	119.70±5.03

(continued)

SNPs	Genotypes	Red blood cell count ($10^{12} \cdot L^{-1}$)	Hemoglobin concentration ($g \cdot L^{-1}$)
g. 83065 G>A	GG	4. 31±0. 61	113. 62±6. 08 a
	GA	4. 12±0. 69	17. 66±7. 93 a
	AA	4. 55±0. 72	129. 38±10. 21 b
g. 83067 C>A	AA	4. 32±0. 83	118. 62±7. 33
	AC	4. 52±0. 75	120. 29±8. 71
	CC	4. 13±0. 71	121. 75±7. 68

Note: Values with lowercase letters within the same column in the same locus denote $P<0.05$.

3 DISCUSSION

Hypoxia-sensing gene pathway is the key regulators of erythropoiesis, vasculogenesis, energy metabolism, cardiopulmonary regulation and iron metabolism in mammals (Warnecke et al., 2004; Mastrogiannaki et al., 2009; Kapitsinou et al., 2010; Ge et al., 2012). As a key member of hypoxia-sensing gene pathways, EPAS1 regulates many variety of genes encoding hypoxia-mediated genes, such as erythropoietin (*EPO*), vascular endothelial growth factor (*VEGF*), adrenomedullin (*ADM*) and glucose transporter type 1 (*GLUT-1*), etc. (Xia et al., 2001; Chavez et al., 2006). Under normoxic conditions, EPAS1 is hydroxylated by prolyl hydroxylases (PHDs), polyubiquitinated by the von Hippel-Lindau protein (PVHL) and then undergoes proteosomal degradation. Under hypoxic conditions, EPAS1 translocates into the nucleus and binds to HIF-1β. The resulting HIF dimer then binds to the hypoxic response element (HRE) which locating on hypoxia-mediated genes, promoting their transcription. Therefore, *EPAS1* plays an important role in response to hypoxia.

Understanding the tissue distribution of *EPAS1* may help identifing its impact on organ-specific tolerance to hypoxia. Previous studies concerning the tissue distribution of *EPAS1* were focused on human and mice. Therefore, to our knowledge, this study represents the first attempt to characterize the mRNA expression profile of *EPAS1* in yaks. This study has provided important information regarding mRNA levels in different yak tissues as well as differences between native high-altitude and low-altitude animal genetics. *EPAS1* mRNA was expressed in all the tissues examined. Consistent with previously published reports, *EPAS1* mRNA expression is the highest in the lung which contain plenty blood vessels (Tian et al., 1997). Tissue hypoxia can upregulate a series of local factors that contribute to angiogenesis and the growth of new capillary vessels, which increase delivery of oxygen. *EPAS1* expression in the lung suggests that *EPAS1* maybe lead to neovascularization that is important for lung function.

Significant differences in the allele and genotype frequencies of the g. 83065 G>A polymorphism among the three yak breeds were detected. High-altitude hypoxia is one environmental stress that determines the survival of yak. Research has established that when altitude increases and the partial pressure of oxygen decreases, the oxygen content in the blood of yaks decreases. Tissue hypoxia is thought to upregulate a series of local factors that promote RBC count, HGB concentration and HCT increase to meet bodily oxygen demand. Jiang et al., (1991) have suggested that the number of alveoli, RBC count and HGB concentration are higher in high-altitude yaks than those in yaks living at lower altitude. We speculate that high-altitude yaks have better adaptability in a hypoxic environment. SNP 3065 G>A was specific to yak, cattle was homozygous with G allele at the same locus. The significant difference among the three populations may be caused by the differences in the breeds, but the significantly higher frequency of the A allele of g. 83065 G>A polymorphism in PL and GN yaks, which have lived at high altitude for many generations, is likely the result of a long period of natural selection for environmental adaptation. Our results further indicated that the g. 83065G>A polymorphism in yak is located within the noncoding regions, which is consistent with the observations in a previous study involving Tibetan individuals (Yi et al., 2010). Yak, which was domesticated in China from wild yak about 4 000 years ago, provides food (meat and milk), transport, shelter and fuel allowing Tibetans to live at high altitudes. Yaks and Tibetans both experienced similar environments in the recent past. Hypoxic environment might work on a similar set of genes in the two genomes, such as genes involved in the hypoxia response and energy metabolism (Qiu et al., 2012). We speculated that parallel evolution on the molecular level may extensively exist between yaks and Tibetans. Considerable evidence showed that introns can lead to alternative splicing which influences almost every step in gene expression from transcription to mRNA export, localization, translation and decay (Cheong et al., 2006). Together, these effects might add up to the alteration of the *EPAS1* gene expression profiles, thus influence its function.

Several genome-wide studies investigating hypoxia adaptation in Tibetans have identified SNPs in *EPAS1* gene represent association to HGB concentration (Beall et al., 2010; Simonson et al., 2010; Peng et al., 2011; Xu et al., 2011). This study showed that *EPAS1* gene is associated with hematologic parameters in PL yaks. Genotypes of g. 83065 G>A were significantly associated with increased HGB concentration. The genotype AA had higher HGB concentration, and would impart better adaptation capability in the hypoxic environment of the plateau. This result was in accordance with the function of *EPAS1* gene on HGB concentration and supported our hypothesis that *EPAS1* could be a candidate gene for hypoxia adaptation in yaks. Many function genes, such as hypoxia inducible factor 1 (*HIF-1*) (Wang et al., 2006) and lactate dehydrogenase b (*LDHB*) (Wang et al., 2013) have been reported to contribute to the hypoxia adaptability of yak. However, to our knowledge, this is the first report to investigate the SNP of the *EPAS1* gene and its association of genotypes with hematologic parameters in yaks.

4 CONCLUSION

It is the first report to study the tissue distribution of the *EPAS1* gene and its association of genotypes with hematologic parameters in yaks. *EPAS1* mRNA had the higher expression level in lung and pancreas. The association analysis of single SNP with two hematologic parameters revealed that the g. 83065 G>A was associated with increased HGB concentration. Although our results were based on a limited number of population and further studies in more yak populations with large sample size are needed to confirm our data. However, these results suggested that *EPAS1* gene could be directly or indirectly involved in adaptation mechanism to hypoxia of yak.

5 MATERIALS AND METHODS

5.1 DNA samples and data source

The animals used in this work were from three yak breeds living at various altitudes areas on the Qinghai-Tibetan Plateau of Yadong County (altitude, 4 500m), Luqu Country (altitude, 3 500 m) and Tianzhu County (altitude, 2 700m), China (Fig. 2). These breeds included Pali (PL; $n=56$), Gannan (GN; $n=187$) and Tianzhu White (TZW; $n=288$). The PL and DT breeds live at higher altitude, whereas the TZW breed live at lower altitude. Blood samples were collected and stored at −20℃. The 56 PL blood samples in eight batches were used for the association study. RBC count and HGB concentration were analyzed using the automated hematology analyzer (PERLONG, Beijing, China). Different tissues including heart, liver, lung, spleen, pancreas, kidney, muscles and ovary were collected from eight healthy adult GN yaks. Fresh tissues were immediately frozen in liquid nitrogen until analysis. All yaks were grazed on natural pasture without feeding supplementation. The whole procedure for collection of bloods and the slaughter of yaks was carried out in strict accordance with the recommendations in the Guide for the Care and Use of Laboratory Animals of the National Institutes of Health, USA.

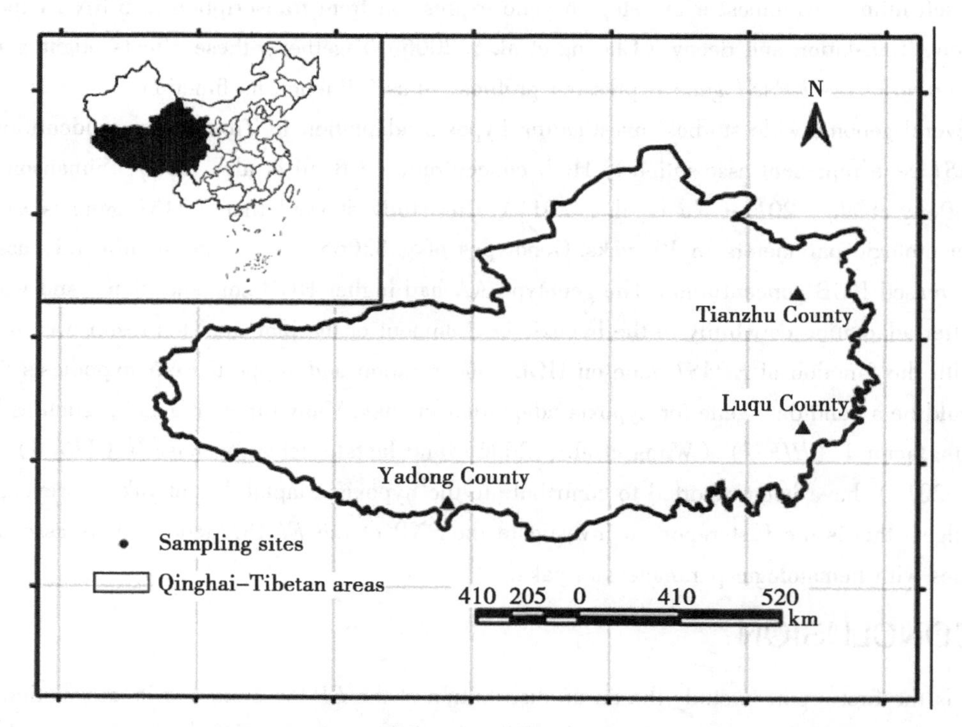

Fig. 2 Geographic locations of the three yak breeds (Pali, Gannan and Tianzhu White)

5.2 Total RNA extraction and DNA isolation

Total RNA was extracted from yak tissues using the TRIzol reagent (Invitrogen, Carlsbad, Calif, USA) according to the manufacturer's instructions. The electrophoresis on ethidium bromide-stained 1. 5% agarose gels was applied to check RNA integrity. The complementary DNA (cDNA) was synthesized from 1μg total RNA by PrimeScript™ RT Master Mix (TaKaRa Biotechnology Co. Ltd., Dalian, China) following the manufacturer's protocol. Blood genomic DNA was extracted with a Genomic DNA Isolation Kit (Tiangene, Beijing, China) according to the manufacturer's instructions.

5. 3 Tissue expression profile analysis

The *EPAS1* mRNA expression levels in different tissues were determined by real-time RT-PCR. The *GAPDH* was used as the control housekeeping gene. Real-time PCR was performed in a 25μL reaction agent comprising 12. 5μL of SYBR® premix *Ex Taq*™ (TaKaRa Biotechnology Co. Ltd., Dalian, China), 1μL of each forward and reverse primer ($10\mu mol\ L^{-1}$) (Table 6), 1μL cDNA template and 9. 5μL ddH_2O. PCR was performed using the following program: 95℃ for 30s, 40 cycles of 95℃ for 30 s, and 60℃ for 30s, and then the melting curve analysis was carried out to determine the specificity of PCR products. Three PCR reactions were performed for each sample and then averaged. The $2^{-\Delta\Delta CT}$ method was used to evaluate the fold changes of *EPAS1* expression levels.

Table 6 Primer pairs used in this study

Primer name	Primer sequence (5′→3′)	Annealing temperature (℃)	Amplified region (bp)
Primer pairs for measuring yak *EPAS*1 gene expression			
EPAS1-S	CGTGGTGACCCAAGATGGTG	60	573-691
EPAS1-A	GGTCACAGGGATGAGTGAAGTCAA		
GAPDH-S	CCACGAGAAGTATAACAACACC	60	422-542
GAPDH-A	GTCATAAGTCCCTCCACGAT		
Primer pairs for screening yak *EPAS1* gene polymorphisms			
E1-S	GCGTTTGACAGCGTCCTTC	61	1-415
E1-A	AAATCGTCCCTGGGTCCTC		
E2-S	TGATGTTCTGGAGACTAACCCCT	61	53 067-53 439
E2-A	GGTGGAAGGTGTGCAGAGTG		
E3-S	TGTCCCGACGACTCCTGTAAC′	60	61 984-62 357
E3-A	CACGGCAATGAAACCCTCC		
E4-S	CTGTGTGTCTATGGATCTGCCG	62	62 691-63 053
E4-A	CTATGGGTGTCTGTTTTTTCTGG		
E5/6-S	ACACACAGACCACACACACGGA	62	66 177-66 828
E5/6-A	AGCACTGGTGGGATGAAACTGT		
E7-S	GAATGTGTTTCCTTGAATGTGTGC	58	75 520-75 875

(continued)

Primer name	Primer sequence (5′→3′)	Annealing temperature (℃)	Amplified region (bp)
E7-A	TGAGGACATGAGTCTTGTATGCC		
E8-S	GCTAACCTCTCTGGACCACCTTC	64	82 753–83 087
E8-A	TGTCAGCCTCAGGAACACCC		
E9-S	GGTTCTGATGGAAAATGTGGAG	60	83 630–83 998
E9-A	GATGATTAGCAAGACTCAGAGAAGG		
E10-S	GACCCTTTTAGGAACTCTTGTGC	64	84 843–85 247
E10-A	GAAACCCGAGGAAGAGGAGC		
E11-S	AGCCTTAACTTGGTCACCTTCC	64	85 625–85 993
E11-A	GGATTTGACACGGGTAAAGCAG		
E12-S	ACAGCGTCCTTGAGGCAGTG	64	87 020–87 329
E12-A	CTGCTCAGGCTCCGTCTTCT		
E13/14-S	TCTCACGCCCTCTCTCTCTG	60	88 501–89 208
E13/14-A	TACCCCTCACGGGATTCT		
E15-S	ATGAGTTCATCCAGAATCCCGT	64	89 179–89 450
E15-A	CCTTGTGCTGAGATTTGCCA		
E16-S	GGAGTCAGGAAAACTGGAAGC	64	91 249–91 449
E16-A	ACAGTCATATCTGGTCAGTTCCG		

5.4 Polymorphisms detection

To identify single-nucleotide polymorphisms (SNP) variations in the sequence of the *EPAS1* gene, DNA poolsequencing was used in the present study. First, three pools were constructed with each pool containing 50 individual DNA samples that were randomly chosen from each yak breed. Then, using these pools as PCR templates, amplifications were carried out. The *EPAS1* gene of yak is consist of 16 exons and 15 introns. Based on the nucleotide sequence of the bovine *EPAS1* gene (GenBankaccession no. NC_0073091), we designed 16 primer pairs to amplify the segments of the *EPAS1* exon regions including exon-intron boundaries (Table 6).All PCR fragments were purified with a Gel Extraction Mini Kit (Tiangene, Beijing, China) and then sequenced in both forward and reverse directions by using an ABI 3 730 DNA analyzer (Applied Biosystems, Foster City,CA, USA).According to the sequencing maps, variations of the sequences can be easily identified. In this study, all 531 samples from three yak breeds were genotyped by directly sequencing method of PCR products respectively.

5.5 Statistical analysis

Allele and genotype frequencies among groups were compared using the chi-square test. SHEsis was used to assess Hardy-Weinberg equilibrium (HWE) and haplotypes for each pair of segregating sites (Yong and Lin 2005). An association test between the genotypes of SNPs and two hematologic parameters was performed by SAS software (ver. 9. 1. 3), based on the following mixed model:

$$Y_{ijkl}=\mu+G_i+S_k+B_j+e_{ijkl}$$

Where, Y_{ijkl} is the phenotypic value of the target trait, μ is the population mean, G_i is the effect of the ith genotypes, S_k is the effect of kth sex, gender, B_j is the effect of the jth batch, and e_{ijkl} is the random residual (Tao et al., 2013).

ACKNOWLEDGEMENTS

Research supported by the Special Fund for Agro-scientific Research in the Public Interest, China (201003061), the Key Technologies R&D Program of China during the 12th Five-Year Plan period (2012BAD13B05), the Great Project of Science and Technology of Gansu Province in China (1102NKDA027), and the National Natural Science Foundation of China (31101702).

References

[1] Beall C M, Cavalleri G L, Deng L, Elston R C, Gao Y, Knight J, Li C, Li J C, Liang Y, McCormack M, Montgomery H E, Pan H, Robbins P A, Shianna K V, Tam S C, Tsering N, Veeramah K R, Wang W, Wangdui P, Weale M E, et al., 2010. Natural selection on *EPAS*1 (*HIF*2α) associated with low hemoglobin concentration in Tibetan highlanders. *Proceedings of the National Academy of Sciences of the United States of America*, 107, 11459-11464.

[2] Chavez J C, Baranova O, Lin J, Pichiule P. 2006. The transcriptional activator hypoxia inducible factor 2 (HIF-2/EPAS-1) regulates the oxygen-dependent expression of erythropoietin in cortical astrocytes. *The Journal of Neuroscience*, 26, 9471-9481.

[3] Cheong H S, Yoon D H, Kim L H, Park B L, Choi Y H, Chung E R, Cho Y M, Park E W, Cheong I C, Oh S J, Yi S G, Park T, Shin H D. 2006. *Growth hormone-releasing hormone* (*GHRH*) polymorphisms associated with carcass traits of meat in Korean cattle. *BMC Genetics*, 7, 35.

[4] Ge R L, Simonson T S, Cooksey R C, Tanna U, Qin G, Huff C D, Witherspoon D J, Xing J, Bai Z Z, Prchal J T, Jorde L B, McClain D A. 2012. Metabolic insight into mechanisms of high-altitude adaptation in Tibetans. *Molecular Genetics and Metabolism*, 106, 244-247.

[5] Gou X, Wang Z, Li N, Qiu F, Xu Z, Yan D, Yang S, Jia J, Kong X, Wei Z, Lu S, Lian L, Wu C, Wang X, Li G, Ma T, Jiang Q, Zhao X, Yang J, Liu B, et al., 2014. Whole-genome sequencing of six dog breeds from continuous altitudes reveals adaptation to high-altitude hypoxia. *Genome Research*, doi: 10. 1101/gr. 171876. 113.

[6] Jiang J C, Gama R Z, He M L. 1991. Comparison of the tibetan plateau yak blood physiology value from altitude. *Acta Ecologiae Animalis Domastic*, 22, 20-26.

[7] Kapitsinou P P, Liu Q, Unger T L, Rha J, Davidoff O, Keith B, Epstein J A, Moores S L, Erickson-Miller C L, Haase V H. 2010. Hepatic HIF-2 regulates erythropoietic responses to hypoxia in renal anemia. *Blood*, 116, 3039-3048.

[8] Li L, Shen M, Yu H. 2006. Significance and determination of RBC, Hb and Mb in yak of various ages. *Acta Ecologiae Animalis Domastici*, 27, 51-54. (in Chinese)

[9] Li Y, Wu D D, Boyko A R, Wang G D, Wu S F, Irwin D M, Zhang Y P. 2014. Population variation revealed high-altitude adaptation of Tibetan mastiffs. *Molecular Biology and Evolution*, 31, 1200-1205.

[10] Ma X, Cui Y, He J, Yang B. 2011. Observation of the histological structure of adult yak bronchial arteries. *Chinese Veterinary Science*, 39, 261-265. (in Chinese)

[11] Mastrogiannaki M, Matak P, Keith B, Simon M C, Vaulont S, Peyssonnaux C. 2009. HIF-2α, but not HIF-1α, promotes iron absorption in mice. *Journal of Clinical Investigation*, 119, 1159-1166.

[12] Maxwell P H. 2005. Hypoxia-inducible factor as a physiological regulator. *Experimental Physiology*, 90, 791-797.

[13] Van Patot M C, Gassmann M. 2011. Hypoxia: Adapting to high altitude by mutating *EPAS-1*, the gene encoding HIF-2α. *High Altitude Medicine & Biology*, 12, 157-167.

[14] Peng Y, Yang Z H, Zhang H, Cui C Y, Qi X B, Luo X J, Tao X A, Wu T Y, Ouzhuluobu, Basang, Ciwangsangbu, Danzengduojie, Chen H, Shi H, Su B. 2011. Genetic variations in Tibetan populations and high-altitude adaptation at the Himalayas. *Molecular Biology and Evolution*, 28, 1075-1081.

[15] Qing G, Simon M C. 2009. Hypoxia inducible factor-2α: a critical mediator of aggressive tumor phenotypes. *Current Opinion in Genetics & Development*, 19, 60-66.

[16] Qiu Q, Zhang G, Ma T, Qian W, Wang J, Ye Z, Cao C, Hu Q, Kim J, Larkin D M, Auvil L, Capitanu B, Ma J, Lewin H A, Qian X, Lang Y, Zhou R, Wang L, Wang K, Xia J, et al., 2012. The yak genome and adaptation to life at high altitude. *Nature Genetics*, 44, 946-949.

[17] Simonson T S, Yang Y, Huff C D, Yun H, Qin G, Witherspoon D J, Bai Z, Lorenzo F R, Xing J, Jorde L B, Prchal J T, Ge R. 2010. Genetic evidence for high-altitude adaptation in Tibet. *Science*, 329, 72-75.

[18] Tao C, Wang W, Zhou P, Xia T, Zhou X, Zeng C, Zhang Q, Liu B. 2013. Molecular characterization, expression profiles, and association analysis with hematologic parameters of the porcine *HPSE* and *HPSE*2 genes. *Journal of Applied Genetics*, 54, 71-78.

[19] Tian H, Mcknight S L, Russell D W. 1997. Endothelial PAS domain protein 1 (EPAS1), a transcription factor selectively expressed in endothelial cells. *Genes & Development*, 11, 72-82.

[20] Wang D P, Li H G, Li Y J, Guo S C, Yang J, Qi D L, Jin C, Zhao X Q. 2006. *Hypoxia-inducible factor* 1α cDNA cloning and its mRNA and protein tissue specific expression in domestic yak (*Bos grunniens*) from Qinghai-Tibetan plateau. *Biochemical and Biophysical Research Communications*, 348, 310-319.

[21] Wang G, Zhao X, Zhong J, Cao M, He Q, Liu Z, Lin Y, Xu Y, Zheng Y. 2013. Cloning and polymorphisms of yak *lactate dehydrogenase b* gene. *International Journal of Molecular Sciences*, 14, 11994-12003.

[22] Warnecke C, Zaborowska Z, Kurreck J, Erdmann V A, Frei U, Wiesener M, Eckardt K U. 2004. Differentiating the functional role of hypoxia-inducible factor (HIF) -1alpha and HIF-2alpha (EPAS-1) by the use of RNA interference: Erythropoietin is a HIF-2alpha target gene in Hep3B and Kelly cells. *FASEB Journal*, 18, 1462-1464.

[23] Xia G, Kageyama Y, Hayashi T, Kawakami S, Yoshida M, Kihara K. 2001. Regulation of vascular endothelial growth factor transcription by endothelial PAS domain protein 1 (EPAS1) and possible involvement of EPAS1 in the angiogenesis of renal cell carcinoma. *Cancer*, 91, 1429-1436.

[24] Xu S, Li S, Yang Y, Tan J, Lou H, Jin W, Yang L, Pan X, Wang J, Shen Y, Wu B, Wang H, Jin L. 2011. A genome-wide search for signals of high-altitude adaptation in Tibetans. *Molecular Biology and Evolution*, 28, 1003-1011.

[25] Yi X, Liang Y, Huerta-Sanchez E, Jin X, Cuo Z X, Pool J E, Xu X, Jiang H, Vinckenbosch N, Korneliussen T S, Zheng H, Liu T, He W, Li K, Luo R, Nie X, Wu H, Zhao M, Cao H, Zou J, et al., 2010. Sequencing of 50 human exomes reveals adaptation to high altitude. *Science*, 329, 75-78.

[26] Yong Y, Lin H. 2005. SHEsis, a powerful software platform for analyses of linkage disequilibrium, haplotype construction, and genetic association at polymorphism loci. *Cell Research*, 15, 97-98.

(Managing editor ZHANG Juan)

(发表于《Journal of Integrative Agriculture》，院选 SCI，IF：0.833)

High Gene Flows Promote Close Genetic Relationship among Fine-wool Sheep Populations (*Ovis aries*) in China

HAN Ji-long[1], YANG Min[2], GUO Ting-ting[1], LIU Jian-bin[1],
NIU Chun-e[1], YUAN Chao, YUE Yao-jing[1], YANG Bo-hui[1]

(1. Lanzhou Institute of Husbandry and Pharmaceutical Sciences, Chinese Academy of Agricultural Sciences, Lanzhou 730050, China; 2. Institute of Animal Science, Chinese Academy of Agricultural Sciences, Beijing 100193, China)

Abstract: The aim of our present study was to construct genetic structure and relationships among Chinese fine-wool sheep breeds. 46 individuals from 25 breeds or strains were genotyped based on the Illumina *Ovine* 50K SNP array. Meanwhile, genetic variations among 482 individuals from 9 populations were genotyped with 10 microsatellites. In this study, we found high genetic polymorphisms for the microsatellites, while 7 loci in the Chinese superfine Merino strain (Xinjiang Types) (CMS) and 5 loci in Gansu alpine superfine-wool sheep strain (GSS) groups were found deviated from HWE. Genetic drift $F_{ST}=0.019$ ($P<0.001$) and high gene flows were detected in all the 7 fine-wool sheep populations. Phylogenetic analysis showed fine-wool sheep populations were clustered in a group independent from the Chinese indigenous breeds such that the 7 fine-wool sheep clustered distinct from Liangshan semi-fine wool sheep (LS) and Hu Sheep (HY) reflected by different population differentiation analyses. Overall, our findings suggested that all fine-wool sheep populations have close genetic relationship, which is consistent with their breeding progress. These populations, therefore, can be regarded as open-breeding populations with high levels of gene flows. Furthermore, the two superfine-wool strains, *viz.* CMS and GSS, might be formed by strong artificial selection and with frequent introduction of Australian Merino. Our results can assist in breeding of superfine-wool sheep and provide guidance for the cultivation of new fine-wool sheep breeds with different breeding objectives.

Key words: Chinese fine-wool sheep; indigenous sheep breeds; genetic relationship; gene flow; microsatellites

1 INTRODUCTION

Merino is a wool-type sheep which is certainly one of the most widely used breeds in upgrading programs worldwide by Food and Agriculture Organization of the United Nations (FAO 2007). The recent statistics shows that this fine-wool sheep alone contributes more than 50 percent of the world's sheep population. Merino wool is popular as a natural fiber which provides excellent breathability. Wool quality traits, such as fiber diameter, length and strength, are important criteria in Merino breeding programs throughout the world. Chinese fine-wool sheep were derived by political and market demand in recent decades by crossing indigenous Chinese sheep as female parent with European and/or Australian Merino and finally accompanied by the progressive hybridization with Australian Merino after a complex breeding progress. In China, fine-wool sheep industry is fast growing and contributes more than one third of total world wool production (FAO 2011-2012). Furthermore, China has also become the largest wool textile and garment producing and exporting country. In order to fulfill the high demand of fine-wool, more quantity of wool with better quality is required to be produced in the country itself. These fine-wool sheep is adapting well and produce more wool and meat in the difficult situation brought about by various climatic condition of the country such as scarcity of arable land and pasture deterioration along with the dry and cold climate of northern China which were once considered as immense hindrances for this industry.

National Fine-Wool Sheep Association Breeding Program was formed in China when Australian Merino was introduced in country. This program used best linear unbiased prediction (BLUP) statistical software for selection of breeding stock especially for superior rams. Moreover, embryo transfer and artificial insemination played a vital role to increase the population size of high-quality Chinese fine-wool sheep. Even though Australian Merino has very important role in upgrading Chinese fine-wool sheep breeds, indigenous sheep might lose their advanced genes, such as disease resistance and adaptability in the local environment, after frequent introgression with Australian Merino. However, the second national survey of the genetic resources of livestock and poultry in China revealed that indigenous pure breeds have been reduced drastically due to substitution or crossbreeding with commercial breeds. The loss of diversity in livestock species has important economic, ecological, and scientific implications, as well as social considerations (Lancioni et al., 2013). Genetic characterization of breeds is the pre-requisite for genetic conservation programs and developing breeding strategies (Yilmaz et al., 2014). Understanding genetic diversity and population structure of those populations is essential for fine-wool sheep breeding. Genomic variation such as mitochondrial genome mutations and microsatellites were always being used to study the genetic history and genetic diversity of domestic animals (Groeneveld et al., 2010). Recently, many molecular researches have been performed placing emphasis on wool and growth performance indexes (Zhang et al., 2013; Liu et al., 2014; Wang et al., 2014). However, very little work has been performed on the adaptation of these fine-wool sheep in different climates. One region near TRPM8 gene detected which might be related to the adaptation to cold climate (Fariello et al., 2014) and an allele OAR22_ 18929579-A was observed to be correlated with climatic variation by genome scan for selection based on the Sheep HapMap dataset (Lv et al., 2014). Nevertheless, almost no study was on the adaptability of those Chinese fine-wool sheep. Very little is known on the genetic diversity and population structure of those populations, especially the heredity contribution of Australian Merino to Chinese fine-wool sheep.

In our present study, we focus to characterize the levels of genetic diversity, phylogenetic relationships and patterns of Australian Merino introgression and admixture among the 7 main fine-wool sheep populations with their elaborate crossbreeding history. Therefore, the genome wide *Ovine* 50K SNP array and 10 microsatellites have been used to analyze the population structure and relationship between Chinese fine-wool sheep and indigenous breeds.

2 RESULTS

To investigate the population structure, we have firstly constructed a neighbor-joining (NJ) tree basing on genome-wide allele sharing among 46 sheep from 25 populations (Fig. 1). The NJ tree separated out fine-wool sheep from indigenous Chinese sheep. For better understanding the population structure, we further performed a principal component analysis (PCA) using a subset of 47816 SNPs in all selected breeds/strains of sheep. The analysis showed two principal components (Fig. 2) that revealed different patterns between Chinese indigenous breeds and fine-wool sheep. This finding was consistent to NJ tree results. Furthermore, four Mongolian lineage sheep (Mongolian, Hu sheep, Small-Tailed Han and Tan sheep) clustered together; Thirteen Tibetan lineages sheep from the Qinghai Tibet Plateau (Minxian black fur, Guide black fur, Ganjia, Tianjun, Qilian, Qiaoke, Awang, Duoma, Oula, Gangba, Jiangzi, Langkazi, Huoba sheep) were clustered in the same branch, and among these, only Minxian black fur sheep showed close genetic relationship with Mongolian lineage sheep. The NJ-tree showed clearly that all seven main fine-wool sheep populations clustered together distinct from other indigenous breeds. This result was similar to that of the PCA analysis. We also found Kazakh sheep were in a separated branch close with Mongolian lineage sheep. In our study, we found four groups, namely: Kazakh, Tibetan, Mongolian and fine-wool sheep and the genetic relationships among these populations are consistent with their breeding progress. This result was well-supported by the traditional classification (Zheng 1988).

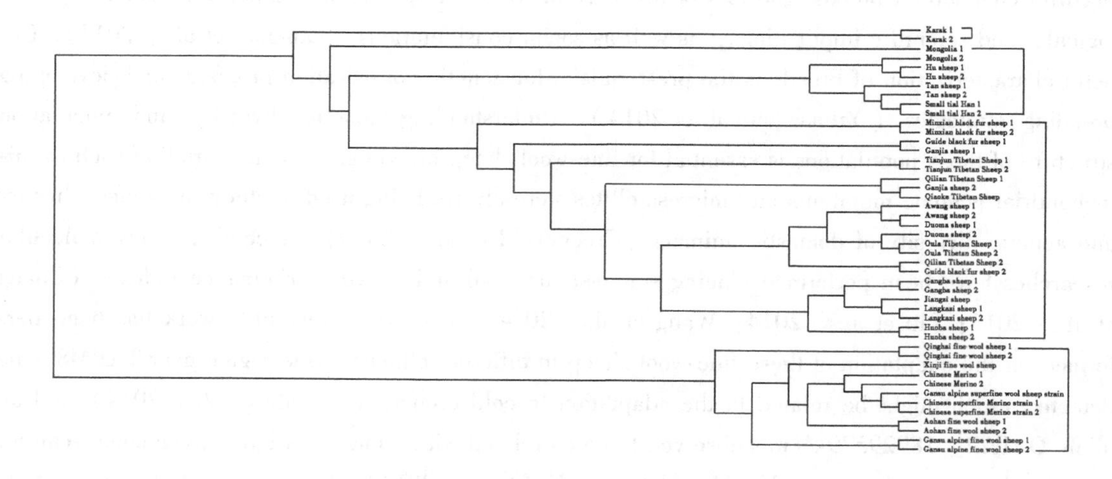

Fig. 1 Neighbour-joining trees of Chinese fine-wool sheep and indigenous sheep breeds

Note: Four branches showed from up to down: Kazak, Mongolia, Tibetan lineages sheep and fine-wool sheep

A total of 153 alleles were observed from the 10 microsatellites in the 482 genotyped individu-

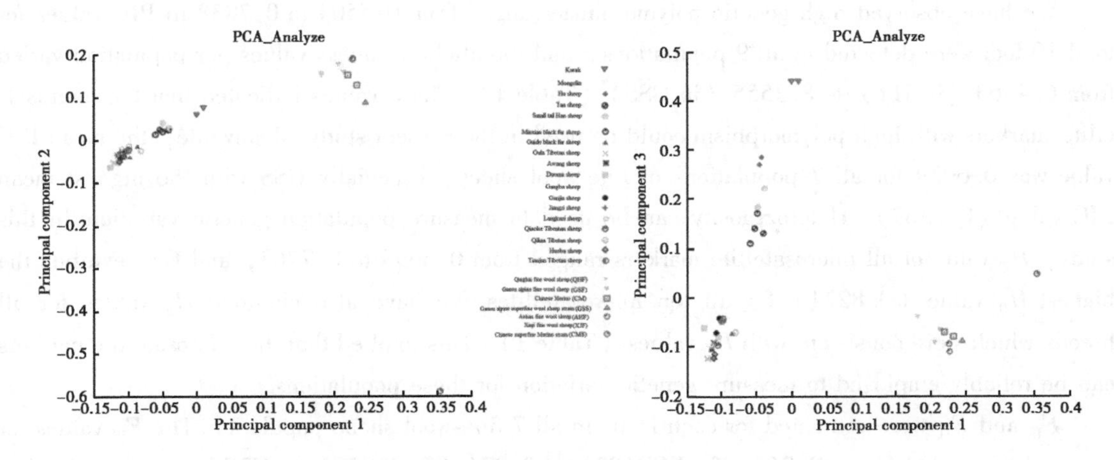

Fig. 2 Population structures of Chinese fine-wool sheep and indigenous sheep breeds were revealed by principal component analysis

als. The alleles number of each locus were from CP34 (7) to CB226 (24). There was no linkage disequilibrium between all the 10 loci at 95% confidence intervals which illustrated that the 10 sites with independent inheritance. Two superfine-wool sheep, viz., CMS and GSS, were with high mean number of alleles per locus which were 10.50 and 9.60 respectively whereas LS and HY had the lowest mean numbers of alleles per locus which were respectively 6.80 and 6.90 respectively (Table 1). Hardy-Weinberg equilibrium (HWE) for each locus and population were observed for 10 loci (Table 1). Consistent to our expectation, significant deviations from HWE in all the ten loci were observed in superfine-wool sheep. When all the breeds were combined to detect the loci deviated from HWE, 7 loci in the CMS group and 5 loci GSS were found.

Table 1 Genetic diversity parameters estimated in the 9 populations

Populations	Abbreviation	Sample sizes	MNA	AR	H_E	H_O	Mean PIC	HWE
Chinese Merino (Xinjiang types)	CM	73	9.10	7.2187	0.6993	0.7527	0.6603	4
Chinese superfine Merino strain (Xinjiang Types)	CMS	113	10.50	7.7555	0.7511	0.7864	0.7182	7
Gansu alpine fine-wool sheep	GSF	32	7.00	6.6659	0.7343	0.7645	0.6855	1
Gansu alpine superfine-wool sheep strain	GSS	43	9.60	8.3555	0.7763	0.8224	0.7369	5
Hu Sheep	HY	33	6.90	6.4363	0.7088	0.7889	0.6573	3
Aohan fine-wool sheep	AHF	78	8.60	6.599	0.7005	0.7444	0.6547	4
Qinghai fine-wool sheep	QHF	40	8.80	7.9545	0.7694	0.8179	0.7298	4
Liangshan semi-fine wool sheep	LS	30	6.80	6.5143	0.7126	0.7470	0.6598	4
Xinji fine-wool sheep	XJF	40	8.50	8.2705	0.7162	0.7999	0.6712	0

Noet: Parameters estimated using 10 microsatellites in 9 populations. *MNA* mean number of alleles/locus, *AR* allelic richness, H_E mean expected heterozygosity, H_O mean observed heterozygosity, *PIC* polymorphic information content, *HWE* number of loci with Hardy-Weinberg equilibrium deviations.

We have observed high genetic polymorphism ranged from 0.5504 to 0.7838 in PIC values for total 10 loci were detected in all 9 populations, and the allelic richness values per population varied from 6.4363 (in HY) to 8.3555 (in GSS) (Table 1). These values indicated that the microsatellite markers with high polymorphism could be used in the present study. Meanwhile, the mean PIC value was 0.6939 for all 7 populations of fine-wool sheep, especially GSS with the highest mean PIC value (0.7369). Heterozygosity can be used to measure population genetic variation. In this study, H_E value of all microsatellite markers ranged from 0.6693 to 0.7763, and GSS exhibits the highest H_O value (0.8224) for all ten microsatellites. We have also obtained H_E values for all breeds which were consistent with H_O values (Table 1). This implied that the microsatellite markers can be reliably employed to measure genetic variation for these populations.

F_{IS} and F_{IT} were estimated for each locus in all 7 fine-wool sheep population. The F_{IS} values for the markers ranged from −0.264 (for FCB193) to 0.074 (for ILSTS11) (Table 2), showing low level of inbreeding in all 7 fine-wool sheep populations, and F_{IT} values ranged from −0.239 (for FCB193) to 0.064 (for JMP58) (Table 2). Correspondingly, levels of apparent breed differentiation were considerable with F_{ST} values. The least pairwise F_{ST} value was 0.0079 between CM and XJF. Mean F_{ST} (0.019) of all 7 fine-wool sheep at the 95% confidence intervals indicated that 98.1% genetic diversity caused by genetic variation within breed. The value of G_{ST} for all 10 loci of 7 fine-wool sheep was 0.017 (Table 2), which suggested that low genetic diversity among breeds and little genetic variation resulted from the differences between the populations. Furthermore, we found high gene flow between fine-wool sheep population and all the N_m values were greater than 8 (Table 3). The highest gene flow was estimated for the CM-XJF pair (N_m = 31.30). Meanwhile, the greatest pairwise F_{ST} estimates were 0.0302 between AHF and QHF, and the lowest gene flow was detected for this pair (N_m=8.02).

Table 2 The results from F-statistics analysis of 7 fine-wool sheep populations

Loci	F_{IT}	F_{ST}	F_{IS}	G_{ST}
MCM140	−0.183*	0.008**	−0.193*	0.009
FCB193	−0.239*	0.020**	−0.264*	0.014
MAF214	0.026*	0.015**	0.011*	0.019
JMP58	0.064*	0.020*	0.044*	0.023
CP34	0.012*	0.016**	−0.004*	0.015
CB226	0.000*	0.035*	−0.036*	0.022
VH72	−0.113*	0.019**	−0.135*	0.015
ILSTS11	−0.100*	0.028**	0.074*	0.018
SPCPSP9	−0.053*	0.034*	−0.090*	0.020
FCB304	−0.057*	0.018*	−0.078*	0.017
ALL	−0.045*	0.019**	−0.066*	0.017
95% confidence intervals	−0.112*	0.016*	−0.131**	

Noet: * Significant P value<0.05, ** Extremely significant P value<0.01.

Table 3 Genetic differentiation parameters estimated in the 9 sheep populations

Group	CM	CMS	GSF	GSS	HY	AHF	QHF	LS	XJF
CM	—	25. 20	10. 44	21. 46	3. 25	11. 15	10. 32	4. 34	31. 30
CMS	0. 0084	—	10. 90	19. 85	3. 56	9. 14	13. 93	4. 85	25. 44
GSF	0. 0234	0. 0224	—	19. 85	4. 15	9. 36	13. 61	5. 25	13. 30
GSS	0. 0115	0. 0124	0. 0124	—	4. 85	14. 54	20. 63	6. 54	18. 55
HY	0. 0713	0. 0656	0. 0569	0. 0490	—	3. 71	3. 05	3. 82	3. 52
AHF	0. 0219	0. 0266	0. 0260	0. 0169	0. 0632	—	8. 02	5. 33	11. 79
QHF	0. 0236	0. 0176	0. 0180	0. 0120	0. 0757	0. 0302	—	3. 95	17. 15
LS	0. 0544	0. 0489	0. 0385	0. 0368	0. 0625	0. 0449	0. 0595	—	4. 28
XJF	0. 0079	0. 0097	0. 0185	0. 0133	0. 0663	0. 0208	0. 0144	0. 0552	—

Noet: Pairwise genetic distances (F_{ST}), and number of effective migrants per generation (Nm) are presented below and above the diagonal, respectively. Population abbreviations are as reported in Table 1.

As expected, small pairwise distances were observed among 7 fine-wool sheep and the highest genetic distances were observed between HY and fine-wool sheep (Appendix A). Five populations (CMS, CM, XJF, AHF and GSS), had a small pairwise distance whereas GSF stays close with QHF. The trees were constructed with UPGMA, using the classical genetic distance (DA)

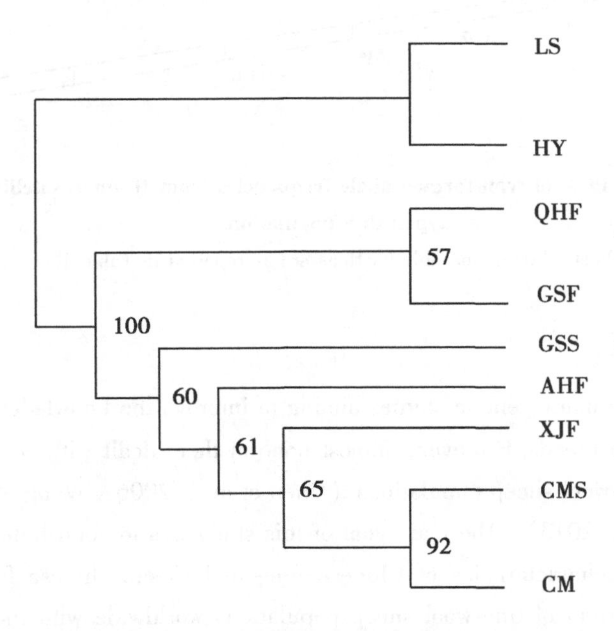

Fig. 3 UPGMA tree of genetic relationships among 9 populations using DA genetics distances

Noet: The numbers on nodes are percentage bootstrap values in 1000 replications. Population abbreviations are as reported in Table 1.

(Fig. 3). The tree suggested two main groups such as fine-wool sheep where 7 fine-wool sheep clustered close together, and rest where other two breeds namely LS and HY, made a separate branch. The PCA results stayed consistent with the topology of the phylogenetic trees, which was constructed to examine the intergroup relationships. The first three principal coordinates in our analysis explained 75. 98% of the total variation. The first divided the populations into two major groups, LS and HY far away from 7 fine-wool sheep (Fig. 4). In order to confirm the findings of PCA, population structure analysis was performed. When $K=5$, the diagram clearly exhibited a great genetic background between fine-wool sheep breeds and rest (LS and HY), among which CMS and AHF showed a little different component content (Fig. 5).

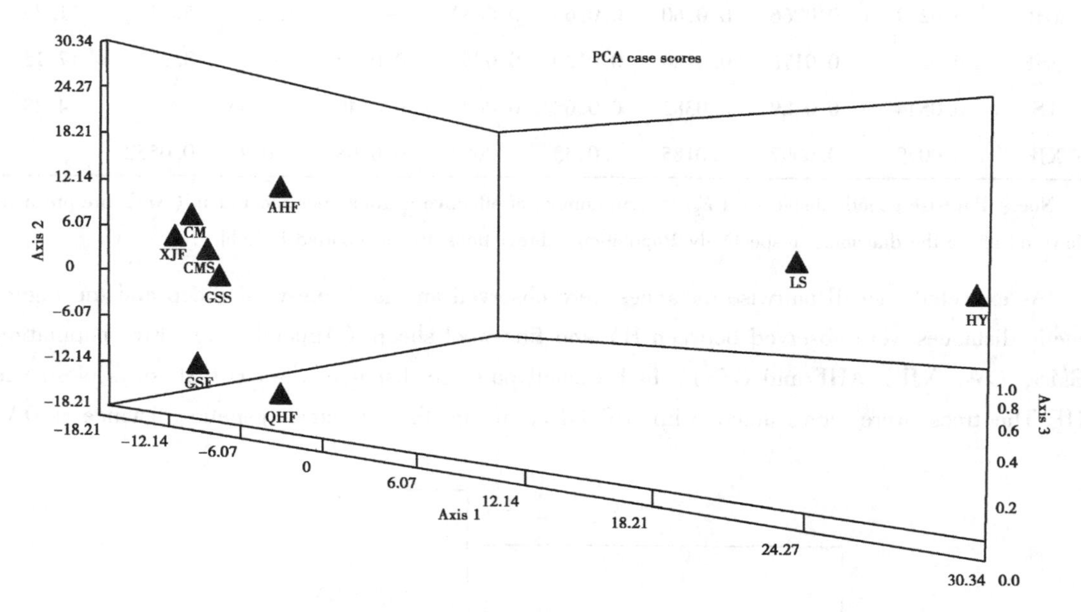

Fig. 4 PCA of transformed allele frequencies from 10 microsatellites typed in 9 populations

Noet: Population abbreviations are as reported in Table 1.

3 DISCUSSION

Currently, there are many genetic studies aiming to improve the knowledge of the genetic composition of Chinese sheep breeds. However, almost none of them dealt with comprehensive and systematic study on the fine-wool sheep populations (Chen et al., 2006; Wang et al., 2006; Niu et al., 2012; Zhao et al., 2013). The main goal of this study was to contribute to the knowledge of population structure and characteristics of Chinese fine-wool sheep. Chinese fine-wool sheep have constituted a large proportion of fine-wool sheep populations worldwide with the increasing demand for fine-wool. Even though Chinese fine-wool sheep breeds are found in different geographical distribution and exhibit different phenotypic traits, but still phenotypically similar to Merino, so they can also be called as Merino-typed breeds. When tracing back the ancestry history of these breeds, all were having same breeding background crossed between Australian Merino and Chinese indigenous breeds. Nonetheless, each breed or strain with special phenotypic character for the geographical iso-

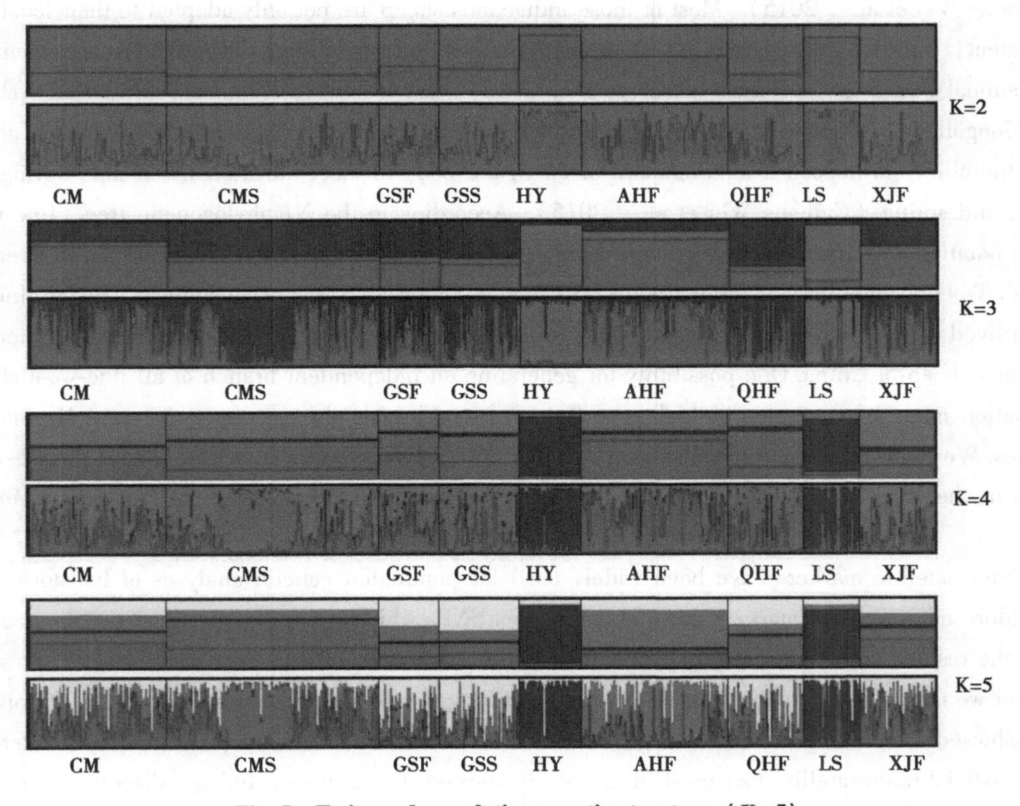

Fig. 5 Estimated population genetic structure ($K=5$)

Noet: Population abbreviations are as reported in Table 1.

lation. This study provides new insight into the genetic structure of fine-wool sheep which was separated from the indigenous sheep that once as their ancestors. Since gene flow occurred frequently among fine-wool sheep populations, high gene flow may reduce the original rich genetic variation among those breeds. Based on the results we thus suggest the selection within breed is a good strategy to upgrade those populations.

Many genome wide scan studies were conducted specifically to shed light on the population structure in sheep with the advantage for plentiful SNP makers that could be used to characterize genetic diversity, identify breeds and detect population structure (Kijas et al., 2012; Ciani et al., 2014; Fariello et al., 2014). In this study, 47816 SNPs were applied to analyze the population structure between fine-wool sheep and Chinese indigenous sheep. We analyzed the population by using NJ-tree and to better understand the population structure of all 46 sheep from 25 populations, we further performed a principal component analysis (PCA) where four clusters were formed, including Kazak, Mongolia, Tibetan lineages sheep and fine-wool sheep. Tibetan sheep defines separate group except for Minxian black fur sheep which were produced by crossing Mongolian-type sheep. Considering all the findings of our studies and based on archaeological, morphological, historical and available molecular genetic information (Zheng, 1988 and Zhong et al., 2010), we recommended three different sheep lineage groups such as Mongolian, Kazakh and Tibetan for Chinese sheep in contrary to the recent study which includes Mongolian and Kazakh sheep in one group

(Caihong Wei et al., 2015). Most of those indigenous sheep are not only adapted to their local environment, but also are regarded as important genetic resources and one of the major components of agro-animal husbandry societies (Zhao et al., 2013). For instance, Tibetan (thin-tailed) sheep and Mongolian (fat-tailed) sheep are abundant in high latitudes; Kazakh (fat-rumped) sheep have the ability to deposit a large amount of fat in the body to meet nutritional demands during the winter and spring (Caihong Wei et al., 2015). According to the NJ phylogenetic tree, fine-wool sheep population defined an independent branch separated from Chinese indigenous sheep breeds, and all Tibetan sheep breeds formed a cluster together. One reason was because those Tibetan lineage sheep lived in Qinghai-Tibetan Plateau, which is the naturally geographical isolation from other indigenous sheep of China. One possibility for generating an independent branch of all fine-wool sheep population might be generated by high gene flows between breeds also may have the same Merino ancestries. We found Chinese superfine-wool Merino strain stay far away from the cluster of fine-wool sheep in the PCA result (Fig. 2), which may have increased introgression of Australian Merino gene.

Microsatellite markers have been widely used for population genetic analyses of livestock species. More microsatellite markers mean abundance of SNPs which is a condition to ensure the accuracy of the results. *Ovine* SNP50 chip and microsatellite markers were simultaneously used in our study, and we chose the 10 loci which have wide genome coverage. Our results suggested that 9 populations showed a relatively high level of genetic variation as estimated by allelic diversity and heterozygosity. All 10 microsatellite loci used in this study showed no significant linkage disequilibrium and almost all the loci with high allele number, which can provide an evidence to select the loci for the future genetic diversity study of the fine-wool sheep. Most of the loci displayed in HWE in all the 9 population except 2 superfine strains CMS and GSS, which may be caused by strong artificial selection just for achieving finer wool fiber. In addition, strong artificial selection might be an important driving force for the formation of species diversity (Caihong Wei et al., 2015). It is found that significant genetic improvements have happened in fiber diameter of CMS (Di et al., 2014). Deviations from HWE are generally caused by non-random mating and inbreeding (Hartl and Clark 1997). The two strains, namely CMS and GSS, may be non-random mating with the flock with superfine-wool. This analysis revealed that the PIC values show strong variation and high mean value of allelic richness in all 9 populations. The level of PIC is generally used to measure the polymorphism for a marker locus (Botstein et al., 1980) and the mean of allelic richness is relevant for estimating the value of polymorphism (Leberg 2002). In conclusion, the microsatellite markers used in this study have many alleles; and a high level of gene diversity and sufficiently polymorphic which suggest that the 10 microsatellite markers used in this study are suitable.

Generally, genetic variation within populations was larger than that among populations. Genetic drift among all 7 fine-wool sheep populations ($F_{ST}=0.019$), only approximately 2% of the genetic variation is distributed among populations. Wright (1978) proposed that there is almost no genetic differentiation when the F_{ST} values are between 0 and 0.05 among populations. The highest gene flow was estimated for the CM-XJF pair ($N_m=31.30$), because of the background of the breeding practice which the superior rams from Chinese Merino (Xinjiang types) were used for progressive hybridization with Xinji fine-wool sheep during the breeding progress. The high gene flows in all 7 fine-wool sheep populations also might seriously affect the differentiation and genetic structure of these fine-wool sheep populations.

The PCA grouped 9 populations into two different genetic clusters which is consistent with the microarray analysis. Meanwhile, the fine-wool sheep with a closer distance was also clearly reflected in the constructed phylogenetic trees. Our results show that genetic relationships of the semi-fine wool sheep LS with all the Chinese fine-wool sheep are not close even though once it were introduced by Chinese Merino (Xinjiang types). One study about the genetic diversity of six semi-fine wool sheep breeds showed that high gene flows or close relationships existing between Liangshan semi-fine wool sheep and other five breeds including Yunnan semi-fine wool sheep, Guizhou semi-fine wool sheep, Pengbo semi-fine wool sheep and foreign breeds Corriedale sheep and Romney sheep, and always Corriedale and Romney were respectively used for introgression in those semi-fine wool breeds in China (Zhu et al., 2013). The five fine-wool sheep populations (CMS, CM, XJF, AHF and GSS) are with a closer pairwise distance, which is consistent with the breeding process fact that the Chinese Merino is widely bred in other 3 breeds (CMS, XJF and AHF), at the same time because of excessive introgression Australian Merino leads to superfine-wool strain GSS with a close distance. GSF and QHF clustered together to the other five fine-wool sheep. We can easily understand it for the populations of GSF and QHF with geographical proximity presenting a good adaptability to alpine pasture. The gene flow occurred accompanying with superior rams exchanged, thus caused close genetic relationship between them. The wide gene flow seriously affects differentiation and genetic structure of the seven fine-wool sheep population, which can also be observed in the genetic structure. Therefore, it supports the historical origin of the Australian Merino derived breed.

4 CONCLUSIONS

In this study, the major finding is that the genetic relationship between the 7 fine-wool sheep population have small pairwise distances exhibiting high gene flow across these populations. Subsequently, there is a long distance for the fine-wool sheep's genetic differentiation from indigenous breeds which once was their important ancestral population. National Fine-Wool Sheep Association Breeding Program plays a vital role on the formation of different breeds with high quality fine-wool, therefore the production systems have been strongly influenced by genetic resources of Australian Merino. As we all know, high genetic diversity is a valuable resource of genetic variation in sheep, along with the introgression of exotic breeds, the important economic traits within those breeds should be enhanced and strengthened during the breeding programs. Besides, adaptability and disease resistance characteristics of the indigenous breed should also be addressed. However, for ensuring concrete conclusion, the findings of these current studies need to be validated with a larger number of individuals per populations.

5 MATERIALS AND METHODS

5.1 Animals and sample preparation

In this study, 46 sheep belonging to 25 populations (We have selected one or two individuals from each sheep population) were collected from a wide geographic range of China. In addition, a total of 482 animals from 9 sheep populations were used consisting 7 main Chinese fine-wool sheep populations (Fig. 6) and two other sheep breeds, namely, Hu sheep with high reproductive performance and Liangshan semi-fine wool sheep. The latter two breeds were taken as controls.

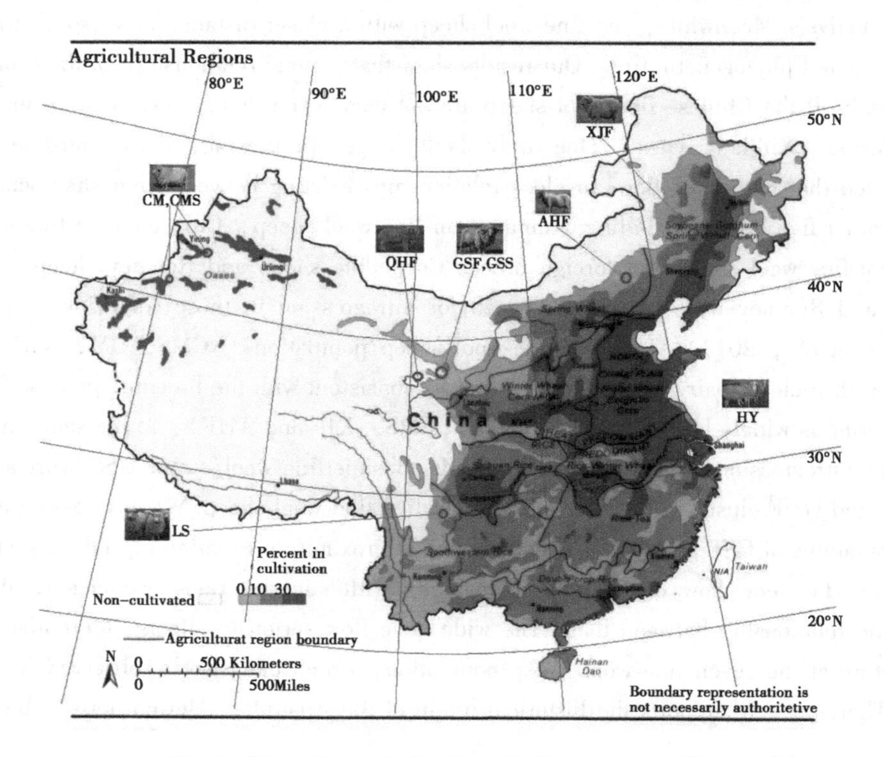

Fig. 6 Map of sample locations in the present study

Note: Our samples were collected from 9 sheep populations, with sample size for each population shown in table 1, the locations were highlighted in red color. Phenotypes of the 9 sheep populations showed in the figure. This map was downloaded from the website www. worldofmaps. net and Population abbreviations are as reported in Table 1. Chinese superfine Merino strain derived from Chinese Merino almost with the same phenotypes, and they come from two sheep breeding farms (Xinyuan and Baicheng), so do the Gansu alpine fine-wool sheep and Gansu alpine superfine-wool sheep strain with almost the same phenotypes and come from the same farm.

5.2 DNA extraction

Blood was collected via jugular venipuncture into EDTA coated vacutainer tubes. All efforts were made to minimize any discomfort for sheep during blood collection. Genomic DNA was extracted from the collected blood samples using the TIANGEN DNA Mini Kit (TIANGEN Bio., Beijing, China). We used NanoDrop 2000 (NanoDrop Technologies, Wilmington, DE, USA) to detect the concentration and purity of the nucleic acids, finally put them in the ultra-low temperature freezer.

5.3 Genetic population structure

Genotyping of the *Ovine* 50K SNP chips from Illumina *Inc.* which was performed by BGI Tech Infinium Genotyping platform using 55mL of approximately 70ng · mL^{-1} genomic DNA, before raw signal intensities were converted into genotype calls using the Genome Studio software. Following quality control, 47816 SNPs with minor allele frequencies (MAF) ≥5% and call rates ≥95% and Hardy Weinberg equilibrium test $P \geq 0.999$, A total of 6425 were removed, Genotypes are available formatted for analysis in PLINK (Purcell et al., 2007). Principal component analysis (PCA) was implemented in the program EIGENSTRAT which was performed for individual basing genotype data and applied the admixture (Alexander et al., 2009). Software and neighbor-joining method was used to construct the phylogenetic tree using Neighbor in the PHYLIP version 3.69 packages (Plotree and Plotgram 1989).

5.4 Microsatellite marker typing

Ten of the microsatellites had been selected from the sheep diversity list recommended by FAO and International Society of Animal Genetics (ISAG). They were amplified by polymerase chain reaction (PCR) where the forward primer 5'end of each microsatellite was labeled either with fluorescent dye FAM, Tet or Rox, TAMRA according to the expected allele size range (Appendix B).

PCR amplified products were detected by ABI3730XL Genetic Analyzer (Applied Biosystems, Foster City, CA, USA). The amplified fragments were analyzed and sized with 3730 Data Collection software and GENEMARKER to declare alleles after the purification of the PCR products.

5.5 Microsatellite marker data analyses

Allele frequencies and mean number of alleles (MNA) were estimated by direct counting. Parameters of loci diversity were estimated for all microsatellite markers in all populations using the Excel Microsatellite Toolkit software (Park, 2001), including observed heterozygosity (H_O), expected heterozygosity (H_E) and polymorphic information content (PIC). Allelic richness (AR) and linkage disequilibrium (LD) were estimated using FSTAT 2.9.3.3 software (Goudet et al., 2002). An exact test was used to determine deviations from Hardy Weinberg equilibrium (HWE) proportions and heterozygosity deficiency by GENEPOP (Raymond and Rousset, 1995) and the method of Guo and Thompson (1992). The same software was used to evaluate gene flow by calculating the number of effective migrants (*Nm*) using private alleles. Wright's F-statistics for each locus was calculated using Weir and Cockerman's method inbuilt in the ARLEQUIN package (Weir and Cockerham 1984; Excoffier et al., 2005). A significance test on the estimates of Wright's *F*-statistics (F_{IT}, F_{IS} and F_{ST}) for each microsatellite locus was obtained by constructing 95% confidence intervals.

To detect genetic relationship among 9 populations, the genetic distances were estimated according to Nei's (1978) unbiased genetic distance by DA using DISPAN (Ota 1993). Bootstrap values were computed over 1000 replicates and additionally a Neighbor-Net graph was constructed based on DA distances. Based on genotypes at the 10 marker loci, principal component analysis (PCA) was performed using MVSP 3.1 (Kovach, 2008). The model-based program STRUCTURE 2.3.3 (Pritchard et al., 2000) was used to analyze the population structure of the 482 sheep. The tests were done on the basis of an admixture model with correlated allele frequencies. The length of the burn-in period was set to 5000 iterations followed by 10000 iterations for Markov chain Monte Carlo sampling. Individuals were grouped into a predefined number of clusters with 100 independent runs for each K value. The optimal K was determined by the modal distribution of ΔK (Evanno et al., 2005). The results were recorded into the DISTRUCT program to provide a graphic display (Rosenberg 2004), and the results for $K=2$ to 5 are shown.

ACKNOWLEDGEMENTS

This work was financially sponsored by Earmarked Fund for Modern China Wool & Cashmere Technology Research System (CARS-40-03) and National Natural Science Foundation for Young Scholars of China (31402057). Project support was provided by the ASTIP (Agricultural Science and Technology Innovation Program) for Genetic Resource and Breeding of Fine-Wool Sheep, CAAS. The authors wish to thank Shuru cheng for technical assistance.

References

[1] Alexander D H, Novembre J, Lange K. 2009. Fast model-based estimation of ancestry in unrelated individuals. *Genome Research*, 19, 1655-1664.

[2] Botstein D, White R L, Skolnick M, Davis R W. 1980. Construction of a genetic linkage map in man using restriction fragment length polymorphisms. *American Journal of Human Genetics*, 32, 314-331.

[3] Chen S, Duan Z, Sha T, Xiangyu J, Wu S, Zhang Y. 2006. Origin, genetic diversity, and population structure of Chinese domestic sheep. *Gene*, 376, 216-223.

[4] Ciani E, Crepaldi P, Nicoloso L, Lasagna E, Sarti F M, Moioli B, Napolitano F, Carta A, Usai G, D'Andrea M, Marletta D, Ciampolini R, Riggio V, Occidente M, Matassino D, Kompan D, Modesto P, Macciotta N, Ajmone-Marsan P, Pilla F. 2014. Genome-wide analysis of Italian sheep diversity reveals a strong geographic pattern and cryptic relationships between breeds. *Animal Genetics*, 45, 256-266.

[5] Di J, Ainiwaer L, Xu X, Zhang Y, Yu L, Li W. 2014. Genetic trends for growth and wool traits of Chinese superfine Merino sheep using a multi-trait animal model. *Small Ruminant Research*, 117, 47-51.

[6] Excoffier L, Laval G, Schneider S. 2005. Arlequin (version 3.0): an integrated so ftware package for population genetics data analysis. *Evolutionary Bioinformatics Online*, 1, 47.

[7] Evanno G, Regnaut S, Goudet J. 2005. Detecting the number of clusters of individuals using the software STRUCTURE: a simulation study. *Molecular Ecology*, 14, 2611-2620.

[8] Fariello M I, Servin B, Tosser-Klopp G, Rupp R, Moreno C, San C M, Boitard S. 2014. Selection signatures in worldwide sheep populations. *PLoS One*, 9, e103813.

[9] Goudet J. 2002. FSTAT, a Statistical Program to Estimate and Test Gene Diversities and Fixation Indices (Verison 2.9.3.3). Available at: http://www2.unil.ch/popgen/softwares/fstat.htm.

[10] Groeneveld L F, Lenstra J A, Eding H, Toro M A, Scherf B, Pilling D, Negrini R, Finlay E K, Jianlin H, Groeneveld E, Weigend S. 2010. Genetic diversity in farm animals-a review. *Animal Genetics*, 411, 6-31.

[11] Guo S W, Thompson E A. 1992. Performing the exact test of Hardy-Weinberg proportion for multiple alleles. *Biometrics*, 361-372.

[12] Hartl D L, Clark A G. 1997. Principles of population genetics. In: p. 542.
[13] Kijas J W, Lenstra J A, Hayes B, Boitard S, Neto L R P, San Cristobal M, Servin B, McCulloch R, Whan V, Gietzen K, Paiva S, Barendse W, Ciani E, Raadsma H, McEwan J, Dalrymple B. 2012. Genome-Wide Analysis of the World's Sheep Breeds Reveals High Levels of Historic Mixture and Strong Recent Selection. *PLoS Biology*, 10, e1001258.
[14] Kovach W L. 2008. Kovach Computing Services: MVSP Version 3. 1. Anglesey, Wales.
[15] Lancioni H, Di Lorenzo P, Ceccobelli S, Perego U A, Miglio A, Landi V, Antognoni M T, Sarti F M, Lasagna E, Achilli A. 2013. Phylogenetic relationships of three Italian merino-derived sheep breeds evaluated through a complete mitogenome analysis. *PLoS One*, 8, e73712.
[16] Leberg P L.2002.Estimating allelic richness: effects of sample size and bottlenecks.*Molecular Ecology*, 11, 2445-2449.
[17] Liu N, Li H, Liu K, Yu J, Bu R, Cheng M, De W, Liu J, He G, Zhao J. 2014. Identification of skin-expressed genes possibly associated with wool growth regulation of Aohan fine wool sheep. *BMC Genetics*, 15, 144.
[18] Lv F, Agha S, Kantanen J, Colli L, Stucki S, Kijas J W, Joost S, Li M, Marsan P A.2014. Adaptations to Climate-Mediated Selective Pressures in Sheep. *Molecular Biology and Evolution*, 31, 3324-3343.
[19] Nei M. 1978. Estimation of average heterozygosity and genetic distance from a small number of individuals. *Genetics*, 89, 583-590.
[20] Niu L L, Li H B, Ma Y H, Du L X. 2012. Genetic variability and individual assignment of Chinese indigenous sheep populations (Ovis aries) using microsatellites. *Animal Genetics*, 43, 108-111.
[21] Ota T. 1993. DISPAN: genetic distance and phylogenetic analysis. Pennsylvania State University, University Park, PA.
[22] Park S. 2001. Microsatellite toolkit. Department of Genetics, Trinity College, Dublin, Ireland.
[23] Plotree D, Plotgram D. 1989. PHYLIP-phylogeny inference package (version 3.2). *cladistics*, 5, 163-166.
[24] Pritchard J K, Stephens M, Donnelly P. 2000. Inference of population structure using multilocus genotype data. *Genetics*, 155, 945-959.
[25] Purcell S, Neale B, Todd-Brown K, Thomas L, Ferreira M A R, Bender D, Maller J, Sklar P, de Bakker P I W, Daly M J, Sham P C. 2007. PLINK: A tool set for whole-genome association and population-based linkage analyses. *American Journal of Human Genetics*, 81, 559-575.
[26] Raymond M, Rousset F. 1995. GENEPOP (Version 1.2): Population Genetics Software for Exact Tests and Ecumenicism. *Journal of Heredity*, 86, 248-249.
[27] Rosenberg N A.2004.DISTRUCT: a program for the graphical display of population structure.*Molecular Ecology Notes*, 4, 137-138.
[28] Wang X, Ma Y, Chen H. 2006. Analysis of the genetic diversity and the phylogenetic evolution of Chinese sheep based on Cyt b gene sequences. *Acta Genetica Sinica*, 33, 1081-1086. (in Chinese).
[29] Wang Z, Zhang H, Yang H, Wang S, Rong E, Pei W, Li H, Wang N. 2014. Genome-wide association study for wool production traits in a Chinese Merino sheep population. *PLoS One*, 9, e107101.
[30] Wei C H, Wang H H, Liu G, Wu M M, Cao J X, Liu Z, Liu R Z, Zhao F P, Zhang L, Lu J, Du L X. 2015. Genome-wide analysis reveals population structure and selection in Chinese indigenous sheep breeds. *BMC Genomics*, 16, 194.
[31] Weir B S, Cockerham C C.1984.Estimating F-statistics for the analysis of population structure.*Evolution*, 38, 1358-1370.

[32] Wright S. 1978. Evolution and the genetics of populations. A treatise in four volumes. Volume 4. *Variability within and among natural populations*. In: p.580.

[33] Yilmaz O, Cemal I, Karaca O. 2014. Genetic diversity in nine native Turkish sheep breeds based on microsatellite analysis. *Animal Genetics*, 45, 604-608.

[34] Zhang L, Mousel M R, Wu X, Michal J J, Zhou X, Ding B, Dodson M V, El-Halawany N K, Lewis G S, Jiang Z. 2013. Genome-wide genetic diversity and differentially selected regions among Suffolk, Rambouillet, Columbia, Polypay, and Targhee sheep. *PLoS One*, 8, e65942.

[35] Zhao E, Yu Q, Zhang N, Kong D, Zhao Y. 2013. Mitochondrial DNA diversity and the origin of Chinese indigenous sheep. *Tropical Animal Health and Production*, 45, 1715-1722.

[36] Zheng P L. 1988. Sheep and Goat Breeds in China. *Shanghai Scientific and Technology Publishers*. (in Chinese).

[37] Zhong T, Han J L, Guo J, Zhao Q J, Fu B L, He X H, Jeond J T, Guan WJ, Ma Y H. 2010. Genetic diversity of Chinese indigenous sheep breeds inferred from microsatellite markers. *Small Ruminant Research*, 90, 88-94.

[38] Zhu L, Lan R, Yang J, Munaier S, Yang H, Shen X, Hong Q. 2013. Study on genetic relationship among six semi-fine wool sheep breeds. *China Herbivores*, 33, 5-9. (in Chinese)

（发表于《journal of integrative agriculture》，院选 SCI，IF：0.833）

Analgesic and Anti-inflammatory Effects of Hydroalcoholic Extract Isolated from Semen Vaccariae

WANG Lei, CUI Don-gan, WANG Xu-rong, ZHANG Jing-yan, YANG Zhi-qiang, QIN Zhe, KONG Xiao-jun, WANG Xue-zhi, LI Jian-xi

(Engineering & Technology Research Center of Traditional Chinese Veterinary Medicine of Gansu Province/Lanzhou Institute of Husbandry and Pharmaceutical Sciences of Chinese Academy of Agricultural Sciences, Lanzhou 730050, China)

Abstract: Semen vaccariae, the seeds of *Vaccaria segetalis* (Neck.) Garcke, is usually used as an important medication for female mammary gland diseases; it has also been used to promote lactation for centuries in China.The purpose of this work was to evaluate the analgesic and anti-inflammatory effects of hydroalcoholic extract from semen vaccariae (HESV) with oral doses of 50, 100 and 200mg/kg · bw in mice and rats.We observed that the HESV could effectively inhibit acetic acid-induced abdominal contraction and could elevate the latency time to thermal stimuli in the hot-plate test in mice.In the xylene-induced ear-swelling test in mice, HESV could suppress the ear swelling. Additionally, HESV could significantly decrease the peritoneal capillary permeability and leukocyte infiltration in mice induced by the intraperitoneal injection of acetic acid.HESV also significantly reduced paw thickness 2–4 hours after the injection of carrageenan in the carrageenan-induced rat paw edema test.This study was the first to demonstrate that the oral administration of HESV might play an important role in the process of analgesia and anti-inflammation, supporting its use for female mammary gland diseases in traditional medicine.

Key words: *Semen vaccariae*; Hydroalcoholic extract; Analgesic activity; Anti-inflammatory activity

INTRODUCTION

Semen vaccariae, the ripe seeds of *Vaccaria segetalis* (Neck.) Garcke, is a traditional herb described in Compendium of Materia Medica and is considered as an important medication for treating female mammary gland diseases and for promoting lactation.Semen vaccariae is usually used to activate blood circulation, regulate menstrual disturbances, dispel edema, and promote lactation for centuries in China (Li and Liang, 2007; Sang et al., 2003a).

Recent studies have shown that semen vaccariae is rich in flavonoids, triterpene saponins, alkaloids, cyclic peptide, phenolic acid and steroids cyclopeptides (Sang et al., 2003b); it also possesses various bioactivities.Zhang et al. (2013) have demonstrated that polysaccharide isolated from semen vaccariae has an inhibitory effect on the benign prostatic hyperplasia in mice induced by Pule'an.Segetalin A and segetalin E from semen vaccariae strongly inhibit the proliferation of human microvascular endothelial cells (Hua et al., 2009), and segetalin A and segetalin B have an estrogen-like activity (Itokawa et al., 1995).Triterpenoid saponin, isolated from semen vaccariae, has an inhibitory effect on luteal cells (Sang et al., 2000).Tong et al. (2012) have demonstrated that semen vaccariae positively impacts mammary gland development by promoting the proliferation and secretion functions of bovine mammary epithelial cells and modulating the expression of E-cadherin and β-catenin via the Wnt signaling pathway *in vitro*, supporting the traditional use of semen vaccariae for promoting lactation.The aim of this study were to evaluate the analgesic and anti-inflammatory effects of hydroalcoholic extract from semen vaccariae (HESV) in mice and rats and to present evidence for the traditional use of semen vaccariae to treat female mammary gland diseases, such as mastitis.

MATERIALS AND METHODS

Reagents and drugs

Carrageenan, indomethacin, Evans blue and crystal violet were obtained from Sigma-Aldrich Chemical Co. (St.Louis, MO, USA).Acetic acid was purchased from Tianjin Damao Chemical Reagent Factory (Tianjin, China).Xylene was purchased from Tianjin Fuchen Chemical Reagent Factory (Tianjin, China).The control medicinal material of semen vaccariae and vaccarin reference substance were purchased from the National Institutes for Food and Drug Control (Beijing, China).

HESV preparation

Semen vaccariae was purchased from the local market in Lanzhou city, China.Dongan Cui, Ph.D., of the Lanzhou Institute of Husbandry and Pharmaceutical Sciences of the Chinese Academy of Agricultural Sciences, China performed botanical identification and thin layer chromatography analysis according to the protocols described by the Chinese pharmacopoeia (2010).In particular, heavy metals, residual pesticides and microbial contamination of semen vaccariae were also detected, and the results met the safety standards of traditional Chinese medicine in China.

First, washing, drying, pulverizing and screening with a 3-mm mesh, treated semen vaccariae; then, the powder (200g) was refluxed with 70% ethanol (2 000mL) for 4h at 7℃ according to an established extraction possess (results not published).The HESV (20. 21g) was obtained by freeze-drying with a lyophilizer (Beijing Songyuan Huaxing Technology Develop CO.Ltd., Beijing, China) after extraction with hydroalcoholic liquid evaporation under reduced pressure at 45℃ and the yield ratio was 10. 1%.The marker compound of semen vaccariae was vaccarin.The content of vaccarin was 60. 19mg/g determined with high-performance liquid chromatography (HPLC) according to the method described by the Chinese pharmacopoeia (2010).Karl Fischer analysis was used to determine the water content, which was between 3. 50% and 4. 11%.

Animal care and handling

Balb/C mice (18–22g) and Sprague-Dawley (SD) rats (160–200g) were purchased from the experimental animal center of Lanzhou University and were housed in plastic cages with free access to food and water at a temperature of 22±1℃ and a relative humidity of 50%±10%.This study was approved by the Institutional Animal Care and Use Committee of Lanzhou Institute of Husbandry and Pharmaceutical Sciences of the Chinese Academy of Agricultural Sciences (SCXK20008-0003).The animal protocols were in compliance with the ethical guidelines for the treatment of animals of the International Association for the Study of Pain (Zimmermanm, 1983).

Acetic acid-induced abdominal constriction in mice

The acetic acid-induced abdominal constriction test in mice was conducted according to a described protocol (Cho et al., 2013; Young et al., 2005).The mice were randomly divided into one of five groups ($n=10$) and were given intraperitoneal injections of 0. 7% acetic acid at a dose of 10mL/kg · bw 1h after oral administration of HESV (50mg/kg · bw, 100mg/kg · bw and 200mg/kg · bw) in the treatment groups, oral administration of indomethacin (3mg/kg · bw) in the reference group, or oral administration of normal saline (10mL/kg · bw) in the control group. The number of writhing movements, abdominal musculature contraction followed by the extension of hind limbs, was registered within 30min after the injection of acetic acid.The inhibition ratio was calculated using the following equation:

$$\text{Inhibitionratio} = \frac{Nc - Nt}{Nc} \times 100\%$$

Nc represents the number of writhing movements in the control group, and Nt represents the number of writhing movements in the treatment group.

Hot-plate test in mice

The hot-plate test was performed according to a previously described protocol (Shinde et al., 1999).Groups of 10 female Balb/C mice were treated orally with HESV (50, 100 and 200mg/kg · bw) or indomethacin (3mg/kg · bw) as a reference; normal saline (10mL/kg · bw) was used as a control.The latency times to thermal stimuli were recorded before (0min) and 30, 60 and 90 min following treatment with a cut-off time of 60sec to avoid paw lesions.

Xylene-induced ear swelling in mice

The xylene-induced ear-swelling test was performed according to a previously described method (Kou et al., 2005).Male mice were randomly divided into five groups (n=10 in one group), and ear swelling was induced by smearing 30μL of xylene on the anterior and posterior surfaces of the right ear 1h following treatment with HESV (50mg/kg · bw, 100mg/kg · bw and 200mg/kg · bw, p.o.), indomethacin (3mg/kg · bw, p.o.) or normal saline (10mL/kg · bw, p.o.).The left ear was as control on the same mouse.One hour later, all mice were sacrificed by cervical dislocation. Circular sections of both ears were collected with an 8 mm diameter punch and were immediately weighed.The swelling degree and inhibition ratio were calculated using the following equations:

$$\text{Swelling degree (SD)} = \frac{Wr - Wl}{Wl} \times 100\%$$

$$\text{Inhibitionratio} = \frac{SDc - SDt}{SDc} \times 100\%$$

Wr represents weight of the right ear, and Wl represents weight of the left ear of the same mouse; SDc represents the swelling degree in the control group, and SDt represents the swelling degree in the treatment group.

Acetic acid-induced leukocyte infiltration and capillary permeability in mice

The test was performed according to a previously described method (Lucena et al., 2007). The mice were randomly divided into five groups ($n=10$ in one group), and 0.5% Evans blue solution (10mL/kg · bw) was intravenously injected to the tail veins of mice 1h following treatment with HESV (50mg/kg · bw, 100mg/kg · bw and 200mg/kg · bw, p.o.), indomethacin (3mg/kg · bw, p.o.) or normal saline (10mL/kg · bw, p.o.); 10 min later, leukocyte infiltration and capillary permeability were induced by the intraperitoneal injection of 0.7% acetic acid at a dose of 10mL/kg · bw. Twenty minutes after the intraperitoneal injection of 0.7% acetic acid, all mice were sacrificed by cervical dislocation, and the peritoneal cavity was immediately opened; the peritoneal fluid was then collected after washing with 5mL sterile saline. Finally, the total leukocyte counts were recorded in the peritoneal fluid (50μL of peritoneal fluid was taken and diluted in 450μL of Türk's solution) according to microscopic analysis. The remaining peritoneal fluid was centrifuged at 3 000 rpm for 15 min and the supernatant was collected. The absorbance of the supernatant was determined at 606 nm using an Evolution 300 UV-VIS spectrophotometer (Thermo Scientific, USA). The concentration of Evans blue in peritoneal cavity indicated the peritoneal capillary permeability induced by acetic acid (Lucena et al., 2007). The inhibition ratios of leukocyte infiltration (IRLI) and capillary permeability (IRCP) in the treatment groups were calculated using the following equations:

$$IRLI=\frac{Lc-Lt}{Lc}\times 100\%$$

$$IRCP=\frac{Ec-Et}{Ec}\times 100\%$$

Lc represents the number of leukocytes in the control group, and Lt represents the number of leukocytes in the treatment group; Ec represents the concentration of Evans blue in the control group and Et represents the concentration of Evans blue in the treatment group.

Carrageenan-induced paw edema in rats

The carrageenan-induced paw edema test was conducted in rats according to a previously described method (Mandegary et al., 2012; Posadas et al., 2004). The rats were randomly divided into five groups ($n=10$ in one group), and edema of the right hind paw was induced by hypodermic injection with 100μL of 1% carrageenan suspension in normal saline 1h after treatment with HESV (50mg/kg · bw, 100mg/kg · bw and 200mg/kg · bw, p.o.), indomethacin (3mg/kg · bw, p.o.) or normal saline (10mL/kg · bw, p.o.). The right hind paw thickness was determined using a vernier caliper before and 1, 2, 3 and 4h after the injection of carrageenan. The swelling degree and inhibition ratio were calculated with the following equations:

$$\text{Swelling degree (SD)} = \frac{PTt-PT0}{PT0}\times 100\%$$

$$\text{Inhibition ratio}=\frac{SDc-SDt}{SDc}\times 100\%$$

PT0 represents the right hind paw thickness before carrageenan injection, and PTt represents the right hind paw thickness 1h, 2h, 3h, or 4h after carrageenan injection; SDc represents the swelling degree in the control group, and SDt represents the swelling degree in the treatment group.

STATISTICAL ANALYSIS

The data are expressed as means ± SE. Data were analyzed with SPSS (version 17.0, IBM SPSS Statistics; USA) using a one-way analysis of variance (ANOVA) followed by least significant difference (LSD) as the post hoc test. The results were considered statistically significant at $P<0.05$.

RESULTS

Analgesic activity of HESV

The results of the analgesic effects of HESV are presented in tables 1 and 2. In the acetic acid-induced abdominal constriction test, HESV caused 37.54%, 53.50% and 57.83% inhibition of constrictions at oral doses of 50mg/kg · bw, 100mg/kg · bw, and 200mg/kg · bw. There was a significant and dosedependent analgesic effect of HESV at 50–200mg/kg · bw compared to the control. In the hot-plate test, there was no significant effects were observed at 60 min after treatment at an oral dose of 50mg/kg · bw. However, there was a significant increase in the latency to thermal stimuli at 100mg/kg · bw at 60 min after drug administration and the effect of HESV (200mg/kg · bw) was significantly higher. In both cases, HESV increased the latency time, at 90 min after drug administration, at doses of 50mg/kg · bw, 100mg/kg · bw and 200mg/kg · bw.

Table 1 Effect of HESV on acetic acid-induced abdominal constriction in mice

Groups	Dose (mg/kg · bw)	Number of writhing	Inhibition (%)
Control	—	$41.10±5.69^{A}$	—
Indomethacin	3	$16.20±3.94^{B}$	60.58
HESV	50	$25.67±4.97^{B}$	37.54
	100	$19.11±3.86^{B}$	53.50
	200	$17.33±5.89^{B}$	57.83

Table 2 Effect of HESV on the hot-plate test in mice

Groups	Dose (mg/kg · bw)	Latency period (s)			
		0min	30min	60min	90min
Control	—	22.23±4.53	$18.59±3.87^{a}$	$18.16±4.98^{A}$	$17.14±5.85^{A}$
Indomethacin	3	21.89±3.44	23.32±8.47	$26.84±6.14^{B}$	$33.41±7.77^{B}$
HESV	50	21.91±4.13	20.54±3.36	22.31±3.47	$24.35±5.06^{B}$
	100	20.18±6.35	20.03±5.99	$23.86±4.18^{B}$	$27.25±5.29^{B}$
	200	21.85±3.98	$24.44±5.81^{b}$	$25.69±4.66^{B}$	$28.16±4.58^{B}$

Anti-inflammatory activity of HESV

In this study, HESV significantly suppressed xylene-induced ear-swelling in mice in a dose-dependent manner (See table 3). Additionally, the inhibition ratio of HESV (200mg/kg · bw, p.o.) exceeded that of indomethacin at 3mg/kg · bw. After the intraperitoneal injection of acetic acid, there were significant increases in total leukocyte numbers and the Evans blue content extruded into the mouse peritoneal cavity in the control group (tables 4 and 5). Additionally, compared to the control, HESV could effectively decrease the total leukocyte numbers and the Evans blue content in the peritoneal fluid. Furthermore, in the acetic acid-induced leukocyte infiltration test, HESVmediated inhibition at a dose of 200mg/kg · bw exceeded that of indomethacin at a dose of 3mg/kg · bw. In the carrageenan-induced paw edema, the hind paws of rats exhibited marked swelling at 3h after stimulation by carrageenan in the control and treatment groups; the degrees of swelling in the indomethacin and HESV (50mg/kg · bw, 100mg/kg · bw, and 200mg/kg · bw, p.o.) treatment groups were significantly decreased in the endpoint 4h of observation compared to the control (table 6).

Table 3 Effect of HESV on xylene-induced ear-swelling in mice

Groups	Dose (mg/kg · bw)	Swelling (%)	Inhibition (%)
Control	—	64.16 ± 5.20^{A}	—
Indomethacin	3	30.76 ± 4.29^{B}	52.06
	50	53.68 ± 3.70^{B}	16.33
HESV	100	40.80 ± 4.62^{B}	36.41
	200	25.89 ± 3.27^{B}	59.65

Note: Values within the same row that are marked with different superscript letters differ significantly: a, b = $P<0.05$; A, B = $P<0.01$.

Table 4 Effect of HESV on acetic acid-induced leukocyte infiltration in mice

Groups	Dose (mg/kg · bw)	Total leukocytes ($\times 10^{6}$)	Inhibition (%)
Control	—	5.13 ± 0.72^{A}	—
Indomethacin	3	3.27 ± 0.36^{B}	36.26
	50	4.07 ± 0.62^{B}	20.66
HESV	100	3.49 ± 0.43^{B}	31.97
	200	3.14 ± 0.45^{B}	38.79

Table 5 Effect of HESV on the acetic acid-induced Evans blue leakage test in mice

Groups	Dose (mg/kg · bw)	Evans blue (μg/mL)	Inhibition (%)
Control	—	7.13 ± 0.99^{A}	—
Indomethacin	3	4.68 ± 0.20^{B}	34.36
	50	5.32 ± 0.40^{B}	25.39
HESV	100	4.81 ± 0.32^{B}	32.54
	200	4.61 ± 0.16^{B}	35.34

Table 6 Anti-inflammatory effect of HESV on carrageenan-induced paw edema in rats

Groups	Dose (mg/kg · bw)	1h		2h		3h		4h	
		Swelling (%)	Inhibition (%)	Swelling (%)	Inhibition (%)	Swelling (%)	Inhibition (%)	Swelling (%)	Inhibition (%)
Control	—	32.94± 5.75a	—	58.59± 3.75A	—	71.61± $9.92^{A,a}$	—	62.61± 4.66^{A}	—
Indom-ethacin	3	30.94± 3.53	6.07	44.13± 4.80^{B}	24.68	45.12± 3.43^{B}	36.99	26.95± 7.29^{B}	56.96
	50	32.86± 2.11	0.24	54.64± 5.07	6.74	65.43± 2.61^{b}	11.42	41.67± 4.48^{B}	33.45
HESV	100	32.10± 2.09	2.55	44.40± 4.73^{B}	24.22	52.84± 3.63^{B}	26.21	31.55± 5.72^{B}	49.61
	200	28.76± 3.92^{b}	12.69	39.01± 3.40^{B}	33.42	42.29± 5.89^{B}	40.94	25.10± 4.64^{B}	59.91

Note: Values within the same row that are marked with different superscript letters differ significantly: a, b= $P<0.05$; A, B = $P<0.01$.

DISCUSSION

Semen vaccariae has been widely available for centuries in China as an important medication for female mammary gland diseases. The present results provided the first evidence of the analgesic and anti-inflammatory activities of HESV in mice and rats.

In the acetic acid-induced abdominal constriction test and the hot-plate test, HESV significantly inhibited the number of mouse writhing and increased the latency to thermal stimuli. Thus, HESV might possess a strong analgesic effect. It exerted analgesic activity through peripheral mechanisms and central mechanisms in part.

The inflammatory reaction is a combination of a number of overlapping reactions within the body. In the early stages of this process, many types of inflammatory mediators, such as prostaglandins, bradykinin and histamine are produced by the inflamed tissue during the vascular reaction stage. Under the effect of these mediators, the endothelial cells of blood vessels shrink and endothelial cell gaps form. Additionally, local vasopermeability was enhanced by leukocyte-mediated endothelial cell injury (Santos et al., 2012). Peritoneal capillary permeability and leukocyte infiltration in the abdominal cavity are aggravated by intraperitoneal injection of acetic acid in mice (Li et al., 2010). HESV significantly suppressed xylene-induced ear-swelling in mice and decreased the total leukocyte numbers and the Evans blue content in the peritoneal fluid induced by intraperitoneal injection of acetic acid. Given these results, it was suggested that HESV might exhibit antiinflammatory effects through modulating the cascade reaction of inflammatory mediators and decreasing leukocyte-mediated endothelial cell injury. Carrageenan-induced paw edema, including edema and hyperalgesia, is a reliable and repeatable model for evaluating the antiinflammatory activities of natural products (Huang et al., 2011). In this study, the degree of swelling stimulated by carrageenan were significantly decreased by HESV compared to the control. It is a well-established fact that non-steroidal anti-inflammatory drugs exert their analgesic and anti-inflammatory activities by inhibiting cyclooxygenase activity to decrease the inflammatory reaction (Rosa et al., 1971; Vane, 1971). We tentatively hypothesized that HESV could decrease inflammatory reactions by inhibiting cyclooxygenase activity in rats.

CONCLUSION

The present work is the first report on the analgesic and anti-inflammatory activities of HESV. We demonstrated that the oral administration of HESV might play an important role in the process of analgesia and anti-inflammation, supporting its traditional use to treat female mammary gland diseases, such as mastitis. Future investigations will focus on the broader involvement of the chemical constituents and mechanism (s) responsible for the pharmacological activities.

ACKNOWLEDGEMENTS

This work was funded by the Special Fund for Agroscientific Research in the Public Interest (201303040-01) and the Special Fund of the Chinese Central Government for Basic Scientific Research Operations in Commonwealth Research Institutes (1610322014004).

References

[1] Cho IJ, Lee CW, Lee MY, Kang MR, Yun J, Oh SJ, Han SB, Lee K, Park SK, Kim HM, Jung SH and Kang JS (2013). Differential anti-inflammatory and analgesic effects by enantiomers of zaltoprofen in rodents. *Int.Immunopharmacol.*, 16: 457-460.

[2] Hua H, Feng L, Zhang XP, Zhang LF and Jin J (2009). Studies on active substance of antiangiogenesis in vaccaria segetalis. *Lishizhen Med.and Mat.Med.Res.*, 20: 698-700.

[3] Huang MH, Wang BS and Chiu CS (2011). Antioxidant, antinociceptive and anti-inflammatory activities of xanthii fructus extract. *J.Ethnopharmacol.*, 135: 545-552.

[4] Itokawa H, Yun Y, Morita H, Takeya K and Yamada K (1995). Estrogen-like activity of cyclic peptides from vaccaria segetalis extracts. *Planta Med.*, 61: 561-562.

[5] Kou JP, Sun Y, Lin YW, Cheng ZH, Zheng W, Yu BY and Xu Q (2005). Anti-inflammatory activities of aqueous extract from radix ophiopogon japonicus and its two constituents. *Biol.Pharm.Bull.*, 28: 1234-1238.

[6] Li F and Liang JY (2007). Research progress of vaccaria segetalis. *Strait Pharm.J.*, 19: 1-4.

[7] Li MX, Shang XF, Zhang RX, Jia ZP, Fan PC, Ying Q and Wei LL (2010). Antinociceptive and anti- inflammatory activities of iridoid glycosides extract of Lamiophlomis rotata (Benth.) Kudo. *Fitoterapia*, 81: 167-172.

[8] Lucena GMRS, Gadotti VM, Maffi LC, Silva GS, Azevedo MS and Santos ARS (2007). Antinociceptive and anti-inflammatory properties from the bulbs of Cipura paludosa Aubl. *J.Ethnopharmacol.*, 112: 19-25.

[9] Mandegary A, Pournamdari M, Sharififar F, Pournourmohammadi S, Fardiar R and Shooli S (2012). Alkaloid and flavonoid rich fractions of fenugreek seeds (Trigonella foenum-graecum L.) with antinociceptive and anti-inflammatory effects. *Food Chem.Toxicol.*, 50: 2503-2507.

[10] National pharmacopoeia committee (2010). Chinese pharmacopoeia. Chinese Medical Science and Technology press, Beijing, China, pp.49-50.

[11] Posadas I, Bucci M, Roviezzo F, Rossi A, Parente L, Sautebin L and Cirino G (2004). Garrag-

eenan- induced mouse paw oedema is biphasic, age-weight dependent and displays differential nitric oxide cyclooxygenase-2 expression.*Brit.J.Pharmacol.*, 142: 331–338.

[12] Rosa DM, Papadimitriou JM and Willoughby DA (1971).A histopathological and pharmacological analysis of the mode of action of nonsteroidal anti-inflammatory drugs.*J.Pathol.*, 105: 239–256.

[13] Sang SM, Lao AN and Chen ZL (2003a).Chemistry and bioactivity of the seeds of vaccaria segetalis. *Oriental Foods and Herbs*, 21: 279–291.

[14] Sang SM, Lao AN, Leng Y and Gu ZP (2000). Segetoside F, new triterpenoid saponin with inhibition of luteal cell from the seeds of vaccaria segetalis.*Tetrahedron Lett.*, 41: 9205-9207.

[15] Sang SM, Xia ZH, Lao AN, Cao L, Chen ZL, Jun U and Yasuo F (2003b).Studies on the constituents of the seeds of vaccaria segetalis.*Heterocycles*, 59: 811–821.

[16] Santos P, Watkinson AC and Hadgraft J (2012).Influence of penetration enhancer on drug permeation fromvolatile formulations.*Int.J.Pharm.*, 439: 260–268.

[17] Shinde UA, Phadke AS, Nair AM, Mungantiwar AA, Dikshit VJ and Saraf MN (1999).Studies on the anti- inflammatory and analgesic activity of Cedrus deodara (Roxb.) Loud.wood oil.*J.Ethnopharmacol.*, 65: 21–27.

[18] Tong JJ, Li Y, Liu R, Gao XJ and Li QZ (2012).Effect of Semen vaccariae and Taraxacu mogono on cell adhesion of bovine mammary epithelial cells.*J.Integr.Agr.*, 11: 2043–2050.

[19] Vane JR (1971).Inhibition prostaglandine synthesis as a mechanism of action for aspirin-like drugs. *Nature New Biol.*, 231: 232–235.

[20] Young HY, Luo YL, Cheng HY, Hsieh WC, Liao JC and Peng WH (2005).Analgesic and anti-inflammatory activities of 6-gingerol.*J.Ethnopharmacol.*, 96: 207–210.

[21] Zhang HJ, Jing Y and Wu GT (2013). Inhibitory effects of crude polysaccharides from semen vaccariae on benign prostatic hyperplasia in mice.*J.Ethnopharmacol.*, 145: 667–669.

[22] Zimmermanm M (1983). Ethical guidelines for investigations of experimental pain in conscious animals.*Pain*, 16: 109–110.

（发表于《Pakistan Journal of Pharmaceutical Sciences》，SCI：0. 682）

A Method for Multiple Identification of Four β_2-Agonists in Goat Muscle and Beef Muscle Meats Using LC-MS/MS Based on Deproteinization by Adjusting pH and SPE for Sample Cleanup

XIONG Lin, GAO Ya-qin, LI Wei-hong,
GUO Tian-feng, YANG Xiao-lin

(Lanzhou Institute of Husbandry and Pharmaceutical Sciences, Chinese Academy of Agricultural Sciences/Laboratory of Quality & Safety Risk Assessment for Livestock Product (Lanzhou), Ministry of Agriculture, Lanzhou 730050, China)

Abstract: An LC-MS/MS method was established for simultaneous identification of levels of bilpaterol, cimbuterol, clenproperol, and bambuterol in goat and beef muscle. Meat samples were subjected to extraction using an aqueous acidic solution and cleansed using MCX solid phase extraction. β_2-Agonists residues were extracted using an aqueous acidic solution. Proteins in extraction liquids was removed based on adjusting pH using perchloric acid and a sodium hydroxide solution, followed by SPE for clean-up. Linearity values of the method based on R^2 values were 0.9976–0.9997 with 70.1%–108.8% recovery and relative standard deviation values of 3.5%–13.3%. Low limits of detection (LOD) and quantitation (LOQ) of 0.01–0.02 and 0.02–0.08μg/kg, respectively, were achieved for spiked goat and spiked beef muscle. The method was sensitive and specific and is an improvement over other currently available technologies.

Key words: β_2-agonists; LC-MS/MS; Feed additive; Goat muscle meat; Beef muscle meat

INTRODUCTION

β_2-Agonists are compounds normally used in symptomatic treatment of asthma and chronic bronchitis and also in prevention of exercise-induced asthma. Moreover, because β_2-agonists can induce production of leaner meat, they are widely used as feed supplements to increase feed efficiency and the rate of weight gain in food-producing animals (1). However, these compounds have also been misused as nutrient repartitioning agents in livestock where they serve to divert nutrients from fat deposition in animals to production of muscle tissue. Misapplication or illegal use of β_2-agonists can result in meat residues that are harmful to both animal and human health due to inducing heart palpitations, diarrhea, muscle tremors, and even malignant tumors (2, 3). For this reason, β_2-

agonists have been banned as growth promoters for rearing of livestock in many European Union (EU) countries and in China. Therefore, for monitoring of food safety, sensitive and specific analytical methods are needed to determine levels of β_2-agonists.

SDetection techniques for identification of β_2-agonists at trace levels include HPLC (4), GC-MS (5, 6), and LC-MS or LC-MS/MS (7-9). HPLC is a semi-qualitative method that can produce false-positive results. The GC-MS method has demonstrated excellent sensitivity, but sample treatment is time-consuming and requires derivatization, so application of GC-MS for detection of β_2-agonists decreases year by year. LC-MS or LC-MS/MS is considered to be the best method of identification of β_2-agonist residues due to simplicity, high qualitative capacity, and high sensitivity in quantitative determination. Before detection of analytes in animal derived food, pre-treatment and concentration steps, which often influence analytical accuracy, are inevitable and critical. For extraction and purification, techniques such as immunoaffinity chromatography (IAC) (10), liquid-liquid extraction (LLE) (11), and solid phase extraction (SPE) (12, 13) are used for testing of β_2-agonist residues. Compared with other methods, SPE is a more effective tool for concentration of trace elements and separation of constituents and has been developed as the main analytical method for use with β_2-agonists. Mixed-mode solid-phase extraction cartridges, such as the mixed-mode cation exchanger (MCX) (14) and the strong cation exchanger (SCX) (13), are commonly used in the cleanup procedure for β_2-agonists in animal tissue.

Biological specimens for use in determination of β_2-agonists include human and bovine urine (15, 16), swine urine (17), pig feed (18, 19), human plasma (20), porcine muscle (21-23), pig liver and kidney (9), infant formula (24), rat urine (25), drinking water (2), and bovine retina (26). Matrices for determination of β_2-agonists are mainly focused on porcine tissue and urine. No previous reports exist regarding analysis of β_2-agonist residues in goat muscle or beef muscle meats, which are the leading animal-derived foods worldwide with ever increasing numbers of cattle and goats. Misapplication or illegal use of β_2-agonists in feeding of cattle and goats is inevitable.

There are more than 30 kinds of β_2-agonist species, only a dozen of which, including clenbuterol (27, 28), ractopamine (29, 30), salbutamoland, and terbutaline (12), have established detection methods. There are no reports on testing methods for β_2-agonist zilpaterol, cimbuterol, clenproperol, and bambuterol residues in goat and beef muscle meats. For these reasons, sensitive analytical methods for detection of these substances are needed for control of illegal use.

This study focused on development of a rapid multi-residue method for identification of levels of zilpaterol, cimbuterol, clenproperol, and bambuterol in goat and beef muscle meats. Meat proteins were removed based on disruption of pH, and extraction was performed using an SPE technique with MCX solid-phase extraction cartridges, followed by direct LC-MS/MS analysis.

MATERIALS AND METHODS

Chemicals and reagents Analytical drug standards for zilpaterol hydrochloride (purity>99.3) and cimbuterol (purity>99.6) were obtained from Witega Laboratorien Berlin-Adlershof GmbH (Berlin, Germany). Clenproperol (purity>98.0) and bambuterol hydrochloride (purity>98.0) standards were obtained from Dr.Ehrenstorfer GmbH (Augsburg, Germany).A β-glucuronidase/arlsulfatase standard was obtained from Merck KgaA (Darmstadt, Germany).HPLC-grade methanol, acetonitrile, and formic acid were obtained from the Tedia Company Inc. (Fairfield, OH, USA). A Milli-Q ultrapure water system (Molsheim, France) was used to obtain HPLC-grade water.*tert*-Butyl methyl ether, ammonium hydroxide, and other chemicals and solvents used were of analytical grade, unless otherwise stated, and were obtained from Sinopharm (Shanghai, China).

Instruments and apparatus An Agilent 1200-6410 LC-MS/MS apparatus (Agilent, Palo Alto, CA, USA), equipped with an ESI ion source was used.LC separation was performed using an Agilent Eclipse Plus C_{18} column (100×3.0 mm, 1.8 μm).Detection was performed on a triple quadrupole mass-spectrometric detector (TQD) with an electrospray ionization (ESI) interface.An electronic analytical balance (accurate to 0.1mg and 0.01 g, BSA224S-CW; Sartorius, Göttingen, Germany), a Labinco L24 vortex mixer (Labinco Beheer B.V., Breda, Netherlands), an Omnifuge 2.ORS centrifuge (Heraeus, Osterode, Germany), a Haier refrigerator (Haier, Qingdao, China), a Sping constant temperature oscillator (Oahua, Changzhou, China), a DP-01 vacuum pump (Autoscience, Shanghai, China), a Waters Oasis MCX (3mL, 60mg) SPE column (Waters, Milford, MA, USA), and N100DR nitrogen evaporators (Peak Scientific, Inchinnan, UK) were used in the sample treatment process.

Sample collection and preparation Frozen samples of goat and beef muscle meats were obtained from farm animals in Lanzhou, China in June of 2014 and kept frozen at −20℃ in refrigerator until analysis.β_2-Agonists were not used in animal feeding.Meat samples were cut into pieces by kitchen knife and homogenised in a blender.Stock solutions of individual standard compounds were prepared in methanol (HPLC grade) at 100 μg/L and were stored at −20℃ in refrigerator in the dark.The expiry date of standard stock solutions was 6 months.A 1mg/L mixed standard solution of the 4 β_2-agonists zilpaterol, cimbuterol, clenproperol, and bambuterol was prepared using individual stock solutions, diluted with a mobile phase solution of methanol-water (10/90, v/v) containing 0.1% formic acid, and the expiry date was 1 week at 4℃ in refrigerator.This mixed-standard solution was used for spiking in recovery experiments and for preparation of calibration standards.

Extraction A 5.0 g comminuted meat sample (goat muscle or beef muscle) was transferred to a 50mL centrifuge tube.Some agonists in animal tissues are present as conjugates after metabolism. Therefore, for detection, they must be restored to the form of β_2-agonists via enzymatic hydrolysis, which is generally performed using β-glucuronidase/arylsulphatase.Therefore, 10mL of an acetate buffer (pH 5.2) and 20μL of β-glucoronidase/arylsulphatase were added to the centrifuge tube. Enzymatic hydrolysis of conjugated metabolites was carried out using constant temperature oscillator of the mixture for 6 h at 42℃.

After cooling to room temperature, crude extracts were centrifuged at 10 000×g for 10 min at 25℃.The supernatant was transferred to another tube and mixed with 5mL of 0.1 M perchloric acid. The mixture was vortexed by vortex mixer for 10 min and adjusted to pH 1.0±0.2 using perchloric acid, followed by centrifugation at 10 000×g for 10 min.The supernatant was transferred to another tube and adjusted to pH 9.5±0.2 using a 10 M sodium hydroxide solution.Then, 15mL of ethyl acetate was added, followed by vortexing for 5 min and centrifugation at 5 000×g for 5 min.The supernatant organic solution was then poured into a tube and mixed with 10mL of tert-butyl methyl ether, then vortexed again for 5 min and centrifuged at 5 000×g for 5 min.The supernatant organic solution was poured into the same tube used in the last step, then dried under nitrogen at 50℃ and reconstituted to 5. 0mL using water containing 2% formic acid.

Purification An MCX cartridge (3. 0mL, 60mg) was conditioned sequentially using 3. 0mL of methanol and 3. 0mL of water containing 2% formic acid.The solution was applied to an SPE cartridge placed on a vacuum manifold device.Subsequently, the cartridge was washed with 3mL of water containing 2% formic acid and 3mL of methanol, then dried under a strong vacuum by vacuum pump for 2–3 min, followed by washing with 5mL of methanol containing 5% ammonium hydroxide and collecting the eluate.Finally, the eluate was evaporated to dryness under nitrogen evaporators and the residue was dissolved in 0. 4mL of methanol/water containing 0. 1% formic acid (1/9, v/v).A clean-up step was completed using filtration of samples through 0. 22 μm Millipore Millex-GV filters and collection of filtrates in LC vials.

LC operating conditions LC separation was performed using an Agilent Eclipse Plus C_{18} column (100×3. 0 mm) with a 1. 8 μm particle size.The injection volume was 10μL and the eluent flow rate was 0. 30mL/min.The mobile phase consisted of (A) methanol containing 0. 1% formic acid and (B) water containing 0. 1% formic acid.The gradient elution protocol used was initially 80% B, decreased to 40% B in 4 min, decreased to 20% B in 1 min, and increased to 80% B in 1 min.The column was then equilibrated for 4 min before the next injection.The column and the ambient temperature were maintained at 30±1℃.

Mass spectrometry Ionisation parameters for the heated ESI source were optimised based on increasing the possible intensities of all analytes.The capillary voltage was 3. 2 kV and the source block and desolvation temperatures were 110 and 350℃, respectively.Nitrogen was used as the cone gas and the desolvation gas at flow rates of 50 and 660 L/h, respectively.The monitoring mode used was multiple reaction monitor (MRM).Ionisation mode, precursor ion, product ions, cone voltage, and collision energies of each compound and retention times are shown in Table 1.

Table 1 Parameters of MRM conditions and retention times of 4 β_2-agonists

No.	Compound	ESI	Retention time (min)	Parent ion (m/z)	Product ions (m/z)	Cone voltage (V)	Collision energy (eV)
1	zilpaterol	+	2. 02	262. 0	244. 0, 185. 1	20	12, 25
2	cimbuterol	+	2. 40	234. 8	160. 0, 216. 5	18	15, 8
3	clenproperol	+	3. 71	263. 5	245. 2, 203. 1	20	10, 20
4	bambuterol	+	4. 66	368. 0	294. 0, 72. 0	25	30, 20

RESULTS AND DISCUSSION

The effect of deproteinization for disruption of pH Meat samples contain a large amount of protein and other polar compounds, the chemical properties of which are similar to β_2-agonists to some extent. These compounds also can be adsorbed and desorbed in mixed-mode SPE cartridges as β_2-agonists, so interferents such as protein and polar compounds must be removed from extracts before SPE. Proteins are amphoteric compounds that exhibit characteristics of the isoelectric point, so they can precipitate out from an extraction solution by disruption pH to the isoelectric point. The developed method adjusted pH twice, first to 1.0±0.2 using perchloric acid, and then to 9.5±0.2 using a 10 M sodium hydroxide solution. These pH adjustment eliminated almost all proteins.

Comparison of SPE after deproteinization and direct SPE is shown in Fig.1 using a spiked beef muscle sample at 0.1μg/kg. Under SPE conditions after deproteinization, cleaner chromatograms (1) were obtained for qualitative detection of the 4 β_2-agonists of interest in beef meat samples. All 4 β_2-agonists target peaks and peaks of impurities were isolated from the baseline, and peak areas of impurites were small enough that matrix effects could be ignored. Disruption of pH was, therefore, effective for elimination of the influence of impurites. By contrast, chromatograms (2) contained many peaks of impurities using direct SPE without deproteinization. Zilpaterol and bambuterol were not completely separated from impurites, which can cause low sensitivity and bad selectivity. Thus, the deproteinization step for disruption of pH before SPE was necessary.

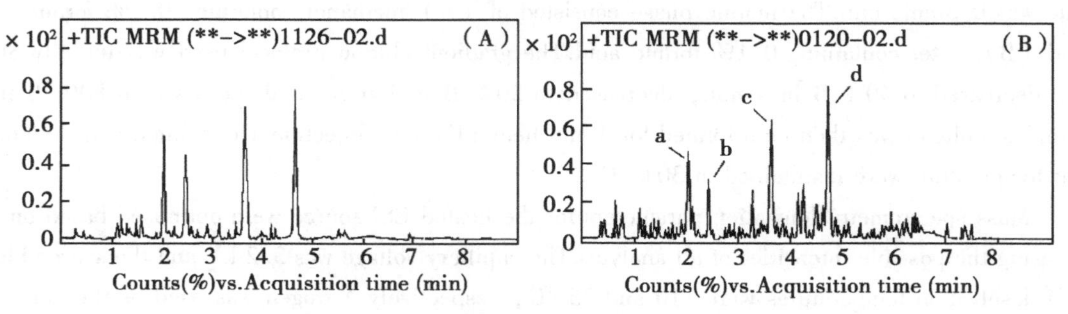

Fig.1 Chromatograms of spiked beef muscle samples treated using SPE after deproteinization (A) and direct SPE (B)

Note: spiked beef muscle at 0.1 μg/kg for zipaterol (a), cimbuterol (b), clenproperol (c), and bambuterol (d).

Optimization of the eluent dosage In order to develop a useful detection method, a sample preparation procedure is required for cleaning of β_2-agonists. Beef muscle meat samples were used as the initial matrix for method validation. The 4 β_2-agonists of interest were spiked at 0.5μg/kg to determine the extraction efficiency, and the dosage of eluent methanol containing 5% ammonium hydroxide for the target compound was the primary factor affecting the cleaning result. Insufficient dosages failed to wash absorbed β_2-agonists from the SPE column. Excessive dosages wasted eluents and polluted the environment. An eluent to ammonia-methanol ratio of 5/95 (v/v) at levels of 0.5, 1.0, 1.5, and 2.0mL/g was used. Better recovery was achieved for the 4 β_2-agonists at 1mL/g. Recovery did not exhibit any obvious variation when eluent dosages were increased to 1.5 and 2.0mL/g, and an eluent dosage of 0.5mL/g led to a lower recovery of less than 70%. A cleansing agent dosage of 1.0mL/g was used in the final optimized method with recovery levels of the 4 β_2-agonists between 86%-90%.

Method validation Beef and goat muscle meats were obtained from β_2-agonist-free farm animals.

Method validation experiments were performed using optimized experimental conditions. Chromatograms for zilpaterol, cimbuterol, clenproperol, and bambuterol spiked at 0.1μg/kg in goat muscle samples prepared using the optimised treatment conditions and instrument conditions are shown in Fig.2. Good chromatographic resolution was achieved.

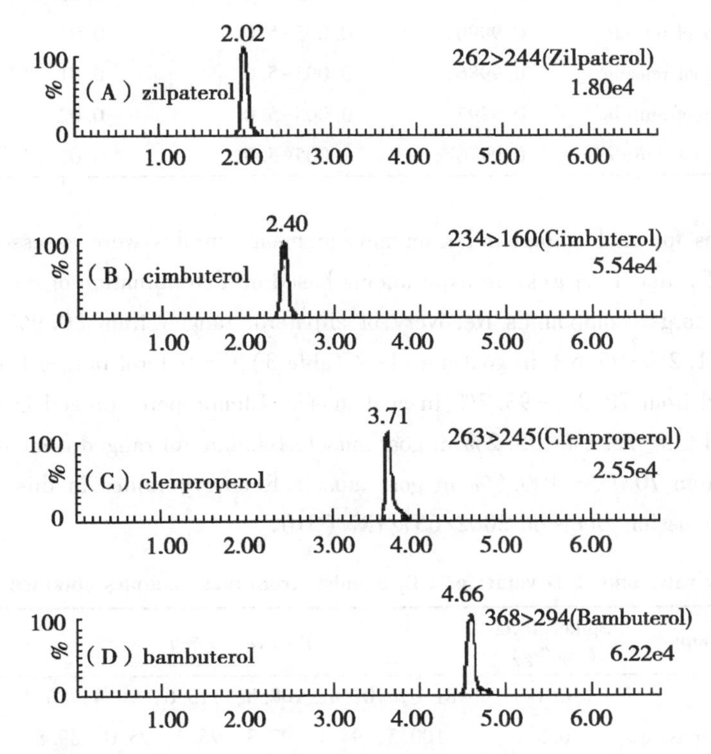

Fig.2 Extract chromatograms for a mixture of 4 β_2-agonists obtained from a blank beef-muscle sample spiked at 0.5μg/kg

Linearity Analysis of experimental data validated the linearity of the method for detection of β_2-agonists in goat and beef muscle samples over a range that depended on the target compound. A multipoint calibration was used in recovery experiments. Calibration curves were obtained based on linear regression analysis of experimental data for spiking using matrix-matched calibration curves for blank goat and blank beef muscle samples with 6 calibration points at 0.005, 0.05, 0.1, 0.2, 0.5, and 5.0μg/kg. Calibration curves were constructed, and squared regression coefficients (R^2) were calculated for each compound (Table 2). Statistical analysis of analytical data for standards demonstrated a satisfactory relationship between the concentration and the response, and the correlation coefficients (R^2) obtained using this method ranged from 0.9976 to 0.9997, which contained the stipulation that $R^2>0.99$ for drug residue analysis.

Table 2 Linear ranges, regression coefficients, and LOD and LOQ values of 4 β_2-agonists

Compound	Sample	R^2	Linear range (μg/kg)	LOD (μg/kg)	LOQ (μg/kg)
Zilpaterol	beef muscle	0.9991	0.005-5.0	0.02	0.08
	goat muscle	0.9981	0.005-5.0	0.02	0.06

(continued)

Compound	Sample	R^2	Linear range (μg/kg)	LOD (μg/kg)	LOQ (μg/kg)
Cimbuterol	beef muscle	0.9988	0.005-5.0	0.01	0.03
	goat muscle	0.9995	0.005-5.0	0.01	0.03
Clenproperol	beef muscle	0.9990	0.005-5.0	0.01	0.03
	goat muscle	0.9986	0.005-5.0	0.01	0.03
Bambuterol	beef muscle	0.9997	0.005-5.0	0.02	0.02
	goat muscle	0.9976	0.005-5.0	0.01	0.02

Recovery rates for each compound from muscle meat samples were assessed at concentration levels of 0.1, 0.5, and 1.0μg/kg in experiments based on determination of ratios of measured and added amounts of target compounds. Recovery of zilpaterol ranged from 80.9% - 108.8% in beef muscle and from 71.2%-95.6% in goat muscle (Table 3). Cimbuterol ranged from 71.2%-99.2% in beef muscle and from 70.9%-93.7% in goat muscle. Clenproperol ranged from 75.8%-97.6% in beef muscle and from 72.1%-95.2% in goat muscle. Bambuterol ranged from 76.8%-100.4% in beef muscle and from 70.1%-100.5% in goat muscle. Recovery values in this study meet the requirements of Commission Decision 2002/657/EC (31).

Table 3 Recovery rates and RSD values of 4 β_2-agonists from meat samples obtained using LC-MS/MS

Compound	Sample	Spiked level (μg/kg)	Recovery (%)	RSD (%)
Zilpaterol		0.1	81.6, 102.4, 108.8, 113.6, 90.4, 86.4	13.3
	beef muscle	0.5	100.5, 94.1, 97.3, 96.2, 95.0, 89.8	3.7
		1	80.64, 92.4, 87.0, 96.1, 83.6, 80.9	7.3
		0.1	95.1, 85.0, 92.0, 78.1, 74.2, 76.3,	10.6
	goat muscle	0.5	96.5, 77.2, 97.2, 77.1, 94.4, 82.5	10.9
		1	95.6, 85.2, 71.2, 75.6, 78.6, 82.1	10.4
Cimbuterol		0.1	99.2, 96.0, 112.8, 71.2, 76.0, 84.0	11.4
	beef muscle	0.5	86.7, 84.9, 97.6, 92.0, 80.3, 77.9	8.4
		1	87.0, 79.0, 72.1, 91.4, 80.1, 75.2,	8.9
		0.1	87.1, 91.4, 86.2, 78.6, 74.2, 75.8	8.1
	goat muscle	0.5	84.2, 73.6, 95.0, 92.6, 70.9, 72.5	13.1
		1	93.7, 92.6, 91.5, 80.2, 75.6, 72.0	11.3
Clenproperol		0.1	83.2, 88.8, 78.4, 80.8, 99.2, 95.2	9.4
	beef muscle	0.5	97.6, 89.3, 86.9, 93.8, 98.7, 95.7,	5.0
		1	75.8, 83.6, 94.2, 76.5, 77.1, 80.9	8.5
		0.1	72.1, 84.6, 78.6, 95.2, 84.7, 92.1	10.0
	goat muscle	0.5	80.5, 79.1, 82.5, 71.6, 77.3, 82.5	5.2
		1	87.5, 92.4, 87.5, 86.3, 85.1, 92.6	3.5

(continued)

Compound	Sample	Spiked level (μg/kg)	Recovery (%)	RSD (%)
Bambuterol		0. 1	76. 8, 108. 4, 88. 0, 100. 4, 86. 4, 105. 6	11. 5
	beef muscle	0. 5	87. 7, 97. 8, 96. 2, 95. 7, 100. 0, 93. 4	4. 4
		1	78. 8, 83. 6, 91. 4, 82. 0, 75. 8, 80. 2	6. 5
		0. 1	87. 5, 82. 2, 74. 2, 95. 6, 87. 6, 82. 3	8. 4
	goat muscle	0. 5	75. 3, 90. 4, 82. 4, 100. 5, 77. 9, 80. 2	11. 1
		1	92. 4, 84. 2, 82. 4, 74. 1, 70. 1, 82. 4	9. 7

Precision The precision of the method was determined using repeated spiked samples and expressed as a relative standard deviation (RSD). Method precision was evaluated based on analysis of spiked samples ($n=6$) (Table 3). RSD values of zilpaterol ranged from 3. 7%-3. 3% in beef muscle and from 10. 4%-10. 9% in goat muscle. Cimbuterol ranged from 8. 4%-10. 4% in beef muscle and from 8. 1%-11. 3% in goat muscle. Clenproperol ranged from 5. 0%-9. 4% in beef muscle and from 3. 5%-10. 0% in goat muscle. Bambuterol ranged from 4. 4%-11. 5% in beef muscle and from 8. 4%-11. 1% in goat muscle. Thus, all RSD values were less than 15% at the 3 different concentration levels of 0.1μg/kg, 0.5μg/kg, and 1μg/kg.

LOD and LOQ Sensitivity values of the method were evaluated based on limit of detection (LOD) and LOQ values. The limit of quantitation (LOQ) was determined using spiked samples with the minimum detectable amount of analyte set at a signal-to-noise ratio (S/N) of 10. LOD and LOQ values (Table 2) of target compounds achieved using the method were within the range of 0. 1-0. 2μg/kg and 0. 02-0. 08μg/kg for spiked beef muscle, and 0. 1-0. 2μg/kg and 0. 02-0. 06μg/kg for spiked goat muscle. According to CRL guidance paper (2007) (32), the maximum detection limits of target compounds are zilpaterol, muscle 5μg/kg; clenproperol, muscle 0. 5μg/kg; and cimbuterol, muscle 0. 5 ppb. LOQ values for the method reported herein were 1 ppb zilpaterol in beef muscle and 1 ppb zilpaterol in goat muscle, 0. 5 ppb clenproperol in beef muscle and 0. 5μg/kg clenproperol in goat muscle, 0. 3 ppb cimbuterol in beef muscle and 0. 4μg/kg cimbuterol in goat muscle, and 0. 3μg/kg bambuterol in beef muscle and 0. 3 ppb bambuterol in goat muscle. LOQ values reported in this study meet the requirements of Commission Decision.

Matrix effects Signal suppression is commonly encountered in LCESI-MS analysis for complex sample matrices. Optimization of sample preparation methods can reduce the level of matrix-induced interference, but interference can not be eliminated. Therefore, matrix effects must be reduced using a pre-treatment cleaning step, for delivery of better sensitivity. Using the same pre-treatment method and instruments, samples of goat and beef muscle meats were cleansed and tested. Matrix effects in these meat samples were evaluated based on LC-MS/MS analysis of blank and spiked samples at 0. 1μg/kg (Fig.3). After cleaning by deproteinization based on adjustment of pH and SPE (MCX column), chromatograms of both blank beef (1) and blank goat samples (2) exhibited few impurity-containing peaks. The area of impurity-containing peaks was small enough to be ignored. The 4 target compound peaks in chromatograms for spiked goat (3) and spiked beef samples

(4) were separated from the baseline within 5 min and the peak shape was well defined with no interfering peaks observed at the retention times of the 4 β_2-agonists. Thus, the pre-treatment method was effective for purification of muscle samples.

Real samples method application The method was applied for analysis of 25 beef muscle and 25 goat muscle samples collected from different markets (Lanzhou and Zhangye) in China. All meat samples were processed according to the method described, and no detectable residues of zilpaterol, cimbuterol, clenproperol, or bambuterol were identified in any meat sample, probably due to stringent regulations of the government of China. This article does not contain any studies with human or animal subjects performed by the any of the authors.

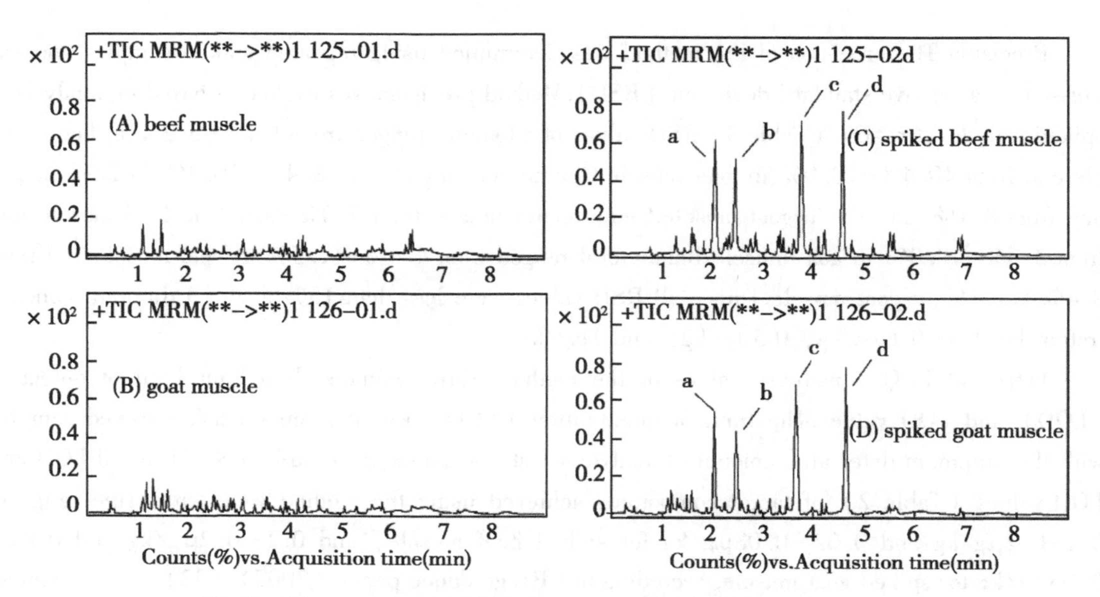

Fig.3 Chromatograms of blank meat samples and spiked samples

Note: at 0. 1μg/kg for zilpaterol (a), cimbuterol (b), clenproperol (c), and bambuterol (d)

An LC-MS/MS method was used for identification of zilpaterol, cimbuterol, clenproperol, and bambuterol in beef and goat muscle samples at trace levels. A combination of deproteinization based on adjustment of pH and solid phase extraction was used for extraction and purification of β_2-agonist residues from meat samples. This new sample purification approach exhibited good specificity, linearity (R^2 = 0.9976-0. 9997), accuracy (recovery, 70. 1% – 108. 8%; RSD, 3. 5% – 13. 3%), and precision (LOD = 0. 01–0. 02μg/kg; LOQ = 0. 02–0. 08μg/kg). An advantage was high sensitivity, compared with other published methods. The achieved decision limit (0. 3μg/kg) and detection capability (1μg/kg) proved the efficiency of the methodology for identification of ultratrace levels of β_2-agonist residues. Good validation results were obtained taking into account difficulties of a multi-residue analytical method for 4 different β_2-agonists. The method can be a valuable tool in regulatory laboratories for control of illegal use of drugs in meat production.

Acknowledgments The research was funded by a research grant from the Chinese Academy of Agricultural Science (1610322014014).

Disclosure The authors declare no conflict of interest.

References

[1] Magda C, Piotr S, Marek K, Natalia M, Jolanta K.Determination of β-blockers and β-agonists using gas chromatography and gas chromatography-mass spectrometry-A comparative study of the derivatization step.J.Chromatogr.A 1218: 8110-8122 (2011).

[2] Juan C, Igualada C, Moragues F, Leóna N.Development and validation of a liquid chromatography tandem mass spectrometry method for the analysis of β_2-agonists in animal feed and drinking water.J.Chromatogr.A 1217: 6061-6068 (2010).

[3] Oraphan A, Wolfgang B, Markus H, Christian WK, Leena S.A sensitive nonaqueous capillary electrophoresis-mass spectrometric method for multiresidue analyses of β-agonists in pork. Biomed. Chromatogr.24: 588-599 (2010).

[4] Morales-Trejo F, León SV, Escobar-Medina A, Gutiérrez-Tolentino R.Application of high-performance liquid chromatography-UV detection to quantification of clenbuterol in bovine liver samples.J.Food Drug Anal.21: 414-420 (2013).

[5] Corcia DD, Morra V, Pazzi M, Vincenti M.Simultaneous determination of β_2 agonists in human urine by fast-gas chromatography/mass spectrometry: Method validation and clinical application.Biomed.Chromatogr.24: 358-366 (2010).

[6] Gallo P, Brambilla G, Neri B, Fiori M, Testa C, Serpe L.Purification of clenbuterol-like β_2-agonist drugs of new generation from bovine urine and hair by α_1-acid glycoprotein affinity chromatography and determination by gas chromatography-mass spectrometry.Anal.Chim.Acta 587: 67-74 (2007).

[7] González-Antuna A, Rodríguez-González P, Centineo G, Alonsoa JIG.Simultaneous determination of seven β_2-agonists in human and bovine urine by isotope dilution liquid chromatography-tandemmass spectrometry using compound-specific minimally ^{13}C-labelle danalogues.J.Chromatogr.A 1372: 63-71 (2014).

[8] Du W, Zhang SR, Fu Q, Zhao G, Chang C.Combined solid-phase microextraction and high-performance liquid chromatography with ultroviolet detection for simultaneous analysis of clenbuterol, salbutamol and ractopamine in pig samples.Biomed.Chromatogr.27: 1775-1781 (2013).

[9] Shao B, Jia XF, Zhang J, Meng J, Wu YN, Duan HJ, Tu XM.Multi-residual analysis of 16 β-agonists in pig liver, kidney and muscle by ultra performance liquid chromatography tandem mass spectrometry.Food Chem.114: 1115-1121 (2009).

[10] Shen JZ, Zhang Z, Yao Y, Shi WM, Liu YB, Zhang SX.Time-resolved fluoroimmunoassay for ractopamine in swine tissue.Anal.Bioanal.Chem.387: 1561-1564 (2007).

[11] Keskin S, Özer D, Temizer A.Gas chromatography-mass spectrometric analysis of clenbuterol from urine.J.Pharmaceut.Biomed.18: 639-644 (1998).

[12] Wang GM, Zhao J, Peng T, Chen DD, Xi CX, Wang X, Zhang JZ.Matrix effects in the determination of β-receptor agonists in animal-derived foodstuffs by ultra-performance liquid chromatography tandem mass spectrometry with immunoaffinity solid-phase extraction.J.Sep.Sci.36: 796-802 (2013).

[13] Wang LQ, Zeng ZL, Wang XF, Yang JW, Chen ZH, He LM.Multiresidue analysis of nine β-agonists in animal muscles by LC-MS/MS based on a new polymer cartridge for sample cleanup.J.Sep.Sci. 36: 1843-1852 (2013).

[14] Fan S, Miao H, Zhao YF, Chen HJ, Wu YN.Simultaneous detection of residues of 25 β_2-agonists and 23 β-blockers in animal foods by high-performance liquid chromatography coupled with linear ion trap mass spectrometry.J.Agr.Food Chem.60: 1898-1905 (2012).

[15] González-Antuña A, Domínguez-Romero JC, García-Reyes JF, Rodríguez-González P, Centineoc G,

Alonso JIG, Molina-Díazb A.Overcoming matrix effects in electrospray: Quantitation of β_2-agonists in complex matrices by isotope dilution liquid chromatography-mass spectrometry using singly ^{13}Clabeled analogues.J.Chromatogr.A 1217: 3612-3618 (2013).

[16] Jenny C, Jack H.Quantitative multi-residue determination of β_2-agonists in bovine urine using on-line immunoaffinity extraction-coupled column packed capillary liquid chromatography-tandem mass spectrometry.J.Chromatogr.B 691: 357-370 (1997).

[17] Zhang GJ, Fang BH, Liu YH, Wang XF, Xu LX, Zhang YP, He LM.Development of a multi-residue method for fast screening and confirmation of 20 prohibited veterinary drugs in feedstuffs by liquid chromatography tandem mass spectrometry.J.Chromatogr.B 936: 10-17 (2013).

[18] Gao F, Wu M L, Zhang Y, Wang G, Wang QJ, He PG, Fang YZ.Sensitive determination of four β_2-agonists in pig feed by capillary electrophoresis using on-line sample preconcentration with contactless conductivity detection.J.Chromatogr.B 973: 29-32 (2014).

[19] Thi AHN, Thi NMP, Thi TD, Thi TT, Jorge S, Thi Q.Hoa N, Peter CH, Thanh DM.Simple semi-automated portable capillary electrophoresis instrument with contactless conductivity detection for the determination of β_2-agonists in pharmaceutical and pig-feed samples.J.Chromatogr.A 1360: 305-311 (2014).

[20] Kramer S, Blaschke G.High-performance liquid chromatographic determination of the β-selective adrenergic agonist fenoterol in human plasma after fluorescence derivatization.J.Chromatogr.B 751: 169-175 (2001).

[21] Du W, Zhao G, Fu Q, Sun M, Zhou HY, Chang C.Combined microextraction by packed sorbent and high-performance liquid chromatography-ultraviolet detection for rapid analysis of ractopamine in porcine muscle and urine samples.Food Chem.145: 789-795 (2014).

[22] Jelka P, Nina P, Ana V, Dinka M, Nada V.Determination of residual ractopamine concentrations by enzyme immunoassay in treated pig's tissues on days after withdrawal.Meat Sci.90: 755-758 (2012).

[23] Ki HP, Jeong-Heui C, Soon-Kil C, Jong-Hyouk P, Ki SK, Hee RP, Hyung SK.Development of QuEChERS-based extraction and liquid chromatography-tandem mass spectrometry method for quantifying flumethasone residues in beef muscle.Meat Sci.92: 749-753 (2012).

[24] Zhan J, Zhong YY, Yu XJ, Peng JF, Chen SB, Yin JY, Zhang JJ, Zhu Y.Multi-class method for determination of veterinary drug residues and other contaminants in infant formula by ultra performance liquid chromatography-tandem mass spectrometry.Food Chem.138: 827-834 (2013).

[25] Juan CD-R, Juan FG-R, Rubén M-R, Esther M-L, María, LD, Antonio, M-D.Detection of main urinary metabolites of β_2-agonists clenbuterol, salbutamol and terbutaline by liquid chromatography high resolution mass spectrometry.J.Chromatogr.B 923-924: 128-135 (2013).

[26] Lee DW, Mona IC, Daniel RD.Multiresidue confirmation of β-agonists in bovine retina and liver using LC-ES/MS/MS.J.Chromatogr.B 813: 35-45 (2004).

[27] Wang L, Li YQ, Zhou YK, Yang Y.Determination of Four β_2-agonists in meat, liver and kidney by GC-MS with dual internal standards.Chromatogr. 71: 737-739 (2010).

[28] Qiao FX, Du JJ.Rapid screening of clenbuterol hydrochloride in chicken samples by molecularly imprinted matrix solid-phase dispersion coupled with liquid chromatography.J.Chromatogr.B 923-924: 136-140 (2013).

[29] Eshaq (Isaac) S, Sin CC, Sami J, Gene A, Michael KH.Determination of ractopamine in animal tissues by liquid chromatography-fluorescence and liquid chromatography/tandem mass spectrometry.Anal.Chim.Acta 483: 137-145 (2003).

[30] Jean-Philippe A, Philippe M, Bruno LB, Francois A.Identification of ractopamine residues in tissue

and urine samples at ultra-trace level using liquid chromatography-positive electrospray tandem mass spectrometry.J.Chromatogr.B 774：59-66（2002）.

[31] Commission Decision.Implementing Council Directive 96/23/EC concerning the performance of analytical methods and the interpretation of results，14，August，2002，7-8（2002）.

[32] Commission of the European Communities.CRLs view on state of the art analytical methods for national residue control plans.CRL guidance paper，7，December，2007，4-5（2007）.

（发表于《Food Science and Biotechnology》，SCI：0.653）

Application of Orthogonal Design to Optimize Extraction of Polysaccharide from *Cynomorium songaricum* Rupr (Cynomoriaceae)

WANG Xiao-li[1], WANG Fan[2], JING Yan-jun[2],
WANG Yong-gang[2], LIN Peng[2], YANG Lin[2] *

(1. Lanzhou Institute of Husbandry and Pharmaceutical Sciences of Chinese Academy of Agricultural Sciences, Lanzhou 730050; 2. School of Life Science and Engineering, Lanzhou University of Technology, Lanzhou 730050, China)

Abstract: Purpose: To optimize the extraction technology of polysaccharides from Cynomorium songaricum Rupr by ultrasonic-assisted extraction (UAE).

Methods: Four parameters including ultrasonic power, ratio of raw material to water, extraction temperature, and extraction time were optimized by orthogonal design. The effects of the factors on the yield of polysaccharides were also studied. The hydroxyl and 1, 1-diphenyl-2-picryl-hydrazyl (DPPH) radical scavenging activities were determined in vitro by spectrophotometry.

Results: The optimal conditions were as follows: 1 : 30 as ratio of raw material to water, extraction for 70 min at 80℃ with ultrasonic power being 420 W. Under these conditions, the yield of polysaccharides was up to 4.51%, which was significantly higher than that obtained under the initial conditions (3.82%). DPPH radical scavenging activity reached 64.82% at 0.012mg/mL, while hydroxyl radical scavenging activity was 18.36% at 0.5mg/mL.

Conclusion: Ultrasonic-assisted extraction technology is a useful tool for the extraction of bioactive components from biological materials.

Key words: *Cynomorium songaricum*; Polysaccharide; Ultrasonic-assisted extraction; Orthogonal design

Tropical Journal of Pharmaceutical Research is indexed by Science Citation Index (SciSearch), Scopus, International Pharmaceutical Abstract, Chemical Abstracts, Embase, Index Copernicus, EBSCO, African Index Medicus, JournalSeek, Journal Citation Reports/Science Edition, Directory of Open Access Journals (DOAJ), African Journal Online, Bioline International, Open-J-Gate and Pharmacy Abstracts

INTRODUCTION

Cynomorium songaricum Rupr. (Cynomoriaceae) is an achlorophyllous holoparasite with distribution in northwestern China. Among its chemical constituents, steroids, triterpenes, flavonoids and lignans have been reported previously [1,2]. Stems of *Cynomorium songaricum* Rupr. are known as suo yang in China. According to the ancient Chinese medical literature, *C. songaricum* is effective in regulating endocrinopathy and heightening sexual function and anti-aging such as dementia [3,4]. Because it works usually together with other Chinese herbs, its individual bioactivity has been scarcely studied.

Polysaccharide is one of common ingredients in many fruits with formidable health care functions in anti-tumor[5,6], antioxidation [7,8], antiinflammatory[9] and human immunity improvement [10]. The research which emphases on the modern medicine and functionalized food chemistry mainly focuses onthe physiological activities and structures of polysaccharides. So far, a large number of researches on extraction technologies of various polysaccharides have been reported, including solvent extraction [11], biological enzymatic methodand ultrasonic extraction [13,14].

The aim of this study was to apply ultrasonicassisted extraction (UAE) to isolate polysaccharides from *C. songaricum*, and to optimize the extraction parameters using orthogonal design which is an effective method to evaluate the effects of multiple factors and their interaction on one or more response variables.

EXPERIMENTAL

Material and reagents

Cynomorium songaricum Rupr. (Cynomoriaceae) used in this study was provided by Yide biological Technology Co, Ltd (Guazhou, PR China). One hundred grams of C. *songaricum* was dried, ground to powder and then sieved through a mesh screen with size of aperture 370μm. All solvents were analytical grade and obtained from Beijing Solarbio Co, Ltd (Beijing, P. R China).

Extraction and determination of polysaccharide

Polysaccharide was extracted from the dried samples (1. 0g) by ultrasonic-assisted treatment in an ultrasonic processor (KQ-250DB, Kunshan Ultrasonic Instruments Co, Ltd, Jiangsu, China). Then, impurity proteins, vitamins, lipids and other pulp components were separated by centrifugation in a 50mL centrifuge tube. The supernatant was collected and concentrated in a vacuum concentrator and precipitated with 80% ethanol, and then the precipitate was washed sufficiently with ethanol, and suspended by 100mL of water. The solution was filtered later. The content of polysaccharide extracted from *C. songaricum* was measured by phenol-sulfuric acid method using glucose as standard [15]. The optical density of reaction solution was measured at 485nm. The content of polysaccharide in this study was calculated according to the equation of linear regression ($Y = 12.6308x-0.0144$, $R^2=0.9997$) based on the standard curve of which horizontal coordinate and vertical coordinate denoted the concentration of glucose (mg/mL) and OD485, respectively. The yield of polysaccharide could be calculated as in Eq 1. Polysaccharide yield (%) = (CNV) / 1000W) 1000 ···. (1) where C is the concentration of polysaccharide calculated by the calibrated regression equation (mg/mL), N is the dilution factor, V is the total volume of extraction solution (mL) and W is the weight of raw material (g).

SINGLE-FACTOR TEST

Effect of ultrasonic power on the yield of polysaccharide

The ultrasonic power was set at 240, 360, 480 and 600 W, respectively, meanwhile, ratio of raw material to water, extraction temperature, extraction times and extraction duration were fixed at 1; 30, 80℃, twice and 60 min. Under these conditions, the content of polysaccharide extracted was determined by phenol-sulfuric acid method and calculated.

Effect of ratio of raw material to water on the yield of polysaccharide

The ultrasonic power, extraction temperature, extraction duration and extraction times were fixed at 480 W, 70°C, 60 min and twice; in addition, the ratio of raw material to water was set at 1 : 6, 1 : 10, 1 : 20, 1 : 30, 1 : 40 and 1 : 50 g/mL, respectively. Under these conditions, the content of polysaccharide extracted was determined by phenol-sulfuric acid method and calculated according to equation (1) .

Effect of extraction temperature on the yield of polysaccharide

The extraction temperature was set at 60℃, 70℃, 80℃, and 90℃, with the ratio of raw material to water, ultrasonic power, extraction times, and extraction duration at 1 : 30, 480 W, twice and 60 min, respectively. The content of polysaccharide extracted was determined by phenol-sulfuric acid method and calculated as in Eq 1 above.

Effect of extraction time on the yield of polysaccharide

The extraction duration was set at 30min, 50min, 70min and 90min, respectively, and ratio of raw material to water, ultrasonic power, extraction times, and extraction temperature were fixed at 1 : 30, 480 W, twice and 80°C. The content of polysaccharide extracted was determined by phenol-sulfuric acid method and calculated as in Eq 1 above.

Optimization of extraction conditions by orthogonal design

According to the results of the single factor experiment, orthogonal array [L_9 (3^4)] with four factors at three levels was chosen to seek the optimum conditions for the maximal yield of polysaccharide extracted from *C. songaricum*. Factors and its levels of orthogonal design were shown in Table 1 and the results of orthogonal test are shown in Table 2 and Table 3. Each of the 9 experiments was performed in triplicate.

Analysis of variance

Analysis of variance (ANOVA) is one of the most important statistical tools, which is used to uncover the main factor and interaction effects of variables. It is also used to identify the procedure parameters that are statistically significant. In ANOVA, F ratio is employed to recognize these significant parameters from others. F value is the ratio of the variance estimation of the treatment effect to the variance estimation of the error. The large value of F ratio means that the selected parameter has a significant effect on the evaluation index compared with the error variation. The sum of squares and degrees of freedom corresponding to the eliminated terms are added into the residual sum of squares and degree of freedom.

Determination of antioxidant activity in *vitro Scavenging* of DPPH radical assay

The DPPH scavenging assay was performed according to the method of Shimada et al[16]. 0. 2 mM DPPH in 60% ethanol was prepared before UV measurements. Then, 3. 0mL of polysaccharide extracted from Cynomorium songaricum Rupr (0. 002, 0. 004, 0. 008 and 0. 012mg/mL) was added into 1. 0mL DPPH, and kept at room temperature for 30 min in the dark. The absorbance of the solution was measured at 525 nm later. Ascorbic acid (0. 002, 0. 004, 0. 008 and 0. 012mg/mL) was used as positive controls.

The scavenging activity of DPPH radical (%) was calculated according to the method of Shimada et al [16].

Hydroxyl radical scavenging assay

Hydroxyl radical scavenging activity was measured according to the method of Jin et al [17]. 10mL reaction mixture is inclusive of 2mL of 200 mM sodium phosphate buffer (pH 7. 4), 1. 5mL of 5. 0 mM 1, 10-phenanthroline aqueous solution, 1. 0mL of 7. 5 mM $FeSO_4$ aqueous solution, 1. 0mL of 0. 1% H_2O_2 aqueous solution, 0. 1mL of the polysaccharide aqueous solution with different concentrations (0. 2mg/mL, 0. 30mg/mL, 0. 40mg/mL and 0. 50mg/mL) and 4. 4mL of distilled water. After incubated at 37℃ for 1 h, the absorbance of the mixture was measured at 510 nm. The scavenging activity of hydroxyl radical production was calculated according to the method of Jin et al [17].

Statistical analysis

The values are expressed as means ± standard deviation. Statistical analysis was done by Origin Lab software (version 8. 0). ANOVA was used to analyze significant differences among the trails in the orthogonal design. The criterion for statistical significance was $p<0.05$, and extreme significance was $p<0.01$.

RESULTS

Effect of ultrasonic power on yield of polysaccharides

The effects of different ultrasonic powers on the yield of polysaccharides were investigated in the commended study. As shown in Fig. 1a, the extraction yield increased as the increase of ultrasonic power ranged from 240 to 480 W, and reached the maximum value at 480 W.

Effect of ratio of raw material to water on yield of polysaccharides

The effects of different ratios of raw material to water, on the yield of polysaccharides were studied. As shown in Fig. 1b, the extraction yield continued to increase evidently with the growth in the ratio of raw material to water until reaching at 1 : 30 g/mL.

Table 1 Orthogonal factors and levels

Level	Factors			
	A (Ultrasonic power, W)	B (ratio of raw material to water, g/mL)	C (extraction Temperature, ℃)	D (extraction time, min)
1	420	1 : 25	75	60
2	480	1 : 30	80	70
3	540	1 : 35	85	80

Effect of extraction temperature on yield of polysaccharides

The effects of different extraction temperature on the yield of polysaccharides were evaluated in this research. As was indicated in Fig. 1c, the yield of polysaccharides was increasing all the time with the increase of extraction temperature range of 60-80℃ and reached the peak value at 80℃. The yield of polysaccharides decreased when the extraction temperature exceeded 80°C.

Effect of extraction time on extraction yield of polysaccharides

The effect of extraction time on yield of polysaccharides from *C. songaricum* was shown in Fig. 1d. It could be found that the extraction yield increased as extraction time ascended from 30 to 70 min, peaked at 70 min, and then decreased.

Optimization of extraction conditions of polysaccharides

Orthogonal array [L_9 (3^4)] was used to optimize the extraction parameters of polysaccharides. Results were shown in Table 2 and Table 3. There were nine experiments corresponding to the nine rows and four columns. Each experiment was repeated three times under the same conditions to eliminate the effects of noise sources in the process. The responses of all tests were given in Table 2. The significance of the factors was studied by ANOVA. The ANOVA summary was shown in Table 3. The F-value was used for qualitative analysis of whether effective factors existed. Factors A and C are significant for the polysaccharide yield from *C. songaricum* at 99% confidence level. Meanwhile, factors B and D show less effect on the polysaccharide yield from *C. songaricum* at the 90% confidence level.

Table 2 Results of orthogonal experiment

No.	A (Ultrasonic power, W)	B (ratio of raw material to water, g/mL)	C (extraction Temperature, ℃)	D (extraction time, min)	Yield (%, n=3)
1	1	1	1	1	3.84
2	1	2	2	2	4.37
3	1	3	3	3	3.73
4	2	1	2	3	3.88
5	2	2	3	1	3.67
6	2	3	1	2	3.95
7	3	1	3	2	3.08
8	3	2	1	3	3.34
9	3	3	2	1	4.01
k1	3.980	3.600	3.710	3.840	
k2	3.831	3.792	4.087	3.798	
k3	3.479	3.898	3.493	3.652	
R	0.501	0.298	0.594	0.188	

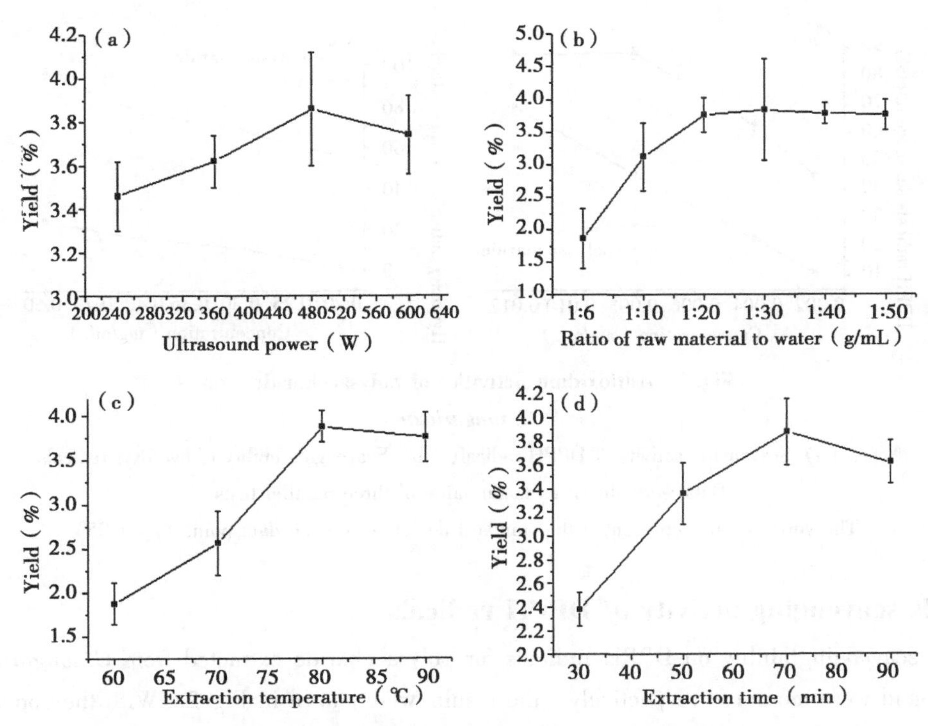

Fig. 1 Effects of different factors on extraction yield of polysaccharide

Note: ultrasound power (a), ratio of raw material to water (b), extraction temperature (c), and extraction time (d). All data were mean values of triplicate. The vertical error bars represented the standard deviation of each data point ($P<0.05$)

Table 3 ANOVA data of the orthogonal experiment

Factor	SS	Df	MS	*F*-value	*P*-value
A (ultrasonic power)	1.192	2	0.596	7.584	0.004
B (ratio of raw material to water)	0.410	2	0.205	2.610	0.101
C (extraction temperature)	1.623	2	0.811	10.323	0.001
D (extraction time)	0.175	2	0.087	1.111	0.351
Error (%)	1.415	18	0.079		

Note: SS=sum of squares; DF=degree of freedom; MS=mean square

The results indicated that factor C was the most significant factor contributing to the extraction efficiency of polysaccharides, followed by factor A, factor B, and factor D. The optimum condition was obtained by orthogonal design, and was recommended as follows: ultrasonic power 420W, ratio of raw material to water 1 : 30 g/ml, extraction temperature 80℃, and extraction time 70 min. To test the reliability of the commended method, The polysaccharides was extracted for five times under above conditions, and results showed that experimental value 4.51% ($n=5$, RSD= 1.98%) was consistent with the predictive values.

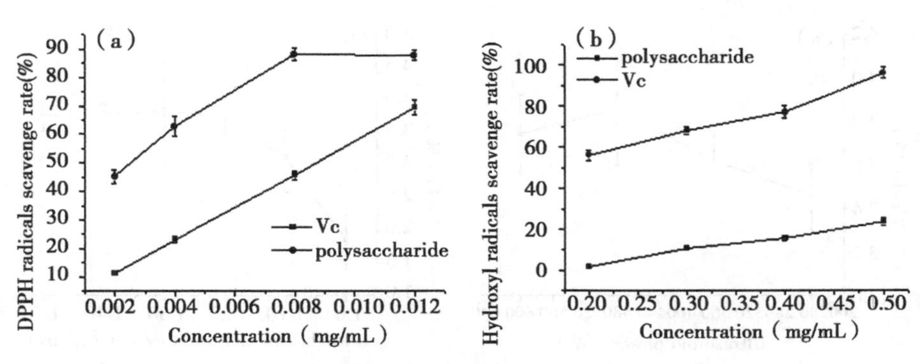

Fig. 2 Antioxidant activities of polysaccharide from *C. songaricum*

Note: (a) Scavenging activity of DPPH radical, (b) Scavenging ability of hydroxyl radicals. Data were shown as mean value of three parallel tests. The vertical bars represented the standard deviation of each data point ($p<0.05$)

Radicals scavenging activity *of* DPPH radicals

The scavenging ability on DPPH radicals for polysaccharide extracted from *C. songaricum* and ascorbic acid were measured respectively, the results were shown in Fig. 2a. With the concentration of polysaccharide increasing from 0. 002mg/mL to 0. 012mg/mL, the DPPH radical scavenging activity increased all the time until reached to 64. 82%, and the maximum scavenging rate of ascorbic acid could reach 88. 61% at 0. 008mg/mL. This manifested that the polysaccharide from *C. songaricum* had astrong scavenging ability on DPPH radicals.

Scavenging activity of hydroxyl radicals

The result of hydroxyl radical scavenging activities of the polysaccharide was given in Fig. 2b, which showed the difference of scavenging activity between the polysaccharide and ascorbic acid. Ascorbic acid was used as positive control. The scavenging activity increased significantly in a concentrationdependent way with polysaccharide concentration range of 0. 2-0. 5mg/mL, and achieved maximum value of 18. 36% at 0. 5mg/mL, however the maximum scavenging activity of ascorbic acid was 91. 71% at 0. 5mg/mL, which indicated that hydroxyl radicals scavenging activities of polysaccharide were not obvious compared with the ascorbic acid.

DISCUSSION

Instead of conventional extraction methods including fluid extraction, Soxhlet extraction, supercritical or subcritical extraction and pressurized liquid extraction, polysaccharides were extracted from *C. songaricum* Rupr using ultrasonic-assisted extraction in this study. Because ultrasonic extraction could obtain higher efficiency of extraction and more healthy extractive by reducing the handling time and use of solvent; meanwhile, its low operated temperatures can reduce the heat loss caused by high temperatures; in addition, it also prevents the vaporization of solvent during boiling and preserves bioactive substance. It is very useful for the extraction of thermolabile compounds[18,19]. In the present study, the best extraction technological parameters for polysaccharides from *C. songaricum* using ultrasonic-assisted extraction procedure is being reported for the first time. The ultrasound-assisted extraction parameters were optimized (ultrasonic power, extraction temperature, extraction time, and ratio of raw material to water) for increase extraction efficiency of the polysaccharides from *C. songaricum* by orthogonal experimental design. We found that the ultrasonic power and extraction temperature were significant for the polysaccharide extraction from *C. songaricum* at the 99% confidence level. While Ajaz Ahmad [19] found that the extraction temperature was an important factor for extraction efficiency of thepolysaccharide at the 95% confidence level in optimal conditions of ultrasonic-assisted extraction of polysaccharides from Paeonia *emodi*. The p-value was used as evaluation index to check the significance of each coefficient and the interaction strength between variables [20].

Polysaccharides have been studied profoundly as additives in food and pharmaceutical applications due to their unique bioactivity[21,22]. We just determined the scavenging ablity on hydroxyl and DPPH radicals *in vitro*. Several studies have also reported that polysaccharides obtained by ultrasound treatment have different antioxidant activities depending on the extraction conditions (e. g., ultrasound power, extraction time, and extraction temperature)[23]. Some reports indicated that polysaccharide extracted by ultrasonic-assisted extraction had higher antioxidant activities than those polysaccharide extracted by enzyme-assisted extraction [24]. But the extraction technology of polysaccharides from *C. songaricum* and its bioactivities need to be further studied.

CONCLUSION

The four parameters (ultrasonic power, ratio of raw material to water, extraction temperature, and extraction time) tested in UAE were optimized by orthogonal design. The yield of polysaccharides was as high as 4.51% under the optimized extraction conditions as follows: 1:30 as ratio of raw material to water, extraction for 70 min at 80℃ with ultrasonic power being 420 W. The polysaccharides extracted from *C. songaricum* showed significant antioxidant activity.

ACKNOWLEDGEMENT

This research was supported by Agricultural Scientific Research Projects: Forage Feed Resources Development and Utilization of Technology Research and Demonstration, China (no. 201203042), Chinese National Natural Science Foundation (31460032) and Natural Science Foundation of Gansu Province, China (nos. 1212RJYA008 and 1308RJZA287).

References

[1] Chu Q, Tian X, Lin M, Ye J. Electromigration profiles of Cynomorium songaricum based on capillary electrophoresis with amperometric detection. J Agric Food Chem, 2006, 54: 7979-7983.

[2] Jiang Z. H., Tanaka T, Sakamoto M, Jiang T, Kouno I. Studies on a medicinal parasitic plant: Lignans from the stems of Cynomorium songaricum. Chem Pharm Bull, 2001, 49: 1036-1038.

[3] Qi Y, Su G. E. Progress in the research of Cynomorium (Cynomorium songaricum). Zhongcaoyao, 2000, 31: 146-148.

[4] Lu Y, Wang QG, Melzig MF, Jenett SK. Extracts of Cynomorium songaricum Protect SK-N-SH Human Neuroblastoma Cells against Staurosporine-induced Apoptosis Potentially through their Radical Scavenging Activity. Phytother. Res, 2009, 23: 257-261

[5] Chen XP, Wang WX, Li SB, Xue JL, Fan LJ, Sheng ZJ, Chen YG. Optimization of ultrasound-assisted extraction of Lingzhi polysaccharides using response surface methodology and its inhibitory effect on cervical cells. Carbohyd Polym, 2010, 80: 944-948.

[6] Zhu ZY, Liu RQ, Si CL, Zhou F, Wang YX, Ding LN, Jing C, Liu AJ, Zhang YM. Structural analysis and antitumor activity comparison of polysaccharides from Astragalus. Carbohyd Polym, 2011, 85: 895-902.

[7] Shi YL, Yang LH, Cai DH. The effect of pleurotus eryngii polysaccharide on exhausted mice's resistance to oxidation and injury. J Phys Educ, 2005, 12: 56-58. (in Chinese)

[8] Chen J, Zhang T, Jiang B, Mu W, Miao M. Characterization and antioxidant activity of Ginkgo biloba exocarp polysaccharides. Carbohyd Polym, 2012, 87: 40-45.

[9] Chen W, Wang WP, Zhang HS, Huang Q. Optimization of ultrasonic-assisted extraction of water-soluble polysaccharides from Boletus edulis mycelia using response surface methodology. Carbohyd Polym, 2012, 87: 614-619.

[10] Yi Y, Zhang MW, Liao ST, Zhang RF, Deng YY, Wei ZC, Tang XJ, Zhang Y. Structural features and immunomodulatory activities of polysaccharides of longan pulp. Carbohydrate Polymers, 2012, 87: 636-643.

[11] Wang LC, Zhang K, Di LQ, Liu R, Wu H. Isolation and structural elucidation of novel homogenous polysaccharide from Mactra veneriformis. Carbohyd Polym, 2011, 86 (2): 982-987.

[12] Yi Y, Zhang MW, Liao ST, Tang XJ, Zhang RF, Wei ZC. Optimization of ultrasonic-enzyme-assisted extraction technology of polysaccharides from Longan pulp. T Chinese Soc Agr Mach, 2010, 41: 131-136.

[13] Hromadkova Z, Ebringerova A. Ultrasonic extraction of plant materials investigation of hemicellulose release from buck wheat hulls. Ultrason Sonochem, 2003, 10: 127-133.

[14] Xie JH, Shen MY, Xie MY, Nie SP, Chen Y, Li C. Ultrasonic-assisted extraction, antimicrobial and antioxidant activities of *Cyclocarya paliurus* (Batal.) Iljinskaja polysaccharides. Carbohyd Polym, 2012, 89: 177-184.

[15] Zhang Q, Zhang TM. Determination of polysaccharide content by phenol-sulfuric acid method. Shandong Food Sci Technol, 2004, 13: 17-18. (in Chinese)

[16] Jin M, Cai Y X, Li J R, Zhao H. 1, 10-Phenanthroline-Fe2+ oxidative assay of hydroxyl radical produced by H_2O_2/Fe^{2+}. Prog. Biochem. Biophys, 1996, 23: 553-555.

[17] Shimada K, Fujikawa K, Yahara K, Nakamura T. Antioxidative properties of xanthan on the autoxidation of soybean oil in cyclodextrin emulsion, J Agric Food Chem, 1992, 40: 945-948.

[18] Wen H, An X, Hai N, Zhen J, Jiawen W. Optimised ultrasonic-assisted extraction of flavonoids from Folium eucommiae and evaluation of antioxidant activity in multi-test systems in vitro. Food Chem, 2009, 114: 1147-1154.

[19] Ahmad A, Alkharfy K M, Wani T A, Mohammad R. Application of Box-Behnken design for ultrasonicassisted extraction of polysaccharides from Paeonia emodi. Int J Biol Macromol, 2015, 72: 990-997.

[20] Karacabey E, Mazz G. Optimisation of antioxidant activity of grape cane extracts using response surface methodology. Food Chem, 2010, 119: 343-348.

[21] Bendjeddou D, Lalaoui K, Satta D. Immunostimulating activity of the hot water-soluble polysaccharide

extracts of Anacyclus pyrethrum, Alpinia galanga andCitrullus colocynthis. J Ethnopharmacol, 2003, 88: 155-160.

[22] Tian Z, Yong C. S, Ming Q. F, Ren X. T. Cloning, expression and biochemical characterization of UDPglucose 6-dehydrogenase, a key enzyme in the biosynthesis of an anti-tumor polysaccharide from the marine fungus Phoma herbarum YS4108. Process Biochem, 2011, 46: 2263-2268.

[23] Bao Y, Mouming Z, John SH, Ning Y, Yueming J. Effect of ultrasonic treatment on the recovery and DPPH radical scavenging activity of polysaccharides from longan fruit pericarp. Food Chem, 2008, 106: 685-690.

[24] Ningbo L, Jianjun Z, Xingqian Y, Shan L, Wenjun W, Ronghua ZH, Jian X, Shiguo CH, Donghong L. Ultrasonic-assisted enzymatic extraction of polysaccharide from Corbicula fluminea: Characterization and antioxidant activity. LWT-Food Sci Tech, 2015, 60: 1113-1121.

(发表于《Tropical Journal of Pharmaceutical Research》)

A New Pleuromutilin Derivative: Synthesis, Crystal Structure and Antibacterial Evaluation*

YI Yun-peng[1], YANG Guan-zhou[2], GUO Zhi-ting[1],
AI Xin[1], SHANG Ruo-feng[1**], LIANG Jian-pinga[1**]

(1. Key Laboratory of New Animal Drug Project of Gansu Province/ Key Laboratory of Veterinary Pharmaceutical Development, Ministry of Agriculture, China/Lanzhou Institute of Husbandry and Pharmaceutical Sciences of CAAS, Lanzhou 730050, China; 2. School of Pharmacy, Lanzhou University, Lanzhou 730020, China)

Abstract: A new pleuromutilin derivative, 14-O- [(4, 6-diaminopyrimidine-2-yl) thioacetate] mutilin, was synthesized and structurally characterized by IR, NMR spectra and single-crystalX-ray diffraction. This compound contains a 5-6-8 tricyclic carbon skeleton and a pyrimidine ring. Its crystal is of orthorhombic system, space group $P2_1$ with $a = 10.0237$ (6), $b = 12.6087$ (7), $c = 10.3749$ (8) Å, $\beta = 101.48$ (1)°, $V = 1284.99$ (14) $Å^3$, $Z = 2$, F (000) = 540, D_c (Mg/m^3) = 1.299, μ = 0.165 mm^{-1}, R = 0.0649 and wR = 0.0797. The *in vitro* antibacterial activity study using Oxfordcup assay showed this compound displayed more potent activity than pleuromutilin and similarantibacterial activity to that of tiamulin.

Key words: Pleuromutilin; Single-crystal structure; Antibacterial activities

1 INTRODUCTION

Pleuromutilin (1) was first isolated in a crystalline form from cultures of two species of basidiomycetes, *Pleurotus mutilus* and *P. passeckerianus* in 1951[1]. Pleuromutilin is a diterpene, constituted of a rather rigid 5-6-8 tricyclic carbon skeleton with eight stereogenic centers[2,3]. The derivatives of pleuromutilin have received much investigative attention due to their high activities against drugresistant Gram-positive bacteria and mycoplasmas *in vitro* and *in vivo*[4], pharmacodynamic properties[5], and no target-specific cross-resistance to other antibiotics[6,7]. The molecular modifications of pleuromutilin were focused essentially on the C^{-14} glycolic acid chain, especially with a thioether group and a basic group together[8]. Further alterations within this group led to the develop-

---

* Supported by Basic Scientific Research Funds in Central Agricultural Scientific Research Institutions (No. 1610322014003) and the Agricultural Science and Technology Innovation Program (ASTIP).

** Corresponding authors. Shang Ruo-Feng, born in 1974, doctor in medicinal chemistry. E-mail: shangrf1974@163.com; Liang Jian-Ping, born in 1963, doctor in animal drugs. E-mail: liangjp100@sina.com.

ment of three drugs: tiamulin, valnemulin, and retapamulin. The first two drugs were used as veterinary drugs[9], while retapamulin became the first pleuromutilin approved for human use in 2007[10]. Footprinting analysis showed that pleuromutilin derivatives inhibited the bacterial protein synthesis via a specific interaction with the 23S rRNA of the 50S bacterial ribosome subunit[4,11].

Herein, we prepare the title compound, 14-O-[(4, 6-diaminopyrimidine-2-yl) thioacetate] mutilin (3), which was characterized by IR, ^1H NMR and ^{13}C NMR spectral analysis, and report its single-crystal structure. The synthesis route is depicted in Scheme 1.

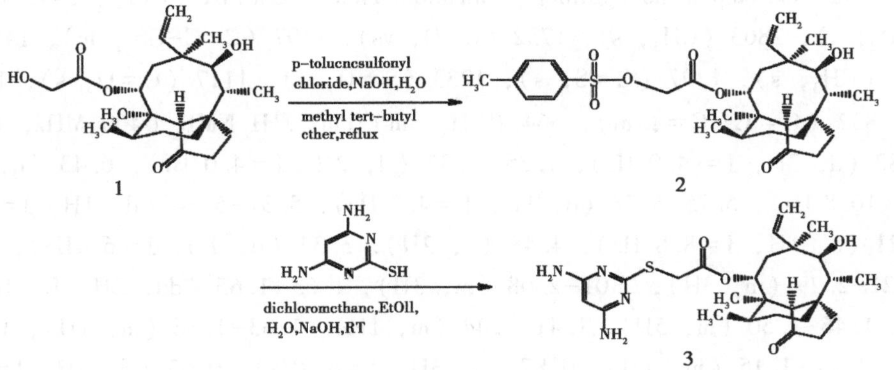

p-toluenesulfonyl chloride, NaOH, H_2O; methyl tert-butyl ether, reflux

dichloromethane, EtOH, H_2O, NaOH, RT

Fig. 1 Synthetic route of the title compound

2 EXPERIMENTAL

2.1 Reagents and instruments

All reagents obtained from commercial sources were of AR grade and used without further purification. Melting points were determined on a Tianda Tianfa YRT-3 apparatus (China) with open capillary tubes and were uncorrected. IR spectra were obtained on a Thermo Nicolet NEXUS-670 spectrometer and recorded as KBr thin film and the absorptions were reported in cm-1. NMR spectra were recorded on Bruker-400 MHz spectrometers in appropriate solvents. Chemical shifts (δ) in ^1H NMR were expressed in parts per million (ppm) relative to the tetramethylsilane. Multiplicities of signals are designated as s (singlet), d (doublet), t (triplet), q (quartet), m (multiplet), br (broad), etc. ^{13}C NMR spectra were recorded on 100 MHz spectrometers. The single-crystal structure of the title compound was determined on an Agilent SuperNova X-diffractometer. All reactions were monitored by TLC on 0.2 mm thick silica gel GF254 pre-coated plates. After elution, plates were visualized under UV illumination at 254 nm for UV active materials. Further visualizations were achieved by staining with 0.05% KMnO4 aqueous solution. Column chromatography was carried out on silica gel (200-300 mesh). The products were eluted in appropriate solvent mixture under air pressure. Concentration and evaporation of the solvent after reaction or extraction were carried out on a rotary evaporator.

2.2 Synthesis and characterization of the title compound

2.2.1 Synthesis of intermediate 2

Forty percent NaOH solution (5mL, 50 mmol) was added dropwise to a mixture of pleuromutilin (7.57 g, 20 mmol) and *p*-toluenesulfonyl chloride (4.2 g, 22 mmol) in t-butyl methyl ether (20mL) and water (5mL). The mixture was stirred vigorously under reflux for 1 h, then diluted with water (50mL) and stirred under an ice bath for 15 min, followed by washing with water (50mL) and cold t-butyl methyl ether (20mL). Filtration afforded compound 2 as white solid. It was used in the next step without further purification. Yield: 93%. IR (KBr): 3446 (OH, s), 2924 (CH_2, s), 2863 (CH_3, s), 1732 (C=O, vs), 1597 (C=C-C=, m), 1456 (CH_2, s), 1371 (CH_3, s), 1297 (C=S, s), 1233 (S=O, s), 1117 (C=O, s), 1035 (C-OH, s), 832 (C=C-C=, m), 664 (CH_2, m) cm^{-1}. ^1H NMR (400 MHz, CDCl3) δ 7.80-7.82 (d, 2H, J= 4.0 Hz), 7.35-7.37 (d, 2H, J=4.0 Hz), 6.43 (q, 1H, J= 17.2 Hz, 10.8 Hz), 5.75-5.78 (d, 1H, J =4.2 Hz), 5.31-5.34 (d, 1H, J=6.4 Hz), 5.17-5.21 (d, 1H, J=8.8 Hz), 4.48 (s, 2H), 3.34 (d, 1H, J=6.4Hz), 2.45 (s, 3H), 2.21-2.29 (m, 3H), 2.01-2.08 (m, 3H), 1.63-1.65 (dd, 2H, J_1=10Hz, J_2= 7.2 Hz), 1.46-1.50 (m, 5H), 1.41-1.44 (m, 1H), 1.33-1.36 (m, 1H), 1.22-1.26 (s, 5H), 1.11-1.15 (m, 1H), 0.87 (d, 3H, J=6.8Hz), 0.63 (d, 3H, J=6.8Hz); 13^C NMR (100 MHz, $CDCl_3$) δ 216.7, 164.8, 145.2, 138.6, 132.5, 129.9, 127.9, 117.2, 74.4, 70.2, 64.9, 57.9, 45.3, 44.4, 43.9, 41.7, 36.4, 35.9, 34.3, 30.2, 26.7, 26.3, 24.7, 21.6, 16.4, 14.6, 11.4.

2.2.2 Synthesis of the title compound 3

A solution of 30% aqueous NaOH (2.2 mmol) was added to a stirred mixture of 4,6-diamino-2-mercaptopyrimidine hydrate (2.8mg, 2.0 mmol) in 5mL CH_3OH, and the resulting reaction mixture was stirred at room temperature for 30 min. Compound 2 (1.17 g, 2.2 mmol) in 15mL CH_2Cl_2 was added dropwise to the reaction mixture and stirred at 0°C for 36 h. Then the reaction mixture was concentrated in vacuo, followed by the addition of EtOAc and washing with brine and water. The organic layer was dried over anhydrous $MgSO_4$, filtered, and concentrated in vacuo to give the residue that was purified by column chromatography using silica gel to give the desired product[12]. White solid; yield: 95%, IR (KBr): 3475 (NH, m), 3377 (NH, m), 2933 (CH_2, s), 1728 (C=O, s), 1619 (C=O, vs), 1582 (C=C-C=, s), 1547 (C=C-C=, s), 1465 (CH_2, s), 1309 (OH, s), 1153 (C-S, m), 1117 (C=O, m), 1019 (C-O, m) cm^{-1}. 1H NMR (400 MHz, DMSO) δ ^1H NMR (400 MHz, DMSO) δ 6.14 (d, *J*=11.2 Hz, 1H), 6.11-5.97 (m, 3H), 5.52 (d, *J*=7.8 Hz, 1H), 5.12 (d, *J*=15.7 Hz, 1H), 5.01 (dd, J=26.9, 15.7 Hz, 1H), 4.50 (d, J =5.8 Hz, 1H), 4.17-3.92 (m, 1H), 3.80 (q, J=16.0Hz, 1H), 3.42 (s, 1H), 2.39 (s, 1H), 2.29-2.14 (m, 1H), 2.12-1.88 (m, 4H), 1.75-1.54 (m, 2H), 1.54-1.11 (m, 9H), 1.12-0.89 (m, 4H), 0.82 (d, *J*= 6.5 Hz, 3H), 0.62 (d, *J*=6.4 Hz, 3H). ^{13}C NMR (101 MHz, DMSO) δ 217.6, 168.3, 167.3, 163.8, 141.2, 115.8, 79.6, 73.1, 70.0, 60.2, 57.7, 45.4, 44.5, 44.0, 42.0, 36.8, 34.5, 33.3, 30.6, 29.0, 27.1, 24.9, 21.2, 16.6,

15.0, 12.0.

2.3 Crystal data and structure determination

The crystals of the title compound suitable for X-ray structure determination were obtained by slowly evaporating a mixed solvent of ethanol and acetone for about twenty days at room temperature. A colorless single crystal with dimensions of 0.31mm × 0.35mm × 0.36mm was selected and-mounted in air onto thin glass fibers. X-ray intensity data were collected at 291.73 (10) K on an Agilent SuperNova - CCD diffractometer equipped with a mirror - monochromatic MoKa (λ = 0.71073 Å) radiation. A total of reflections were collected in the range of $3.55 \leqslant \theta \leqslant 26.02°$ (index ranges: $-12 \leqslant h \leqslant 12$, $-14 \leqslant k \leqslant 15$, $-7 \leqslant l \leqslant 12$) by using an ω scanmode with 4161 independent ones (R_{int} = 0.0493), of which 2223 with $I > 2\sigma(I)$ were considered as observed and used in the succeeding refinements. The structure was solved by direct methods using SUPERFLIP program[13] and refined with SHELXL program[14,15] by full - matrix least - squares techniqueson F^2. The non-hydrogen atoms were refined anisotropically, and hydrogen atoms were determined with theoretical calculations. A full - matrix least - squares refinement gave the final R = 0.0649, wR = 0.0797 ($w = 1/[\sigma^2(Fo^2) + (0.000P)^2]$, where $P = (Fo^2 + 2Fc^2)/3$), $(\Delta/\sigma)_{max}$ = 0.000, S = 0.947, $(\Delta\rho)_{max}$ = 0.213, and $(\Delta\rho)_{min}$ =. 0.276 e/Å^3.

2.4 Antibacterial activity measurement

Oxford cup assay was performed to evaluate the inhibition rate of the title compound in the growth of bacteria. Inoculum was prepared in 0.9% saline using McFarland standard and spread uniformly on nutrient agar plates. The title compound, as well as pleuromutilin and tiamulin fumarate used as reference drugs, was weighed 12800 μg accurately and dissolved with 3mL ethanol, followed by diluting to 10mL with distilled water. Then all the solutions were diluted with distilled water by two folds. The resulting solutions were added individually to the Oxford cups which were placed at equidistance on the above agar surfaces. The zone of inhibition for each concentration was measured after 24 h incubation at 37℃. The same procedure was repeated in triplicate.

3 RESULTS AND DISCUSSION

The title compound was prepared according to Scheme 1. Almost all pleuromutilin derivatives were synthesized from 22-O-tosylpleuromutilin (Compound 2) which was obtained by the reaction of pleuromutilin and p-toluenesulfonyl chloride to activate the 22-hydroxyl of pleuromutilin. The title compound was then obtained in good yield by the nucleophilic attack of 4,6-diamino-2-mercaptopyrimidine hydrate on compound 2 under alkaline conditions. The IR, 1^H NMR and ^{13}C NMR for the obtained compounds are all in good agreement with the assumed structure. A single crystal of the title compound was cultured for X-ray diffraction analysis to confirm the configuration. The selected bond lengths and bond angles are shown in Table 1, and hydrogen bonding parameters in Table 2.

Table 1 Selected Bond Lengths (Å) and Bond Angles (°)

Bond	Dist.	Bond	Dist.	Bond	Dist.
S (1) -C (16)	1.799 (6)	C (1) -C (2)	1.559 (7)	C (7) -C (8)	1.571 (7)
S (1) -C (17)	1.784 (6)	C (1) -C (8)	1.539 (6)	C (7) -C (24)	1.514 (8)
O (1) -C (1)	1.469 (5)	C (2) -C (3)	1.538 (7)	C (7) -C (26)	1.551 (8)
O (1) -C (15)	1.317 (6)	C (2) -C (9)	1.557 (7)	C (9) -C (10)	1.531 (8)
O (2) -C (15)	1.216 (6)	C (2) -C (21)	1.552 (7)	C (9) -C (22)	1.535 (7)
O (3) -C (12)	1.208 (6)	C (3) -C (4)	1.556 (7)	C (10) -C (11)	1.527 (7)
O (4) -C (6)	1.445 (5)	C (3) -C (12)	1.545 (7)	C (10) -C (13)	1.517 (7)
N (3) -C (18)	1.349 (7)	C (4) -C (5)	1.541 (6)	C (15) -C (16)	1.494 (7)
N (4) -C (20)	1.375 (7)	C (4) -C (11)	1.547 (7)	C (18) -C (19)	1.394 (7)
N (1) -C (17)	1.311 (7)	C (4) -C (14)	1.544 (6)	C (19) -C (20)	1.369 (8)
N (1) -C (18)	1.370 (7)	C (5) -C (6)	1.569 (7)	C (24) -C (25)	1.298 (7)
N (2) -C (17)	1.331 (6)	C (5) -C (23)	1.559 (7)		
N (2) -C (20)	1.350 (6)	C (6) -C (7)	1.542 (7)		

Angle	(°)	Angle	(°)	Angle	(°)
C (17) -S (1) -C (16)	100.1 (3)	C (14) -C (4) -C (3)	102.4 (4)	O (3) -C (12) -C (3)	128.5 (5)
C (15) -O (1) -C (1)	121.0 (4)	C (14) -C (4) -C (11)	108.2 (4)	O (3) -C (12) -C (13)	123.2 (5)
C (17) -N (1) -C (18)	113.4 (5)	C (4) -C (5) -C (6)	112.3 (5)	C (13) -C (12) -C (3)	108.3 (5)
C (17) -N (2) -C (20)	113.2 (5)	C (4) -C (5) -C (23)	110.7 (5)	C (12) -C (13) -C (14)	105.5 (4)
O (1) -C (1) -C (2)	104.2 (4)	C (23) -C (5) -C (6)	112.4 (4)	C (13) -C (14) -C (4)	106.1 (4)
O (1) -C (1) -C (8)	106.5 (4)	O (4) -C (6) -C (5)	111.6 (4)	O (1) -C (15) -C (16)	112.6 (5)
C (8) -C (1) -C (2)	116.1 (4)	O (4) -C (6) -C (7)	108.9 (4)	O (2) -C (15) -O (1)	125.4 (5)
C (3) -C (2) -C (1)	109.6 (4)	C (7) -C (6) -C (5)	115.8 (4)	O (2) -C (15) -C (16)	121.9 (5)
C (3) -C (2) -C (9)	108.1 (4)	C (6) -C (7) -C (8)	113.9 (4)	C (15) -C (16) -S (1)	117.0 (4)
C (3) -C (2) -C (21)	110.8 (5)	C (6) -C (7) -C (26)	107.2 (5)	N (1) -C (17) -S (1)	113.1 (4)
C (9) -C (2) -C (1)	111.2 (5)	C (24) -C (7) -C (6)	111.2 (5)	N (1) -C (17) -N (2)	131.1 (5)
C (21) -C (2) -C (1)	106.8 (5)	C (24) -C (7) -C (8)	108.3 (5)	N (2) -C (17) -S (1)	115.8 (5)
C (21) -C (2) -C (9)	110.4 (4)	C (24) -C (7) -C (26)	111.9 (5)	N (3) -C (18) -N (1)	115.1 (6)
C (2) -C (3) -C (4)	118.0 (4)	C (26) -C (7) -C (8)	104.2 (5)	N (3) -C (18) -C (19)	122.7 (6)
C (2) -C (3) -C (12)	117.0 (4)	C (1) -C (8) -C (7)	120.8 (5)	N (1) -C (18) -C (19)	122.2 (6)
C (12) -C (3) -C (4)	101.9 (4)	C (10) -C (9) -C (2)	111.9 (4)	C (20) -C (19) -C (18)	116.5 (6)
C (5) -C (4) -C (3)	115.4 (5)	C (10) -C (9) -C (22)	109.9 (5)	N (2) -C (20) -N (4)	114.4 (6)
C (5) -C (4) -C (11)	112.3 (4)	C (22) -C (9) -C (2)	116.6 (5)	N (2) -C (20) -C (19)	123.5 (5)
C (5) -C (4) -C (14)	111.2 (4)	C (11) -C (10) -C (9)	113.7 (5)	C (19) -C (20) -N (4)	122.0 (6)
C (11) -C (4) -C (3)	106.6 (4)	C (10) -C (11) -C (4)	112.7 (5)	C (25) -C (24) - (7)	129.0 (7)

Table 2 Hydrogen Bond Lengths (Å) and Bond Angles (°)

D-H...A	d (D-H)	d (H...A)	d (D...A)	∠DHA
N (3) -H (1) ...O (4)[a]	0.86	2.17	2.938	149
N (3) -H (3B) ...O (2)[b]	0.86	2.27	3.069	155
O (4) -H (4) ...N (1)[c]	0.82	2.18	2.872	142

(continued)

D-H... A	d (D-H)	d (H... A)	d (D... A)	∠DHA
C (21) -H (21A) ... O (1)	0.96	2.22	2.593	167
C (21) -H (21B) ... O (3)	0.96	2.26	2.940	127
C (22) -H (22C) ... O (1)	0.96	2.53	2.915	106
C (26) -H (26A) ... O (4)	0.96	2.48	2.807	104

Noet: Symmetry codes: (a) $1+x, y, z$; (b) $-1-x, -0.5+y, -1-z$; (c) $-1+x, y, z$

The crystal structure in Fig. 2 shows that the whole molecule of the complex consists of a 5-6-8 tricyclic carbon skeleton and a pyrimidine ring, in which all bond lengths are in normal ranges. Five chiral carbons can be found in the molecule and the absolute configurations of C (2), C (3), C (4), C (5), C (9) are S, R, R, R and S. The five-membered ring C (3), C (4), C (11), C (12), C (14) is not planar and the dihedral angles formed by C (3) -C (12) -C (13) -C (14) and C (1) -C (4) -C (14) is 38.133°. The bond lengths of C (3) -C (12) and C (12) -C (13) are shorter than that of C (3) -C (2) and C (13) -C (14), which may be caused by the conjugation with carbonyl (C=O). The eight-membered ring C (2), C (3), C (4), C (5), C (6), C (7), C (8), C (1) exhibits a boat conformation, while the six-membered ring C (3), C (4), C (11), C (10), C (9), C (2), C (3) exhibits a chair conformation and the dihedral angles formed by C (3) -C (4) -C (11), C (3) -C (11) -C (10) -C (2) and C (10) -C (9) -C (2) are 46.674 and 47.847°, respectivel.

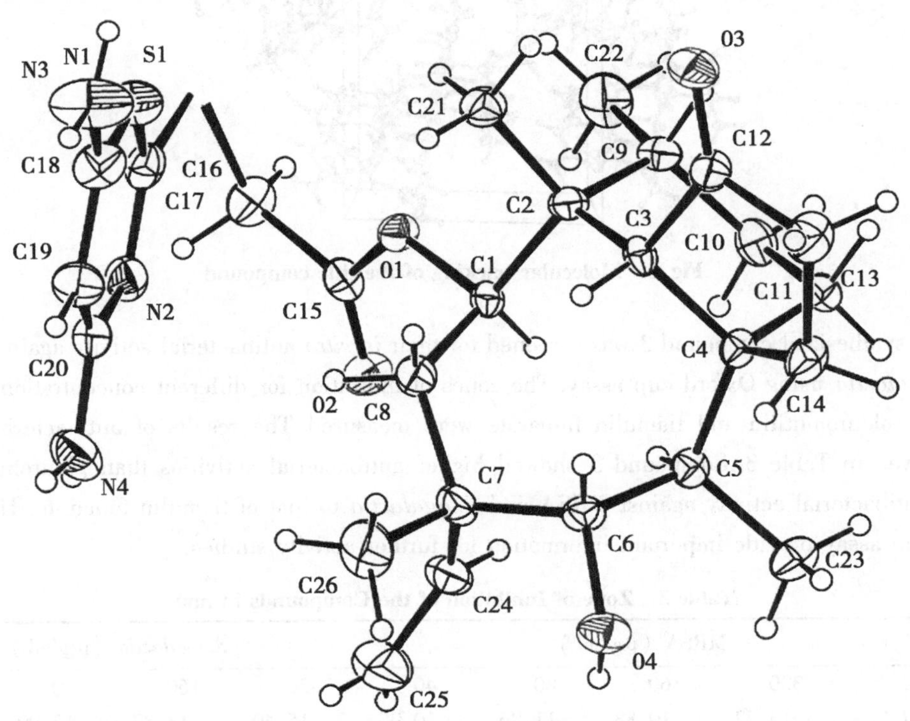

Fig. 2 Molecular structure of the title compound (Hydrogen atoms were omitted for clarity)

The side chain of C (14) exhibits a zig-zag conformation. The bond length of O (2) -C (15) is shorter than that of O (1) -C (1) which may be caused by the conjugation with carbonyl (C=O). Four intramolecular hydrogen bonds are formed in the structure at C (21) -H (21A) ···O (1), C (21) -H (21B) ···O (3), C (22) -H (22C) ···O (1) and C (26) -H (26A) ···O (4) via C atoms of tricyclic skeleton and the O atoms of ester, hydroxyl group and carbonyl group, respecttively. At the same time, three intermolecular hydrogen bonds are formed between two molecules of the title compound at N (3) -H (1) ···O (4), N (3) -H (3B) ···O (2), and O (4) -H (4) ···N (1) . One of the amino groups at the terminal side chain takes part in two hydrogen bonds and its N atom acts as two hydrogen-bond donors to the O atom of hydroxyl group and carbonyl group 3. Meanwhile, the O atom of hydroxyl group also acts as a hydrogen-bond donor to the N atom of the other amino group at the terminal side chain. These intermolecular hydrogen bonds link the molecular network structure and play key roles in stabilizing the crystal packing (Fig. 3) .

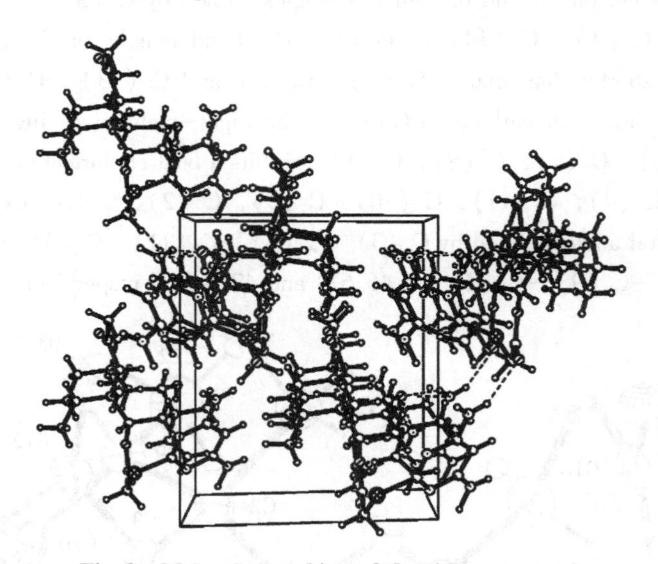

Fig. 3 Molecular packing of the title compound

The synthesized compound 3 was screened for their *in vitro* antibacterial activity against MRSA, and *S. agalactia* using Oxford cup assay. The zones of inhibition for different concentrations of compound 3, pleuromutilin and tiamulin fumarate were measured. The results of antibacterial activities aregiven in Table 3. Compound 3 showed higher antibacterial activities than pleuromutilin but similar antibacterial activity against MRSA and *S. agalactia* to that of tiamulin fumarate. The data of Oxford cup assay provide important information for further activity studies.

Table 3 Zone of Inhibition of the Compounds in mm

	MRSA (μg/mL)				*S. agalactia* (μg/mL)			
Compound	320	160	80	40	320	160	80	40
Compound 3	13. 22	10. 83	11. 86	10. 95	15. 40	13. 87	12. 00	11. 37
Pleuromutilin	11. 77	10. 88	10. 07	9. 52	12. 36	12. 25	11. 56	10. 18
Tiamulin	13. 33	12. 31	11. 44	10. 18	14. 32	13. 43	12. 67	11. 89

References

[1] Kavanagh, F.; Hervey, A.; Robbins, W.J.Antibiotic substances from basidiomycetes: VIII.pleurotus multilus (Fr.) sacc.and pleurotus passeckerianus pilat.Proceedings of the National Academy of Sciences of the United States of America, 1951, 37, 570-574.

[2] Arigoni, D.Structure of a new type of terpene.Gazz.Chim.Ital.1962, 92, 884-901.

[3] Birch, A.; Holzapfel, C.W.; Rickards, R.W.The structure and some aspects of the biosynthesis of pleuromutilin.Tetrahedron, 1966, 22, 359-387.

[4] Poulsen, S.M.; Karlsson, M.; Johansson, L.B.; Vester, B.The pleuromutilin drugs tiamulin and valnemulin bind to the RNA at the peptidyl transferase centre on the ribosome. Molecular Microbiology, 2001, 41, 1091-1099.

[5] Berner, H.; Schulz, G.; Fischer, G. Chemie der Pleuromutiline, 3. Mitt.: Synthese des 14-o-acetyl-19, 20-dihydro-a-nor-mutilins. Monatshefte fur Chemie/Chemical Monthly, 1981, 112, 1441-1450.

[6] Frank, S.; Paola, F.; Ada, Y.; Jo rg, M.; Harms, S.Inhibition of peptide bond formation by pleuromutilins: the structure of the 50S ribosomal subunit from deinococcus radiodurans in complex with tiamulin.Mol.Microbiol., 2004, 54, 1287-1294.

[7] Ling, Y.; Wang, X.; Wang, H.; Yu, J.; Tang, J.; Wang, D.; Chen, G.; Huang, J.; Li, Y.; Zheng, H.Design, synthesis, and antibacterial activity of novel pleuromutilin derivatives bearing an amino thiazolyl ring.Arch.Pharm. (Weinheim, Ger.), 2012, 345, 638-646.

[8] Egger, H.; Reinshagen, H.New pleuromutilin derivatives with enhanced antimicrobial activity.I.Synthesis.The Journal of Antibiotics, 1976, 29, 915-922.

[9] Tang, Y.Z.; Liu, Y.H.; Chen, J.Y.Pleuromutilin and its derivatives-the lead compounds for novel antibiotics.Mini Reviews in Medicinal Chemistry, 2012, 12, 53-61.

[10] Daum, R.S.; Kar, S.; Kirkpatrick, P.Retapamulin. Nature Reviews Drug Discovery, 2007, 6, 865-866.

[11] Long, K.S.; Hansen, L.H.; Jakobsen, L.; Vester, B.Interaction of pleuromutilin derivatives with the ribosomal peptidyl transferase center.Antimicrob Agents Chemother., 2006, 50, 1458-1462.

[12] Spicer, T.P.; Jiang, J.; Taylor, A.B.; Choi, J.Y.; Hart, P.J.; Roush, W.R.; Fields, G.B.; Hodder, P.S.; Minond, D.Characterization of selective exosite-binding inhibitors of matrix metalloproteinase 13 that prevent articular cartilage degradation in vitro.Journal of Medicinal Chemistry, 2014, 57, 9598-9611.

[13] Palatinus, L.; Chapuis, G.SUPERFLIP-a computer program for the solution of crystal structures by charge flipping in arbitrary dimensions.Journal of Applied Crystallography, 2007, 40, 786-790.

[14] Sheldrick, G.M.SADABS, Program for Bruker Area Detector Absorption Correction. University of Götingen, Germany, 1997.

[15] Sheldrick.G.M.SHELXS-97, Program for Crystal Structure Solution. University of Götingen, Germany, 1997.

（发表于《结构化学》）

Belamcanda Chinensis (L.) DC: Ethn-opharmacology, Phytochemistryand Pharmacology of an Important Traditional Chinese Medicine

XIN Rui-hua, ZHENG Ji-fang, CHENG Long, PENG Wen-jing, LUO Yong-jiang

(Key Laboratory of New Animal Drug Project, Gansu Province/Engineering & Technology Research Centre of Traditional Chinese Veterinary Medicine of Gansu Province/Key Laboratory of Veterinary Pharmaceutics Development, Ministry of Agriculture/Lanzhou Institute of Husbandry and Pharmaceutical Sciences of CAAS, Lanzhou 730050, China)

Abstract: Background: *Belamcanda chinensis* (L.) DC (Iridaceae), a widely used traditional Chinese medicine known as *She Gan* (Chinese: 射干), is a flowering perennial herb native to East Asia. For thousands of years in China, the rhizome of *Belamcanda chinensis* has been used to treat inflammation, oxyhepatitis, mumps, acute mastitis, and asthma, as well as throat disorders such as cough, tonsillitis and pharyngitis. *Belamcanda chinensis* is now listed in the Pharmacopoeia of the People's Republic of China. The present paper reviews the advancements in the investigation of botany, ethnopharmacology, phytochemistry, pharmacology and toxicology of *Belamcanda chinensis*.

Materials and Methods: Information on *Belamcanda chinensis* was collected from scientific journals, books, theses and reports via library and electronic search (PubMed, CNKI, Elsevier, ACS, Google Scholar, Baidu Scholar, Web of Science and Science Direct).

Results: A number of chemical compounds have been isolated from *Belamcanda chinensis*, and the major isolated compounds have been identified as isoflavonoids, flavonoids and iridal-type triterpenoids. Among these active compounds, the effects of tectoridin and tectorigenin have been widely investigated. The primary active components in *Belamcanda chinensis* possess a wide range of pharmacological activities, including anti-inflammatory, anti-oxidative, anti-tumour, anti-alcohol injury, cardiovascular and oestrogenic activities.

Conclusions: As an important traditional Chinese medicine, *Belamcanda chinensis* has been demonstrated to have marked bioactivity, especially in the respiratory system and as an oestrogenic and hepatoprotective agent. This activity is related to its traditional use and provides opportunities for the development of novel drugs and therapeutic products for various diseases. However, the toxicity of *Belamcanda chinensis* will require further study, and more attention should be devoted to its better utilization.

Key words: *Belamcanda chinensis*; Ethnopharmacology; Phytochemistry; Pharmacology; Toxicology

INTRODUCTION

Belamcanda chinensis (L.) DC, which belongs to the family Iridaceae, has been used worldwide as a traditional medicine for thousands of years (Qin et al., 1999). *Belamcanda chinensis* is cold-natured and bitter in taste (Liu et al., 2011a). Due to its effects of heat-clearing, detoxifying and reducing pharyngeal swelling (Qin et al., 1998; Ni et al., 2012), *Belamcanda chinensis* is used mainly in the clinical treatment of respiratory diseases (Yang, 2012; Shi, 2011) such as upper respiratory tract infection, acute and chronic pharyngitis, tonsillitis, chronic sinusitis, bronchitis, asthma, emphysema, pulmonary heart disease and sore throat (Wu et al., 2014; Zhao et al., 2013; Yang, 2013; Wang, 2005; Zhang, 2010c; Shi et al., 2012), treatments that have been listed in the "Chinese Pharmacopoeia" (Committee for the Pharmacopoeia of P. R. China, 2010).

Belamcanda chinensis includes only one species in China. The genus is widespread in Jilin, Liaoning, Hebei, Shanxi and Shandong provinces (Huang, 2010). In some districts, *Iris dichotoma* Pall (also known as the wild iris) and *Iris tectorum* maxim (also known as Chuan Iris) are substituted for *Belamcanda chinensis* (Qin et al., 2003a). These herbs are referred to as *Belamcanda chinensis* herbs. Studies have shown that they are similar in chemical composition and pharmacological activities but belong to different species (Huang et al., 1997; Wu et al., 1990). In recent years, numerous studies have reported the chemical compositions of *Belamcanda chinensis*. A series of flavonoids, triterpenoids, quinones, steroids and other volatile components have been found in the genus *Belamcanda chinensis*, with isoflavone and flavonoid compounds being the two main groups of constituents (Qin et al., 2000; Wang et al., 2006; Xiao et al., 2008; Li, 2003; Meng et al., 2004; Zhong et al., 2001). Pharmacology studies have shown that these flavonoids have a wide range of pharmacological activities such as anti-inflammatory, antibacterial, oestrogen-like, anti-oxidation, anticancer, hepatoprotective and hypoglycaemic lipid-lowering effects (Zhang et al., 2010b).

To better understand the chemical composition and the pharmacological effects of *Belamcanda chinensis*, the recent advances in ethno-pharmacology, phytochemistry, biological and pharmacological activities of *Belamcanda chinensis* are summarized in this review. Considering that *Belamcanda chinensis* has many synonyms (http: //www.theplantlist.org), we use *Belamcanda chinensis* as the name of the plant throughout this review.

BOTANY AND ETHNO-PHARMACOLOGY

Botany

Belamcanda chinensis (L.) DC, first listed in Shen Nong Ben Cao Jing, is also known as butterfly flower and flat bamboo (Fig. 1). *Belamcanda chinensis* is a perennial herbaceous plant that belongs to the Iridaceous family and grows on hillsides, on arid hillsides, and in meadows, ditches, thickets, forest-edge open fields, valleys, grassy slopes and other places (http: //

www. zysj. com. cn).

Fig. 1 Belamcanda chinensis

Native to East Asia, *Belamcanda chinensis* has naturalised in North America but is distributed mainly in the Himalayas, Indochina, and the Russian Far East. In China, *Belamcanda chinensis* is widely distributed in Heilongjiang, Jilin, Liaoning, Inner Mongolia, Hebei, Shanxi, Ningxia, Gansu and other provinces (Ma et al., 1995).

Belamcanda chinensis grows from a fleshy, knobby and usually orange or pale brown rhizome that creeps just below ground level. The root system of *Belamcanda chinensis* consists of a thickened crown (approximately 4-8cm long) at the base of the plant, which has fibrous roots underneath; spreading rhizomes are also produced. Both the crown and the rhizomes have an orange interior. Su Song from the Northern Song dynasty said, "Its root is like the long rod of ancient hunters;" thus, in Chinese, it is named Shegan. The erect central stalk is 50-120cm tall and either branched or unbranched; it is terete, fairly stout, glabrous, glaucous, and pale green. This stalk terminates in a cyme or compound cyme of flowers. There are pairs of small linear-lanceolate bracts at each fork of the stalk; these bracts are slightly membranous and tend to wither away. Each flower spans approximately 4cm across, consisting of 6 spreading tepals, 3 distinct stamens, a style with a tripartite stigma, and an inferior ovary. The tepals are orange with purple dots and are elliptical-oblong in shape, while the ovary is green, glabrous, and narrowly ovoid. Each cyme usually produces several flowers. The blooming period occurs from mid-to-late summer and lasts approximately 1-2 months. There is no noticeable floral scent. Each flower is replaced by an oblong seed capsule approximately 1″long; the 3 sides of this capsule become strongly recurved, revealing a mass of shiny black seeds that resembles a blackberry. *Belamcanda chinensis* has sword-shaped alternate leaves approximately 30-60cm long and 2.5-4cm wide; the leaves originate primarily towards the bottom of the flowering stalk. These leaves are often grouped together into the shape of a fan; the leaves are green to grey-blue, linear in shape, glabrous, and glaucous. Their margins are smooth, while their veins are parallel. This plant can spread either by rhizomes or by seeds (Li et al., 2010; http://frps.eflora.cn.).

ETHNO-PHARMACOLOGY

With special biological and pharmacological effects, *Belamcanda chinensis* has played an important role in traditional Chinese medicine (TCM) and for thousands of years was considered a unique and effective medicine for treating respiratory disease, such as cough, sore throats and asthma, and 3 000 years ago, *Belamcanda chinensis* was listed in a variety of ancient Chinese medical documents such as the "Shen Nong Ben Cao Jing", the oldest classical medicinal book, and was also listed in "Wu Pu Ben Cao", "Guang Ya", "Ben Cao Shi Yi", "Guang Zhou Flora", "Gang Mu", "Zhong Yao Zhi", and other books. In these monographs, *Belamcanda chinensis* was described as effective in stimulating the pharynx to expel phlegm, relieve cough, clear heat and detoxify. Therefore, it is used to treat a productive cough, wheezing, aphthous stomatitis, pharyngeal swelling, black skin and other symptoms. According to the authoritative textbook of Science of Chinese Materia Medica, *Belamcanda chinensis* is cold-natured, bitter in taste and attributive to the liver and lung meridians. *Belamcanda chinensis* has the power to clear heat from the throat and to cure people from the affliction of exogenous wind-heat. In the context of clinical practice, *Belamcanda chinensis* is mainly applied for coughs with an abundance of phlegm, pharyngalgia, gum swelling, hoarse throat, thirst, dry stool and dark urine. Meanwhile, *Belamcanda chinensis* is slightly toxic, so pregnant women and people with spleen deficiency are not suitable candidates for its use (http://www.satcm.gov.cn).

Due to their distinctive clinical effects, many classic prescriptions developed by the ancient famous doctors were handed down from generation to generation and received repeated clinical verification for thousands of years. Moreover, in modern clinical practice, these classicprescriptions have been employed in a more extensive and intensive way. For example, She Gan Ma Huang Tang was used to cure cough in ancient times (Table 1). However, currently, many prescriptions are used to treat bronchitis, bronchial asthma, pneumonia, emphysema, cor pulmonale, allergic rhinitis, and itchy skin. These prescriptions have also been applied clinically in the forms of pills, granules, and capsules for the sake of convenience.

Along with the development of pharmacology, pharmacy and other disciplines, *Belamcanda chinensis* is used as a key ingredient in combination with other Chinese herbs to treat a variety of diseases in traditional Chinese medicine. *Belamcanda chinensis* is documented in the 2010 edition of the Chinese Pharmacopoeia, in which it is classified as a heat-clearing and detoxifying drug. It is believed to be the most important Chinese medicine for treating pharyngitis and sore throat (Committee for the Pharmacopoeia of P. R. China, 2010, State Administration of Traditional Chinese Medicine,

1996). Approximately 9 preparations in which *Belamcanda chinensis* was the main and active component were listed in the Chinese Pharmacopoeia, including "Qing Yan Li Ge Wan" and "Qing Ge Wan", which have been widely used for clearing heat from the throat, reducing swelling and alleviating pain, curing dry throat and thirst, and reducing pharyngeal swelling. *Belamcanda chinensis* also has remarkable effects on respiratory disease in children, such as "Xiao Er Yan Bian Ke Li", "Xiao Er Qing Fei Zhi Ke Pian" and "She Gan Li Yan Kou Fu Ye", which were considered as conventional drugs to clear heat of lung; reduce pharyngeal swelling, coughs with abundance of phlegm, thirst and dry stool (Table 1).

In China, *Belamcanda chinensis* is cultivated mainly for medicinal uses, with its rhizome being the medicinal part, while its seeds have been extracted with ethanol for treating pain in the bones and muscles. Because the flowers of *Belamcanda chinensis* are attractive in colour and beautiful in shape, it is also cultivated as an ornamental flower in gardens and courtyards (Gong et al., 2004).

Main components

Many compounds have been isolated from *Belamcanda chinensis*, including isoflavones, flavonoids, iridoid glycosides, diterpenes, triterpenoids, alkaloids, phytosterols, polysaccharides, etc. (Table 2). The chemical structures of the primary compounds are shown in Fig. 2.

ISOFLAVONES AND FLAVONOIDS

The chemical constituents of *Belamcanda chinensis* have been investigated, and various compounds, especially isoflavones, flavonoids and flavonols, have been reported (Ji et al., 2001). Most of the compounds reported belong to the isoflavonoid family, including tectorigenin (1), tectorigenin-7-*O*-β-glucosyl (1→6) glucoside (2), tectoridin (3), irigenin (4), iridin (5), irisflorentin (6), dichotomitin (7), 3', 5'-dimethoxyirisolone-4-O-β-D-glucoside (8), nonirisflorentin (9), iristectorigenin A (10), iristectorigenin B (11), 5, 6, 7, 3'-tetrahydroxy-4'-methoxyisoflavone (12), 6'-O-vanilo-riridin (13), iristectrigenin (17), iristectrigenin (17), iristectrigenin A-7-glucoside (18), 8-hydroxytectrigenin (19), 8-hydroxyiristectrigenin A (20), 8-hydroxyirigenin (21), tectrigenin-4'-glucoside (22), irilin D (23), isotectrigenin (24), astragalin (25), 5, 7, 4'-dihydroxy-6, 3', 5'-trimethoxyisoflavone (26), genistein (27), dimethyltectorigenin (28), isoirigenin (29), isoirigenin-7-O-β-D-glucoside (30), irisolon (31), 3'-hydroxytectoridin (32), and tectoruside (33). Compounds that belong to the flavonoid family (Jin et al., 2008a) include 5, 4' dihydroxy-6, 7-methylenedioxy-3' methoxyflavone (38), iristectogenin (39), hispidulin (40) and acetovanillone (41). Flavonol compounds include isorhamnetin (42), rhamnazin (43) and rhamnocitrin (44). Other flavonoids include mangiferin (45), 7-O-methylmangiferin (49), 3'-hydroxyltectoridin (50), iristectorin (51), and isoiridin (54).

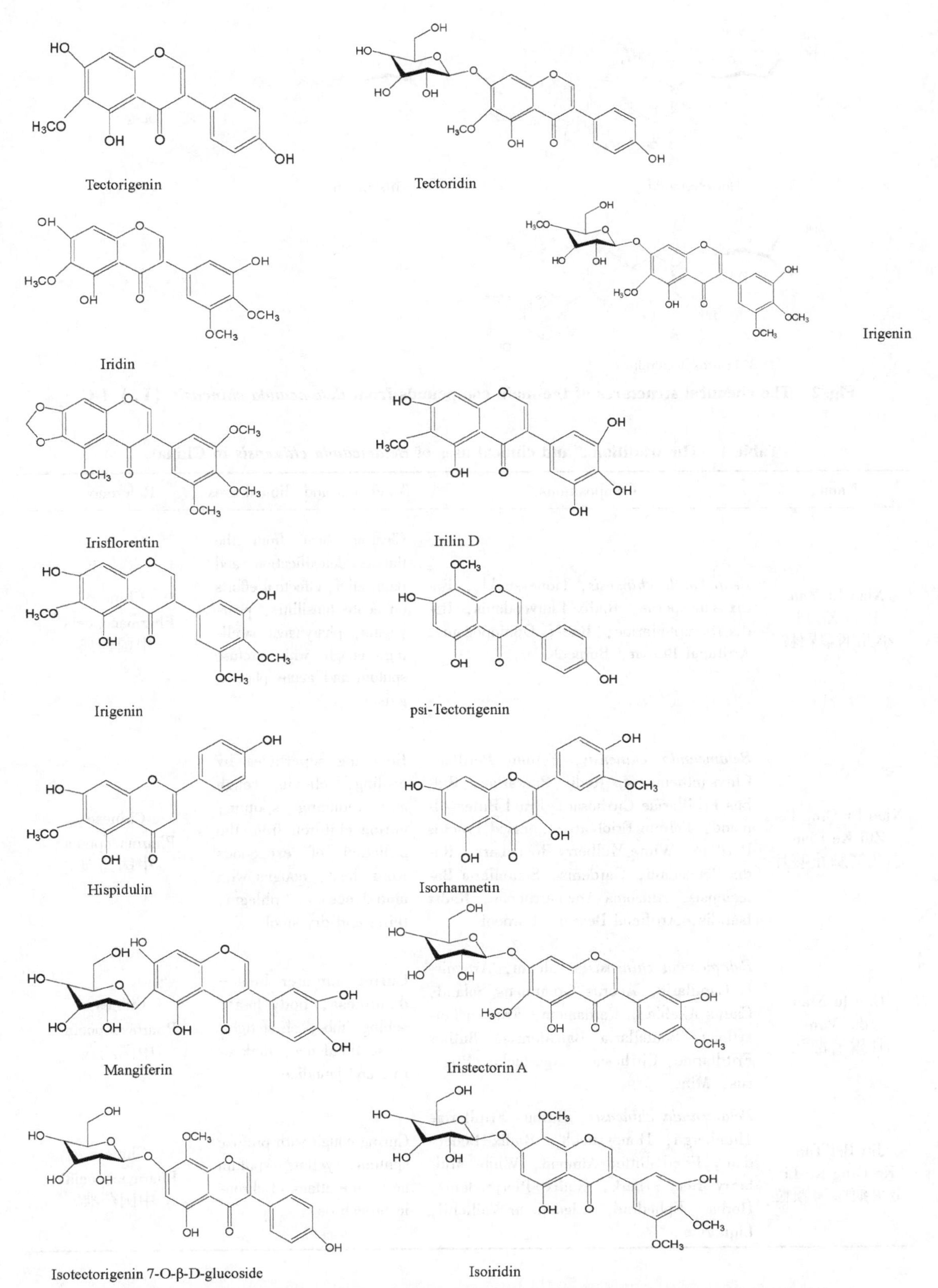

The chemical structures of the main compounds from *Belamcanda chinensis* (L.) DC

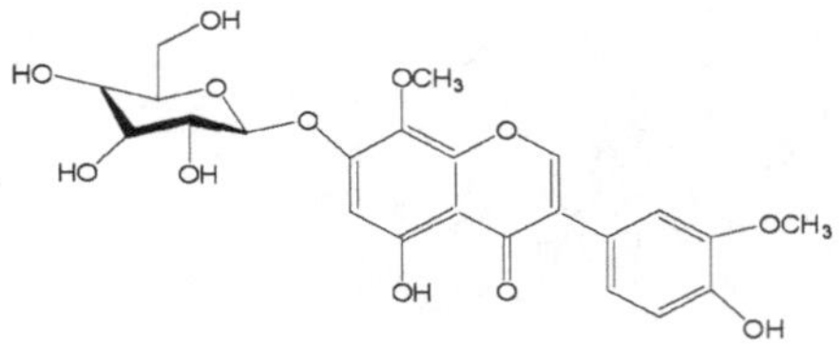

Homotectoridin

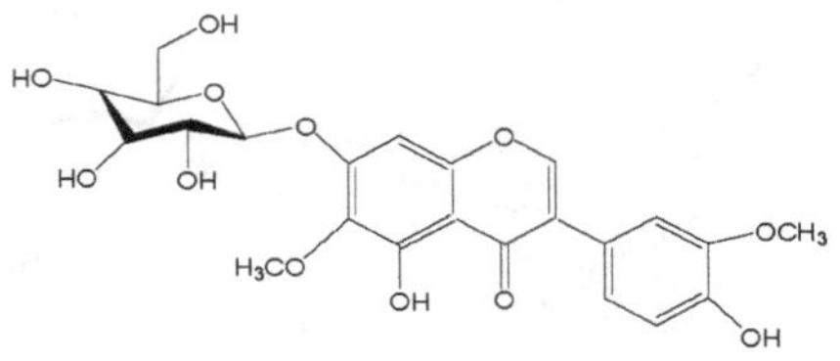

Iristectorin

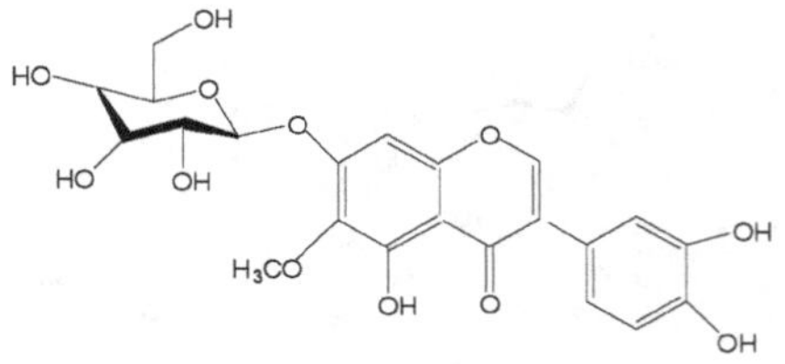

3'-Hydroxytectoridin

Fig. 2 The chemical structures of the main compounds from *Belamcanda chinensis* (L.) DC

Table 1 The traditional and clinical uses of *Belamcanda chinensis* in China

Name	Compositions	Traditional and clinical uses	References
Xiao Er Yan Bian Ke Li 小儿咽扁颗粒	*Belamcanda chinensis*, Honeysuckle, Radix Tinosporae, Radix Platycodonis, Radix Scrophulariae, Radix Ophiopogonis, Artificial Bezoar, Borneol	Clearing heat from the throat; detoxification and pain relief; effectual efforts on acute tonsillitis, pharyngitis, pharyngeal swelling, cough with profuse sputum and acute pharyngitis	Chinese Pharmacopoeia[a] 中国药典
Xiao Er Qing Fei Zhi Ke Pian 小儿清肺止咳片	*Belamcanda chinensis*, Folium Perillae, Chrysanthemums, Radix Puerariae, Bulbus Fritillariae Cirrhosae, Fried Bitter Almond, Folium Eriobotryae, Fried Fructus Perillae, White Mulberry Root-bark, Radix Peucedani, Gardenia, Scutellaria Baicalensis, Rhizoma Anemarrhenae, Radix Isatidis, Artificial Bezoar, Borneol	Relieving superficies by cooling, relieving cough and reducing sputum; curing children from the affliction of exogenous wind-heat, coughs with abundance of phlegm, thirst and dry stool	Chinese Pharmacopoeia[a] 中国药典
Gan lu Xiao du Wan 甘露消毒丸	*Belamcanda chinensis*, Talcum, Artemisia Capillaris, Acorus Gramineus Soland, Caulis Akebiae, Cardamom, Fructus Forsythiae, Scutellaria Baicalensis, Bulbus Fritillariae Cirrhosae, Agastache Rugosus, Mint	Curing summer heat-dampness, body heat, aching limbs, chest tightness, flatulence, dark urine and jaundice	Chinese Pharmacopoeia[a] 中国药典
Jin Bei Tan Ke Qing Ke Li 金贝痰咳清颗粒	*Belamcanda chinensis*, Bulbus Fritillariae Thunbergii, Honeysuckle, Radix Peucedani, Fried Bitter Almond, White Mulberry Root - bark, Radix Platycodonis, Herba Ephedrae, LigusticumWallichii, Liquorice	Curing cough with profuse sputum, yellow sputum and acute attack of chronic bronchitis	Chinese Pharmacopoeia[a] 中国药典

(continued)

Name	Compositions	Traditional and clinical uses	References
Gui Lin Xi Gua Shuang 桂林西瓜霜	*Belamcanda chinensis*, Mirabilitum Praeparatum, Borax, Cortex Phellodendri, Chinese Goldthread, Radix Sophorae Subprostratae, Bulbus Fritillariae Thunbergii, Indigo Naturalis, Borneol, Rheum Officinale, Scutellaria Baicalensis, Liquorice, Mentha-camphor	Heat-clearing and detoxifying; reducingswelling and alleviating pain; curing aphthous stomatitis; reducing pharyngeal swelling, swelling gums and acute and chronic pharyngitis	Chinese Pharmacopoeia[a] 中国药典
Qing Yan Li Ge Wan 清咽利膈丸	*Belamcanda chinensis*, Fructus Forsythiae, Gardenia, Scutellaria Baicalensis, Prepared Radix et Rhizoma Rhei, Stir-baked Fructus Arctii, Mint, Radix Trichosanthis, Radix Scrophulariae, Schizonepeta Spica Radix Sileris, Radix Platycodonis, Liquorice	Clearing heat from the throat; reducing swelling and alleviating pain; curing pharyngalgia, inhibited chest and diaphragm, bitter taste, dry stool and dark urine	Chinese Pharmacopoeia[a] 中国药典
Qin Yan Run HouWan 清咽润喉丸	*Belamcanda chinensis*, Radix Sophorae Subprostratae, Radix Platycodonis, Fried Bombyx Batryticatus, Gardenia, Moutan Radicis Cortex, Chinese Olive, Radix Tinosporae, Radix Ophiopogonis, Radix Scrophulariae, Rhizoma Anemarrhenae, Radix Rehmanniae, Radix Paeoniae Alba, Bulbus Fritillariae Thunbergii, Liquorice, Borneol, Pulvis Cornus Bubali Concentratus	Clearing heat from the throat; reducing swelling and alleviating pain, thirst, cough with profuse sputum; reducing pharyngeal swelling,hoarse throat	Chinese Pharmacopoeia[a] 中国药典
Qing Ge Wan 清膈丸	*Belamcanda chinensis*, Honeysuckle, Fructus Forsythiae, Radix Scrophulariae, Radix Sophorae Subprostratae, Chinese Goldthread, Radix Rehmanniae Praeparata, Radix Gentianae, Gypsum, Compound of Glauber-salt and Liquorice, Radix Platycodonis, Radix Ophiopogonis, Mint, Radices Rehmanniae, Borax, Liquorice, Bezoar, Borneol, Pulvis Cornus Bubali Concentratus	Clearing heat from the throat; reducing swelling and alleviating pain; curing dry throat and thirst; reducing pharyngeal swelling, hoarse throat and dry stool	Chinese Pharmacopoeia[a] 中国药典
Lu Si Ge Wan 鹭鸶咯丸	*Belamcanda chinensis*, Herba Ephedrae, Bitter Almond, Gypsum, Liquorice, Herba Asari, Fried Fructus Perillae, Fried Mustard, Stir-baked Fructus Arctii, Pericarpium Trichosanthis, Indigo Naturalis, Clamshell, Radix Trichosanthis, Gardenia, Artificial Bezoar	Curing cough and hoarse throat	Chinese Pharmacopoeia[a] 中国药典
She Gan Gao 射干膏	*Belamcanda chinensis*, Rhizoma Cimicifugae, Gardenia Florida, Radix Scrophulariae, Phaseolus Angularis, Cortex Phellodendri, Honey, Rehmannia Juice, Date	Effective for treatment of dry throat, aphtha and swelling gingiva	"Sheng Ji Zong Lu", Vol. 124[b] 《圣济总录》卷一二四

(continued)

Name	Compositions	Traditional and clinical uses	References
She Gan Yin 射干饮	*Belamcanda chinensis*, Radix Aucklandiae, Indian Bread Pink Epidermis, Ginseng, Dried Tangerine Peel	Effective for treatment of indigestion and vomiting	"Sheng Ji Zong Lu", Vol. 83[b] 《圣济总录》卷八十三
She Gan San I 射干散 I	*Belamcanda chinensis*, Concretio Silicea Bambusae, Mirabilite, Cornu Rhinoceri, Radix Scrophulariae, Rhiizoma Cimicifugae Foetidae, Alum, Xanthate, Prepared Radix Glycyrrhizae	Curing pharyngitis and irritable feverish sensation in chest	"Qi Xiao Liang Fang", Vol. 61[b] 《奇效良方》卷六十一
She Gan San II 射干散 II	*Belamcanda chinensis*, Radix Paeoniae Rubra, Rhiizoma Cimicifugae Foetidae, Almond, Fructus Arctii, Sweet gum, Radix Puerariae, Herba Ephedrae, Liquorice	Curing wind-heat attacking upward and reducing pharyngeal swelling	"Tai Ping Sheng Hui Fang", Vol. 35[b] 《太平圣惠方》卷三十五
She Gan Wan I 射干丸	*Belamcanda chinensis*, Pinellia Ternata, Rhizoma Zingiberis, Flos Farfarae, Chinese Honey Locust, Dried Tangerine Peel, Radix Stemonae, Schisandra Chinensis, Herba Asari, Bulb of Fritillary, Poria Alba, Semen Pruni	Curing cough and cough with profuse sputum.	"Sheng Ji Zong Lu", Vol. 65[b] 《圣济总录》卷六十五
She Gan Ma Huang Tang 射干麻黄汤	*Belamcanda chinensis*, Herba Ephedrae, Ginger, Flos Farfarae, Schisandra Chinensis, Date, Pinellia Ternata	Curing cough	"Jin Gui Yao Lue"[b] 《金匮要略》卷上**
Luo Shi She Gan Tang 络石射干汤	*Belamcanda chinensis*, Trachelospermum Jasminoides, Chinese Herbaceous Peony, Rhizoma Cimicifugae, Nidus Vespae, Fructus Tribuli	Reducing pharyngeal swelling	"Sheng Ji Zong Lu", Vol. 122[b] 《圣济总录》卷一二二
She Gan Xiao Du Yin 射干消毒饮	*Belamcanda chinensis*, Radix Scrophulariae, Fructus Forsythiae, Schizonepeta, Fructus Arctii, Liquorice	Curing measles and cough; reducing pharyngeal swelling	"Zhang Shi Yi Tong", Vol. 15[b] 《张氏医通》卷十五
She Gan Shu Nian Zi Tang 射干鼠粘子汤	*Belamcanda chinensis*, Fructus Arctii, Rhizoma Cimicifugae, Liquorice	Curing aphthous stomatitis and dry stool; reducing pharyngeal swelling	"Xiao Er Dou Zhen Fang Lun"[b] 《小儿痘疹方论》
She Gan Tang I 射干汤 I	*Belamcanda chinensis*, Orange Osmanthus, Rhizoma Cimicifugae, Radix Angelicae Dahuricae, Liquorice, Cornu Rhinoceri, Almond	Reducing pharyngeal swelling	"Wai Tai Mi Yao", Vol. 23[b] 《外台秘要》卷二十三
She Gan Tang II 射干汤 II	*Belamcanda chinensis*, Pinellia Ternata, Cortex Cinnamomi, Herba Ephedrae, Radix Asteris, Liquorice, Ginger, Date	Curing choking cough in children	"Bei Ji Qian Jin Yao Fang", Vol. 5[b] 《备急千金要方》卷五

(continued)

Name	Compositions	Traditional and clinical uses	References
She Gan Tang III 射干汤 III	*Belamcanda chinensis*, Chinese Herbaceous Peony, Semen Coicis, Cortex Cinnamomi, Concha Ostreae, Gypsum	Curing stroke and sweating	"Qi Xiao Liang Fang", Vol. 1[b] 《奇效良方》卷一
She Gan Tang IV 射干汤 IV	*Belamcanda chinensis*, Gardenia Florida, Indian Bread Pink Epidermis, Rhizoma Cimicifugae, Radix Paeoniae Rubra, Bighead Atractylodes Rhizome	Treatment of gastritis	"Sheng Ji Zong Lu", Vol. 129[b] 《圣济总录》卷一二九
She Gan Tang V 射干汤 V	*Belamcanda chinensis*, Sophora Subprostrata Root, Fructus Forsythiae, Fructus Arctii, Radix Scrophulariae, Schizonepeta, Radix Sileris, Radix Platycodonis, Liquorice, Artichoke Hearts	Curing pharyngeal tumefaction	"Yi Xue Ji Cheng", Vol. 2[b] 《医学集成》卷二
Huang Qin She Gan Tang 黄芩射干汤	*Belamcanda chinensis*, Scutellaria Baicalensis, Fructus Aurantii Immaturus, Pinellia Ternata, Liquorice, Rhizoma Cimicifugae, Cassia Bark	Curing throat obstruction	"Sheng Ji Zong Lu", Vol. 124[b] 《圣济总录》卷一二四
Han Hua She Gan Wan 含化射干丸	*Belamcanda chinensis*, Rhizoma Cimicifugae, Borax, Liquorice, Fermented Blank Bean, Almond	Reducing pharyngeal swelling and pain in the root of tongue	"Tai Ping Sheng Hui Fang", Vol. 18[b] 《太平圣惠方》卷十八
Da She Gan Tang 大射干汤	*Belamcanda chinensis*, Angelica Sinensis, Herba Ephedrae, Cinnamomum Cassia, Cortex Mori, Fructus Aurantii, Radix Asteris, Radix Angelicae Pubescentis, Almond, Pinellia Ternata, Liquorice	Curing cough and vomiting	"Xing Yuan", Vol. 3[b] 《杏苑》卷三
Jia Wei She Gan Tang 加味射干汤	*Belamcanda chinensis*, Radix Rehmanniae, Radix Platycodonis, Fructus Forsythiae, Scutellaria Baicalensis, Fritillary, Radix Scrophulariae, Liquorice, Schizonepeta, Great Burdock	Reducing pharyngeal swelling	"Nang Mi Hou Shu"[b] 《囊秘喉书》卷下

Table 2 The compounds isolated from *Belamcanda chinensis* herbs

(The structures of the main compounds are illustrated in Fig. 2)

No.	Chemical compounds	Plant	Part of plant	Reference
	Isoflavones			
1	tectorigenin	*Belamcanda chinensis* (L.) DC, *Iris tectorum* Maxim, *Iris dichotoma* Pall.	rhizomes	Li et al. (1986), Qiu et al. (2006b), Hideyuki et al. (2001), Jin et al. (2008), Zhou et al. (2000), Masataka et al. (2007), Yeon et al. (2011), Li et al. (2009), Zhou et al. (1997), Qiu et al. (2009), Qiu et al. (2006), Wu et al. (2008a)

(continued)

No.	Chemical compounds	Plant	Part of plant	Reference
2	tectorigenin -7-O-β-glucosyl	*Belamcanda chinensis* (L.) DC.	rhizomes	Li et al. (2009)
3	tectoridin	*Belamcanda chinensis* (L.) DC, *Iris tectorum* Maxim.	rhizomes	Li et al. (1986), Qiu et al. (2006b), Xu et al. (1999), Jin et al. (2008), Zhou et al. (2000), Masataka et al. (2007), Yeon et al. (2011), Li et al. (2009), Zhou et al. (1997), Qiu et al. (2009), Zhang et al. (2011)
4	irigenin	*Belamcanda chinensis* (L.) DC, *Iris dichotoma* Pall., *Iris tectorum* Maxim.	rhizomes	Ji et al. (2001), Li et al. (1986), Xu et al. (1999), Hideyuki et al. (2001), Jin et al. (2008), Masataka et al. (2007), Yeon et al. (2011), Li et al. (2009), Qiu et al. (2006), Wu et al. (2008a), Liu et al. (2005), Liu et al. (2011b), Zhang et al. (2011)
5	iridin	*Iris tectorum* Maxim, *Belamcanda chinensis* (L.) DC.	rhizomes	Qiu et al. (2006b), Hideyuki et al. (2001), Jin et al. (2008), Zhou et al. (2000), Yeon et al. (2011), Li et al. (2009), Zhou et al. (1997), Zhang et al. (2011)
6	irisflorentin	*Belamcanda chinensis* (L.) DC, *Iris dichotoma* Pall., *Iris tectorum* Maxim.	rhizomes	Ji et al. (2001), Li et al. (1986), Xu et al. (1999), Hideyuki et al. (2001), Jin et al. (2008), Yeon et al. (2011), Li et al. (2009), Qiu et al. (2006), Wu et al. (2008a), Liu et al. (2005), Zhang et al. (2011)
7	dichotomitin	*Iris dichotoma* Pall., *Belamcanda chinensis* (L.) DC	rhizomes	Li et al. (1986), Jin et al. (2008), Li et al. (2009), Wu et al. (2008a), Zhang et al. (2011)
8	3', 5'-dimethoxyirisolone-4-O-β-D-gl ucoside	*Belamcanda chinensis* (L.) DC	rhizomes	Jin et al. (2008), Li et al. (2009)
9	nonirisflorentin	*Belamcanda chinensis* (L.) DC	rhizomes	Jin et al. (2008)
10	iristectorigenin A	*Iris tectorum* Maxim, *Belamcanda chinensis* (L.) DC	rhizomes	Xu et al. (1999), Yeon et al. (2011), Li et al. (2009), Qiu et al. (2009), Zhang et al. (2011)
11	iristectorigenin B	*Belamcanda chinensis* (L.) DC	rhizomes	Li et al. (2009)
12	5,6,7,3'-tetrahydroxy-4'-methoxyisofl avone	*Belamcanda chinensis* (L.) DC.	rhizomes	Hideyuki et al. (2001), Jin et al. (2008)

(continued)

No.	Chemical compounds	Plant	Part of plant	Reference
13	6" -O-vaniloyiridin	*Belamcanda chinensis* (L.) DC	rhizomes	Hideyuki et al. (2001), Jin et al. (2008)
14	6" -O-*p*-hydroxybenzoyliridin	*Belamcanda chinensis* (L.) DC.	rhizomes	Hideyuki et al. (2001), Jin et al. (2008)
15	2, 3-dihydroirigenin	*Belamcanda chinensis* (L.) DC	rhizomes	Hideyuki et al. (2001)
16	tectoirigenin	*Belamcanda chinensis* (L.) DC.		Hideyuki et al. (2001)
17	iristectrigenin	*Belamcanda chinensis* (L.) DC	rhizomes	Masataka et al. (2007)
18	iristectrigenin A-7-glucoside	*Belamcanda chinensis* (L.) DC	rhizomes	Masataka et al. (2007)
19	8-hydroxytectrigenin	*Belamcanda chinensis* (L.) DC	rhizomes	Masataka et al. (2007)
20	8-hydroxyiritectrigenin A	*Belamcanda chinensis* (L.) DC	rhizomes	Masataka et al. (2007)
21	8-hydroxyirigenin	*Belamcanda chinensis* (L.) DC	rhizomes	Masataka et al. (2007)
22	tectrigenin-4'-glucoside	*Belamcanda chinensis* (L.) DC	rhizomes	Masataka et al. (2007)
23	irilin D	*Belamcanda chinensis* (L.) DC	rhizomes	Masataka et al. (2007), Yeon et al. (2011), Li et al. (2009), Qiu et al. (2006)
24	isotectrigenin	*Belamcanda chinensis* (L.) DC	rhizomes	Masataka et al. (2007)
25	astragalin	*Belamcanda chinensis* (L.) DC	rhizomes	Masataka et al. (2007)
26	5, 7, 4'-dihydroxy-6, 3', 5'-trimethoxyis oflavone	*Belamcanda chinensis* (L.) DC *Iris tectorum* Maxim.	rhizomes	Qiu et al. (2009), Liu et al. (2011b)
27	genistein	*Belamcanda chinensis* (L.) DC	rhizomes	Ji et al. (2001)

(continued)

No.	Chemical compounds	Plant	Part of plant	Reference
28	dimethyltec-torigenin	*Belamcanda chinensis* (L.) DC	rhizomes	Ji et al. (2001)
29	isoirigenin	*Belamcanda chinensis* (L.) DC	rhizomes	Liu et al. (2011b)
30	isoirigenin 7-O-β-D-glucoside	*Belamcanda chinensis* (L.) DC	rhizomes	Qiu et al. (2006)
31	irisolon	*Belamcanda chinensis* (L.) DC	rhizomes	Ji et al. (2001), Zhang et al. (2011)
32	3'-hydroxy-tectoridin	*Belamcanda chinensis* (L.) DC	rhizomes	Qiu et al. (2006b), Li et al. (2009)
33	tectoruside	*Belamcanda chinensis* (L.) DC	rhizomes	Li et al. (2009)
34	5-hydroxy-isoflavone	*Belamcanda chinensis* (L.) DC	rhizomes	Liu et al. (2011b)
35	5-hydroxy-6, 7-methyl-enedioxy-3', 4', 5'-tri-methoxy-isoflavone	*Belamcanda chinensis* (L.) DC	rhizomes	Sikc et al. (1993)
36	6-methoxy-5, 7, 8, 4'-tetra-hydryoxyisof lavone	*Belamcanda chinensis* (L.) DC	rhizomes	Song et al. (2007)
37	4'-methoxy-5, 6-di-hydroxyisoflavone-7-O-β-D-glucopyr-anoside Fla-vonoids	*Belamcanda chinensis* (L.) DC	rhizomes	Song et al. (2007)
38	5, 4'-di-hydroxy-6, 7-methyl-enedioxy-3'-methoxy-flavone	*Belamcanda chinensis* (L.) DC	rhizomes	Jin et al. (2008)

(continued)

No.	Chemical compounds	Plant	Part of plant	Reference
39	iristectogenin	*Belamcanda chinensis* (L.) DC	rhizomes	Jin et al. (2008)
40	hispidulin	*Belamcanda chinensis* (L.) DC	rhizomes	Jin et al. (2008), Wang et al. (2011)
41	acetovanilnone	*Belamcanda chinensis* (L.) DC	rhizomes	Jin et al. (2008)
	Flavonols			
42	isorhamnetin	*Belamcanda chinensis* (L.) DC	rhizomes	Jin et al. (2008)
43	rhamnazin	*Belamcanda chinensis* (L.) DC	rhizomes	Jin et al. (2008)
44	rhamnocitrin	*Belamcanda chinensis* (L.) DC	rhizomes	Hideyuki et al. (2001)
	Other flavonoids			
45	mangiferin	*Belamcanda chinensis* (L.) DC	rhizomes	Wang et al. (2011), Li et al. (2009), Zhang et al. (2011)
46	neomangiferin	*Belamcanda chinensis* (L.) DC	rhizomes	Zhang et al. (2011)
47	isomangiferin	*Belamcanda chinensis* (L.) DC	rhizomes	Li et al. (2009)
48	wogonin	*Iris dichotoma* Pall.	rhizomes	Li et al. (1986)
49	7-O-methylmangiferin	*Belamcanda chinensis* (L.) DC	rhizomes	Wang et al. (2011), Li et al. (2009)
50	3'-hydroxyltectoridin	*Belamcanda chinensis* (L.) DC	rhizomes	Wang et al. (2011)
51	iristectorin	*Belamcanda chinensis* (L.) DC	rhizomes	Li et al. (2009)
52	iristectorin A	*Belamcanda chinensis* (L.) DC	rhizomes	Qiu et al. (2006b), Li et al. (2009), Qiu et al. (2009), Zhang et al. (2011)
53	iristectorin B	*Belamcanda chinensis* (L.) DC	rhizomes	Wang et al. (2011), Qiu et al. (2009), Zhang et al. (2011)
54	isoiridin	*Belamcanda chinensis* (L.) DC	rhizomes	Wang et al. (2011)
	Triterpenoids			

(continued)

No.	Chemical compounds	Plant	Part of plant	Reference
55	(6R, 10S, 11S, 14S, 26R) -26-hydroxy-1-5-methycidone-spiroirid-16-enal	*Belamcanda chinensis* (L.) DC	rhizomes	Kunihiko et al. (2000)
56	(6R, 10S, 11R) -26-ξ-hydroxy-13R - oxa spiroirid	*Belamcanda chinensis* (L.) DC	rhizomes	Kunihiko et al. (2000)
57	iso - iridogermana	*Belamcanda chinensis* (L.) DC	rhizomes	Kunihiko et al. (2000)
58	belamcandal	*Belamcanda chinensis* (L.) DC		
59	dibelamcandal A	*Belamcanda chinensis* (L.) DC	rhizomes	Song et al. (2011)
60	28-deacetyl belamcandal	*Belamcanda chinensis* (L.) DC	rhizomes	Kunihiko et al. (2000), Abe et al. (1991)
61	16-O-acety lisoiridogermanal	*Belamcanda chinensis* (L.) DC	rhizomes	Kunihiko et al. (2000)
62	iritectol A	*Iris tectorum* Maxim.	rhizomes	Rui et al. (2007)
63	iritectol B	*Iris tectorum* Maxim.	rhizomes	Rui et al. (2007)
64	isoiridogermanal	*Iris tectorum* Maxim.	rhizomes	Rui et al. (2007)
65	6R,10S,11R-26-hydroxy-(13R) -oxasp iroirid-16-enal	*Belamcanda chinensis* (L.) DC	rhizomes	Kunihiko et al. (2000)
66	iridobelamal A	*Belamcanda chinensis* (L.) DC *Iris tectorum* Maxim.	rhizomes	Kunihiko et al. (2000), Rui et al. (2007)
	Others			
67	shegansu A	*Belamcanda chinensis* (L.) DC	rhizomes	Li et al. (1998), Zhou et al. (1997)

(continued)

No.	Chemical compounds	Plant	Part of plant	Reference
68	shegansu B	*Belamcanda chinensis* (L.) DC	rhizomes	Zhou et al. (2000)
69	shegansu C	*Belamcanda chinensis* (L.) DC	rhizomes	Li et al. (1998)
70	belamcandaquinone A	*Belamcanda chinensis* (L.) DC	rhizomes	Yoshiyasu et al. (1993)
71	belamcandaquinone B	*Belamcanda chinensis* (L.) DC	rhizomes	Yoshiyasu et al. (1993)
72	rhamnazin	*Iris dichotoma* Pall.	rhizomes	Li et al. (1986)
73	belamcandones A-D	*Belamcanda chinensis* (L.) DC	seeds	Kstsura et al. (1995)
74	stigmasterol	*Belamcanda chinensis* (L.) DC	rhizomes	Ji et al. (2001)
75	β-sitosterol	*Belamcanda chinensis* (L.) DC	rhizomes	Ji et al. (2001), Lin et al. (1998), Wu et al. (2008a), Liu et al. (2005)
76	daucosterol	*Belamcanda chinensis* (L.) DC	rhizomes	Zhou et al. (2000), Zhou et al. (1997), Liu et al. (2005)
77	β-daucosterol	*Belamcanda chinensis* (L.) DC	rhizomes	Wu et al. (2008a)
78	isorhapontigenin	*Belamcanda chinensis* (L.) DC	rhizomes	Zhou et al. (2000), Li et al. (1998), Zhou et al. (1997)
79	resveratrol	*Belamcanda chinensis* (L.) DC	rhizomes	Zhou et al. (2000), Li et al. (2009), Li et al. (1998), Zhou et al. (1997)
80	*p*-hydroxybenzoic acid	*Belamcanda chinensis* (L.) DC	rhizomes	Zhou et al. (2000), Li et al. (1998), Zhou et al. (1997)
81	uridine	*Belamcanda chinensis* (L.) DC	rhizomes	Wu et al. (2008a)
82	cycloartanol	*Belamcanda chinensis* (L.) DC	rhizomes	Wu et al. (2008a)
83	apocynin	*Belamcanda chinensis* (L.) DC	rhizomes	Wang et al. (2011), Li et al. (1998), Wu et al. (2008a)
84	isoferulic acid	*Belamcanda chinensis* (L.) DC	rhizomes	Qiu et al. (2006b)

Abe et al. (1991) isolated 9 bicyclic triterpenoid compounds and 4 iris types of fatty acid esters from *Belamcanda chinensis*. The triterpenoid compounds include (6R, 10S, 11S, 14S, 26R) -26-hydroxy-15-methylidenespiroirid-16-enal (55), iso-iridogermana (57), belamcandal (58), 28-deacetylbelamcandal (60), and 16-O-acetylisoiridogermanal (61). Later, Fang et al. (2007) isolated iritectol A (62), iritectol B (63) and isoiridogermanal (64).

Kunihiko et al. (2000) isolated 6R, 10S, 11R-26-hydroxy- (13R) -oxaspiroirid-16-enal (65) and iridobelamal A (66).

OTHER CONSTITUENTS

Quinone compounds that have been isolated from *Belamcanda chinensis* include belamcandaquinone A (70) and belamcandaquinone B (71) (Yoshiyasu et al., 1993). Several ketones have also been isolated from the seeds of *Belamcanda chinensis*, including rhamnazin (72) and belamcandones A-D (73) (Katsura et al., 1995). Three steroid compounds that have been isolated are stigmastan (74), β-sitosterol (75) and daucosterol (76). In addition, isorhapontigenin (78), resveratrol (79), hydroxybenzoic acid (80), uridine (81), and cycloartanol (82) have also been isolated.

PHARMACOLOGICAL EFFECTS OF THE CHEMICAL CONSTITUENTS IN *BELAMCANDA CHINENSIS*

Pharmacological studies have demonstrated that many of the chemical compounds in *Belamcanda chinensis*, especially isoflavones, show various types of bioactivity. The various chemical components of *Belamcanda chinensis* will provide a good foundation for further development of new drugs. In Table 3, we list the active compounds and extracts of *Belamcanda chinensis* and their bioactivity.

Antiinflammatory activity

Many studies show that the isoflavones extracted from *Belamcanda chinensis* affect the inflammatory process of the mammalian system and possess anti-inflammatory as well as immune-modulatory activity both *in vitro* and *in vivo* (Wu et al., 1990). An *in vivo* study using the inflammatory reaction model of rat paw swelling and mouse ear oedema showed that *Belamcanda chinensis* extracts had the potential to prevent inflammation (Abe et al., 1991). Hyaluronidase is a proteolytic enzyme that specifically degrades extracellular hyaluronic acid. As early as 1968, studies have shown that tectoridin and tectorigenin from iris exhibited excellent anti-inflammatory activity by markedly inhibiting hyaluronidase activity *in vitro* and *in vivo*, and these chemicals were more potent than the non-steroidal anti-inflammatory drug phenylbutazone sodium salt (Esaki et al., 1968). More recently, Japanese scholars reported that one of the mechanisms of the anti-inflammatory activity of the rhizomes of *Belamcanda chinensis* is the inhibition of PGE_2 production by tectorigenin and tectoridin, with tectorigenin being more potent than tectoridin, due to the decreased expression of cyclooxygenase (COX) -2 in the inflammatory cells (Shin et al., 1999; Kim et al., 1999a). They further investigated the structure-activity relationship of various isoflavones in the inhibition of PGE_2 production and found that 6-methoxylation and 5-hydroxylation increased the potency of PGE_2 inhibition. Production and that 7- O -glycosylation decrease the inhibitory activity (Yamaki et al., 2002). Another anti-inflammatory mechanism was brought up. Because nitric oxide (NO) produced by inducible nitric oxide synthase (iNOS) is a mediator of inflammation, studies reported that tectorigenin showed dose-dependent inhibitory effects on the expression of inducible nitric oxide synthase (iNOS), the production of nitric oxide (NO), the secretion of interleukin (IL) - 1βly, the expression of cyclooxygenase (COX) - 2 and the production of prostaglandin E_2 (PGE_2), which was caused by the blocking of nuclear factor kappa-B (NF-κB) activation (Kim et al., 1999b; Cheol et al., 2008; Ahn et al., 2006; Park et al., 2004).

Table 3 Pharmacological effects of *Belamcanda chinensis*

Effect	Extract or compounds	*In vivo*	*In vitro*	Reference
Anti - inflammatory activity	The ethanol extracts of *Belamcanda chinensis*, *Iris dichotoma Pall.*, and *Iris tectorum Maxim.*	At 8 - 22 g/kg in rats and mice, the extracts markedly inhibited both early and late inflammation.		Wu et al. (1990)
	Extracts of *Belamcanda chinensis*	At 0.32 g/kg, 0.64 g/kg, 1.28 g/kg extracts of *Belamcanda chinensis* clearly inhibited the extent of rat paw swelling; 0.46 g/kg, 0.92 g/kg, and 1.84 g/kg decreased mouse oedema and reduced the stretching times significantly.		Li et al. (2008)
	Tectorigenin and tectoridin	Tectoridin and tectorigenin exhibited excellent anti-inflammatory activity by markedly inhibiting the hyaluronidase activity in rats.		Esaki et al. (1968)
	Tectorigenin and tectoridin		These two chemical compounds suppressed prostaglandin E^2 production by rat peritoneal macrophages stimulated by a protein kinase C activator, 12-O-tetradecanoylphorbol 13-acetate (TPA), or an endomembrane Ca^{2+}-ATPase inhibitor, thapsigargin. Tectorigenin inhibited prostaglandin E^2 production more potently than tectoridin.	Kim et al. (1999a)
	Tectorigenin		Tectorigenin at 10 - 100 μM showed dose - dependent inhibitory effects on the expression of inducible nitric oxide synthase (iNOS) in LPS-activated RAW264.7 cells.	Kim et al. (1999b)

(continued)

Effect	Extract or compounds	*In vivo*	*In vitro*	Reference
Anti - tumour activity	Tectorigenin and genistein		Both tectorigenin and genistein exhibited cytotoxicity against various human cancer cells, induced differentiation of human promyelocytic leukaemia HL - 60 cells to granulocytes and monocytes/macrophages, caused apoptotic changes to DNA in the cells, inhibited autophosphorylation of epidermal growth factor (EGF) receptor by EGF and decreased the expression of Bcl-2 protein.	Lee et al. (2001)
	Tectorigenin		Tectorigenin (up to 400 μM) showed cell - cycle - specific inhibition and arrested cells at G2/M phase.	Fang et al. (2008)
	Tectorigenin		Tectorigenin could inhibit proliferation of human hepatoma cells (SMMC-7721) in a concentration - dependent manner from 1.00 $\mu g \cdot mL^{-1}$ to 8.00 $\mu g \cdot mL^{-1}$. The 48-hour maximal inhibition rate reached 76.57%; IC_{50} was (3.71 ± 1.17) $\mu g \cdot mL^{-1}$.	Wu et al. (2008b)
	Tectorigenin and tectoridin	When tectorigenin was administered subcutaneously at a dose of 30mg/kg for 20 days to mice implanted with murine Lewis lung carcinoma (LLC), significant (30.8%) inhibition of tumour volume resulted. Tectorigenin and tectoridin, when administered i.p. at the same dosage for 10 days to ICR mice bearing sarcoma 180, caused significant reductions in tumour weight, by 44.2% and 24.8%, respectively.	Both compounds decreased angiogenesis of both chick embryos in the chorioallantoic membrane assay and basic fibroblast growth factor - induced vessel formation in the mouse Matrigel plug assay. They also reduced the proliferation of calf pulmonary arterial endothelial (CPAE) cells and were found to possess relatively weak gelatinase/collagenase inhibitory activity in vitro.	Jung et al. (2003)

(continued)

Effect	Extract or compounds	*In vivo*	*In vitro*	Reference
	Tectorigenin		Tectorigenin suppressed the proliferation of HSC-T6 cells. Tectorigenin at the concentration of 100 μg/mL greatly inhibited the viability of HSC-T6 cells in a time- and dose-dependent manner and induced the condensation of chromatin and fragmentation of nuclei. When applied for 48 h, the percentage of cell growth and apoptosis reached 46.3% ± 2.37% ($P=0.004$) and 50.67% ± 3.24%, respectively.	Wu et al. (2010)
	Extract of *Belamcanda chinensis* (BCE)		BCE at 100, 400 or 1400 μg/mL decreased LNCaP tumour cell proliferation and downregulated the expression of AR, PDEF, NKX3.1 and PSA.	Thelen et al. (2007)
	Tectorigenin and *B. chinensis* extract	Tectorigenin at 400mg/mL led to the downregulation of PDEF, PSA hTERT and IGF-1 receptor mRNA expression *in vitro* and caused reductions in PSA secretion and IGF-1 receptor protein expression.	The extract (6.7mg/g) markedly inhibited the development of tumours on male athymic nude BALB/c - nu mice.	Thelen et al. (2005)
	Tectorigenin		Treatment with tectorigenin (10-100 μmol/L) caused the upregulation of oestrogen receptor β (ERβ), the preferred receptor for phytoestrogens, resulting in antiproliferative effects.	Stettner et al. (2007)
Estrogenic activity	Irigenin, tectorigenin and tectoridin.		The oestrogenic activity of these isoflavones was tested using Ishikawa cells. Irigenin, tectorigenin and tectoridin were highly oestrogenic (EC_{50} = 0.75, 0.42 and 0.81 μg/mL, respectively.	Lee et al. (2005a)

(continued)

Effect	Extract or compounds	*In vivo*	*In vitro*	Reference
	Tectorigenin	Tectorigenin administered intravenously to ovariectomised (ovx) rats inhibited pulsatile pituitary LH secretion.	Tectorigenin (7mg) bound to both receptor subtypes had a strong hypothalamotropic and osteotropic effect on human breast cancer MCF7 cells.	Seidlová et al. (2004)
	The ethanol extract of *Belamcanda chinensis*.		The recombinant yeast system with both a human oestrogen receptor expression plasmid and a reporter plasmid was used to confirm that the extract had high oestrogenic activities with a relative potency (RP) of 1.26×10^{-4}.	Zhang et al. (2005)
	Antioxidative activity	Tectorigenin	Tectorigenin could scavenge superoxide radicals, hydroxyl radicals and hydrogen peroxide in the system.	Qin et al. (2003b)
	Tectorigenin		Tectorigenin at 5mg/kg and 10mg/kg had antioxidant effects on 1,1-diphenyl-2-picrylhydrazyl (DPPH) radical and xanthine-xanthine oxidase superoxide anion radical on rats.	Lee et al. (2000)
	Isorhapontigenin		The results showed that ISOR (10^{-4}, 10^{-5} and 10^{-6} mol/L) significantly inhibited MDA formation in liver microsomes, brain mitochondria and synaptosomes induced by Fe^{2+}-Cys. ISOR markedly prevented the decrease in GSH in mitochondria and synaptosomes induced by H_2O_2 and the increase of ultra-weak chemiluminescence during lipid peroxidation induced by VitC-ADP-Fe^{2+} as well as oxidative DNA damage induced by $CuSO_4$-Phen-VitC-H_2O_2.	Wang et al. (2001)

(continued)

Effect	Extract or compounds	*In vivo*	*In vitro*	Reference
Anti-alcohol injury	Tectorigenin and tectoridin		Studies have reported that both tectorigenin and tectoridin inhibited the lipopolysaccharide-induced nitric oxide release from primary cultured rat cortical microglia (IC_{50}: 1.3-9.3 μM). These results indicate that both compounds have therapeutic potential in alcoholism.	Yuan et al. (2009)
Hepatoprotective effect	Tectoridin	Tectoridin (25, 50 and 100mg/kg) given intragastrically five times on three consecutive days to mice was found to significantly attenuate the increase in alanine aminotransferase, aspartate aminotransferase and triglyceride levels and hepatic mitochondria dysfunction that were induced by acute ethanol exposure. These results showed that tectoridin has hepatoprotective effects.	Tectoridin inhibited the decrease of PPARα expression and its target genes at the mRNA level and inhibited the decrease in enzyme activity levels, suggesting that tectoridin protected against ethanol-induced liver steatosis mainly through modulating the disturbance of the PPARα pathway and ameliorating mitochondrial function.	Xiong et al. (2010)
	Tectorigenin	Tectorigenin (100mg/kg) administered intraperitoneally for CCl_4-induced liver injury in mice significantly inhibited the increase in plasma ALT, AST and LDH activity. Tectoridin may be a prodrug that is transformed to tectorigenin.		Lee et al. (2003)
Hypoglycaemic effects	Tectorigenin and tectoridin	Oral administration at 100mg/kg for 10 consecutive days to streptozotocin-induced diabetic rats significantly inhibited sorbitol accumulation in the tissues such as the lens, sciatic nerves and red blood cells. Tectorigenin showed stronger inhibitory activity than tectoridin. Tectorigenin may be a promising compound for the prevention and/or treatment of diabetic complications.		Jung et al. (2002)

(continued)

Effect	Extract or compounds	*In vivo*	*In vitro*	Reference
	Tectorigenin	Intraperitoneal administration of tectorigenin (5-10mg/kg) for seven days to streptozotocin-induced rats significantly reduced the blood glucose, total cholesterol, LDL- and VLDL-cholesterol and triglyceride levels.		Lee et al. (2000)
Cardiovascular and cerebrovascular effects	Tectorigenin	Tectorigenin (50, 100, 200 μmol/L^{-1}) had significant protective effects on the damage of the vascular endothelial cells induced by low-density lipoprotein. It also inhibited the LDL-induced mRNA overexpression of monocyte chemoattractant protein-1 and intercellular adhesion molecule-1.		Wang et al. (2010)
Antithrombotic effect	Ethanol extract of *Belamcanda chinensis*	Intragastric administration of a 75% ethanol extract of *Belamcanda chinensis* to rats could significantly delay carotid throm bosis after electrical stimulation.		Zhang et al. (1997a)
	Tectorigenin		Tectorigenin inhibited arachidonic acid and collagen-induced platelet aggregation.	Lo et al. (2003)
Antifungal activity	Tectorigenin	Tectorigenin showed marked antifungal activity against dermatophytes of the genera trichophyton, and the minimum inhibitory concentration (MIC) was in the range of 3.12-6.25mg/ml.		Oh et al. (2001)

Anti-tumour activity

Lee et al. (2001) studied the cytotoxic effects of six isoflavonoids that were isolated from flowers of *Pueraria thunbergiana* Benth. They concluded that tectorigenin may be a possible therapeutic agent for leukaemia because both tectorigenin and genistein exhibited cytotoxicity against various human cancer cells, induced differentiation of human promyelocytic leukaemia HL-60 cells to granulocytes and monocytes/macrophages, caused apoptotic changes to DNA in the cells, inhibited auto-phosphorylation of epidermal growth factor (EGF) receptor by EGF and decreased the expression of Bcl-2 protein. Supporting those findings, another group found that tectorigenin showed cell-cycle-specific inhibition and arrested cells at the G2/M phase (Fang et al., 2008). In addition, *in vitro* studies showed that tectorigenin could dose-dependently inhibit the proliferation of SMMC-7721 human hepatoma cells, which may be related to the promotion of apoptosis (Wu et al., 2008). Furthermore, Jung's group found that both tectorigenin and tectoridin inhibited angiogenesis *in vitro*. *In vivo* studies showed that subcutaneous and intra-peritoneal administration of either tectorigenin or tectoridin significantly inhibited tumour growth, indicating that both compounds had significant anti-tumour activities (Jung et al., 2003; Wu et al., 2010).

Recent advances have indicated that protein tyrosine kinases have a role in the pathophysiology of cancer, which prompted researchers to systematically synthesise tyrosine phosphorylation inhibitors as potential drugs (Levitzki et al., 1994; Levitzki et al., 1998; Manash et al., 2004; Tony et al., 1997). Ohuchi et al. (1999) found that Isoflavones could inhibit tyrosine kinase activity, which further impaired the transport of nuclear factor-κB, leading to anti-tumour effects. However, additional evidence demonstrated that Ψ-tectorigenin 6, an isomer of the tectorigenin, was a tyrosine kinase inhibitor with greater activity (Himpens et al., 1994; Umezaawa et al., 1992; Johnson et al., Filipeanu et al., 1995). These results indicate that the modification of tectorigenin would aid in the development of more effective anticancer drugs. Some of the *Belamcanda chinensis* extracts have been evaluated to determine their potential as anticancer drugs. Among them, tectorigenin was found to be useful for the prevention or treatment of human prostate cancer by *in vitro* and *in vivo* methods (Thelen et al., 2007). Tectorigenin treatment resulted in the down-regulation of PDEF, PSA, hTERT and IGF-1 receptor mRNA expression *in vitro* as well as the reduction of PSA secretion and IGF-1 receptor protein expression, suggesting that it has anti-proliferative potential. Moreover, animal experiments demonstrated that *Belamcanda chinensis* markedly inhibited the development of tumours *in vivo* (Thelen et al., 2005). Another group obtained similar results, demonstrating via *in vitro* studies that tectorigenin and irigenin regulate prostate cancer cell number by regulating the cell cycle to inhibit proliferation (Morrissey et al., 2004). Later, Stettner et al. (2007) found that tectorigenin treatment caused the up-regulation of oestrogen receptor β (ERβ), the preferred receptor for phytoestrogens, resulting in anti-proliferative effects, which may be the mechanism underlying the beneficial effects of tectorigenin on prostate cancer.

In 2001, tectorigenin was shown to have anti-mutagenic activity. Tectorigenin had suppressive effects on umu gene expression of the SOS response in *Salmonella* typhimurium against the mutagens (Miyazawa et al., 2001). Zhang et al. (2010a) investigated the effects of tectorigenin on pulmonary fibroblasts in the Idiopathic Pulmonary Fibrosis (IPF) animal model and the underlying molecular mechanisms. Tectorigenin was found to inhibit the proliferation of pulmonary fibroblasts *in vitro* and to enhance miR-338* expression, which might in turn down-regulate LPA1, indicating a potential inhibitory role for tectorigenin in the pathogenesis of IPF. The effects of tectorigenin on proliferation and apoptosis of hepatic stellate cells (HSC-T6 cells) were also investigated, and tectorigenin was shown to suppress the proliferation of HSC-T6 cells and to induce their apoptosis in a time-and dose-dependent manner (Park et al., 2002).

Estrogenic activity

Tectoridin, a major isoflavone isolated from the rhizome of *Belamcanda chinensis*, is known as a phytoestrogen. Phytoestrogens are the natural compounds isolated from plants that are structurally similar to animal oestrogen, 17-β-oestradiol. Four isoflavones that were isolated from the rhizome of *Belamcanda chinensis*, iristectorigenin A, irigenin, tectorigenin and tectoridin, were tested for their oestrogenic activity using Ishikawa cells. The results showed that irigenin, tectorigenin and tectoridin were highly oestrogenic, while iristectorigenin A exhibited weak oestrogenic activity (Lee et al., 2005a). The oestrogenic activity of kakkalide and tectoridin has been compared with their metabolites. The data showed that all compounds expressed oestrogenic effects by increasing the proliferation of MCF-7 cells and inducing oestrogen-response c-fos and pS2 mRNA expression, with the metabolites being more potent (Shin et al., 2006). The human oestrogen receptor (hER) exists as two subtypes, hERα and hERβ. Seidlová-Wuttke et al. found that tectorigenin bound to both receptor subtypes and had a strong hypothalamotropic and osteotropic effect but no effect in the uterus or the mammary gland (Seidlová et al., 2004). In contrast, isoflavone glycoside was found to bind weakly to both receptors (Keiko et al., 2002). The oestrogenic activity of 70% ethanol extracts of *Belamcanda chinensis* was also confirmed using a recombinant yeast system with both a human oestrogen receptor expression plasmid and a reporter plasmid (Zhang et al., 2005). Certain compounds isolated from the rhizomes of *Belamcanda chinensis*, including resveratrol, iriflophenone, tectorigenin, tectoridin and belamphenone, had been shown to stimulate MCF-7 and T-47D human breast cancer cell proliferation (Monthakantirat et al., 2005).

Kyungsu et al. (2009) investigated the molecular mechanisms underlying the oestrogenic effect of tectoridin and found that this effect occurred mainly via the GPR30 and ERK-mediated rapid nongenomic oestrogen signalling pathway, while genistein exerted its oestrogenic effects via both an ER-dependent genomic pathway and a GPR30-dependent nongenomic pathway. Genistein was found to reduce the luteinising hormone-releasing hormone (LHRH)-induced release of luteinising hormone (LH) and follicle-stimulating hormone (FSH) from pro-oestrous rat hemi-pituitaries incubated in vitro (Melanie et al., 1995). When given intravenously to ovariectomised (ovx) rats, genistein inhibits pulsatile pituitary LH secretion. Upon chronic application to ovx rats, a *Belamcanda chinensis* extract containing 5% *Belamcanda chinensis* at a daily dose of 33mg or 130mg had no effect on uterine weight or on oestrogen-regulated uterine gene expression, but oestrogenic effects in the bone (i.e., effects on the bone mineral density of the metaphysis of the tibia) were established (Seidlová et al., 2004).

Scavenging free radicals and anti-oxidative activity

The poly-phenolic compounds from plants are by far the most frequently reported as antioxidants. Using a biochemiluminescence technique, Qin et al. (2003b) found that the Isoflavones isolated from *Belamcanda chinensis*, including irigenin, tectorigenin, tectoridin and 5,6,7,4'-tetrahydroxy-8-methoxyisoflavone, could scavenge superoxide radicals, hydroxyl radicals and hydrogen peroxide in the system, with tectorigenin being the strongest oxygen free radical scavenger of the four. *In vitro* studies showed that tectorigenin had antioxidant effects on 1,1-diphenyl-2-picrylhydrazyl (DPPH) radical, xanthine-xanthine oxidase superoxide anion radical, and lipid peroxidation in rat microsomes induced by enzymatic and non-enzymatic methods (Lee et al., 2000). The isoflavonoid fractions of *Belamcanda chinensis* have the capability to scavenge free radicals, to reduce transition-metal ions and to protect polyunsaturated fatty acids from peroxidation (Dorota et al., 2010). Studies have shown that tectorigenin has a significant protective effect on the damage to VECs induced by ox-LDL, and tectorigenin significantly inhibited the MCP-1 and ICAM-1 mRNA overexpression in VECs induced by ox-LDL, which was thought to play an important role in anti-atherosclerosis (Wang et al., 2010).

Isorhapontigenin, isolated from *Belamcanda chinensis*, is a derivative of stilbene and is structurally similar to resveratrol. *In vitro* studies have shown that isorhapontigenin significantly suppressed various types of oxidative damage induced in rat liver microsomes, brain mitochondria and synaptosomes respectively, showing a more potent anti-oxidative activity than the classical antioxidant vitamin E (Wang et al., 2001). In addition, isorhapontigenin was found to dose-dependently inhibit the production of superoxide anion and hydrogen peroxide in phorbol myristate acetate (PMA)-activated rat neutrophils. Scanning electron microscopy results showed that isorhapontigenin protected against surface changes in rat neutrophils and inhibited the release of β-glucuronidase from the activated neutrophils. Electron-spin resonance (ESR) spectra demonstrated that isorhapontigenin could scavenge oxygen free radicals generated in the PMA-activated neutrophils, resulting in the inhibition of respiratory burst in PMA-activated rat neutrophils (Fang et al., 2002).

Anti-alcohol injury and hepatoprotective effects

Microglial cells are the primary immune cells in the central nervous system, and they play an important role in the inflammatory process of brain damage. Increasing evidence shows that microglial activation is related to neurological dysfunction and can regulate alcoholism (Crews et al., 2006). Studies have reported that both tectorigenin and tectoridin inhibit the microglial activation, shown as the inhibition of the lipopolysaccharide-induced nitric oxide release from primary cultured rat cortical microglia, with tectorigenin showing a stronger inhibitory effect. The results indicate that both compounds have therapeutic potential in the treatment of alcoholism (Yuan et al., 2009). Later, the hepatoprotective effects and the mechanisms of tectoridin hepatoprotective effects were investigated. Tectoridin was found to significantly attenuate the increases in the levels of alanine aminotransferase, aspartate aminotransferase and triglyceride and the hepatic mitochondrial dysfunction that were induced by acute ethanol exposure. Furthermore, tectoridin inhibited the decrease in expression of PPARα and its target genes at the mRNA level and inhibited the decrease in enzyme activity levels, suggesting that tectoridin protected against ethanol-induced liver steatosis mainly through modulating the disturbance of the PPARα pathway and retaining mitochondrial function (Xiong et al., 2010). Several groups found that tectoridin may be a pro-drug transformed to tectorigenin, as only intra-peritoneal administration of tectorigenin and oral but not intra-peritoneal administration of tectoridin exhibited hepatoprotective activities in CCl_4-intoxicated model animals, and this hepatoprotective effect may be related to the inhibition of β-glucuronidase, the increase in GSH content and GST activity and the inhibition of apoptosis (Lee et al., 2003; Jung et al., 2004; Lee et al., 2005b).

Hypoglycaemic and hypolipidaemic effects

Studies have shown that tectorigenin and tectoridin isolated from the rhizomes of *Belamcanda chinensis* could inhibit aldose reductase, which plays important roles in diabetic complications. Moreover, oral administration of either compound significantly inhibited tissue sorbitol accumulation in streptozotocin-induced diabetic rats, suggesting that the compounds may be candidates for the prevention and/or treatment of diabetic complications (Jung et al., 2002). *In vitro* studies also showed that tectorigenin had potent hypoglycaemic activity (Bae et al., 1999). Intra-peritoneal administration of tectorigenin significantly reduced the blood glucose, total cholesterol, LDL-cholesterol, VLDL-cholesterol and triglyceride levels in the streptozotocin-induced diabetic rats, thus showing potent hypoglycaemic and hypolipidaemic effects in vivo (Lee et al., 2000).

Cardiovascular and cerebrovascular effects

Tectorigenin had significant protective effects against damage to the vascular endothelial cells induced by low-density lipoprotein. It also inhibited the mRNA overexpression of monocyte chemoattractant protein-1 and intercellular adhesion molecule-1 induced by LDL, which may be one of the mechanisms of atherosclerosis (Wang et al., 2010). One study reported that the anticoagulant effect of *Belamcanda chinensis* that is due to one of its components, acidic polysaccharide (MW of 10,000). Intra-gastric administration of the 75% ethanol extract of *Belamcanda chinensis* to rats could significantly delay carotid thrombosis after electrical stimulation (Zhang et al., 1997a). Tectorigenin was also found to inhibit arachidonic acid and collagen-induced platelet aggregation, suggesting that tectorigenin may be one of the active ingredients that resulted in the antithrombotic effect of *Belamcanda chinensis* (Lo et al., 2003).

Antibacterial and antifungal activity

i) Antibacterial activity

Previous studies have shown that the *Belamcanda chinensis* had different degrees of antibacterial effects on many bacteria, including *Staphylococcus aureus*, influenza A, Group B streptococcus, pneumococcus, meningococcus, *E. coli*, typhoid and paratyphoid bacillus, and *Haemophilus influenzae* bacilli (Yu et al., 2001). An *in vitro* study showed that isoflavone glycosides did not inhibit the growth of *Helicobacter pylori*. However, their aglycones, irisolidone, tectorigenin and genistein, inhibited *Helicobacter pylori* growth (Bae et al., 2001).

ii) Antifungal activity

The antifungal activity of *Belamcanda chinensis* against 17 strains of fungi and 6 strains of bacteria was investigated, and the results showed that tectorigenin had marked antifungal activity against dermatophytes of the genus *Trichophyton* (Oh et al., 2001). A water decoction of *Belamcanda chinensis* had no inhibitory effects on normal ocular pathogenic fungi (Wei et al., 1994) including *Aspergillus phyllotreta*, variegated song, earth song, Japanese song, *Fusarium moniliforme*, and pear Fusarium yeast, but it could inhibit dermatophytes, which are superficially pathogenically fungal, including *Trichophyton rubrum*, Trichophyton, wool-like spores, *Epidermophyton floccosum*, Xu Lanshi Trichophyton plaster-like spores, Trichophyton violaceum, and Trichophyton canis (Wang et al., 1997; Liu et al., 1998).

Electron microscopy results showed that when the concentration or the duration of treatment with the ether extract of *Belamcanda chinensis* was increased, the cell wall of *Trichophyton rubrum* became roughened and developed disruptive hollows that eventually collapsed, while the mycelium gradually swelled, the cell wall gradually thickened, organelle swelling increased, intracellular particles with a high electron density emerged, and *Trichophyton rubrum* eventually disintegrated (Liu et al., 1999).

iii) Antiviral activity

Previous *in vitro* studies showed that a water decoction of *Belamcanda chinensis* showed inhibitory effects against influenza virus, adenovirus, echovirus, Coxsackievirus, herpes virus, and wild iris aglycones and was found to be the active antiviral ingredient. Mangiferin showed strong inhibition of the *in vitro* replication of type II herpes simplex virus (Liu et al., 2000). A recent study reported that a 60% ethanol extract of *Belamcanda chinensis* significantly delayed the onset of the cytopathic effect of influenza virus FM1, adenovirus III and herpes simplex virus, but it showed no antiviral activity against enterovirus Cox B3. The *in vivo* study showed that treatment of mice with a 60% ethanol extract of *Belamcanda chinensis* could significantly inhibit the increase of the lung/body weight ratio caused by influenza virus, indicating that *Belamcanda chinensis* played a role in the inhibition of viral pneumonia (Han et al., 2004).

Effects on the digestive system

Intra-gastric administration of the leaching solution of ethanol and ethanol-water extract of *Belamcanda chinensis* to rabbits can promote saliva secretion, and tectoridin may be one of the active ingredients. Studies also showed that injection of the 75% ethanol extract of *Belamcanda chinensis* into the duodenum of anesthetised rats persistently accelerated bile secretion, possibly due to the presence of mangiferin in the extract (Zhang et al., 1998). Moreover, the 75% ethanol extract of *Belamcanda chinensis* also showed weak antiulcer effects because intra-gastric administration of the extract inhibited the ulceration induced by flooding stress, hydrochloric acid and indomethacin ethanol by 26%-48% (Zhang et al., 1997b). In addition, intra-gastric administration of a 75% ethanol extract of *Belamcanda chinensis* to mice caused significant inhibition of small intestine diarrhoea caused by castor oil but caused weak inhibition of the large intestine diarrhoea caused by folium sennae and no inhibition of ink gastrointestinal propulsive motility in mice (Zhang et al., 1997c).

Other pharmacological effects

Components of *Belamcanda chinensis* have been found to have ichthyotoxic activity. Japanese scholars (Ho et al., 1999) isolated eleven iridal-type triterpenoids from the hexane and ether extracts of *Belamcanda chinensis*. Among them, three compounds showed potent ichthyotoxic activity against killie-fish (*Oryzias latipes*), but others did not.

Toxicity and side effects

i) Skin allergies

It has been reported that four days after treatment with *Belamcanda chinensis*, a patient presented with red rashes of various sizes and blisters on the skin of the neck and back, possibly due to an allergy to the antiviral injection. The symptoms gradually disappeared after medicine withdrawal and anti-allergy treatment (Gao et al., 2006). Because the *Belamcanda chinensis* antiviral injection has a very complex composition, it is difficult to identify the allergy-causing substance. Therefore, it is recommended that care should be exercised in the clinical use of *Belamcanda chinensis*, and the injection should be used with caution in patients with drug allergies.

ii) Systemic muscle rigidity

One report indicated that patients had muscle rigidity of the neck, limbs and abdominals after using *Belamcanda chinensis*. Symptoms included masseter tension, speech and language impairment, physical impairment and plate-like abdominal muscles. However, the underlying mechanism is not yet clear. (Li, 2005).

CONCLUSION

Belamcanda chinensis is a traditional Chinese medicine commonly used in China, mainly in the clinical treatment of respiratory diseases. The roots of this plant are used alone or in combination with other Chinese herbs as a key ingredient to treat bronchial asthma, tonsillitis and cough in children in traditional Chinese medicine. However, further studies are required to understand the mechanisms of its positive roles in the respiratory system. In addition, a relationship between pharmacological effects and traditional uses of *Belamcanda chinensis* must also be scientifically verified.

Extracts of *Belamcanda chinensis* are rich in isoflavone and triterpenoid compounds. Pharmacology studies have confirmed that these compounds have anti-inflammatory, antibacterial, antioxidant, anti-tumour, free radical scavenging, hepatoprotective, hypoglycaemic and oestrogen-like effects. Among these effects, its hepatoprotective and oestrogen-like effects are most promising. *Belamcanda chinensis* could be used to treat menopausal syndrome and prostate cancer as well as in health product development. In addition, isoflavone and its glycosides (primarily tectoridin and tectorigenin) have been identified as the two main active chemical constituents in *Belamcanda chinensis*, with the glycosides possessing more potent activity. Furthermore, studies have shown that tectoridin may be a prodrug transformed to tectorigenin (Park et al., 2004; Kyoung et al., 2005; Bae et al., 1999).

In this review, current pharmacological data is limited to studies on just a few chemical compositions (tectorigenin and tectoridin) in many cases, showing that effort is required to isolate more biologically active compounds. Although the use of in *vitro* test systems remains popular, there is a clear need for more *in vivo* research and eventually clinical trials. Additionally, some of the studies provide *in vivo* data only, without identifying the underlying mechanism.

The toxicity of *Belamcanda chinensis* has been noted in ethno-medicine and has been validated by toxicological studies. However, the toxicity of medicinal preparations, including doses and safe limits for administration, is not well characterized and requires future attention. Thus, prolonged and high-dose intake of traditional formulations containing *Belamcanda chinensis* should be avoided until the results of more in-depth toxicity studies become available. In China, *Iris dichotoma* Poll and *Iris tectorum* Maxim are substituted for *Belamcanda chinensis* in some districts (Qin et al., 2003a). Studies have shown that these three extracts have similar chemical compositions and pharmacological activities (Huang et al., 1997; Wu et al., 1990), but there is a strong demand for detailed evaluation of their pharmacological and toxicological differences.

In traditional medicine, *Belamcanda chinensis* is commonly used not only in China but also in Korea and Japan. According to the literature reviewed, numerous studies have reported the chemical composition, pharmacological activity and underlying mechanisms of activity of *Belamcanda chinensis*, based on new drugs that have been developed and the number of patent applications. However, the pathway of its distribution, absorption, metabolism and excretion must be clarified by future pharmacokinetic studies. More knowledge of *Belamcanda chinensis* will enhance our understanding of the material basis of treatment with Chinese herbs and will significantly improve the clinical use and effectiveness of these herbs. Taken together, the importance of *Belamcanda chinensis* has been highlighted based on its prominent usage in traditional medicine as well as its potential for use in beneficial therapeutic remedies. Nevertheless, there is clearly a need for further studies focusing on the mechanism, pharmacokinetics and toxicity of *Belamcanda chinensis*.

ACKNOWLEDGEMENTS

This work was financed by the Special Fund for Agro-Scientific Research in the Public Interest (No. 201303040-18).

References

[1] Abe, F., Chen, R.F., Yamauchi, T., 1991. Iridals from *Belamcanda chinensis* and *Iris japonica*. Phytochemistry, 30, 3379-3382.

[2] Ahn, K.S., Noh, E.J., Cha, K.H., 2006. Inhibitory effects of irigenin from the rhizomes of *Belamcanda chinensis* on nitric oxide and prostaglandin E_2 production in murine macrophage RAW264.7 cells. Life Sci. 78, 2336-2342.

[3] Bae, E.A., Han, M.J., Kim, D.H., 2001. In vitro anti-helicobacter pylori activity of irisolidone isolated from the flowers and rhizomes of *Puerariathun bergiana*. Plantamedica. 67, 161-163.

[4] Bae, E.A., Han, M.J., Lee, K.T., Choi, J.W., Park, H.J., Kim, D.H., 1999. Metabolism of 6" -O-xylosyltectoridin and tectoridin by human intestinal bacteria and their hypoglycemic and in vitro cytotoxic activities. Department of Food and Nutrition, Kyung Hee University. 22, 1314-1318.

[5] Cheol, H.P., Eun, S.K., Sang, H.J., 2008. Tectorigen ininhibits IFN-γ/LPS-induced inflammatory responses in murine macrophage RAW264.7 cells. Arch Pharm Res. 31, 1447-1456.

[6] Committee for the Pharmacopoeia of P.R.China, 2010. Pharmacopoeia of P.R.China, Part I. China Medical Science and Technology Press, P.R.China (in Chinese).

[7] Crews, F.T., Bechara, R., Brown, L.A., Guidot, D.M., Mandrekar, P., Oak, S., Qin, L., Szabo, G., Wheeler, M., Zou, J., 2006. Cytokines and alcohol. Alcohol Clin Exp Res. 30, 720-730.

[8] Dorota, W., Bogdan, J., Ireneusz, K., Wiesław, O., Adam, M., 2010. Antimutagenic and antioxidant activities of isoflavonoids from *Belamcanda chinensis* (L.) DC. Mutation Research/Genetic Toxicology and Environmental Mutagenesis. 696, 148-153.

[9] Esaki, S., Nihon, Y.Z., 1968. Pharmacological studies of tectoridin and tectorigenin. Folia Pharmacologica Japonica. 64, 186-198. 130-131.

[10] Fang, R., Houghton, P.J., Hylands, P.J., 2008. Cytotoxic effects of compounds from *Iris tectorum* on human cancer cell lines. J Ethnopharmacol. 118, 257-263.

[11] Fang, R., Houghton, P.J., Luo, C., Hylands, P.J., 2007. Isolation and structure determination of triterpenes from *Iris tectorum*. Phytochemistry. 68, 1242-1247.

[12] Fang, Y.N., Liu, G.T., 2002. Effect of isorhapontigenin on respiratory burst of rat neutrophils. Phytomedicine. 9, 734-738.

[13] Feng, C., Zhou, T.T., Fan, G.R., Wei, H., 2009. HPLC-DAD/ESI-MS in analyzing chemical constituents of Rhizoma *belamcanda*. Academic Journal of Second Military Medical University. 30, 817-

820.
[14] Filipeanu, C. M., Brailoiu, E., Huhurez, G., Slatineanu, S., Baltatu, O., Branisteanu, D.D., 1995. Multiple effects of tyrosine kinase inhibitors on vascular smooth muscle con traction. Eur J Pharm.281, 29-35.
[15] Gao, J.L., Chen, S., 2006. A case of skin allergy caused by *Belamcanda chinensis* antiviral injection. Qilu Pharmaceutical Affaris.25, 508 (in Chinese).
[16] Gong, W., Li, L., Chen, F., 2004. Beautiful flowering plant of perennial roots-Iridaceous plants. Terrtory & Natural Resources Study.4, 96-98 (in Chinese).
[17] Han, Y., Kong, H., Li, Y.P., 2004. Antiviral experimental study on *Belamcanda chinensis*. Chinese Traditional and Herbal Drugs.35, 306-308 (in Chinese).
[18] Hideyuki, I., Satomi, O., Takashi, Y., 2001. Isoflavonoids from *Belamcanda chinesis*. Chem. Pharm.Bull.49, 1229-1231.
[19] Himpens, B., De, S. H., Bollen, M., 1994. Modulation of nucleocytosolic [Ca^{2+}] gradient in smooth muscle by protein phosphorylation. FASEB J.11, 879-883.
[20] Ho, H., Onous, S., Miyake, Y., 1999. Iridal-type triterpenoids with ichthyotoxic activity from *Belamcanda chinensis*. Journal of Natural Products.1, 89-93.
[21] http://frps.eflora.cn. Institute of Botany, The Chinese Academy of Science (in Chinese).
[22] http://www.satcm.gov.cn. State Administration of Traditional Chinese Medicine of the People's Republic of China (inChinese).
[23] http://www.theplantlist.org. The Plant List, version1. Published on the Internet (accessed 01.01.10).
[24] Huang, L., 2010. Study on the chemical constituents of *Iris dichotoma* P. and Pharmaphylogenetic investigations on *Iris* in China (Doctoral thesis). Peking Union Medical College Hospital (in Chinese).
[25] Huang, M.S., 1997. Qualitative research of wild *Iris*, Chuan *Iris* and *Belamcanda chinensis* by thin layer chromatography and high performance liquid chromatography. Chin J Pharm Anal.17, 112-115 (in Chinese).
[26] Jin, L., Chen, H.S., Jin, Y.S., Liang, S., Xiang, Z.B., Lu, J., 2008a. Chemical constituent from *Belamcanda chinensis*. Journal of Asian Natural Products Research.10, 89-94.
[27] Johnson, M.S., Wolbers, W.B., Noble, J., Fennell, M., Mitchell, R., 1995. Effect of tyrosine kinase inhibitors on luteinizing hormone-releasing hormone (LHRH) -induced gonadotropin release from the anterior pituitary. Molecular and Cellular Endocrinology.109, 69-75.
[28] Jung, S.H., Lee, Y.S., Lee, S., Lim, S.S., Kim, Y.S., Ohuchi, K., Shin, K.H., 2003. Anti-angiogenic and anti-tumour activities of isoflavonoids from the rhizomes of *Belamcanda chinensis*. Planta Med.69, 617-622.
[29] Jung, S.H., Lee, Y.S., Lee, S., Lim, S.S., Kim, Y.S., Shin, K.H., 2002. Isoflavonoids from the rhizomes of *Belamcanda chinensis* and their effects on aldose reductase and sorbitol accumulation in streptozotocin induced diabetic rat tissues. Arch Pharm Res.25, 306-312.
[30] Jung, S.H., Lee, Y.S., Lim, S.S., Lee, S., Shin, K.H., Kim, Y.S., 2004. Antioxidant activities of isoflavones from the rhizomes of *Belamcanda chinensis* on carbon tetrachloride-induced hepatic injury in rats. Arch Pham Res.27, 184-188.
[31] Katsura, S., Kazuo, H., Ryohei, k., 1995. Belamcandones A-D, dioxotetrahydrodibenzofurans from *Belamcanda chinensis*.38, 703-709.
[32] Keiko, M., Tohru, A., Toshiharu, H., 2002. Interaction of Phytoestrogens with Estrogen Receptors α and β (II). Biol.Pharm.Bull.25, 48-52.

[33] Kim, H.K., Cheon, B.S., Kim, Y.H., 1999b.Effects of naturally occurring flavonoids on nitric oxide production in the macrophage cell line RAW264.7 and their structure-activity relationships. Biochem Pharmacol.58, 759-765.

[34] Kim, Y.P., Yamada, M., Lim, S.S., 1999a.Inhibition by tectorigenin and tectoridin of prostaglandin E_2 production and cyclooxygenase-2 induction in rat peritoneal macrophages.Biochim Biophys Acta. 1438, 399-407.

[35] Kunihiko, T., Yoji, H., Sumiko, S., Yoshio, H., Taro, N., 2000.Iridals from *Iris tectorum* and *Belamcanda chinensis*.Phytochemistry.53, 925-929.

[36] Kyoung, A.K., Kyoung, H.L., Sungwook, C., Rui, Z., Myung, S.J., So, Y.K., Hee, S.K., Dong, H.K., Jin, W.H., 2005.Cytoprotective effect of tectorigenin, a metabolite formed by transformation of tectoridin by intestinal microflora, on oxidative stress induced by hydrogen peroxide.European Journal of Pharmacology.519, 16-23.

[37] Kyungsu, K., Saet, B.L., Sang, H.J., Kwang, H.C., Woo, P., Young, C.S., Chu, W.N., 2009.Tectoridin, a poor ligand of estrogen receptor α, exerts its estrogenic effects via an ERK-dependent pathway.Mol and Cells.27, 351-357.

[38] Lee, H.U., Bae, E.A., Kim, D.H., 2005b.Hepatoprotective effect of tectoridin and tectorigenin on tert-butyl hyperoxide-induced liver injury.College of Pharmacy, Kyung Hee University.97, 541-544.

[39] Lee, H.W., Choo, M.K., Bae, E.A., Kim, D.H., 2003.Beta-glucuronidase inhibitor tectorigenin isolated from the flower of *Puerariathun bergiana* protects carbon tetrachloride-induced liver injury. Liver Int.23, 221-226.

[40] Lee, K.T., Sohn, I.C., Kim, D.H., Choi, J.W., Kwon, S.H., Park, H.J., 2000.Hypoglycemic and hypolipidemic effects of tectorigenin and kaikasaponin Ⅲ in the streptozotocin-induced diabetic rat and their antioxidant activity in vitro.Arch Pharm Res.23, 461-466.

[41] Lee, K.T., Sohn, I.C., Kim, Y.K., 2001.Tectorigenin, an isoflavone of Pueraria thunbergiana Benth., induces differentiation and apoptosis in human promyelocytic leukemia HL-60 cells. Biol Pharm Bull.10, 1117-1121.

[42] Lee, S.H., Jin, J.L., Yoo, H.H., Jin, Y., 2005a.Evaluation of the estrogenic activity of isoflavones from the rhizome of *Belamcanda chinensis*.Food Science and Biotechnology.14, 39-41.

[43] Levitzki, A., 1998.Protein tyrosine kinase inhibitorsas therapeutic agents.Top Curr Chem.211, 1-15.

[44] Levitzki, A., 1994.Signal-transduction therapy.A novel approach to disease management.Eur J Biochem.226, 1-13.

[45] Li, C.J., 2005.A case report of whole body muscle stiffness caused by *Belamcanda chinensis*. New Chin Med.36, 609 (in Chinese).

[46] Li, G.X., Qin, W.Y., Qi, Y., 2008.The anti-inflammatory and analgesia experiment research on the *Belamcanda chinsnsis* extract.J Prac Trad Chin Int Med.22, 3-4.

[47] Li, H., Jiang, Y.L., Yang, C.H., 2010.The Resources and Ornamental Use of Iridaceae Plants in Guizhou Province.Guizhou Forestry Science and Technology.38, 27-31 (in Chinese).

[48] Li, J., Li, W.Z.M., Huang, W., Cheung, A.W.H., Bi, C.W.C., Duan, R., Guo, A.J.Y., Dong, T.T.X., Tsim, K.W.K., 2009. Quality evaluation of Rhizoma Belamcandae (*Belamcanda chinensis* (L.) DC.by using high-performance liquid chromatography coupled with diode array detector and mass spectrometry.Journal of Chromatography A.1216, 2071-2078.

[49] Li, X.L., 2003.The research situation of Chinese medicine *Belamcanda chinensis* (L.) DC.Strait Pharmaceutical Journal.15, 72-74 (in Chinese).

[50] Li, Y.Q., Lu, Y.R., Wei, L.X., 1986.Study on flavonoids of *Iris* dichotoma Pall.Acta Pharmaceu-

tica Sinica.21, 836-841 (in Chinese).

[51] Lin, M., Zhou, L.X., He, W.Y., Cheng, G.F., 1998.Shegansu C, a novel phenylpropanoid ester of sucrose from *Belamcanda chinensis*.JANPR.1, 67-75.

[52] Liu, B.G., Ma, Y.X., Gao, H., Wu, qiong., 2008.Tectorigenin monohydrate: an isoflavone from *Belamcanda chinensis*.ActaCryst.64, 2056.

[53] Liu, C.P., Nan, G.R., Wang, F.R., Si, R.L., Wang, G.S., Wang, L., Liu, G.S., Zhao, Y.Z., 1999.Electron microscopy observation of *Belamcanda chinensis* ether extract on red hair versicolor bacterium.Chinese Journal of Dermatology.32, 341-342 (in Chinese).

[54] Liu, C.P., Wang, F.R., Nan, G.R., Si, R.L., Wang, G.S., Guo, W.Y., 1998.Bacteriostatic action of *Belamcanda chinensis* extract on skin ringworm fungi. Chinese Journal of Dermatology. 31, 310-311 (in Chinese).

[55] Liu, J., Chen, H.S., Wang, J.J., 2005.Studies on chemical constituents of the roots of *Belamcanda chinensis* (L.) DC.Journal of Chinese Medicinal Materials.28, 29 (in Chinese).

[56] Liu, S.L., Chen, G., Liu, P., Tian, D.Z., Wang, P., 2011a.Statistical analysis of 146 traditional Chinese medicine properties which belong to the lung channel.Lishizhen Medicine and Materia Medica Research.22, 2528-2530 (in Chinese).

[57] Liu, W.N., Luo, J.G., Kong, L.Y., 2011b.Application of complexation high-speed counter-current chromatography in the separation of 5-hydroxyisoflavone isomers from *Belamcanda chinensis* (L.) DC.Journal of ChromatographyA.1218, 1842-1848.

[58] Liu, X.H., Pan, J.H., Wang, Y.X., 2000.Quantitative determination of mangiferin in Rhizoma *Belamcandae* and its substitute of *Iris* L..Chinese Traditional and Herbal Drugs.10, 739-740 (in Chinese).

[59] Lo, W.L., Wu, C.C., Chang, F.R., Wang, W.Y., Khalil, A.T., Lee, K.H., Wu, Y.C., 2003. Antiplatelet and anti-HIV constituents from Euchresta formosana.Nat Prod Res.17, 91-97.

[60] Ma, M., Yan, P., Li, X.Y., 1995.The research of *irides* plants in Xinjiang.Shihezi agricultural Journals.4, 9-16 (in Chinese).

[61] Manash, K.P., Anup, K.M., 2004.Tyrosine kinase-role and significance in cancer.Int.J.Med.Sci.1, 101-115.

[62] Masataka, M., Yukari, I., Momoyo, I., Kinuko, I., Junko, K., Fumiko, S.N., Yoshizumi, M., Chiaki, N., 2007.New isoflavones from *Belamcandae* rhizome.J Nat Med.61, 329-333.

[63] Melanie, S., Johnson, W., Bart, W., Jillian, N., Myles, F., Rory, M., 1995.Effect of tyrosine kinase inhibitors on luteinizing hormone-releasing hormone (LHRH) -induced gonadotropin release from the anterior pituitary.Molecular and Cellular Endocrinology.109, 69-75.

[64] Miyazawa, M., Sakano, K., Nakamura, S., 2001.Antimutagenic activity of isoflavone from *Puerarialobata*.J Agric Food Chem.49, 336-341.

[65] Monthakantirat, O., De, E.W., Umehara, K., Yoshinaga, Y., Miyase, T., Warashina, T., Noguchi, H., 2005.Phenolic constituents of the rhizomes of the Thai medicinal plant *Belamcanda chinensis* with proliferative activity for two breast cancer cell lines.J Nat Prod.68, 361-364.

[66] Morrissey, C., Bektic, J., Spengler, B., 2004.Phytoestrogens derived from *Belamcanda chinensis* have an antiproliferative effect on prostate cancer cells in vitro.Journal of Urology.172, 2426-2433.

[67] Ni, P.M., Cheng, X.N., 2012.Treatment clinical experience of mucousy.Shandong J Tradit Chin Med.31, 138-139 (in Chinese).

[68] Oh, K.B., Kang, H., Matsuoka, H., 2001.Detection of antifungal activity in *Belamcanda chinensis* by a single-cell bioassay method and isolation of its active compound, tectorigenin.Bioscience, Biotechnology, and Biochemistry.65, 939-942.

[69] Ohuchi, k., Kim, Y.P., Lim, S.S., Lee, S., Ryu, N.A., 1999.Inhibition of cyclooxygenase-2 induction in rat peritoneal macrophages by tectorigenin isolated from the rhizomes of *Belamcanda chinensis*, and its mechanism of action.Proceedings of the International Symposium on Recent Advances in Natural Products Research, 3rd, Seoul, Republic of Korea.19, 12-24.

[70] Park, E.K., Shin, Y.W., Lee, H.U., Lee, C.S., Kim, D.H., 2004.Passive cutaneous anaphylaxis-inhibitory action of tectorigenin, a metabolite of tectoridin by intestinal microflora. Biol Pharm Bull.27, 1099-1102.

[71] Park, K.Y., Jung, G.O., Choi, J., 2002.Potent antimutagenic and their anti-lipid peroxidative effect of kaikasaponin III and tectorigenin from the flower of *Pueraria thunbergiana*.

[72] Qin, M.J., Huang, Y., Yang, G., Xu, L.S., Zhou, K.Y., 2003a.RbcL sequence analysis of *Belamcanda chinensis* and related medicinal plants of *Iris*.Acta Pharmaceutica Sinica.38, 147-152 (in Chinese).

[73] Qin, M.J., Ji, W.L., Liu, J., Zhao, J., Ju, G.D., 2003b.Scavenging effects on radicals of isoflavones from rhizome of *Belamcanda chinensis*.Chinese Traditional and Herbal Drugs.7, 640-641 (in Chinese).

[74] Qin, M.J., Xu, G.J., Xu, L.S., Wang, Q., 1998.Character identification research of *irides* Medicinal roots.Chinese herbal medicine.29, 695-697 (in Chinese).

[75] Qin, M.J., Xu, L.S., Tian, Z.J.H, Wang, Q., Xu, G.J., 2000.A preliminary study on the distribution pattern of isoflavones in rhizomes of *Iris* from China and its systematic significance.Journal of Systematics and Evolution.38, 343-349 (in Chinese).

[76] Qin, M.J., Yu, G.D., Wang, X.H., 1999.Biology research of *Belamcanda chinensis* (L.) DC. (1) taxonomic comprehensive characteristics and the group of investigation. Chinese wild plant resources.18, 30-31 (in Chinese).

[77] Qiu, Y.K., Gao, B.Y., Xu, B.X., Liu, K., 2006b.Study on Chemical Consituents of *Belamcanda chinensis*.Chin Pharm J.41, 1133-1135 (in Chinese).

[78] Qiu, Y.K., Gao, Y.B., Xu, B.X., Liu, K., 2006a.Isolation and identification of isoflavonoids from *Belamcanda chinensis*.Chinese Journal of Medicinal Chemistry.3, 175-177 (in Chinese).

[79] Qiu, Z.H., Zhang, Z.G., Wang, J.H., Lv, T.S., 2009.Study on the isoflavonoids of *Iris tectorum*. Journal of Chinese Medicinal Materials.32, 1392-1394 (in Chinese).

[80] Seidlovά, W.D., Hesse, O., Jarry, H., Rimoldi, G., Thelen, P., Christoffel, V., Wuttke, W., 2004.*Belamcanda chinensis* and the thereof purified tectorigenin have selective estrogen receptor modulator activities.Phytomedicine.11, 392-403.

[81] Shi, G.B., 2011.She Gan Ma Huang Tang in treating the cold cough clinical observation (Master thesis). Guangzhou University of Chinese Medicine (in Chinese).

[82] Shi, L., Ji, Z.Q., Kang, W.Y., 2012.The clinical application of She Gan antiviral injection.J China Pharm.23, 3359-3360 (in Chinese).

[83] Shin, J.E., Bae, E.A., Lee, Y.C., Ma, J.Y., Kim, D.H., 2006.Estrogenic effect of main components kakkalide and tectoridin of *Puerariae Flos* and their metabolites.Biol Pharm Bull.29, 1202-1206.

[84] Shin, K.H., Kim, Y.P., Lim, S.S., 1999.Inhibition of prostaglandin E2 production by the isoflavones tectorigenin and tectoridin isolated from rhizomes of *Blamcanda chinensis*.Planta Med.65, 776-777.

[85] Sick, W.W., Hee, W.E., 1993.An isoflavone noririsflorentin from *Belamcanda chinensis*. Phytochemistry.33, 939-940.

[86] Song, Z.J., Luo, F., Zhou, Y., Bai, B.R., Peng, S.L., Ding, L.S., 2007.Two new isofla-

vonoids from the rhizomes of *Belamcanda chinensis*.Chinese Chemical Letters.18, 694-696 (in Chinese).

[87] Song, Z.J., Xu, X.M., Deng, W.L., Peng, S.L., Ding, L.S., Xu, H.H., 2011.A new dimeric Iridal triterpenoid from *Belamcanda chinensis* with significant molluscicidea ctivity.Organic Letters.13, 462-465.

[88] State Administration of Traditional Chinese Medicine, 1996.Chinese Materia Medica.Shanghai Scientific and Technical Publishers.2122-2126 (in Chinese).

[89] Stettner, M., Kaulfuss, S., Burfeind, P., Schweyer, S., Strauss, A., Ringert, R.H., Thelen, P., 2007.The relevance of estrogen receptor-β expression to the antiproliferative effects observed with histonedeacetylase inhibitors and phytoestrogensin prostate cancer treatment.Mol Cancer Ther.3, 2626-2633.

[90] Tony, H., 1997.Oncoprotein networks.Cell.88, 333-346.

[91] Umezaawa, K., 1992.Biological activity of oncogene function inhibitors isolated from microorganisms. Seikagaku.64, 160-171.

[92] Thelen, P., Peter, T., Hunermund, A., Kaulfuss, S., Seidlová, W.D., Wuttke, W., Ringert, R.H., Seseke, F., 2007.Phytoestrogens from *Belamcanda chinensis* regulate the expression of steroid receptors and related cofactors in LNCaP prostate cancer cells.BJU Int.100, 199-203.

[93] Thelen, P., Scharf, J.G., Burfeind, P., 2005.Tectorigenin and other phytochemicals extracted from leopard lily *Belamcanda chinensis* affect new and established targets for therapies in prostate cancer.Carcinogenesis vol.26, 1360-1367.

[94] Wang, G., 2005.Clinical study on treating children's asthmatic bronchitis wheezing with cold-heat jumble with the She Gan Ma Huang Chai Hu particles (Master thesis).Shandong Traditional Chinese Medicine University (in Chinese).

[95] Wang, H., Yang, F.Q., Pan, M., Ming, J.L., 1997.Bacteriostatic action of 10 kinds of traditional Chinese medicine on shallow pathogenic fungi.Journal of Traditional Chinese Medicine.38, 431-432 (in Chinese).

[96] Wang, J.F., Zhang, Y.P., Yang, C.Y., Hou, M.X., Wang, F., Wang, G.Y., Wang, S.H., 2010.Effect of tectorigenin on MCP-1 and ICAM-1 mRNA expression ininjured vascular endothelial cells.China J Chin Mater Med.35, 2001-2003 (in Chinese).

[97] Wang, Q.L., Lin, M., Liu, G.T., 2001.Antioxidative activity of natural isorhapontigenin.Japanese Journal of Pharmacology.87, 61-66.

[98] Wang, X.H., Liang, Y., Peng, C.L., Xie, H.C., Pan, M., Zhang, T.Y., Ito, Y., 2011. Preparative isolation and purification of chemical constituents of *Belamcanda* by MPLC, HSCCC, and PREP-HPLC.Journal of Liquid Chromatography Related Technologies.34, 241-257.

[99] Wang, X., Qin, M.J., Wu, G., Su, P., 2006.Chemical constituents of the leaves of *Iris* songarica.Journal of China Pharmaceutical University.37, 222-225 (in Chinese).

[100] Wei, W., Lu, M.M., Wu, J.F., Ye, L.L., 1994.Bacteriostatic observation of 68 Chinese herbal medicine on eight kinds of eye common pathogenic fungi.Journal of Nanjing College of Traditional Chinese Medicine.5, 60 (in Chinese).

[101] Wu, J.L., Wang, X.L., Zhang, R.X., 2008b.Study on anti-hepatocarcinoma activity of tectorigenin from *Pueraria Flos* in vitro.Journal of Shenyang Pharmaceutical University.1, 79-80 (in Chinese).

[102] Wu, J.H., Wang, Y.R., Huang, W.Y., 2010.Anti-proliferative and pro-apoptotic effects of tectorigenin on hepatic stellate cells.World J Gastroenterol.31, 3911-3918.

[103] Wu, S.H., Zhang, G.G., Zuo, T.T., Li, Y.N., 2008a.Isolation and identification of chemical constituents from the rhizome of *Belamcanda chinensis* (L.) DC.Journal of Shenyang Pharmaceutical University.25, 796-799 (in Chinese).

[104] Wu, Y.L., 2014.She Gan Ma Huang Tang in treating theinfantile capillary bronchitis clinical observation.Pharmaceutical journal of north.11, 28-29 (in Chinese).

[105] Wu, Z.F., Xiong, C.M., (1990).The pharmacological research of wild *Iris*, Chuan *Iris* and *Belamcanda chinensis*.6, 28-30 (in Chinese).

[106] Xiao, Z.G., Zheng, D.P., 2008.The research progress of tectorigenin.Journal of Shangrao Normal College.28, 41-49 (in Chinese).

[107] Xiong, Y., Yang, Y.Q., Yang, J., Chai, H.Y., Li, Y., Yang, J., Jia, Z.M., Wang, Z. R., (2010).Tectoridin, an isoflavone glycoside from the flower of *Puerarialobata*, prevents acute ethanol-induced liver steatosis in mice.Toxicology.276, 64-72.

[108] Xu, Y.L., Ma, Y.B., Xiong, J., (1999).Isolfavoonoids of *Iris* tectorrum.Acta Botanica Yunnanica.21, 125-130 (in Chinese).

[109] Yamaki, K., Kim, D.H., Ryu, N., (2002).Effects of naturally occurring isoflavones on prostaglandin E_2 production.Planta Med.68, 97-100.

[110] Yang, L., (2013).The clinical research of JWSGMH prescription in treating bronchial asthma (Master thesis).Hubei University of Chinese Medicine (in Chinese).

[111] Yang, T.H., (2012).She Gan Ma Huang Tang combined with acupoint applicationin treating the postinfectious cough clinical observation (Master thesis).Guangzhou University of Chinese Medicine (in Chinese).

[112] Yeon, S.L., Seon, H.K., Jin, K.K., Sanghyun, L., Sang, H.J., Soon, S.L., (2011). Preparative isolation and purification of seven isoflavones from *Belamcanda chinensis*.PhytochemAnal.22, 468-473.

[113] Yoshiyasu, F., Yuuko, K., Junko, O., Mitsuaki, K., (1993).Belamcanda quinones A and B, novel dimeric 1, 4-benzoquinone derivatives possessing cyclooxyenase inhibitory activity.34, 7633-7636.

[114] Yu, J., Xu, L.H., Wang, Y., Xiao, Y., Yu, H., (2001).Experimental research on antibacterial effect of *Belamcanda Chinensis* and *portulaca Oleracece* L.on P.aeruginosa *in vitro*.Journal of Norman Bethune University of Medical Science.27.

[115] Yuan, D., Xie, Y.Y., Bai, X., Wu, X., Yang, J.Y., Wu, C.F., (2009).Inhibitory activity of isoflavones of *Pueraria* flowers on nitric oxide production from lipopolysaccharide activated primary rat microglia.Journal of Asian Natural Products Research.11, 471-481.

[116] Zhang, C.Z., Wang, S.X., Zhang, Y., Chen, J.P., Liang, X.M., (2005).*In vitro* estrogenic activities of Chinese medicinal plants traditionally used for the management of menopausal symptoms.J Ethnopharmacol.98, 295-300.

[117] Zhang, H., Liu, X.F., Chen, S., Wu, J.H., Ye, X., Xu, L.Z., Chen, H.M., Zhang, D. P., Tan, R.X., Wang, Y.P., (2010a).Tectorigenin inhibits the *in vitro* proliferation and enhances miR-338* expression of pulmonary fibroblasts in rats with idiopathic pulmonary fibrosis.J Ethnopharmacol.131, 165-173.

[118] Zhang, L., (2010b).The effects of She Gan Ma Huang Tang on airway inflammation and remodeling in chronic asthma mice (Doctoral thesis).Heilongjiang University of Chinese Medicine (in Chinese).

[119] Zhang, L., Zhang, Y.K., Dai, R.J., Deng, Y.L., (2010c0.The Pharmacological effect of C-glycoside flavones in the leaves of *Belamcanda chinensis*.Nat Prod Res Dev.22, 728-730 (in Chi-

nese).
[120] Zhang, M.F., Shen, Y.Q., Zhu, Z.P., Wang, H.W., Mi, C.F., (1997c).Researches on traditional Chinese medicine medicinal of XinWen (Hot) together to Pi Wei jing (V) -antidiarrhea effect.Chinese Journal of Basic Medicine in Traditional Chinese Medicine.5, 2-5 (in Chinese).
[121] Zhang, M.F., Shen, Y.Q., Zhu, Z.P., Wang, H.W., Yang, Z.F., (1997a).Researches on traditional Chinese medicine medicinal of XinWen (Hot) together to Pi Wei jing (Ⅵ) -antithrombotic and anticoagulant.China J Chin Mater Med.11, 691-693 (in Chinese).
[122] Zhang, M.F., Shen, Y.Q., Zhu, Z.P., Yang, Z.F., Mi, C.F., (1997b).Researches on traditional Chinese medicine medicinal of XinWen (Hot) together to PiWeijing (Ⅱ) -antiulcer effect. Pharmacology and Clinics of Chinese Materia Medica.4, 1-5 (in Chinese).
[123] Zhang, M.F., Zhu, Z.P., Shen, Y.Q., Wang, H.W., Gu, Y., Liu, X.P., (1998).Researches on traditional Chinese medicine medicinal of XinWen (Hot) together to Pi Wei jing (I) -cholagogic effect.Chinese Journal of Basic Medicine in Traditional Chinese Medicine.8, 16-19 (in Chinese).
[124] Zhao, Y., Wang, Q., Liu, H.P., Zhou, X.M., He, H.L., Cao, Z.D., (2013).Meta analysis of She Gan Ma Huang Tang in treating the infantile cough variant asthma.19, 324-328 (in Chinese).
[125] Zhong, M., Guan, X.J., Huang, B.S., Qiu, Y.X., (2001).Modern research progress on Chinese medicine rhizome *belamcandae*. Traditional Chinese medicine journal.24, 904-907 (in Chinese).
[126] Zhou, L.X., Lin, M., (2000).A new stilbene dimer-Shegansu B from *Belamcanda chinensis*.JANPR.2, 169-175.
[127] Zhou, L.X., Lin, M., (1997).Studies on chemical constituents of *Belamcanda chinensis* (L.) DC. Chinese Chemical Letters.8, 133-134.

(发表于《African Journal of Traditional, Complementary and Alternative medicines》, SCI: 0.500)

Evaluation of Analgesic and Anti-inflammatory Activities of Compound Herbs Puxing Yinyang San

WANG Lei, WANG Xu-rong, ZHANG Jing-yan, YANG Zhi-qiang, LIU You-bin, CUI Don-gan, QIN Zhe, MENG Jia-ren, KONG Xiao-jun, WANG Xue-zhi*, LI Jian-xi*

(Engineering & Technology Research Center of Traditional Chinese Veterinary Medicine of Gansu Province/Key Laboratory of Veterinary Pharmaceutics Discovery, Ministry of Agriculture/Key Laboratory of New Animal DrugProject, Gansu Province/Lanzhou Institute of Husbandry and Pharmaceutical Sciences of Chinese Academy of Agricultural Sciences, Lanzhou 730050, China)

Abstract:

Background: Bovine mastitis is one of the most relevant and problematic diseases to treat and control in practice. Puxing Yinyang San (PYS) is a compound of herbs to treat bovine mastitis in China. This study was performed to evaluate the analgesic and anti-inflammatory activities of PYS in mice and rats.

Materials and methods: The analgesic and anti-inflammatory activities of PYS were determined using acetic acid-induced writhing response, hot plate test, xylene-induced ear swelling test, carrageenan-induced paw edema test, and acetic acid-induced capillary permeability andleukocyte infiltration test with oral doses of 155, 310 and 620mg/kg · bw in mice or rats.

Results: The acetic acid-induced writhing response was dose-dependently inhibited by oral administration of PYS and the latency time tothermal stimuli was increased in the hot plate test, especially 90 minutes after treatment. In the xylene-induced ear swelling, PYS significantly decreased swelling degree in a dose-dependent manner. Additionally, PYS significantly suppressed the peritoneal capillary permeability and leukocyte infiltration in mice induced by intraperitoneal injection of acetic acid. PYS also significantly reduced the carrageenan-induced rat paw edema at 2, 3, and 4 h after the carrageenan injection. The results suggested that PYS possessed significant analgesic and anti-inflammatory activities.

Conclusion: This study was the first to demonstrate that oral administration of PYS might play an important role in the process of analgesia and anti-inflammation, supporting its treatment for mastitis. Future investigations will focus on the broader involvement of the ingredients andmechanisms responsible for pharmacological activities of PYS.

Key words: Puxing Yinyang San; Analgesic activity; Anti-inflammatory activity; Herb

INTRODUCTION

Mastitis is the most common disease in dairy cows and the principal cause of economic losses in the dairy industry (Demon et al., 2013). Itundermines udder health, reduces milk quality, loses milk production and entails prohibitive costs (Piepers et al., 2007). The incidence ofsubclinical mastitis in dairy cows was 22%-65% in the world (Biiragohin and Dutta, 1999; Gitau et al., 2014; Lam et al., 2013; Mukesh et al., 2014; Ramírez et al., 2014), and it was 27%-55% in China (Xiao, 2011). Although it has been studied for more than one hundred years, bovine mastitis is still one of the most relevant and problematic diseases to treat and control in practice because of the complexity and mutability of the condition (Green and Bradley, 2013). Recently, it is common to cure mastitis in dairy cows with intra-mammary antimicrobials in practice (Hill et al., 2009; Pol and Ruegg, 2007). However, antimicrobial treatment increases the risk of antibiotic residues in milk and antimicrobial resistance, which may endanger public security. Therefore, it is still important to search for more effective and safer medicine to treat or prevent dairy cows with mastitis. Many Chinese herbs possess bacteriostatis or anti-inflammatory ability, and provide holistic therapy for interrelated diseases. Puxing Yinyang San (PYS), pharmaceutical product named "Ru Ning San", consists of three Chinese herbal medicines: *Taraxacum mongolicum* Hand. -Mazz., *Vaccaria segetalis* (Neck.) Garcke, and *Epimedium* brevicornu Maxim (Table 1). *Taraxacum mongolicum* is a broad-spectrum herb (Li et al., 2008; Song et al., 2006). The polysaccharides and total flavonoids are main antibacterial components of *Taraxacum mongolicum* (Gu and Wang, 2007; Song et al., 2010); taraxasterol and phenolic acids show significant anti-inflammatory ability (Liu et al., 2012; Park et al., 2011). *Vaccaria segetalis* possesses abilities of activating blood circulation, regulating menstrual disturbances, dispelling edema, and promoting lactation for centuries in China (Li and Liang, 2007). Polysaccharide isolated from *Epimedium* has inhibitory effects on *Bacillus subtilis*, *Escherichia coli* and *Staphylococcus aureus* (Su et al., 2011). Total Flavonoids of *Epimedium* has anti-inflammatory activity (Zhang et al., 1999), and icariin is a safe and effective natural anti-inflammatory drug (Wu et al., 2009). PYS has been shown effective to treat subclinical mastitis in our previous studies and the effective rate was 88.89% (Wang, 2013). It has also been demonstrated that PYS is not toxic if orally administered in acute toxicity tests in mice (maximum daily dose=40g/kg · bw) and subchronic toxicity test at doses of 750, 1 500, and 3 000g/kg · bw (Wang et al., 2013). However, its mechanism to treat subclinical mastitis has not been studied.

The aim of the present study was to evaluate the analgesic and anti-inflammatory effect of PYS through five animal models, including aceticacid-induced mouse writhing, hot plate test, xylene-induced mouse ear swelling, acetic acid-induced mouse capillary permeability and leukocyte infiltration and carrageenan-induced rat hind paw edema.

MATERIALS AND METHODS

Reagents and drugs

Carrageenan, indometacin, Evans blue and Crystal violet were purchased from Sigma-Aldrich

Chemical Co. (St. Louis, USA). Acetic acidwas obtained from Tianjin Damao Chemical Reagent Factory (Tianjin, China). Xylene was purchased from Tianjin Fuchen Chemical ReagentFactory (Tianjin, China). The control medicinal materials of *Taraxacum mongolicum*, *Vaccaria segetalis*, and *Epimedium* were purchased from National Institutes for Food and Drug Control (Beijing, China). The reference substances of vaccarin (95.1%), icariin (100%), and caffeic acid (≥ 98%) were also purchased from National Institutes for Food and Drug Control (Beijing, China).

Preparation of PYS

PYS was composed of three herbs–*Taraxacum mongolicum*, 110 g; *Vaccaria segetalis*, 60 g; *Epimedium*, 30 g. Plant parts and origin usedin the formula had been shown in Table 1. The herbs were pre–treated by washing and drying. Following that, herbs were mixed as previouslydescribed and crushed through 120 mesh sieve.

Table 1 Component herbs of PYS

Herb name	Plant part	Origin	Grams, g	Content (%)
Pu gong ying (*Taraxacum mongolicum* Hand. –Mazz.)	whole plant	Gansu, China	110	55
Wang bu liu xing (*Vaccaria segetalis* (Neck.) Garcke)	Seed	Hebei, China	60	30
Yin yang huo (*Epimedium* brevicornu Maxim)	Leaf	Sichuan, China	30	15

The TLC was carried out according to the protocol described by Chinese pharmacopoeia (2010). Briefly, 2.50 g PYS was dissolved in 40mL 5% formic acid, extracted for 20 minutes with ultrasonic, filtered and the filtrate was volatilized until dry. Following that, the residue was dissolved in 10mL distilled water and filtered. The filtrate was extracted with 10mL ethyl acetate for two times. Then the ethyl acetate extraction was volatilized until dry and the residue was dissolved in 1mL methanol and spotted on a silica G thin layer plate (Taizhou Luqiao Sijia Biochemical Plastics Factory, China) with 2μL, and developed at room temperature with butyl acetate–formic acid–water (7 : 2.5 : 2.5). The solution of *Taraxacum mongolicum* and caffeic acid were used as the reference substance. The TLC spots were visualized by spraying with alchlor reagent and the color stripes were displayed at 366 nm (CabUVIS, DESAGA GmbH, Germany). 3.00 g PYS was dissolved in 40mL 70% methanol, extracted for 30minutes with ultrasonic, filtered and spotted on a polyamide film plate (Taizhou Luqiao Sijia Biochemical Plastics Factory, China) with 2μL, and water–methanol (6 : 4) was used as the developing reagent. The solution of *Vaccaria segetalis* and vaccarin were used as the reference substance. The TLC spots were visualized by spraying with alchlor reagent (2%) dissolved in ethanol solvent and the color stripes were displayed at 366 nm. 2.30 g PYS was macerated in 20mL ethanol for 30 minutes, then filtered and the filtrate was volatilized until dry. Following that, the residue was dissolved in 10mL distilled water and the filtrate was extracted with 10mL ethyl acetate for two times. Then the ethyl acetate extraction was volatilized until dry. The residue was dissolved in 1mL ethanol and spotted on a silica H plate (Taizhou Luqiao Sijia Biochemical Plastics Factory, China) with 2μL, and developed with ethyl acetate–butanone–formic acid–water (10 : 1 : 1 : 1). The solution of *Epimedium* and icariin were used as the reference substance. The TLC spots were visualized by spraying with alchlor reagent and the color stripes were displayed at 366 nm.

Animal care and handling

Male or female Balb/C mice (18-22 g) and SD rats (160-200 g) were obtained from the experimental animal center of Lanzhou University. They were housed in plastic cages at a temperature of 22±1℃, relative humidity 50±10% with free access to food and water. This study was approved by the Institutional Animal Care and Use Committee of Lanzhou Institute of Husbandry and Pharmaceutical Sciences of the Chinese Academy of Agricultural Sciences (SCXK20008-0003). The animal protocols were in compliance with the ethical guidelines for the treatment of animals of the International Association for the Study of Pain (Zimmermanm, 1983).

Analgesic tests

Acetic acid-induced writhing response

The writhing response in mice was carried out by using the method of previous study (Young et al., 2005). PYS was administrated orallyto mice at doses of 155, 310, and 620mg/kg · bw, respectively. Mice received indometacin (3mg/kg · bw) in reference group and normal saline (10mL/kg · bw) in control group. After 60 minutes treatment, each mouse was intraperitoneally administered 0.7% acetic acid (10mL/kg · bw). The number of writhing movements was recorded in a period of 0-30 minutes after the injection of acetic acid. The inhibition ratio wascalculated using the following equation:

$$\text{Inhibition ratio} = (Nc - Nt)/Nc \times 100\%$$

Nc represents the number of writhing in the control group, and *Nt* represents the number of writhing in the drug treated group.

Hot plate test

This test was carried out according to previously described method (Shinde et al., 1999). Mice whose forepaw licking or jumping on theheated plate (55±1°C) within 30 s was used in this experiment. Mice were treated orally with PYS (155, 310, and 620mg/kg · bw) or indometacin (3mg/kg · bw) as reference. Normal saline was used in control animals (10mL/kg · bw). Before and 30, 60, and 90 min after treatment, mice were individually put on the heated plate and the time for forepaw licking or jumping was registered as the latency time to thermal stimuli with a cut-off time of 60 s to avoid paw lesions.

Anti-inflammatory Tests

Xylene-induced ear swelling

Experiments were carried out according to previously described method (Kou et al., 2005). Mice were pre-treated with PYS (155, 310, and 620mg/kg · bw) 60 min prior to the induction of ear swelling. Ear swelling was induced by smearing 30μL of xylene on the anterior and posterior surfaces of the right ear. The positive control received indometacin (3mg/kg · bw) and the control group received the same volume of normal saline. One hour later, the mice were sacrificed by cervical dislocation and circular sections were taken from both ears at the same place with a diameter of 8 mm punch, and weighed. The swelling degree and inhibition ratio were calculated using the following equations:

$$\text{Swelling degree(SD)} = (Wr-Wl)/Wl\times100\%$$
$$\text{Inhibition ratio} = (SDc-SDt)/SDc\times100\%$$

Wr represents weight of the right ear, and *Wl* represents weight of the left ear of the same mouse; *SDc* represents the swelling degree in the control group, and *SDt* represents the swelling degree in the drug treated group.

Leukocyte infiltration and capillary permeability caused by intraperitoneal injection of 0.7% acetic acid

Experiments were carried out according to previously described method (Lucena et al., 2007). PYS (155, 310, and 620mg/kg · bw), indometacin (3mg/kg · bw) and normal saline (10mL/kg · bw) was administered orally to mice, respectively. After 60 min, 0.5% Evans blue solution (10mL/kg · bw) was intravenously injected to the tail veins of mice; 10 min later, mice were injected with 0.7% acetic acid solution in normal saline by intraperitoneal route. 20 min after the intraperitoneal injection, mice were sacrificed by cervical dislocation and the peritoneal cavity was opened and washed with 5mL sterile saline. The peritoneal washes were collected. Finally, 50μL of peritoneal fluid was taken and diluted in 450μL of Türk's solution for total leukocyte counts under microscope (Olympus Company, Japan). The remainder of the peritoneal fluid was centrifuged at 3000 rpm for 15 min. The absorbance of the supernatant was read at 606 nm using an Evolution 300 UV-VISspectrophotometer (Thermo Scientific, USA). The concentration of Evans blue leaked into the peritoneal cavity was calculated according to the standard curve of Evans blue, which indicated the peritoneal capillary permeability induced by acetic acid (Lucena et al., 2007). The inhibition ratios of leukocyte infiltration (IRLI) and capillary permeability (IRCP) in drug treated groups were calculated using the following equations:

$$IRLI = (Lc-Lt)/Lc\times100\%$$
$$IRCP = (Ec-Et)/Ec\times100\%$$

Lc represents the number of leukocytes in the control group, and *Lt* represents the number of leukocytes in the drug treated group; *Ec* represents the concentration of Evans blue in the control group, and *Et* represents the concentration of Evans blue in the drug treated group.

Carrageenan-induced paw edema

This test was based on the method of the previous study (Kou et al., 2005). PYS (155, 310, and 620mg/kg · bw), indometacin (3mg/kg · bw) and normal saline (10mL/kg · bw) was administered orally to rats, respectively. One hour later, the edema of the right hind paw was induced by hypodermic injection with 100μL of 1% carrageenan suspension in normal saline. Paw thickness was determined with a vernier caliper before and 1, 2, 3 and 4 h after the injection of carrageenan. The swelling degree and inhibition ratio were calculated using the following equations:

$$\text{Swelling degree (SD)} = (PTt-PT_0)/PT_0\times100\%$$
$$\text{Inhibition ratio} = (SDc-SDt)/SDc\times100\%$$

PT_0 represents the right hind paw thickness before carrageenan injection, and *PTt* represents the right hind paw thickness 1, 2, 3, or 4 h after carrageenan injection; *SDc* represents the swelling degree in the control group, and *SDt* represents the swelling degree in the drug tested group.

Statistical analysis

The data were expressed as mean ± SE. Data using SPSS software program version 17. 0 by a one-way ANOVA followed by LSD as the post hoc test. The results were considered statistically significant at $P<0.05$.

RESULTS AND DISCUSSION

Owing to the complexity and mutability of the condition, bovine mastitis is still one of the most relevant and problematic diseases to treatand control in practice (Green and Bradley, 2013). PYS has been shown to have obvious effect on preventing and treating subclinical mastitis of dairy cows in our previous studies (Wang, 2013). In this present study, it demonstrated that PYS had analgesic and anti-inflammatory activities according to five animal models.

TLC Chromatograms of PYS

PYS is a mixture of *Taraxacum mongolicum*, *Vaccaria segetalis*, and *Epimedium.* The results of TLC analyses were shown in Fig. 1, Fig. 2, and Fig. 3, which were in accord with the requirement of the detection of traditional Chinese herbal medicine in Chinese Pharmacopoeia. PYS contains several kinds of compounds and only the three marked components , caffeic acid, vaccarin, and icariin,

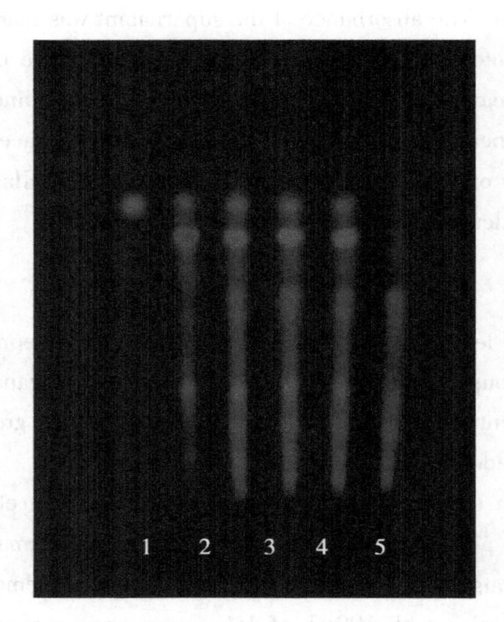

Fig. 1 The TLC Chromatograms of Caffeic Acid (1), *Taraxacum mongolicum* Hand. -Mazz (2), PYS (3, 4, and 5), and negative control (6)

were detected using TLC, because the low concentration of other compounds or lack of single component standards. In addition, the preparation of PYS was only through physical process, smash and mixture and the chemical components would not be changed between PYS and the threeherbs.

Analgesic Activity of PYS

Two animal models, the acetic acid-induced writhing response and the hot plate test, were employed to investigate the analgesic activity ofPYS in this study. Acetic acid-induced writhing test is believed to indicate the involvement of peripheral mechanisms (Deraedt et al., 1980), whereas the hot plate test is believed to investigate the central mechanisms (Ferreira et al., 2004). The acetic acid-induced writhing model is a visceral pain model. In the process, arachidonic acid releases by cyclooxygenase and prostaglandins biosynthesis increases such as prostaglandins, serotonin, and histamine, which play an important effect on the nociceptive mechanism (Duarte et al., 1988). Table 2 showed the analgesic effect of PYS on the acetic acid-induced writhing test in mice. PYS (155, 310, and 620mg/kg · bw) and indometacin significantly restrained the number of abdominal writhing compared with the control group. It indicated the good peripheral analgesia of PYS and the mechanism of the analgesic effect may be associated with blockage of arachidonic acid metabolite synthesis.

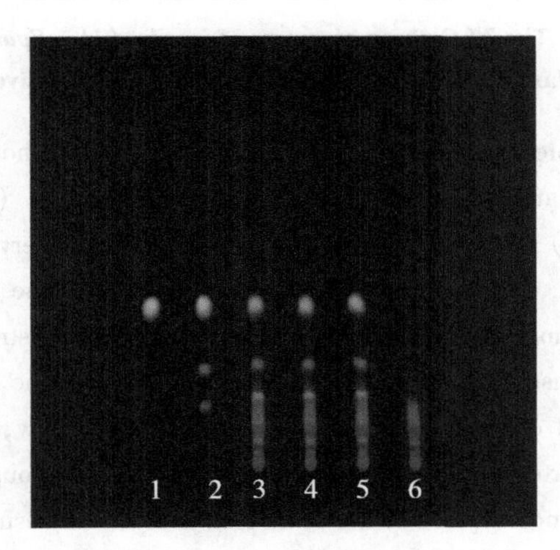

Fig. 2 The TLC Chromatograms of Vaccarin (1), *Vaccaria segetalis* (Neck.) Garcke (2), PYS (3, 4, and 5), and negative control (6)

Table 2 Effect of PYS on the acetic acid-induced writhing test in mice.

Groups	Dose (mg/kg · bw)	Number of writhing	Inhibition (%)
Control	—	41.10±5.69	—
Indometacin	3	16.20±3.94**	60.58
	155	29.00±2.55**	29.44
PYS	310	26.33±6.60**	35.94
	620	21.11±3.66**	48.64

Note: Each value represents mean ± SE. (n=10).

** P<0.01 significantly different from the control group.

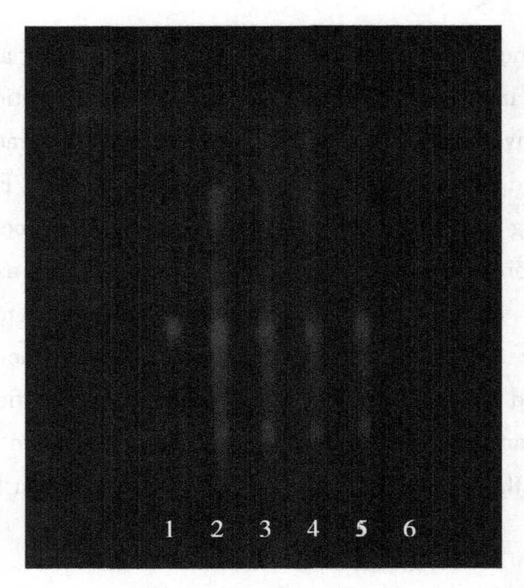

Fig. 3 The TLC chromatograms of icariin (1), *Epimedium* brevicornu Maxim (2), PYS (3, 4, and 5), and negative control (6)

To check for the possible central analgesic activity of PYS, the hot-plate test was carried out, as it was sensitive to strong analgesics and had limited tissue damage (Ferreira et al., 2004). In the hot-plate test (Table 3), although no significant effect was observed after 60 min of drug administration at doses of 155, and 310mg/kg · bw, significant increase in the latency to the thermal stimuli was observed at 620mg/kg · bw after 60 min of drug administration. Interestingly, in both cases, the effect was long lasting, and PYS increased the latency time could be observed at 90 min after drug administration, at doses of 155, 310, and 620mg/kg · bw, respectively. In this study, it was possible that PYS exerted an analgesic effect at least in part through central mechanisms, because of the specific central nociception of hot plate test. Given both results of the two tests, we suggested that PYS had strong analgesic activity.

Table 3 Effect of PYS on the hot plate test in mice

Groups	Dose (mg/kg · bw)	Latency period (s)			
		0 min	30 min	60 min	90 min
Control	—	22.23±4.53	18.59±3.87	18.16±4.98	17.14±5.85
Indometacin	3	21.89±3.44	23.32±8.47	26.84±6.14**	33.41±7.77**
	155	22.16±4.06	18.43±4.56	20.81±5.66	27.83±6.14**
PYS	310	22.54±4.24	19.11±5.04	23.38±5.91	27.57±6.37**
	620	22.37±6.05	24.23±8.30	24.79±5.97*	29.71±5.52**

Note: Each value represents mean ± SE. (n=10). * $P<0.05$, ** $P<0.01$ significantly different from the control group at each corresponding time.

Anti-inflammatory Activity of PYS

The process of acute inflammatory resulted from vasodilation, capillary permeability accentuation, and edema formation at the inflammatory site. The xylene-induced ear swelling was a common inflammatory model to evaluate vasodilation and substance P partially participated in the process (Luber-Narod et al., 1997). Administration of xylene on ears brought about a significant increase of ear weight in the control group (Table 4). Compared with the control group, PYS significantly suppressed xylene-induced ear swelling in mice and it was dose-dependent. The results indicated that PYS might reduce the release of substance P or antagonize ear swelling induced by xylene.

Peritoneal capillary permeability and leukocyte infiltration in the abdominal cavity could be aggravated by intraperitoneal injection of acetic acid (Li et al., 2010). Injection of acetic acid led to a significant increase of total leukocytes and Evans blue content extruded into mice peritoneal cavity in the control group (Table 5 and Table 6). Compared with the control group, PYS as well as indometacin significantly reduced total leukocytes and Evans blue content induced by acetic acid in mice. These results indicated that peritoneal leukocyte infiltration and capillary permeability induced by acetic acid were suppressed by oral administration of PYS in a dose-dependent manner.

Table 4 Effect of PYS on the xylene-induced ear swelling test in mice

Groups	Dose (mg/kg · bw)	Swelling (%)	Inhibition (%)
Control	—	64.16±5.20	—
Indometacin	3	30.76±4.29**	52.06
	155	55.48±5.59**	13.53
PYS	310	43.60±4.70**	32.04
	620	38.92±5.83**	39.34

Note: Each value represents mean ±SE. (n=10). ** P<0.01 significantly different from the control group.

The carrageenan-induced paw edema, including edema and hyperalgesia, is a reliable and repeatable model to evaluate the anti-inflammatory activity of natural products (Huang et al., 2011). The hind paw edema is a biphasic phase of various mediators release in sequence to generate the inflammatory response (Vinegar et al., 1969). The first phase is correlated with the release of histamine, serotonin and bradykinin (Di Rosa et al., 1971). The late phase, a complement-dependent reaction, contributes to elevated production of prostaglandins in tissues (Di et al., 1972; Garćia et al., 2004; Posadas et al., 2004). Injection of carrageenan brought about a significant increase of paw thickness in the control group and the inflammation approximately reached its maximum at 3 h after the carrageenan injection (Table 7). Compared withthe control group, the edema was significantly decreased by administering PYS at 2, 3, and 4 h after the carrageenan injection. The resultsdemonstrated that the anti-inflammatory effect of PYS mainly played in the late phase in the carrageenan-induced rat paw edema test. Theanti-inflammatory mechanism of PYS might inhibit the synthesis of prostaglandins. According to all the anti-inflammatory tests, we suggestedthat PYS displayed a significant anti-inflammatory effect.

Table 5 Effect of PYS on the acetic acid-induced leukocyte infiltration test in mice

Groups	Dose (mg/kg · bw)	Total leukocytes ($\times 10^6$)	Inhibition (%)
Control	—	5.13±0.72	—
Indometacin	3	3.27±0.36 **	36.26
	155	4.52±0.55 **	11.89
PYS	310	3.76±0.24 **	26.71
	620	3.66±0.31 **	28.65

Note: Each value represents mean ± SE. (n=10). ** $P<0.01$ significantly different from the control group.

Table 6 Effect of PYS on the acetic acid-induced Evans blue leakage test in mice

Groups	Dose (mg/kg · bw)	Evans blue (ug/mL)	Inhibition (%)
Control	—	7.13±0.99	—
Indometacin	3	4.68±0.20 **	34.36
	155	5.86±0.72 **	17.81
PYS	310	5.33±0.37 **	25.25
	620	5.10±0.24 **	28.47

Note: Each value represents mean ± SE. (n=10). ** $P<0.01$ significantly different from the control group.

Table 7 Effect of PYS on the carrageenan-induced paw edema test in rats

Groups	Dose (mg/kg · bw)	1 h		2 h		3 h		4 h	
		Swelling (%)	Inhibition (%)	Swelling (%)	Inhibition (%)	Swelling (%)	Inhibition (%)	Swelling (%)	Inhibition (%)
Control	—	32.94±5.75	—	58.59±3.75	—	71.61±9.92	—	62.61±4.66	—
Indometacin	3	30.94±3.53	6.07	44.13±4.80 **	24.68	45.12±3.43 **	36.99	26.95±7.29 **	56.96
	155	32.29±6.10	1.97	58.42±4.67	0.29	70.89±6.50	1.01	49.82±4.98 **	20.43
PYS	310	31.29±5.14	5.01	47.14±7.85 **	19.54	52.90±2.97 **	26.13	31.56±3.49 **	49.59
	620	31.19±2.92	5.31	47.73±4.04 **	18.54	49.76±5.68 **	30.51	31.62±5.21 **	49.50

Note: Each value represents mean ± SE. ($n=10$). ** $P<0.01$ significantly different from the control group at each corresponding time.

CONCLUSION

This study suggested that PYS possessed significant analgesic and anti-inflammatory activities, supporting its use to treat mastitis in dairy cows; and more investigations for ingredients and mechanisms responsible for the pharmacological activities of PYS should be studied further.

ACKNOWLEDGEMENTS

This work was funded by the Special Fund for Agro-scientific Research in the Public Interest

(201303040-01), the Special Fund of Chinese Central Government for Basic Scientific Research Operations in Commonweal Research Institutes (1610322014004) and 948 project (2014-Z9).

References

[1] Biiragohin, J. and Dutta, G. N. (1999). Treatment of bovine subclinical mastitis. Indian Vet. J., 76 (7): 646-649.

[2] Chinese Pharmacopoeia Committee. (2010). Chinese pharmacopoeia, 2010 ed. China medical science press, Beijing, China.

[3] Demon, D., Breyne, K., Schiffer, G. and Meyer, E. (2013). Short communication: Antimicrobial efficacy of intramammary treatment with a novel biphenomycin compound against Staphylococcus aureus, Streptococcus uberis, and Escherichia coli-induced mouse mastitis. J. Dairy Sci., 96: 7082-7087.

[4] Deraedt, R., Jouquey, S., Delevallee, F. and Flahaut, M. (1980). Release of prostaglandins E and F in an algogenic reaction and its inhibition. Eur. J. Pharmacol., 61 (1): 17-24.

[5] Di Rosa, M., Papadimitriou, J. M. and Willoughby, D. A. (1971). A histopathological and pharmacological analysis of the mode of action of nonsteroidal anti-inflammatory drugs. J. Pathol., 105: 239-256.

[6] Di Rosa, M., Sorrentino, L. and Parente, L. (1972). Non-steroidal anti-inflammatory drugs and leucocyte emigration. J. Pharm. Pharmacol., 24 (7): 575-577.

[7] Duarte, I. D., Nakamura, M. and Ferreira, S. H. (1988). Participation of the sympathetic system in acetic acid-induced writhing in mice. Braz. J. Med. Biol. Res., 21: 341-343.

[8] Ferreira, M. A. D., Nunes, O. D. R. H., Fontenele, J. B., Pessoa, O. D. L., Lemos, T. L. G. and Viana, G. S. B. (2004). Analgesic and anti-inflammatory activities of a fraction rich in oncocalyxone A isolated from Auxemma oncocalyx. Phytomedicine, 11 (4): 315-322.

[9] Garüa, M. D., Fern. ndez, M. A., Alvarez, A. and Saenz, M. T. (2004). Antinociceptive and anti-inflammatory effect of the aqueous extract from leaves of Pimenta racemosa var. ozua (Mirtaceae). J. Ethnopharmacol., 91 (1): 69-73.

[10] Gitau, G. K., Bundi, R. M., Vanleeuwen, J. and Mulei, C. M. (2014). Mastitogenic bacteria isolated from dairy cows in Kenya and their antimicrobial sensitivity. J S Afr Vet Assoc., 85 (1): 950.

[11] Green, M. and Bradley, A. (2013). The changing face of mastitis control. Vet. Rec., 173 (21): 517-521.

[12] Gu, Y. J. and Wang, L. J. (2007). Extraction of total flavonoids from Taraxacum mongolicum and its bacteriostatic performance. J. Northeast Forestry University, 35 (8): 43-45.

[13] Hill, A. E., Green, A. L., Wagner, B. A. and Dargatz, D. A. (2009). Relationship between herd size and annual prevalence of and primary antimicrobial treatments for common diseases on dairy operations in the United States. Prev. Vet. Med., 88: 264-277.

[14] Huang, M. H., Wang, B. S. and Chiu, C. S. (2011). Antioxidant, antinociceptive, and anti-inflammatory activities of Xanthii Fructus extract. J. Ethnopharmacol., 135 (2): 545-552.

[15] Kou, J.P., Sun, Y., Lin, Y.W., Cheng, Z.H., Zheng, W., Yu, B.Y. and Xu, Q. (2005). Anti-inflammatory activities of aqueous extract from radix ophiopogon japonicus and its two constituents. Biol. Pharm. Bull., 28: 1234-1238.

[16] Lam, T.J.G.M., van den Borne, B.H.P., Jansen, J., Huijps, K., van Veersen, J.C.L., van

Schaik, G.and Hogeveen, H. (2013). Improving bovine udder health: A national mastitis control program in the Netherlands.J.Dairy Sci., 96 (2): 1301-1311.

[17] Li, F. and Liang, J. Y. (2007). Research progress of vaccaria segetalis. Strait Pharm. J., 19: 1-4.

[18] Li, L. S., Shi, W. J., Guan, M., Li, X. J., Fang, T. and Zheng, J. L. (2008). Study on active component and bacteriostasis of tetraploid Plant of Herba Taraxaci. Chinese J. Experimental Traditional Medical Formulae, 14 (6): 55-58.

[19] Li, M. X., Shang, X. F., Zhang, R. X., Jia, Z. P., Fan, P. C., Ying, Q. and Wei, L. L. (2010). Antinociceptive and anti-inflammatory activities of iridoid glycosides extract of Lamiophlomis rotate (Benth.) Kudo. Fitoterapia, 81: 167-172.

[20] Liu, J. T., Han, P., Liu, L. B., Cheng, Y. and Zhang, X. M. (2012). Study on anti-inflammatory effects of taraxasterol crude extract. Progress in Veterinary Medicine, 33 (12): 104-106.

[21] Luber-Narod, J., Austin-Ritchie, T., Hollins, C., Menon, M., Malhotra, R. K., Baker, S. and Carraway, R. E. (1997). Role of substance P in several models of bladder inflammation. Urol Res, 25 (8): 395-399.

[22] Lucena, G. M. R. S., Gadotti, V. M., Maffi, L. C., Silva, G. S., Azevedo, M. S. and Santos, A. R. S. (2007). Antinociceptive and anti-inflammatory properties from the bulbs of Cipura paludosa Aubl. J. Ethnopharmacol., 112: 19-25.

[23] Mukesh, Kr. S., Thombare, N. N. and Biswajit, M. (2014). Subclinical mastitis in dairy animals: Incidence, economics, and predisposing factors. Sci. World J., . doi: 10. 1155/2014/523-984.

[24] Park, C. M., Jin, K. S. and Lee, Y. W. (2011). Luteolin and chicoric acid synergistically inhibited inflammatory responses via inactivation of PI3K-Akt pathway and impairment of NF-κB translocation in LPS stimulated RAW 264. 7 cells. Eur. J. Pharmacol, 660 (2-3): 454-459.

[25] Piepers, S., De Meulemeester, L., De Kruif, A., Opsomer, G., Barkema, H. W. and De Vliegher, S. (2007). Prevalence and distribution of mastitis pathogens in subclinically infected dairy cows in Flanders, Belgium. J. Dairy Res., 74: 478-483.

[26] Pol, M. and Ruegg, P. L. (2007). Treatment practices and quantification of antimicrobial drug usage in conventional and organic dairy farms in Wisconsin. J. Dairy Sci., 90: 249-261.

[27] Posadas, I., Bucci, M., Roviezzo, F., Rossi, A., Parente, L., Sautebin, L. and Cirino, G. (2004). Carrageenan-induced mouse paw oedema is biphasic, age-weight dependent and displays differential nitric oxide cyclooxygenase-2 expression. Brit. J. Pharmacol., 142: 331-338.

[28] Ramírez, N. F., Keefe, G., Dohoo, I., Sánchez, J., Arroyave, O., Cerón, J., Jaramillo, M. and Palacio, L. G. (2014). Herd- and cow-level risk factors associated with subclinical mastitis in dairy farms from the high plains of the northern Antioquia, Colombia. J. Dairy Sci., 97 (7): 4141-4150.

[29] Shinde, U. A., Phadke, A. S., Nair, A. M., Mungantiwar, A. A., Dikshit, V. J. and Saraf, M. N. (1999). Studies on the anti-inflammatory and analgesic activity of Cedrus deodara (Roxb.) Loud. wood oil. J. Ethnopharmacol., 65: 21-27.

[30] Song, X. Y., Liu, Q., Zhang, Y. Z., Liu, Y. X., Yang, L. and Wu, D. Y. (2010). Study on extraction technology and antibacterial activity of polysaccharides from Herba Taraxaci. China Pharmacy, 21 (47): 4453-4455.

[31] Song, Y. M., Li, D. G., Zhang, Y., Zhang, W. G., Jin, Y. P., Zhou, L. and Tang, J. M. (2006). The bacteriostasis of the extracts from Herba taraxaci on bacteria derived from bovine hidden mastitis. Acta Agriculture Boreali-occidental is Sinica, 15 (4): 55-57.

[32] Su, Y. , Liu, P. J. , Li, L. M. and Zheng, X. C. (2011). Extraction of Epimedium polysaccharides by microwave-assisted technology and the antibacterial activity. Food Science and technology, 36 (11): 158-161.

[33] Vinegar, R. , Schreiber, W. and Hugo, R. (1969). Biphasic development of carrageenan edema in rats. J. Pharmacol. Exp. Ther. , 166: 96-103.

[34] Wang, G. Q. (2013). Study on the quality control, toxicology and pharmacodynamics of "Ru Ning San" for treatment and prevention of subclinical mastitis in dairy cows. Chinese Academy of Agricultural Sciences Dissertation, Beijing, China.

[35] Wang, G.Q., Wang, X.R., Zhang, K., Zhang., J.Y., Wang, X.Z., Meng, J.R., Chang, R. X., Yang, Z.Q.and Li, J.X. (2013). Study on acute toxicity test in mice and sub-acute toxicity test in rats of "Ru Ning San" . China Animal Husbandry & Veterinary Medicine, 40 (5): 129-133.

[36] Wu, J. F. , Dong, J. C. , Xu, C. Q. , Zhang, X. J. and Zeng, L. X. (2009). Effects of icariin on inflammation model stimulated by lipopolysaccharide in vitro and in vivo. Chinese J. Integrated Traditional and Western Medicine, 29 (4): 330-334.

[37] Xiao D. H. (2011). Diseases of dairy cattle. China Agricultural university press, Beijing, China.

[38] Young, H.Y., Luo, Y.L., Cheng, H.Y., Hsieh, W.C., Liao, J.C.and Peng, W.H. (2005). Analgesic and anti- inflammatory activities of [6]-gingerol.J.Ethnopharmacol., 96: 207-210.

[39] Zhang, Y. F. and Yu, Q. H. (1999). The anti-inflammatory effects of total flavonoids of Epimedium. J. Shenyang Pharmaceutical University, 16 (2): 122-124.

[40] Zimmermanm, M. (1983). Ethical guidelines for investigations of experimental pain in conscious animals. Pain, 16: 109-110.

（发表于《African Journal of Traditional, Complementary and Alternative Medicines》）

Study on Extraction and Antioxidant Activity of Flavonoids from *Cynomorium songaricum* Rupr.

WANG Yong-gang[1,2], WANG Fan[2], SUN Shang-chen[2],
ZHU Xin-qiang[1*], WANG Chun-mei[1], YANG Lin[2],
LIN Peng[2], WANG Xiao-Li[1*]

(1. Lanzhou Institute of Husbandry and Pharmaceutical Sciences, Chinese Acadeny of Agricultural Sciences, 730050 Lanzhou, China; 2. Experiment Education Centre, School of Life Science and Engineering, Lanzhou University of Technology, 730050 Lanzhou, China)

Abstract: Response surface methodology (RSM) was applied to predict optimum conditions for extraction offiavonoids from *Cynomorium songaricum* R u p r. A central composite design was used to monitor the effect of extraction temperature, extraction time, ratio of water to raw material on yield of total flavanoids. The optimum extraction condi-tions were obtained as ratio of water to raw material of 30 : 1, extraction temperature of 86℃ and extraction time of 2. 5 h. The results showed that the experimental values of 27. 12% were consistent with the predictive values. The antioxidaant experimental results showed the DPPH radical scavenging activity increased with the flavonoids concentration until a maximum value of 89. 06% at 0. 006mg/mL and the maximum scavenging rate of ascorbic acid could reach 89. 04% at 0. 006mg/mL. The hydroxyl radical scavenging activity increased significantly in a concentration-dependent way at the flavanoids concentration range of 0. 1-0. 4mg/mL, and achieved maximum scav-enging activity of 20. 48% at 0. 4mg/mL, the maximum scavenging activity of ascorbic acid was 91. 71% at 0. 5mg/ml, which indicated that flavonoids of *C songaricum* had a strong DPPH radical scavenging ability compared with the positive control *Vc*, and hydroxyl radical scavenging ability was much weaker than the positive control V_c。 The extraction method was applied successfully to extract total flavonoids from *C. songaricum*.

Key words: *Cynomorium songaricum* Rupn; Flavonoid; Ethanol reflux extraction; Response surface methodology

AIMS AND BACKGROUND

Cynomorium songaricum Rupr. (Cynomoriaceae) is an achlorophyllous holoparasite that is distributed in the north-western part of China. Among its chemical constituents, steroids, triterpenes, fiavonoids and liguans have been reported previously[1,2]. Stems of *Cynomorium songaricum*

* *Received* 16 *April* 2015
Revised 19 *June* 2015

Rupr. axe known in Chinese as suo - yang. According to the ancient Chinese medical literature, C. songaricum is effective in regulating endocrinopathy and improving sexual fimction and also against symptoms of aging, including dementia[3,4]. Because it works usually together with other Chinese herbal medicaments, its individual bioactivity has been scarcely studied.

More and more studies showed that fiavonoids have many biological activities, such as antibacterial, antiviral, anticancer and antioxidant effects. In fact, flavonoids consist of fiavones, fiavanone, fiavanols, fiavonols and fiavanonols that comprise a large group of secondary metabolites in plants[5,6], for example vegetables, fruits, flowers, roots, stems and herbs[7]. Present studies have shown that fiavonoids from *Coreopsis tinctoria* have a lot ofbioactive components[8]. Response surface methodol-ogy (RSM) is a very useful tool for this purpose, which was first introduced by Box and Wilson[1,9-11]. As a package of statistical and mathematical techniques employed for developing, improving, and optimising process, RSM can be effectively used to evaluate the effects of multiple factors and their interaction on one or more response variables[12-15]. In this study, RSM was used to optimise the extraction conditions (ratio of water to material, extraction temperature and extraction time). The extraction ef-ficiency was validated by RSM.

EXPERIMENTAL

Plant material and chemicals

Cynomorium songaricum Rupr. (Cynomoriaceae) used for this study was provided by Yide biological technology Co., Ltd. (Guazhou, P. R. China). 100 g of *C songaricum* were dryed, grinded to powder and then sieved through 40-mesh screen. Allanalytical grade solvents were from Beijing Solarbio Co., Ltd. (Beijing, P. R. China) and Standard of Rutin (Batch No F20080313, > 95%) was obtained from Nationalmedicine group chemical reagent Co., Ltd. (Beijing)

Extraction of flavonids and determination of extraction yield

The dried flowers samples (1.0 g) were extracted for fiavonoids by 70% ethanol reflux extraction with different temperature of the water bath (70 to 95℃) with a ratio of water to material (mL/g) (ranging from 15 : 1 to 40 : 1) for a given extraction time (ranging from 1.0 to 3.0 h). The extracts were added to a defined volume by water (100mL) and then been filtered. The content of fiavonoids extracted from *C. songari-cum* was measured according to the method. The optical density of reaction solution was measured at 510 nm. The content of fiavonoids in this sample was calculated according to equation of linear regression ($Y = 11.66x + 0.0001$, $R^2 = 0.9996$) based on the standard curve whose horizontal coordinate and vertical coordinate denoted the concentration ofrutin (mg/mL) and OD510, respectively. The flavonoids yield could be calculated as follows:

$$\text{yield}\ (\%) = \frac{C \times N \times V}{W \times 1\ 000} \times 100 \qquad (1)$$

Where C is the concentration of fiavonoids calculated by the calibrated regressionequation (mg/ml); N-the dilution factor; V- the total volume of extraction solution (ml), and W-the weight of raw material (g).

Experimental design

In this study, single factor experiment was employed to guide the preliminary range of the variables including X_1 (ratio of water to raw material), X_2 (extraction tempera-ture), mad X_3 (extraction time). Then a central composite design was used to investigate the effects of three independent variables, X_1 (ratio of water to raw material), X_2 (extraction temperature), mad X_3 (extraction time) on the yield of flavonoids (19. The independent variables were coded at three levels El, 0, mad 1), and the complete design consisted of 20 experimental points including six replications of the centre points (all variables were coded as zero). In detail, X_1 (ratio of water to raw material, 25 : 1, 30 : 1, 35 : 1), X_2 (extraction temperature 80, 85, 90℃), mad X_3 (extraction time1.5, 2.0, 2.5 h) were investigated.

Statistical analysis

Experimental data showed that response variables were fitted to a quadratic polynomial model. The general form of the quadratic polynomial model was as follows:

$$Y = \beta_0 + \sum_{i=1}^{3} \beta_i x_i + \sum_{i=1}^{3} \beta_i x_i^2 + \sum_{i<j}^{3} \sum_{j=1}^{3} \beta_y x_i x_j \tag{2}$$

Where Yis the measured response associated with each factor level combination; β_0, β_i, β_{ii} and β_{ij} – the regression coefficients for imercept, linear, quadratic and interaction terms, respectively; x_i and x_j – the coded independent variables. Design expert software (Trial version 7.1.6.) was used to estimate the response of each set of experimental design and optimised conditions. The fitness of the quadratic polynomial model was inspected by the regression coefficient R^2. F-value and p-value were used to check the significances of the regression coefficient.

Determination of antioxidant activities in vitro

Effect of scavenging of (DPPH) radicals. The DPPH scavenging assay was performed according to Shimada et alp The solution of 0.2 mM DPPH in 60% ethanol wasprepared before UV measurements. Then, 3.0mL of the flavonoids (0.002, 0.004, 0.006, 0.008, 0.012, 0.016mg/mL) were added to 1.0mL DPPH, and kept at room temperature for 30 min in the dark, and the absorbance at 525 nm was then measured against a blank. Ascorbic acid was used as positive controls. The scavenging activity of DPPH radical (%) was calculated according to Shimada et al.[16]

Hydroxyl radical scavenging assay. Hydroxyl radical scavenging activity was meas-ured according to Jin et al.[17] with appropriate modifications. 10mL of reaction mixture contained 2mL of 200 mM sodium phosphate buffer (pH 7.4), 1.5mL of 5.0 mM 1, 10-phenanthroline aqueous solution, 1.0mL of 7.5 mM $FeSO_4$ aqueous solution, 1mL of 0.1% H202 aqueous solution, 0.1mL of the polysaccharide aqueous solution wit& different concentration (0.1, 0.2, 0.30, 0.40, and 0.50mg/mL, respectively) and distilled water. After incubating at 37℃ for 1 h, the absorbance of the mixture was measured at 510 mn.

RESULTS AND DISCUSSION

Effect of ratio of water to raw material on yield of flavonoids

At a temperature of 80℃, extraction time of 2 h, the extraction efficiency of fiavo-noids with the increase of ratio of water to raw material extraction increased gradually (Fig. 1a). When the value of ratio of water to raw material was more than 35mL/g, extraction yield increased very slow and stable. Considering the cost actually, the value of 35mL/g was chosen.

Effect of extraction temperature on extraction yield of flavonoids

At a fixed ratio of water to material of 30 : 1, extraction time of 2 h, the extraction ef-ficiency of fiavonoids will increase gradually and tended to be genfie wit & the increase of the extraction temperature (Fig. lb), which is due to molecular diffusion rate increasing gradually wit & the rise of temperature. But fiavonoids are heat-sensitive material, the temperature is too high, it was easy to cause degradation, and a low temperature extraction will save energy consumption, so 90-C was a better extraction temperature.

Effect of extraction time on extraction yield of flavonoids

At a fixed ratio of water to material of 30 : 1, extraction temperature of 90℃, the extraction efficiency of fiavonoids increased with the extension of extracting time and tended to stable (Fig. 1*c*). The extraction efficiency was highest in 2. 5 h.

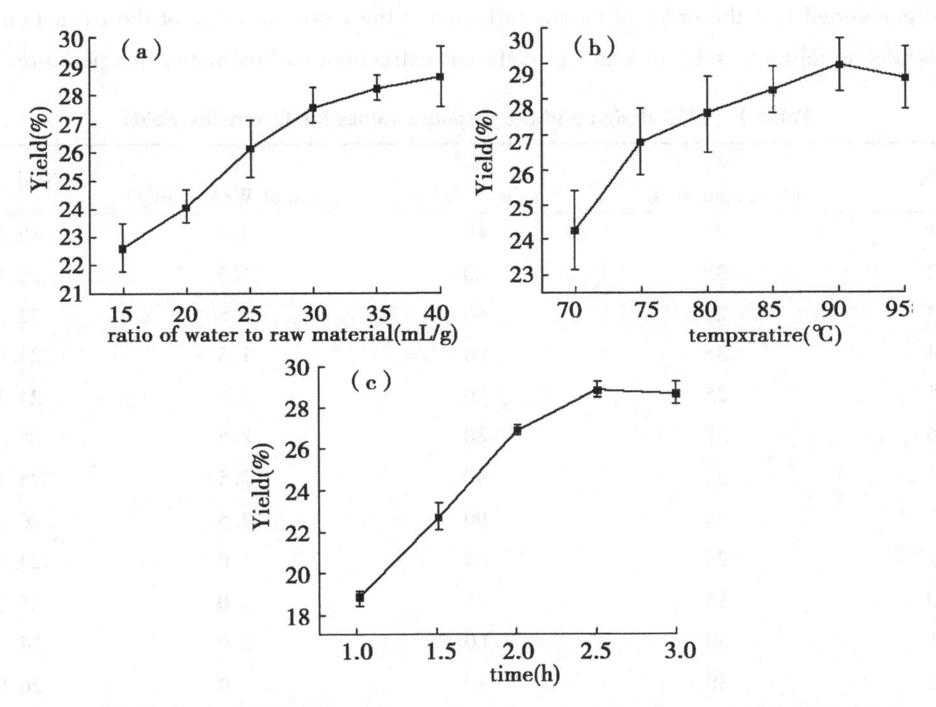

Fig. 1 Effects of different ratio of water to raw material (a), extraction temperature (b), and extraction time (c) on extraction yield offlavonoids

Note: all data were mean values of triplicate. The vertical error bars represented the standard deviation of each data point ($P<0.05$)

Optimisation of extraction conditions of flavonoids

According to the values obtained in the single factor experiment and method of central composite designed experiment, RSM was applied to optimise the extraction conditions of flavanoids from

C. songaricum. Table 1 showed the experiments design matrix with the response values obtained. Multiple regression analysis of the experimental data yielded the following secondorder polynomial equation:

$$Y- 358.61+ 3.30X_1+ 6.48X_2+ 35.39X_3+ 0.0025X_1X_2\ 0.15X_1X_3 \\ 0.27X_2X_3\ 0.05X12\ 0.03X_2^2\ 3.65X_3^2. \quad (3)$$

The ANOVA results for the fitted quadratic polynomial model of extraction offiavanoids axe shown in Table 2. The model F-value of 28.79 with a low probability p-value indicates high significance of the model. The coefficient of determination (R^2) of 0.9628 was the proportion of variability in the data explained or accountedfor by the model. The significance of each coefficient was determined using F-value and p-value. The results are given in Table 3. It could be seen that the independent variables (X_2 X_3), and the quadratic terms (X_2X_3, X_1) significantly affected the yield of fiavanoids ($P<0.01$), the other independent variables (X_1) and quadratic terms (X_2^2, $X_3{}^2$) were significant too ($P<0.05$) and the two-vaxiable interaction X_1X_2, X_1X_3 had no significant influence ($P>0.01$) on the extraction yield of flavanoids. By observing linear and quadratic coefficients we concluded that the order of factros influencing the response value of the extraction yield of flavonoids was as follows: ratio of water to material>extractiontime>extraction temperature.

Table 1 CCD design and the response values for flavonoids yields

No.	X_1 (temperature, ℃)	X_2 (time, rain)	X_3 (ratio of W/M, mFg)	Yield (%)
1	25	80	1.5	19.22
2	35	80	1.5	20.66
3	25	90	1.5	22.26
4	35	90	1.5	24.68
5	25	80	2.5	24.79
6	35	80	2.5	25.46
7	25	90	2.5	25.87
8	35	90	2.5	26.06
9	25	85	2.0	24.87
10	35	85	2.0	25.28
11	30	80	2.0	24.18
12	30	90	2.0	26.86
13	30	85	1.5	24.28
14	30	85	2.5	26.64
15	30	85	2.0	26.82
16	30	85	2.0	26.68
17	30	85	2.0	25.89
18	30	85	2.0	27.01
19	30	85	2.0	27.42
20	30	85	2.0	26.91

Table 2 Analysis of variance for fitted quadratic polynomial model

Source	SS[a]	DF[b]	MS[c]	*F*-value	*p*-value
Model	86. 39351	9	9. 599278	28. 79023	< 0. 0001
X_1	2. 63169	1	2. 63169	7. 892986	0. 0185
X_2	13. 04164	1	13. 04164	39. 11459	<0. 0001
X_3	31. 39984	1	31. 39984	94. 17465	< 0. 0001
X_1X_2	0. 03125	1	0. 03125	0. 093725	0. 7658
X_1X_3	1. 125	1	1. 125	3. 374109	0. 0961
X_2X_3	3. 61805	1	3. 61805	10. 85128	0. 0081
X_1^2	4. 621536	1	4. 621536	13. 86095	0. 0040
X_2^2	1. 993255	1	1. 993255	5. 978187	0. 0346
X_3^3	2. 284105	1	2. 284105	6. 850506	0. 0257
Residual	3. 334214	10	0. 333421	—	—
Lack of fit	2. 05153	5	0. 410306	1. 599405	0. 3094
Pure error	1. 282683	5	0. 256537	—	—
Cor. total	89. 72772	19	—	—	—
	R^2-0. 9628	R^2 adj-0. 9294			

Note: a-sums of squares; b-degree of freedom; c-mean square.

Response surface analysis

The relationship between independent and dependent variables was illustrated by the tridimensional representation of the response surfaces by the model. They are presented in Fig. 2a, b, c for the independent variables (The ratio of water to material, extraction temperature and extraction time) and were obtained by keeping one of the variables constant, which indicated the changes in extraction yield under different conditions.

Fig. 2a show s the 3D response surfaces, the combined effect of the ratio of water to raw material and extraction temperature on the extraction yield, and it reveals that the extraction yield was minimum at low and high levels of the ratio of water to raw material and extraction temperature. When ratio of water to material was at a certain value, the extraction yield increased with the increase of the extraction temperature. Howev er, it was not significant that the increase of the extraction temperature affected the extraction yield at a certain ratio of water to material. From Fig. 2b, the results indicate that the interactions between ratio of water to raw material and extraction time were significant when the other variables were fixed at a constant. As shown in Fig. 2c, extraction temperature, extraction time display a quadratic effect on the response. Thek interaction effecting on the yield was significant.

Optimisation and verification

The optimum condition was obtained by Design Expert 7. 1. 6 software, and were recommended

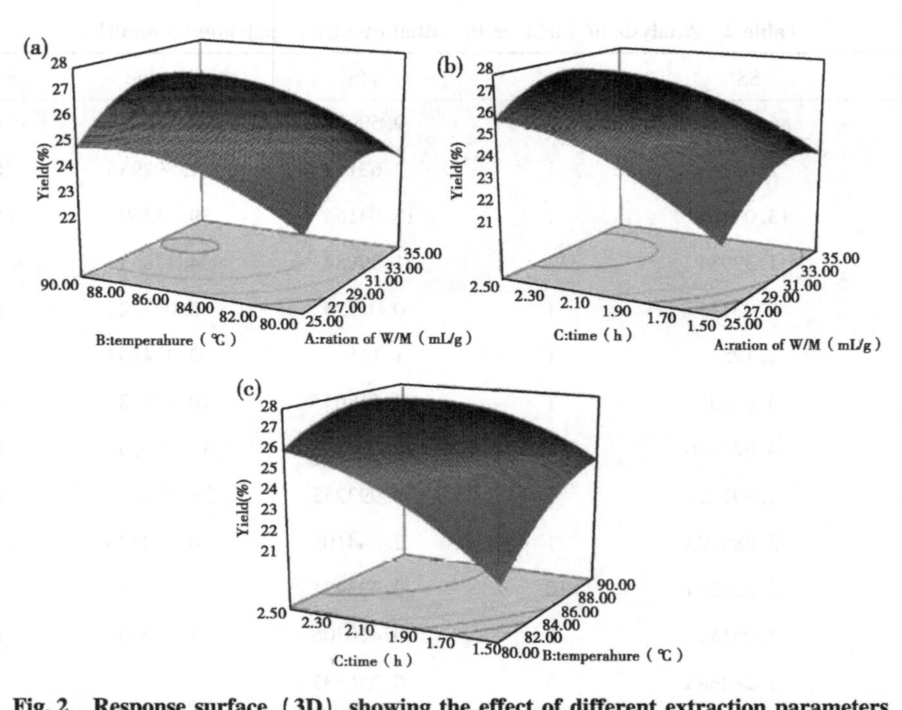

Fig. 2 Response surface (3D) showing the effect of different extraction parameters and their interac tions on the yield

Note: X_1 extraction temperatme, ℃; X_2 extraction time, h; X_1-the ratio of water to raw matelial, mL/g.

as follows: ratio of waler to raw material of 30.78, extraction temperature of 86.59℃ and extraction time of 2.29 min. The maximum predicted yield of 27.51% was obtainedunder those conditions. To further test the reliability of the experimental method. The extraction conditions were adjusted according to the actually production as follows: ratio of water to raw material of 30, extraction temperature of 86℃ and extraction time of 2.5 h. The results showed that the experimental values of 27.12% were not only consistent with the predictive values, but were also better than any single factor experiments. The response model was adequate for the optimisation of extraction process, and the model of equation (3) was accurate.

Antioxidant activity

Scavenging activity of DPPH radicals, in this work, DPPH free-radical scavengingeffect of flavonoids and ascorbic acid were measured respectively as shown in Fig. 3a. At a concentration ranging from 0.002 to 0.016mg/mL, the DPPH radical scavenging activity increased with the flavonoids concentration until a maximum value of 89.06% at 0.006mg/mL, whereas the maximum scavenging rate of ascorbic acidcouldreach 89.04% at 0.006mg/mL. Flavanoids scavenging activity of DPPH radicals was as strong as ascorbic acid[18].

Scavengirlg activity of hydroxyl radical. The results of hydroxyl radical scavengingactivities of the fiavonoids are given in Fig. 3b, which showed that the difference of scavenging activity between

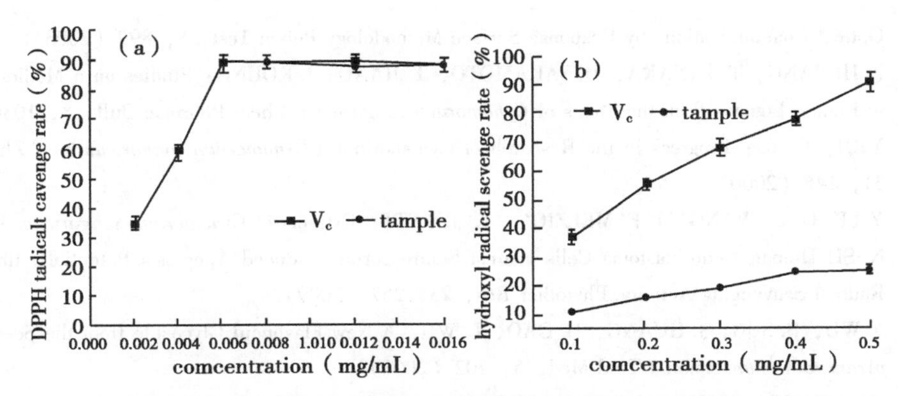

Fig. 3 Antioxidant activities offlavonoids from *C songaricum*

Note: *a*− scavengingactivity of DPPH radical; *b* scavenging ability of hydroxyl radicals (data were shown as mere ($n=3$) The vertical bars represented me standard deviation of each dam point ($P<0\ 05$))

the fiavonoids with ascorbic acid as a positive control. The scavenging activity increased significantly in a concentration-dependent way at the fiavonoids concentration range of 0. 1 0. 4mg/mL, and achieved maximum scav-enging activity of 20. 48% at 0. 4mg/mL, however, the maximum scavenging activity of ascorbic acid was 91. 71% at 0. 5mg/mL, which indicated that hydroxyl radicals scavenging activities offlavonoids were not obviously compared with the ascorbic acid.

CONCLUSIONS

In this study, an efficient technology has been developed for rapid extraction offlavonoids from *C songaricum* by RSM. The optimum extraction conditions for thepolysaccharides were as follows: ratio of water to raw material of 30 : 1, extractiontemperature of 86℃ and extraction time of 2. 5 h. The results showed that the experi-mental values of 27. 12% were not only consistent with the predictive values, but were also better than any single factor experiments. The work provided a novel and efficient method for the extraction of flavonoids from *C songaricum*. The antioxidant activities measurement *in vitro* demonstrated that flavonoids can scavenge free radicals to some extent. However, further investigation of structural identifications is required to elucidate the antioxidant mechanism. We hope this research will promote the use of fiavonoids from *C. songaricum* in food and drug industry.

ACKNOWLEDGEMENTS

This research was supported by Special Scientific Research Fund of Agricultural Public Welfare Profession of China (No 201203042, 201403048-8), Science and Technology Innovation Projects of CAAS (No CAAS-ASTIP-2014-LIHPS-08), Chinese National Natural Science Foundation (No 31460032), Natural Science Foundation of Gansu Province (No 1212RJYA008, 1308RJZA287), and the Foundation of Excellent Young Teachers of LUT (No 10-061406).

References

[1] H. KIM, J. KIM, J. CHO, J. HONG: Optimization and Characterization of UV-curable Adhesives for

Optical Communications by Response Surface Methodology Polym Test, 8, 899 (2003).

[2] Z. H. JIANG, T. TANAKA, M. SAKAMOTO, T. JIANG, I. KOUNO: Studies on a Medicinal Parasitic Plant: Lignans from the Stems of *Cynomorium songaricum*. Chem Pharmac Bull, 8, 1036 (2001).

[3] Y QI, G. SU: Progress in the Research of Cynomorium (*Cynomorium songaricum*) 'Zhongcaoyao' 31, 146 (2000).

[4] Y LU′ Q′ G′ WANG′ M′ F′ MELZIG′ S′ K′ JENETT: Extracts of *Cynomorium songaricum* Protect SK-N-SH Human Neuroblastoma Cells against Staurosporine-induced Apoptosis Potentially through Their Radical Scavenging Activity. Phytother Res, 23, 257 (2009).

[5] Y WU, B. SUN, J. HUANG, H. GAO, L. WU: A New Flavonoid Glycoside from the Seeds of *Fagopyrum tataricum*. Asian J Trad Med, 5, 202 (2007).

[6] Y ZHANG, S. SHI, M. ZHAO, Y. JIAN G, P. TU: A Novel Chalcone from *Coreopsis tinctoria* Nutt. Biochem Syst and Ecol, 10, 766 (2006).

[7] Y YANG, B. Z. DUAN: An Elementary introduction to the Reform of the Experimental Teaching of Pharmacognosy. Res Explor Lab, 25, 984 (2006).

[8] T. CUSHNIE, A. J. LAMB: Antimicrobial Activity of Flavonoids. hat J Antimicrob AG, 5, 343 (2005).

[9] B. ALIAKBARIAN, A. FATHI, P. PEREGO, F. DEHGHANI: Extraction of Antioxidants from Winery Wastes Using Subcritical Water J Supercrit Fluid, 65, 18 (2012).

[10] G. HANRAHAN, K. LU: Application of Factorial and Response Surface Methodology in Modem Experimental Design and Optimization. Crit Rev Anal Chem, 3 4, 141 (2006).

[11] W. XIAO, L. HAN, B. SHI: Optimization of Microwave-assisted Extraction of Flavonoid from Radix Astragali using Response Surface Methodology. Sep Sci Technol, 3, 671 (2008).

[12] M.AMENI, H.YOUNESI, N.BAHRAMIFAR, A.A.Z.LORESTANI, F.GHORBANI, A.DANESHI, M.SHARIFZADEH: Application of Response Surface Methodology for Optimization of Lead Biosorption in an Aqueous Solution by *Aspergillus niger*.J Hazard Mater, 1, 694 (2008).

[13] M. MUTHUKUMAR, D. MOHAN, M. RAJENDRAN: Optimization of Mix Proportions of Mineral Aggregates Using Box Behnken Design of Experiments. Cement Concrete Comp, 7, 751 (2003).

[14] Z. L. SHENG, P. F. WAN, C. L. DONG, Y H. LI: Optimization of Total Flavonoids Content Extracted from *Flos populi* Using Response Surface Methodology. had Crop Prod, 43, 778 (2013).

[15] X. Ir. ZHU, 55 L. MANG, J. XIE, P. WANG, W. K. SU: Response Surface Optimization of Mechanochemical-assisted Extraction of Flavonoids and Telrpene Trilactones from Ginkgo Leaves. had Crop Prod, 11, 1041 (2010).

[16] K. SHIMADA, K. FUJIKAWA, K. YAHARA, T NAKAMURA: Antioxidative Propelties of Xanthan on the Autoxidation of Soybean Oil in Cyclodextrin Emulsion. J Agr Food Chem, 6, 945 (1992).

[17] M. J [N, 55 X. CAI, J. R. LI, H. ZHAO: 1, 10-Phenanthroline-Fe^{2+} Oxidative Assay of Hydroxyl Radical Produced by H_2O_2/Fe^{2+}. Prog Biochem. Biophys, 23, 553 (1996).

[18] S. S. MITIC, M. B. STOJKOVIC, J. LJ. PAVLOVIC, M. N. MITIC, B. T. STOJANOVIC: Antioxidant Activity, Phenolic and Mineral Content of *Stachys germanica* L. (Lamiaceae). Oxid Commun, 35 (4), 1011 (2012).

[19] S. L. XIONG, A. L. LI, F. REN, H. J [N]. Study on Extraction of Total Flavonoids from Powder or Husks of Different Cultivars of Buckwheat and Analysis on Their Free Radical Scavenging Activities. Food Sci, 30, 118 (2009).

（发表于《Oxidation Communications》）

Study on the Extraction and Oxidation of Bioactive Peptide from the *Sphauercerpus grailis*

LENG Fei-fan [1,2], SUN Shang-chen[2], ZHU Xin-qiang[1],
LIU Lin[2], LIU Xiao-feng[2], WANG Yong-gang[2*], WANG Xiao-Li[1*]

(1. aLanzhou Institute of Husbandry and Pharmaceutical Sciences, Chinese Academy of Agricultural Sciences, 730050 Lanzhou, China;
2. Experiment Education Centre, School of Life Science and Enginee ring, Lanzhou University of Technology, 730050 Lanzhou, China)

Abstract: In this work, a single factor and orthogonal experimental design was utilised to optimise extraction condition of bioactive peptide from the *Sphauerocerpus grailis* by neutral protease hydrolysis method and the degree of hydrolysis was taken as evaluation index of the extraction yield. Oxidation activity ofbioactive peptide was studied. The results showed that the optimum extraction condition was the ratio of enzyme to substrate concentration of 5% and enzymolysis for 1. 5 h at 37℃. Under this condition, the degree of hydrolysis reached up to 30%. The antioxidant experimental results showed the DPPH radical scavenging activity increased with the bioactive peptide concentration until a maximum value of 53. 28% at 1. 0mg/mL and the maximum scavenging rate of hydroxyl radical could reach 75. 12% at 1. 0mg/mL.

Key words: Sphauerocerpus grailis; Bioactive peptide; Orthogonal experimental design

AIMS AND BACKGROUND

Sphauerocerpus grailis is a perennial herb and is mainly distributed in Neimenggu, Heilongjiang, Gansu, Qinghai, and Tibet alpine areas. It is rich in protein and various kinds of mineral elements, commonly known as little ginseng, and can be used to treat rheumatism, tetanus, cranker and other diseases in accordance with historical records. At present, the study of the concentrates is focused mainly on their botanical characteristics, cultivation and distribution, simple processing technology, the extraction of essential nutrient, chemical component and polysaccharide[1-3].

Peptides axe compounds whose strucmre is between amino acids and protein molecule. Some office low molecular bioactive peptides have a wide variety of biological fimctions, such as immune regulation, antihypertensive, lower cholesterol, anticancer, antioxidation, promote growth, adjust the food fiavour, etc[3-5] Recently, the research of bioactive peptides represents a boom. The

* *Received* 16 *April* 2015
Revised 17 *May* 2015

development goal of peptide products in China is mainly concentrated on soybean and casein, the study about peptides of the *S. grailis* is less. In this paper, the optimum extraction conditions of bioactive peptide from the S. grailis were studied using a single factor and orthogonal experimental design, and the biological activity of bioactive peptide was investigated, thus providing theoretical foundation for the development and application ofbioactive peptides.

EXPERIMENTAL

Materials

S. grailis was provided from Minle County in Zhangye City of Gansu Province.

Methods

Extraction ofbioactive peptide. Complete, no pests, clean S. *grailis* were shattered with a mill at constant temperature and drying for 2 h. An extracting solution of peptide was prepared by degreasing, alkalisolution and acid-isolation separating protein, enzymolysis and inactivating, bleaching and filtrating[3,4].

Measure of bioactive peptide. The extracting solution of peptide was centrifuged for 10 min at 5000 r/min and 2mL of supematant were put in a triangle bottle with 0. 1mol/L NaOH solution titrated to pH 8. 2, and 4mL of 37% formaldehyde solution (adjust pH to 8. 2 ahead of time) were added, then pH was adjusted to 9. 2 to record alkaline at this time. The amino nitrogen was calculated using the following formula:

$$h = ((V_1 - V_0) \times N_{NaOH} \times 14.008)/2 \qquad (1)$$

Where h is the amino acid content in the samples, mg/mL; V_1 the volume of the consumed sodium hydroxide by sample, ml; V_0 - the volume of the consumed sodium hydroxide by blank control, mL, and N_{NaOH} -the concentration of sodium hydroxide solution, mol/L.

The degree of hydrolysis (DH) was calculated using the following formula:

$$DH = (h_1 - h_0) / h_{tot} \qquad (2)$$

Where h_1 is the amino nitrogen content in the hydrolysate, mg/mL; h_0 the amino nitrogen content in the suspension of S. *grailis* protein powder, mg/mL; h_{tot} - the total amino nitrogen content in the hydrolysate, mg/mL.

Single factor experiment. Three factors including enzyme to substrate concentration ratio, enzymolysis temperature and time were selected to conduct single factor experiment[4-7].

Orthogonal experimental design. According to single factor test results, a multiple factor multiplelevel orthogonal experimental design was applied to optimise the extraction conditions of bioactive peptide from the S. *grailis*[7].

Oxidation resistance of bioactive peptide

Activity of DPPH radicals removal. A solution of 0. 04mg/mL DPPH in 60% ethanol was prepared and 3. 0mL of the extraction solution (0. 05mg/mL, 0. 1mg/mL, 0. 25mg/mL, 0. 5mg/mL, 1. 0mg/mL) were added to 1. 5mL DPPH, kept at room temperature for 30 min in the dark, and the absorbance at 517 nm was then measured against a blank. Ascorbic acid was used as positive control. The scavenging activity of DPPH radical (%) was calculated according to the following formula:

$$\text{clearance rate}\ (\%) = \left(1 - \frac{A_1 - A_2}{A_0}\right) \times 100 \qquad (3)$$

Where A_1 is the absorbance of DPPH solution after added extract solution; A_2 the absorbance of extracting solution; A_0 the absorbance without added extracting solution. Activity of hydroxyl radicals removal. A mixture of 0. 3mL of 6 mmol/L $FeSO_4$ aqueous solution, 0. 3mL of 6 mmol/L H_2O_2 aqueous solution, 1. 5mL of 6 mmol/L salicylic acid, 0. 1mL of the extraction solution with different concentration (0. 05mg/mL, 0. 1mg/mL, 0. 25mg/mL, 0. 5mg/mL, 1. 0mg/mL) and 7. 8mL distilled water was prepared. After incubating at 37℃ for 1 h, the absorbance of the mixture was measured at 510 nm (Refs 9–12). The absorbance of the blank control was determined simultaneously with the following formula:

$$\text{clearance rate}\ (\%) = \frac{A_0 - A_1 + A_2}{A_0} \times 100 \qquad (4)$$

Where A_0 is the absorbance without added extracting solution; A_1– the absorbance of DPPH solution with added extract solution, and A_2 the absorbance of extracting solution.

RESULTS AND DISCUSSION

Single factor experimental design

Determination of the enzyme to substrate concentration ratio. It can be seen from Fig. 1a, when the enzyme to substrate concentration ratio was 5%, DH reached maximum. DH is not in direct proportion to the increase along with the increase of enzyme to substrate concentration ratio, which means that the effect is not significant. It can be noted that the best enzyme to substrate concentration ration is 5%. It can be seen from Fig. 1 b that DH increased obviously with the extension of time at the beginning of the reaction, the change of DH is not obvious after 1. 5 h, however, hydrolysis reached maximum. The optimum hydrolysis time was determined as 1. 5 h. Fig. 1c revealed that DH increased significantly with the increase in extraction temperature from 24 to 37℃ and achieves the maximum value at 37℃, and then DH has a downward trend with the increase of temperature. The reason may be that high temperature can lead to protease inactivation. In addition, the extraction at high temperature can cause thermal denaturation of the peptide. The best extraction temperature was determined as 37℃ (Refs 10 and 11).

Orthogonal experimental design. According to the results of single factor experiment, 3 factors 3 levels orthogonal experimental design was applied to optimise the extraction conditions of bioactive peptide9. Factors and the levels of orthogonal design and the orthogonal experiment results are shown in Tables 1 and 2.

Table 1 Factors and levels of orthogonal design

Levels	Factors		
	enzyme to substrate concenlration ratio (%)	enzymolysis temperature (℃)	enzymolysis time (h)
1	4	30	1
2	5	37	1. 5
3	6	44	2

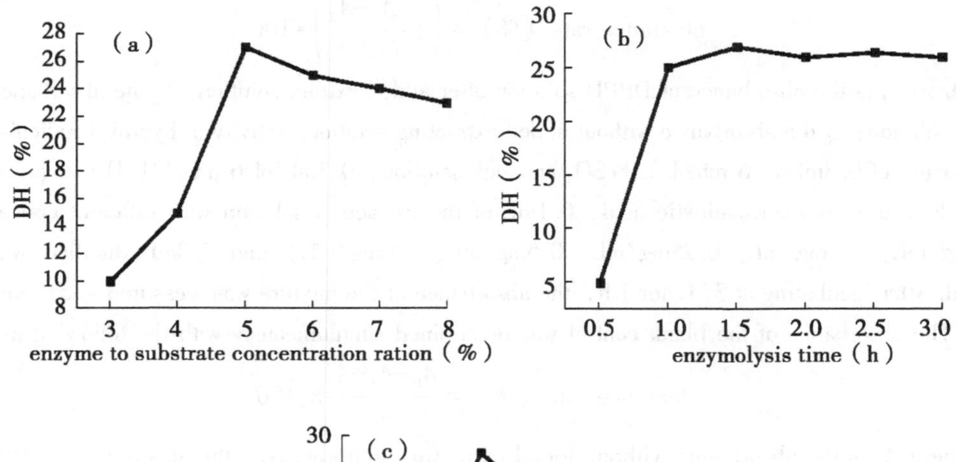

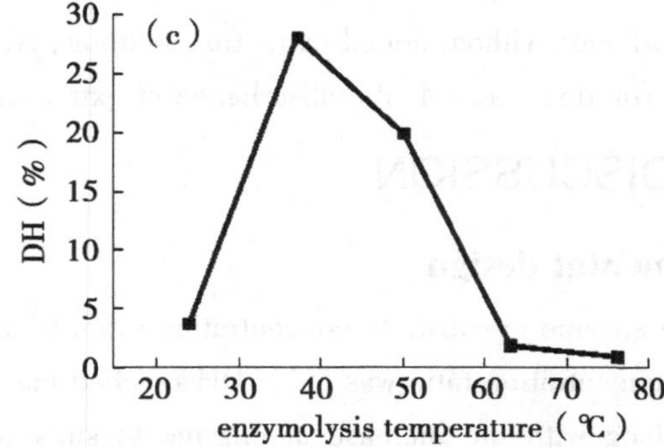

Fig. 1 Effect of different factors on DH

Note: *a*-enzyme to substrate concentration ratio; *b*-enzymolysis time, and *c*-enzymolysis temperature

Table 2 Orthogonal experiment results

Number	A (%)	B (℃)	C (h)	Degree of hydrolysis (%)
1	1	1	1	16.0
2	1	2	2	30.0
3	1	3	3	26.0
4	2	1	2	19.0
5	2	2	3	28.5
6	2	3	1	27.5
7	3	1	3	17.5
8	3	2	1	26.5
9	3	3	2	27.0
K_1	24.0	17.5	23.3	
K_2	25.0	28.3	25.3	
K_3	23.7	26.8	24.0	
R	1.3	10.8	2.0	

Note: *K value is the average of the sum of DH with the same number in the same column*; R *value-the measure of variation.*

Orthogonal experiment results showed that the best combination of the *S. grailis* protein levels of neutral protease hydrolysis were $A_1B_2C_2$, namely, enzyme to substrate concentration ratio of 5%, enzymolysis time of 1. 5 h, enzymolysis temperature of 37℃, and under these conditions, DH of bioactive peptide was 30%. The order of the effect of the factors on DH value was: B (enzymolysis time), C (enzymolysis temperature) and A (ration of enzyme to substrate concentration) in turn according to the *R* value. *Antioxidant activities ofbioactive peptide.* DPPH freeradical scavenging effect of bioactive peptide and ascorbic acid were measured, respectively as shown in Fig. 2a. The DPPH radical scavenging activity increased with the bioactive peptide concentration until a maximum value of 53. 28% at 1. 0mg/mL, whereas the maximum DPPH radical scavenging rate of ascorbic acid reached 95. 61% at 1. 0mg/mL. The results of hydroxyl radical scavenging activities of the bioactive peptide are given in Fig. 2b, and the maximum scavenging rate of hydroxyl radical reached 75. 12% at 1. 0mg/mL. The maximum hydroxyl radical scavenging activity of ascorbic acid was 93. 66% at 1. 0mg/mL, which indicated that DPPH radicals and hydroxyl radicals scavenging activities of bioactive peptide were not obviously compared with those of ascorbic acid.

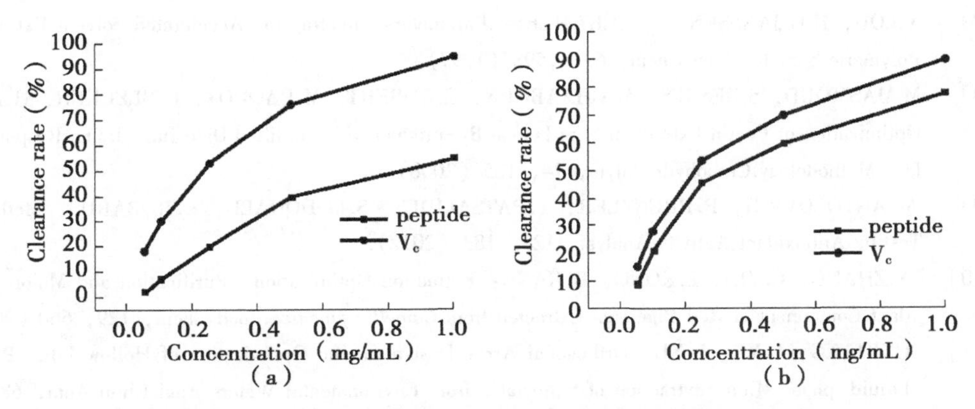

Fig. 2 Antioxidant activities of bioactive peptide from *S. grailis*

Note: *a*- scavenging activity of DPPH radicals, and *b*-scavenging ability of hydroxyl radicals

CONCLUSIONS

The experimental results show that the best combination of tJae *S. grailis* protein levels for neutral protease hydrolysis were enzyme to substrate concentration ratio of 5%, enzymolysis time of 1. 5h and enzymolysis temperature of 37℃. Under these conditions, the degree of hydrolysis of bioactive p eptide was 30%. The factors of primary and secondary order axe enzymolysis time > enzymolysis temperature > enzyme to substrate concentration ratio. Biological activity test results show that *S. grailis* active peptides have the ability to remove DPPH free radical and hydroxyl free radical.

ACKNOWLEDGEMENTS

This research was supported by Special Scientific Research Fund of Agricultural Public Welfare Profession of China (No 201203042, 20140304 8-8), Science and Technology Innovation Projects of CAAS (No CAAS-ASTIP-2014-LIHPS-08), Chinese National Natural Science Foundation (No 31460032), Natural Science Foundation of Gansu Province (No 1212RJYA008,

1308RJZA287), and the Foundation of Excellent Young Teachers of LUT (No 10-061406).

References

[1] T R.CHEN, G.H.LUO, T X.ZU, S.H.YANG, D.X.WANG: Study on the Processing of Sphauericepus Gracilis Spirulina Healthy Beverage.Food Sci, 25 (11), 121 (2004).

[2] D.AGYEI, M.K.DANQUAH: hadustrial-scale Manufacturing of Pharmaceutical-grade Bioactive Peptides.Biotechnol Adv, 29, 272 (2011).

[3] N.BENKERROUM: Amimicrobial Peptides Gener ate d from Milk Proteins: A Smrvey and Prospects for Application in the Food Industry.Dairy Technol, 63, 320 (2010).

[4] B.LAPORNIK, K.M.PROSEK, A.GOLC WONDRA: Comparison of Extracts Prepared from Plant By-products Using Different Solvents and Extraction Time.Food Eng, 71, 214 (2005).

[5] P.LIN, X.P.JIA, 55 Z.WANG, H.B.ZHU, Q.M.CHEN: Ultrasonically Assisted Extraction of Rutin from *Artemisia selengensis* Turcz: Comparison with Conventional Extraction Techniques.Food Anal Method, 5, 261 (2009).

[6] Z.Y.JU, L.R.HOWARD: Effects of Solvent and Temperature on Pressurized Liquid Extraction of Amhocyanins and Total Phenolics from Dried Red Grape Skin.Food Chem, 51, 5207 (2003).

[7] X.LOU, H.G.JANSSEN, C.A.CRAMERS: Parameters Affecting the Accelerated Solvent Extraction of Polymeric Samples.Anal Chem, 69, 1598 (1997).

[8] M.MASMOUD, S.BESBES, M.CHAABOUN, C.ROBERT, M.PAQUOT, C.BLECKER, H.ATTIA: Optimization of Pectin Extraction from Lemon By-product with Acidified Date Juice Using Response Surface Methodology.Carbohydr Polym, 74, 185 (2008).

[9] M. ANTOLOVICH, P. PRENZLER, E. PATSALIDES, S. McDONALD, K. ROBARDS: Methods for Testing Antioxidant Activity.Analyst, 127, 183 (2012).

[10] Y.ZHANG, C.YIN, L.KONG, D.JIANG: Extraction Optimisation, Purification and Major Antioxidant Component of Red Pigments Extracted from *Camellia japonica*.Food Chem, 129, 660 (201la).

[11] C.ZHANG, L.YE, L.XU: Orthogonal Array Design for the Optimization of Hollow Fiber Protected Liquid-phase Micro-extraction of Salicylates from Environmental Waters.Anal Chim Acta, 689, 219 (201lb).

[12] S.S.MITIC, M.B.STOJKOVIC, J.LJ.PAVLOVIC, M.N.MITIC, B.T.STOJANOVIC: Antioxidant Activity, Phenolic and Mineral Content of *Stachys germanica* L. (Lamiaceae). Oxid Commun, 35 (4), 1011 (2012).

(发表于《Oxidation Communications》)

The Two Dimensional Electrophoresis and Mass Spectrometric Analysis of Differential Proteome in Yak Follicular Fluid

GUO Xian, PEI Jie, CHU Min, WANG Hong-bo, DING Xue-zhi, YAN Ping

(Key Laboratory of Yak Breeding Eagineering of Gansu Province/
Lanzhou Institute of Husbandry and Pharmaceutical Sciences,
Chinese Academy of Agricultural Sciences, 730050 Lanzhou, China)

Abstract: Proteome refers to the entire protein set encoded by a single genome. To understand file mechanisms underlying the seasonal reproduction of yak at the protein level and to examine file changes of protein composition in the yak follicular fluid and plasma, file differential protein and its composition in the yak follicular fluid and plasma have been identified by two dimensional electrophoresis (2DE) and Mass Spectrometry (MS). The follicular fluid and plasma from yak on Qinghai Plateau were subjected to 2DE and silver staining. The gel image was analyzed by image master 2D Platinum Software followed by Matrix Assisted Laser Desorption/Iomzation Time of Flight (MALDI-TOF/TOF)$_{TM}$ MS identification. The proteins of high abundance were removed using Proteo Extract Albumin/Immunoglobulin G (IgG) Removal kit. Gel images with high resolution, cleat´ background and good reproducibility from follicular fluid and plasma were obtained by 2DE. Comparison between images from follicular fluid and plasma showed 24 differentially expressed proteins, of which two were up-regulated and twenty two were down-regulated. MS analysis identified eight proteins and five unknown proteins. The proteomic images were successfully separated and identified differential proteins between follicular fluid and plasma in the yak which provided experimental evidence for the mechanisms of yak follicular development and the understanding of microenvironment for oocyte development.

Key words: Yak; Follicular fluid; plasma; Differential proteome; Two dimensional electrophoresis; Mass spectrometrv identification

INTRODUCTION

Proteome refers to the entire protein set encoded by a single genome (Nowak, 1995). Proteomics is the study of proteome which includes the study of proteincomposition and its changing panems from cellular and systematic levels that leads to better understanding of physiological and pathological processes (Blackstock and Weir, 1999). Proteomics contains a series of high throughput methods that can simultaneously examine all proteins in cells, organisms or biological fluid including protein sample preparation, two dimensional electrophoresis (2DE), image analysis, protein separation and identification by Mass Spectrometry (MS) and database search. Proteomics is the milestone indicating that life science research has entered the post genome era and is the key content. Follicular fluid is composed of the plasma components across the blood-ovary barrier and the metabolic products of the granulosa cells which serves as the microenvironment during oocyte maturation. It contains materials involved in oocyte mitosis, ovulation, differentiation of ovarian cells into functional corpus luteum and fertilization. Therefore, the content of follicular fluid reflects the developmental stage of oocytes and the maturation level of follicles (Eppig et al., 2005; Sugiura et al., 2005). Follicular fluid is the culture medium for oocyte development and differentiation which directly or indirectly affects the development potential of the oocytes (Coleman et al., 2007). Growth factors and cytokines such as the growth hormone, macrophage inflammatory protein and macrophage colony stinmlating factor may serves as the markers for oocyte maturation (Kawano et al., 2004). Spitzer et al. (1996) compared the protein expression between the matured and inunatta-efollicular fluid in human by 2DE and found that the differential expression panems during different stages reflected the physiological condition to some extent which could be considered as critical markers for follicular fluid maturation. Huang et al. (2013) successfully separated and identified several differentially expressed proteins between matured and immature follicular fluid in buffalo by 2DE-MS.

Yaks (*Boa grunniens*) reach late sexual maturity and their reproduction is obviously seasonal. Tile reproductive mechanism is distinct from other bovine species (Sarkar et al., 2008). Yaks have low fecundity and their reproductive performnance is mainly determined by follicle growth and development in the ovaries and is also affected by nutrition, feeding and management. Although, people tried to improve yak reproductive performance by controlled reproduction (Zi et al., 2006), embryological technology (Yu et al., 2010) and molecular biology methods (Ding et al., 2012; Qiu et al., 2012), in addition to tile reinforcement of feecling and management, tile resuks are unsatisfactory which is mainly because the regulatory mechanisms of yak follicular development are still poorly understood. Moreover, it has been reported that 6 months old yaks cliff er from adult yaks in the ovarian and follicular development including the weight and size of the ovary and number of follicles and oocytes (Xu et al., 2012). The development potential of in vitro yak oocytes is different dming the reproductive and nonreproductive season (Guo et al., 2012). Hence, the protein composition change dming follicular development in yak needs further investigation.

Researcher planned to study the follicular fluid and plasma from yaks dming the reproductive season by proteomic approaches. The comparison of proteincomposition between follicular fluid and plasma and the establishment of 2DE protein images could help us to screen the key proteins involved in follicular development and oocyte maturation which could provide reference for the research on the expression, quantification and localization of these proteins during follicular development. The study findings could elucidate the mechanisms underlying the seasonal reproduction in yak which could technically increase the efficiency of yakreproduction and accelerate yak breeding as well as theoretically improve the *in vitro* culture system for yak oocytes.

The study examined tile changes of protein composition in the yak follicular fluid and plasma based on tile proteomic strategy to screen tile key proteins involved in follicular development and oocyte maturation. The study aimed to elucidate tile mlderlying regulatory mechanisms and understand tile pattern of seasonal reproduction in yak from proteomic perspective. Tile findings would shed light on the improvement of ovarian reset-ce and reproductive performance in yak.

MATERIALS AND METHODS

Collection and purification of follicular fluid: The ovaries and blood samples were collected from tile yak slaughter in Qinghai Plateau. Follicular fluid was drawn front follicles with a diameter > 8 nun using sterile syringes followed by centrifugation twice (lst, 5 000 rpm for 15 rain at room temperatrue, collect supematant; 2nd, 1 2000 rpm for 10 rain at room temperatrue, collect supematant) to completely remove granulosa cells, oocytes and debris. The blood samples which contained ethylenediaminetet-raacetic acid anticoagulant were centrfuged at 5 000 rpm for 20 min. Tile supernatant plasma was collected. Both follicular fluid and plasma samples were then aliquoted into 1. 5mL EP tubes and stored at −80℃ for future use.

Protein extraction and quantification: Follicular fluid and plasma samples (200μL each) were placed in sample lubes with added protein extraction buffer for grinding. Then, samples were placed on ice for 1 h and centrifuged at 13 000 rpm for 20 rain at 4℃. Supematant was collected. Proteo Extract Albumin/IgG Removal kit (Calbiochem Company, Massachusetts and Urfited States of America) was used to remove proteins of high abundance per inanufacturer's manual. Protein concentration was measured by Bradford Method (Bradford, 1976). Samples were stored at −80℃ and total protein extraction was tested by Sodium Dodecyl Sulfate Polyacrylamide Gel Electrophoresis (SDS-PAGE).

TWO DE

1st dimensional Isoelectric Focusing (IEF): Protein samples (100 μg) were loaded into the 13 cra, pH 3-10 non-linear gel strips [hnmobilized pH Gradient (IPG), Amersham Bioscience Inc., Amersham and Urfited Kingdom (UK)] to perform the 1 st dimensional IEF in tile solid phase with pH gradient [Ettan IPGphor IEF System, GE Amersham (GE Healthcare Life Sciences) Buckinghamslm'e and UK]. Tile electrophoresis condition was: 30 V for 12 h, 500 V for 1 h, 1 000 V for 1 h, 8 000 V for 8 h and 500 V for 4 h.

2nd dimensional SDS-PAGE: After IEF, the IPG gel strip was transferred to tile 2nd dimensional SDS-PAGE (12.5%). The electrophoresis was run under constant current 15 mA/gel for 30 rain and then 30 mA/gel until tile bromophenol blue reached 0. 5cm from the bottom of tile gel [Hofer SE 600, GE Amersham (GE Healthcare Life Sciences, Buckinghamslfire and UK)].

Gel staining and image analysis: The gel was stained with silver nitrate and scanned by hnage Scanner [Umax Powerlook 2110XL, GE Amersham (GE Healthcare LifeSciences), Buckinghamslfire and UK] (Chevallet et al., 2006). Image Master Software was used to analyze tile images and match tile protein spots. Protein spots withfold change in light density >1. 5 were considered as differentially expressed proteins.

MS identification: Tile differentially expressed protein spots were cut out from file gel and analyzed by Matrix Assisted Laser Desorption/Ionization Time of Flight (MALDI) -TOF/TOF™ MS (4 800proteomics analyzer, Applied Biosystems Inc., California and USA). The sequence reads were searched in the primary andsecondary MS database using Mascot Software.

RESULTS

The 2DE images from yak follicular fluid and plasma after proteins of high abundance were removed: Both plasma and follicular fluid samples contained protein ofhigh abundance at 66.2 ku as there were dark bands in the SDS-PAGE (Fig. 1). Hence, to obtain 2DE images with high resolution, good quality and reproducibility, theProteoExtract Albmnin/IgG Removal kit was used to remove the protein of high abundance in plasma and follicular fluid. After the highly abundant protein was removed, the band at 66.2 ku became significantly lighter in both plasma and follicular fluid SDS-PAGE (Fig. 1). The efficiency was ideal.

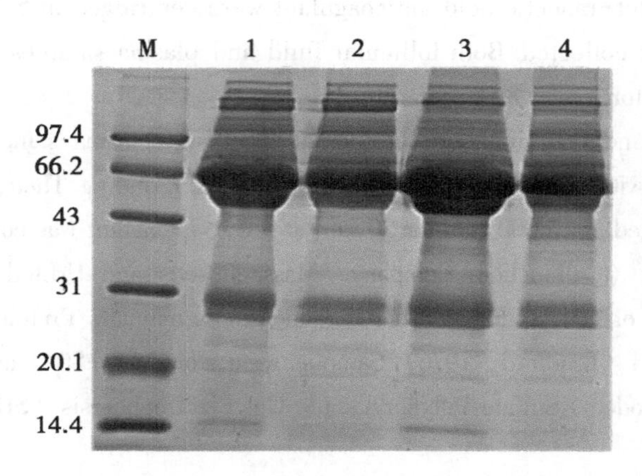

Fig. 1 SDS-PAGE of yak follicular fluid and plasma samples

Note: M: Marker; 1: plasma sample before highly abundant protein was removed;
2: plasma sample after highly abundant protein was removed;
3: follicular fluid sample before highly abundant protein was removed;
4: follicular fluid sample after highly abundant protein was removed

The establishment of 2DE images from yak follicular fluid and plasma samples: After the protein of high abundance was removed, yak plasma and follicular fluid samples were subjected to 2DE (Fig. 2).

Analysis of differentially expressed proteins in yak follicular fluid and plasma: The 2DE were perfornedthree times. Twenty four protein spots with fold change >1.5and 16 protein spots with fold change >2 were obtained. Using plasma samples as control, two protein spots were found inthe follicular fluid samples with fold change >1.5 were up-regulated and the rest 22 were downregulated. Part of the enlarged differentially expressed protein spots was shown (Fig. 3).

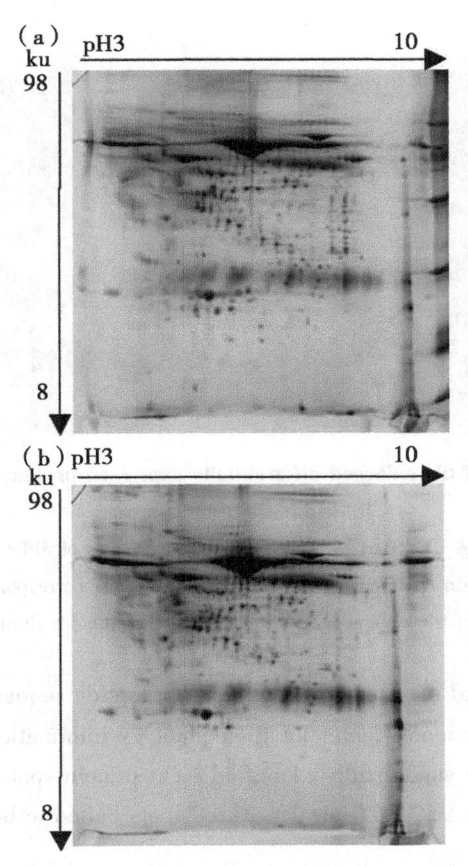

Fig. 2 Comparison between yak plasma and follicular fluid 2DE plasma image after highlyabundant protein was removed

Note: (a) plasma 2DE image after highly abundant protein was removed and (b) follicular fluid 2DE image after highly abundant protein was removed

Table 1 Identification of differentially expressed proteins by MS

Protein names	NCBI accession No.	Molecular weight/ (u)	PI	Protein score	Protein score C.I. (%)	DLfi'erences multiples
Fibrinogen alphachain precursor	gil75812954	67484. 2	6. 73	122	100	2. 92477
PR domain zinc finger protein 5	gil300798423	75021. 4	8. 95	68	98. 01	2. 80851
C4b-binding protein alpha chain	gil76677514	22393. 3	6. 34	215	100	2. 72175
Fibrinogen gamma-B chain precursor	gil27806893	50839. 5	5. 54	213	100	2. 65529
C－type lectin domain family 9 member A-like isoform X_2	gil528950702	28015. 1	9. 27	68	99. 31	2. 50686
Transfenin	gil602117	79869. 5	6. 75	79	100	2. 49628
Protein gamma, GTP binding	gil224267	8119. 1	4. 76	68	98. 27	2. 39829
TPA: keratin 6A-like	gil296487880	63341. 2	8. 41	85	99. 96	2. 14482

Identification of differentially expressed proteins by MS: The 16 differentially expressed protein

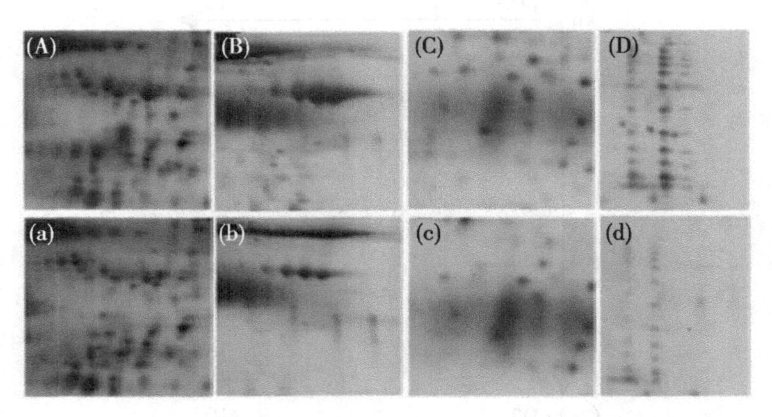

Fig. 3 Part of tile enlarged differentially expressed protein spots between follicular fluid and plasma

Note: A-D) Part of tile enlarged 2DE images of differentially expressed protein spots on tile yak plasma; a-d) tile corresponding positions of tile enlarged 2DE images of yak follicular fluid

spots were cut from the gel and digested by enzymes. The peptide sequences were analyzed and identified by MS. By searching National Center for Biotechnology Information (NCBI) *Box taurus* database using Mascot, the study successfully identified eight protein spots whose protein scores were > 65 and confidence were above 95% (Table 1). Three spots Failed to be identified and five were unknown proteins.

DISCUSSION

Proteomics is an effective method to study follicular fluid and plasma, among which 2DE/MS is a classic strategy of follicular fluid and plasma protein analysis andidentification (Cnanier, 1988; Huang et al., 2014). Protein sample preparation is the basis and prerequisite for proteomic research and is the key step of 2DE whichdirectly affects the quality of image and the reliability and reproducibility of the experiment (Simpson, 2003). With the development of follicles, the follicular fluid, made up by the secretion of granulosa cells in the lumen and the substances selectively across the bloodovary barrier, gradually accumulates. Its composition is similar to plasma which mainly includes proteins, enzymes, ammo polysaccharides, steroid hormones and other small molecule metabolites (Eppig et al., 2005; Fortune, 1 994).

Follicular fluid and plasma contain large amount of impurity and highly abundant proteins which could inevitably affect protein IEF. Proteo Extract Albmnin/IgG Removal kit was used to remove proteins of high abundance in plasma and follicular fluid and obtained protein samples of high quality which was tested by SDS-PAGE. The experimental platform composed of IEF (pH 3-10; 13cm IPG gel strip) and SDS-PAGE (12.5%) for the analysis of follicular fluid and plasma was established. The 2DE images with high resolution and clear background from yak follicular fluid and plasma were obtained when the sample loading quantity was 100pg.

The analysis of follicular fluid and plasma proteins by proteomic approaches not only can screen for differentially expressed proteins between the two samples but also can validate or con'ect the previous conclusions. Angelucci et al (2006) analyzed human follicular fluid by proteomic approaches and identified 695 protein spots between 10-200 kDa and pH 3-10 of which only 625 were present in the plasma (Angelucci et al., 2006). MALDI-TOF-MS identified 183 proteins shared by follicular fluid and plasma and 27 specifically expressed proteins in either sample. Schweigert et al. (2006) analyzed protein and peptides in human follicular fluid and plasma by surface enhanced laser desorption/iomzation TOF-MS and found that haptoglobin (α1 and α2 chares), haptoglobin 1 and transtilyretin could serve as tile candidate markers foi oocyte matmation and quality diagnosis (Schweigert et al., 2006). Tile comparison of yak follicular fluid and plasma by 2DE showed compositionalsimilarity as well as differentially expressed protein spots between tile two samples. Hence, tile study found 24 differentially expressed protein spots whose fold change was >1.5, of which 16 had fold change >2. Of which, 2 were upregulated and 22 were downregulated in follicular fluid. Eight protein spots were successfully identified by MS which played regulatory roles in follicular fluid and plasma and greatly affected oocyte maturation (Table 1). The expression of the differential proteins was in accordance with its function. However, the detailed mechanisms require further investigation.

Novel proteins in follicular fluid can be identified by proteomic approaches and database search. These proteins might be relevant to the development and maturation of follicles or oocytes. Anahory et al. (2002) analyzed the protein content in the matured human follicular fluid and identified new proteins in it. They ism 2D PAGE using matured human follicular fluid and obtained around 600 protein spots on the gel. According to the expression levels, fora spots were picked and analyzed by direct sequencing or MALDI-MS of which three were identified and for the first time they were reported as the components of follicular fluid. Lee et al. (2005) identified novel proteins in the matured follicular fluid by proteomic approaches (2DE and MALDI-MS). All together 180 protein spots were fomld and of which 10 were identified. Database search suggested that six proteins had already been reported by reverse tramcription-polymerase chain reaction, genes that encoded the fora newly identified proteins were found to be expressed in human primary granulosa cells. In tiffs study, 16 differentially expressed protein spots were picked for MS analysis of which eight were successfully identified by Mascot search (NCB_ *IBos taurus*) and three failed to be identified. The rest five proteins could not be matched by Mascot search which suggested that there were novel yak-specific proteins dm'rug follicular development. However, its structure and function need to be further investigated and validated.

CONCLUSION

The proteins of high abundance were removed by Proteo Extract Albumin/IgG Removal kit and then the 2DE images were obtained with plentiful protein spots, clear background, high resolution and good reproducibility fi'om yak follicular fluid and plasma. These images were analyzed by hnage-Master 2D Platmum Soft ware and 24 differentially expressed proteins were found of which eight were successfully identified by MS. These proteins might be relevant to the regulatory mechanismsmlderlying the development of follicular fluid in yak.

ACKNOWLEDGEMENTS

The research was supported or partly supported by grants from National Natural Science Foundation of China (31301976), China Agriculture Research System (CARS-38), Central Public-interest Scientific Institution Basal Research Fmld (1610322014010) and National Science and Technology Support Project (2013BAD16B09).

References

[1] Anshory, T., H. Dechaud, R. Beimes, P. Marin, N. J. Lamb and D. Laoudj, 2002. Identification of new proteins in follicular fluid of mature human follicles. Electrophoresis, 23: 1197-1202.

[2] Angelucci, S., D. Ciavardelli, F. ch Giuseppe, E. Eleuterio and M. Sulpizio et al., 2006. Proteome analysis of human follicular fluid. Bioclffm. Biophys. Acta (BBA) -Protems Proteomics, 1764: 1775-1785.

[3] Blackstock, W. P. and M. P. Weir, 1999. Proteomics: Quantitative and physical mapping of cellular proteins. Trends Biotechnol., 17: 121-127.

[4] Bradford, M. M., 1976. A rapid and sensitive method foi' the quantitation of microgram quantities of protein utilizing the principle of protein-dye binding. Anal. Biochem., 72: 248-254.

[5] Chevalier, M., S.Luche and T.Rabilloud, 2006.Silver staining of proteim in polyacrylamide gels. Nat. Prot., 1: 1852-1858.

[6] Coleman, N. V., G. A. Shagiakhmetova, I. Y. Lebedeva, T. I. Kuzmma and A. K. Golubev, 2007. In vitro maturation and early developmental capacity of bovine oocytes cultured in pure follicular fluid and supplementation with follicular wall. Theriogenology, 67: 1053-1059.

[7] Ding, X. Z., C. N. Liang, X. Guo, C. F. Xing and P. J. Bao et al., 2012. A novel single nucleotide polymorphism in exon 7 of LPL gene and its association with carcass traits and visceral fat deposition in yak (Bos grunniens) steers. Mol. Biol. Rep., 39: 669-673.

[8] Eppig, J. J., F. L. Pendola, K. Wigglesworth and J. K. Pendola, 2005. Mouse oocytes regulate metabolic cooperativity between granulosa cells and oocytes: Ammo acid transport. Biol. Reprod., 73: 351-357.

[9] Fortune, J. E., 1994. Ovarian follicular′ growth and development in marrmlals. Biol. Reprod., 50: 225-232.

[10] Gramer, F., 1988.Extraction of plant proteins for two-dimensional electrophoresis.Electrophoresis, 9: 712-718.

[11] Guo, X., X. Ding, J. Pei, P. Bao, C. Liang, M. Chu and P. Yah, 2012. Efficiency of/n vitro embryo production of yak (Bos grunniens) cultured in different maturation and culture conclitions. J. Applied Anim. Res., 40: 323-329.

[12] Husng, Y. L., D. L. Husng, J. L. Gusng, Q. Z. Pan and Q. Fu et al., 2013. Establishment of two dimensional electrophoresis method and mass spectrometry analysis of the differential proteins of buffalo follicular fluid. Acta Veterinaria Zootechnica Smica, 44: 1244-1250.

[13] Huang, Y. L., Q. Fu, D. L. Husng, F. M. Chen and F. L. Husng et al., 2014. 2-DE clifferential analysis of ALB/IgG removal buffalo follicular fluid. Acta Veterinaria Zootechnica Sinica, 45: 1113-1119.

[14] Kawano, Y., J. Fukuda, K. Nasu, M. Nishida, H. Narahara and I. Miyakawa, 2004. Production of macrophage hfflammatory protein - 3α in human follicular fluid and cultured granulosa

cells. Fertil. Steril. , 82: 1206-1211.

[15] Lee, H.C., S.W.Lee, K.W.Lee, S.W.Lee, K.Y.Cha, K.H.Kim and S.Lee, 2005.Identification of new proteins in follicular fluid from mature human follicles by direct sample rehycaration method of two-dimensional polyacrylamide gel electrophoresis.J.Korean.Med.Sci., 20: 456-460.

[16] Nowak, R., 1995.Entering the postgenome era.Science, 270: 368-369.

Qiu, Q. , G. J. Zhang, T. Ma, W. Qian and J. Wang et al. , 2012. The yak genome and adaptation to life at high altitude. Nat. Genet. , 44: 946-949.

[17] Sarkar, M. , D. H. Sengupta, B. D. Bora, J. Rajkhoa and S. Bora et al. , 2008. Efficacy of Heat-synch protocol for reduction of estrus, synchronization of ovulation and timed artificial insemination in yaks (Poephagus grunniens L.) Anim. Reprod. Sci. , 104: 299-305.

[18] Schweigert, F. J. , B. Gericke, W. Wolfram, U. Kaisers and J. W. Dudenhausen, 2006. Peptide and protein profiles in serum and follicular fluid of women undergomg IVF. Hum. Reprod. , 21: 2960-2968.

[19] Simpson, R. J. , 2003. Proteins and Proteomics: A Laboratory Manual. Cold Spring Harbor Laboratory Press, Cold Spring Harbor, NY. , USA. , ISBN-13: 9780879695545, Pages: 926.

[20] Spitzer, D. , K. F. Mmrach, F. Lottspeich, A. Staudach and K. Ilhnensee, 1996. Different protein patterns derive from follicular fluid of mature and immature human follicles. Hum. Reprod. , 11: 798-807.

[21] Sugiura, K. , F. L. Pendola and J. J. Eppig, 2005. ocyte control of metabolic cooperativity between oocytes and companion granulosa cells: Energy metabolism. Dev. Biol. , 279: 20-30.

[22] Xu, J. T. , Y. G. Sun, C. Dongzhi, Z. J. Ms, S. H. Chen and Y. C. Feng, 2012. Research on the ovarian and follicular development in 6 months of age and adult yaks. CNn. Heilongjiang Anita. Sci. Vet. Med. , 3: 41-44.

[23] Yu, S. J. , Y. H. Yong mid Y. Cui, 2010. Oocyte morphology from primordial to early tertiary follicles of yak. Repro & Domest. Arran. , 45: 779-785.

[24] Zi, X.D., S.M.He, H.Lu, J.A.Feng, J.Y.Lu, S.Chang and X.Wang, 2006.Induction of estrus in suckled female yaks (Bos grunniens) and synchromzation of ovulation in the non-sucklers for timed artificial insemination using progesterone treatments and Co-Synch regimens. Arum.Reprod.Sci., 92: 183-192.

(发表于《Journal of Animal and Veterinary Advances》)

Prevalence of *blaZ* Gene and other Virulence Genes in Penicillin-Resistant *Staphylococcus aureus* Isolated from Bovine Mastitis Cases in Gansu, China

YANG Feng[1,*], WANG Qi[2], WANG Xu-rong[1], WANG Ling[1], XIAO Min[1], LI Xin-pu[1], LUO Jin-yin[1], ZHANG Shi-dong[1], LI Hong-sheng[1]

(1. Key Lab of New Animal Drug Project, Key Laboratory of Veterinary Pharmaceutical Development, Ministry of Agriculture/Lanzhou Institute of Husbandry and Pharmaceutical Sciences of Chinese Academy of Agricultural Science, Lanzhou 730050, China;
2. State Key Laboratory of Applied Organic Chemistry, Lanzhou University, Lanzhou 730000, China)

Abstract: Staphylococcus aureus is a major etiological agent of bovine mastitis worldwide. In this study, 37 strains of S. aureus resistant to penicillin were isolated from bovine mastitis cases in Gansu Province for investigating *blaZ* and virulence-related genes, including *tst*, *eta*, *etb*, *lukPV*, *lukED*, *lukM*, *hla*, *hlb*, *hld*, and *edin*. Antibiotic resistance was based on disk diffusion method and *blaZ* and virulenceassociated genes were studied by polymerase chain reaction. Penicillin resistance gene *blaZ* was detected in 35/37 (94.6%) of penicillinresistant *S. aureus isolates*. *tst*, *lukPV*, *lukED*, *hla*, *hlb*, and *hld* were observed in 5.4%, 2.7%, 89.2%, 70.3%, 73.0%, and 70.3% of the penicillin-resistant isolates, respectively, while *eta*, *etb*, *lukM*, and *edin* were not detected in any isolates. *blaZ* carried by penicillinresistant *S. aureus* isolates may be the main reason for phenotypic penicillin resistance. Virulence determinants encoded by *lukED*, *hla*, *hlb*, and *hld* genes may play important roles in bovine mastitis pathogenesis of penicillin-resistant *S. aureus* in Gansu Province.

Key words: Bovine mastitis; *Staphylococcus aureus*; antibiotic resistance; *blaZ*; virulence-related genes

Staphylococcus aureus is the primary contagious pathogen in bovine mastitis, causing severe economic losses to the dairy industry worldwide [1]. Penicillin is the most commonly used drug in the treatment of mastitis[2], which has led to an increase in the number of resistant strains [3].

The pathogenic basis of *S. aureus* and its response to antibiotic therapy depend on various antibiotic resistance-and virulence-associated genes carried by the pathogen[4]. Resistance gene *blaZ* is responsible for resistance to penicillin[5]. Currently, over 40 virulence-associated genes have been reported among various *S. aureus strains*[6]. Although some but not all of the virulence-related genes in *S. aureus strains* from bovine mastitis have been reported in South and East China[7,8], little is known yetabout the virulence genes in penicillin-resistant *S. aureus strains* recovered from bovine mastitis in Northwest China. Therefore, the aims of this work were first to determine the genetic basis of penicillin resistance and second to investigate virulence-related genes in penicillinresistant *S. aureus strains* from bovine mastitis cases in Gansu Province.

Thirty-seven bacterial isolates from bovine mastitis isolates from clinical cases and 11 isolates from subclinical cases) were collected at 3 farms located in Gansu Province in China during 2014. Mastitis infection was confirmed by the California Mastitis Test. The isolates were identified as penicillin-resistant *S. aureus* strains by morphological characterization, biochemical testing, and disk diffusion method according to Clinical Laboratory Standards Institute standards[9]. *S. aureus* ATCC 29213 was used as control isolate. The penicillin-resistant gene *blaZ* and several virulence-related genes encoding toxic shock syndrome (*tst*), exfoliatins (*eta*, *etb*), leukotoxins (*lukPV*, *lukED*, *lukM*), hemolysins (*hla*, *hlb*, *hld*), and EDIN (*edin*) were detected by simplex PCR according to Olsen et al.[10] and Jarraud et al.[11], respectively.

As shown in the Table, 94.6% of S. aureus isolates resistant to penicillin were shown to have the expected penicillin resistance gene *blaZ*. *lukED* was the most prevalent virulence gene (89.2%), followed by *hlb* (73.0%), *hla* (70.3%) *hld* (70.3%), *tst* (5.4%), and *lukPV* (2.7%). *eta*, *etb*, *lukM*, and *edin were* not detected in any strains. In the *blaZ*-positive isolates, *lukED*, *hla*, *hlb*, and *hld were* the most commonly occurring virulence-related genes, detected in 86.5%, 67.6%, 70.3%, and 67.6% of isolates, respectively. In contrast, *tst* and *lukPV* were found in only two isolates and one isolate, respectively. For the two *blaZ* negative *S. aureus* isolates, the detection rates of *lukED*, *hla*, *hlb*, and *hld* were all 2.7%.

Table 1 Distribution of *blaZ* and virulence-related genes in penicillin-resistant *S. aureus* from cows with mastitis in Gansu

Resistance genotype	No. (%) of strains	No. (%) of virulence-related genes									
		tst	*eta*	*etb*	*lukPV*	*lukED*	*lukM*	*hla*	*hlb*	*hld*	*edin*
$blaZ^+$	35(94.6)	2(5.4)	0(0)	0(0)	1(2.7)	32(86.5)	0(0)	25(67.6)	26(70.3)	25(67.6)	0(0)
$blaZ^-$	2(5.4)	0(0)	0(0)	0(0)	0(0)	1(2.7)	0(0)	1(2.7)	1(2.7)	1(2.7)	0(0)
Total	37(100)	2(5.4)	0(0)	0(0)	1(2.7)	33(89.2)	0(0)	26(70.3)	27(73.0)	26(70.3)	0(0)

Note: $blaZ^+$: Strains positive for blaZ; $blaZ^-$: Strains negative for *blaZ*.

The activated *blaZ* could encode β-lactamase enzyme (penicillinase), which inactivates the antibiotic through hydrolysis of the peptide bond in the β-lactam ring [5]. In this study, not all the penicillin-resistant *S. aureus* isolates exhibited genotypic resistance to penicillin, corresponding well to the findings of previous studies [12]. This is in agreement with data from Gao et al. [13], showing that no resistance genes could be determined in some phenotypically resistant *Streptococcus agalactiae* isolates. In some isolates, phenotypic resistance may have been caused by point mutations rather than gene acquisition. Additionally, except for the general resistance mechanisms[14], other pathways such as biofilm formation may play a major role in the resistance mechanisms [15].

The penicillin-resistant *S. aureus* isolates in the present study showed a high frequency of *lukED*. The *lukED* gene could encode bicomponent leukotoxins with which *S. aureus* can target and kill innate immune cells critical for defense against bacterial infections [16]. A high prevalence of *hla*, *hlb*, and *hld* was also observed in the *S. aureus* isolates, in agreement with findings reported in China and other countries[7,17-19]. Hemolysins encoded by *hla*, *hlb*, or *hld* aid the *S. aureus* population in the invasion of the host cells and cause damage [20]. The observation of frequent *lukED*, *hla*, *hlb*, and *hld* suggest that bicomponent leukotoxins and hemolysins encoded by these corresponding virulence-related genes may have crucial roles in pathogenesis of bovine mastitis caused by penicillin-resistant *S. aureus* in Gansu Province.

In conclusion, our results suggest that the *blaZ* carried by penicillin-resistant *S. aureus* may play a major role in the penicillin-resistant phenotype, but the resistance gene cannot be used alone as a diagnostic indicator for penicillin resistance. Further studies should be performed to develop accurate molecular indicators of antibiotic resistance. In addition, the detection of genes encoding virulence determinants suggests that *lukED*, *hla*, *hlb*, and *hld* are the most prevalent virulence-related genes in penicillin-resistant *S. aureus* from bovine mastitis cases in Gansu. These results emphasize the need for monitoring the genetic basis of antimicrobial resistance and virulence determinants in *S. aureus*.

ACKNOWLEDGMENTS

This study was supported by the Special Fund of the Chinese Central Government for Basic Scientific Research Operations in Commonweal Research Institutes (No. 1610322015007), the National Key Science and Technology Support Project in the 12th Five-Year Plan of China (No. 2012BAD12B03), and the Natural Science Foundation of Gansu Province (No.145RJYA311).

References

[1] SupréK, Lommelen K, De Meulemeester L. Antimicrobialsusceptibility and distribution of inhibition zone diameters of bovine mastitis pathogens in Flanders, Belgium. Vet Microbiol 2014; 171: 374-381.

[2] Thomson K, Rantala M, Hautala M, PyöräläS, Kaartinen L. Cross-sectional prospective survey to study indication-based usage of antimicrobials in animals: results of use in cattle. BMC Vet Res 2008; 4: 15.

[3] Gentilini E, Denamiel G, Betancor A, Rebuelto M, Rodriguez Fermepin M, De Torres RA. Antimicrobial susceptibility of coagulase-negative staphylococci isolated from bovine mastitis in Argentina. J Dairy Sci 2002; 85: 1913-1917.

[4] Peacock SJ, Moore CE, Justice A, Kantzanou M, Story L, Mackie K, O'Neill G, Day NPJ. Virulent combinations of adhesin and toxin genes in natural populations of Staphylococcus aureus. Infect Immun 2002; 70: 4987-4996.

[5] Jensen SO, Lyon BR. Genetics of antimicrobial resistance in Staphylococcus aureus. Future Microbiol 2009; 4: 565-582.

[6] Stotts SN, Nigro OD, Fowler TL, Fujioka RS, Steward GF. Virulence and antibiotic resistance gene combinations among *Staphylococcus aureus* isolates from coastal waters of Oahu, Hawaii. J Young Invest 2005; 12: 1-8.

[7] Yang, FL, Li, XS, Liang XW, Zhang XF, Qin GS, Yang BZ. Detection of virulence-associated genes in *Staphylococcus aureus* isolated from bovine clinical mastitis milk samples in Guangxi. Trop Anim Health Pro 2012; 44: 1821-1826.

[8] Memon J, Yang Y, Kashifa J, Yaqoob M, Buriroa R, Soomroa J, Liping W, Hongjie F. Genotypes,

virulence factors and antimicrobial resistance genes of *Staphylococcus aureus* isolated in bovine subclinical mastitis from Eastern China.Pak Vet J 2013；33：486-491.

[9] CLSI.Performance Standards for Antimicrobial Disk and Dilution Susceptibility Tests for Bacteria Isolated from Animals；Approved Standard，Third Edition.CLSI Document M31-A3.Wayne，PA，USA：Clinical and Laboratory Standards Institute，2008.

[10] Olsen JE，Christensen H，Aarestrup FM.Diversity and evolution of *blaZ* from *Staphylococcus aureus* and coagulase-negative staphylococci.J Antimicrob Chemoth 2006；57：450-460.

[11] Jarraud S，Mougel C，Thioulouse J，Lina G，Meugnier H，Forey F，Nesme X，Etienne J，Vandenesch F.Relationships between *Staphylococcus aureus* genetic background，virulence factors，agr groups (alleles)，and human disease.Infect Immun 2002；70：631-641.

[12] Frey Y，Rodriguez JP，Thomann A，Schwendener S，Perreten V.Genetic characterization of antimicrobial resistance in coagulase-negative staphylococci from bovine mastitis milk.J Dairy Sci 2013；96：2247-2257.

[13] Gao J，Yu FQ，Luo LP，He JZ，Hou RG，Zhang HQ，Li SM，Su JL，Han B.Antibiotic resistance of *Streptococcus agalactiae* from cows with mastitis.Vet J 2012；194：423-424.

[14] Pantosti A，Sanchini A，Monaco M.Mechanisms of antibiotic resistance in *Staphylococcus aureus*.Future Microbiol 2007；2：323-334.

[15] Croes S，Deurenberg RH，Boumans MLL，Beisser PS，Neef C，Stobberingh EE.*Staphylococcus aureus* biofilm formation at the physiologic glucose concentration depends on the S.aureus lineage.BMC Microbiol 2009；9：229.

[16] Reyes-Robles T，Alonzo F 3rd，Kozhaya L，Lacy DB，Unutmaz D，Torres VJ.*Staphylococcus aureus* leukotoxin ED targets the chemokine receptors CXCR1 and CXCR2 to kill leukocytes and promote infection.Cell Host Microbe 2013；14：453-459.

[17] Fueyo JM，Mendoza MC，Rodicio MR，Muñiz J，Alvarez MA，Martín MC.Cytotoxin and pyrogenic toxin superantigen gene profiles of *Staphylococcus aureus* associated with subclinical mastitis in dairy cows and relationships with macrorestriction genomic profiles.J Clin Microbiol 2005；43：1278-1284.

[18] Haveri M，Roslöf A，Rantala L，PyöräläS.Virulence genes of bovine *Staphylococcus aureus* from persistent and nonpersistent intramammary infections with different clinical characteristics.J Appl Microbiol 2007；103：993-1000.

[19] Ote I，Taminiau B，Duprez JN，Dizier I，Mainil JG.Genotypic characterization by polymerase chain reaction of *Staphylococcus aureus* isolates associated with bovine mastitis.Vet Microbiol 2011；153：285-292.

[20] Lowy FD.Is Staphylococcus aureus an intracellular pathogen？Trends Microbiol 2000；8：341-343.

（发表于《Turkish Journal of Veterinary and Animal Sciences》）

Hematologic, Serum Biochemical Parameters, Fatty Acid and Amino Acid of *Longissimus Dorsi* Muscles in Meat Quality of Tibetan Sheep

WANG Hui, HUANG Mei-zhou, LI Sheng-kun , WANG Sheng-yi,
DONG Shu-wei, CUI Dong-an, QI Zhi-ming, LIU Yong-ming

Abstract:

Background: Reference values are very important in clinical management of diseased animal. The mutton quality characteristics not only could reflect the breed of sheep, but also allow for a selection of improved meat quality in the breed analyzed with more attractive sensory attributes. The aim of this study was to establish reference values for commonly used hematologic and biochemical parameters and evaluating the amino acid and fatty acid composition of meat quality traits of Tibetan sheep.

Materials, Methods & Results: A total of 80 Tibetan sheep were randomly selected. Blood samples were collected used for hematological and biochemical analysis. The contents of mineral elements in serum were also measured. The animals were slaughtered according to commercial procedures. The mutton samples were removed from the longissimus dorsi (LD) muscles used for amino acid and fatty acid composition analyzed. We have established a locally relevant reference parameters for commonly used hematological and biochemical tests. Most hematological and serum biochemical values were similar to those of yaks and camels, but the Hemoglobin (HGB) concentration was a little higher. The concentrations of Cu, Zn and Se were remarkably lower than critical values in the present study. LD muscle in Tibetan sheep had a higher proportion of total PUFA. We found a large amount of Glu, Lys, Asp, Leu and Arg in the LD muscle of Tibetan sheep.

Discussion: Hematological and biochemical parameters are useful tools in measuring the physiological status of animals because they may provide information for diagnosis and prognosis of diseases. Tibetan sheep has survived for millions of generations in this low oxygen condition on the plateau. Tibetan sheep must have evolved exceptional mechanisms to adapt to this extremely inhospitable habitat. The study about Tibetan sheep will undoubtedly facilitate the discovery of potential molecular mechanisms of high-altitude adaptation. In the present study of the Tibetan sheep, all the hematological parameters referring to the erythrocyte and leukocyte series and also to the blood platelet counting were within the reference intervals considered as normal for the ovine species. The hematological results for Tibetan sheep were basically within the reference ranges for other ruminants, including yaks, Przewalski's, Tibetan gazelles and camels. Trace element deficiency and unbalanced distribution is a major problem in the QTP, which has caused enormous economic loss to the herdsmen. The concentrations of Cu, Zn and Se were remarkably lower than critical values in the present study. Fatty acid composition plays an important role in meat quality, not only in the flavor quality, but also in the nutritional value of meat. Meanwhile, fatty acid analysis can be also used to evaluate the quality of animal meat. LD muscle in Tibetan sheep had a higher proportion of total PUFA, which were considered as flavor precursors of meat. This provides an evidence for that Chinese indigenous sheep show better mutton flavor intensity. Amino acids are the fundamental units of protein, and amino acid content plays an important role in meat quality by providing nutritive value and flavour characteristics. We found a large amount of Glu, Lys, Asp, Leu and Arg in the LD muscle of Tibetan sheep. The fatty acid and amino acid of LD muscles not only could reflect the character of Tibetan sheep mutton, but also allow for a selection of improved meat quality in the breed analyzed with more attractive sensory attributes.

Key words: Tibetan sheep; Hematologic; Biochemical parameters; Meat quality traits

INTRODUCTION

The Qinghai–Tibetan Plateau (QTP) is a large elevated plateau (mean elevation 4 500m) mainly across the Qinghai Province and Tibet Autonomous Region of China (Fig. 1). The QTP is an important biogeographical area and has recently become a focus for biodiversity studies. Having been living in such a harsh environment for several thousand years, yak has evolved the special mechanism of nutrient digestion and metabolism in response to a lack of nutrition[22,31]; also Tibetan sheep seem to have developed some special physiological features for their survival. Tibetan sheep is one of the dominant livestock species on the QTP, with a population of over 50 million animals providing meat, milk and major source of income for most nomadic and seminomadic peoples in these regions[26].

Reference ranges of physiological parameters can be useful for the evaluation of the state of health in specimens of the species as well as diagnostics and prevention of diseases. Locally relevant reference ranges for commonly used hematologic and biochemical parameters are essential for screening and safety follow up of trial participants as well as for routine clinical management[8]. As an increased emphasis on meat quality among producers and consumers, the interest for meat quality traits and the possibility for implementing these traits into breeding programmes have attracted interest among sheep breeders and sheep breeding companies[17]. Thus, the objective of this study was aimed at establishing reference values for commonly used hematological and biochemical parameters and evaluating the amino acid and fatty acid composition of meat quality traits of Tibetan sheep.

MATERIALS AND METHODS

Study area

Rangelands of the QTP, although sparsely populated and contributing little to China's overall economy, play an important environmental role throughout Asia. They contain high biodiversity values and can also potentially provide China with a source of cultural and geographic variety in the future.

Tibetan sheep is one of the finest local sheep breeds in China. They have formed abundant variation and strong adaptability to environment in morphology, structure, behavior and physiology, etc, and have contained a great number of favorable genes due to the long period of natural selection and artificial selection. Tibetan sheep is one of the mostimportant livestock species found on the QTP, with a population of over 50 million animals; it's also the main source of all human's production and means of subsistence in these regions. The stable development of the Tibetan sheep industry relates our country's economic boom in national minority area and social stability.

Sampling

Two types of blood samples were collected from the jugular vein into vacutainer tubes from 80 (40 male and 40 female) Tibetan sheep (6-10 mo of age, weighing 10.0 to 20.0kg), each of 10mL (Fig. 1); the first blood sample was collected in plain vacutainer tube and used for obtaining serum. The second blood sample was collected in vacutainertubel containing EDTA as anticoagulant and used for hematological analysis. Blood samples in plain tubes were centrifuged at 1008 g for 15 min, after which serum was harvested, stored at -20°C, and were used for measuring serum biochemical constituents.

The animals were slaughtered according to commercial procedures. The mutton samples of Tibetan sheep were removed from the Longissimus dorsi (LD) muscle, placed in zipped plastic bags, and frozen for instrumental quality analysis. Samples for amino acid and fatty acid composition determination were stored in -80°C until analyzed.

Analysis of hematologic and biochemical parameters

White blood cell (WBC), Neutrophils percentage (NEU%), Lymphocytes percentage (LYM%), Monocytes percentage (MON%), Eosinophils percentage (EOS%), Basophils percentage (BAS%), Neutrophils (NEU), Lymphocytes (LYM), Monocytes (MON), Eosinophils (EOS), Basophils (BAS), red blood cell (RBC), Hemoglobin (HGB), hematocrit (HCT), mean corpuscular volume (MCV), mean corpuscular hemoglobin (MCH), mean corpuscular hemoglobin concentration (MCHC), red cell distribution width-coefficient of variation (RDW-CV), red cell distribution width-standard deviation (RDWSD), platelets (PLT), mean platelet volume (MPV), platelet distribution width (PDW) and plateletcrit (PCT) were determined using an automatic hematology analyzer (UniCel DxH800)[2].

The serum content of total protein (TP), albumin (ALB), alanine aminotransferase (ALT), aspartate aminotransferase (AST), alkaline phosphatase (ALP), creatine kinase (CK- NAC), amylase (Amy), Lipase, γ - glutamyl transferase (GGT), blood glucose (GLU), total bilirubin (T. BILI), direct bilirubin (D. BILI), blood urea nitrogen (BUN), creatinine (CRE), cholesterol (CHOL), calcium (Ca) and phosphorus (P) were determined on an automatic analyzer (UniCel DxC800)[2] using commercial test kits[2].

Amino acid analysis

About 100mg of minced muscle samples of meat used for amino acid analysis were hydrolyzed with 10mL of HCl (6mol/L) at 110°C for 22 h in sealed evacuated tubes[1]. Amino acid composition of the muscle powder was analyzed using ion-exchange chromatography with an automatic amino acid analyzer (L-8900)[3].

Fatty acid composition analysis

Fatty acid methyl esters were prepared following the direct method of O' Fallon et al.[16], and identified and quantified using gas chromatography (GC-450)[4], equipped with a flame ionization detector (FID) and a fused silica capillary column (120 m×250μm×0.25 μm) 5. The temperature program was as follows: the carrier gas (helium) flow rate was 0.81mL/min, the column was equilibrated at 45℃ for 3 min, then it was increased to 175°C at 13°C/min where it was maintained for 27 min, further it was increased to 215℃ at 4°C/min and kept for 5 min. Carrier gas was helium (2mL/min), and the split ratio was 30: 1. The identification of fatty acids was accomplished by comparison with external standards. The amount of each fatty acid was calculated as peak area percentage of total fatty acids.

Analysis of trace elements

Copper (Cu), iron (Fe), zinc (Zn), manganese (Mn) were determined by atomic absorption spectrophotometry (AAS ZEEnit 700)[6][26]. The hydride generation atomic fluorescence spectrometry was used for Selenium (Se) determination [14].

Statistical Analysis

The SPSS procedure (Version 19.0) 7 was used for all statistical analyses of data. We performed initial descriptive statistics, including mean and standard deviation (SD).

RESULTS

The minimum and maximum values, arithmetic means and standard deviation of means of hematologic values and serum biochemical parameters are presented in Tables 1 and 2, respectively. The results obtained from the QTP area demonstrated that HGB (90.8-160.1 g/L) and RBC (8.636×10^{12}/L) were a little higher than values set as standards (90.0-145.0 g/L and 8.0×10^{12}/L, respectively) on the clinical hematology machines being used for clinical trials assessments in the study area, but there was no significant difference. Such variations are expected for animals in different geographical/ ecological locations.

To investigate the relationships between blood physiological and biochemical indexes in Tibetan sheep, Pearson correlation analyses wereperformed for the accessions (Table 3). Among the chemical composition and trace elements, closely positive associations were recognized between the contents of HGB and TP, EOS and ALT, BAS and ALT, EOS and ALP ($P<0.01$). The relationship between MON% and ALP was indicated that there was a very significantly negative association ($P<0.01$). The relationships between blood physiological and biochemical indexes will provide reference date for breeding, disease prevention, feeding and management.

The mean concentration of Cu, Zn and Se, observed in the present study were 0.259, 1.915 and 0.04μg/mL, respectively (Table 4). The levels of serum Fe and Mn were in normal range. The concentrationsof Cu, Zn and Se were remarkably lower than critical values in the present study.

Table 5 and Fig. 2 shows results of fatty acid composition of LD muscle from Tibetan sheep. The proportion of polyunsaturated fatty acids (PUFA) (C16: 1, C18: 1, C18: 2, C18: 3) was similar to saturated fatty acids (SFA) (C14: 0, C16: 0, C18: 0); there was no significant difference. LD muscle in Tibetan sheep had a higher proportion of total PUFA.

Table 6 shows the amino acid composition in LD muscle of Tibetan sheep. We found a large amount of Glu, Lys, Asp, Leu and Arg in LD muscle of Tibetan sheep.

Table 1 Normal values for some hematological parameters inTibetan sheep

Blood index	Mean±SD	Range
WBC (10^9/L)	9. 395±4. 418	3. 50–18. 0
NEU%	49. 59±11. 60	26. 3–70. 1
LYM%	46. 61±10. 19	35. 2–71. 0
MON%	2. 115±1. 056	0. 20–4. 50
EOS%	2. 933±1. 757	0. 90–5. 10
BAS%	0. 071±0. 085	0–0. 40
NEU (10^9/L)	4. 758±1. 824	2. 42–6. 92
LYM (10^9/L)	3. 761±1. 908	2. 10–8. 35
MON (10^9/L)	0. 168±0. 118	0. 01–0. 49
EOS (10^9/L)	0. 565±0. 819	0. 02–2. 9
BAS (10^9/L)	0. 016±0. 026	0–0. 10
RBC (10^{12}/L)	8. 636±2. 484	3. 75–12. 6
HGB (g/L)	117. 0±16. 70	90. 8–160. 1
HCT (%)	36. 88±9. 548	16. 6–53. 2
MCV (fL)	43. 39±8. 186	33. 2–70. 4
MCH (pg)	14. 58±4. 638	10. 0–27. 2
MCHC (g/L)	311. 2±38. 72	250. 0–456. 2
RDW–CV (%)	18. 20±3. 098	15. 0–25. 6
RDW–SD (fL)	31. 46±6. 166	21. 6–43. 2
PLT (109/L)	407. 6±285. 1	72. 4–840. 0
MPV (fL)	5. 453±1. 327	4. 59–17. 2
PDW (fl)	15. 25±0. 768	13. 30–17. 20
PCT (%)	0. 230±0. 141	0. 04–0. 50

Note: WBC, White blood cell; NEU%, Neutrophils percentage; LYM%, Lymphocytes percentage; MON%, Monocytes percentage; EOS%, Eosinophils percentage; BAS%, Basophils percentage; NEU, Neutrophils; LYM, Lymphocytes; MON, Monocytes; EOS, Eosinophils; BAS, Basophils; RBC, red blood cell; HGB, Hemoglobin; HCT, hematocrit; MCV, mean corpuscular volume; MCH, mean corpuscular hemoglobin; MCHC, mean corpuscular hemoglobin concentration; RDW–CV, red cell distribution width–coeffi–cient of variation; RDW–SD, red cell distribution width–standard deviation; PLT, platelets; MPV, mean platelet volume; PDW, platelet distribution width; PCT, plateletcrit. SD: Standard deviation.

DISCUSSION

Hematological and biochemical parameters are useful tools in measuring the physiological status of animals because they may provide information for diagnosis and prognosis of diseases[25,15]. This study aimed at establishing hematologic and biochemical reference values to serve as standards for the interpretation of laboratory results during screening and follow–ups in clinical trials and routine healthcare in the Tibetan plateau area. Tibetan plateau has a low partial pressure of oxygen and a high level of ultraviolet radiation[9]. Non-native animals such as humans that visit such high-altitude regions may experience life-threatening acute mountain sickness. In contrast, the Tibetan sheep, which has survived for millions of generations in this low oxygen condition on the plateau. Tibetan sheep must have evolved exceptional mechanisms to adapt to this extremely inhospitable habitat. The study about Tibetan sheep will undoubtedly facilitate the discovery of potential molecular mechanisms of high-altitude adaptation. In the present study of the Tibetan sheep, all the hematological parameters referring to the erythrocyte and leukocyte series and also to the blood platelet counting were within the reference intervals considered as normal for the ovine species[5].

Biochemical tests commonly used during screening/enrollment and safety monitoring of trial participants in animal study area are ALT, AST, Bilirubin (Total and Direct), Urea and Creatinine. Liver damage induces an increase in the serum total bilirubin, and the haemic compound is considered as a sensitive indicator for liver injury. As bilirubin concentrations, high serum activities of some enzymes highly expressed in liver in ruminants such as AST, GGT and LDH are observed in liver injury and highly contribute to evaluate the degree of tissue damage [13]. The values of the general clinical biochemistry parameters that were selected to help in the characterization of the metabolic and nutrition state and renal, hepatic, and thyroid functions, also situated within the reference intervals published for the ovine species [7].

Table 2 Normal values for some biochemical parameters inTibetan sheep serum

Parameter	Mean±SD	Range
TP (g/L)	63. 50±9. 49	42. 0-74. 0
ALB (g/L)	31. 78±7. 24	17. 1-45. 0
ALT (U/L)	31. 19±21. 50	10. 0-85. 2
AST (U/L)	541. 6±728. 9	40. 2-2 160. 0
ALP (U/L)	299. 1±181. 4	100. 1-480. 1
CK-NAC (U/L)	57. 89±46. 51	10. 2-140. 5
Amy (U/L)	119. 0±63. 57	40. 4-200. 4
Lipase (U/L)	150. 1±31. 23	105. 1-195. 0
GGT (U/L)	10. 05±6. 16	1. 20-20. 6
GLU (mmol/L)	3. 353±1. 64	1. 51-7. 00
T. BILI (μmol/L)	6. 433±5. 03	0. 41-16. 2
D. BILI (μmol/L)	1. 657±1. 21	0. 10-3. 00
BUN (mmol/L)	26. 64±22. 10	6. 9-84. 7
CRE (μmol/L)	111. 5±65. 23	31. 2-210. 0
CHOL (mmol/L)	6. 140±1. 369	3. 1-8. 0
Ca (mmol/L)	2. 944±0. 553	1. 9-3. 6
P (mmol/L)	2. 375±1. 174	1. 0-4. 3

Noet: TP, total protein; ALB, albumin; ALT, alanine aminotransferase; AST, aspartate aminotransferase; ALP, alkaline phosphatase; CK-NAC, creatine kinase; Amy, amylase; GGT, γ-glutamyl transferase; GLU, blood glucose; T. BILI, total bilirubin; D. BILI, direct bilirubin; BUN, blood urea nitrogen; CRE, creatinine; CHOL, cholesterol; Ca, calcium; P, phosphorus. SD: Standard deviation.

Table 3 Pearson phenotypic correlations between blood physiological and biochemical indexes in Tibetan sheep

Index	TP	ALB	ALT	AST	ALP	CK-NAC	Amy	Lipase	GGT	GLU	T. BILI	D. BILI	BUN	CRE	CHOL	Ca	P
WBC	0.381	0.140	-0.171	0.331	-0.019	-0.002	-0.006	-0.200	0.207	-0.045	0.040	-0.177	0.000	-0.260	-0.087	-0.177	-0.360
NEU%	0.103	0.222	0.220	-0.036	-0.172	0.289	0.141	0.301	0.166	0.220	0.072	-0.060	0.061	0.029	0.320	-0.042	-0.238
LYM%	-0.171	0.016	0.225	-0.380	-0.134	0.305	0.011	0.248	0.306	-0.081	-0.240	0.060	0.264	0.188	0.201	-0.362	-0.192
MON%	-0.296	-0.090	-0.162	-0.308	-0.550**	0.347	0.107	0.385	0.156	-0.020	-0.123	0.042	-0.204	0.312	0.288	0.082	-0.170
EOS%	-0.037	0.070	0.491*	-0.390	0.032	0.137	-0.017	0.033	-0.232	0.008	-0.185	0.061	-0.023	0.117	0.287	-0.169	0.154
BAS%	0.210	0.173	0.446*	-0.067	0.226	0.102	0.037	-0.028	-0.287	0.033	-0.242	-0.198	-0.016	0.182	0.186	-0.083	0.100
NEU	0.233	0.176	-0.131	0.030	-0.120	0.354	0.122	0.104	0.263	-0.144	-0.180	-0.011	0.213	0.018	0.040	-0.285	-0.318
LYM	0.128	0.061	-0.391	0.223	-0.131	0.180	0.129	0.002	0.501*	0.018	0.106	0.150	0.220	-0.149	0.153	-0.179	-0.057
MON	-0.108	0.000	-0.192	-0.092	-0.241	0.421*	0.193	0.449*	0.355	-0.100	-0.262	0.279	0.045	0.441*	0.259	0.144	-0.313
EOS	0.086	0.131	0.742**	0.239	0.517**	0.127	0.094	-0.097	-0.228	0.129	0.077	0.106	-0.013	-0.104	0.089	0.137	-0.143
BAS	0.086	0.035	0.587**	-0.126	0.338	0.070	-0.031	-0.098	-0.341	-0.004	-0.149	-0.029	0.199	-0.014	0.081	-0.221	0.194
RBC	0.260	0.167	0.439*	0.225	0.511*	0.117	0.231	0.168	0.103	-0.426*	-0.293	-0.183	0.364	-0.167	0.202	-0.275	-0.242
HGB	0.684**	-0.079	0.115	0.175	0.276	-0.034	0.340	0.114	0.157	-0.284	-0.317	0.079	0.165	-0.432*	0.117	-0.083	-0.282
HCT	0.511*	0.006	0.247	0.291	0.289	-0.108	0.233	-0.124	-0.164	-0.389	-0.291	-0.423*	0.054	-0.358	-0.127	-0.199	-0.142
MCV	0.417*	0.328	-0.215	0.320	0.355	0.028	-0.098	0.128	0.062	-0.182	0.242	0.213	0.084	-0.326	0.106	0.128	-0.173
MCH	0.361	0.191	-0.286	0.123	0.078	0.269	0.049	0.271	-0.096	-0.151	0.206	0.342	-0.022	-0.183	0.220	0.126	-0.341
MCHC	0.460*	0.203	0.297	0.125	0.301	0.090	0.224	0.096	0.098	-0.183	-0.300	0.123	0.267	-0.132	0.173	0.137	-0.150
RDW-CV	0.344	0.186	0.118	0.301	0.429*	0.179	0.027	0.163	0.261	-0.368	-0.040	0.144	0.424*	-0.343	0.169	-0.250	-0.320
RDW-SD	0.378	0.155	0.136	0.116	0.061	0.229	0.178	0.305	0.428*	-0.018	-0.084	-0.067	0.269	-0.211	0.294	-0.197	-0.428*
PLT	0.292	0.263	-0.222	0.037	0.140	-0.317	-0.438*	-0.188	0.068	-0.274	-0.019	-0.207	0.144	-0.252	-0.233	-0.411	0.353
MPV	0.159	-0.049	0.319	-0.057	0.507*	-0.139	0.082	-0.035	0.099	-0.304	-0.261	0.133	0.270	-0.208	0.001	-0.112	0.132
PDW	0.312	-0.199	0.112	0.120	0.193	0.252	0.269	0.184	-0.369	-0.120	0.064	0.346	-0.060	-0.219	0.154	0.135	-0.356
PCT	0.325	0.364	-0.514*	0.115	0.119	-0.343	-0.546*	-0.462	-0.089	-0.022	0.296	-0.168	0.258	-0.325	-0.546*	-0.484	0.516*

Note: ** Significant at 0.01 probability level. * Significant at 0.05 probability level.

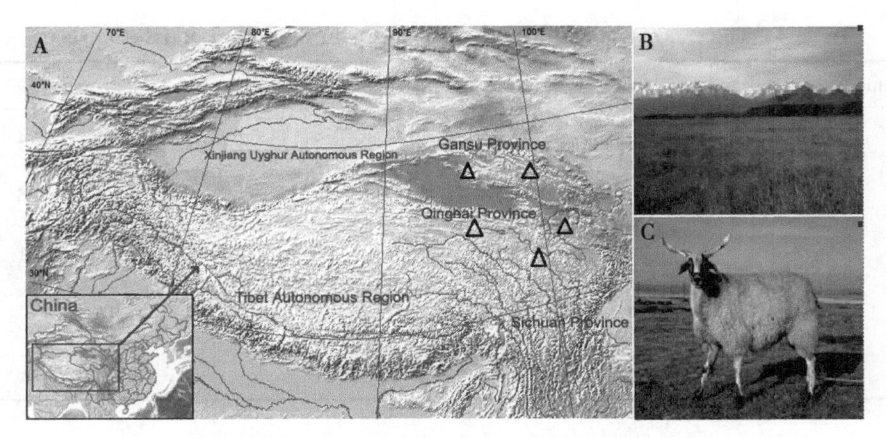

Fig. 1　The location of the sampling sites (A), blank triangle, The rangelands of Qinghai–Tibetan plateau (B), The Tibetan sheep (C)

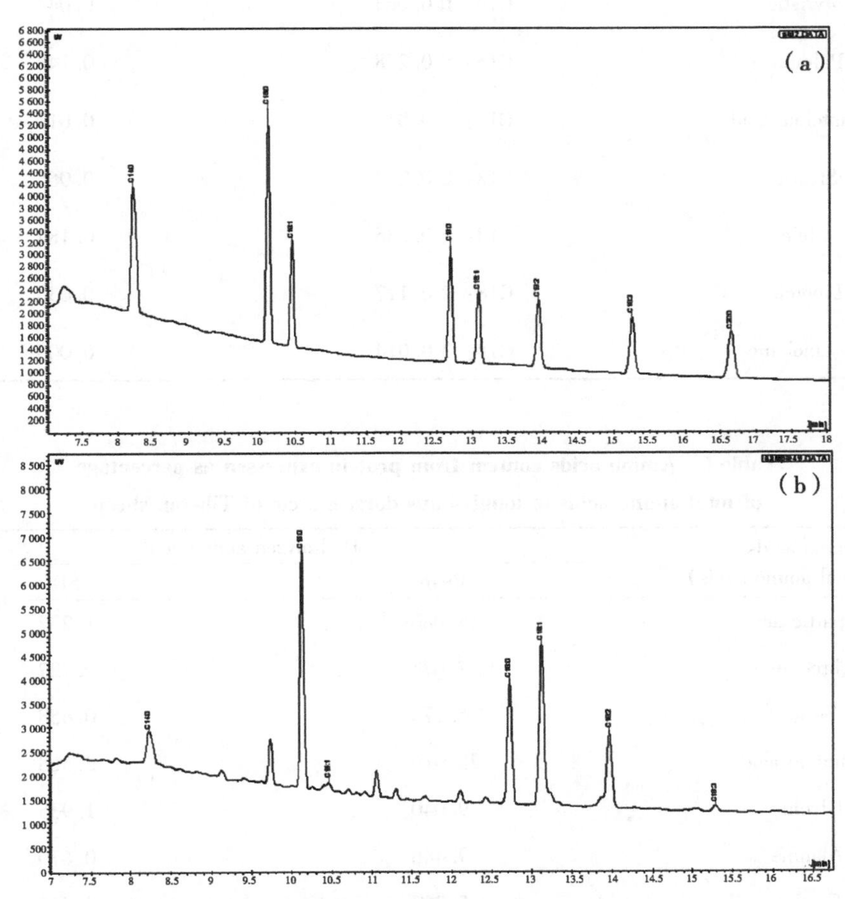

Fig. 2　Gas chromatograms of fatty acid for standard (a) and Longissimus dorsi muscle of Tibetan sheep (b)

Table 4　The concentrations of mineral elements in the serum of Tibetan sheep

Element (μg/mL)	Mean ± SD	Range
Se	0.040±0.013	0.018-0.064
Cu	0.259±0.207	0.038-0.555
Mn	0.193±0.066	0.100-0.301
Fe	9.678±4.794	4.10-20.1
Zn	1.915±0.285	1.41-2.47

Table 5　Fatty acid composition in LD muscle of Tibetan sheep

Fatty acid (%)	Mean	SD
Myristic	C14: 0 0.064	0.044
Palmitic	C16: 0 0.278	0.148
palmitoleic acid	C16: 1 0.018	0.011
Stearic	C18: 0 0.212	0.065
Oleic	C18: 1 0.386	0.187
Linoleic	C18: 2 0.117	0.036
α-Linolenic	C18: 3 0.014	0.005

Table 6　Amino acids content from protein expressed as percentage of total amino acids in longissimus dorsi muscle of Tibetan sheep

Amino acids (% of total amino acids)	Hydrolyzed amino acids	
	Mean	SD
Aspartic acid	15.086	1.232
Threonine	7.078	0.722
Serine	6.173	0.458
Glutamin acid	28.016	2.568
Glycine	9.040	1.935
Alanine	9.486	0.670
Cystine	5.067	0.463
Valine	8.186	0.494
Methionine	1.496	1.583

(continued)

Amino acids (% of total amino acids)	Hydrolyzed amino acids	
	Mean	SD
Isoleucine	7.083	0.795
Leucine	12.930	1.170
Tyrosine	6.377	0.368
Phenylalanine	7.421	0.726
Lysine	16.243	2.435
Histidine	4.122	0.608
Arginine	10.499	0.734
Proline	6.691	1.010

As far as the characterization of the energetic state of the ruminants is concerned, the glucose and the total cholesterol were evaluated. The level of glycaemia could contribute to the assessment of energy status diagnosis in association with insulin which has a close relationship with the nutritional state of the animal and with respect to the free fatty acids, high levels generally indicate a negative energy balance, conversely low levels indicate a positive balance [2].

The protein state and the kidney function can be estimated in ewes by the simultaneous determination of the total protein and urea, associated with HCT and HGB, and to the serum creatinine in the kidney function. Creatinine is mainly filtered by the kidney and there is little-to-no tubular reabsorption of creatinine [7].

The hematological results for Tibetan sheep were basically within the reference ranges for other ruminants, including yaks [18], Przewalski's [19], Tibetan gazelles[19] and camels [11]. Some physical adaptations have induced specific changes in HGB with increasing altitude. The HGB concentration was a little higher than the value for yaks (108 ± 6.51) g/L[18] and camel (109 ± 24) g/L[11], so Tibetan sheep have an improved capability for accommodating the low-oxygen circumstances without having an increased concentration of hemoglobin. However, most of the non-Tibetans in the region were immigrants whom from low-lying provinces and their physiological adaptation for environment of high altitude might be different from the Tibetan, which could contribute tothe ethnic differences in HGB level [28].

Trace elements are integral components of certain enzymes and of other biologically important compounds that have major physiological and bio-chemical roles, and micronutrient malnutrition is a major healthproblem in China[29]. Trace element requirements vary with age and production level-young, pregnant and lactating animals have the greatest need. QTP is also China's important base of animal husbandry. Every year a great number of sheep died in the area, which is attributed to malnourishment and mineral deficiency [23]. Trace element deficiency and unbalanced distribution is a major problem in the region[18,27], which has caused enormous economic loss to the herdsmen.

Deficiencies are more accurately diagnosed from blood or tissue tests. Copper and zinc are the most important essential minerals necessary for the normal functioning of animals' reproduction functions. Investigators[10,32] reported that the levels of serum Cu in sheep should be between 0. 8 and 1. 2 μg/mL, Zn should be between 6. 9 and 14. 86 μg/mL, and Se should be between 0. 06 and 0. 20 μg/mL. In ruminants, average blood Cu values of<0. 5 μg/mL are a sign of severe Cu deficiency[33]. The concentrations of Cu, Zn and Se were remarkably lower than critical values in the present study. The results indicated that most of the sheep were deficient in Cu, Zn and Se. Tibetan sheep have some physiological peculiarities in trace element metabolism maybe due to geographic (altitude, latitude, climate) and dietary factors.

Fatty acid composition plays an important role in meat quality, not only in the flavor quality, but also in the nutritional value of meat[30]. Meanwhile, fatty acid analysis can be also used to evaluate the quality of animal meat[20]. Fatty acids, especially polyunsaturated fatty acids (PUFA), are major flavor precursors of meat. C18: 2 was one of polyunsaturated fatty acids, which were a positive correlation with flavor of meat [3]. Furthermore, the amount and composition of fatty acids in the food are also associated with human health. The percentage and total amount of saturated fatty acids (SFA) have been identified as dietary risk factors[16], which may related to various cancers and especially coronary heart disease. PUFA are required for the normal composition of sperm, retina and brain lipids and for the optimal maturation of visual and cortical function in preterm infants[21]. Therefore, increasing PUFA content and reducing SFA content of meat has become a tendency of meat quality improvement.

LD muscle in Tibetan sheep had a higher proportion of total PUFA, which were considered as flavor precursors of meat. This provides an evidence for that Chinese indigenous sheep show better mutton flavor intensity. On the other hand, meat of Tibetan sheep is good for human health because these PUFA are essential to the human body, involved in many biological functions.

Protein is important resources for human with essential amino acids as one of the most important nutritional qualities [4]. Amino acids are the fundamental units of protein, and amino acid content plays an important role in meat quality by providing nutritive value and flavour characteristics[1]. We found a large amount of Glu, Lys, Asp, Leu and Arg in the LD muscle of Tibetan sheep. Analyses of the amino acid compositions of some red meats showed that glutamic acid was present at the highest concentration in camel meat[6] and buffalo meat[34], which were similar to our study. The results from those studies found that the glutamic acid contents of camel and buffalo were 16. 35-17. 25 (g/16 g N) and 12. 32-12. 69 (g/100g protein), respectively.

In conclusion, the Tibetan sheep in the study were selected and management strictly. The results of hematologic and serum biochemical parameters of Tibetan sheep have important reference value for researchers, veterinarians and herdsman. The fatty acid and amino acid of *Longissimus Dorsi* muscles of Tibetan sheep not only could reflect the character of Tibetan sheep mutton, but also allow for a selection of improved meat quality in the breed analyzed with more attractive sensory attributes.

MANUFACTURERS

[1]Becton Dickinson vacutainer. Franklin Lakes, NJ, USA.

[2]Beckman Coulter. Brea, CA, USA.

[3]Hitachi High-Technologies Corporation. Tokyo, Japan.

[4]Varian. Palo Alto. CA, USA.

[5]SGE-FFAP. Melbourne, Australia.

[6]Analytik Jena. Jena, Germany.

[7]SPSS Inc. Chicago, IL, USA.

Funding. Financial assistance for this research was provided by Central Public-interest Scientific Institution Basal Research Fund (NO. 1610322013003), National Key Technology Research and Development Program of the Ministry of Science and Technology of China (NO. 2012BAD12B03), and Special Fund for Agroscientific Research in the Public Interest (NO. 201303040-17).

Ethical approval. All animal protocols have been reviewed and approved by the Institutional Animal Care and Use Committee of Lanzhou Institute of Husbandry and Pharmaceutics Sciences of Chinese Academy of Agricultural Sciences (Animal use permit: SCXK20008-0003).

Declaration of interest. The authors report no conflicts of interest. The authors alone are responsible for the content and writing of the paper.

References

[1] Cai Z. W., Zhao X. F., Jiang X. L., Yao Y. C., Zhao C. J., Xu N. Y. & Wu C. X. 2010. Comparison of muscle amino acid and fatty acid composition of castrated and uncastrated male pigs at different slaughter ages. *Italian Journal of Animal Science*. 9 (2): e33.

[2] Caldeira R. M., Belo A. T., Santos C. C., Vazques M. I. & Portugal A. V. 2007. The effect of long-term feed restriction and over-nutrition on body condition score, blood metabolites and hormonal profiles in ewes. *Small Ruminant Research*. 68 (3): 242-255.

[3] Cameron N. D., Enser M., Nute G. R., Whittington F. M., Penman J. C., Fisken A. C., Perry A. M. & Wood J. D. 2000. Genotype with nutrition interaction on fatty acid composition of intramuscular fat and the relationship with flavour ofpig meat. Meat Science. 55 (2): 187-195.

[4] Chen D. W. & Zhang M. 2007. Non-volatile taste active compounds in the meat of Chinese mitten crab (Eriocheir sinensis). Food Chemistry. 104 (3): 1200-1205.

[5] Chen J. 2004. Domestic Animals Physiology. Nanjing Agricultural University. 4th edn. Beijing: China Agricultural Press, 98-105.

[6] Dawood A. & Alhankal M. A. 1995. Nutrient composition of Najdi camel meat. Meat Science. 39 (1): 71-78.

[7] Dias I. R., Viegas C. A., Silva A. M., Pereira H. F., Sousa C. P., Carvalho P. P., Cabrita A. S., Fontes P. J., Silva S. R. & Azevedo J. M. T. 2010. Haematological and biochemical parameters in Churra-da-Terra-Quente ewes from the northeast of Portugal. Arquivo Brasileiro de Medicina Veterinária e Zootecnia. 62 (2): 265-272.

[8] Dosoo D. K., Kayan K., Adu-Gyasi D., Kwara E., Ocran J., Osei-Kwakye K., Mahama E., Amenga-Etego S., Bilson P., Asante K. P., Koram K. A. & Owusu-Agyei S. 2012. Haematological and biochemical Rreference values for healthy adults in the middle belt of Ghana. PLoS One. 7 (4): e36308.

[9] Ge R. L., Cai Q., Shen Y. Y., San A., Ma L., Zhang Y., Yi X., Chen Y., Yang L., Huang Y., He R., Hui Y., Hao M., Li Y., Wang B., Ou X., Xu J., Zhang Y., Wu K., Geng C., Zhou W., Zhou T., Irwin D. M., Yang Y., Ying L., Bao H., Kim J., Larkin D. M., Ma J.,

Lewin H. A. , Xing J. , Platt R. N. , Ray D. A. , Auvil L. , Capitanu B. , Zhang X. , Zhang G. , Murphy R. W. , Wang J. , Zhang Y. P. & Wang J. 2013. Draft genome sequence of the Tibetan antelope. Nature Communications. 4: 1858.

[10] Kargin F. , Seyrek K. , Bulduk A. & Aypak S. 2004. Determination of the levels of zinc, copper, calcium, phosphorus and magnesium of Chios ewes in the Aydin region. Turkish Journal of Veterinary and Animal Sciences. 28 (3): 609-612.

[11] Liu Z. P. 2003. Studies on the hematological and trace element status of adult bactrian camals in China. Veterinary research communications. 27 (5): 397-405.

[12] Lu C. Y. , Yan X. P. , Zhang Z. P. , Wang Z. P. & Liu L. W. 2004. Flow injection on-line sorption preconcentration coupled with hydride generation atomic fluorescence spectrometry using a polytetrafluoroethylene fiber-packed microcolumn for determination of Se (IV) in natural water. Journal of Analytical Atomic Spectrometry. 19 (2): 277-281.

[13] LubojacákV. , PechováA. , Dvoák R. , Drastich P. , Kummer V. & Poul J. 2005. Liver steatosis following supplementation with fat in dairy cow diets. Acta Veterinaria Brno. 74 (2): 217-224.

[14] O'allon J. V. , Busboom J. R. , Nelson M. L. & Gaskins C. T. 2007. A direct method for fatty acid methyl ester synthesis: application to wet meat tissues, oils, and feedstuffs. Journal of Animal Science. 85 (6): 1511-1521.

[15] Oliveira-Junior A. A. , Tavares-Dias M. & Marcon J. L. 2009. Biochemical and hematological reference ranges for Amazon freshwater turtle, Podocnemis expansa (Reptilia: Pelomedusidae), with morphologic assessment of blood cells. Research in Veterinary Science. 86 (1): 146-151.

[16] Pascual J. V., Rafecas M., Canela M. A., Boatella J., Bou R., Barroeta A. C. & Codony R. 2007. Effect of increasing amounts of a linoleic-rich dietary fat on the fat composition of four pig breeds. Part II: Fatty acid composition in muscle and fat tissues. Food Chemistry. 100 (4): 1639-1648.

[17] Pethick D. W. , Ball A. J. , Banks R. G. & Hocquette J. F. 2010. Current and future issues facing red-meat quality in a competitive market and how to manage continuous improvement. Animal Production Science. 51 (1): 13-18.

[18] Shen X. Y. , Du G. Z. & Li H. 2006. Studies of a naturally occurring molybdenum induced copper deficiency in the yak. Veterinary Journal. 171 (2): 352-357.

[19] Shen X. & Jiang Z. 2012. Serum biochemical values and mineral contents of tissues in Przewalski and Tibetan gazelles. African Journal of Biotechnology. 11 (3): 718-723.

[20] Skonberg D. I. & Perkins B. L. 2002. Nutrient composition of green crab (Carcinus maenus) leg meat and claw meat. Food Chemistry. 77 (4): 401-404.

[21] Tapiero H. , Ba G. N. , Couvreur P. & Tew K. D. 2002. Polyunsaturated fatty acids (PUFA) and eicosanoids in human health and pathologies. Biomedicine and Pharmacotherapy. 56 (5): 215-222.

[22] Wang H. , Long R. , Zhou W. , Li X. , Zhou J. & Guo X. 2009. A comparative study on urinary purine derivative excretion of yak (Bos grunniens), cattle (Bos taurus), and crossbred (Bos taurus x Bos grunniens) in the Qinghai-Tibetan plateau, China. Journal of Animal Science. 87 (7): 2355-2362.

[23] Wang H. , Liu Y. , Qi Z. , Wang S. , Liu S. , Li X. , Wang H. , Wang X. , Xia X. & Zhu X. 2014. The estimation of soil trace elements distribution and soil-plant-animal continuum in relation to trace elements status of sheep in Huangcheng area of Qilian mountain grassland, China. Journal of Integrative Agriculture. 13 (1): 140-147.

[24] Wang X. , Wang H. , Li J. , Yang Z. , Zhang J. , Qin Z. , Wang L. & Kong X. 2014. Evaluation of

bioaccumulation and toxic effects of copper on hepatocellular structure in mice. Biological trace element research. 159 (1-3): 312-319.

[25] Whiting S. D., Guinea M. L. & Limpus C. J. 2007. Blood chemistry reference values for two ecologically distinct populations of foraging green turtles, eastern Indian Ocean. Comparative Clinical Pathology. 16 (2): 109-118.

[26] Wu S. M., Danba C., Huang S. Y., Zhang D. L., Chen J., Gong G., Xu M. J., Yuan Z. G. & Zhu X. Q. 2011. Seroprevalence of Toxoplasma gondii infection in Tibetan sheep in Tibet, China. Journal of Parasitology. 97 (6): 1188-1189.

[27] Xin G. S., Long R. J., Guo X. S., Irvine J., Ding L. M., Ding L. L. & Shang Z. H. 2011. Blood mineral status of grazing Tibetan sheep in the Northeast of the Qinghai-Tibetan Plateau. Livestock Science. 136 (2): 102-107.

[28] Xing Y., Yan H., Dang S., Zhuoma B., Zhou X. & Wang D. 2009. Hemoglobin levels and anemia evaluation during pregnancy in the highlands of Tibet: a hospital-based study. BMC Public Health. 9 (1): 336.

[29] Yang X., Chen W. & Feng Y. 2007. Improving human micronutrient nutrition through biofortification in the soil-plant system: China as a case study. Environmental Geochemistry and Health. 29 (5): 413-428.

[30] Yu K., Shu G., Yuan F., Zhu X., Gao P., Wang S., Wang L., Xi Q., Zhang S., Zhang Y., Li Y., Wu T., Yuan L. & Jiang Q. 2013. Fatty acid and transcriptome profiling of longissimus dorsi muscles between pig breeds differing in meat quality. International Journal of Biological Sciences. 9 (1): 108-118.

[31] Zhang Y., Guo X., Long R., Zhou J., Zhu Y. & Mi J. 2011. Effects of dietary nitrogen level on ruminal fermentation, digestibility and metabolism of nutrients in Yaks. Chinese Journal of Animal Nutrition. 23: 956-964.

[32] Zhong J.F., Zhang L., Zhou X.H., Chao S.Y., Zhang B.Y., Wei B.D.& Xiao X.S.2007.Sufficiency or lack of minerals in the soil, forage and animal ecosystem in Hainan area in Qinghai provence.Acta Ecologiae Animalis Domastici.28: 44-48.

[33] Zhou L. Y., Long R. J., Pu X. Y., Qi J. & Zhang W. W. 2009. Studies of a naturally occurring sulfur-induced copper deficiency in Przewalski's gazelles. Canadian Veterinary Journal. 50 (12): 1269-1272.

[34] Ziauddin K. S., Mahendrakar N. S., Rao D. N., Ramesh B. S. & Amla B. L. 1994. Observations on some chemical and physical characteristics of buffalo meat. Meat Science. 37 (1): 103-113.

（发表于《Acta Scientiae Veterinariae》）

Optimization Extracting Technology of *Cynomorium songaricum* Rupr. Saponins byUltrasonic and Determination of Saponins Content in Samples with Different Source

WANG Xiao-li[1], WEI Qing-wei[2], ZHU Xin-qiang[1],
WANG Chun-mei[2], WANG Yong-gang[2], LIN Peng[1], YANG Lin[1]
(1. Lanzhou Institute of Animal Husbandry and Veterinary Medicine, China Academy of Agricultural Sciences, Lanzhou 730050, China; 2. School of Life Science and Engineering, Lanzhou University of Technology, Lanzhou 730050, China)

Abstract: Extraction process was optimized by single factor and orthogonal experiment (L_9 (3^4)). Moreover, thecontent determination was studied in methodology. The optimum ultrasonic extraction conditions were: ethanol concentration of 75%, ultrasonic power of 420 w, the solid–liquid ratio of 1 : 15, extraction duration of 45 min, extraction temperature of 90°C and extraction for 2 times. Saponins content in Guazhou samples was significantly higher than those in Xinjiang and Inner Mongolia. Meanwhile, Guazhou samples harvested in April and May were higher than that in Guazhou *Cynomorium songaricum* numbered "the three – nine" . In addition, saponin content in various samples grown in separate years is different. This ultrasonic extraction process can significantly improve the saponins extraction efficiency. Determination by this method is fast, easy – to – operate. The result of this method is reliable. The conclusion of study could provide scientific basis for rational development and quality control of *Cynomorium songaricum Rupr.*

Key words: Content determination; *Cynomorium songaricum Rupr*; Process optimization; saponins; ultrasonic extraction

INTRODUCTION

In recent years, *Cynomorium songaricum* Rupr. (cs) has been playing core role in health – care and has become a new material for developing and producing health foods and medicines (Zhao et al. , 2010). Up tonow, there have been a few study of *Cynomorium songaricum Rupr.* saponins (*cs* saponins) ; in addition, literature concerning extraction as well as determination content in *cs* saponins has not been reported (Zhao et al. , 2010; Wang et al. , 2014; Yoo et al. , 2014; Capote and de Castro, 2007). Therefore, in this study, ultrasonic extraction was applied for extraction of *cs* saponins and the optimal conditions of extractionprocess were investigated by orthogonal experiment. The *cs* saponins content in different samples from various origins also were determined in this study to provide basis for establishment extraction methods, content determination together with quality standard of *cs* saponins (Nickrent et al., 2005).

MATERIALS AND METHODS

Reagents

Ursolic acid, Vanillic aldehyde, glacial acetic acid, perchloric acid and other chemicals were analytical grade reagents and the purity of Ursolic acid was 98%.

Sample

Cs provided by GuaZhou YiDe biotechnology. Ltd. Samples dried to constant weight at 50°C, crushing and through 100 mesh sieve.

Apparatus

A KQ-600DE ultrasonic equipment from Kun Shan Ultrasonic Instrumants co., Ltd, a AB104-N electronic scales from Mettle-Toledo, a CARIN Austrilia RTY, Ltd and a Exceed-Da-20 Aike Barnstead from Chengdu Kangning Experimental water Factory.

The preparation of the standard solution

Ursolic acid Standard solution (0.10192mg/mL) was prepared by dissolving 2.6mg which dried to constant weight in methanol and made to volume in a 25mL volumetric flask.

The preparation of the test sample solution

Cs powder was sophisticated weighted 1.0000g, reflux extracted with ethyl alcohol and obtained *cs* extracting solution, then ethanol was recycled by reduced pressure and distilled with 40mL hot water and extracted with 40mL pertroleum ether 2 times. The above solution extracted with 40mL water saturation n-butyl alcohol, then combined the n-butyl alcohol liquid and evaporated to dryness. The residue dissolved in methanol and transferred to 50mL volumetric flask with methanol constant volume and shook well (Jin et al., 2012). Further precision took 1mL to10mL volumetric flask with methanol constant volume and shook well again as the test sample solution. Drawn the standard curve with absorbance value as the ordinate and ursolic acid concentration as the abscissa. Obtained regression curve equation: $Y = 52.5318X - 0.0872$ ($r = 0.9990$) at the determination wavelength at 548 nm, showed that ursolic acid solution was on 8.5-23.5 μg/mL (Yoo et al., 2014; Capote and de Castro, 2007).

RESULTS AND DISCUSSION

In this study, the single factor experimental was employed to guide the preliminary range of the variables including the ethanol concentration, ultrasonic power, the solid-liquid ratio and extraction time. In detail, the ethanol concentration was set respectively for 50%, 60%, 70%, 80%and 90%, the ultrasonic power was set respectively for 240, 360, 480 and 600 w, the solidliquid ratio was set respectively for 1 : 6, 1 : 10, 1 : 20, 1 : 30, 1 : 40, the temperature was set respectively for 50, 60, 70, 80, 90 and 95°C, extraction time was set respectively for 30, 60, 90 and 120 min and extraction times was set respectively for 1, 2, 3, 4 times, the nexamined the influence of their to *cs* saponins extraction rate.

Orthogonal experiment and the results

According to the values obtained in the single factor experiment, orthogonal experiment was applied to optimum the extraction conditions of saponins from cs. Choose the ethanol concentration (A), ultrasonic power (B), the solid-liquid ratio (C) and extraction time (D) as examine factors and designed 4 factors 3 levels orthogonal experiment. Factors and levels were shown in Table 1. The results of orthogonal experiment were shown in Table 2. The results of variance analysis were shown in Table 3.

Table 1 Factors and levels

Levels	Factors			
	A (The ethanol concentration)	B (Ultrasonic power)	C (The solid-liquid ratio)	D (Extraction time)
1	65%	300w	1 : 15	45 min
2	70%	360w	1 : 20	60 min
3	75%	420w	1 : 25	75 min

Table 2 Results of orthogonal experiment

Number	A	B	C	D	Extract ratio (%) (n=3)
1	1	1	1	1	6. 18
2	1	2	2	2	6. 09
3	1	3	3	3	6. 03
4	2	1	2	3	5. 87
5	2	2	3	1	5. 78
6	2	3	1	2	6. 24
7	3	1	3	2	5. 81
8	3	2	1	3	6. 28
9	3	3	2	1	6. 29
k_1	6. 099	5. 954	6. 236	6. 082	
k_2	5. 966	6. 052	6. 084	6. 049	
k_3	6. 130	6. 188	5. 874	6. 063	
R	0. 164	0. 234	0. 362	0. 033	

Table 3 The results of variance analysis

Soruce of variation (SV)	Sum of squares (SS)	Free (f)	Squares (S)	F value	Significant P
The ethanol concentration	0. 137	2	0. 069	11. 370	0. 001
Ultrasonic power	0. 247	2	0. 124	20. 456	2.321×10^{-5}
The solid-liquid ratio	0. 592	2	0. 296	49. 001	5.215×10^{-8}
Extraction time	0. 005	2	0. 003	0. 416	0. 666
Error	0. 109	18	0. 006		

Table 4 Different *cs* saponins samples content from different sources

Sample number	Sample origin	Acquisition time	Saponins content (n=3)
The three-nine	Guazhou	2011. 01	5. 70%
001	Guazhou	2009. 05	3. 69%
002	Guazhou	2009. 05	4. 93%
003	Guazhou	2011. 05	6. 25%
004	Guazhou	2011. 05	6. 69%
005	Guazhou	2011. 04	6. 48%
006	Guazhou	2010. 04	6. 40%
Xinjiang	Jimusaer	2012. 03	4. 47%
Inner Monglia	Ejinaqi	2012. 03	5. 15%

Orthogonal experiment data (Table 3 and 4) were shown various factors affecting cs saponins extraction rate sequence was: the solid-liquid ratio > ultrasonic power > ethanol concentration > extraction time. Got the best conditions was A3B3C1D1, namely the ethanol concentration of 75%, ultrasonic power 420 w and the solid-liquid ratio 1 : 15 and extraction time of 45 min.

Verified experiment

According to the best extraction conditions what were determined by orthogonal experiment, the ethanol concentration of 75%, ultrasonic power of 420 w, the solid-liquid ratio of 1 : 15, extraction time of 45 min and extraction temperature of 90°C, extracting 2 times, cs saponinshave been measured the average extraction rate was 6. 40% (n=5, RSD=1. 13%), The result shown that a higher content than orthogonal experimental results and the extraction technology can effectively improve the saponins extraction rate and good repeatability.

Determination of saponins content in the samples (Yoo et al., 2014)

Cs powder was sophisticated weighted 1. 0000 g in the 100mL round bottom flask, obtained *cs* saponins extract solution which used extraction conditions by orthogonal experiment, then ethanol was recycled by reduced pressure and residue was dissolved with 40mL distilled hot water and extracted 3 times with 40mL pertroleum ether. Different *cs* saponins samples content from different sources shown as Table 4.

The experimental results showed that Guazhou samples saponins content were significantly higher than Xinjiang and Inner Mongolia and Guazhou samples which harvested in April and May were higher than Guazhou three-nine *Cynomorium songaricum*. In addition, the different years samples saponin content is different (Luan and Li, 2010).

CONCLUSION

Ultrasonic extraction have time saving, high efficiency, energy saving etc., sometimes even than modern extraction method, such as supercritical fluid extraction, microwave extraction, has been widely used extraction of effective ingredients in plant and can effectively increase the extraction rate. This study results show that the ultrasound power, the solid-liquid ratio, the alcohol concentration has a significant effect on *cs* saponins extraction rate, screened extraction process can obviously increase *cs* saponins extraction rate.

Ursolic acid as standard, this study adopts the vanillin-perchlorate chromogenic method measured cs saponins contents, precision, repeatability, stability and recovery experiment results show that this method is stable and reliable, repeatability well, can be used to cs saponins content determination (Zhao et al. , 2010; Wang et al. , 2014).

Different sources cs saponins results show that the content of saponins with Guazhou cs saponins content was significantly higher in Inner Mongolia, Xinjiang samples. *Guazhou cs saponins* is *number one*, *cs* saponins content and medicinal materials quality have a certain degree of relation, but need combination cs saponins active research to clarify. Guazhou samples which harvested in April and May were higher than Guazhou three-nine *Cynomorium songaricum.*, suggested that cs Saponins content has a tendency to increased gradually at the early stage of the unearthed, unearthed period and exuberant plant growth period. In addition, the different years samples saponin content is different. This may be related to that the growing environment of soil nutrients, water and climate factors (Zhao et al. , 2010; Wang et al. , 2014; Yoo et al. , 2014; Capote and de Castro, 2007). The research results will provide the basis for the establishment of cs saponins extraction, content determination and quality standard.

ACKNOWLEDGMENT

This research was supported by Agricultural Scientific research projects. Forage feed resources development and utilization of technology research and demonstration (201203042).

References

[1] Capote, F. P. and M. D. L. de Castro, 2007. Ultrasound inanalytical chemistry. Anal. Bioanal. Chem. , 387: 249-257.

[2] Jin, S., Eerdunbayaer, A.Doi, T.Kuroda, G.Zhang, T.Hatano and G.Chen, 2012.Polyphenolicconstituents of cynomorium songaricum rupr.and antibacterial effect of polymeric proanthocyanidin on methicillin-resistant staphylococcus aureus.J.Agr.Food Chem., 60: 7297-7305.

[3] Luan, N. and D. Li, 2010. Study on supercritical CO_2 extraction of flavonoids from Cynomorium songaricum. Zhong Yao Cai, 33: 1167-1171.

[4] Nickrent, D. L. , J. P. Der and F. E. Anderson, 2005. Discovery of the photosynthetic relatives of the "Maltese mushroom" Cynomorium. BMC Evol. Biol. , 5: 1-11.

[5] Wang, X. L. , Y. N. Wang, X. X. Yang and M. Xu, 2014. Optimization of ultrasonic extraction process oftotal flavonoids from mulberry leaves by orthogonal experiment. Agric. Biotechnol. , 3 (4): 12-13.

[6] Yoo, D.Y., J.H.Choi, W.Kim, H.Y.Jung, S.M.Nam, J.W.Kim, Y.S.Yoon, K.Y.Yoo, M.H.Won and I.K.Hwang, 2014.Cynomorium songaricum extract enhances novel object recognition, cell proliferation and neuroblast differentiation in the mice via improving hippocampal environment.BMC Complem. Altern.M., 14 (5): 2-8.

[7] Zhao, G. , J. Wang, G. W. Qin and L. H. Guo, 2010. Cynomorium songaricum extracts functionally modulate transporters of γ-aminobutyric acid and monoamine. Neurochem. Res. , 35: 666-676.

（发表于《Advance Journal of Food Science and Technology》）

Optimization of Ultrasound-Assisted Extraction of Tannin from *Cynomorium songaricum*

WANG Xiao-li [1], JING Yan-jun[2], ZHU Xin-qiang[1],
WANG Chun-mei[2], WANG Yong-gang[2], LIN Peng[1], YANG Lin[1]

(1. Lanzhou Institute of Animal Husbandry and Veterinary Medicine, China Academy of AgriculturalSciences, Lanzhou 730050, China;
2. School of Life Science and Engineering, Lanzhou University of Technology, Lanzhou 730050, China)

Abstract: To optimize ultrasound-assisted extraction technology of tannin from the succulent stem of *Cynomoriumsongaricum*. The content of tannin in *C. songaricum* samples of different source was comparatively studied. Extraction rate of tannin as indexes, extraction process was optimized by single factor experiment and L9 (3^4) orthogonal experiment. The best extraction technology for ultrasonic extraction were 75% as the extraction solventethanol concentration, 600 w as the ultrasonic power, 1 : 65 as the solid-liquid ratio, 105 min as the extraction time and extraction 2 times. According to the results of content determination, *C. songaricum* samples from Guazhou hasthe highest tannin content than samples from Xinjiang and Inner Mongolia and in the samples of Guazhou which were harvested in April and May has more content than three-nine *C. songaricum* (harvested in the third nine-day period after the winter solstice). This study could provide scientific basis for rational development and qualitycontrol of *C. songaricum*.

Key words: Content determination; *Cynomorium songaricum*; Process optimization; Tannin; Ultrasonic extraction

INTRODUCTION

Cynomorium songaricum, referred to as Elixir of Life, is dried fleshy stems from *Cynomorium songaricum* Rupr. *Cynomorium songaricum* Rupr is succulent perennial herbs as well as achlorophyllous holoparasite belonging to *Cynomoriaceae* in conjunction with Cynomorium Songaria genera. *C. songaricum* always parasitize root of plants which is Zygophyllaceae together with Natraria L (Yoo et al., 2014; Zhao et al., 2010; Yu et al., 2010). Reports show that *C. Songaria* tannin is one of effective constituents among its compositions, steroids, triterpenes and so on. Meanwhile, in addition to effect of hemostatic convergence, it plays important role in anti-oxidation, reducing hematic fat, falling blood pressure as well as anti-tumor. Currently, there are few study about *C. songaria* tannin and no report on ultrasound-assisted extraction technology of tannin from the succulent stem of *C. songaricum*. As a consequence, our study uses ultrasound-mediated method (Capote and de Castro, 2007) to extract C. Songaria tannin and optimize conditions by single-factor and orthogonal test (Wang et al., 2014). At the same time, we measure tannin content in *C. songaricum* sample from different sources. The recommended study established a precise and effective extraction method of tannin's extraction, ulteriorly, provided references for extraction of *C. songaria* tannin as well as formulating Quality standards. Moreover, it can be supplement to scientific foundations for exploitation and utilization of *C. songaricum*.

MATERIALS AND METHODS

Apparatus and reagents

Numeric control ultrasonic cleaning machine for this study was purchase in Kunshan ultrasonic instrument Co., LTD (KQ-600DE). All the reagents were weighed by electronic balance (AB104-N, Mettle-Toledo). Eco pure water meter was provided by Chengdu corning experimental water factory (Exceed-Da-20). All the reagents in the experiments were analytical grade. The solutions were prepared with distilled water.

Sample: *Cynomorium songaricum* was provided by Guazhou Yide biological technology co., LTD. Sample was dried at 50° C to constant weight, was subsequently crushed through 100 mesh sieve. Afterwards, the crushed material is sealed to preserve.

Preparation for reagents

0.6g indicarminum (Tianjin Guangfu fine chemical industry, China) was precisely weighed and dissolved by sulfuric acid (Baiyin Liangyou chemical reagent, China) with 50% volume fraction. Then the solution was diluted to 100mL with distilled water. 25g gelatin (Shanghai gelatin factory, China) was soaked in saturated sodium chloride (Tianjin Baishi Chemicals, China) and dissolved completely by heating. Acidic sodium chloride was prepared by adding 25mL concentrated sulfuric acid into 975mL saturated sodium chloride. 0.79g potassium permanganate (Tianjin Baishi Chemicals, China) was accurately weighed and dissolved, whereafter, diluted to 500mL in volumetric flask.

Titration with potassium permanganate

10mL extract of *C. songaricum* together with 5mL indigo carmine indicator was diluted to 100mL with distilled water. The solution was titrated with 0.01mol/L potassium permanganate standard solution until its color changed from green to drab yellow. The letter "a" was used to be volume of potassium permanganate consumed.

10mL extract of *C. songaricum* and 10mL of Gelatin solution and acidic sodium chloride, respectively. In addition, 1.0g kaolin (Zhongqin in the Shanghai chemical reagent co., LTD) was added into the solution. Afterwards, the mixture was shaken for many minutes and filtered. The residue was washed for two or three times with distilled water. 5mL indigo carmine indicator was added in the mix containing filtrate as well as washings. Then solution was diluted to 100mL with distilled water. The solution was titrated with 0.01mol/L potassium permanganate standard solution until its color changed from green to drab yellow. The letter "b" was used to be dosage of potassium permanganate consumed.

Calculation of tannin content:

$$\text{Tannin content }(\%) = \frac{(a-b)\times N\times 0.042}{W\times D} = \times 100\%$$

In this formula: "a" represents the total volume of potassium permanganate standard solution consumed by sample. "b" is dosage of potassium permanganate standard solution used by non-tannin in the sample. "N" is on behalf of the concentration of potassium permanganate standard solution. The number "0.042" is the weight of tannin per gram (g). "W" stands for the weight of sample (g). "N" is on behalf of the dilution factor.

RESULTS AND DISCUSSION

Single factor test

Extraction solvents, extraction frequency, ultrasound power, ratio of raw material to water and ultrasonic duration were selected to be testing factors on extraction efficiency. Our study indicated that the optimum condition of single factor was: 70% ethanol as extraction solvent, extracting for two times, at 800 W ultrasonic power for 120 min and ratio of raw material to water being 1 : 70.

Orthogonal test

Comprehensive evaluation of technical parameters (extraction solvents, extraction frequency, ultrasound power, ratio of raw material to water and ultrasonic duration) of ultrasonic extraction process of tannin from *Cynomorium songaricum* were investigated by orthogonal experiment. Extraction efficiency of tannin was used to be assessment index. There were nine experiments in accord with four columns (A, B, C, D) as well as nine rows (1 to 9). As revealed in Table 1 to 3 the effects of various factors presented an overall order of B (ultrasonic power) >A (ethanol concentration) >D (ultrasonic duration) >C (ratio of raw material to water). Analysis of variance manifested that the four factors have great significance on extraction efficiency. The optimal conditions match was $A_3B_3C_1D_1$ (ethanol concentration = 80%, ultrasonic power = 600 W, ratio of raw material to water = 1 : 65, ultrasonic duration = 105 min).

Table 1 Orthogonal factors and levels

Level factor	A (ethanol concentration)	B (ultrasonic power)	C (ratio of raw material to water)	D (ultrasonic duration)
1	60%	480 w	1 : 65	105 min
2	70%	540 w	1 : 70	120 min
3	80%	600 w	1 : 75	135 min

Table 2 Results of orthogonal test

Number	A	B	C	D	Extraction efficiency of tannin (%) (n=3)
1	1	1	1	1	2.94
2	1	2	2	2	2.98
3	1	3	3	3	3.13

(continued)

Number	A	B	C	D	Extraction efficiency of tannin (%) (n=3)
4	2	1	2	3	2. 80
5	2	2	3	1	3. 34
6	2	3	1	2	3. 40
7	3	1	3	2	2. 86
8	3	2	1	3	3. 43
9	3	3	2	1	3. 56
K1	3. 018	2. 866	3. 258	3. 279	
K2	3. 182	3. 252	3. 116	3. 081	
K3	3. 283	3. 366	3. 110	3. 123	
R	0. 265	0. 500	0. 148	0. 198	

Table 3 Analysis of variance

Factor (SV)	Sum of square of deviation (SS)	Degree of freedom (f)	Mean square (S)	F value	P
Ethanol concentration	0. 323	2	0. 162	70. 182	*
Ultrasonic power	1. 237	2	0. 619	268. 495	**
Ratio of raw material to water	0. 126	2	0. 063	27. 410	*
Ultrasonic duration	0. 195	2	0. 098	42. 386	*
Error	0. 041	18	0. 002		
$F_{0.01}$ (2, 2) = 99. 00, $F_{0.05}$ (2, 2) = 19. 00					

Table 4 Results of tannin content in various samples

Samples number	Source region of samples	Acquisition time	Tannin content (n=3)
Three and nine	Guaprefecture	2011. 01	3. 35%
01	Guaprefecture	2009. 05	1. 88%
02	Guaprefecture	2009. 05	3. 35%
03	Guaprefecture	2011. 05	3. 47%
04	Guaprefecture	2011. 05	3. 55%
05	Guaprefecture	2011. 04	3. 51%
06	Guaprefecture	2010. 04	3. 64%
Sinkiang	Jimsar County	2012. 03	3. 27%
Neimenggu	Ejin Banner	2012. 03	3. 18%

Verification test

Cynomorium songaricum from different origin were determined in three replications, as shown in Table 4. Result suggested that content oftannin in all samples except number "01" was more than

3. 3%. Our study showed that the optimal conditions were reasonable, feasible and stable.

CONCLUSION

Ultrasound-mediated extraction is time-saving, effective and energy-saving. Meanwhile, it can increase extraction efficiency. It has been widely used in determination of plants' active ingredient. Study indicated that ethanol concentration, ultrasonic power, ratio of raw material to water together with ultrasonic duration had markedly effect on extraction efficiency of tannin. Moreover, the optimal condition was as follows: Ethanol concentration was 80%. Ultrasonic power is equal to 600 W. Ratio of raw material to water was 1 : 65. Ultrasonic duration was the same with 105 min. *Cynomorium songaricum* tannin content was measured by titration with potassium permanganate. The titration method was easy to operate and with great repeatability. Verification test showed that tannin concentration in *Cynomorium songaricum* differed because of their origin as well as growing period: *C. songaricum* samples from Guazhou has highest tannin content than others; tannin content in the samples of Guazhou which were harvested in April and May was highest. This study provided a precise and effective extraction method for tannin's exploitation and utilization (Capote and de Castro, 2007; Yoo et al., 2014; Wang et al., 2014; Yu et al., 2010; Zhao et al., 2010).

ACKNOWLEDGMENT

This research was supported by Agricultural Scientific research projects. Forage feed resources development and utilization of technology research and demonstration (201203042).

References

[1] Capote, F.P.and M.D.L.de Castro, 2007. Ultrasound in analytical chemistry. Anal. Bioanal. Chem., 387: 249-257.

[2] Wang, X.L., Y.N.Wang, X. X. Yang and M. Xu, 2014. Optimization of ultrasonic extraction process of total flavonoids from mulberry leaves by orthogonal experiment. Agric. Biotechnol., 3 (4): 12-13.

[3] Yoo, D.Y., J.H.Choi, W.Kim, H.Y.Jung, S.M.Nam, J.W.Kim, Y.S.Yoon, K.Y.Yoo, M.H.Won and I.K.Hwang, 2014. *Cynomorium songaricum* extract enhances novel object recognition, cell proliferation and neuroblast differentiation in the mice via improving hippocampal environment. BMC Complem. Altern M., 14 (5): 2-8.

[4] Yu, F.R., Y.Liu, Y. Z. Cui, E. Q. Chan, M. R. Xie, P. P. McGuire and F. H. Yu, 2010. Effects of a flavonoid extract from *Cynomorium songaricum* on the swimming endurance of rats. Am. J. Chinese Med., 38 (01): 65-73.

[5] Zhao, G., J. Wang, G. W. Qin and L. H. Guo, 2010. *Cynomorium songaricum* extracts functionally modulate transporters of γ-aminobutyric acid and monoamine. Neurochem. Res., 35: 666-676.

(发表于《Advance Journal of Food Science and Technology》)

航天诱变航苜1号紫花苜蓿兰州品种比较试验*

杨红善**，常根柱，周学辉***

（中国农业科学院兰州畜牧与兽药研究所，兰州　730050）

摘　要：航天育种是近十几年来快速发展的农业高科技新领域，已经变成一种主要的诱变育种方法。本研究以航天搭载的紫花苜蓿种子为基础材料，选育出我国第一个航天诱变多叶型紫花苜蓿新品种“航苜1号”。为验证该品种的生产性能，在兰州进行生产性能的品种比较试验，试验结果表明：多叶性状使航苜1号叶量比对照增加5.72%；干草产量（14237.5kg/hm^2）比对照提高13.26%；第一茬和第二茬草开花初期粗蛋白质含量分别为20.08%和18.42%，分别高于对照2.97%和5.79%；18种必需氨基酸为12.32%，高于对照1.57%；微量元素明显优于对照品种。多叶率与其他各指标间的相关性分析表明，多叶率和草产量、粗蛋白、氨基酸总量呈显著正相关。由此可见，该品种具有多叶率高、产草量高和营养含量高的特性。

关键词：航天育种；紫花苜蓿；多叶性状；品比实验

发达的草业是农业现代化的重要标志，紫花苜蓿（*Medicago sativa*）是世界上种植面积最大的牧草种类。就牧草品种而言，目前我国优质的牧草育成品种缺乏，选育具优质、丰产、抗逆性强的新品种迫在眉睫，牧草大多数为多年生、异花授粉植物，遗传背景复杂，传统的育种方法周期长，转基因育种经费高、难度大等问题[1]。将高科技的空间技术与常规农业育种相结合，形成航天诱变育种技术，其目的是利用空间的特殊环境，如空间宇宙射线、微重力、高真空、弱磁场等因素，共同作用使种子产生变化，引起生物体变异，地面种植获得有益变异体，并选育新种质、新材料、培育新品种的植物育种技术[2]。该育种技术也是近十几年来快速发展的农业高科技新领域，航天诱变育种与传统育种相比较具有变异频率高、变异幅度大、有益变异多、稳定性强、优势明显等特点[3-17]。我国多叶型紫花苜蓿常规育种的研究已经开展多年，是目前国内无登记的新品种，航苜1号是我国第一个审定登记的多叶型紫花苜蓿省级育成新品种。目前有专家对紫花苜蓿多叶性状的遗传稳定性、受气候条件的影响、对产量和营养是否具有明显的贡献率等存在质疑，有关研究指出在开展多叶苜蓿的常规育种中多叶型苜蓿的产草量较低，多叶率有较大差异[13-14]，这与我们研究的航天诱变多叶型紫花苜蓿所表现的多叶性状具有稳定、高度遗传的特性，叶量显著提高，产草量高、营养成分高等特点相反。本研究以航天育种选育的多叶型紫花苜蓿为主试品种，以亲本未搭载原品种、国外多叶型紫花苜蓿和国产优质品种为对照，在黄土高原

* 基金项目：甘肃省农业生物技术研究与应用开发项目（GNSW-2014-19），中央级公益性科研院所基本科研业务费专项资金项目（1610322014022），甘肃省青年科技基金计划项目（145RJYA273）和寒生、旱生灌草新品种选育（CAAS-ASTIP-2014-LIHPS-08）资助。

** 作者简介：杨红善（1981—　），男，甘肃兰州人，助研。E-mail：Yanghsh123@126.com.

*** 通讯作者：Corresponding author.

半干旱区（兰州）进行品种比较试验，进一步研究该品种的生产性能。

1 材料与方法

1.1 试验材料

1.1.1 主试品种

航苜 1 号紫花苜蓿（*M. sativa* cv. *Hangmu No.* 1）牧草育成品种，利用航天诱变育种技术选育而成。2014 年 3 月通过甘肃省草品种审定委员会审定登记为育成品种（登记号：GCS014）；2014 年 4 月 25 日通过国家草品种审定委员会评审，批准参加国家草品种区域试验。该品种叶以 5 叶为主，复叶多叶率达 41.5%，叶量为总产量的 50.36%；干草产量 15 529.9kg/hm^2，高于对照 12.8%；粗蛋白质含量 20.08%，高于对照 2.97%；18 种氨基酸总量为 12.32%，高于对照 1.57%；种子千粒重 2.39g，牧草干鲜比 1：4.68。

航苜 1 号亲本材料种子航天搭载处理：2002 年 3 月将精选的三得利紫花苜蓿种子分为 2 份，1 份按要求密封搭载于神舟 3 号飞船，搭载时间为 2002 年 3 月 25 日至 2002 年 4 月 1 日，飞行高度为 198~338km，倾角 42.4°，绕地球 108 圈；另 1 份作为地面对照（CK），称为亲本未搭载原品种，地面低温储存[4-5]。

1.1.2 对照品种

三得利紫花苜蓿（*M. sativa* cv. *Sanditi*）亲本未搭载原品种；中兰 1 号紫花苜蓿 *M. sativa* cv. *Zhonglan* No. 1）国内优质、丰产品种；先行者紫花苜蓿（*M. sativa* cv. *Concett*）加拿大多叶型紫花苜蓿品种。

1.2 试验地概况

试验区设在中国农业科学院兰州畜牧与兽药研究所大洼山试验站。东经 103°44′36″，北纬 36°02′20″，海拔 1697m，年均降水量 324.5 mm；年均温 9.3℃，极端最高温 39.1℃，最低温 -23.1℃，≥0℃ 的年活动积温 3 700℃·d，≥10℃ 的年活动积温 1 900~2 300℃·d。无霜期 169d；相对湿度 58%，蒸发量 1 450.0 mm，日照时数 2751.4h；无霜期 179d。为我国典型的黄土高原半干旱区，属于兰州盆地黄河南岸三级阶地，土质为 III 级自重湿陷性黄土。

1.3 试验方法

1.3.1 试验田间种植

2011 年，参试品种重复 4 次，小区随机区组排列。小区代码：航苜 1 号（HM1）、三得利（S）、中兰 1 号（ZH1）、先行者（C）。小区面积：2.7m×7.0m = 18.9m^2，行距：30cm，播种量：1.5g/m^2（每小区 28.35g）。小区四周隔离带 1.5m，小区间距 1m。

1.3.2 多叶率指标定义及测定方法

结合航苜 1 号紫花苜蓿选育过程及专家现场测试，本研究首次提出了 3 项多叶率衡量指标，包括多叶枝率、复叶多叶率和品种多叶率。指标的定义和测定方法如下。

（a）多叶枝率：指具有多叶的枝条数占所测定总枝条数的百分比，多叶枝条指枝条上具多叶性状的枝条；测定方法：每试验小区以 1m 为 1 个样段，随机选择 3 个样段，距地面 6cm 刈割，将具有多叶的枝条和非多叶枝条分开并计数，计算多叶枝条占总枝条数的百分比。

（b）复叶多叶率：指 1 个枝条上 3 片以上的复叶数占总复叶数的百分比（复叶数指 1 个叶柄上长有 3 片以上的叶片）；测定方法：在多叶枝率测定中确定的多叶枝条中随机取 20 个枝条，测定每一枝条上 3 片以上复叶数占总复叶数的百分比。

（c）品种多叶率=多叶枝率×复叶多叶率。

1.3.3 生育期、株高测定、草产量、种子产量、干鲜比、茎叶比等指标的测定方法

参考中华人民共和国农业行业标准《草品种区域试验规程，禾本科牧草》执行[6]。

（a）生育期：分为播种、出苗、分枝、孕蕾、开花、乳熟、成熟7个阶段。（b）株高测定：测量从地面至植株最高部位的绝对高度，分枝期、孕蕾期、初花期、结荚期测定。（c）草产量测定：包括第1次刈割的产量和再生草产量，在开花初期测定，测产时先割去小区两侧边行及前后各50cm，并移出小区（本部分不计入产量），将余下部分刈割测产、称重，按实际面积计算产量。（d）种子产量：种子成熟后，先割去小区两侧边行及前后各50cm，并移出小区（本部分不计入产量），将余下部分刈割、自然风干、打碾、称重，按实际面积计算产量。（e）茎叶比测定：测草产量的同时进行，从每小区随机取3~5把草样，将4个重复的草样混合均匀，取鲜草0.5kg样品，将茎和叶、花序按两部分分开，待风干后称其重量，求其百分数。（f）干鲜比：每次刈割测产后，从每小区随机取3~5把草样，将4个重复的草样混合均匀，取约1 000g的样品，剪成3~4cm长，编号称重，然后在干燥气候条件下自然风干后称重。计算干草产量和干鲜比。

1.4 数据处理

采用国际通用软件SPSS13.0进行统计分析，用Excel2007进行数据处理和制表。

2 结果与分析

2.1 物候期

航苜1号的物候期与对照基本一致，相对个别对照略有提前（表1）。4月中下旬播种，4月底出苗，6月中上旬分枝，7月上旬现蕾，7月中旬开花，9月中下旬种子成熟；第二年，4月中旬返青，8月中下旬种子成熟。

表1 物候期（日/月）

试验区	年度	品种名称	播种期	出苗/返青	分枝期	现蕾期	开花盛期	结荚期	成熟期
兰州	2011	HM1	22/4	30/4	8/6	3/7	18/7	15/8	20/9
		S	22/4	30/4	7/6	5/7	20/7	17/8	22/9
		ZH1	22/4	30/4	9/6	3/7	17/7	19/8	23/9
		C	22/4	30/4	6/6	1/7	15/7	13/8	24/9
	2012	HM1	—	16/4	27/4	30/5	16/6	15/7	18/8
		S	—	16/4	28/4	2/6	18/6	17/7	16/8
		ZH1	—	16/4	27/4	28/5	14/6	15/7	21/8
		C	—	19/4	29/4	3/6	20/6	20/7	22/8
	2013	HM1	—	12/4	25/4	26/5	14/6	12/7	14/8
		S	—	12/4	23/4	28/5	16/6	14/7	16/8
		ZH1	—	12/4	24/4	26/5	14/6	12/7	12/8
		C	—	16/4	25/4	29/5	17/6	15/7	17/8

注：HM1：航苜1号 Hangmu No.1；S：三得利 Sanditi；ZH1：中兰1号 Zhonglan No.1；C：先行者 Concett。下同。

2.2 多叶率、茎叶比、干鲜比

多叶率和叶茎比指标取 3 年试验平均值进行分析，航苜 1 号多叶枝率为 83.49%，复叶多叶率为 41.5%，品种多叶率为 34.64%，多叶率指标极显著高于先行者（P<0.01）。航苜 1 号叶茎比显著高于 3 个对照品种（P<0.05）；叶片占干草总产量的百分比航苜 1 号为 50.36%、三得利为 47.97%、中兰 1 号为 47.67%、先行者为 47.27%，航苜 1 号平均高于对照 5.72%，其叶总量明显高于其他对照品种（表 2）。

表 2　多叶率、茎叶比、干鲜比

品种名称	多叶枝率（%）	复叶多叶率（%）	品种多叶率（%）	叶茎比	干鲜比
HM1	83.49aA	41.5aA	34.65aA	1.014aA	1∶4.68
S	0	0	0	0.922bAB	1∶4.27
ZH1	0	0	0	0.911bAB	1∶4.50
C	30.58bB	10.7bB	3.27bB	0.896bB	1∶4.45

注：同列不同小写字母表示差异显著（P<0.05），不同大写字母表示差异极显著（P<0.01）。下同。

2.3 产草量

草产量每年于开花初期测定，同时测定株高（表 3）。干草产量取 3 年平均值分析比较，航苜 1 号为14 237.5kg/hm^2，对照品种三得利为 12 407.1 kg/hm^2，中兰 1 号为 13 384.4 kg/hm^2，先行者为 11 919.0 kg/hm^2，航苜 1 号比三得利高产 14.75%，中兰 1 号高产 6.37%，先行者高产 19.45%，比对照品种平均高产 13.26%。

表 3　株高及干草产量

试验区	年度	品种名称	第一次刈割		第二次刈割		第三次刈割		总产量
			株高（cm）	干草产量（kg/hm^2）	株高（cm）	干草产量（kg/hm^2）	株高（cm）	干草产量（kg/hm^2）	干草产量（kg/hm^2）
兰州	2011	HM1	74.9	5 720.4	73.6	2860.2	46.7	2 012.8	10 593.4
		S	71.2	4 684.7	70.1	2 342.3	45.5	1 648.3	8 675.3
		ZH1	75.5	5 496.2	70.5	2 748.1	39.1	1 933.8	10 178.1
		C	69.2	4 399.3	68.0	2 199.6	31.3	1 547.9	8 146.8
	2012	HM1	96.0	7 301.7	92.1	4 867.8	75.3	3 042.3	15 211.8
		S	85.4	6 562.8	83.5	4 375.2	68.6	2 734.6	13 672.6
		ZH1	94.7	7 139.3	90.7	4 759.6	72.2	2 974.7	14 873.6
		C	73.3	6 326.1	71.6	4 217.4	60.7	2 635.8	13 179.3
	2013	HM1	98.3	7 946.4	97.1	5 579.4	79.3	3 381.5	16 907.3
		S	86.5	6 990.5	84.5	4 908.2	71.5	2 974.7	14 873.4
		ZH1	96.7	7 097.7	93.7	4 983.5	76.2	3 020.3	15 101.5
		C	76.2	6 782.5	74.6	4 762.2	63.9	2 886.2	14 430.9
	3 年平均	HM1	89.7	6 989.5	87.6	4 435.8	67.1	2 812.2	14 237.5
		S	81.0	6 079.3	79.4	3 875.2	61.9	2 452.5	12 407.1
		ZH1	88.9	6 577.7	84.9	4 163.7	62.5	2 642.9	13 384.4
		C	72.9	5 836.0	71.4	3 726.4	51.9	2 356.6	11 919.0

2.4 种子产量

种子产量取3年平均值分析比较（表4），航苜1号为289.6kg/hm²，三得利为288.8kg/hm²，中兰1号为298.1kg/hm²，先行者为265.6kg/hm²，航苜1号种子产量比对照品种平均高产1.90%。

表4 种子产量

品种名称	2011年	2012年	2013年	3年平均
HM1	202.5	328.7	337.6	289.6bA
S	201.7	327.3	337.3	288.8bA
ZH1	207.5	339.4	347.4	298.1aA
C	184.6	303.7	308.4	265.6cB

2.5 营养成分

2012年，分别于第一茬草和第二茬草开花初期采样，自然风干后测定8大营养成分，包括粗蛋白质、粗纤维、粗脂肪、粗灰分、钙、磷、无氮浸出物和水分（表5）。粗蛋白质含量分析比较：第一茬草，航苜1号为20.08%，对照品种三得利为19.72%，先行者为19.9%，中兰1号为18.89%，航苜1号平均高于对照2.97%，差异不显著。第二茬草，航苜1号为18.42%，对照品种三得利为16.89%，先行者紫为18.18%，中兰1号为17.22%，航苜1号平均高于对照5.79%，差异显著。

表5 营养成分

	品种	粗蛋白质（%）	粗纤维（%）	粗脂肪（%）	粗灰分（%）	钙（%）	磷（%）	无氮浸出物（%）	水分（%）
第1茬	HM1	20.08±0.05aA	27.49	3.11	10.15	1.62	0.40	29.74	9.43
	S	19.72±0.04abAB	22.23	1.70	9.90	1.37	0.31	37.24	9.21
	ZH1	18.89±0.06bcBC	23.37	2.09	9.61	1.39	0.31	36.85	9.16
	C	19.90±0.09aA	18.83	3.12	10.72	1.55	0.34	37.03	10.40
第2茬	HM1	18.42±0.15aA	22.78	1.68	9.31	0.87	0.53	37.99	9.82
	S	16.89±0.13bB	26.00	1.75	8.91	0.76	0.36	38.02	8.43
	ZH1	17.22±0.09bB	26.31	1.76	9.42	0.84	0.45	36.29	9.00
	C	18.18±0.11aA	23.92	1.94	9.16	0.84	0.43	37.49	9.31

2.6 必需氨基酸

2013年，第一茬草开花初期采样，自然风干后测定18种必需氨基酸（表6）。以18种必需氨基酸总量进行分析：航苜1号为12.32%，三得利为11.67%，中兰1号为12.52%，先行者为12.19%，航苜1号比三得利高5.57%，平均高于对照1.57%。其中，天门冬氨酸平均高于对照7.78%，苏氨酸平均高于对照2.43%，丝氨酸平均高于对照5.88%，谷氨酸平均高于对照2.16%，缬氨酸平均高于对照5.22%，组氨酸平均高于对照4.34%，其他各氨基酸与对照基本持平或略低于对照（如脯氨酸）。

2.7 微量元素

2012 年，第一茬草开花初期采样，风干后测定 6 种常见微量元素，包括全锌、全锰、全铜、全铁、全镁和全硼，各参试品种的 6 种微量元素（表 7）。分析比较结果：航苜 1 号全锌为 15.59%，平均高于对照 8.6%；全锰为 4.65%，平均高于对照 23.3%；全铜为 1.57%，平均和对照持平；全铁为 39.97%，平均高于对照 23.2%；全镁为 0.60%，平均高于对照 7.64%；全硼为 5.15%，平均高于对照 14.4%。其中，航苜 1 号的全锌、全锰、全铁、全硼与各对照品种差异极显著（摩<0.01）。

2.8 各指标间的相关性分析

相关分析表明，多叶率和草产量、粗蛋白、氨基酸总量及全锰呈显著正相关；与全锌极显著正相关；与粗纤维、全铁和全镁呈正相关；与种子产量和全铜呈负相关（表 8）。

表 6　18 种必需氨基酸含量

氨基酸	HM1	S	ZH1	C	氨基酸	HM1	S	ZH1	C
天门冬氨酸 Asp	2.17	2.00	2.04	2.00	蛋氨酸 Met	0.18	0.20	0.18	0.18
苏氨酸 Thr	0.56	0.53	0.57	0.54	异亮氨酸 Ile	0.50	0.48	0.53	0.50
丝氨酸 Ser	0.60	0.56	0.58	0.56	亮氨酸 Leu	0.78	0.73	0.82	0.79
谷氨酸 Glu	1.26	1.16	1.30	1.24	酪氨酸 Tyr	0.30	0.33	0.37	0.35
脯氨酸 Pro	1.22	1.22	1.32	1.33	苯丙氨酸 Phe	0.70	0.69	0.74	0.71
甘氨酸 Gly	0.56	0.52	0.56	0.58	赖氨酸 Lys	0.80	0.74	0.80	0.80
丙氨酸 Ala	0.61	0.56	0.62	0.60	组氨酸 His	0.32	0.30	0.31	0.31
胱氨酸 Cys	0.26	0.26	0.28	0.24	精氨酸 Arg	0.50	0.45	0.51	0.50
缬氨酸 Val	0.74	0.68	0.73	0.70	色氨酸 Trp	0.26	0.26	0.26	0.26
总氨基酸 TotalAA	12.32aAB	11.67bB	12.52aA	12.19aAB					

表 7　常见微量元素（%）

品种名称	全锌	全锰	全铜	全铁	全镁	全硼
HM1	15.59±0.18aA	4.65±0.69aA	1.57±0.05bB	39.97±0.76aA	0.60±0.03aA	5.15±0.04aA
S	14.46±0.16bBC	4.01±0.87bB	1.71±0.03aA	39.87±0.71aB	0.59±0.02aA	4.55±0.04cC
ZH1	14.57±0.10bB	4.13±0.36bB	1.58±0.03aB	30.15±0.34bB	0.59±0.02aA	4.10±0.06dD
C	14.01±0.22cC	3.20±0.89cC	1.43±0.03cC	26.72±0.52cC	0.52±0.04bA	4.90±0.05bB

表 8　航苜 1 号各指标间的相关性

项目	多叶率	草产量	种子产量	粗蛋白	粗纤维	氨基酸	全锌	全锰	全铜	全铁
草产量	0.656*									
种子产量	−0.058	0.674*								
粗蛋白	0.601*	−0.006	−0.406							
粗纤维	0.676	0.956**	0.678*	0.116						

（续表）

项目	多叶率	草产量	种子产量	粗蛋白	粗纤维	氨基酸	全锌	全锰	全铜	全铁
氨基酸	0.237*	0.496	0.309	-0.142	0.275					
全锌	0.790**	0.925**	0.524	0.270	0.961**	0.317				
全锰	0.522*	0.915**	0.791**	0.020	0.977**	0.202	0.902**			
全铜	-0.232	0.221	0.715**	-0.089	0.398	-0.316	0.285	0.559		
全铁	0.401	0.488	0.488	0.370	0.706*	-0.349	0.685*	0.758**	0.757	
全镁	0.166	0.623*	0.790**	0.037	0.712**	0.165	0.606*	0.786**	0.710	0.649*

注：** 表示 0.01 水平上显著相关，* 表示 0.05 水平上显著相关

3 讨论

航天育种将高科技的空间技术与常规农业育种相结合是近 10 多年来快速发展的高创新性育种方法，已成为空间生命科学的重要研究内容之一。航天育种在农作物、蔬菜和花卉已开展多年研究，并培育出多个品种应用于生产中，最显著的变异特性是生物量提高、抗逆性增强、品质提高和早熟性，这对紫花苜蓿育种而言极为重要[1,3]。

与谷类作物不同，紫花苜蓿的经济产量主要以草产量和品质来衡量，其中，草产量主要由叶、茎组成，另外 60%的总可消化养分、70%的粗蛋白、90%的维生素存在于叶片中，因此多叶性状利于苜蓿优质。多叶型紫花苜蓿具有羽状三出复叶变异为由 4 个以上单叶组成的复叶[13-14]。常规育种认为紫花苜蓿中多叶性状不能稳定遗传，但航天诱变过程中高能重粒子能导致种子细胞中 DNA 双链断裂，高真空、微重力可以阻延或抑制断裂链的修复，使得断裂的碱基在 DNA 双链周围游动，当返回地面后获得重力，断裂的碱基在双链的其他部位发生重叠，从而产生 DNA 结构中碱基的缺失和重组，属于内源基因的改良，因此产生的多叶性状变异的遗传稳定较高[2,8-10]。本研究团队通过航天诱变育种技术选育出的多叶型紫花苜蓿新品种“航苜 1 号”，其多叶枝率为 83.49%，复叶多叶率为 41.5%，品种多叶率为 34.64%，而国外多叶型紫花苜蓿先行者，多叶枝率为 30.58%，复叶多叶率为 10.7%，品种多叶率为 3.27%，可见航天育种技术选育的品种其多叶率指标明显高于常规育种选育品种。多叶性状使航苜 1 号叶量增多，叶片总量平均高于对照 5.72%，而叶量增加使光合作用增强，草产量显著提高。因 70%左右的营养物质存在于叶片中，因此叶量增多也有效提高了品种的营养成分、总氨基酸和微量元素，航苜 1 号蛋白质含量平均高于对照 2.97%，18 种必需氨基酸总量平均高于对照 1.57%，微量元素全锌平均高于对照 8.72%、全锰平均高于对照 22.97%、全铁平均高于对照 23.42%、全镁、平均高于对照 4.74%和全硼平均高于对照 14.01%。多叶率与其他各指标间的相关性分析可以看出，多叶率和草产量、粗蛋白、氨基酸总量和全锰呈正相关，且差异显著；与全锌呈正相关，且差异极显著；与粗纤维、全铁和全镁呈正相关，但差异不显著。多叶率与种子产量呈负相关，但差异不显著，因叶量增多使得植物光合作用增强，生物量显著提高，一定程度影响了营养生长，使种子产量降低，这与植物生长规律相一致。

4 结论

利用航天诱变育种技术选育出我国第一个多叶型紫花苜蓿新品种“航苜 1 号”，该品种

特性是优质、丰产，表现为多叶率高、产草量高和营养含量高。航苜 1 号的品种多叶率为 34.65%，显著高于国外多叶型紫花苜蓿先行者，多叶性状使得叶量比对照增加 5.72%，干草产量比对照增产 13.26%，粗蛋白质（两茬）平均高于对照 4.38%，18 种必需氨基酸总量高于对照 1.57%，微量元素明显优于对照品种。多叶率和草产量、营养成分、18 种氨基酸总量及微量元素呈正相关，并且与多数指标差异显著，多叶性状可以有效提高紫花苜蓿品种的产量和品质。

参考文献

[1] 杨红善，常根柱，包文生．紫花苜蓿的航天诱变．草业科学，2013，30（2）：253-258.

[2] 刘纪元．中国航天诱变育种．北京：中国宇航出版社，2007.

[3] 杨红善，常根柱，柴小琴，等．紫花苜蓿航天诱变田间形态学变异研究．草业学报，2012，21（5）：222-228.

[4] 范润钧，邓波，陈本建，等．航天搭载紫花苜蓿连续后代变异株系选育．山西农业科学，2010，38（5）：7-9.

[5] 范润钧．空间搭载紫花苜蓿种子第一代植株表型变异及基因多态性分析．兰州：甘肃农业大学，2010.

[6] 负旭疆，苏加楷，齐晓，等．NY/T2322-2013．草品种区域试验规程，禾本科牧草．北京：中国农业出版社，2013.

[7] 王密．紫花苜蓿种子空间诱变变异效应的研究．呼和浩特：内蒙古农业大学，2010.

[8] 杜连莹．实践八号搭载 8 个苜蓿品种生物学效应研究．哈尔滨：哈尔滨师范大学，2010.

[9] 张月学，刘杰琳，韩微波，等．空间环境对紫花苜蓿的生物学效应．核农学报，2009，23（2）：266-269.

[10] 李红，李波，李雪婷，等．卫星搭载对苜蓿突变株蛋白表达的影响．草业科学，2013，30（11）：1749-1754.

[11] 郭慧慧，任卫波，解继红，等．卫星搭载后紫花苜蓿 DNA 甲基化变化分析．中国草地学报，2013，35（5）：29-33.

[12] 马学敏，张治安，邓波，等．不同含水量紫花苜蓿种子卫星搭载后植株叶片保护酶活性的研究．草业科学，2011，28（5）：783-787.

[13] 王雯玥，韩清芳，宗毓铮，等．多叶型和三叶型紫花苜蓿产量与相关性状的回归分析．中国农业科学，2010，43（14）：3044-3050.

[14] 杜红梅．多叶型紫花苜蓿生物学特性的初步研究．长春：东北师范大学，2010.

[15] 康俊梅，张铁军，王梦颖，等．紫花苜蓿 QTL 与全基因组选择研究进展及其应用．草业学报，2014，23（6）：304-312.

（发表于《草业学报》）

丹翘液对脂多糖诱导 RAW264.7 细胞炎症相关因子的抑制效应分析*

魏立琴[1,2**]，王东升[1]，董书伟[1]，
邝晓娇[1]，张世栋[1***]，严作廷[1***]

（1. 中国农业科学院兰州畜牧与兽药研究所/农业部兽用药物创制重点实验室/甘肃省中兽药工程技术研究中心，兰州 730050；2. 甘肃农业大学动物医学院，兰州 730070）

摘　要：为研究丹翘液的体外抗炎作用及其抗炎机制，利用 LPS 诱导巨噬细胞建立炎症模型，MTT 法检测不同浓度丹翘对 RAW264.7 细胞活力影响；NO 测试盒检测丹翘液对 LPS 诱导的 NO 释放量的影响；ELISA 方法检测丹翘液对 LPS 诱导的 TNF-α，IL-6 和 PGE_2 分泌的影响；RT-PCR 方法检测丹翘液对 LPS 诱导的 TNF-α、IL-6、COS-2 和 iNOS 基因转录的影响；Western blot 检测对细胞核内 NF-κB p65 蛋白表达的影响。结果显示丹翘液小于 700μg·mL^{-1} 对细胞无毒性作用；丹翘液各剂量组（100、300、600μg·mL^{-1}）能不同程度地抑制 LPS 诱导的 NO、PGE_2、TNF-α 和 IL-6 的分泌；能显著抑制 iNOS、COX-2、TNF-α 和 IL-6 基因转录和细胞核内 NF-κB p65 蛋白表达。丹翘液的抗炎作用机制可能与抑制 NF-κB 通路激活，进而抑制炎症介质和炎性细胞因子的转录和表达有关。

关键词：丹翘液；抗炎机制；RAW264.7；LPS；炎症模型

炎症是机体对各种致炎因素及损伤所产生的具有防御意义的应答性反应，其在一定时期内对机体是有利的，可以提高机体抵抗力，但剧烈或持久的炎症反应会对机体造成损失，影响动物机体的健康，炎症反应是一些疾病的发病基础[1-2]。近年来，中药制剂对炎症有较好的抑制效果，由于其毒副作用小，对机体不良反应较少，其影响炎症免疫反应作用的机制受到了广泛的关注[3-4]。中药影响炎症免疫反应的机制包括影响炎性介质和炎性细胞因子的作用、核因子的功能、活性氧的生成、神经内分泌激素的生成等[4-6]。

丹翘液是以丹参和连翘等中草药为主要成分，依据活血化瘀缩宫排脓的治疗原则研制而成的中药子宫灌注剂，临床上主要用于防治奶牛子宫内膜炎[7-8]。动物药理学研究显示，丹翘液具有良好的抗炎、镇痛的药理效应[9]，但是丹翘液的抗炎作用机制尚不清楚。因此，

---

* 基金项目：“十二五”国家科技支撑计划项目（20012BAD12B03）；中国农业科学院科技创新工程——奶牛疾病研究。

** 魏立琴（1986—　），女，甘肃兰州人，硕士，主要从事中兽医临床和奶牛疾病防治的研究，E-mail：24weiliqin@163.com.

*** 通讯作者：张世栋（1983—　），男，甘肃张掖人，助理研究员，硕士，主要从事药理毒理学研究，E-mail：xhangshidong@caas.cn；严作廷（1962—　），男，甘肃武威人，研究员，博士，主要从事中兽医临床和奶牛疾病防治的研究，E-mail：yanzuoting@caas.cn.

本试验利用脂多糖（lipopolysaccharide，LPS）诱导巨噬细胞 RAW 264.7，建立炎症模型，并研究了丹翘液对炎症细胞模型的影响，旨在探讨丹翘液的体外抗炎作用及其抗炎机制，为该新药的研发提供理论基础和实验证据。

1 材料与方法

1.1 材料

1.1.1 细胞株

小鼠单核/巨噬细胞株 RAW264.7（购自中国科学院上海生命科学研究院细胞资源中心）添加含 10%胎牛血清、100U · mL^{-1}青霉素、100μg · mL^{-1}链霉素的 DMEM 培养基，置于 37℃、5%CO_2、饱和湿度的培养内常规培养。取对数生长期的细胞用于试验。

1.1.2 试剂和仪器

能多唐（*E. coli* 0111：B4）、噻唑蓝（MTT）购自 Sigma 上海公司；胎牛血清、DMEM、胰蛋白酶购自美国 GIBCO 公司；小鼠 TNF-α、IL-6、PGE_2 的 ELISA 检测试剂盒购自上海酶联生物科技有限公司；NO 试剂盒购自南京建成生物工程研究所；TRIzol 购自美国 Invitrogen 公司；RT-PCR 试剂盒、SYBR Green PCR Master Mix 试剂盒购自日本 TaKaRa 公司；引物由北京六合华大基因合成；HRP 羊抗兔 IgG 购自西安壮志；兔多克隆抗体 NF-κB p65 购自 Abcam；Dmi6000b 倒置相差显微镜（德国 Leica 公司）；SpectraMaxM2/M2e 多功能酶标仪（美国 Molecular Devices 公司）；CFX96 型定量 PCR 仪（美国 Bio Red 公司）；HeaL Force HF90 型 CO_2 培养箱（中国上海力康有限公司）。

1.1.3 中药丹翘液的制备

称取丹参 1 200g、连翘 800g，粉碎成粗粉，加乙醇加热回流提取 3 次，滤过，合并滤液，减压回收乙醇，干燥，加入 800mL 聚乙二醇-200，4 000r · min^{-1}离心 200min，不清液合并后再加入聚乙二醇-200 定容至 1 000mL，灭菌，即得每 1mL 相当于原生药 2g 的药液，4℃保存备用。

1.2 方法

1.2.1 细胞实验分组及处理

取对数生长期 RAW264.7 细胞接种于培养板，细胞密度为 8×10^4 · mL^{-1}，预培养 24h 后进行分组处理。①正常对照组：不做任何处理，正常培养 24h；②炎症模型组（LPS）：1 μg · mL^{-1}LPS 持续刺激细胞 6h；③丹翘组：600μg · mL^{-1}丹翘液作用细胞 24h；④试验组：1μg · mL^{-1}LPS 持续刺激 RAW264.7 细胞 6h 后，加入不同浓度的丹翘液（100μg · mL^{-1}、300μg · mL^{-1}、600μg · mL^{-1}）共同作用 24h。

1.2.2 细胞毒性试验

细胞毒性评价采用 MTT 试验，取对数生长期 RAW364.7 细胞接种于 96 孔板中，细胞分组及处理方法同 1.2.1，培养结束后，每孔加入 20μL 的 MTT（5mg · mL^{-1}）继续培养 4h，弃去培养液，每孔加入 100μL 的 DMSO 溶解甲瓒，均匀振荡溶解完全后在 570nm 处检测各孔的吸光度。试验重复 3 次以上。

1.2.3 NO 含量测定

取对数生长期 RAW364.7 细胞接种于 24 孔板中，细胞分组处理方法同 1.2.1，培养结束后，收集各组细胞上清液，按照 NO 试剂盒说明书检测细胞上清中 NO 的释放。酶标仪测定550nm处的光吸收值，以亚硝酸钠作为对照品测定并计算NO的浓度。

1.2.4 ELISA检测TNF-α、IL-6、PGE_2含量

取对数生长期RAW264.7细胞接种于24孔板中，细胞分组及处理方法同1.2.1，培养结束后，收集各组细胞培养上清液，ELISA法检测各组TNF-α、IL-6、PGE_2含量，操作步骤按照试剂盒说明书。

1.2.5 RT-PCR法检测TNF-α、IL-6、OCX-2和iNOS基因转录水平

取对数生长期RAW264.7细胞接种于6孔板中，细胞分组及处理方法同1.2.1，培养结束后，Trizol法提取各组细胞总RNA，按试剂盒说明书操作，将mRNA逆转录为cDNA，得到的cDNA用于RT-PCR检测TNF-α、IL-6、COX-2和iNOS基因的相对转录水平。PCR引物见表1。RT-PCR反应条件是95℃ 3min；95℃ 15s，60℃20s，循环40次；55~95℃循环81次，mRNA的相对转录量采用$2^{-\Delta\Delta Ct}$方法计算，并以*β-actin*为内参基因。

表1 RT-PCR引物序列

引物		序列	PCR产物/bp	退火温度/℃
β-actin	Forward	5′-GCCACCAGTTCGCCATGGAT-3′	70	58
	Reverse	5′-GCTTTGCACATGCCGGAGC-3′		
TNF-α	Forward	5′-TCGAGTGACAAGCCCGTAG-3′	180	58
	Reverse	5′-CAGCCTTGTCCCTTGAAGAG-3′		
IL-6	Forward	5′-CTGATGCTGGTGACAACCAC-3′	143	59
	Reverse	5′-TCCACGATTTCCCAGAGAAC-3′		
iNOS	Forward	5′-GGAACCCAGTGCCCTGCTTT-3′	178	65
	Reverse	5′-CACCAAGCTCATGCGGCCT-3′		
COX-2	Forward	5′-TGAGTACCGCAAACGCTTCTC-3′	151	65
	Reverse	5′-TGGACGAGGTTTTTCCACCAG-3′		

1.2.6 Western blot检测NF-κB蛋白表达

取对数生长期RAW364.7细胞接种于6孔板中，细胞分组及处理方法同1.2.1，培养结束后，根据试剂盒说明是提取总蛋白质、细胞核蛋白，BCA法则定各组蛋白质浓度，各组取40αμg蛋白质样品与蛋白质上样缓冲液混匀，沸水中变性10min后，在10%分离胶、5%浓缩胶中100V恒压进行SDS-PAGE电泳2.5h，结束后半干转移法12V恒压转印1h到PVDF膜上，用5%的BSA封闭1h，加入一抗过液孵育（4℃），TBST洗膜4次，每次10min，加入HRP标记的二抗在室温缓慢摇动孵育2h，TBST洗涤10min×4次，加ECL，暗室胶片曝光、显影、定影后显示各组细胞中目标蛋白质表达水平。

1.2.7 统计分析

炎性因子释放数据与RTPCR采用GraphPad Prism 5软件分析，试验重复3次以上，数据以$\bar{x}\pm s$表示，组间采用单因素方差（One-Way ANOVA）分析，$P<0.05$为差异具有统计学意义。

2 结果

2.1 丹翘液对RAW264.7细胞活力的影响

图1A为MTT法检测细胞经不同剂量的丹翘液处理后的细胞活力变化。结果显示，与空

白对照组相比，丹翘液≤700μg·mL^{-1}时对细胞无毒性作用（P>0.05），当丹翘液大于70μg·mL^{-1}时，细胞活力明显下降（P<0.01）。图1B为丹翘液和LPS共同处理后细胞的活力变化，结果与丹翘液单独处理的一致。

2.2 丹翘液对LPS诱导的细胞炎症介质的影响

NO检测结构显示，与正常对照组相比，LPS处理组的NO释放量极显著升高（P<0.01），丹翘液处理组的NO释放量不明显，与LPS处理组相比，丹翘液各组均能极显著抑制NO的释放量（P<0.01）。各剂量丹翘液抑制后NO含量与正常对照组差异显著（P>0.01或P>0.05）（图2A）。ELISA检测PGE_2结果显示，与正常对照组相比，LPS处理组的PGE_2释放量极显著升高（P<0.01），与LPS处理组相比，丹翘液高、中剂量组能显著抑制PGE_2的释放量（P>0.05或P>0.01），低剂量差异不显著（P<0.05）。高剂量丹翘液抑制后PGE_2含量与正常对照组无显著差异（P<0.05），中、低剂量丹翘液抑制后PGE_2含量与正常对照组显著差异（P>0.01或P>0.05）（图2B）。

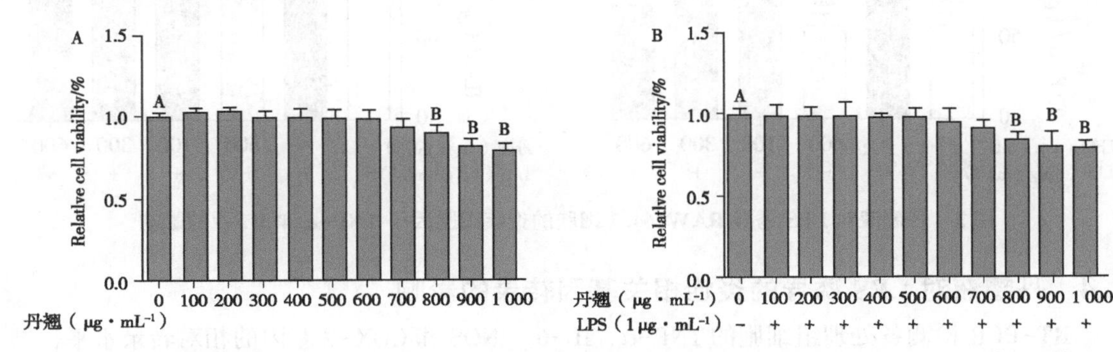

A. 不同浓度的丹翘液（0~1 000μg·mL^{-1}）作用24h时细胞的活力变化；B. 1μg·mL^{-1}LPS持续刺激RAW364.7细胞6h后，不同浓度的丹翘液（0~ 1 000μg·mL^{-1}）作用24h时细胞的活力变化。标注不同大写字母表示组间差异极显著（P<0.01），标相同或未标字母表示组间差异不显著（P>0.05）

图1 丹翘液对RAW364.7细胞的毒活性作用

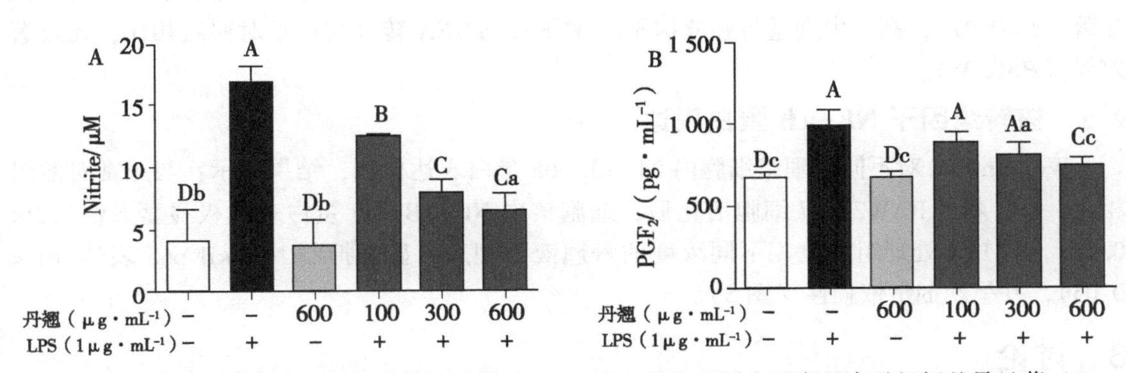

标注不同大写字母表示组间差异极显著（P<0.01），标注不同小写字母表示组间差异显著（P>0.05），标相同小写字母表示组间差异不显著（P>0.05）。下图同

图2 丹翘液对LPS诱导RAW264.7细胞分泌NO和PGE_2的影响

2.3 丹翘液对 LPS 诱导的炎症细胞因子分泌的影响

图 3A、B 为 ELISA 法检测丹翘液对 LPS 诱导的 RAW264.7 细胞上清液中 TNF-α 和 IL-6 含量的影响，结果显示，与正常对照组相比，LPS 处理组的 TNF-α、IL-6 释放量仍显著升高（$P<0.01$），丹翘液组的释放量变化不明显（$P>0.05$），与 LPS 处理组相比，高剂量丹翘液能显著抑制 TNF-α、IL-6 的释放量（$P<0.01$ 或 $P<0.05$），但中、低剂量对 TNF-α、IL-6 的抑制无统计学意义（$P>0.05$）。高剂量丹翘液抑制后 TNF-α、IL-6 含量与正常对照组无显著差异（$P>0.05$），低剂量丹翘液抑制后 TNF-α、IL-6 含量与正常对照组极显著差异（$P<0.01$）。

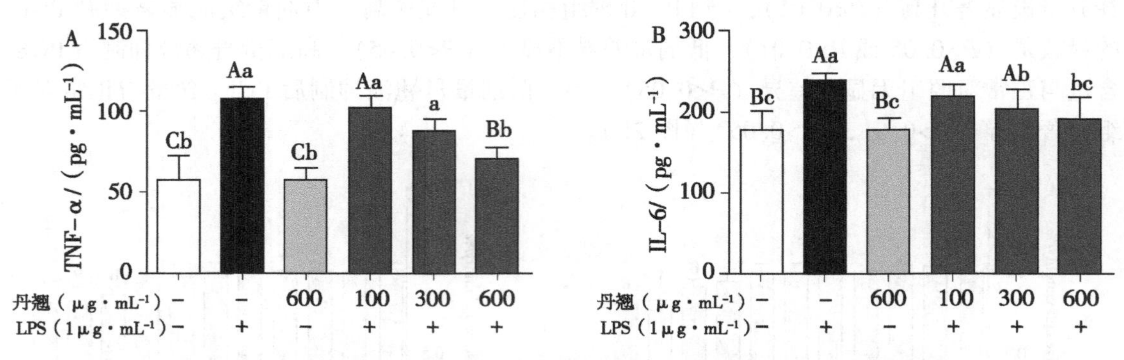

图 3 丹翘液对 LPS 诱导 RAW264.7 细胞的炎症细胞因子 TNF-α 和 IL-6 的影响

2.4 丹翘液对 LPS 诱导的炎症相关基因转录的影响

RT-PCR 检测各处理组细胞的 TNF-α、IL-6、iNOS 和 COX-2 基因的相对转录水平，结果显示，与正常组相比，LPS 处理组的 TNF-α、IL-6、iNOS 和 COX-2 mRNA 转录均极显示升高（$P<0.01$），与 LPS 处理组相比，丹翘液各剂量组均能够显著抑制 LPS 诱导巨噬细胞 iNOS 和 COX-2 mRNA 转录的升高（$P<0.01$ 或 $P<0.05$）（图 4A、B），也能够显著抑制 TNF-α、IL-6 mRNA 转录的升高（图 4C、D），且对 TNF-α、iNOS 和 COX-2 具有一定的剂量效应。各剂量丹翘液抑制后 IL-6、iNOS 和 COX-2 mRNA 转录与正常对照组相比差异极显著（$P<0.01$），高、中剂量丹翘液抑制后 TNF-α mRNA 转录与正常对照组相比，无显著差异（$P>0.05$）。

2.5 核转录因子 NF-κB 蛋白表达

Western blot 对不同处理组细胞内 NF-κB p65 蛋白表达检测，结果显示：与正常对照组相比，LPS 刺激 RAW264.7 细胞活化后，细胞核内 NF-κB p65 蛋白表达极显著升高（$P<0.01$），与 LPS 处理组相比，不同浓度的丹翘液处理后，显著抑制 NF-κB p65 表达（$P<0.05$），并呈现剂量依赖性（图 5）。

3 讨论

炎症是机体对于刺激的一种防御反应，表现为红肿热痛和功能障碍。炎症反应是损伤与抗损伤的过程，损伤因子一方面可以直接或间接破坏组织和细胞，另一方面可以通过炎症充血和渗出反应，杀伤和包围损伤因子，通过实质和间质细胞的再生修复和愈合受损的组织[10-11]。如损伤过程占优势，则炎症加重，并向全身扩散；若抗损伤过程占优势，则炎症

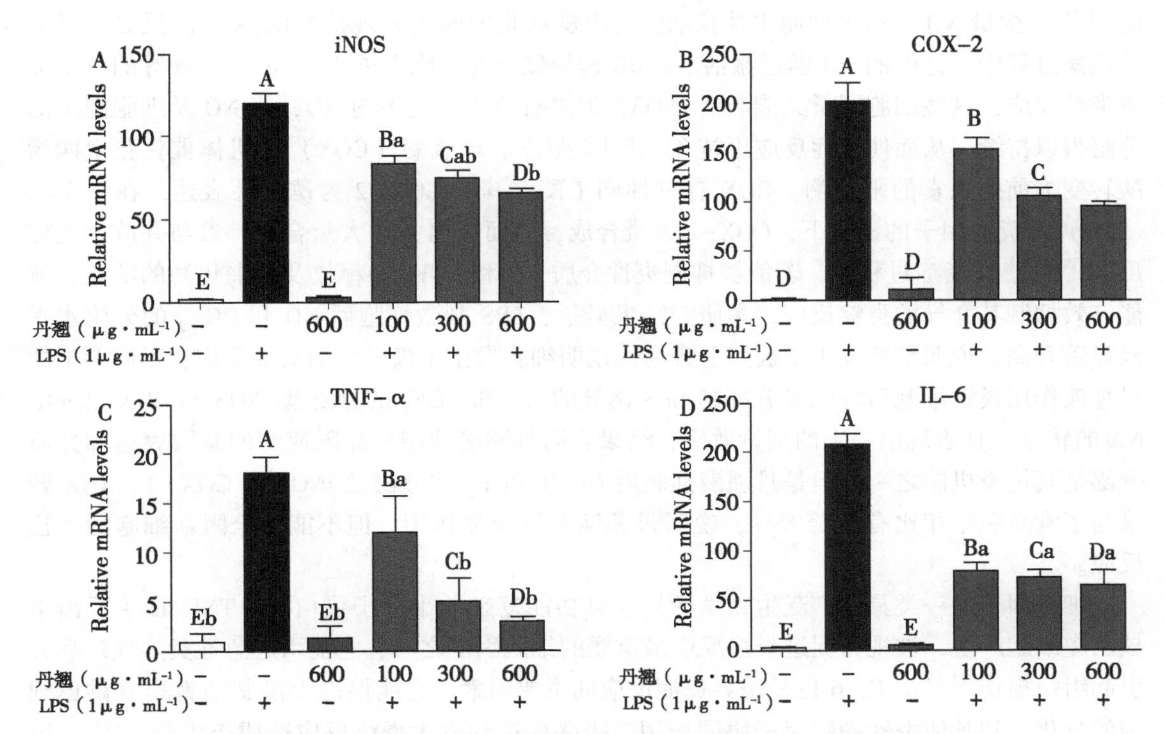

图 4　丹翘液对 LPS 诱导 RAW264.7 细胞的炎症相关基因转录的影响

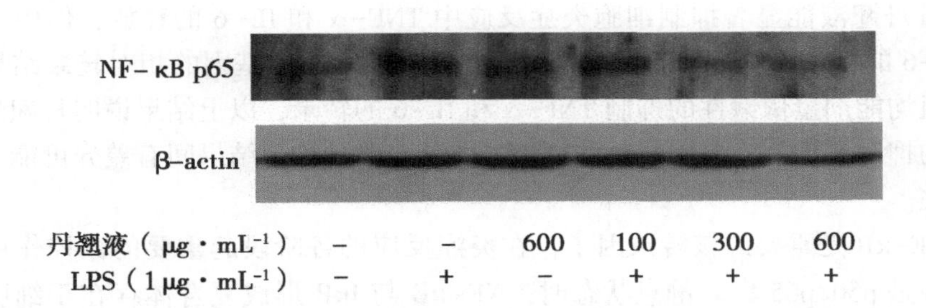

图 5　丹翘液对 LPS 诱导 RAW264.7 细胞的 NF-κB p65 蛋白表达的影响

逐渐趋向痊愈。所以，控制损伤、加强抗损伤过程是机体维持正常生理功能的重要环节。巨噬细胞参与吞噬、免疫调节、组织修复、新陈代谢和炎症等多种生理过程。微生物刺激巨噬细胞产生多种促炎性细胞因子和趋化因子，这些促炎因子反过来引起细胞渗出和激活其他类型的免疫细胞（如 T 细胞），这些因素共同介导形成免疫炎症反应[12-13]。因此本研究利用 LPS 诱导巨噬细胞炎症反应模型对丹翘液的体外抗炎作用及机制进行深入的探讨。本试验首先检测丹翘液对 RAW264.7 细胞活力的影响，MTT 试验结果，药物浓度在 700μg·mL⁻¹以内对 RAW264.7 细胞无明显毒性作用。因此，选择 0~700μg·mL⁻¹的丹翘液进行下一步的抗炎活性研究。

NO 既能参与机体生理过程的调节及免疫应答反应，其过度产生又可导致组织损伤和炎性反应。体内 NO 产生由一氧化氮合酶（NOS）催化 L-精氨酸（L-Arg）产生。NOS 可以分为结构型（eNOS）和诱导型（iNOS）。eNOS 只产生少量 NO，发挥正常生理作用，病原微生物感染及组织损伤都可诱导 iNOS 表达，促进 NO 大量产生。NO 具有抗炎和促炎双重功能[14-15]。少量 NO 可通过抑制中性粒细胞与内皮细胞的黏附表现抗炎活性[16]。反之，炎症性疾病过程中，过度的 NO 通过激活 NF-κB 诱导促炎症细胞因子 TNF-α、IL-6 等的产生促进炎性反应，这些细胞因子又能激活 iNOS，促进机体产生更多的 NO，使 NO 及细胞因子的分泌得以持续，从而使炎症反应更持久，更剧烈[17]。环氧酶（COX）是机体催化花生四稀酸转变为前列腺素的限速酶，COX 有三种同工酶，其中 COX-2 为诱导性表达，在脂多糖（LPS）等炎症因子的诱导下，COX-2 大量合成，进而催化 PGs 大量合成，参与并放大炎症反应[18-21]。在炎症过程中，若能够抑制炎性介质 NO 和 PGE_2 的释放及其催化酶的活化，就能有效抑制其介导的炎症反应。本研究结果显示，LPS 刺激细胞后 NO 和 PGE_2 的分泌水平极显著升高，说明细胞发生了极显著升高，说明细胞发生了极显著的炎症反应，不同剂量的丹翘液作用炎性细胞后均能显著抑制 LPS 诱导的 NO 和 PGE_2 的分泌及 iNOS 和 COX-2 mR-NA 的转录，且表现出一定的剂量效应，结果表明丹翘液抑制炎症限速酶的基因表达和介质分泌是其抗炎机制之一。但是丹翘液抑制后 NO 和 PGE_2 的分泌及 iNOS 和 COX-2 mRNA 转录与正常对照组相比有显著差异，说明丹翘液有抗炎症作用，但不能完全阻断细胞的炎症反应。

细胞因子是一类具有广泛生物学活性、且功能复杂的小分子蛋白质。TNF-α 主要由单核巨噬细胞产生，被视为促进炎症反应最重要的细胞因子之一，它能引起急性炎症性疾病并引起组织损伤[22-23]。IL-6 也是诱导炎症反应的主要因素，它能促进 T 淋肥细胞和 B 淋巴细胞的分化，和其他炎性细胞因子协同作用，级联促进和放大炎症反应造成组织损伤[24]。因此，TNF-α 和 IL-6 作为炎症标志物，对药物的抗炎作用研究具有指示作用[25]。试验结果显示，高剂量丹翘液能显著抑制细胞炎症反应中 TNF-α 和 IL-6 的释放，但中、低剂量对 TNF-α 和 IL-6 的释放抑制效果不显著。同时 TNF-α 和 IL-6 基因的相对转录结果显示，丹翘液各剂量组均能剂量依懒性的抑制 TNF-α 和 IL-6 的转录，以上结果说明丹翘液对促炎细胞因子分泌的抑制是抑制其基因转录的结果。此外，两种检测结果间有差异可能与不同方法的灵敏度有关。

此外，NF-κB 是重要的核转录因子，在炎症反应的各阶段起重要的调控作用，其主要的诱导型亚基是 p50/p65[26]。静息状态时，NF-κB 与 IκB 形成复合体存在于细胞质中。当受到特定刺激后，IκB 被活化的激酶复合体（IκB kinase，IKK）磷酸化，NF-κB 与 IκB 解离，游离的 NF-κB p65 迅速转移到细胞核，与 κB 结合位点结合，启动炎性介质及促炎症细胞因子的转录与表达[27-31]。本试验 NF-κB 蛋白表达结果显示，LPS 刺激细胞后，细胞核内 NF-κB p65 蛋白表达升高，说明 NF-κB 被激活并转移至细胞核内，丹翘液能抑制细胞内 NF-κB p65 蛋白表达升高，说明丹翘液可抑制 NF-κB 的活化与核转移，进而抑制炎症相关基因的表达与炎症介质的产生。

4 结论

丹翘液能够抑制 LPS 刺激巨噬细胞产生性介质 NO 和 PGE_2；也能抑制炎性因子 TNF-α 和 IL-6 的分泌及其基因转录；亦对炎症关键酶 iNOS、COX-2 的基因转录具有抑制作用。丹翘液能抑制 LPS 对巨噬细胞内 NF-κB 信号通路的激活，抑制 NF-κB p65 活化因子的入核转移，从而抑制炎症过程中关键酶和炎症因子的基因表达及炎症介质的产生。因此，丹翘液对 LPS 诱导巨噬细胞炎症反应具有抗炎保护作用。

参考文献

[1] MATHUR N, PEDERSEN B K.Exercise as a mean to control low-grade systemic inflammation [J/OL].*Mediators Inflamm*, 2008, 2008: 109502. [2015-10-23].http: //www.hindawi.com/journals/mi/2008/109502/.

[2] RAMOS G C.Inflammation as an animal development phenomenon [J/OL].*Clin Dev Immunol*, 2012, 2012: 983203. [2015-10-23].http: //www.hindawi.com/journals/jir/2012/983203/.

[3] 郭醉元.溶菌酶与中药治疗奶牛子宫内膜炎的比较研究 [D].哈尔滨：东北农业大学，2008.

[4] 岑玉文，谭行华，张坚生，等.中西医结合疗法改善艾滋病合并肺部感染患者中医症候的随机对照研究 [J].中国中药杂志，2013，38（15）：2448-2452.

[5] 周松，刘永刚，张国祥，等.中药影响炎症免疫反应的机制研究进展 [J].医学综述，2012，18（2）：183-185.

[6] 刘红，孙伟，万毅刚，等.慢性肾病肾组织炎症信号通路 NF-κB 的调节机制及中药的干预作用 [J].中国中药杂志，2013，37（24）：4246-4251.

[7] 王洪海，王国卿，郭梦尧，等，中药治疗奶牛子宫内膜炎研究进展 [J].中国兽医杂志，2010，46（8）：64-66.

[8] SHELDON I M, PRICE S B, CRONIN J, et al.Mechanisms of infertility associated with clinical and subclinical endometritis in high prducing dairy cattle [J].*Reprod Domest Anim*, 2009, 44 (Suppl 3): 1-9.

[9] 魏立琴，王东升，苗小楼，等.丹翘灌注液抗炎镇痛作用的研究 [J] 中国农学通报，2015，31（2）：75-79.

[10] ARROYO A G, IRUELA-ARISPE M L.Extracellular matrix, inflammation, and the angiogenic response [J].*Cardiovasc Res*, 2012, 86 (2): 226-235.

[11] MCCARTHY N.Inflammation: An innate response [J].*Nat Rev Cancer*, 2015, 15 (4): 197.

[12] MANTOVANI A, BISWAS S K, GALDIERO M R, et al.Macrophage plasticity and polarization in tissue repair and remodelling [J].*J Pathol*, 2013, 229 (2): 176-185.

[13] BISWAS S K, CHITTEZHATH M, SHALOVAIN, et al. Macrophage polarization and plasticity in health and disease [J].*Immunol Res*, 2012, 53 (1-3): 11-24.

[14] 王强，瘳清奎.一氧化氮在炎症反应中的作用 [J].医学综述，2002，8（4）：198-200.

[15] ZHOU L, ZHU D Y.Neuronal nitric oxide ynthase: structure, subcellular localization, regulation, and clinical implications [J].*Nitric Oxide*, 2009, 20 (4): 223-230.

[16] MEDZHITOV R.Toll like receptors and innate immunity [J].*Nat Rev Immunol*, 2001, 1 (2): 135-145.

[17] DEBPRASAD C, HEMANTA M, PAROMITA B, et al.Inhibition of NO_2, PGE_2, TNF-α, and iNOX expression by *shorea robusta* L.: An ethnomedicine used for anti inflammatory and amalgesic activity [J/OL].*Evid Based Complement Alternat Med*, 2012, 2012: 254849. [2015-10-23].http: //www.hindawi.com/journals/ecam/2012/254849/.

[18] WANG D, DUBOIS R N.The role of COX-2 in intestinal inflammation and colorectal cancer [J].*Oncogene*, 2010, 29 (6): 781-788.

[19] 徐艺荣.九种黄酮类化合物对 LPS 诱导的 RAW264.7 细胞 PGE_2、COX-2 表达的影响 [D].天津：天津科技大学，2012.

[20] CHEN J, JIANG X H, CHEN H, et al.CFTR negatively regulates cyclooxygenase 2 PGE2 positive feed back loop in inflammation [J].*J Cell Physiol*, 2012, 227 (6): 2759-2766.

[21] LI H, BRADBURY J A, DACKOR R T, et al. Cyclooxygenase 2 regulates Th17 cell differentiation

during allergic lung inflammation [J].*Am J Respir Crit Care Med*, 2011, 184 (1): 37-49.

[22] AHMAD S F, ANSARI M A, ZOHEIR K M, et al.Regulation of TNF-αand NF-κB activation through the JAK/STAT signaling pathway downstream of histamine 4 receptor in a rat model of LPS induced joint inflammation [J].*Immunobiology*, 2015, 220 (7): 889-898.

[23] YAMANAKA H.TNF as a target of inflammation in rheumatoid arthritis [J].Endocr Metab *Immune Disord Drug Targets*, 2015, 15 (2): 129-134.

[24] HONG S H, ONDREY F G, AVIS I M, et al.Cycollxygenase regulates human oropharyhgeal carcinomas via the proinflammatory cytokine IL-6: a general role for inflammation? [J].*FASEB J*, 2000, 14 (11): 1499-1507.

[25] MANJEET K R, GHOSH B.Quercetin inhibits LPS induced nitric oxide and tumor necrosis factor alpha production in murine macrophages [J].*Int J Immunopharmacol*, 1999, 21 (7): 435-443.

[26] 邢飞跃，赵克森，姜勇.NF-κB的信号通路与阴断策略 [J].中国病理生理杂志，2003，19 (6): 849-855.

[27] 乔静.葫芦素E对脂多糖诱导的小鼠巨噬细胞RAW264.7炎症反应的抑制作用及其机制研究 [D].广州：暨南大学，2013.

[28] OH Y C, JEONG Y H, HAJ H, et al.Oryeongsan in hibits LPS induecd production of inflammatory mediators via blockade of the NF-kappaB, MAPK pathways and leads to HO 1 induction in macrophage cells [J].*BMC Complement Altern Med*, 2014, 14: 242.

[29] GUGASYAN R, GRUMONT R, GROSSMANN M, et al.Rel/NF-κB transcription factors: key mediators of B cell activation [J].*Immunol Rev*, 2000, 176: 134-140.

[30] BAWADEKAR M, DE ANDREA M, LO CIGNO I, et al.The extracellular IFI16 protein propagates inflammation in endothelial cell via p38 MAPK and NF κB p65 activation [J].*J Interferon Cytokine Res*, 2015, 35 (6): 441-453.

[31] ZHAO W, SUN Z, WANG S, et al.Wntl participates in inflammation induced by lipopolysaccharide through upregulating scavenger receptor A and NF-κB [J].*In flammation*, 2015, 38 (4): 1700-1706.

（发表于《畜牧兽医学报》）

牦牛 *Ihh* 基因组织表达分析、SNP 检测及其基因型组合与生产性状的关联分析*

李天科**[1,2,3]，赵娟花[1,2]，裴　杰[1,2]，
梁春年[1,2]，郭　宪[1,2]，秦　文[1,2,3]，阎　萍[1,2,3]***

（1. 中国农业科学院兰州畜牧与兽药研究所，兰州　730050；
2. 甘肃省牦牛繁育工程重点实验室，兰州　730050；
3. 甘肃农业大学动物科学技术学院，兰州　730070）

摘　要：为了研究牦牛 *Ihh* 基因多态位点基因型组合与生产性状间的相关性，发现与牦牛生产性状相关的分子标记，同时进一步研究 *Ihh* 基因在牦牛和黄牛各个组织的分布和表达，试验采用高分辨率溶解曲线分析技术（Highresolution melting curve，HRM）进行基因型分型和统计等位基因频率，同时，运用荧光定量 PCR 分析 *Ihh* 基因在牦牛和黄牛不同器官的表达差异。试验结果表明，*Ihh* 基因在牦牛和黄牛的所有组织中均表达。然后采用 SHEsis 和 PHASE 软件对 *Ihh* 基因多态位点进行配对连锁不平衡和单倍型分析，采用 SPSS17.0 进行多态位点单倍型组合与生产性状关联性分析。结果检测到牦牛 *Ihh* 基因外显子 3 的 2 个多态位点 5855（C/T）和 6383（G/A）。群体遗传学分析显示，2 个多态位点均表现为高度多态（$PIC>0.25$）；χ^2 检验表明，甘南牦牛和大通牦牛群体在两个突变位点处于 Hardy-Weinberg 平衡状态（$P>0.05$），而天祝牦牛在 2 个突变位点处未达到 Hardy-Weinberg 平衡（$P<0.05$）；多态位点配对连锁不平衡分析发现，2 个突变位点之间存在强连锁平衡，有 4 种单倍型组合；基因型组合主要发现了 3 种。关联分析表明，*Ihh* 基因 2 个突变位点的 3 种基因型组合对牦牛体斜长、体高、胸围、管围和体重有显著差异（$P<0.05$），具有 CTGA 基因型组合的牦牛个体在体斜长、体高、胸围、管围和体重方面显著高于 CCGG 和 TTAA 型（$P<0.05$）。因此，综上推断 *Ihh* 基因基因型组合与牦牛体斜长、体高、胸围、管围和体重存在相关性。

关键词：牦牛；*Ihh* 基因；多态性；高分辨率熔解曲线；生产性状；荧光定量；基因表达

骨骼是脊椎动物整个身体的支架，它们不仅保护邻近器官，而且保证了机体的自由运动，它们的位置、形状和尺寸基本决定了动物体格的大小。Indian hedgehog（印度豪猪因

* 基金项目：甘肃牧区生产生态生活优化保障技术集成与示范（2012-2016）（2012BAD13B05）；甘肃省科技重大专项计划（1102NKDA027）。

** 作者简介：李天科（1987—　），男，硕士生，主要从事动物遗传育种与繁殖研究，E-mail：litianke1987@163.com.

*** 通讯作者：阎萍，研究员，博士生导师，主要从事动物遗传育种与繁殖研究，E-mail：pingyan@sohu.com.

子，*Ihh*）是一个主要的骨发育调节因子，它对于协调软骨细胞的增殖，软骨细胞的分化和造骨细胞的分化等软骨内骨发育过程是必须的[1-6]。研究发现，*Ihh* 属于一个非常保守的 Hedgehog（Hh）分泌信号家族，在脊椎动物中存在其 2 种同源基因：Sonic Hedgehog（音速豪猪因子，Shh）和 Desert Hedgehog（沙漠豪猪因子，Dhh）[7]。Hedgehog 信号最早是在 1980 年从果蝇体内分离得到[8]，Hedgehog 的突变可使果蝇胚胎发育成毛团状，酷似刺猬，故又被称为刺猬基因[9]。

Ihh 是细胞分泌的一种重要的多肽，由 2 个结构域组成：氨基端结构域（Hh-N）和羧基端结构域（Hh-C），而且高度保守，具有独特的生物活性[10-11]。回顾与 *Ihh* 基因相关的一些研究，可以了解到，*Ihh* 在软骨细胞和成骨细胞的增殖和分化中起调节作用[12]，进一步研究发现，*Ihh* 在成骨细胞中也有表达[13]；K. K. Mak 等[14-15]研究发现，*Ihh* 既可以单独作用于软骨分化，又可以协同 PTHrP（甲状旁腺素相关肽）形成负反馈轴机制，从而维持骨的稳定生长。另也有研究表明，*Ihh* 也在妊娠过程中介导胚胎的植入，敲除 *Ihh* 基因后小鼠不能受孕[16]。表达试验表明，*Ihh* 在小鼠妊娠第 3.5~5.5 天的子宫内膜腔上皮和腺上皮呈高水平表达[17]，且分泌期子宫内膜 *Ihh* 基因表达明显高于增殖期[18]。*Ihh* 基因突变方面的研究主要集中在人和小鼠上，*Ihh* 基因突变的小鼠表现出个体矮小和典型的 A1 型短指/趾症表型[19]；在人上 *Ihh* 的突变主要造成并趾和短指[20]，这都说明 *Ihh* 基因与指骨关节发育密切相关；另外，G. Lettre 等研究发现 *Ihh* 基因的单核苷酸多态性也是人身高差异的主要因素[21]。

目前，对 *Ihh* 基因的相关研究主要集中在人和小鼠上，而在家畜方面，关于 *Ihh* 基因的研究尚未见报道，鉴于 *Ihh* 基因在人和小鼠中具有重要的功能，在牦牛中开展对该基因的相关研究就显得尤为重要。鉴于此，笔者从分子水平入手，初步筛查出 *Ihh* 基因多态位点，并进行多态位点基因型组合与生产性状的相关性分析，找到与生产性状相关的 DNA 标记，为牦牛新品种的选育提供科学依据；同时，以荧光定量 PCR 法分析了 *Ihh* 基因在牦牛和黄牛不同器官的表达差异，为进一步研究 *Ihh* 基因在牦牛体内的分布、表达奠定基础。

1 材料与方法

1.1 试验材料

用于表达谱分析的组织样品采自于 6 头成年甘南牦牛和 6 头成年黄牛的心、肝、脾、肺、肾、肌肉、小肠和胰，取样后立即投入液氮中，回实验室后-80℃保存。用 GIBCO 公司的 TRIZOL Reagent 提取组织总 RNA，用分光光度计测定其浓度，将 RNA 稀释至1μg · μL^{-1} 备用，cDNA 第一链的合成参考 TaKaRa 反转录试剂盒说明书进行。

用于遗传多样性分析的全血样品采自 3 个中国地方牦牛品种（包括 6 月龄的甘南牦牛 200 头、6~9 月龄的大通牦牛 203 头和 2~3 岁的天祝牦牛 208 头），颈静脉采血 10mL，ACD 抗凝（V（血液）：V（ACD）= 6：1），轻微震荡混匀后，-20 ℃保存备用。利用 Relax Gene 血液基因组 DNA 提取系统（天根，北京）提取基因组 DNA，采用 1%的琼脂糖凝胶电泳检测 DNA 的质量，使用 NANO DROP2000（Thermo，美国）检测提取的 DNA 浓度并抽取 20μL 稀释到 18~22ng · μL^{-1}作为 PCR 扩增模板，其余母液保存于-80℃。实地测量 4 项体尺指标（体斜长、体高、胸围、管围）和体重。

1.2 组织表达分析

1.2.1 组织表达谱引物

根据 NCBI 网站中黄牛 *Ihh* 基因的 mRNA（登录号：NM_ 001076870），利用 Pimer5 设计定量 PCR 引物（表 1），内参引物选用文献中牦牛 *GAPDH* 引物[22]，引物由 TaKaRa 公司合成。

表 1 *Ihh* 基因定量分析引物信息

引物 P	引物序列（5′-3′）	退火温度（℃）	片段长度（bp）
Ihh	F：CCAACTACAATCCAGACATCATCTT R：AGATGGCCAGCGAGTTCAG	60.5	102
GAPDH	F：CCACGAGAAGTATAACAACACC R：GTCATAAGTCCCTCCACGAT	54.4	120

1.2.2 荧光定量 PCR

反应体系（10μL）：2μL RT 产物，5μL SYBR Green 荧光染料，上下游引物各 1μL，1μL ddH_2O。反应程序：95℃预变性 30s；95℃5s，退火温度退火 30s，62℃30s，40 个循环；于 60℃采集荧光，采用默认设置自动生成 Ct 值。采用 $2^{-\Delta\Delta Ct}$ 法[22-23]，用 *GAPDH* 作为内参基因对各样本用实时定量 PCR 结果进行校正处理，并以胰腺表达量作为校正者，计算各样本 *Ihh* 基因的相对表达水平。每个样本 3 个重复，取平均值。

1.3 PCR 扩增

1.3.1 SNP 检测

参考 NCBI 数据库牛的 *Ihh* 基因（登录号：NP_ 001070338），并结合牦牛全基因组测序的序列，采用交叉重叠原理[24]设计 4 对引物对 *Ihh* 基因外显子序列进行扩增（表 2），引物由上海生物工程股份有限公司合成。

PCR 扩增体系（25μL）：包括约 20ng · μL^{-1} 的基因组 DNA 1μL，10pmol 的上下游引物各 1μL，2×*Taq* PCR MasterMix（内含 *Taq* DNA 聚合酶、Mg^{2+}、dNTPs 等）（北京，天根）12.5μL，灭菌超纯水 9.5μL。PCR 反应程序：95℃预变性 3min；94℃变性 30s，复性 30s（退火温度见表 2），72℃延伸（时间见表 2），35 个循环；72℃延伸 10min；4℃保存。

表 2 鉴定牦牛 *Ihh* 基因突变位点所用引物信息

引物	引物序列（5′-3′）	退火温度（℃）	片段长度（bp）	延伸时间（s）	扩增位置
Ihh-1	F：AAGGACAGGAGAACACC R：GGAGGCAGTCGGGTAGAGG	54.3	348	20	外显子 1
Ihh-2	F：TACTGCGTGAGTTTGGAT R：GCCTCTTCACCTTCTTGG	59.0	659	30	外显子 2
Ihh-3	F：GGAATCCAAACTCCACCCAG R：CACAGCCGCAAAGCAAGA	56.7	525	30	外显子 3
Ihh-4	F：TCTTGCTTTGCGGCTGTG R：GGGCTTTCCCTACTTATTC	55.3	1 255	50	外显子 3

1.3.2 DNA 测序分析和 HRM 小片段法基因分型[24-25]

PCR 扩增产物通过琼脂糖凝胶回收试剂盒（天根，北京）回收纯化送往大连宝生工生物工程有限公司进行测序，测序结果通过 DNASTAR 软件比对分析。针对所发现的多态位点，利用 Light Scanner Primer Design Software（Idaho 公司，美国）设计出用于 HRM（High Resolution Melting，高分辨率熔解曲线）分析多态位点的引物（表 3），对试验群体样本进行基因型分析。合成 1 对高低温内标[26]（表 4）对熔解曲线进行校正，提高分型的精确度。

HRM-PCR 扩增体系（11 μL）：包括 20ng · μL^{-1}的基因组 DNA 1μL，10pmol 的上下游引物各 0.2μL，2×Taq PCR MasterMix（内含 *Taq* DNA 聚合酶、Mg^{2+}、dNTPs 等）（北京，天根）5μL，灭菌超纯水 3.6μL，10×LC Green 饱和染料（美国，Idaho）1μL。HRM-PCR 反应程序：95℃预变性 5min；94℃变性 20s，Tm（℃）（表 3）退火 20s，72℃延伸 20s，35 个循环；72℃延伸 10min。

高低温内标稀释方法：1OD 内标单链加 400μL 水，退火体系：饱和氯化钠 1μL，内标互补双链各 1μL，补水至 10μL，95℃水浴 3min，然后缓慢降至室温。

HRM 荧光信号采集：将稀释好的高低温内标各 1μL 加入 HRM-PCR 产物单孔中，95℃变性 30s，放入 Light Scanner 96（美国，Idaho）中收集荧光信号。

表 3 HRM 分析多态位点的引物信息

检测位点	引物序列（5′-3′）	退火温度（℃）	片段长度（bp）
5855（C/T）	F：GCAGCTGTCTCCACGCAC F：ACCACATCCTCCACCACC	64.4	112
6383（G/A）	F：GTCCCAACCCAGCCAGCC F：GCAAGCCCACCCAAGAGG	65.5	96

表 4 高低温内标序列信息

高低温内标	Tm（℃）	内标序列（5′-3′）
Low2	69.85	GCGGTCAGTCGGCCTAGCGGTAGCCAGCTGCGGCACTGCGTGACGCTCAG-C3
High2	87.92	ATCGTGATTTCTATAGTTATCTAAGTAGTTGGCATTAATAATTTCATTTT-C3

1.4 数据统计

各组织中表达量的差异用独立样本 *t* 检验进行分析。多态性根据群体遗传学理论直接计算得到基因型和等位基因频率，统计牦牛群体 *Ihh* 基因多态位点的纯和度 *Ho*、杂合度 *He*、有效等位基因数 *Ne*、多态信息含量 *PIC*[27-29]，并进行 Hardy-Weinberg 平衡检验。采用 SPSS17.0 软件对 *Ihh* 基因多态性位点在不同牦牛群体中的基因型分布进行卡方检测，考虑到品种、性别、环境和年龄因素的影响，采用固定模型分析基因型效应对生产性状的影响。

固定模型：$Y_{ijklm}=u+G_i+S_j+A_k+B_l+E_m+\varepsilon_{ijklm}$。

式中，Y_{ijklm}为个体表型记录，u 为群体均值，G_i 为标记基因型效应，S_j 为性别因素，A_k 为年龄因素，B_l 为品种效应，E_m 为环境因素，ε_{ijklm}为随机误差。

2 结果

2.1 牦牛 *Ihh* 基因的组织表达谱分析

利用实时定量 PCR 对牦牛和黄牛 *Ihh* 基因在心、肌肉、小肠、肺、肾、肝、卵巢和胰各个组织的表达情况进行检测，结果见图 1。*Ihh* 基因在牦牛和黄牛所研究的组织中均表达，其中在牦牛肝中的表达量显著高于黄牛（$P<0.05$）；在牦牛和黄牛的卵巢、肺、肾和胰组织中中度表达，在心、脾和肌肉组织中牦牛和黄牛表达量都相对低。

2.2 牦牛 *Ihh* 基因 HRM 方法的分型

通过 DNA 混合池对牦牛 *Ihh* 基因测序后，应用 MEGA.5 软件对拼接得到的 *Ihh* 基因序列与黄牛序列进行比对，并结合测序峰图可知，牦牛 *Ihh* 基因外显子 3 上存在 5855（C/T）和 6383（G/A）的多态位点（图 2），应用高分辨率熔解曲线对多态位点进行基因型分析，结果如图 3 所示。

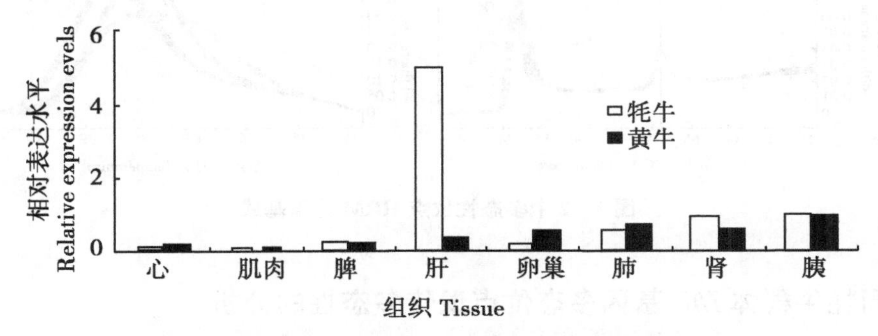

图 1　*Ihh* 基因在牦牛和黄牛不同组织器官的表达模式

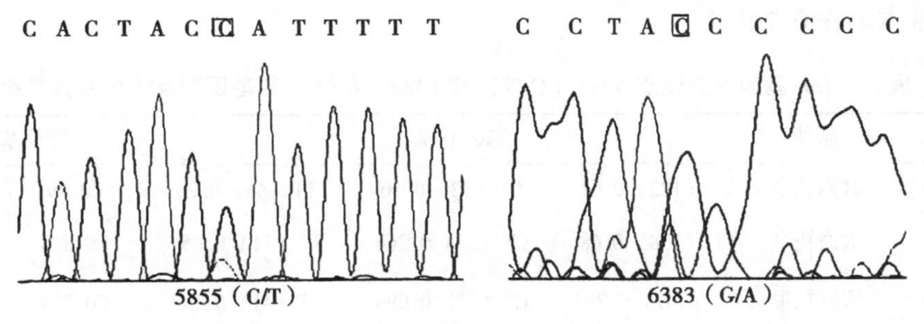

图 2　2 个多态位点的测序峰

5855（C/T）基因型判定：扩增包含多态性位点在内的 112bp 的 DNA 片段，然后将扩增通过 Light Scanner（美国，Idaho）扫描分型，其基本原理是通过 LC Green 饱和染料和 DNA 进行结合，然后通过 Light Scanner 高分辨率分析系统升温，并收集荧光信号得到熔解曲线，在熔解曲线形成过程中，由于核酸分子物理性质的不同，从而导致熔解曲线的不同，达到区分不同样品的目的。结果表明，熔解曲线分析有 3 种基因型：CC、CT 和 TT（图 3A）。统计分析发现，CC 基因型属于优势基因型，个体数明显多于其他基因型。

6383（G/A）基因型判定：扩增包含多态性位点在内的 98bp 的 DNA 片段，然后将扩增通过 LightScanner（美国，Idaho）扫描，通过熔解曲线分析出有 3 种基因型，GG、GA 和 AA（图 3B）。统计分析发现，GG 基因型属于优势基因型，个体数明显多于其他基因型。

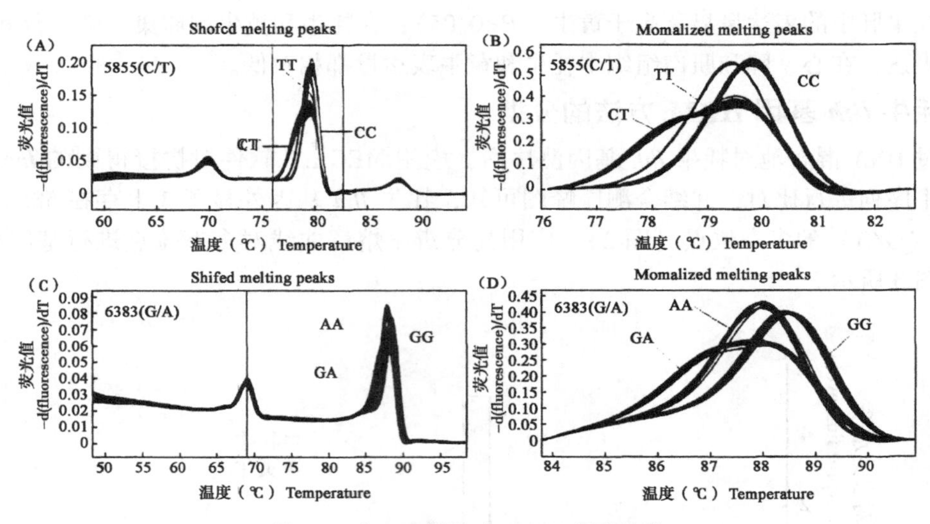

图3　2个多态性位点HRM分型基线

2.3　不同牦牛群体 *Ihh* 基因多态位点遗传多态性的分析

2.3.1　不同牦牛群体 *Ihh* 基因突变位点的基因型频率和等位基因频率

在 *Ihh* 基因5855（C/T）突变位点处，3个牦牛品种的优势等位基因为C（表5）；在 *Ihh* 基因6383（G/A）突变位点处，3个牦牛品种的优势等位基因为G（表5），它们的等位基因频率基本均在0.7以上。

表5　*Ihh* 基因突变位点5855（C/T）和6383（G/A）的基因型及等位基因频率

位点	品种	基因型频率			等位基因频率	
5855（C/T）	甘南牦牛	CC（122）0.61	CT（32）0.16	TT（46）0.23	C0.69	T0.31
	大通牦牛	CC（138）0.68	CT（54）0.266	TT（11）0.054	C0.813	T0.187
	天祝牦牛	CC（147）0.707	CT（20）0.096	TT（41）0.197	C0.755	T0.245
6383（G/A）	甘南牦牛	GG（149）0.745	GA（18）0.09	AA（33）0.165	G0.79	A0.21
	大通牦牛	GG（133）0.655	GA（42）0.207	AA（28）0.138	G0.759	A0.241
	天祝牦牛	GG（124）0.596	GA（38）0.183	AA（46）0.221	G0.688	A0.312

2.3.2　不同牦牛群体 *Ihh* 基因群体遗传特性

多态位点5855（C/T）在甘南牦牛和大通牦牛群体中，表现为高度多态（*PIC*>0.5），天祝牦牛表现为中度多态（0.5<*PIC*<0.25）（表6）；多态位点6383（G/A）在大通牦牛和天祝牦牛群体中表现为高度多态（*PIC*>0.5），在甘南牦牛群体中表现为中度多态（*PIC*<0.5）（表6）。另外Hardy-Weinberg平衡检验发现，在多态位点5855（C/T）和（6383（G/A）处，甘南牦牛和大通牦牛均未达到Hardy-Weinberg平衡（*P*<0.05），天祝牦牛达到了Hardy-Weinberg平衡（*P*>0.05）。

表 6 *Ihh* 基因突变位点 5855（C/T）和 6383（G/A）的遗传多态性

位点	品种	纯合度	杂合度	有效等位基因数	多态信息含量	HWE	
						卡方 χ^2	*P* 值 *P* value
5855（C/T）	甘南牦牛	0.425	0.575	2.353	0.536	0.073	$P<0.01$
	大通牦牛	0.465	0.535	2.151	0.532	0.436	$P<0.05$
	天祝牦牛	0.539	0.461	1.855	0.422	2.404	$P>0.05$
6383（G/A）	甘南牦牛	0.582	0.418	1.718	0.388	1.679	$P<0.05$
	大通牦牛	0.448	0.552	2.232	0.536	0.657	$P<0.05$
	天祝牦牛	0.404	0.596	2.475	0.561	2.674	$P>0.05$

2.4 不同牦牛群体 *Ihh* 基因不同基因型组合与体尺性状的关联性分析

对各牦牛群体 *Ihh* 基因各多态位点进行 HRM 分型统计后，运用 SHEsis 软件进行位点间的连锁不平衡分析，2 个位点间可能存在强连锁不平衡（D'>0.75，$r^2>0.33$）；同时，在 3 个牦牛群体中发现了 4 种单倍型组合（表 7）。参照鲍斌等[30]研究方法从基因型组合角度出发，选取 3 个牦牛群体中共同的基因型组合 CCGG、TTAA 和 CTGA（表 8）运用 SPSS17.0 软件分析这 3 种基因型组合与生长性状的相关性。由表 9 可知，甘南牦牛基因型组合对体重有显著影响（$P<0.05$），对体长、体高、胸围和管围差异不显著（$P>0.05$）；由表 9 可知，大通牦牛基因型组合对体长、体高、胸围、管围和体重都有显著影响（$P<0.05$）；由表 9 可知，天祝牦牛基因型组合对胸围和体重有显著影响（$P<0.05$），而在体长、体高和管围方面差异不显著（$P>0.05$）。总体比较发现 CTGA 基因型组合在各个牦牛群体中都具有较高的生产性状。

表 7 牦牛 *Ihh* 基因多态位点的分布及频率

品种	单倍型组合			
	CG	TA	CA	TG
甘南牦牛	286（0.688）	71（0.177）	26（0.064）	0
大通牦牛	303（0.748）	102（0.245）	28（0.067）	4（0.010）
天祝牦牛	274（0.685）	82（0.205）	2（0.005）	42（0.105）

表 8 牦牛 *Ihh* 基因突变位点 5855（C/T）和 6383（G/A）基因型组合

品种	基因型组合								
	CCGG	CCAA	CCGA	TTGG	TTAA	TTGA	CTAA	CTGA	CTGG
甘南牦牛	122	1	0	1	32	12	0	5	27
大通牦牛	119	0	23	0	41	0	5	15	0
天祝牦牛	139	2	5	2	7	0	17	36	0

表9 甘南牦牛、大通牦牛和天祝牦牛 *Ihh* 基因不同基因型组合与生产性状关联分析

品种	基因型	体长/cm	体高/cm	胸围/cm	管围/cm	体重/kg
甘南牦牛	CCGG（122）	83.91±0.819	85.54±0.998	106.79±1.099	11.66±0.137	86.74±3.133a
	CTGA（5）	83.60±0.668	85.57±0.727	107.49±1.741	11.59±0.112	85.66±1.622a
	TTAA（32）	82.77±1.138	84.46±0.090	104.18±1.791	11.78±0.139	78.05±1.715b
大通牦牛	CCGG（119）	85.86±0.971^{a}	85.32±0.725^{a}	113.97±1.186^{a}	11.88±0.047^{a}	72.71±1.238^{ab}
	CTGA（15）	86.96±0.621^{a}	86.61±0.604^{b}	109.87±1.006^{b}	11.91±0.066^{ab}	76.60±2.023^{a}
	TTAA（41）	84.10±0.758^{b}	85.77±0.691^{a}	109.56±1.145^{b}	12.06±0.057^{b}	70.92±1.531^{b}
天猪牦牛	CCGG（139）	113.50±0.391	110.4±0.654	145.42±0.558^{A}	13.64±0.393	210.48±1.455^{ab}
	CTGA（36）	117.57±1.426	114.08±0.753	156.86±0.791^{C}	14.07±0.701	223.42±1.126^{a}
	TTAA（7）	115.87±0.419	109.43±1.485	150.75±0.8865^{B}	13.55±0.202	196.37±0.871^{b}

注：同列相同字母表示差异不显著（$P>0.05$），不同字母表示差异显著（$P<0.05$）

3 讨论

在长骨发育过程中，*Ihh* 主要表达在肥大软骨细胞区域。*Ihh* 基因敲除的小鼠显示出软骨细胞增殖减少，肥大软骨细胞增加和造骨细胞缺失等多方面的骨发育异常[31-32]。本试验研究的牦牛和黄牛的8个组织中，均检测到 *Ihh* 基因 mRNA 的表达，说明 *Ihh* 基因是黄牛和牦牛很多组织细胞功能所必需。

牛的生长性状属微效多基因控制的数量性状，常规的表型选择，精确性较差，遗传进展缓慢。而且关于调控牛生长的机理目前还不清楚，主效基因还没有确定，目前可能与生产性状有关的候选基因主要有 *GH*[33]、*GHR*[34-35]和 *IGF*[36-37]等，而牛的 *Ihh* 基因位于2号染色体107722665~107728939位置，通过与牛2号染色体上的 QTL 数据（http：//www.animalgenome.org/QTLdb/pubs.php）库进行比对发现，在包括 *Ihh* 基因在内的106.5~115.4cm 的 QTL 区域主要与体尺性状、产犊、结核病和乳房炎等性状有关，而且研究发现 *Ihh* 基因突变的小鼠（E95K 突变）表现出个体矮小和典型的 A1 型短指/趾症表型—第2~第4指中指节严重缩短，第5指中指节缺失[19]，在人上 *Ihh* 基因的突变主要造成并趾和短指[20]，这都证明 *Ihh* 基因与指骨关节发育密切相关。因此把 *Ihh* 基因作为影响牛生产性状的候选基因进行研究。

本试验通过 DNA 混合池测序得到甘南牦牛 *Ihh* 基因2个多态位点，均位于外显子3上，在群体中均存在3种基因型和2种等位基因，TT 和 AA 分别为其优势基因型，其优势等位基因频率都达到了0.7以上，这可能与样本的品种、家系和自然环境有关系。按照 D. Botstein 等[29]提出衡量基因变异程度高低的多态信息含量指标（*PIC*）的标准，本试验中所检测到的2个多态位点杂合度及 *PIC* 值都在高等水平，故这2个多态位点均为高度多态，遗传变异高，选择余地大。Hardy-Weinberg 平衡检测表明，在2个突变位点处，甘南牦牛和大通牦牛均未达到哈代—温伯格平衡（$P<0.05$），这可能是因为甘南牦牛和大通牦牛较长期的人工选育，人为的因素为其不处于 Hardy-Weinberg 平衡的主要因素；而2个突变位点在天祝牦牛中均达到 Hardy-Weinberg 平衡（$P>0.05$），可能因牦牛种群是一个较为原始的封闭种群，长期处于闭锁繁育状态，从而处于动态平衡。

试验中对不同牦牛品种2个突变位点进行连锁不平衡分析后，发现2个突变位点有强连锁不平衡，统计发现了4种单倍型组合，发现了主要的3种基因型组合，对3种基因型组合与生产性状间的相关性分析表明，CCGG、CTGA和TTAA的3种基因型组合可能与牦牛的生产性状相关，而在这3种组合中，CTGA具有较高的生产性能，但在群体中数目少，需要加大人工选育。同时，这种相关性分析降低了单个突变位点受环境、其他位点和其他相关微效基因的影响[38]，得到较可靠的基因型效应估计值，在评估品种和种群的遗传改良时更准确。

4 结论

Ihh 基因2个突变位点的3种基因型组合对牦牛体斜长、体高、胸围、管围和体重均有不同程度的影响，具有CTGA基因型组合的牦牛个体可能体型更大，因此，*Ihh* 基因可以作为主效候选基因或与主基因紧密连锁的分子标记，将来可以尝试通过人工干预，扩大有利基因型组合个体的数目。

参考文献

[1] 齐晓龙，王玉炯，吴更. Hedgehog信号通路研究进展［J］. 中国细胞生物学学报，2013，35（8）：1211-1220.

[2] VORTKAMP A，LEE K，LANSKE B，et al. Regulation of rate of cartilage differentiation by Indian-hedgehog and PTH-related protein［J］. Science，1996，273：613-622.

[3] ST JACQUES B，HAMMERSCHMIDT M，MCMAHON A P. Indian hedgehog signaling regulates proliferation and differentiation of chondrocytes and is essential for bone formation［J］. Genes Dev，1999，13：2072-2086.

[4] KARP S J，SCHIPANI E，ST JACQUES B，et al. Indian hedgehog coordinates endochondral bone growthand morphogenesis via parathyroid hormone relatedprotein-dependent and independent pathways［J］. Development，2000，127：543-548.

[5] LONG F，ZHANG X M，KARP S，et al. Genetic manipulation of hedgehog signaling in the endochondralskeleton reveals a direct role in the regulation ofchondrocyte proliferation［J］. Development，2001，128：5099-5108.

[6] CHUNG U I，SCHIPANI E，MCMAHON A P，et al. Indian hedgehog couples chondrogenesis to osteogenesis in endochondral bone development［J］. J Clin Invest，2001，107：295-304.

[7] 高耀祖，魏垒，李鹏翠，等. Indian hedgehog在软骨内成骨和骨关节炎中作用的研究进展［J］. 中国矫形外科杂志，2013（17）：1472-1475.

[8] NUSSLEIN VOLHARD C，WIESCHAUS E. Mutations affecting segment number and polarity in Drosophila［J］. Nature，1980，287（30）：795-801.

[9] CHANG D T，LOPEZ A，VON KESSLER D P，et al. Products，genetic linkage and limb patterning activityof a murine Hedgehog gene［J］. Development，1994，120（11）：3339-3353.

[10] EHLEN E C，BUELENS L A，VORTKAM P. Hedgehog signaling in skeletal development［J］. BirthDefects Res Part C，Embryo Today Rev，2006，78（3）：267-279.

[11] WEIR E D，PHILBRIEK W M，AMLING M，et al. Targeted overex pression of parathyroid hormone-related peptide in chondrocytes causes chondrodysplasiaand delayed endochondral bone formation［J］. ProcNatl Acad Sci USA，1996，93（19）：10240-10245.

[12] JEMTLAND R，DIVIETI P，LEE K，et al. Hedgehogpromotes primary osteoblast differentiation and

increases PTHrP mRNA expression and iPTHrP secretion [J]. Bone, 2003, 32 (6): 611-620.
[13] 韩磊，张晓玲，李晅，等. Hedgehog 通路对成骨细胞增殖和分化的作用研究 [J]. 上海口腔医学, 2009, 18 (3): 287-290.
[14] MAK K K, KRONENBERG H M, CHUANG P T, et al. Indian Hedgehog signals independently of PTHrPto promote chondrocyte hypertrophy [J]. Development, 2008, 135 (11): 1947-1956.
[15] KOBAYASHI T, SOEGIARTO D W, YANG Y, etal. Indian hedgehog stimulates periarticular chondrocyte differentiation to regulate growth plate length independently of PTHrP [J]. J Clin Invest, 2005, 115 (7): 1734-1742.
[16] LEE K, JEONG J, KWAK I, et al. Indian Hedgehog is amajor mediator of progesterone signaling in the mouse uterus [J]. Nat Genet, 2006, 38 (10): 1024-1029.
[17] KUBOTA K, YAMAUCHI N, YAMAGAMI K, et al. Steroidal regulation of Ihh and Gli-1expression in the ratuterus [J]. Cell Tissue Res, 2010, 340 (2): 389-395.
[18] WEI Q, LEVENS E D, STEFANSSON L, et al. Indian Hedgehog and its targets in human endometrium: menstrual cycle expression and response to CDB-2914 [J]. J Clin Invest, 2010, 95 (12): 5330-5337.
[19] GAO B, HU J, STRICKER S, et al. A mutation in Ihhthat causes digit abnormalities alters its signalling capacity and range [J]. Nature, 2009, 458 (7242): 1196-1200.
[20] McCREADY M E, SWEENEY E, FRYER A E, et al. A novel mutation in the IHH gene causes brachydactyly type A1: a 95-year-old mystery resolved [J]. HumGenet, 2002 (111): 368-375.
[21] LETTRE G, JACKSON A U, GIEGER C, et al. Identification of ten loci associated with height highlightsnew biological pathways in human growth [J]. NatGenet, 2008, 40 (4): 584-591.
[22] PEI J, LANG X, GUO X, et al. Expression profiles ofgrowth hormone receptor and insulinlike growth factor I in cattle and yak tissues revealed by quantitativereal-time PCR [J]. Arch Anim Breed, 2013, 70: 7482-7491.
[23] 李爱民，马 云，杨东英，等. 鲁西牛 ANGPTL6 基因的 3 个多态位点与其生长性状的关联性分析 [J]. 中国农业科学, 2012, 45 (11): 2306-2314.
[24] MICHAEL L, ROBERT P, ROBERT P. Genotypingof single-nucleotide polymorphisms by high-resolution melting of small amplicons [J]. Mol DiagnostGenet, 2004, 50 (7): 1156-1164.
[25] YE M H, CHEN J L, ZHAO G P. Associations of AFABP and H-FABP markers with the content of intramuscu-lar fat in Beijing-you chicken [J]. AnimBiotechnol, 2010, 21: 14-24.
[26] CAMERON N, STEVEN F. Base-pair neutral homozygotes can be discriminated by calibrated high resolution melting of small amplicons [J]. Nucleic AcidsRes, 2008, 1093 (10): 1-8.
[27] NEI M, ROYCHOUDHURY A K. Sampling variances of heterozygosity and genetic distance [J]. Ge-8 5netics, 1974, 76 (2): 379-390.
[28] NEI M, LI W H. Mathematical model for studying genetic variation in terms of restriction endonucleases [J]. Proc Natl Acad Sci USA, 1979, 76 (10): 5269-5273.
[29] BOTSTEIN D, WHITE R L, SKOLNICK M, et al. Construction of a genetic linkage map in man usingrestriction fragment length polymorphisms [J]. Am JHum Genet, 1980, 32 (3): 314-331.
[30] 鲍斌，房兴堂，陈宏，等. 中国荷斯坦牛 STAT5A 基因遗传多态性与泌乳性状的相关分析 [J]. 中国农业科学, 2008, 41 (6): 1872-1878.
[31] ST JACQUES B, HAMMERSCHMIDT M, MCMAHON A P. Indian hedgehog signaling regulates proliferation and differentiation of chondrocytes and is essential for bone formation [J]. Genes Dev, 1999, 13: 2072-2086.
[32] 赵恒伍，张锦程，李长有. 血管内皮生长因子调节成骨细胞中 Ihh 和 p38MAPK 的表达 [J].

中国医科大学学报，2011，40（2）：113-116.

[33] WU Y，PAN A L，PI J S，et al. One novel SNP ofgrowth hormone gene and its associations withgrowth and carcass traits in ducks［J］. Mol Biol Rep，2012（39）：8027-8033.

[34] OUYANG J H，XIE L，NIE Q，et al. Single nucleotidepolymorphism（SNP）at the GHR gene and its associations with chicken growth and fat deposition traits［J］. Br Poult Sci，2008（49）：87-95.

[35] 赵高锋，陈宏，雷初朝，等．秦川牛 GHR 基因 SNPs 及其与生长性状关系的研究［J］. 遗传，2007，29（3）：319-323.

[36] MULLEN M P，BERRY D P，DAWNJ，et al. Singlenucleotide polymorphisms in the insulin-like growthfactor 1（IGF-1）gene are associated with performancein Holstein-Friesian dairy cattle［J］. Origl Res Article，2011（2）：1-9.

[37] TRUJILLO A I，CASAL A，PEÑAGARICANO F，etal. Association of SNP of neuropeptide Y，leptin，andIGF-1genes with residual feed intake in confinementand under grazing condition in Angus cattle［J］. JAnim Sci，2013，91：4235-4244.

[38] 鞠志花，王洪梅，李秋玲，等．荷斯坦牛 κ-酪蛋白基因两 SNPs 位点基因型组合对产奶性状的影响［J］. 农业生物技术学报，2009，17（4）：587-592.

（发表于《畜牧兽医学报》）

中药治疗牛子宫内膜炎的系统评价和 Meta 分析

董书伟，张世栋，王东升，
王　慧，苗小楼，严作廷*，杨志强*
（中国农业科学院兰州畜牧与兽药研究所农业部兽用药物
创制重点实验室/甘肃省新兽药工程重点实验，兰州　730050）

摘　要：为系统评价中药治疗牛子宫内膜炎的临床疗效，计算机检索 CNKI、Wan Fang Data、CBM、PubMed、Web of Science 和 Science Direct 数据库，全面收集中药治疗牛子宫内膜炎的临床对照试验。按照 Cochrane 方法纳入文献，采用 RevMan 5.3 软件进行 Meta 分析。结果共纳入 24 项试验，2 782 头患病牛，其中，试验组 1 739 头，对照组 1 043头，试验组均采用复方中药治疗，对照组采用土霉素、青霉素、青链霉素联合使用和环丙沙星治疗。Meta 分析显示：复方中药治疗牛子宫内膜炎的总有效率优于抗生素整体组［OR=2.40，95%CI（1.74，2.32），$P<0.01$］，但与青链霉素联合使用治疗亚组的总有效率相当；中药治愈率优于整体抗生素组［OR=1.54，95%CI<1.25，1，89），$P<0.01$），但与土霉素治疗亚组的治愈率相当；受胎率显著高于抗生素组［OR=1.68，95%CI<1.40，2.03），$P<0.01$］。发表偏倚分析显示，本研究纳入的文献有一定的发表偏倚性。根据现有证据显示，复方中药治疗牛子宫内膜炎的疗效优于抗生素，并有助于提高牛的受胎率。

关键词：子宫内膜炎；Meta 分析；中药；牛

奶牛子宫内膜炎是由致病微生物感染子宫黏膜而导致的一种繁殖障碍性疾病，是危害奶牛养殖业的三大疾病之一[1]。奶牛子宫内膜炎在美国、加拿大和法国部分地区的发病率为 28%[2]36%[3]和 32%[4]，瞿自明等报道我国 16 个城市 41 个奶牛场中子宫内膜炎发病率为 17.26%，占不孕牛的 68.34%[5]。子宫内膜炎严重影响奶牛的发情和妊娠，导致屡配不孕，是引起奶牛不孕症的主要原因[4]。据报道，在美国每头子宫内膜炎患病牛就会造成329~386 美元损失[6]，另外，它延长胎间距，降低产奶量，诱发乳房炎，导致奶牛丧失繁殖能力而被迫淘汰，因此，子宫内膜炎给奶牛养殖业造成了巨大的经济损失。

奶牛患子宫内膜炎的主要病因是病原微生物的感染，包括葡萄球菌、大肠杆菌、链球菌、化脓棒状杆菌和支原体等[7-8]，因此，临床上治疗奶牛子宫内膜炎的方法主要是子宫灌注抗生素[9]和激素疗法[10]，常用土霉素、青霉素、青链霉素联合使用或盐酸环丙沙星等抗生素和前列腺素类激素，并且曾经取得良好的控制效果，但是随着抗生素的长期和大肆滥用，造成了耐药菌株的不断出现和牛奶中抗生素残留，降低了药物的疗效，并已严重威胁着人类的健康和乳制品的食品安全问题[11]。

中药是我国传统医学的瑰宝，具有安全、有效、低毒、低残留、零休药期和不易产生耐药性的特点，在治疗人类和动物疾病方面有独特的优势，因此，利用中药治疗奶牛子宫内膜炎符合奶牛绿色健康养殖的行业需求。我国兽医科技人员在防治奶牛子宫内膜炎方面研发了多种中药制剂，并发表了大量的临床疗效试验，但是报道结果差别却很大，质量参差不齐，不能为临床兽医工作者提供实践参考，导致中药的疗效尚未得到业界的广泛认可。

Meta分析是针对同一科学问题而开展的多个独立研究结果进行方差合并及定量分析的统计学方法，其本质是通过增加样本含量提高参数统计功效，从而提高结果的可靠性，是一种客观定量的综述形式，其在人类循证医学上应用十分广泛，为临床决策提供了强有力的依据。近年来，该方法在优化日粮中营养素的添加水平和评价动物生长性能等畜牧领域中已有应用[12-13]，但在我国兽医学领域的应用却鲜有报道。作者拟利用Meta分析方法，对前人发表的关于中药治疗奶牛子宫内膜炎的临床对照试验进行系统评价，为防治奶牛子宫内膜炎提供有价值的参考，也为我国兽医学研究提供新的方法。

1 材料与方法

1.1 检索策略

通过计算机检索CNKI、Wan Fang Data、中国生物医学数据库（Chinese Biomedical Database，CBM）和PubMed、Science Direct、Web of Science数据库，检索时限均为从建库至2015年1月。检索词包括研究对象和干预措施，中文检索词为中药、草药、中草药、子宫内膜炎、子宫炎、牛；英文检索词为Traditional Chinese Medicine、Chinese Herbal Drug、TCM、Endometritis、Metritis、Cow，所有检索采用主题词和自由词相结合的方式，同时手工检索纳入研究的参考文献。并与本领域专家学者，有关作者和兽药厂家联系，收集正在进行的试验和“灰色文献”，收集会议论文集，获取未发表的文献信息。

1.2 文献筛选

首先，用文献管理软件NoteExpress3.0将初步筛选到的文献归类、整理和去重。其次，正式筛选前，随机抽取检索结果中的文献进行预筛选，发现筛选标准中的问题应及时修改。再次，通过阅读标题和摘要进行初筛，再通过阅读全文复筛排除不符合纳入和排除标准的文献。最后，所有文献的初筛和复筛均由2名研究者独立筛选文献，提取资料并由另一方进行检查核对，如有分歧则由第三名研究者介入讨论协调解决。

1.3 文献的纳入标准

1.3.1 研究设计

所有研究均为随机对照试验（randomized controlled trial，RCT）或非随机的临床对照试验（clinical controlled trial，CCT），无论是否使用随机化分组和盲法均应纳入。

1.3.2 研究对象

子宫内膜炎患病牛。

1.3.3 干预措施

试验组药物为中药，对照组药物为土霉素、青霉素、青链霉素联合使用和盐酸环丙沙星；所有病例均先用生理盐水清洗子宫，将分泌物排出后再子宫灌注药物。

1.3.4 结局指标

疗效指标：总有效率、痊愈率和治疗后3个发情周期内的受胎串。药物治疗牛子宫内膜炎后，通过检查阴道、子宫状态的变化及分泌物的数量、色泽、形状来判断药物的疗效。痊愈的判定标准：临床症状消失，子宫大小、质地恢复正常，黏液变透明清亮，黏液量正常，无臭味，发情恢复正常；有效的判定标准：临床症状消失，但是子宫状态、黏液状态、发情状态和3个情期内受胎率，这四项指标中有一项未恢复正常；无效的判定标准：临床症状没有改善甚至恶化；受胎的判定标准：3个发情周期内授精，经人工或B超检查确认妊娠。总有效头数=痊愈头数+有效头数。

1.4 文献的排除标准

文献的排除标准：①综述、评论、单个病例报告和诊疗体会；②重复发表和同次试验；③研究对象不是牛；④干预措施是中药和其他药物或生物制剂联用，对照药物是中药或是和西药联用；⑤评价指标不一致；⑥未设对照组。

1.5 数据提取

①一般资料：题目、作者、发表日期和文献来源；②研究特征：研究对象、治疗组和对照组动物总数，给药后总有效数、治疗组和对照组给药方式。

1.6 方法学质量评价

纳入研究文献质量评价采用修正后JADAD评分量表，包括随机序列的产生、分组隐藏、盲法、退出与失访，若有失访或退出，是否采用意向性分析法（1TT），如果每项仅为提及，记为1分，如果说明具体方法记为2分，≤3分为低质量研究，4~7分为高质量研究。

1.7 统计学方法

采用Cochrane协作网提供的Rev Man 5.3统计软件进行Meta分析。分类变量采用比值比（odds ratio，OR）；计量资料采用加权均数差（weighted 171ean difference，WMD），并计算各效应量的95%可信区间（confidence interval，C1）。各纳入研究结果间的异质性采用X^2检验。当各研究间有统计学同质性（$P>0.1$，$I^2<50\%$），采用固定效应模型进行合并分析；如各研究间存在统计学异质性（$P<0.1$，$I^2>50\%$），分析其异质性来源，采用随机效应模型进行分析，如两组间异质性过大或无法查找数据来源时，采用描述性分析。根据各纳入文献中对照组干预措施的不同，将对照组分为土霉素、青霉素、青链霉素联合使用和盐酸环丙沙星进行亚组分析，并采用倒漏斗图分析的方法检验是否存在发表偏倚[14]。

2 结果

2.1 文献检索与筛选

按照检索策略共得到文献740篇，阅读文献题目及摘要，排除非临床研究，非对照试验、综述、重复发表的文献197篇，进一步阅读全，排除不符合纳入标准的文献123篇，再排除对照药物非土霉素、青霉素、青链霉素和环丙沙星的文献51篇，最终纳入24篇文献[15-37]，发表时间1995—2014年，文献类型包括期刊、专利、学位论文和会议论文，具体筛选流程见图1。

2.2 纳入文献的基本特征及方法学质量评价

纳入的所有文献均在国内进行，纳入研究对象为产后20d以上的子宫内膜炎患病牛，共治疗2 782头，其中试验组1 739头，对照组1 043头，单个研究样本量在10~315头，见表1。试验组均使用中药治疗，虽然具体配方不同，但主要治则均是清热解毒、活血化瘀、祛腐排脓，还有催情促孕，主要药物有当归、丹参、益母草、红花、蒲公英和淫羊藿等。给药途径主要是子宫灌注，剂型主要是中药灌注剂，只有李正团[24]使用散剂经口灌服，对照组均使用土霉素、青霉素、青链霉素和环丙沙星，见表1。

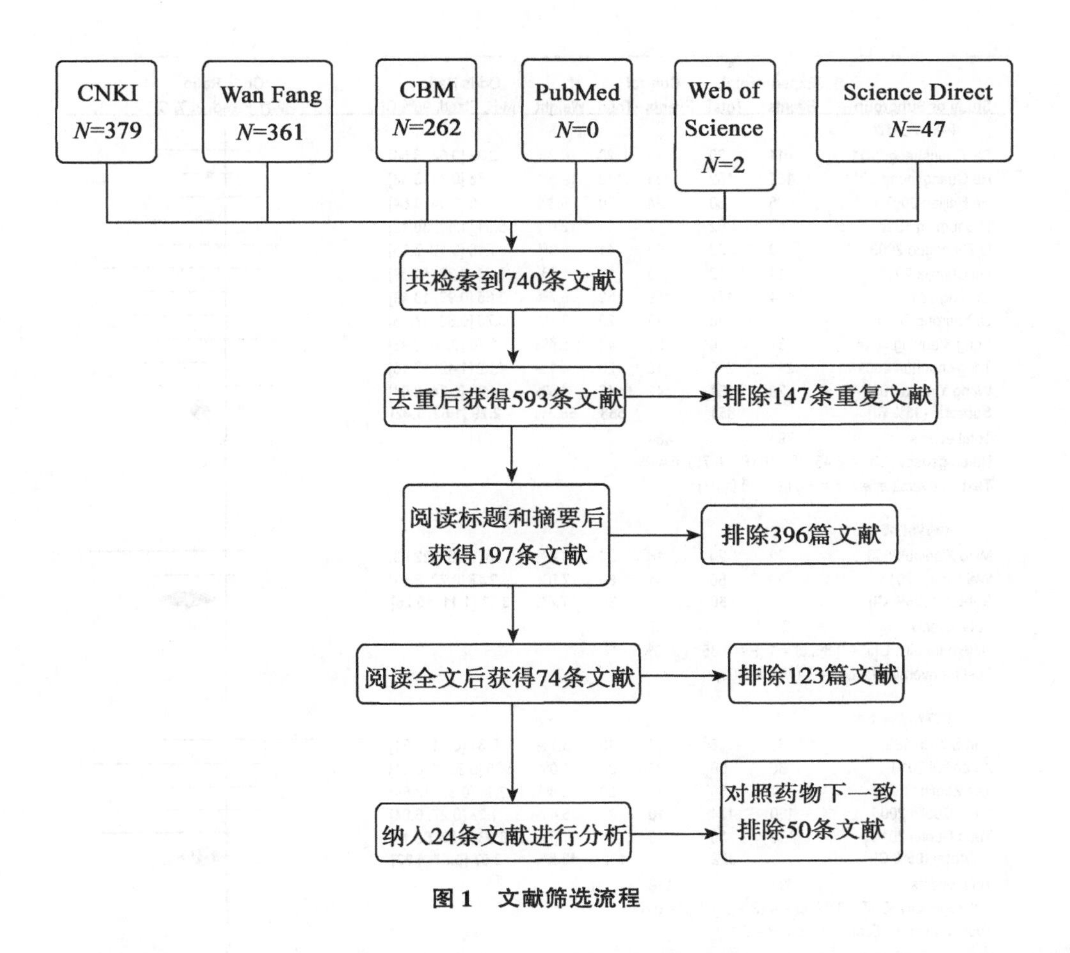

图 1　文献筛选流程

方法学质量评价显示：24 篇文献中只有何海健等[30]提及采用分娩先后顺序的方法随机分组，大部分文献只是提及随机分组，还有文献[24,33-35]未提及随机分组；没有文献提及随机分配方案的隐藏和使用盲法，1 篇文献[36]提及治疗中的脱落或失访情况，但未进行 ITT 分析。1 篇文献[22]提及药物的不良反应，其他研究的评价条目均显示不清楚，根据 JADAD 评分标准，本研究纳入的文献评分大部分只有 1 分，说明证据质量较低。

2.3　治疗牛子宫内膜炎的总有效率 Meta 分析

有 22 项文献提到总有效率，各文献间异质性 $I^2<50\%$，无统计学异质性，故采用固定效应模型，Meta 分析结果显示，中药和抗生素治疗组的总有效率分别为 93.5%（1 318/1 409）和 87.1%（776/ 891），与抗生素治疗组相比，置信区间落在无效线的右侧，则表明中药组的效应量大于抗生素组，即复方中药治疗牛子宫内膜炎的总有效率显著优于抗生素组［OR=2.40，95%CI（1.74，3.32），$P<0.01$］。亚组分析结果显示，中药组的总有效率均显著优于土霉素对照组［OR = 2.29，95% CI（1.53，3.42），$P<0.0001$］、青霉素对照组［OR=3.37，95%CI（1.11，10.26），$P=0.03<0.05$］和环丙沙星对照组［OR = 2.77，95%CI（1.15，6.65），$P=0.02<0.05$］；而中药治疗组的总有效率与青链霉素对照组相当，统计学无差异［OR=2.07，95%CI（0.87，4.92），$P=0.15>0.05$］，见图 2。

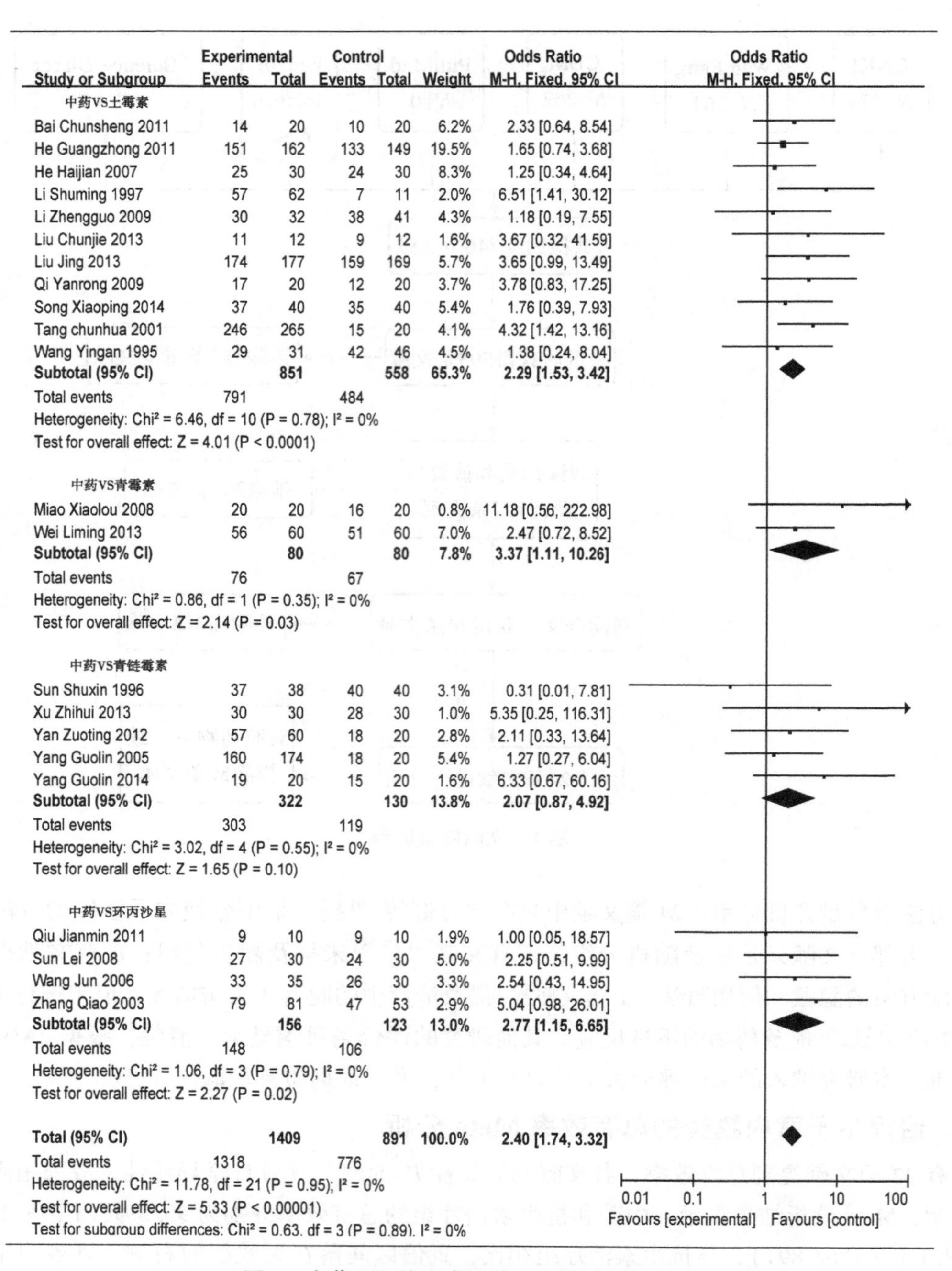

图 2　中药组和抗生素组的总有效率比较森林

2.4　治疗牛子宫内膜炎的治愈率 Meta 分析

有 18 项研究提到治愈率，各研究间异质性 I^2= 44%<50%，有低度异质性，采用固定效应模型分析。Meta 分析结果显示，与抗生素治疗组相比，置信区间落在无效线的右侧，则表明中药组的效应量大于抗生素组，即复方中药治疗牛子宫内膜炎的总治愈率优于抗生素组

[OR=1.54，95%CI（1.25，1.89）]，两组间差异有统计学意义（$P<0.01$），中药和抗生素治疗的治愈率分别为85.6%（1 424/ 1 664）和79.0%（868/1 122）。亚组分析结果显示，中药组的治愈率均显著优于青霉素组 [OR=2.19，95%CI（1.46，3.30），$P=0.000\ 2<0.05$]、青链霉素联合使用组 [OR=3.28，95%CI（1.67，6.42），$P=0.0005\sim0.05$] 和环丙沙星组 [OR=2.29，95% CI（1.08，4.86），$P=0.03<0.05$]，但中药组与土霉素组的治愈率相当 [OR=1.11，95%CI（0.84，1.46），$P=0.39>0.05$]，见图3。

2.5 给药后3个情期受胎率的Meta分析

有17项试验研究了牛子宫内膜炎治疗后的受胎率，各研究间异质性 $I^2=33\%<50\%$，有低度异质性，采用固定效应模型进行分析。Meta分析结果显示，与抗生素治疗组相比，置信区间落在无效线的右侧，则表明中药组的效应量大于抗生素组，即复方中药治疗牛子宫内膜炎的受胎率优于抗生素组 [OR= 1.68，95%CI（1.40，2.03）]，两组间差异有统计学意义（$P<0.01$），中药和抗生素治疗组的受胎率分别为76.8%（1 336/1 740）和66.6%（639/959）。亚组分析结果显示，中药治疗后牛在3个情期内的受胎率均显著优于土霉素组 [OR=1.60，95%CI（1.26，2.04），$P=0.000\ 1\sim0.05$]、青霉素对照组 [OR= 1.58，95% CI（1.04，2.39），$P=0.03<0.05$]、青链霉素对照组 [OR=1.78，95%CI（1.09，2.91），$P=0.02<0.05$] 和环丙沙星对照组 [OR=3.0，95%CI（1.34，6.73），$P=0.008\sim0.05$]，见图4。

表1 纳入文献的基本特征及质量评价

纳入研究		患病牛头严		干预措施		给药途径	疗效指标	质量评价				
作者	时间	试验组	对照组	试验组	对照组			随机方法	分组隐藏	盲法	脱落失访	评分
宋晓平	2014	40	40	中药	土霉素	子宫灌注	①②	1	0	0	0	1
刘春杰	2013	13	13	中药	上霉素	子宫灌注	①③	1	0	0	0	1
刘 镜	2013	177	169	中药	上霉素	子宫灌注	①②	1	0	0	0	1
白春生	2011	20	20	中药	上霉素	子宫灌注	①②	1	0	0	0	1
何光中	2011	162	149	中药	上霉素	子宫灌注	①②③	1	0	0	0	1
祁燕蓉	2009	20	20	中药	上霉素	子宫灌注	①③	1	0	0	0	1
李正国	2009	32	41	中药	上霉素	经口灌服	①②	0	0	0	0	0
何海健	2007	30	30	中药	上霉素	子宫灌注	①②	3	0	0	0	3
汤春华	2001	265	20	中药	上霉素	子宫灌注	①③	0	0	0	0	0
李树明	1997	62	11	中药	土霉素	子宫灌注	①②③	1	0	0	0	1
王应安	1995	31	46	中药	土霉素	子宫灌注	①②	0	0	0	0	0
杨国林	2014	20	20	中药	青链霉素	子宫灌注	①②③	1	0	0	0	1
徐志辉	2013	30	30	中药	青链霉素	子宫灌注	①②③	1	0	0	0	1
严作廷	2012	60	20	中药	青链霉素	子宫灌注	①②③	1	0	0	0	1
杨国林	2005	174	20	中药	青链霉素	子宫灌注	①②③	1	0	0	0	1
张来英	1998	15	15	中药	青链霉素	子宫灌注	③	1	0	0	1	3
孙树新	1995	38	40	中药	青链霉素	子宫灌注	①③	1	0	0	0	1
魏立明	2013	60	60	中药	青霉素	子宫灌注	①②	1	0	0	0	1

（续表）

纳入研究		患病牛头严		干预措施		给药途径	疗效指标	质量评价				
作者	时间	试验组	对照组	试验组	对照组			随机方法	分组隐藏	盲法	脱落失访	评分
苗小楼	2008	20	20	中药	青霉素	子宫灌注	①②	1	0	0	0	1
刘俊平	2008	315	137	中药	青霉素	子宫灌注	②③	1	0	0	0	1
邱建民	2011	10	10	中药	环丙沙星	子宫灌注	①③	1	0	0	0	1
孙 蕾	2008	30	30	中药	环丙沙星	子宫灌注	①③	1	0	0	0	1
王 军	2006	35	30	中药	环丙沙星	子宫灌注	①②③	1	0	0	0	1
张 桥	2003	81	53	中药	环丙沙星	子宫灌注	①②	0	0	0	0	0

疗效指标是①有效率，②治愈率，③3 个情期内的受胎率

2.6 发表偏倚分析

若纳入文献分布呈现倒漏斗形式，且漏斗图结构对称，则表明文献的发表偏倚性较小或无偏倚性[14]。本研究中总有效率的文献均在漏斗图内，但是对称性相对较差，表明本研究中总有效率有一定的发表偏倚性（图 5A），但偏倚程度不大；治愈率和受胎率的文献均有 1 篇不在漏斗图内，但治愈率的文献分布漏斗图对称性较好，说明其发表偏倚性较小（图 5B）；而受胎率文献分布对称性较差，表明受胎率研究发表偏倚性较大（图 5C）。

3 讨论

3.1 中药对奶牛子宫内膜炎的疗效评价

奶牛子宫内膜炎主要由致病微生物感染引起，因此，临床兽医工作者多采用土霉素、青链霉素或环丙沙星等抗生素，促进炎性分泌物排出，但抗生素和激素疗法产生的药物残留和耐药性问题，使得人们不得不寻求绿色、安全的中药疗法。为了系统评价中药和抗生素对奶牛子宫内膜炎的疗效，本研究经筛选后纳入 24 项临床对照试验，共治疗 2 782 头患病牛，试验组均采用复方中药治疗，对照组采用土霉素、青霉素、青链霉素联合使用和环丙沙星治疗，Meta 分析结果显示：中药和抗生素治疗组的总有效率分别为 93.5%（1 318/1 409）和 87.1%（776/ 891），复方中药治疗牛子宫内膜炎的总有效率显著优于抗生素组（$P<0.01$），但和青链霉素联合使用治疗组的总有效率相当；中药和抗生素治疗的总治愈率分别为 85.6%（1 424/1 664）和 79.0%（868/ 1 122），显著优于抗生素组，但和土霉素治疗组的治愈率相当。中兽医学认为，奶牛子宫内膜炎是由于湿热内阻，胞宫血瘀所致，治则清热解毒，活血化瘀”“，辅以祛腐排脓和催情促孕，多选用黄连、连翘、当归、黄芪、丹参、益母草、红花、蒲公英和淫羊藿，主要是在清宫液[17,29,35,38]和促孕灌注液[26-27]的配方基础上加减而成，方中诸药协同，发挥杀菌、抗炎、消肿、去腐生肌、改善子宫微循环和增强机体免疫力的作用，其有效成分（如盐酸小檗碱、连翘苷、有机酸、黄酮类化合物等活性成分）对金黄色葡萄球菌、大肠杆菌、链球菌及化脓性杆菌有显著的抑杀作用[34]；益母草和红花含有的生物碱能增强子宫平滑肌的收缩力；丹参、红花、蒲公英等含有的丹参酮等活性成分可以调节子宫内膜炎患病牛的全血黏度、血浆黏度、红细胞变形指数等血液流变学状态，达到活血祛瘀的疗效[39]；黄芪、当归等含有的多糖能调节机体免疫系统；生物碱、黄酮类、皂苷类和挥发油等成分具有明显的抗炎活性，能有效抑制子宫内膜的炎性损伤[40]。在本研

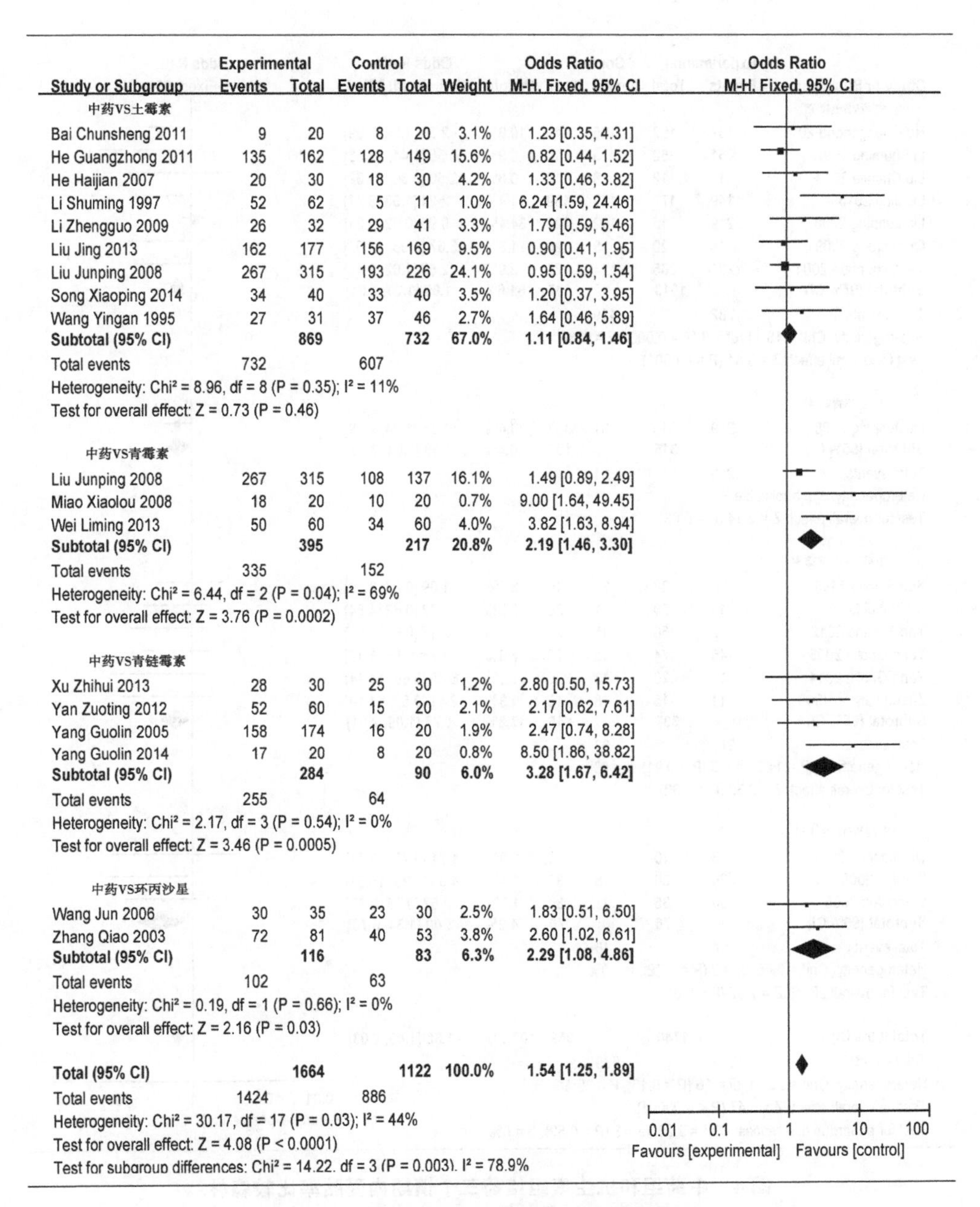

图 3　中药组和抗生素组的治愈率比较森林

究中，还发现中药治疗后奶牛的受胎率（76.8%）显著高于抗生素治疗奶牛（66.6%）（*P*<0.01），说明中药有助于促进牛的发情和胚胎着床，从而提高受胎率。因此，中药正是通过抗菌消炎、免疫调控、活血化瘀和催情促孕等多靶点、多途径的治疗奶牛子宫内膜炎，并促进预后的子宫功能恢复。

根据现有证据显示，复方中药治疗牛子宫内膜炎的疗效显著优于抗生素，未发现不良反应，并且可以促进子宫复旧，缩短产犊间隔，降低淘汰率，是治疗奶牛子宫内膜炎的一种较

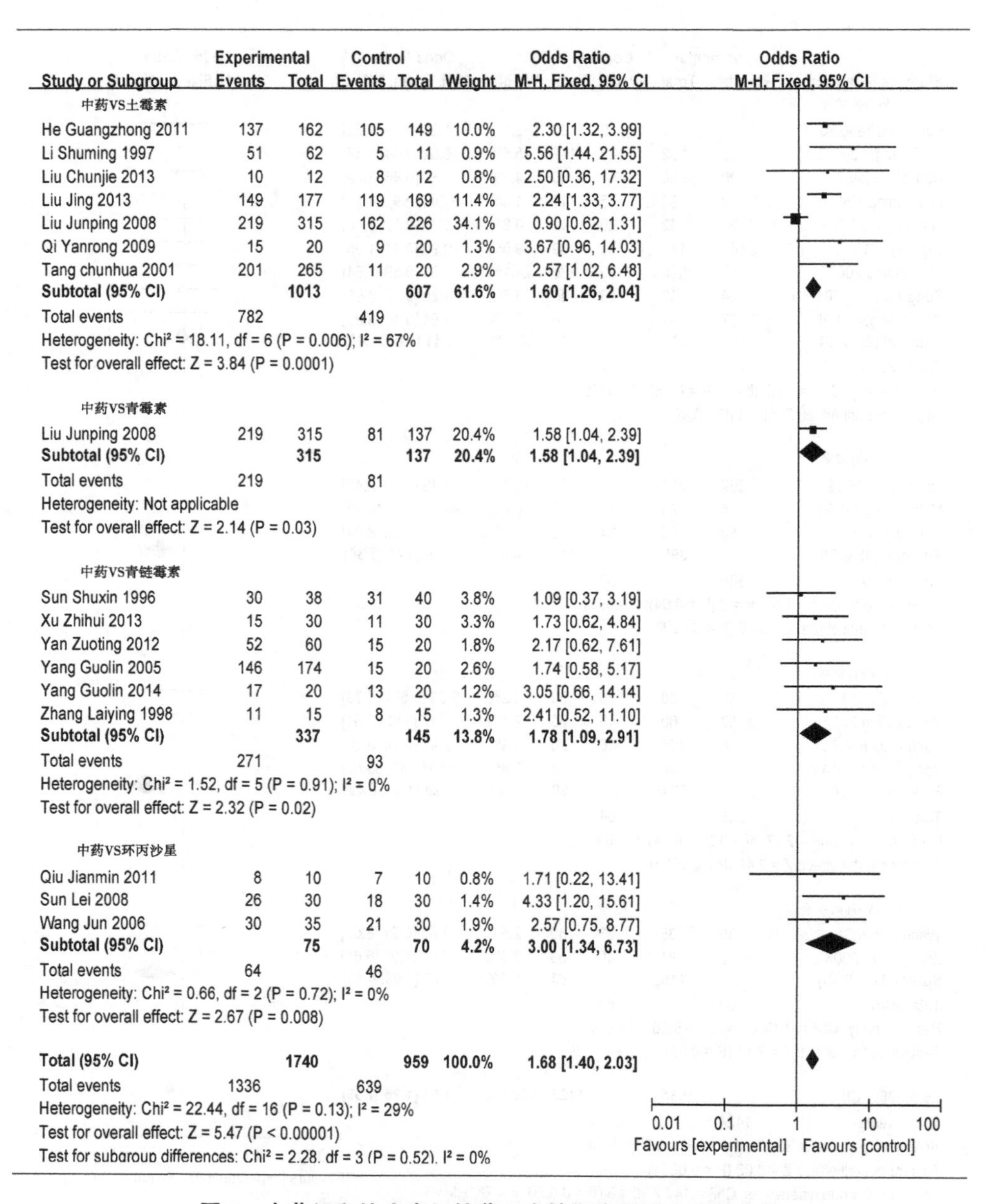

图 4　中药组和抗生素组给药三个情期内受胎率比较森林

好方法，可在临床上推广应用，但在不具备中药的条件下，也可选择青链霉素和土霉素治疗。

3.2　本系统评价的局限性

系统评价是建立在已发表文献的基础之上，结论的可靠性受纳入文献质量的影响较大[41]，由于本论文是评价中药对奶牛子宫内膜炎的疗效，虽然在外文数据库中进行了检索，但纳入的 24 篇文献仍然全部来自国内，且均发表在国内一般专业期刊上，并有部分 RCT 样本量较小，缺乏国外大规模的 RCT[7]。由于兽医学方面的 RCT 尚未有统一标准，临床试验

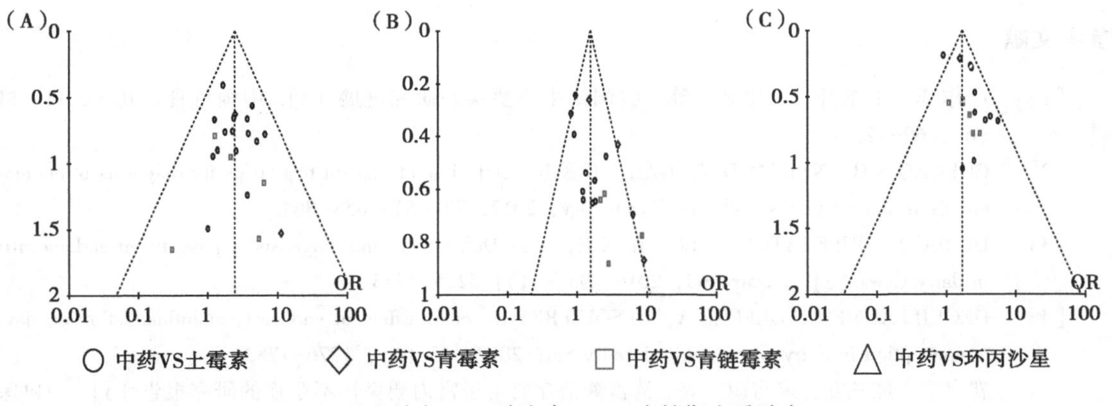

A. 总有效率；B. 治愈率；C. 3 个情期内受胎率

图 5 中药组和抗生素组疗效指标比较的倒漏斗

多由实施者自己决定，所以本次纳入文献的 JADAD 方法学质量评价多为 1 分，属低质量证据，大部分文献未提及具体的随机方法，所有文献均未描述分配方案的隐藏及盲法的具体实施情况，故可能产生选择及测量偏倚，从而影响本系统评价结果。

其次，本评价纳入文献在方法学和临床上存在异质性，如各个研究中所用的中药成分不完全相同，其疗效也有所差别，但由于其组方的治疗原则都是在中兽医指导下确定的，在根本上是一致的，因此，中药成分不同对本结果是有一定影响的。另外，未纳入灰色文献及正在进行的试验，纳入文献的倒漏斗图呈非对称性分布，提示存在一定的发表偏倚性。因此，本系统评价存在一定的局限性，尚需进一步开展更大规模和高质量的 RCT 来支持和验证。

3.3 对今后研究的启示

近年来，随着我国 GDP 增长，对农业科学研究的支持力度，尤其是畜牧兽医领域投入不断增加，兽医科技人员开展了众多的药物治疗动物疾病的临床疗效试验，但试验的方法学质量偏低，把“随机”当成“随便”，不用盲法设计试验，这样取得的结果不能让人信服。建议今后的研究应详细报告随机分配序列的产生和随机分组的隐藏，使用盲法与安慰剂对照，详细报告病例的退出和脱落情况，避免使用主观的复合结局评分指标和自拟的疗效评价标准，临床试验的报告应按照最新的 CONSORT 标准进行，提高试验的质量和结果的真实度。

尽管畜牧兽医领域已有大量的研究结果，但对于单个试验，受制于实验条件、实验动物数量和状态等多方面因素影响，得到的数据或结论可能存在不同甚至相反的情况。因此，研究人员在不同条件下做的临床试验，就可以通过 Meta 分析方法将多个试验的结果纳入到同一个研究中，得到更接近真实的客观的结论，从而提高这些科研数据的应用价值，为临床兽医的决策提供参考。

4 结论

本系统评价通过对 24 项临床对照试验的 Meta 分析，发现复方中药对奶牛子宫内膜炎的疗效显著优于抗生素，但总有效率和青链霉素联合使用相当，治愈率和土霉素相当；并且中药可以显著提高奶牛治疗后 3 个情期内的受胎率。本系统评价为临床治疗奶牛子宫内膜炎提供了一定的决策参考。

参考文献

[1] 严作廷，王东升，王旭荣，等．我国奶牛主要疾病研究进展［J］．中国草食动物，2011，31（6）：69-72.

[2] CHEONG S H，NYDAM D V，GALVAO K N，etal. Use of reagent test strips for diagnosis of endometri tis in dairy COyVS［J］. Theriogenology，2012，77（5）：858-864.

[3] DUBUC J，DUFFIELD T F，LESLIE K E，et al. Deft mhons and diagnosis of postpartum endometritis in dairy Cows［J］. J Dairy Sci，2010，93（11）：5225-5233.

[4] DEGUILLAUME L，GEFFRE A，DESQUILBET L，et al. Effect of endocervical inflammation on days to conception in dairy cowss［J］. J Dairy Sci，2012，95（4）：1776-1783.

[5] 瞿自明，熊三友，宋富源，等．清宫液治疗奶牛子宫内膜炎性不孕症的研究报告［J］．中兽医医药杂志，1989（6）：1-6.

[6] OVERTON M，FETROW J，Economics of Postpar tmn Uterine Health［Z］. Hartland Wisconsin，USA，2008：29-44.

[7] DUBUCJ，DUFFIELD T F，LESLIE K E，et al. Ran domized clinical trial of antibiotic and prostaglandin treatments for utenne health and reproductive per forimance in dairy cows［J］. J Dairy Sci，2011，94（3）：1325-1338.

[8] MACHADO V S，BICALHO M L，MEIRA J E B，et al. Subcutaneous nmnunlzatlon with inactivated bacte rial components and purified protem of Escherichia coli，Fusobacterium necrophorum and Trueperegga pyogenes prevents puerperal metntm in Holstein dair y cows［J］. PLoS One，2014，9（3）：e91 734.

[9] HAIMERL P，HEUWIESER W. Invited revleyv：An tibiotic treatment of metntm in dairy cows：A system atm approach［J］. J Dairy Sci，2014，97（11）：6649-6661.

[10] HAIMERL P，HEUWIESER W，ARLT S. Therapy of bovine endometritis with prostaglandin F2α：a ifle taanalysis［J］. J Dairy Sci，2013，96（5）：2973-2987.

[11] DOLEJSKA M，JURCICKOV A Z，LITERAK I，et al. IncN plasmids carrying bla CTX-M-1 in Escherichia coli isolates on adairyfarm［J］. *Vet* Microbiol，2011，149（34）：513-516.

[12] 李飞，徐明，曹阳春，等．Meta 分析方法优化泌乳奶牛日粮碳水化合物平衡指数［J］．畜牧兽医学报，2014，45（9）：1457-1466.

[13] RABOISSON D，MOUNIE M，MAIGNE Y，Diseases，reproductive performance，and changes in milk pro duction associated with subclinical ketosis in dairy Cows：A meta-analysis and revlew［J］. J Dairy Sci，2014，97（12）：7547-7563.

[14] 熊加川，田茂露，何朝霞，等．霉酚酸酯治疗过敏性紫癜肾炎疗效和安全性的系统评价［J］．中国循证医学杂志，2014，14（2）：184-190.

[15] 宋晓平，智晓艳，崔恩慧，等．一种治疗奶牛子宫内膜炎的中药组合物及制备方法和应用［P］．2014：07-02.

[16] 杨国林．一种治疗牛子宫内膜炎的中药组合物及其制备方法和应用［P］．2014：12-20.

[17] 魏立明，王晓平，赵真，等．中草药制剂灌注治疗奶牛子宫内膜炎效果观察［J］．甘肃畜牧兽医，2013（4）：47-48.

[18] 刘春杰．阿魏酸钠抗炎分子机制及对奶牛子宫内膜炎疗效初步观察［D］．长春：吉林大学，2013.

[19] 刘镜，何光中，张晓可，等．一种治疗奶牛子宫内膜炎的中药组合物及其制备方法［P］．2013：08-14.

[20] 徐志辉，李俊，周山山．纯中药灌注剂治疗奶牛子宫内膜炎的疗效观察［J］．黑龙江动物繁殖，2013：21（3）：34-35.
[21] 白春生，张尚昆，薛兴中．奶牛子宫内膜炎的治疗试验［J］．新疆畜牧业，2011（2）：22-25.
[22] 何光中，刘镜，杨红文，等．复方中草药防治奶牛子宫内膜炎的效果［J］．安徽农业科学，2011，29（25）：15387-15389.
[23] 邱建民．“宫得健泡腾栓”的制备及临床试验［D］．扬州：扬州大学，2011.
[24] 李正国．抗奶牛子宫内膜炎复方中药的药理作用研究［D］．石河子：石河子大学，2009.
[25] 祁燕蓉，何生虎．中药复方治疗奶牛子宫内膜炎的疗效观察［Z］．中国畜牧兽医学会家畜内科学分会2009年学术研讨会论文集，青岛，2009：85-87.
[26] 祁生旺，田志，宋新军，等．奶牛“子宫消炎促孕灌注液”的临床试验［J］．畜牧与饲料科学，2008（2）：92-93.
[27] 刘俊平，刘风军，王勇胜，等．中西药对影响奶牛胚胎移植效率的繁殖疾病疗效的对比［J］．黑龙江畜牧兽医，2008（2）：9-14，
[28] 孙蕾，李琳，王军．复方中药对奶牛子宫内膜炎的疗效观察［J］．上海畜牧兽医通讯，2008（6）：71.
[29] 苗小楼，李芸，苏鹏，等．“宫康”治疗奶牛子宫内膜炎的疗效观察［J］．动物医学进展，2008，29（增）：74-75.
[30] 何海健，兰新财．宫炎消栓防治奶牛子宫内膜炎的疗效观察［J］．中兽医学杂志，2007（3）：3-5.
[31] 王军．治疗奶牛子宫内膜炎新药EDM的研制［D］．长春：吉林农业大学，2006.
[32] 杨国林，巩忠福，严作廷，等．一种治疗牛子宫内膜炎的中药［P］．2005：08-10.
[33] 张桥，彭本英，金巍．奶牛慢性子宫内膜炎的治疗试验［J］．湖北农学院学报，2002（5）：221-222.
[34] 汤春华，张礼中，任建清，等．中草药制剂防治奶牛繁殖障碍疾病的效果［J］．江苏农业科学，2001（1）：61-62.
[35] 王应安，张寿，杜志权，等．盐酸土霉素及清宫液治疗奶牛子宫内膜炎试验［J］．青海畜牧兽医杂志，1995，25（1）：18-19.
[36] 张来英，陈登天，田志军．“清宫液”治疗奶牛不孕症对比试验［J］．草食家畜，1998（2）：46-47.
[37] 孙树新，赵福连，商福文，等．中药“西瓜霜”治疗母牛子宫炎试验［J］．河北畜牧兽医，1996，12（1）：20-21.
[38] 严作廷，王东升，李世宏，等．清宫助孕液治疗奶牛子宫内膜炎临床试验［J］．中国奶牛，2012（4）：26-28.
[39] 吴金节，章孝荣，刘亚，等．奶牛子宫内膜炎防治前后血液流变学参数的变化［J］．南京农业大学学报，2000，22（2）：65-68.
[40] 姜巍，杨国林，杨志强，等．清宫液Ⅱ号抗炎药理学研究［J］．中兽医医药杂志，2005（5）：12-13.
[41] MATHIE R T，CLAUSEN J，Vetermary homeopa thy：meta arialysis of randomised placebo controlled trials［J］．Homeopathy，2015，104（1）：2 8.

（发表于《畜牧兽医学报》）

不同宿主来源的耐甲氧西林金黄色葡萄球菌分子流行病学研究进展*

苏　洋[1,2**]，陈智华[2]，邓海平[1]，李春慧[1,2]，蒲万霞[1***]

（1. 中国农业科学院兰州畜牧与兽药研究所/农业部兽用药物创制重点实验室/甘肃省新兽药工程重点实验室，兰州　730050；
2. 甘肃农业大学动物医学学院，兰州　730070）

耐甲氧西林金黄色葡萄球（Methicillin-resistant *Staphylococcus aureus*，MRSA）是医院内和社区内获得性感染的主要病原菌，也引起动物疾病的发生，而且健康人群的携带率为20%~30%[1]，其携带的mecA耐药基因编码与β内酰胺类抗生素亲和力极低的PBP2a蛋白而导致菌株的耐药。自1940年青霉素被用于治疗因金黄色葡萄球菌（*S. aufeus*）引起的感染后，很快便出现了耐药菌株；1959年，半合成青霉素-甲氧西林首次被用于治疗耐青霉素的*S. aureus*后，1961年既有MRSA的报道。在随后的几十年中其耐药菌株在世界范围内不断地出现，由于其高度耐药性和异质性给临床治疗带来了极大的困难。美国每年因感染MRSA而导致的死亡人数已超过艾滋病，我国MRSA的感染率达到了33.3%~80.4%。MRSA的感染已经成为研究的热点，许多学者将MRSA的感染同艾滋、乙型肝炎列为世界范围内最难解决的3大感染性疾病。近年来，越来越多不同动物源性MRSA的出现，给兽医临床和公共卫生安全带来极大的威胁。本文就近年来国内外对MRSA的分型方法及其在动物种的流行情况作一综述。

1　分型方法

1.1　脉冲场凝胶电泳分型方法（PFGE）

PFGE是20世纪80年代Schwartz和Cantor等建立的，可以用来分离整条染色体这样超大分子量的细菌DNA，被认为是鉴定*S. aureus*的“金标准”。MRSA在溶菌酶和溶葡萄球菌素的作用下获得完整DNA，并通过核酸限制性内切酶*Sma* I或*Csp* I消化后可获得10~20个长约10~700 kb的DNA片段。DNA片段经PFGE分离，根据Tenover等提出的标准[2]分为：①相同（Indistinguishable）：酶切图谱间具有同样的条带数，而且相应条带大小相同，流行病学上则认为相同，可认为是同一型别；②紧密相关（Closely related）：由点突变、插入或缺失DNA等导致的2~3个差异条带，很有可能是同一型别，定为亚型；③可能相关（Possiblyrelated）：两个独立的基因变件可能引起4~6个条带的差异，也许是同一型别，可定为亚型；④不相关（Differentl：3个或更多独立基因变异所致7个或以上的差异条带，可认为在流行病学上不相关性。

---

* 基金项目：中央级公益性科研院所基本科研业务费专项资金项目（161032201 1010）。

** 作者简介：苏洋（1985—　），男，四川绵阳人，硕士研究生，主要从事兽医病原微生物耐药性研究。

*** 通讯作者：E-mail：puwanxia@ caas. ca.

1.2 多位点序列分型（MLST）

该方法是在多位点酶凝胶电泳（MLEE）基础之上改进而来，*S. aureus* 的 MLST 分型研究选择 *arC*、*aroE*、*glpF*、伊放、*pta*、*tpi*、*yqil* 7 个看家基因，PCR 扩增、测序及序列比对，将测序结果与相应基因的标准序列进行比对并截为标准长度，提交数据库（http：llwww. mist. net）获得每株菌株各个位点的等位基因号，并由每株细菌 7 个看家基因的等位基因号再获得序列号（Sequence typing，STs）[3]。

1.3 *SCCmec* 分型

该方法是根据 *SCCmec* 染色体盒的组成元件不同为基础，包括 *mec* 基因复合体、*CCr* 基因复合体及脚 *ec* 和 *ccr* 基因复合体之外的 *J* 非编码区。由于 *SCCmec* 染色体盒携带不同的基因元件，到目前它已经被鉴定出 *SCCmec* Ⅰ~Ⅺ共 11 型（IWG. scc），而其中的 *SCCmec* Ⅳ由于 J 区的不同又被分为Ⅳa、Ⅳb、ⅣC、Ⅳd、Ivg、Ivh、Ⅳi、Ⅳj 8 个亚型[4]。近年来 SCC-mec 已建立起多种分型方法，但广泛使用的是 Zhang 等建立的多重 PCR 方法，该方法能同时检测 I、II、III、IVa、Ⅳb、IVC、IVd 及 V 8 个型[5]。

1.4 spa 基因分型

Spa 蛋白是位于 *S. aureus* 细胞壁的一种蛋白成分，具有种、属特异性而无型的特异性。spa 分型仅涉及 spa 基因序列，该基因长约 2 150 bp，包括 Fc 结合区、x 域和 C 末端 3 个区域。其中 x 域含 2~15 个长 21~27 bp 的重复序列，该区域由于重复序列的数目、特征及排列顺序不同具有高度多态性。根据 spa 基因的高度重复可变序列设计引物对 x 区域进行扩增，将 PCR 产物测序后与数据库中已有的型别比对。目前，已经建立了 spa 重复序列和分型数据库（http：//spa. ridom. de），已有重复序列 563 个、基因型 11 372 个。

2 MRSA 与寄主的关系

为研究不同宿主来源 MRSA 的分子流行病学情况，对其进行 SCCmec、spa、MLST 及 PFGE 等分型研究结果表明，某些菌株具有较强的宿主特异性，但这种特异性又在某些菌株中并非是绝对的。如，目前所知的几种人源性流行菌株 ST5-MRSA. I、ST5. MRSA. II、ST5-MRSA. Ill、ST5-MRSA. IV、ST8. MRSA. I、ST. MRSA. II、ST8. MR-SA. III、ST8. MRSA. IV、ST22. MRSA. IV、ST30. MRSA. Ⅳ、ST45-MRSA-II 及 ST45. MRSA-IV 等未在动物中发现，并且具有地理性分布特点[6-7]。但某些 MRSA 既能感染人类又能感染不同动物，如 ST22. MRSA. IVMRSA 在人类、猪和马中都曾被发现[8]；此外，最初与猪感染最为密切的 ST398 LA-MRSA，后来发现它同样感染奶牛、家禽、犬及马[8]，在欧洲部分地区流行的奶牛乳房炎 MRSA 的 MLST 分型主要为 ST398[9]。研究表明生活在养猪场附近人群中所携带的 MRSA 属于 ST398 家系[10-11]。荷兰 12%人源性 MRSA 为 ST398 型，而 1. 7%人源性 MRSA 在 8 个欧洲国家也检测为 ST398[12]。Sergio 等首次从亚洲地区的养猪场和工人上分离到 ST398 LA-MRSA 菌株[13]。这表明部分 MRSA 的宿主特异性并非是绝对的，有着更大的感染范围。

3 MRSA 在动物中的流行情况

3.1 MRSA 在宠物中的流行情况

在社区获得性 MRSA（CA-MRSA）流行加剧的情况下，驯养动物特别是与人类接触最为密切的宠物不可避免的暴露于 MRSA 中，而宠物中的 MRSA 又将给其他动物和人类健康带来严重的威胁。在英国，首次从宠物中分离出MRSA，研究表明这些菌株属于ST22，推测可能来源于人类。医院内获得性MRSA（HA-MRSA）和CA-MRSA也从犬中分离到，这与其犬主人有关[14]。因此，推测宠物感染或定植MRSA与其主人、临床兽医师及兽医外科手术有关。犬与携带MRSA的人类或其他动物接触后，自身便成为MRSA的寄主，只是不表现出临床症状，在创伤或术后可能会导致MRSA的机会性感染。

3.2 MRSA在家畜中的流行

3.2.1 MRSA在猪中的流行　引起人们关注动物源性

MRSA是源于荷兰某养猪场的意外发现，其饲养员MRSA的携带率是荷兰普通国民的760倍，并且这些菌株主要是MLST ST398 MRSA。此后，ST398 MRSA在新西兰、丹麦、德国、葡萄牙、加拿大和美国等地区的养猪场检出。起初ST398 LA. MRSA的报道仅见于欧洲等国，但最近也在亚洲国家的养猪场分离到ST398 LA. MRSA。虽然ST398在猪中的检出率很高，但它毒力不强，临床致病性感染鲜有报道。此外，其他MRSA也在猪中发现，一种普遍为人所感染的USA100菌株也从猪群中分离到，这可能是从人类传给猪的[15]；ST9 MRSA首次从四川双流地区的养猪场分离到[16]；ST9、ST912及ST1297 MRSA也从中国部分地区的养猪场和猪场员工中分离到[17]。表明不同MRSA可感染或定植于猪群中。

3.2.2 IVlRSA在奶牛中的流行

*S. aureus*是引起奶牛乳房炎的主要病原菌之一，而牛源MRSA自1972年首次报道后，就引起了兽医界的重视。在比利时，从118株乳房炎*S. aurus*中分离出11株MRSA，分型结果表明它们属于ST398 LA. MRSA[18]；在德国，从17家奶牛养殖场临床型乳房炎奶样分离到25株MRSA，牛场员工分离到2株，均属于MLST ST398，spa分型为spa type011（23株）、spatype034（3株）、spa type2576（1株），主要携带SCCmec type Ⅴ。ST398 MRSA为欧洲国家奶牛乳房炎MRSA的一优势菌系，同样ST8、ST1、ST239、ST329、ST8、ST425、ST130和ST1245在土耳其、匈牙利等欧洲国家的奶牛乳房炎奶样中零星发现[9]。亚洲地区关于奶牛乳房炎MRSA的研究报道较少，来自日本和韩国地区的研究报道表明，ST5、ST580、ST1和ST72 MRSA也从牛奶样中检出[9]。最近，Garcia-Alvarez等第一次从奶牛乳房炎奶样中检出携带SCCmecXI染色体盒的MRSAt[19]。表明ST398与其他STs MRSA都能引起奶牛乳房炎，而且存在地区性差异，虽然感染率并不高，但它的威胁仍很大。例如，像ST398这样物种特异性不强且具有传播性的菌株，一旦在牛群中传播开来将会给奶牛养殖业带来严重的威胁。

3.2.3 MRSA在马中的流行

MRSA感染或定植于马主要与环境、创伤和与其接触者等有关。在人群中流行的USA500 ST8 MRSA和ST254 MRSA均在马中发现，推测是源于人类[20,21]；ST398 MRSA也在马中发现，其定植率达10.9%[22]。不同MRSA可感染或定植于马，给马的健康带来潜在性威胁。

3.3 MRSA在家禽中的流行

MRSA感染和定植家禽的报道并不多见。从肉鸡样品中分离出的8株MRSA，分型表明均属于spa type t1456，MLST分型的ST398[23]。从鸡肉、火鸡肉及其加工产品中也分离到37.2%（32/86）的MRSA，其中28株的MLST分型为ST398，spa分型为t011、t034、t899、t2346及t6574，含SCCmec typeⅣ或Ⅴ；2株的MLST分型为ST9，携带*SCCmec*染色体盒和

spatype t1430；剩余 2 株分别属于 MLST 分型的 ST5 和 ST1791，都携带 SCCmec typeⅢ染色体盒，都是 spa 分型的 t002[24]。

4 MRSA 在人与动物之间相互传播的情况

多数关于 MRSA 在不同物种间传播的研究，是对分离自不同动物和与其直接或间接接触人类的 MRSA 做耐药谱型和分子分型。研究资料显示，源于动物和人的 MRSA 无法用基因型和耐药普型加以区分，所以推测是 MRSA 在人与动物之间相互传播，但 MRSA 在不同物种间传播目前仍尚无定论。一种不能用 PFGE 分型，MLST 分型为 ST398，携带 spa-type108 和 *SCCmec* type Ⅴ的 LA-MRSA 从猪、猪场员工及一位猪场员工的妻子和幼女中分离到，表明其在人类和猪之间相互传播[25]。MRSA 在人和奶牛间的传播也有报道，27 株分离自亚Ⅰ临床型奶牛乳房炎和一株牛场员工的 MRSA 无法通过耐药谱型和基因型加以区分，表明存在 MRSA 在奶牛和人之间相互传播的可能[26]。类似情况也在宠物和人之间发生，从一只健康犬和携带 MRSA 主人上分离的两株 MRSA 具有完全一样的 PFGE 和 *SCCmec* 分型结果，表明 MRSA 可能从人传播到犬[27]。种种事实表明，可能存在 MRSA 在人和动物间相互传播的现象，虽然目前不能确定具体的传播方，但 MRSA 在不同物种问的传播必将引起公众对社会公共卫生安全的高度担忧。

5 结语

细菌在抗生素的持续选择压力下，对其产生的抗性也越来越强。虽然目前很难断言 MRSA 给公共卫生安全带来多大的危害，但作为人和动物感染的主要病原菌，它确实已给人类疾病的防治造成极大的困难；同时，对动物健康、兽医临床及食品安全具有很大的威胁挑战，因此，对不同感染源及地域的 MRSA 进行深入的流行病学研究显得尤为重要。

参考文献

[1] Van Belkum A, Melles D C, Nouwen J, et al. Co. evolutionaryaspects of human colonization and infection by Staphylococcusaureus [J]. Infect Genet Evol, 2009, 9: 32-47.

[2] Tenover F C, Arbeit R D, Goering R V, et al. Interpreting chro. mosomal DNA restriction patterns produced by pulsed · field gel electrophoresis: Criteria for bacterial strain typing [J]. J Clin Mi-erobiol, 1995, 33: 2233-2239.

[3] Enright M C, Day N P, Davies C E, et al. Multilocus sequencetyping for characterization of methicillin resistant and methi. cillin-susceptible clones of Staphylococcus aurues [J]. J Clin Microbiol, 2000, 38 (3): 1008-1015.

[4] Milheirico C, Oliveira D C, de Lencastre H. Multiplex PCR strategy for subtyping the staphylococcal cassette chromosome mec type Ⅳ in methicillin-resistant Staphylococcus alureus: SCCmecⅣ multiplex' [J]. J Amimicrob Chemother, 2007, 60: 42-48.

[5] Zhang Kun-yah, McClure J A, Elsayed S, et al. Novel multiplexPCR assay for characterization and concomitant subtyping of staphylococcal cassette chromosome mec types Ⅰ to Ⅴ in me-thicillin-resistant Staphylococcus al//Bus [J]. J Clin Microbiol, 2005, 43 (10): 5026-5033.

[6] Monecke S, Coombs G, Shore A C, et al. A field guid to pandemic, epidemic and sporadic clones of methicillin-resistant Staphylococcus aureus [J]. PIoS One, 2011, 6: e17936.

[7] Enright M C, Robinson A, Randle G, et al. The evolutionary history of methicillin-resistant Staphylo-

coccus aureus (MRSA) [J]. PNAS, 2002, 99: 7687-7692.
[8] Cuny C, Friedrich A, Svetlana K, et al. Emergency of methi-cillin-resistant Staphylococcus aulreus (MRSA) in different animal species [J]. Int J Med Microbiol, 2010, 300: 109-117.
[9] Holmes M A, Zadoks R N. Methicillin resistant S. aufeus in human and bovine mastitis [J]. J Mammary Gland Biol Neoplasia, 2011, 16: 373-382.
[10] Voss A, Loeffen F, Bakker J, et al. Methicillin-resistant Staphylococcus aureus in pig farming [J]. Emerg Infect Dis, 2005, 11: 1965-1966.
[11] Armand-Lefevre L, Ruimy R, Andremont A. Clonal comparison of Staphylococcus aLIFeus isolates firom healthy pig farmers, human controls, and pigs [J]. Emerg Infect Dis, 2005, 11: 711. 714.
[12] Cleef B A V, Monnet D L, Voss A, et al. Livestock-associatedmethicillin-resistant Staphylococcus aureus in humans, Europe [J]. Emerg Infect Dis, 2011, 17: 502-505.
[13] Sergio D M, Koh T H, Hsu L Y, et al. Investigation of methi-cillin-resistant Staphylococcus au/eus in pigs used for research [J]. J Med Microbiol, 2007, 56: 1107-1109.

(发表于《中国预防兽医学报》)

牦牛 KAP3.3 基因的克隆及生物信息学分析*

王宏博**，梁春年，包鹏甲，张良斌，
裴　杰，吴晓云，赵娟花，阎　萍***
（1. 中国农业科学院兰州畜牧与兽药研究所，兰州　730050；
2. 甘肃省牦牛繁育工程重点实验室，兰州　730050）

摘　要：本研究以牦牛的角蛋白关联蛋白 3.3（*KAP*3.3）基因作为研究对象，通过查询 GenBank 中收录的黄牛 *KAP*3.3 基因的 mRNA 序列设计 1 对特异性引物，通过逆 PCR、PCR 以及基因克隆测序的方法首次获得牦牛的 *KAP*3.3 基因完整的 CDS 区序列，并对其进行生物信息学分析。结果表明：牦牛 *KAP*3.3 基因的 CDS 区为 297 bp，编码 98 个氨基酸；编码的蛋白质属于亲水性蛋白。二级结构含有延伸链、β 转角以及无规卷曲 3 种。氨基酸序列与黄牛的完全一致，具有 Keratin，high sulphur matrix protein 家族的完整结构域。

关键词：牛；*KAP*3.3 蛋白；CDS 区

角蛋白关联蛋白（keratin associated protein，KAP）作为绒毛的重要蛋白，是毛基质的主要成分，与角蛋白中间丝（keratin intermediate filament，KIF）共同构成了毛纤维[1]，在毛干的韧性和强度方面具有重要作用。*KAP* 家族基因在决定绒毛的细度、长度、强度、弯曲、光泽度、弹性等起了关键作用[2]。*KAP*6、*KAP*7 和 *KAP*8 与羊毛的直径相关[3]。而 *KAP*1.1 和 *KAP*1.3 有一个多态性位点与羊毛的细度有显著相关性[4]。研究发现，*KAP*3.2 基因与毛长和产绒量有一点相关性[5]。而獭兔 *KAP*3.2 基因与其毛密度有关[6]。*KAP* 家族基因的研究报道相对较多，但是关于 *KAP*3.3 基因的相关报道却在国内外都未发现。且研究 *KAP* 家族基因的对象多是绵羊、山羊、人和小鼠，牦牛方面该基因家族的研究尚未见报道。

牦牛主要分布在青藏高原及其周边，能够适应严寒等恶劣环境。牦牛是唯一产绒的牛种，天祝白牦牛是我国稀有珍贵品种，因其全身被白色绒毛而有“祁连白牡丹”等美称。因天祝白牦牛白色绒毛有利于染色，可在多种商品中使用，具有较高的经济价值，研究其绒毛的生长特性对于提高牦牛的产绒量有至关重要的作用。本实验通过克隆得到天祝白牦牛 *KAP*3.3 基因 CDS 区全长序列，并对其蛋白理化性质、亚细胞定位、结构域及功能预测、系统进化等进行分析，为进一步研究该蛋白生理功能及周期表达差异做铺垫。

---

* 资助项目：甘肃牧区生产生态生活优化保障技术集成与示范（2012BAD13B05）；中国农业科学院牦牛资源与育种创新团队（CAAS-ASTIP-2014-LIHPS-01）；现代农业（肉牛牦牛）产业技术体系（CARS-38）。

** 作者简介：王宏博（1977—　），男，甘肃庄浪人，博士，副研究员，研究方向为动物生产，E-mail：hongbo610@163.com.

*** 通讯作者：阎萍，E-mail：pingyan@sohu.com.

1　材料与方法

1.1　实验材料

1.1.1　样品、克隆载体及感受态细胞

实验样品取自甘肃省天祝藏族自治县，采集天祝白牦牛皮肤组织液氮冻存带回实验室，保存在-80℃冰箱中备用；大肠杆菌（*E. coil*）由大连宝生物公司提供；克隆载体 pGEM-T Easy 采自普洛麦格生物技术有限公司。

1.1.2　分子生物学试剂

RNA 提取试剂盒、2×Taq PCR MasterMix、琼脂糖凝胶回收试剂盒和质粒提取试剂盒采自 TIANGEN 公司；反转录试剂盒由 TaKaRa 公司提供；琼脂糖由上海 YITO 公司提供；琼脂粉、酵母浸出粉和胰蛋白胨采自 OXOID 公司，X-gal、IPTG、Amp 采自上海生工生物公司。

1.1.3　引物设计及合成

以 GenBank 中黄牛 *KAP*3.3 基因的 mRNA 序列（登录号：XM_ 002702188.3）为模板，使用 Primer Primer5.0 设计 1 对引物，序列为 F：5′-AAGCCACTGATGACACCTCA -3′，R：5′-GAGCCACAGTTAGTTGCAGG-3′，委托上海生工生物公司进行合成。

1.2　实验方法

1.2.1　提取皮肤样品中的总 RNA

于 10 月中旬选取 1 只成年的体质健康的雄性天祝白牦牛，将体侧部位绒和毛剪去，用刀片切割皮肤组织，使用 RNA 提取试剂盒提取牦牛皮肤组织 RNA，用 1%琼脂糖凝胶电泳检测后，将浓度稀释至 500 ng/μL，保存于-80℃冰箱中。

1.2.2　RT-PCR

按照 TaKaRa 公司反转录试剂盒的要求进行反转录反应，合成牦牛皮肤组织的 cDNA。主要步骤为在 PCR 管依次加入 5 ×gDNA EraserBuffer 2μL，gDNA Eraser 1μL，RNase Free dH_2O 6μL，500 ng/μL 总 RNA 1μL，混匀后 42℃ 2 min。在以上反应中依次加入 5 ×Prime Script Buffer 4μL，RNase Free dH_2O 4μL，RT PrimerMix 1μL，Prime-Script RT Enzyme Mix 1μL，混合后 42℃ 15 min，85℃5 s 条件下反应，得到 cDNA，采用 NanoDrop2000/2000c 分光光度计检测浓度后，放入-20℃保存备用。

1.2.3　PCR 扩增

根据设计的引物，以 cDNA 为模板，使用 2×Taq PCR MasterMix 进行扩增。反应体系 50μL：模板 2μL，10 mmol/L 引物 F、R 各 1μL，2×TaqPCR MasterMix 25μL，ddH_2O 20μL。PCR 反应体系：预变性 94℃ 4 min；变性 94℃ 30 s，退火 62.8℃ 30 s，延伸 72℃ 50 s，循环 30 次；72℃ 10 min。使用 1%琼脂糖凝胶对 PCR 产物进行电泳检测，根据琼脂糖凝胶回收试剂盒的操作步骤将 PCR 产物回收并取 6μL 加入 1%琼脂糖凝胶进行检测。

1.2.4　克隆 *KAP*3.3 基因

将检测合格的 PCR 产物加入含有 pGEM-T Easy 克隆载体的液体中，并加入 T4DNA 连接酶以及 Buffer，在 16℃水浴锅中进行过夜连接。将过夜连接的质粒加入到 *E. coil* DH5α 感受态细胞中并加入 SOC 培养基，在恒温摇床 37℃，200 r/min 振荡培养 1 h，取适量涂布于含有 X-gal、IPTG 和 Amp 抗性的 LB 固体培养基上，在恒温培养箱 37℃培养 12 h。使用接种环挑取培养基上白色阳性光滑菌落，并将其接种于含有 Amp 的 LB 液体培养基中，在

37℃恒温培养箱中 200 r/min 进行过夜培养。依照质粒回收试剂盒的说明对菌液中的质粒进行提取，进行双酶切鉴定。将鉴定合格的菌液委托上海生工生物公司进行序列测定。

1.2.5　*KAP*3.3 基因的生物信息学分析

使用在线软件 Open Reading Frame Finder（ORF Finder）（http：//www.ncbi.nlm.nih.gov/gorf/orfig.cgi）预测牦牛 *KAP*3.3 基因的开放阅读框；将牦牛 *KAP*3.3 基因编码蛋白的氨基酸序列输入到在线软件 ProtScale（http：//web.expasy.org/protscale/）和 Protparam（http：//web.expasy.org/protparam/）进行蛋白质的疏水性质和理化性质分析；使用 PSORT Ⅱ（http：//psort.nibb.ac.jp/）在线软件预测亚细胞定位；使用在线软件 NetPhos2.0Server（http：//www.cbs.dtu.dk/services/NetPhos/）、TMHMM ServerV2.0（http：//www.cbs.dtu.dk/services/TMHMM/）和 SignalP4.1（http：//www.cbs.dtu.dk/ services/SignalP-4.1/）分析蛋白质的磷酸化位点、跨膜结构域及信号肽位点；使用 FoldIndex（http：//bip.weizmann.ac.il/fldbin/findex）在线软件分析氨基酸序列的无序化特性；使用在线软件 Interpro（http：//www.ebi.ac.uk/interpro/）分析蛋白质的结构域；使用 SOPMA（http：//npsa-pbil.ibcp.fr/cgi-bin/npsa_ automat.pl? page=/NPSA/npsa_ sopma.html）在线软件预测蛋白质的二级结构。使用 ProtFun 2.2（http：//www.cbs.dtu.dk/services/ProtFun/）在线软件进行蛋白质功能预测。在 GenBank 中下载不同物种 *KAP*3.3 基因编码的氨基酸序列，使用 MegAlign 软件进行同源性分析；使用 MEGA 5.1 软件进行 NJ 法构建 *KAP*3.3 基因的系统发育树。

2　结果

2.1　牦牛 *KAP*3.3 基因 PCR 扩增

将扩增产物加入到 1%的琼脂糖凝胶中进行电泳，结果显示 PCR 产物的条带与 Marker 的 700 bp 基本一致，该片段与 696 bp 的目的片段大小相符（图 1）。

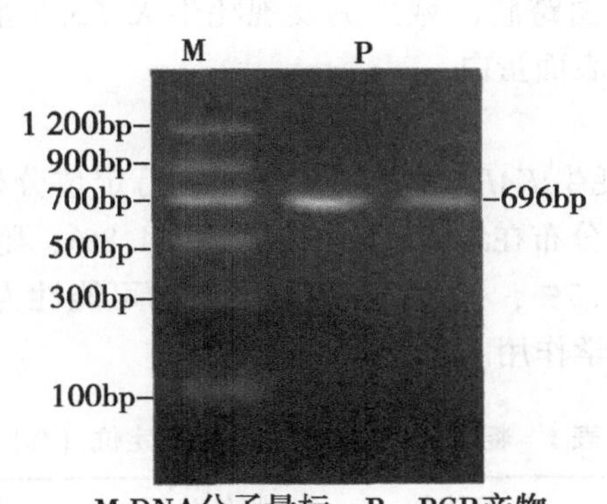

M:DNA分子量标，P：PCR产物

图 1　牦牛 *KAP*3.3 基因的 PCR 电泳

2.2　牦牛 *KAP*3.3 基因的克隆及序列分析

将克隆测序的结果去除载体序列结果与预期一致，为 696 bp。BLAST 分析显示，与黄牛的相似度高达 99%，可以确定该基因是牦牛 *KAP*3.3 基因。ORF Finder 在线分析发现牦牛 *KAP*3.3 基因的开放阅读框为 297 bp，共编码 98 个氨基酸，起始密码子为 ATG，终止密码子为 TGA（图 2）。

atggcttgctgtgctcccctctgctgcagcgcccgtaccagcccc
M A C C A P L C C S A R T S P
gccaccactatctgctcctctgacaaattctgcagatgtggagtc
A T T I C S S D K F C R C C U
tgcctacccagcacctgcccacacacagtctggttactggagcca
C L P S T C P H T U W L L E P
acctgctgtgacaactgccccccaccttgccacattcctcagccc
T C C D N C P P P C H I P Q P
tgtgtgcccacctgcttcctgctcaactcttcccagcccacccca
C U P T C F L L N S S Q P T P
ggcctggaaaccatcaacctcacaacctacactcagcccagctgt
G L E T I N L T T Y T Q P S C
gagccctgcatcccaagctgctgctga
E P C I P S C C *

图 2　牦牛 *KAP*3.3 基因开放阅读框及编码蛋白序列

2.3　牦牛 *KAP*3.3 蛋白的生物信息学分析

2.3.1　理化性质分析

使用 Protparam 在线软件对牦牛 *KAP*3.3 蛋白的理化性质进行分析发现，牦牛 *KAP*3.3 蛋白的分子量为 10.4162 ku，PI 值为 5.37，该蛋白包含 20 种基本氨基酸，含量最高的为 Cys（19.4%），含量最低的为 Lys（1%）、Trp（1%）、Met（1%）和 Tyr（1%）。带负电荷的氨基酸（Asp+Glu）有 5 个，带正电荷的氨基酸（Arg+Lys）有 3 个。

2.3.2　亲疏水性和跨膜结构分析

根据 ProtScale 在线软件分析牦牛 *KAP*3.3 基因氨基酸序列的亲疏水性。由图 3 可知，*KAP*3.3 氨基酸序列第 5 位峰值最高，为 1.967；第 88 位峰值最低为- 1.244，平均峰值为-0.49，属于亲水蛋白质。蛋白跨膜区域预测发现牦牛 *KAP*3.3 蛋白全部氨基酸都在膜的表面，从而证实该蛋白是一种表面蛋白。

2.3.3　亚细胞定位分析

在线软件 PSORT Ⅱ对牦牛 *KAP*3.3 蛋白亚细胞进行定位分析。由表 1 可知，其分布在线粒体的可能性为 21.7%，分布在细胞核的可能性为 34.8%，胞外分泌的可能性为 17.4%，分布在细胞质的可能性为 21.7%，分布在细胞骨架上的可能性为 4.3%。因此可以判断该蛋白主要在细胞核内发挥生物学作用。

表 1　牦牛 *KAP*3.3 蛋白亚细胞定位（%）

亚细胞定位	线粒体	细胞核	胞外分泌	细胞质	细胞骨架
可能性	21.7	34.8	17.4	21.7	4.3

2.3.4　磷酸化和信号肽位点分析

依照 NetPhos2.0Server 软件预测牦牛 *KAP*3.3 基因编码蛋白质磷酸化位点可知（图 4），*KAP*3.3 蛋白共有 5 个磷酸化位点，分别是 Ser 3 个、Thr 2 个。使用 SignalP4.1 预测 *KAP*3.3 蛋白的切割位点以及分泌途径可知该蛋白没有信号肽。

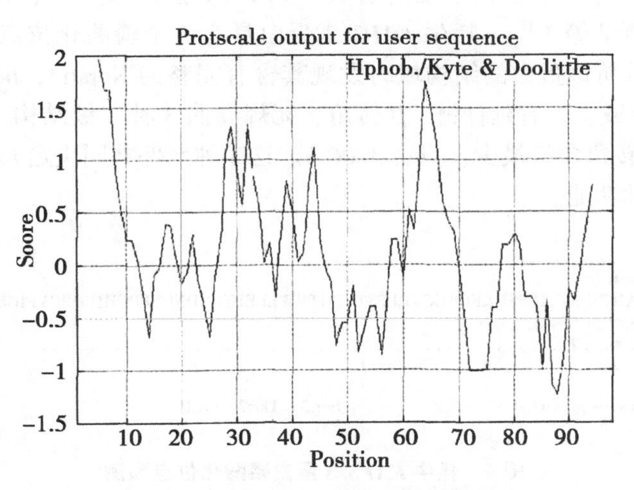

图3　牛 *KAP*3.3 蛋白亲疏水性分析结果

2.3.5　无序化特性及结构域分析

根据 FoldIndex 软件分析牦牛 *KAP*3.3 蛋白氨基酸序列的无序化特性，可知 *KAP*3.3 蛋白中并未发现无序化区域。根据软件 Interpro 对牦牛 *KAP*3.3 蛋白进行结构域预测（图5），可知该序列有完整的 Keratin，high sulphur matrix protein 家族蛋白的结构域。

2.3.6　二级结构预测

使用软件 SOPMA 预测 *KAP*3.3 蛋白的二级结构（图6），可知该蛋白延伸链含 26 个氨基酸，占 26.53%；β 转角含 3 个氨基酸，占 3.06%；无规卷曲含 69 个氨基酸，占 70.41%。2.3.7 功能预测在线软件 ProtFun2.2 预测的结果（表2）可知，*KAP*3.3 是结合转运蛋白，并且都是具有酶功能的蛋白，蛋白加粗的表示功能的指向。

2.3.7　系统进化分析

从 NCBI 数据库中下载以下 *KAP*3.3 蛋白的序列：抹香鲸（*Physetermacrocephalus*）（XM_ 007129082.1）、黄牛（*Bostaurus*）（NM_ 001077103.1）、小须鲸（*Balaenoptera acutorostrata*）（XM_ 007177606.1）、人（*Homo sapiens*）（BC093845.1）、羊驼（*V. pacos*）（XM_ 006214297.1）、埃式象鼩（*E. edwardii*）（XM_ 006889436.1）、大熊猫（*Ailuropoda melanoleuca*（*David*））（XM_ 002924947.1）。根据 MegAlign 软件进行同源性比较分析。图7显示，牦牛 *KAP*3.3 蛋白跟黄牛的完全一致，相似性为 100%。使用 MEGA5.10 软件进行 NJ 法构建系统发育树（图8），结果与同源性分析比较的结果基本一致，牦牛 *KAP*3.3 蛋白与黄牛在同一分支上。

3　讨论

绒毛作为畜牧业重要的副产品，为纺织工业提供重要的天然纤维原料，*KAP* 基因家族是编码绒毛纤维的重要结构蛋白。本实验首次克隆了牦牛 *KAP*3.3 基因完整 CDS 区，并提交到 GenBank（登录号：KM514664）。生物信息学分析结果显示，*KAP*3.3 基因开放阅读框长度为 297 bp，编码 98 个氨基酸。编码蛋白分子量为 10.416 2ku，PI 值 5.37，为亲水性表面蛋白，亚细胞定位结果显示该蛋白主要在细胞核内发挥生物学功能。磷酸化过程是蛋白质翻译后的一个重要过程，与许多生物学功能有关，如 DNA 损伤修复[1]、转录调节[7]、信号转导[8]、细胞凋亡的调节等[9-12]，牦牛 *KAP*3.3 蛋白具有 5 个磷酸化位点。*KAP*3.3 蛋白没有无序化区域，而在分析该蛋白的结构域时发现其含有完整的 Keratin，highsulphur matrix protein 家族蛋白的结构域。含有延伸链、β 转角、无规卷曲 3 种二级结构，无规则卷曲含量最高，占 70.41%，β 转角含量最少，只占 3.06%。这 3 种元件共同决定 *KAP*3.3 蛋白的高级结构，使其具有生物学功能。

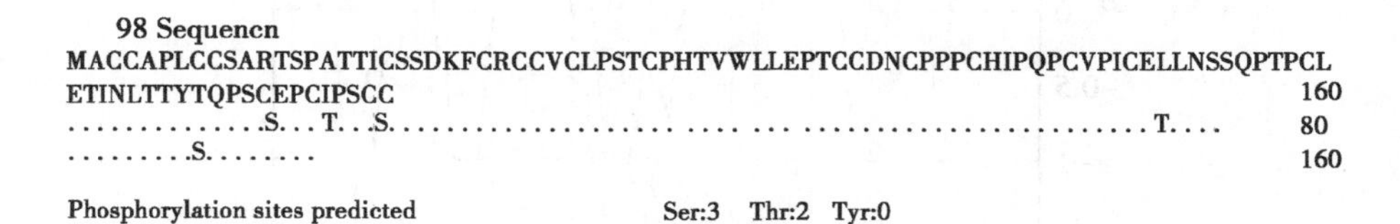

图 4 牦牛 *KAP*3.3 蛋白磷酸化位点预测

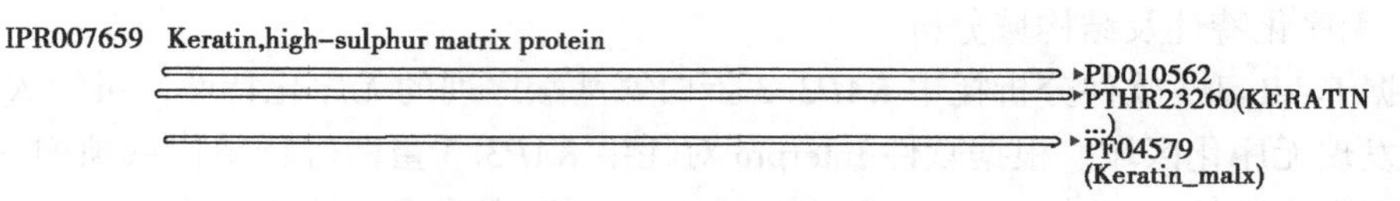

图 5 牦牛 *KAP*3.3 蛋白结构域预测

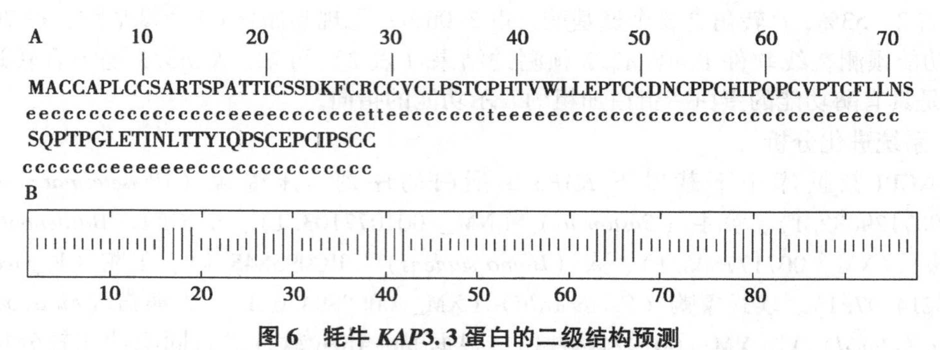

图 6 牦牛 *KAP*3.3 蛋白的二级结构预测

注：A：*KAP*3.3 氨基酸序列中各二级结构单元具体位置：延伸链（e），β 转角（t），无规卷曲（c）；B：*KAP*3.3 二级结构示意图，竖线有长至短分别表示延伸链，β 转角和无规卷曲

表 2 牦牛 *KAP*3.3 蛋白的功能预测

功能范畴	估计概率	先验概率	功能范畴	估计概率	先验概率
氨基酸的生物合成	0.011	0.500	调节功能	0.034	0.211
辅助因子的生物合成	0.210	2.917	复制和转录	0.020	0.075

（续表）

功能范畴	估计概率	先验概率	功能范畴	估计概率	先验概率
细胞膜	0.033	0.541	翻译	0.071	1.614
细胞合成	0.030	0.411	转运和结合	0.773	1.885
中间代谢	0.048	0.762	是否是酶		
能量代谢	0.035	0.389	是	0.368	1.285
脂肪酸代谢	0.017	1.308	否	0.632	0.886
嘌呤和嘧啶	9.331	1.362			

Percent Idenlily

Divergence	1	2	3	4	5	6	7	8		
1		86.9	100.0	85.9	84.8	85.7	84.8	88.9	1	Bos grunniens
2	14.5		86.9	94.9	84.8	80.6	81.8	84.8	2	Plryseter calodon
3	0.0	14.5		85.9	84.8	85.7	84.8	88.9	3	Bos laurus
4	15.7	5.2	15.7		83.8	79.6	84.8	84.8	4	Balaenoptera acutorostrata scammoni
5	17.0	17.0	17.0	18.3		88.8	85.9	83.8	5	Homo sapiens
6	15.9	22.5	15.9	23.9	12.2		86.7	83.7	6	Vicugna pacos
7	17.0	20.9	17.0	20.9	15.7	14.6		87.9	7	Elephantus edwardii
8	12.1	17.0	12.1	17.0	18.3	18.5	13.3		8	Ailuropoda melanoleuca
	1	2	3	4	5	6	7	8		

图 7　不同物种 *KAP*3.3 蛋白相似性分析

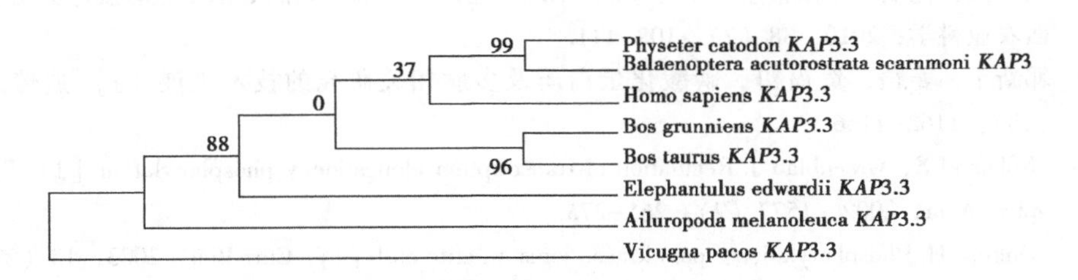

图 8　NJ 法构建 *KAP*3.3 蛋白的系统发育树

牦牛和黄牛 *KAP*3.3 蛋白的系统进化分析发现其保守性达 100%，但编码基因 BLAST 分析结果显示其相似度为仅 99%，说明编码部分蛋白的碱基发生了同义突变。对高原型藏绵羊 *KAP*3.2 的研究发现该基因处于 Hardy-Weinberg 非平衡状态，且等位基因数明显偏低，处于中度多态，推测是由长期迁徙、遗传漂变及本品种选育过程中产毛性状受高强度人工选择导致其高度近交，造成部分等位基因流失有关[13]。系统进化分析发现，牦牛和黄牛的 *KAP*3.3 蛋白处于同一个分支，说明牦牛和黄牛的 *KAP*3.3 亲缘性较近。来自父系遗传[14]、母系遗传[15]及常染色体的研究都显示牦牛与黄牛的遗传距离较近，与本研究结果一致。8 个物种的蛋白相似率达到了 80%以上。说明 *KAP*3.3 蛋白比较保守，吴添文[5]通过比较家兔 *KAP*3.3 基因序列与小鼠、猪、人、黑猩猩、牛等哺乳动物的同源性，发现均达 80%以上，指出该基因在不同物种中存在着一定差异，但也说明该基因在进化过程中相对保守，其结果与本研究一致。

4 结论

牦牛 *KAP*3.3 基因 CDS 区的全长为 297 bp，共编码 98 个氨基酸，该蛋白是亲水蛋白，主要在细胞核内发挥生物学作用，含有 5 个磷酸化位点，有 3 种二级结构，含有完整的 Keratin，high sulphur matrix protein 家族结构域，是一个保守性较高的蛋白。

参考文献

[1] 张桂山，薛景龙，孙俪敏，等．不同品种绵羊皮肤毛囊中角蛋白关联蛋白 8-1 基因差异表达研究［J］. 中国畜牧兽医，2014，41（5）：76-79.

[2] 赵志东．KAP9.2 基因的群体遗传变异及羊绒生长期和休止期基因的表达研究［D］. 杨凌：西北农林科技大学，2012.

[3] Parsons Y，Cooper D，Piper L. Evidence of linkage betweenhighlycineyrosine keratin gene loci and wool fibre diameter in aMerino halfib family［J］. Anim Genet，1994，25（2）：105-108.

[4] 刘桂芬，田可川，张恩平，等．优质细毛羊羊毛细度的候选基因分析［J］. 遗传，2007，29（1）：70-74.

[5] 吴添文．家兔毛囊发育规律及其相关基因的遗传效应、表达规律研究［D］. 扬州：扬州大学，2011.

[6] 宦霞娟，陈玉林．獭兔 KAP 和 FGF5 基因的多态性与经济性状的关系［J］. 西北农业学报，2009，18（6）：12-17.

[7] 贾桦，杨丽娟，李爱华，等．宁夏滩羊 KAP1.1 基因多态性及其与二毛裘皮重要性状关系的研究［J］. 牡丹江医学院学报，2013，34（6）：1-5.

[8] 刘一飞，范瑞文，董常生．羊驼皮肤 KAP3.2 基因 3’端序列的 cDNA 克隆及序列分析［J］. 山西农业科学，2010，38（7）：108-111.

[9] 邓新宇，姜颖，贺福初．磷酸化蛋白质及多肽相关研究的技术进展［J］. 遗传，2007，29（10）：1163-1166.

[10] Kobor M S，Greenblatt J. Regulation of transcription elongationby phosphorylation［J］. Biochim Biophys Acta，2002，1577（2）：261-275.

[11] Ohkura H. Phosphorylation：polo kinase joins an elite club［J］. Curr Biol，2003，13（23）：R912-R914.

[12] Ruvolo P P，Deng X，May W S. Phosphorylation of Bcl2 and regulation of apoptosis［J］. Leukemia，2001，15（4）：515-522.

[13] 王志有，祁全青，陈玉林．高原型藏绵羊 KAP 基因单核苷酸多态性分析［J］. 中国畜牧兽医，2010，37（10）：120-124.

[14] 蔡欣，赵芳芳，孙磊．牦牛与其他家牛属动物 SRY 基因多态性及其父系进化关系分析［J］. 中国畜牧兽医，2014，41（5）：190-195.

[15] 李齐发，李隐侠，赵兴波，等．牦牛线粒体 DNA 细胞色素 b 基因序列测定及其起源、分类地位研究［J］. 畜牧兽医学报，2006，37（11）：1118-1123.

（发表于《遗传育种》）

苦豆子及其复方药对奶牛子宫内膜炎5种致病菌的体外抑菌活性研究*

辛任升**[1,2]，刘　宇[1]，郝宝成[1]，
尚若峰[1]，张　超[1]，梁剑平[1,2]***

（1. 中国农业科学院兰州畜牧与兽药研究所/农业部兽用药物创制重点实验室/甘肃省新兽药工程重点实验室，兰州　730050；
2. 甘肃农业大学动物医学院，兰州　730070）

摘　要：采用酸法提取苦豆子（*Sophoraalopecuroides* L.）的有效成分，并用96孔板微量肉汤稀释法测定苦豆子生物总碱的最小抑菌浓度（MIC）和最小杀菌浓度（MBC）。采用（K-B）纸片扩散法，以盐酸小檗碱为对照，考察苦豆子提取物及其复方灌注剂对致病菌的抑菌效果。结果表明，苦豆子总生物碱对表皮葡萄球菌抑制效果最佳，MIC为16g/L，MBC为32g/L。苦豆子总生物碱、复方灌注剂及硫酸小檗碱的抑菌环平均直径分别为14.28mm、18.61mm和18.02mm。由此得出苦豆子总生物碱及其复方灌注剂有明显的抗菌作用。

关键词：苦豆子总生物碱；体外抑菌试验；最小抑菌浓度；最小杀菌浓度；抑菌环

奶牛子宫内膜炎（puerperal endometritis）是奶牛分娩或产后子宫感染病原微生物引起的急性炎症，是奶牛养殖业中较为常见的一种繁殖障碍性疾病。细菌是导致奶牛子宫内膜炎的主要因素。该病可造成奶牛生理功能紊乱、产奶量下降，甚至终生不孕而被淘汰，给养殖业带来巨大经济损失[1]。有研究发现，引起奶牛子宫内膜炎的主要致病菌有葡萄球菌（*S. aureus*）、链球菌（*Streptococcus*）、大肠埃希杆菌（*Escherichia coli*）、化脓棒状杆菌（*C. pyo-genes*）、绿脓杆菌（*P. Aeruginosa*）等[2]。有研究表明，子宫内致病菌的存在，可引起子宫内膜组织损伤和炎症，降低胚胎存活率[3]。目前，国内外治疗奶牛子宫内膜炎的常用方法有：子宫内疗法、全身疗法、激素疗法、中药疗法、生物学疗法等，其中较常用且有效的方法是子宫灌注抗菌药物或使用激素类药物[4]。随着抗菌药物的大量使用，尤其是抗生素类和磺胺类药物，细菌的耐药性不断增强和动物体内的药物残留已成为亟待解决的问题。因此，研发高效、广谱、低残留、低毒、无公害的药物将成为防治奶牛子宫内膜炎的重中之重。有研究表明，中药或天然植物提取物，可发挥对病原菌的直接杀灭作用而不产生耐药性[5-7]，因此，抗菌中药逐步成为热点。

本试验通过苦豆子总生物碱及复方灌注剂对奶牛子宫内膜炎5种致病菌的抑菌活性研究，以期为奶牛子宫内膜炎的防治提供参考依据。

---

* 基金项目："十二五"农村领域国家科技计划资助项目（2011AA10A214）。

** 作者简介：辛任升（1987—　），男，硕士。

*** 通讯作者，梁剑平，E-mail：liangjp100@ sina. com.

1　材料与方法

1.1　菌种与培养基

表皮葡萄球菌（*S. epidermidis*）CMCC26069，铜绿假单胞菌（*P. Aeruginosa*）ATCC27853，蜡样芽胞杆菌（*Bacillus cereus*）CMCC（B）63302，购自北京北纳创联生物技术研究院；大肠杆菌（*Escherichia coli*）CVCC1570，金黄色葡萄菌（*S. aureus*）CVCC1882购自中国兽医药品监察所；MH 肉汤培养基（meller-Hinton broth，MHB）、MH 琼脂培养基（meller-Hinton Agar，MHA）、营养肉汤培养基（nutrient Broth，NB）均购自北京奥博星生物技术有限责任公司。

1.2　仪器和试剂

ZDX-35BI 型座式自动电热压力蒸汽灭菌器（上海申安医疗器械厂）；PYX-DHS-40X-50 型电热恒温培养箱（上海跃进医疗器械厂）；SW-CJ-IB 型超净工作台（苏州净化设备有限公司）；RE-6000A 旋转蒸发仪（上海亚荣生化仪器厂）；HX-T 电子天平（慈溪市天东衡器厂）；不锈钢数显卡尺（天津市量具厂）；二甲基亚砜（DMSO）购自 Sigma 公司；硫酸小檗碱注射液（10g/L），山西兆益生物有限公司（生产批号：130301）；试验中用到的其他试剂均为分析纯。

1.3　苦豆子有效成分的提取

称取 30℃干燥至恒重的苦豆子粗粉 1 000g，加入 10 倍重量 0.3%的盐酸水溶液浸泡，24h 后，用 8 层纱布过滤。用 8 倍重量 0.3%的盐酸水溶液浸泡，24h 后，用 8 层纱布过滤（同法处理 2 次）。合并 3 次滤液，并用旋转蒸发仪浓缩成浓稠浸膏状。向浸膏中加入无水乙醇，使醇含量达到 60%，充分搅拌静置 12h，滤取上层醇溶液；向下层溶液加入无水乙醇，使醇含量达到 80%，充分搅拌静置 12h，滤取上层醇溶液，合并 2 次滤液，用旋转蒸发仪回收乙醇至浸膏无乙醇味；用氨水碱化至 pH 9~10，加氯化钠至饱和，置于分液漏斗，加氯仿进行萃取，连续萃取 3 次，合并 3 次萃取液，加无水硫酸钠脱水，用旋转蒸发仪回收氯仿，得棕色浸膏。

1.4　水提醇沉法提取复方药有效成分

称取益母草 120g，当归 25g，川芎 30g，黄芩 15g，加 10 倍蒸馏水浸泡 1h，煎煮 1h，8 层纱布过滤；将残渣加 8 倍水量煎煮 1h 后过滤，合并 2 次滤液，用旋转蒸发仪浓缩至浸膏状；加无水乙醇使醇含量达到 60%，充分搅拌静置 12h，滤取上层醇溶液；向下层溶液加入无水乙醇，使醇含量达到 80%，充分搅拌静置 12h，滤取上层醇溶液，合并 2 次滤液，用旋转蒸发仪回收乙醇至浸膏无乙醇味，每毫升约含 6g 生药。

1.5　苦豆子总碱提取液及复方灌注剂的制备

称取苦豆子提取物适量，加少量 DMSO 助溶，用蒸馏水溶解，使其药液质量浓度为 512g/L，0.22μm 微孔滤膜过滤除菌，4℃冰箱保存备用。

称取 20g 苦豆子提取物，10g 复方提取物，0.08g 山梨酸钾，用少量 DMSO 助溶，加 100mL 蒸馏水，121℃灭菌 15min。

1.6　细菌菌悬液的制备

取 9 支盛有 4.5mL 无菌生理盐水（0.9%）的灭菌试管，依次标记 10^{-9}~10^{-1}。精密吸取 0. 5mL 菌液至 10^{-1}试管中，此为 10 倍稀释。将此试管放在旋涡混合器上使菌液混匀，从 10^{-1}试管中吸取 0. 5mL 至 10^{-2}试管混匀，依次稀释到 10^{-9}试管。

取无菌平皿 15 套，分别标注 10^{-9}、10^{-8}、10^{-7}、10^{-6}、10^{-5}各 3 套。每套平皿中加入 15mL MH 琼脂培养基，待冷却凝固。分别吸取 10^{-9} ~ 10^{-5}试管内菌液 100μL 放入到对应平皿中，用 L 棒将菌液涂布均匀，37℃恒温培养箱中培养 18h，对菌落数在 30 ~ 300 的平皿进行计数，求平均值，单位为 CFU/mL。最终配成菌液浓度为 106 ~ 107CFU/mL 菌悬液备用。

1. 7 药敏纸片的制备

用打孔器将新华一号中速定性滤纸打成直径 6mm 的圆片，将滤纸片分为 4 组，分别加入编有 A ~ D 号青霉素小瓶中高压灭菌（121℃，30min），在 60℃条件下烘干，分别加入各供试液浸泡 24h，取出挥干多余滤液，保存备用。

1. 8 苦豆子提取物及制剂对 5 种致病菌的抑菌活性

采用 96 孔聚苯乙烯 U 型微孔板[8-10]测定苦豆子提取物对 5 种致病菌的 MIC、MBC。将 MHB 培养基依次加入到 96 孔板中，每孔 100μL，样品初始质量浓度为 256g/L。第 1 孔中加入样品药液 100μL，采用连续二倍稀释法至第 10 孔，并从第 10 孔吸 100μL 弃去。在每孔加 5μL 菌悬液，第 11 孔为不加药的阴性对照，第 12 孔为不加药也不加菌的阳性对照，每孔重复 3 次。此时 1 ~ 10 孔的药物质量浓度分别为 256g/L、128g/L、64g/L、32g/L、16g/L、8g/L、4g/L、2g/L、1g/L、0. 5g/L。

将微孔板置微型振荡器上震荡 1min，使各孔内溶液混匀，微孔板加盖并用胶纸密封置于湿盒内，35 ~ 37℃孵育 18h 后观察，同法测定复方灌注剂，以肉眼观察无浑浊者为受试菌的 MIC。

另取 1 微孔板，每孔加入 100μL MHB 培养基，再将 MIC 前不同药物浓度的混合液分别取 5μL 加入各孔，每孔 3 个重复。在 37℃条件下孵育 18h 后，在每孔中加入 5g/L 氯化三苯四氮唑（TTC）5μL，35℃孵育 2h 后有细菌生长孔呈红色，无红色者即为该菌的 MBC。

采用（K-B）纸片扩散法[11]测定抑菌圈直径。配制 MHA 平板，琼脂厚度为 4mm，分别吸取 100μL 菌液加到平板内，用 L 棒均匀涂布。将平板均分为 4 个区，分别放入 A ~ D 药敏片，每种细菌重复 3 次。其中，C 药敏片为阳性对照，D 药敏片为空白对照。放置无菌超净台上静置 30min 后，放入 37℃培养箱，24h 后测量抑菌圈。抑菌效果判定：抑菌圈直径 > 20mm 为极敏感，15 ~ 20mm 为高度敏感，10 ~ 15mm 为中度敏感，抑菌圈 < 10mm 为低度敏感[12]。

2 结果

2. 1 苦豆子提取物及复方制剂对 5 种致病菌的 MIC、MBC

苦豆子提取物及复方制剂对供试菌的 MIC、MBC 见表 1。苦豆子提取物及复方制剂对表皮葡萄球菌的 MIC、MBC 最低，分别为 16g/L、32g/L 和 12. 5g/L、25g/L，说明豆子提取物及复方制剂对表皮葡萄球菌具有良好的抑菌效果。

2. 2 苦豆子提取物及复方制剂对 5 种致病菌的抑菌活性

苦豆子提取物及复方制剂对 5 种致病菌的抑菌圈大小见表 2。如表所示，B 与 C 差异不显著，而 A 与 B，A 与 C 差异极显著，说明提高苦豆子提取物的含量，可明显增加其对细菌的抑菌活性。

表 1　苦豆子提取物及复方制剂对 5 种致病菌的 MIC、MBC 值（g/L）

细　菌	MIC		MBC	
	A	B	A	B
金黄色葡萄菌 CVCC1882	32	25	64	50
表皮葡萄球菌 CMCC26069	16	12. 5	32	25
大肠杆菌 CVCC1570	32	25	128	100
蜡样芽胞杆菌 CMCC（B）63302	32	25	64	50
铜绿假单胞杆菌 ATCC27853	32	25	128	100

注：A. 苦豆子提取物；B. 复方苦豆子灌注液

表 2　苦豆子提取物及复方苦豆子灌注液对 5 种致病菌的抑菌活性（mm）

细 菌	n	抑菌圈直径			阴性对照
		A	B	C	
金黄色葡萄球菌 CVCC1882	3	$15.34±1.02^{A}$	$19.02±0.54^{Ba}$	$19.22±1.32^{Ba}$	6. 00±0. 00
表皮葡萄球菌 CMCC26069	3	$16.50±0.64^{A}$	$25.54±1.21^{Ba}$	$24.52±1.35^{Ba}$	6. 00±0. 00
大肠杆菌 CVCC1570	3	$14.12±1.58^{a}$	$15.20±2.06^{a}$	$15.20±2.21^{a}$	6. 00±0. 00
蜡样芽胞杆菌 CMCC（B）63302	3	$13.14±2.32^{A}$	$19.00±2.06^{Ba}$	$19.00±1.71^{Ba}$	6. 00±0. 00
铜绿假单胞杆菌 ATCC27853	3	$12.16±2.28^{a}$	$14.12±3.02^{a}$	$12.02±2.59^{a}$	6. 00±0. 00

注：A. 苦豆子提取物；B. 复方灌注剂；C. 硫酸小檗碱注射液；D. 二甲基亚砜 DMSO。同一行中标有不同小写字母的均数间差异显著（$P<0.05$），标有不同大写字母的均数间差异极显著（$P<0.01$）。

3　讨论

本试验采用纸片扩散法检测苦豆子提取物及复方苦豆子灌注剂对 5 种奶牛子宫内膜炎致病菌的抑菌活性，试验结果表明，金黄色葡萄球菌、表皮葡萄球菌、蜡状芽孢杆菌对苦豆子提取物及其复方制剂的敏感性明显高于大肠杆菌和铜绿假单胞杆菌。因此，革兰阳性菌对苦豆子提取物及复方制剂的敏感性有可能高于革兰阴性菌，确认此猜想有待于进一步试验加以验证。

周娅等[13]报道，苦豆子总碱对金黄色葡萄球菌 ATCC25923、大肠杆菌 25922、铜绿假单胞菌 ATCC27853 的 MIC 分别为 15g/L、15g/L、30g/L；哈丽娜等[14]报道，苦豆子总碱对金黄色葡萄球菌 ATCC2611、大肠埃希菌 ATCC24752、铜绿假单胞菌 ATCC27853 的 MIC 均为 35. 1g/L；本试验得出的 MIC 分别为 16、32、32g/L，说明本试验测得的数据准确有效。

另有报道，石榴皮、五倍子、诃子、黄连、乌梅、艾叶、黄芩、黄芪、黄柏、大黄、马齿苋、白头翁 12 味中药对大肠杆菌的 MIC 为 62.5～500g/L[15]，苍术、金银花、板蓝根、连翘、鱼腥草、大青叶对大肠杆菌的 MIC 为 125～500g/L[16-18]。相比而言，苦豆子提取物及其复方灌注剂的抑菌效果明显优于其他中药提取物，为苦豆子进一步开发提供有力依据。

近年来，盲目滥用抗生素导致细菌耐药性增强，造成畜产品中抗生素残留增加，因此抗生素残留问题已经引起社会的高度重视。寻求绿色、无污染、无公害的药物已迫在眉睫。苦豆子总生物碱是从天然植物苦豆子中提取而来，即不会使细菌产生耐药性也基本无残留，将其制成水溶性灌注剂治疗奶牛子宫内膜炎具有良好的应用前景。

参考文献

[1] 李宏胜，杨峰，王旭荣，等．兰州地区部分奶牛场子宫内膜炎病原菌分离鉴定及抗生素耐药性研究［J］．中国畜牧兽医，2014，41（1）：222-226.

[2] 邓祥华，张志强，范才良，等．奶牛子宫内膜炎病原菌的研究进展［J］．黑龙江畜牧兽医，2013，13（7）：28-30.

[3] Sheldon I M，lewis G S，LeBlanc S，et al. Definingpostpartum uterine disease in cattle［J］．Theriogenology，2006，65（8）：1516-1530.

[4] 赵红霞．奶牛子宫内膜炎致病大肠杆菌的分离、鉴定及耐药性研究［D］．内蒙古呼和浩特：内蒙古农业大学，2008：13.

[5] Tan B K，Vanitha J. Immunomodulatory and antimicrobial effects of some traditional Chinese medicinalherbs：a review［J］．Curr Med Chem，2004，11（11）：1423-1430.

[6] Zuo G Y，Wang G C，Zhao Y B，et al. Screening of Chinese medicinal plants for inhibition against clinicalisolates of methicillin - resistant staphylococcus aureus（MRSA）［J］．J Ethnopharmacol，2008，120（2）：287-290.

[7] 王国强．中兽药应用的现状及发展方向［J］．养殖技术顾问，2012（7）：214.

[8] Georgantelis D，Ambrosiadis I，Katikou P，et al. Effectof rosemary extract，chitosan and α-tocopherol on microbiological parameters and lipid oxidation of freshpork sausages stored at 4℃［J］．Meat Sci，2007，76（1）：172-181.

[9] Georgantelis D，Blekas G，Katikou P，et al. Effect of rosemary extract，chitosan andα-tocopherol on lipid oxidation and colour stability during frozen storage of beef burgers［J］．Meat Sci，2007，75（2）：256-264.

[10] Dong C J，Sung Y L，Ji H Y，et al. Inhibition of pork and fish oxidation by a novel plastic film coated withhorseradish extract［J］．LWT-Food Sci Technol，2009，42（4）：856-861.

[11] 秦晓蓉，张铭金，高绪娜．槲皮素抗菌活性的研究［J］．化学与生物工程，2009，26（4）：55-57.

[12] 房春林，杨光友，杨海涵，等．杨树花提取物及其复方制剂的药理学试验［J］．中国兽医科学，2010，40（2）：205-209.

[13] 周娅，杨志伟，赵建宁，等．苦豆子总碱的体外抗菌活性研究［J］．宁夏医学院学报，2000，22（2）：79-81.

[14] 哈丽娜，杨风琴，肖文婷．宁夏道地药材苦豆子醇提取物体外抑菌作用研究［J］．长治医学院学报，2011，25（5）：324-326.

[15] 陈虹．中草药对 8 种畜禽肠道病原菌的体外抑菌试验［D］．甘肃兰州：甘肃农业大学，2009：13.

[16] 曹翠萍，宁海强，孙丽，等．中药对大肠杆菌抑菌作用及耐药性诱导作用的研究［J］．西南农

业学报，2007，20（5）：1101-1104.
[17] 李少基，陈武，陈足金．12 种中草药的体外抑菌试验［J］．中兽医医药杂志，2004，6（1）：44-46.
[18] 刘富来，冯翠兰，王林川．中草药对肠炎沙门菌的体外抑菌试验［J］．中国兽药杂志，2004，38（11）：28-30.

（发表于《中国兽医学报》）

用主成分分析法研究腹泻仔猪血清生化指标*

黄美州**，刘永明***，王 慧，
王胜义，崔东安，李胜坤，齐志明

（中国农业科学院兰州畜牧与兽药研究所/农业部兽用药物
创制重点实验室/甘肃省新兽药工程重点实验室/甘肃省中
兽药工程技术研究中心，兰州 730050）

摘 要：为了研究腹泻对仔猪血清生化指标的影响，本试验采集了 20 头腹泻仔猪（7~8 日龄）和 10 头健康仔猪（与腹泻仔猪同日龄、同性别）的粪便和血清。荧光定量 PCR 法对粪便进行病原学检测，对采集的血清用血生化分析仪进行检测，并利用主成分分析法对其结果进行分析。结果，①20 头腹泻仔猪粪便的猪流行病腹泻病毒检测结果均为阳性。②十二项血清生化指标被转化成了综合性更强的 4 个主成分，根据这 4 个主成分进行的聚类分析的结果和临床采集样品的分类结果一致。③腹泻仔猪的血清球蛋白、尿素和肌酐水平显著升高（$P<0.05$）。结果表明，由猪流行性腹泻病毒引发的腹泻，可能导致仔猪体内的蛋白质代谢紊乱、肝脏和肾脏的机能受损。

关键词：仔猪腹泻；主成分分析；生化指标

血清生化指标的改变是组织细胞通透性发生改变和机体新陈代谢机能发生改变的反映，通过血清生化指标研究动物机体的疾病，可以为疾病的靶组织或靶器官和疾病的发生机理的研究提供重要的指示和基础[1-2]。近年来哺乳仔猪的腹泻病给整个养猪行业带来重大的损失，已经成为影响仔猪成活率的重要疾病之一。由病毒导致的哺乳仔猪的腹泻由于具有传染性强、致病力强、发病率高、死亡率高等特点，已经成为导致仔猪腹泻的主要病原。虽然近些年来对由病毒导致的哺乳仔猪腹泻病的病原学和流行病学进行了大量的研究[3-4]，对其组织病理学也进行了研究，但发病机理还尚未清楚，对靶器官了解得还不够全面，对血液生化的病理学研究更是稀少[5-6]。因此本试验利用主成分分析对腹泻哺乳仔猪血清生化指标进行研究，了解血液生化病理学，为进一步研究仔猪腹泻的发病机理和作用的靶器官提供了理论基础。

1 材料与方法

1.1 实验动物

于甘肃永靖县玉丰养殖场，选取同日龄（7~8 日龄）发病且具有相同临床症状、同性

* 基金项目：公益性行业专项基金资助项目（20130304-17）；国家科技支撑计划基金资助项目（2012BAD12B03）。

** 作者简介：黄美州（1989— ）男，硕士研究生。

*** 通讯作者，刘永明，E-mail：myslym@ sina. com.

别、体质量相似的20头腹泻哺乳仔猪，另选取与腹泻仔猪同日龄、同性别、体质量相近的10头健康仔猪作为对照。

1.2 粪便和血清的采取

采用腹部挤压法采集健康和腹泻仔猪的粪便，放入粪便采集器中，-20℃贮藏。通过前腔静脉分别对20头腹泻仔猪和10头健康仔猪进行采血，每头10mL，放入一次性采血管中，室温静置2h后，放入离心机中3 000r/min离心10min，收集采血管上部的血清，-20℃贮藏。

1.3 病原的检测

将采集的30份粪便样品，用猪传染性胃肠炎病毒、猪流行性腹泻病毒、猪轮状病毒三重实时荧光RT-PCR检测试剂盒（购自北京世纪元亨动物防疫技术有限公司）进行检测，详细操作按照试剂盒使用说明书进行。最后将所提RNA放入实时荧光定量PCR仪（BIO-RAD CFX96，德国BIO-RAD公司）进行检测。

1.4 血清生化指标的测定

将上述采集的30份血清样品，用全自动生化分析仪（Erba XL-640，德国Erba公司）对血清中的肌酐（CREA）、丙氨酸氨基转移酶（ALT）、碱性磷酸酶（ALP）、门冬氨酸氨基转移酶（AST）、总蛋白（TP）、白蛋白（ALB）、血清胆红素（T-BIL）、尿素氮（BUN）、血糖（GLU）、总胆固醇（TC）、甘油三酯（TG）和肌酸激酶（CK）进行检测，所用生化试剂盒购自宁波美康生物科技股份有限公司。

1.5 数据的统计分析

本试验中所有数据均运用SAS9.2统计分析软件进行分析。随机选择7头健康仔猪和10头腹泻仔猪的血液生化测量数据，采用主成分分析聚类分析方法进行处理。然后通过SAS9.2统计软件中2个样本的t检验的方法对健康仔猪和腹泻仔猪的各项生化指标进行差异性检测。

2 结果

2.1 病原的检测结果

猪传染性胃肠炎病毒、猪流行性腹泻病毒、猪轮状病毒三重实时荧光RT-PCR检测结果见图1，20份哺乳仔猪腹泻粪便的猪流行性腹泻病毒均为阳性，其余的检测结果均为阴性。

2.2 血清生化指标的检测结果

基于主成分对血清生化指标检测分析的结果

对随机选择的7头健康仔猪和10头腹泻仔猪的血清生化指标基于主成分分析的结果见表1，对12项血清生化指标进行主成分分析根据方差累计贡献率达到75%的原则，选择了4个主成分，其特征值分别为：3.918、2.657、2.280、1.142基本反映了原来变量的信息。主成分变量Y1=0.470×ALT+0.363×ALP+0.422× BUN +0.421×CR-0.300GLU+…；主成分变量Y1代表机体肝肾的代谢指征，单独说明原始变量的32.65%。主成分变量Y2=0.449×TP+0.446×ALB+0.424×TG+0.476×TC+…；主成分变量Y2代表脂和血清蛋白代谢指征，单独说明原始变量的22.14%，主成分变量Y3=0.492×CK+…；主成分变量Y3代表肌肉代谢指征，单独说明原始变量的19%。主成分变量Y4=0.412×T-BIL+0.634×AST+…；主成分变

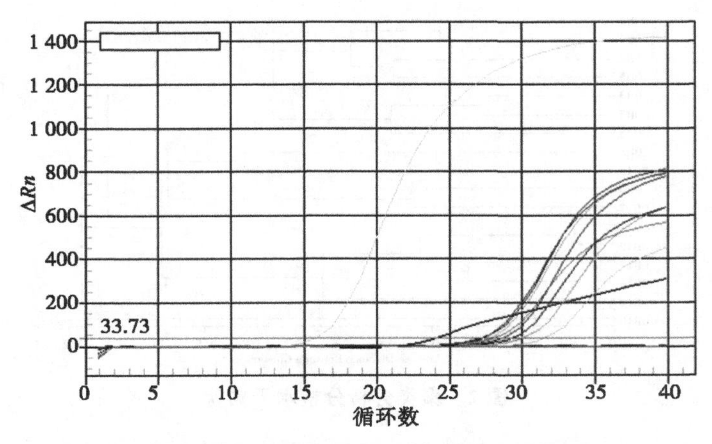

图 1 实时荧光定量 RT-PCR 猪流行性腹泻病毒的扩增曲线

Y4 代表肝胆代谢指征，单独说明原始变量的 9.51%。

根据上述的 4 个主成分对随机选择的 7 头健康仔猪和 10 头腹泻仔猪的血清生进行聚类分析，采用的是类平均的聚类方法。聚类分析的水平树状图见图 2，其中 OB1～OB7 为健康仔猪，OB8～OB17 为腹泻仔猪，从水平树状图，可以将所有的观测样本分为 3 大类：OB1～OB7 为一大类；OB8、OB9、OB11～OB17 为另一大类。OB10 为单独的一类。

除去 OB10 以外，用主成分分析法对检测的 12 项血清生化指标进行聚类分析，所得的分类的结果基本和采集样品的分类结果一致。其中 OB10 单独成了一类。对血清生化 15 个检测指标的测定结果如表 2 所示，表中的数据均为 $\bar{x}±s$。

表 1 血清生化指标主成分矩阵特征值、贡献率、累积贡献率

项　目	肝肾的代谢指 Y1	血脂和血清蛋白代谢指征 Y2	肌肉的代谢指征 Y3	肝胆的代谢指征 Y4
矩阵特征值	3.918	2.657	2.280	1.141
贡献率	0.327	0.221	0.190	0.095
累积贡献率	0.327	0.548	0.738	0.8331

2.3 健康仔猪与腹泻仔猪血清生化指标差异性的检测

结果见表 2 所示腹泻仔猪与健康仔猪相比血清中的 GLB 水平、ALT 水平、BUN 水平极显著的升高（$P<0.01$），肌酐水平和碱性磷酸酶水平显著升高（$P<0.05$），A/G 水平极显著的降低（$P=0.002$），AST/ALT 水平显著地降低（$P=0.04$）。

3 讨论

3.1 主成分的聚类分析

主成分分析能够在保持原资料大部分信息的前提下，有效的压缩变量数，而且生成的新变量解释了样本的某个方面或某些方面的变异情况，对进一步分析和评价样本提供理论依据[7-9]，肖丽等[10]和陈国顺[11]成分别用主成分分析法对不同品种仔猪血液参数差异和不同固

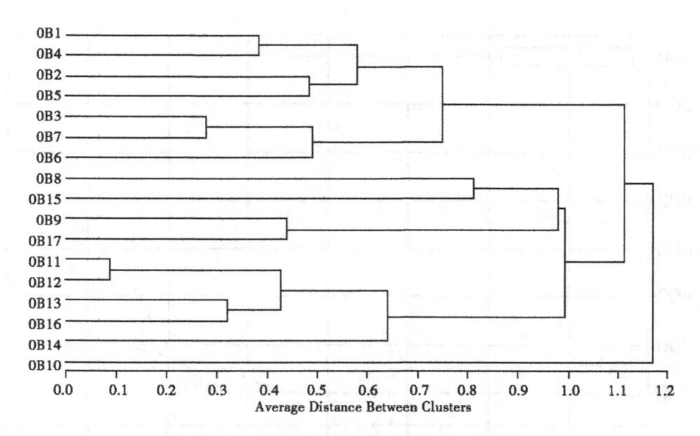

图 2　聚类分析分析水平树状

定效应对血液生化指标的影响进行了研究，在对血常规进行主成分分析时，两者出现了一致的指向相同血液指征的两个主成分，说明主成分分析法具有较强的稳健性。本试验对于观测样本的 12 项血清生化指标经过主成分分析转换成了 4 项能够代表原来样本基本信息的综合性更强的生物学指标，分别为肝肾代谢指征、血脂和血清蛋白代谢指征、肌肉代谢指征、肝胆代谢指征，根据这 4 项主成分进行聚类分析。聚类分析作为一种探索性的分类方法，可将一组数据按照本身的内在规律较合理地分为几类，大大缩小了以往全靠主观判断所造成的误差使数据分析结果更具客观性[12]。本试验采用基于主成分聚类分析法来进一步确认所选择血清生化指标能够有效反映出仔猪健康和腹泻的状态，结果显示，采用基于主成分的聚类分析方法和临床上采集样本时的原始方法一致。基本能将健康和腹泻仔猪聚为两大类，而且将观测样本聚为的两大类和临床上采集的样品的分类结果是一致的，但在主成分聚类分析中，将观测样本 OB10 单独的聚为了一类，而没有聚到腹泻仔猪的一类中，可能的原因，采集的观测样本 OB10 的血清收集或实验过程中发生了一些变化，使血清中的一些生化指标明显高于同组中的其他观测样本的值，如：尿素含量为 12. 3mmol/L，而同组的尿素的平均值为 4. 9mmol/L。

表 2　生化指标的统计分析结果（$\bar{x}$±s）

指标	腹泻仔猪	样本数	健康仔猪	样本数	*P* 值
T-BIL	2. 1±1. 1	20	2. 8±2. 2	10	0. 42
TP	59. 1±14. 8	20	47. 3±24. 2	10	0. 27
ALB	14. 57±4. 9	20	23. 7±11. 0	10	0. 06
GLB	44. 5±10. 9	20	23. 6±14. 6	10	0. 006
A/G	0. 3±0. 1	20	1. 09±0. 4	10	0. 002
ALT	86. 7±18. 3	20	29. 9±13. 6	10	<0. 001
AST	84. 9±29. 9	20	59. 6±49. 5	10	0. 25
AST/ALT	0. 96±0. 2	20	1. 9±1. 2	10	0. 04
ALP	865. 1±324. 6	20	389. 5±240. 9	10	0. 03

（续表）

指标	腹泻仔猪	样本数	健康仔猪	样本数	P 值
CK	668.8±245.3	20	1 188.0±999.5	10	0.14
BUN	39.8±14.5	20	4.9±3.4	10	0.000 6
CR	221.5±36.0	20	76.4±42.6	10	0.017
GLU	2.9±3.4	20	6.1±2.9	10	0.053
TG	0.8±0.5	20	0.5±0.3	10	0.07
TC	3.3±1.6	20	3.1±1.3	10	0.75

3.2 腹泻对仔猪血清生化指标差异性的影响

仔猪机体的代谢是一个动态平衡的过程，血液中的生化指标的变化是仔猪机体代谢平衡状态的一种风向标，它在一定程度上能反映出机体某些物质的代谢状态和某些组织或器官的功能状态，了解血清生化指标的变化，有助于了解一些疾病发生，发展和转轨的过程[13-14]。血清白蛋白主要由肝脏合成，一般急性病毒性肝炎时，肝脏合成蛋白质功能受损较轻，血清白蛋白量略低或正常[2,15-16]。本试验中腹泻仔猪血清中的白蛋白含量略低于健康仔猪血清中白蛋白的含量，但没有达到显著性差异的水平（$P=0.06$），可能是由于猪流行性腹泻病毒（PEDV）发病比较急，对肝脏合成蛋白质功能的损害较轻或没有损害。

血清球蛋白由人体单核-吞噬细胞系统合成，球蛋白增高常见于慢性炎症和感染，当机体内存在病毒等抗原时，机体的免疫器官就会增加对球蛋白的分泌，用于抵抗和消除侵袭的病原体[16-17]。本试验中腹泻仔猪血清中的球蛋白含量显著高于健康仔猪的球蛋白含量，这可能是由于猪流行性腹泻病毒的入侵刺激机体的免疫器官，增加了对球蛋白的分泌。

丙氨酸转氨酶（ALT）和天门冬氨酸转氨酶（AST）是机体内糖和蛋白质互相转变所需的酶。ALT 主要分布于肝细胞的细胞浆中，AST 主要分布在肝细胞的细胞浆和线粒体中，且肝内的丙氨酸转氨酶的活性是血清中的千倍以上，只要有1%的肝细胞损伤，就可使血液中的丙氨酸转氨酶明显升高[17]。本试验中腹泻仔猪血清中的 ALT 和 AST 都高于健康仔猪，但腹泻仔猪血清中的 ALT 显著地高于健康仔猪（$P<0.001$），而 AST 却没有明显的差异（$P=0.25$），使得 AST/ALT 表现出了显著性差异（$P<0.05$），这可能是由于猪流行性腹泻病毒对于肝细胞的损伤较小，ALT 比 AST 对肝细胞的损伤反应更灵敏，因此出现了腹泻仔猪的 ALT 显著性增高而 AST 增高不显著。在由病毒、药物等引起肝内胆汁瘀滞时，由肝脏而来的 ALP 向胆汁中排泄障碍及因胆管内压亢进，致使血中 ALP 显著增高[18-19]。本试验中腹泻仔猪血清中 ALP 显著增高，可能就是 PEDV 病毒引起肝内胆汁瘀滞，使血清中的 ALP 显著增高。

血清尿素（BUN）和血清肌酐（CR）都是作为检查肾小球滤过功能的指标，尿素大部分经肾小球滤过而随尿不断排出体外，肌酐的数值不受食物中蛋白质含量及年龄、性别或运动的影响，经肾小球过滤后，直接从尿中排出[2,16,19]。因此，当肾小球滤过率降低时，就会导致血清中的 BUN 和 CR 增加，本试验中腹泻仔猪血清中的 BUN 和 CR 都显著性增高（$P<0.05$），这一结果预示着 PEDV 病毒可能引起了肾脏的损伤使肾小球的率过滤降低。

综上所述，本试验中所选取的仔猪血液生化指标通过主成分分析能够很好地反映出仔猪健康与患病的情况。由猪流行性腹泻病毒导致的哺乳仔猪的腹泻，不仅仅对哺乳仔猪的肠道造成损伤，通过试验中对血清生化指标的分析表明，猪流行腹泻病毒在导致哺乳仔猪腹泻的同时也对哺乳仔猪的肝脏和肾脏功能造成了损伤。

参考文献

[1] 魏光河，张培晏．运输对荣昌猪生理生化指标的影响［J］．动物医学进展，2011，32（10）：123-127.

[2] 辛亮．不同日粮类型对仔猪生产性能及血清生化指标的影响［D］．南京：南京农业大学，2013.

[3] 高君恺，刘浩飞，杨倩．猪流行性腹泻病毒的研究进展［J］．南京农业大学学报，2014，37（1）：1-5.

[4] 赵梦姣，陈书民，成岩，等．山东省部分地区猪流行性腹泻流行病学调查及其 M 基因遗传变异分析［J］．中国兽医学报，2013，33（10）：1504-1508.

[5] 张雷，董毅．猪传染性胃肠炎病毒感染机理的研究进展［J］．现代畜牧兽医，2009（7）：79-82.

[6] 纪素玲，索良次旦，徐业芬等．西藏仔猪腹泻病的病理学研究［J］．黑龙江畜牧兽医，2001（1）：31-32.

[7] 张鹏．基于主成分分析的综合评价研究［D］．南京：南京理工大学，2004.

[8] 李靖华，郭耀煌．主成分分析用于多指标评价的方法研究——主成分评价［J］．管理工程学报，2002，16（1）：39-43.

[9] Haware R V，Tho I，Bauer B A，et al. Multivariate analysis of relationships between material properties，process parameters and tablet tensile strength for alpha-lactose monohydrates.［J］. Eur J Pharm Biopharm，2009，73（3）：424-431.

[10] 肖丽，屈云龙，季涛，等．基于主成分分析的不同固定效应对种猪血液生化指标的影响［J］．中国畜牧杂志，2014，50（3）：27-31.

[11] 陈国顺．运用聚类分析和主成分分析筛选猪的血清指标［J］．甘肃农业大学学报，2005，40（6）：723-727.

[12] 陈军辉，谢明勇，傅博强，等．西洋参中无机元素的主成分分析和聚类分析［J］．光谱学与光谱析，2006，26（7）：1326-1329.

[13] 伍力，耿梅梅，王文策，等．哺乳藏仔猪发育期血液生化指标动态变化规律研究［J］．西南农业学报，2010，23（2）：570-574.

[14] 赵拴平，王睿琪，唐中林，等．3 个地方品种猪血液生化指标分析［J］．中国畜牧医，2012，39（2）：96-100.

[15] 李华慧．三种补铁剂对哺乳仔猪的生长性能和血液生化指标影响的研究［D］．南宁：广西大学，2011.

[16] 戚应杰．慢性丙型肝炎患者血清球蛋白水平与病毒载量的相关性研究［D］．合肥：安徽医科大学，2012.

[17] 周国华，陈叶青，周红宇，等．血清球蛋白/胆碱脂酶比值的临床价值［J］．南华大学学报：医学版，2007，35（4）：571-574.

[18] 王书凤，龚月生．不同抗生素组合对哺乳仔猪内分泌的调控作用及血液生化指标的影响［J］．安徽农业科学，2014（21）：7037-7040.

[19] 谢秀琼．中药新制剂开发与应用［M］．3 版．北京：人民卫生出版社，2006：750-753.

（发表于《中国兽医学报》）

鱼紫散防治鸡痘有良效*

谢家声，李锦宇，王贵波，
辛蕊华，罗永江，罗超应，郑继方

（中国农业科学院兰州畜牧与兽药研究所，兰州 730050）

鸡痘是由痘病毒引起的一种急性、热性、接触性禽类传染病，家禽中以鸡最为易感，不分年龄、性别和品种均可发生。本病分布广泛，在大中型鸡场更易形成流行，往往从数只发展到整群，延续时间很长，病程3~4周。本病的死亡率虽然不高，但在饲养管理不当而出现并发症时可引起大批死亡，对养鸡业造成严重损失。笔者根据文献报道结合中兽医对本病的认识进行辨证施治，以鱼腥草、紫草为主，配合其他4味中药，应用于2群8 600只患病产蛋鸡群，收到较满意的防治效果，报告如下。

1　临床资料

治疗的2群鸡均为商品蛋鸡，第1群品种为海蓝褐（360日龄）4 320只；第2群为罗曼褐（170日龄）4 280只。鸡舍温度分别为13 ℃和17 ℃，湿度分别约为40%和60%。产蛋率分别为84.0%（3 628/4 320）和80.0%（3 424/4 280），两群鸡均在50日龄时进行过鸡痘疫苗的免疫。查患鸡冠、髯、喙角、眼睑及耳球上均有凹凸不平、灰褐色硬结节或互相连结融合形成的大块厚痂；病鸡数逐日增加，采食量有不同程度下降，产蛋率分别为81.0%（3 500/4 320）和75.0%（3 210/4 280），较患病前分别下降近3.0%和5.0%；日死亡率分别为0.3%（13/4 320）和0.21%（9/4 280）；病变严重病鸡分别为168只和122只，其鼻孔、喙角、冠、髯上，有大量痘痂分布，影响呼吸和采食，隔离饲养并进行治疗和观察。

2　治则及处方

2.1　中药治疗

治疗原则：中兽医理论认为，本病主要由于外感时邪病毒、内有湿热蕴郁所致，故治宜疏风清热、解毒祛湿。

处方组成：鱼腥草70%，紫草10%，蒲公英、牛蒡子、荆芥、防风各5%，诸药共为细粉备用。根据病情，以日粮的1.0%~1.5%拌入饲料，连续饲喂7~10d。

2.2　西药治疗

预防鸡痘诱发葡萄球菌的感染，应用中药的同时，饮水中加入硫氰酸红霉素可溶性粉，按红霉素计，20mg/kg体重。

---

* 基金项目：科技部科研院所技术开发研究专项（2013EG134236）；科技部科技基础性工作专项课题（2013FY110600-8）；甘肃省科技支撑计划（1104NKCA094）；科技部农业科技成果转化资金项目（2012GB23260560）。

3 治疗效果

全群给药 7 d 后，鸡群采食量明显增加，病鸡逐渐减少，疫情得到有效控制。

全群给药 7 d 后，2 群病鸡的产蛋率分别从 81.0%和 75.0%上升到 84.4%和 79.5%。

全群给药 7 d 后，第 1 群总死亡 5 只，总死亡率为 0.12%（5/4 320），日平均死亡率为 0.017%（0.12/7）。

第 2 群总死亡 3 只，总死亡率为 0.07%（3/4 280），日平均死亡率为 0.01%。

4 治疗结果

第一群：病变严重的 168 只，经过 7 d 治疗，135 只病鸡的痘痂完全脱落，仅留有粉红色斑痕，判为治愈，治愈率为 80.4%（135/168）；10 只部分脱落，判为好转，好转率为 6.0%（10/168）；20 只变化不明显，判为无效，无效率为 11.9%（20/168）；死亡 3 只，死亡率为 1.8%（3/168）。

第二群：病变严重的 122 只，经过 7 d 治疗，95 只病鸡的痘痂完全脱落，仅留有粉红色斑痕，判为治愈，治愈率为 77.9%（95/122）；13 只部分脱落，判为好转，好转率为 10.7%（13/122）；9 只变化不明显，判为无效，无效率为 7.4（9/122）%；死亡 5 只，死亡率为 4.1%（5/122）。

5 体会

单味鱼腥草治疗鸡痘的经验曾多次见诸有关文献资料，根据笔者的临床验证，鱼腥草治疗鸡痘确有效果，但仅此一味药，仍嫌“力单味薄”。因此，根据中兽医理、法、方、药原则，在辨证施治基础上“遣方用药”，能使治疗更趋合理，疗效进一步提高。方中鱼腥草、蒲公英清热解毒，紫草凉血、活血、解毒透疹。用于血热毒盛、斑疹紫黑、麻疹不透。其功长于凉血活血，痘疹欲出未出，血热毒盛者皆可用之。荆芥、防风解毒祛风、除湿，牛蒡子宣肺解表、清咽利喉。诸药伍用，共奏疏风清热、解毒祛湿的功效。临床实践证明，以鱼腥草、紫草为主并配合蒲公英、牛蒡子、荆芥、防风以及硫氰酸红霉素中西药结合治疗鸡痘，不仅治疗作用更为显著，同时对未发病鸡群亦有较好的预防作用。

鸡痘是一种流行性极强的接触传染性疾病，养鸡生产中要求分别在 30~40 日龄和 120~140 日龄以鸡痘弱毒疫苗翼下皮肤刺种，可产生坚强的免疫力，但此操作需要逐只抓鸡，不仅耗费人力而且还常常造成极强的应激，影响正常生长和产蛋性能。因此，在一般中小型鸡场，仅在转入蛋鸡笼时，接种一次鸡痘疫苗，甚至不做免疫。有时虽然进行了刺种，但因操作不当，疫苗稀释后污染严重等种种原因，常常导致免疫失败。目前尚无治疗鸡痘的有效药物，因此，通过临床实践，探索治疗鸡痘的有效中药或有效的中西结合治疗药，十分必要。

（发表于《中国兽医杂志》）

不同地区啤酒糟基本成分测定及其分析*

王晓力[1**]，孙尚琛[2]，匡彦蓓[3]，王永刚[2]，朱新强[1]，李秋剑[4]

（1. 中国农业科学院兰州畜牧与兽药研究所，兰州 730050；
2. 兰州理工大学生命科学与工程学院，兰州 730050；
3. 华南师范大学生命科学学院，广州 510630；
4. 深圳市农产品质量安全检验检测中心，深圳 518005）

摘　要：[目的] 测定并分析不同地区啤酒糟的基本成分。[方法] 测定了来自甘肃兰州（2012-B-001）、内蒙古呼和浩特（2012-B-002）和陕西西安（2012-B-003）3个地区啤酒糟的基本成分，并与葡萄渣、甜菜渣、小麦秸秆进行了比较。[结果] 试验表明，3个地区啤酒糟的基本组分中，粗蛋白所占比例均达20%以上，故此类啤酒糟可用于制备蛋白质饲料。2012-B-001、2012-B-002和2012-B-003所含中性洗涤纤维（NDF）和酸性洗涤纤维（ADF）含量相比其他粗饲料较低，对于反刍类家畜具有较好的饲喂效果。此外，3类啤酒糟中，必需氨基酸种类齐全，维生素种类与含量以及矿物质种类与含量均符合国标规定的饲料原料要求。[结论] 以啤酒糟为原料制备动物饲料，尤其是蛋白类饲料有较好的经济效益与开发前景。

关键词：啤酒糟；基本成分；开发；前景

（发表于《安徽农业科学》）

---

* 基金项目：农业部兽用药物创制重点实验室和甘肃省新兽药工程重点实验室开放课题经费，公益性行业（农业）科研专项子课题：甘肃河西走廊荒漠灌区苜蓿高效种植关键技术研究与集成示范；工业副产品的优化利用技术研究与示范项目。

** 王晓力（1965— ），女，甘肃兰州人，副研究员，硕士，从事畜牧方面的研究。

藏绵羊哺乳期羔羊早期补饲培育模式研究*

朱新书**1，王宏博[1]，包鹏甲[1]，
李世红[2]，夏永祺[2]，汪海成[2]，张　功[2]
（1. 中国农业科学院兰州畜牧与兽药研究所/甘肃省牦牛繁育工程重点实验室，
兰州　730050；2. 甘肃省甘南藏族自治州临潭县农牧局畜牧
草原工作站，临潭　747500）

摘　要：[目的] 提高藏绵羊哺乳期羔羊的补饲效果。[方法] 采用单因子设计，探讨藏绵羊哺乳期羔羊的早期补饲培育模式。[结果] 在放牧和补饲粗饲料相同的条件下，同时补馈代乳粉+全价料对藏绵羊哺乳期羔羊的补饲效果明显好于单独补饲玉米或全价料，日增重差异显著（$P<0.05$）或极显著。在有效补饲条件下藏绵羊羔羊生长发育迅速，早期平均日增量可达到140~160g/d，羔羊成活率可达到93.75%。[结论] 合理搭配羔羊代乳粉+全价料是藏绵羊哺乳期羔羊早期补饲培育的优化模式。

关键词：藏绵羊；哺乳羔羊；早期补饲；培育模式

（发表于《安徽农业科学》）

黄酮类化合物的分子修饰与构效关系的研究***

黄　鑫****1,2，梁剑平[1,2]，郝宝成[1]
（1. 中国农业科学院兰州畜牧与兽药研究所，农业部兽用药物
创制重点实验室，甘肃省新兽药工程重点实验室，兰州　730050；
2. 中国农业科学院研究生院，北京　100081）

摘　要：黄酮类化合物是一类从植物中提取出来的、具有广泛药理作用的芳香类天然化

---

* 基金项目：农业部公益性行业（农业）科研专项（201303062）；现代农业（肉牛牦牛）产业技术体系专项（CARS-38）；牦牛资源与育种创新工程项目（CAAS-ASTIP-20140LIHPS-01）。

** 朱新书（1957—　），男，陕西渭南人。制研究员，从事牦牛和藏羊资源利用研究。

*** 基金项目：国家高新技术研究发展计划（863计划）（2011AA10A214）。

**** 黄鑫（1989—　），男，甘肃天水人，硕士研究生，研究方向：兽用天然药物的创制。通讯作者，助理研究员，从事兽用天然药物的创制研究。

合物，其结构主要由母核中 A、B、C 三个环组成。黄酮类化合物由于溶解度、生物利用度差等因素而限制其在临床上的广泛应用。目前国内外学者以天然黄酮类化合物为研究对象，在已知其分子结构信息的基础上，利用不同的分子修饰方法如化学修饰法、酶修饰法、微生物修饰法等对其结构进行改造，综述不同结构修饰对其抗菌、抗病毒、抗氧化、抗炎等药理作用的影响。

关键词：黄酮类化合物；分子修饰；生物活性；构效关系；药理作用

（发表于《安徽农业科学》）

牦牛的肉用特性研究

郭　宪*，裴　杰，包鹏甲，

褚　敏，赵娟花，阎　萍**

（中国农业科学院兰州畜牧与兽药研究所/甘肃省牦牛

繁育工程重点实验室，兰州　730050）

摘　要：[目的] 为了提高牦牛的肉用性能和加快牦牛选育进程，对牦牛的肉用特性进行了研究，为牦牛选育与肉品开发提供基础数据与技术参考。[方法] 以牦牛为研究对象，对牦牛的产肉性能、肉品质、肉脂肪酸进行了测定与分析。[结果] 从产肉性能来看，18 月龄补饲牦牛宰前活重为 137.54kg，胴体重为 66.14kg，净肉重为 52.16kg，屠宰率为 48.09%，净肉率为 37.92%，眼肌面积为 31.53cm^2，各项产肉指标均显著高于当地自然放牧牦牛。从肉品质来看，18 月龄牦牛的肉色、大理石纹评分、失水率、肌纤维直径均低于成年牦牛肉，且肌纤维直径呈显著性差异（$P<0.05$）。18 月龄牦牛肉的系水力、熟肉率均高于成年牦牛肉，且熟肉率呈显著性差异（$P<0.05$）。从肉脂肪酸来看，牦牛补饲后在背最长肌、半腱肌中均检测到肉豆蔻酸、2-甲基-棕榈酸、2-己基-环丙烷辛酸、顺式-11-十八烯酸、反-9-十八碳烯酸、山嵛酸、木蜡酸，而自然放牧条件下均未检测到。牦牛补饲后脂肪酸含量及其成分发生明显变化，表明牦牛肉脂肪酸组成及其含量与饲养条件有关。[结论] 放牧加补饲是提高牦牛产肉性能的主要措施之一，补饲可改善牦牛肉品质。

关键词：牦牛肉；产肉性能；肉品质；脂肪酸

（发表于《安徽农业科学》）

---

* 郭宪（1978—　）男，甘肃环县人，副研究员，博士，从事动物遗传育种与繁殖研究。

** 通讯作者，研究员，博士，博士生导师，从事动物遗传育种与繁殖研究。

牦牛乳氨基酸和脂肪酸的含量研究

郭　宪，裴　杰，褚　敏，
包鹏甲，赵生军，阎　萍
（中国农业科学院兰州畜牧与兽药研究所/甘肃省牦牛繁育
工程重点实验室，兰州　730050）

摘　要：［目的］提高牦牛乳的开发利用价值以及牦牛乳制品的附加值。［方法］对青海高原牦牛乳氨基酸、脂肪酸组分及其含量进行了测定。［结果］在牦牛乳中共检测到16种氨基酸，其谷氨酸、异亮氨酸、酪氨酸和精氨酸含量相对较高，赖氨酸、甘氨酸、蛋氨酸和丙氨酸含量相以较低。牦牛初乳中总氨基酸含量（5.485%）显著高于常乳。在牦牛初乳和常乳中均检测到28种脂肪酸，其中，饱和脂肪酸19种，不饱和脂及酸9种。常乳中的油酸和硬脂酸含量显著高于初乳。［结论］该研究可为牦牛乳的开发利用提供基础数据和科学依据。

关键词：牦牛乳；氨基酸；脂肪酸

（发表于《安徽农业科学》）

绵山羊双羔素提高不同品种绒山羊繁殖率的研究*

冯瑞林**，郭　健，裴　杰，刘建斌，岳耀敬，
郭婷婷，孙晓萍，牛春娥，袁　超，杨博辉
（中国农业科学院兰州畜牧与兽药研究所，兰州　730050）

摘　要：［目的］提高绒山羊繁殖率，降低其养殖成本，增加其养殖效益。［方法］应用绵山羊双羔素（睾酮-3-羧甲基肟·牛血清白蛋白），对辽宁绒山羊、陇东绒山羊、河西绒山羊和新疆绒山羊进行免疫注射，研究绵山羊双羔素对不同品种绒山羊的免疫效果。［结果］辽宁绒山羊试验组产羔率较对照组提高34.31%，差异极显著（$P< 0.01$）；新疆绒山羊试验组产羔率较对照组提高25.49%，差异极显著（$P<0.01$）；陇东绒山羊试验组产羔率较对照组提高24.76%，差异极显著（$P< 0.01$）；河西绒山羊试验组产羔

* 基金项目：国家“十二五”科技支撑项目子课题（2011BAD28B05-1-4）；甘肃省科技重大专项计划项目（1203NKDA023）。

** 冯瑞林（1959—　），男，甘肃兰州人，助理研究员，从事家畜繁殖育种研究。

率与对照组相比提高12.50%，差异显著（$P < 0.05$）。[结论] 绵山羊双羔素可以提高不同品种绒山羊双羔率和产羔率。
关键词：绵山羊双羔素；绒山羊；繁殖率；双羔率；产羔率

（发表于《安徽农业科学》）

绵山羊双羔素提高藏羊繁殖率的研究*

冯瑞林**，郭 健，裴 杰，刘建斌，岳耀敬，
郭婷婷，孙晓萍，牛春娥，袁 超，杨博辉

（中国农业科学院兰州畜牧与兽药研究所，兰州 730050）

摘 要：[目的] 提高藏羊的繁殖率，降低牧区草场的载畜量，减少基础母羊存栏数。[方法] 应用绵山羊双羔素（睾酮-3-羧甲基肟·牛血清白蛋白），对青海省海北州同宝牧场、海北州牧科所羊场和甘肃省甘南州李恰如牧场的藏羊进行免疫试验。[结果] 海北州同宝牧场试验点试验组产羔率为110.00%，对照组产羔率为101.14%，试验组比对照组提高产羔率8.86%（$P < 0.05$）；甘肃李恰如试验点试验组产羔率为133.87%，与对照组相比提高产羔率33.87%（$P < 0.01$）；海北州牧科所羊场试验点试验组产羔率为125.91%，与对照组相比提高产羔率25.91%（$P < 0.01$）。[结论] 只要加强藏羊的饲养管理，应用绵山羊双羔素提高藏羊的繁殖率才成为可能。
关键词：绵山羊双羔素；藏羊；繁殖率；双羔率；产羔率

（发表于《安徽农业科学》）

绵山羊双羔素提高粗毛羊繁率的研究

冯瑞林，郭 健，裴 杰，
刘建斌，岳耀敬，郭婷婷，
孙晓萍，牛春娥，袁 超，杨博辉

（中国农业科学院兰州畜牧与兽药研究所，兰州 730050）

摘 要：[目的] 提高粗毛羊的繁殖率，增加牧户和养殖者的经济效益。[方法] 应用

---

* 基金项目：国家“十二五”科技支撑项目（2011BAD28B05-1-4）；甘肃省科技重大专项（1203NKDA023）。

** 冯瑞林（1959— ），男，甘肃兰州人，助理研究员，从事家畜繁殖育种研究。

绵山羊双羔素（睾酮-3-羧甲基肟·牛血清白蛋白），对滩羊、藏羊、蒙古羊、哈萨克羊和大尾粗毛羊进行免疫注射。［结果］与对照组相比，滩羊试验组产羔率提高19.51%，差异极显著（$P<0.01$）；与对照组相比，藏羊试验组产羔率提高33.87%，差异极显著（$P<0.01$）；与对照组相比，蒙古羊试验组产羔率提高14.33%，差异显著（$P<0.05$）；与对照组相比，哈萨克羊试验组产羔率提高19.35%，差异极显著（$P<0.01$）；与对照组相比，大尾粗毛羊试验组产羔率提高43.35%，差异极显著（$P<0.01$）。［结论］粗毛羊只要加强营养和饲养管理，绵山羊双羔素可以提高粗毛羊的双羔率和产羔率。

关键词：绵山羊双羔素；粗毛羊；繁殖率；双羔率；产羔率

（发表于《安徽农业科学》）

绵山羊双羔素提高东北半细毛羊繁殖率的研究*

冯瑞林**，郭　健，裴　杰，刘建斌，岳耀敬，
郭婷婷，孙晓萍，牛春娥，杨博辉

（中国农业科学院兰州畜牧与兽药研究所，兰州　730050）

摘　要：［目的］挖掘东北半细毛羊的繁殖潜能，提高养殖户的经济效益，增加养殖者的养殖积极性。［方法］在黑龙江省七台河市和牡丹江市，应用绵山羊双羔素（睾酮-3-羧甲基肟·牛血清白蛋白），免疫东北半细毛羊267只。［结果］七台河市试验点试验组产羔率为156.39%，对照组产羔率105.75%，试验组比对照组提高产羔率50.64%；牡丹江市试验点试验组产羔率122.95%，对照组产羔率104.38%，试验组产羔率比对照组提高18.57%。［结论］东北半细毛羊的繁殖潜力可以通过绵山羊双羔素提高其产羔率的方式得到很大提升。

关键词：绵山羊双羔素；东北半细毛羊；繁殖率；双羔率；产羔率

（发表于《安徽农业科学》）

---

* 基金项目：国家“十二五”科技支撑项目（2011BAD28B05－1－4）；甘肃省科技重大专项（1203NKDA023）。

** 冯瑞林（1959—　），男，甘肃兰州人，助理研究员，从事家畜繁殖育种工作。

犬感染犬瘟热病毒后抗体的产生规律及检测方法*

苏贵龙**，李建喜

（中国农业科学院兰州畜牧与兽药研究所，兰州 730050）

摘　要： 由犬瘟热病毒引起的犬瘟热（Canine distemper，CD）疾病，是造成犬及部分食肉目动物的一种具有高度传染性和高死亡率的疾病。犬瘟热病毒感染犬产生抗体的规律在临床上具有非常重要的意义。幼犬在出生后约 35 d 就失去了母源抗体的保护，需要人工对其进行预防免疫，以获得针对该病毒的特异性抗体。犬瘟热病毒在进入机体后，引起以胸腺依赖性的 B2 细胞所介导的体液免疫为主，初次免疫应答产生的 IgM 稍多于 IgG，再次免疫应答时记忆性 B 细胞发挥着主要的作用，在很短时间内使得特异性 IgG 抗体达到较高水平，且在机体内维持很久的时间。通过掌握 IgM 和 IgG 抗体产生的量及变化规律，可以诊断疾病发生的过程，为疾病的治疗和预防提供诊断依据。

关键词： 犬瘟热病毒；IgM；IgG；产生规律；检测方法

（发表于《安徽农业科学》）

提高湖羊羔羊甲级羔皮率的试验研究

孙晓萍***，刘建斌，冯瑞林，郭婷婷，岳耀敬，杨博辉

（中国农业科学院兰州畜牧与兽药研究所，兰州 730050）

摘　要：［目的］提高湖羊羔羊的甲级羔皮率。［方法］将湖羊怀孕母羊分为 2 组，试验组在母羊怀孕后期进行限饲，对照组按照常规方法饲养，探讨提高湖羊羔羊的甲级羔皮率的方法。［结果］试验组羔羊甲级羔皮率达 42.43%，羔羊荐部伸直毛长 1.90cm，被毛的紧贴度达 43.63%。对照组羔羊甲级羔皮率为 22.59%，羔羊荐部伸直毛长 2.08cm，被毛的紧贴度为 24.82%。［结论］在湖羊母羊怀孕后期限制营养供给，可改善湖羊羔羊羔皮品质，提高甲级羔皮率。

关键词： 湖羊；羔羊；甲级羔皮率

（发表于《安徽农业科学》）

* 基金项目：公益性行业专项（201303040-15）。

** 苏贵龙（1988— ），男，甘肃平凉人，硕士研究生，研究方向：中兽医免疫与生物转化技术。* 通讯作者，研究员，博士，硕士生导师，从事中兽医免疫与生物转化技术研究。

*** 孙晓萍（1962— ），女，江苏如皋人，副研究员，从事动物遗传育种研究。

代谢组学在基础兽医学中的应用*

马 宁**，刘希望，李剑勇*，杨亚军，郭沂涛，刘光荣

（中国农业科学院兰州畜牧与兽药研究所兽药重点实验室/
农业部兽用药物创制重点实验室，兰州 730050）

摘 要：代谢组学是利用高通量检测技术对生物体内所有小分子代谢产物同时进行定性和定量分析，结合数据处理和信息建模手段，通过模式识别的方法判断生物体的代谢状态。代谢组学作为系统生物学的重要组成部分，其在药物研发、疾病研究、药物毒理学等领域发挥着越来越大的作用，在基础兽医学研究领域有着巨大的应用空间和研究价值。结合代谢组学相关研究策略及在动物药物代谢、中兽药研发和动物疾病中的应用，论文主要对代谢组学的概况，发展情况和未来趋势等进行论述。

关键词：代谢组学；液质联用；基础兽医学；兽药

（发表于《动物医学进展》）

丹参水提物对山羊子宫内膜上皮细胞炎症模型中***
基质金属蛋白酶-2表达的影响

邝晓娇[1]****，张世栋[1]，董书伟[1]，
王东升[1]，魏立琴[1]，靳亚平[2]，严作廷[1]*

（1. 中国农业科学院兰州畜牧与兽药研究所/
甘肃省中兽药工程技术研究中心，兰州 730050；
2. 西北农林科技大学动物医学院，杨凌 712100）

摘 要：为研究丹参水提物(SME)对子宫内膜上皮细胞炎症模型中基质金属蛋白酶-2(MMP-2)表达的影响，体外培养山羊子宫内膜上皮细胞(EEC)，利用内毒素(脂多糖，LPS)诱导细胞增殖和产生TNF-α，建立EEC炎症模型。然后给予高、中、低不同剂量SME干预。应用

---

* 基金项目：国家自然科学基金项目（31402254）。

** 马宁（1990— ），男，河北保定人，硕士研究生，主要从事药效学代谢组学研究。*通讯作者

*** 基金项目：中央级公益性科研院所基本科研业务费专项资金项目（1610322014001）；中国农业科学院基本科研业务费预算增量项目（2014ZL012）；“十二五”国家科技支撑计划重点项目（2012BAD12B03）；中国农业科学院科技创新工程项目（1610322014028）。

**** 邝晓娇（1986— ），女，河南人，硕士，主要从事生殖与内分泌学研究。*通讯作者

MTT 比色法检测细胞活力,ELISA 试剂盒检测 TNF-α 含量,实时荧光定量 PCR 检测细胞 MMP-2mRNA 表达变化。结果表明,5μg/mL LPS 作用 EEC 12h 时促增殖作用显著(P<0.01);SME 可显著促进 EEC 炎症模型中 MMP-2 的表达,以中、高浓度组的作用较为明显。说明 SME 对 LPS 引起的子宫内膜炎症反应有一定的保护作用。

关键词： 子宫内膜上皮细胞；丹参水提物；内毒素；基质金属蛋白酶-2

（发表于《动物医学进展》）

含查尔酮结构（*E*）-4-[3-(取代苯基)丙烯酰基]苯甲酸钠的制备及抑菌活性研究*

郭沂涛**，刘希望，杨亚军，许春燕，李剑勇*

（甘肃省新兽药工程重点实验室/农业部兽用药物创制重点实验室/中国农业科学院兰州畜牧与兽药研究所，兰州　730050）

摘　要： 由取代苯甲醛和4-乙酰基苯甲酸通过克莱森-斯密特缩合反应制备了12个含有查尔酮结构（*E*）-4-[3-（取代苯基）丙烯酰基]苯甲酸钠类化合物，其化学结构经核磁共振氢谱、高分辨质谱予以确证。体外抑菌试验显示部分化合物对艰难梭菌、产气荚膜梭菌有较好抑制作用，最小抑菌浓度与甲硝唑相当，大部分化合物对大肠埃希菌及金黄色葡萄球菌无明显的抑制作用。

关键词： 查尔酮；制备；抑菌活性；艰难梭菌；最小抑菌浓度

（发表于《动物医学进展》）

* 基金项目：中央公益性科研院所基本科研业务费专项（1610322013005）。

** 郭沂涛（1988—　），男，河北保定人，硕士研究生，主要从事药物合成研究。*通讯作者

苦豆子灌注剂质量的检测*

辛任升**，刘　宇，梁剑平*

（中国农业科学院兰州畜牧与兽药研究所/农业部兽用药物创制重点实验室/甘肃省新兽药工程重点实验室，兰州　730050）

摘　要：为建立苦豆子灌注剂的质量标准，采用薄层色谱法（TLC）法对苦豆子灌注剂进行定性鉴别；采用高效液相色谱法测定槐定碱含量。TLC 中，苦豆子灌注剂与槐定碱对照品在相应位置上显相同颜色的斑点，且专属性强、分离效果好；槐定碱进样量在 0.2mg/mL～1mg/mL（r=0.999 6）范围内线性关系良好，平均加样回收率为 98.84%（RSD=0.338%，n=6）。因此，所建标准测定方法简单，重现性好，可用于该制剂的质量控制。

关键词：苦豆子灌注剂；槐定碱；质量标准；高效液相；薄层色谱

（发表于《动物医学进展》）

奶牛乳房炎致病菌的分离鉴定及耐药性研究***

李新圃****，李宏胜，罗金印，杨　峰，王旭荣

（中国农业科学院兰州畜牧与兽药研究所/农业部兽用药物创制重点实验室/甘肃省新兽药工程重点实验室，兰州　730050）

摘　要：为进一步了解奶牛乳房炎致病菌的菌群分布，以及条件致病菌的致病性和耐药性。采集了 302 份临床型乳房炎乳样，从中分离出 257 株细菌，使用 16SDNA PCR 试剂盒鉴定分析，对其中检出率较高的条件致病菌进行小鼠致病性和耐药性研究。结果显示，检出菌可分为主要致病菌、条件致病菌、乳酸菌及其他菌。条件致病菌化脓隐秘杆菌对小鼠具有明显的致病性，粪肠球菌和屎肠球菌对小鼠没有明显的致病性，但它们对抗生素均有一定的耐药性。粪肠球菌和屎肠球菌具有条件致病性，但它们属于乳酸菌，而乳酸菌一般作为益生菌使用。说明奶牛乳房炎致病菌具有多样性和复杂性，除了常见

---

* 基金项目：“十二五”农村领域国家科技计划项目（2011AA10A214）。

** 辛任升（1987—　），男，山东青岛人，硕士，主要从事中草药开发与利用研究。*通讯作者

*** 基金项目：“十二五”国家科技支撑计划（2012BAD12B03）；甘肃省自然科学基金项目（1308RJZA119）；甘肃省农业科技创新项目（GNCX-2013-59）；甘肃省科技支撑计划（144NKCA240）。

**** 李新圃（1962—　），女，山西万荣人，副研究员，主要从事兽医药理学研究。

的主要致病菌和条件致病菌外，有些益生菌进入乳腺组织也能成为条件致病菌引发乳房炎。

关键词：奶牛乳房炎；致病菌；致病性；耐药性

（发表于《动物医学进展》）

伊维菌素微乳制剂的安全性试验*

邢守叶**[1,2]，周绪正[2]，李　冰[2]，牛建荣[2]，魏小娟[2]，张继瑜[2]*

（1. 中国农业科学院兰州畜牧与兽药研究所/农业部兽用药物创制重点实验室，兰州　730050；2. 甘肃农业大学动物医学院，兰州　730070）

摘　要：按照《中国兽药典》2010版和《兽药及添加剂安全性毒理学评价程序》的要求，对伊维菌素微乳制剂进行肌肉刺激性试验、热原试验和注射途径 LD_{50} 测定，为其临床应用提供安全性依据。肌肉刺激性试验，观验，观察4只家兔注射伊维菌素微乳制剂后48h的变化；热原试验，测家兔的体温变化，判断伊维菌素微乳制剂所含热原的限度是否符合规定；LD_{50} 测定，分别给大鼠、小鼠按组一次腹腔注射不同剂量的伊维菌素微乳，连续观察7~14d，记录各组动物急性毒性反应的症状、病理变化、死亡时间，按简化寇氏法计算其半数致死量（LD50）及95%可信限。结果表明，伊维菌素微乳制剂对家兔股四头肌未见明显刺激作用；热原限度符合规定，无致热源；伊维菌素微乳制剂对大鼠、小鼠腹腔注射的 LD_{50} 为其临床推荐用量（0.2mg/kg）的191.28倍和161.00倍，并且高于伊维菌素原料药的 LD_{50}（24.249 3mg/kg）。伊维菌素微乳制剂的安全性试验结果提示其安全可靠。

关键词：伊维菌素；微乳；肌肉刺激性试验；热原试验；急性毒性试验；安全性评价

（发表于《动物医学进展》）

---

* 基金项目：国家现代农业肉牛牦牛产业技术体系建设专项基金项目（CARS-38）；重点牧区生产生态生活保障技术集成与示范项目（2012BAD13BOO）。

** 邢守叶（1988—　），女，硕士研究生，主要从事兽医药理学与毒理学方面的研究。*通讯作者

银翘蓝芩口服液薄层色谱鉴别方法研究*

许春燕**，刘希望，杨亚军，郭沂涛，李剑勇*

（农业部兽用药物创制重点实验室/甘肃省新兽药工程重点实验室/
中国农业科学院兰州畜牧与兽药研究所，兰州　730050）

摘　要：为建立复方银翘蓝芩口服液的质量标准，采用薄层色谱法对组方中的药材进行定性鉴别，确定了该试剂中金银花、连翘、黄芩及苦参的薄层鉴别方法。在选定的色谱条件下，色谱斑点分离效果较好，该方法能有效的鉴别复方中的金银花、连翘、黄芩及苦参。所建立的方法简便、专属性强、重复性好，可为该口服液的质量控制提供依据。
关键词：银翘蓝芩口服液；薄层色谱；鉴别方法；质量标准

（发表于《动物医学进展》）

仔猪腹泻病原学研究进展

黄美州，刘永明*，王　慧，崔东安，王胜义，李胜坤

（中国农业科学院兰州畜牧与兽药研究所/农业部兽用药物
创制重点实验室/甘肃省新兽药工程重点实验室/甘肃省
中兽药工程技术研究中心，兰州　730050）

摘　要：仔猪腹泻的病原复杂，且有些病原变异速度快，导致近年来仔猪腹泻病的发病率和死亡率居高不下，给仔猪腹泻病的病原学研究带来了极大的困难。为了弄清仔猪腹泻病的主要病原，许多专家和学者从不同的角度、用不同的方法对仔猪腹泻病原学进行了大量的研究，病原的主要功能编码基因和功能蛋白在病原侵袭宿主过程中的作用是近年来的一个研究热点。论文从病原的主要编码基因和功能蛋白的角度，对近年来引起仔猪腹泻的主要病原（细菌、病毒）的研究进展进行了综述，并根据近年来的仔猪腹泻病病原学的研究进展，对未来仔猪腹泻病原学的发展提出了一些看法。
关键词：仔猪腹泻；病原学；流行病学

（发表于《动物医学进展》）

---

* 基金项目：公益性行业（农业）科研专项经费项目（201303040-12）；兰州市科技计划项目（2012-2-73）。

** 许春燕（1988—　），女，河南驻马店人，硕士研究生，主要从事中兽药研究。* 通讯作者

紫菀化学成分及药理作用研究进展*

彭文静**，辛蕊华，任丽花，罗永江，
王贵波，罗超应，谢家声，李锦宇，郑继方*

（中国农业科学院兰州畜牧与兽药研究所/甘肃省新兽药工程
重点实验室/甘肃省中兽药工程技术研究中心，兰州　730050）

摘　要：中药紫菀为菊科多年生草本植物紫菀的干燥根茎，味辛、苦，性温，归肺经，具有润肺下气、止咳化痰平喘的作用，是临床常用的润肺祛痰止咳药。近代药理学研究表明，紫菀中化学成分含量丰富，主要有萜类、肽类、黄酮类、有机酸类等；其药理学作用也十分广泛，有抗菌、抗肿瘤、镇咳祛痰平喘、抗病毒、抗氧化活性等作用。近年来紫菀的研究主要集中在止咳化痰以及抗肿瘤等方面，其他方面的药理作用却有待于进一步研究。通过查阅国内外相关文献，综述了紫菀的化学成分和药理作用的研究进展，整理了紫菀研究的基本信息，以期为紫菀的进一步研究提供参考。

关键词：紫菀；化学成分；药理作用；肝脏毒性

（发表于《动物医学进展》）

青海省某奶牛场顽固性乳房炎主要病原菌的分离鉴定与耐药性研究***

林　杰****，王旭荣，杨志强，王　磊，张景艳，
孔晓军，王学智，秦　哲，王孝武，李建喜*

（中国农业科学院兰州畜牧与兽药研究所/农业部兽用药物
创制重点实验室/甘肃省新兽药工程重点实验室，兰州　730050）

摘　要：为了明确青海省某奶牛场顽固性乳房炎的病原菌，并制定有效的防控措施，本试验采用细菌分离、16SrNDA 序列分析、细菌基因分型等方法对病牛乳样中的病原菌进

---

* 基金项目："十二五"国家科技支撑计划项目（2011BAD34B03）；中央级公益性科研院所基本科研业务费专项资金项目（1610322013015）；甘肃省科技支撑计划项目（1304NKCA155）。

** 彭文静（1989—　），女，湖南人，硕士研究生，主要从事中兽医药学研究。*通讯作者

*** 基金项目：国家奶牛产业技术体系科学家岗位项目（CARS37）。

**** 林杰（1991—　），女，河北唐山人，硕士研究生，主要从事动物疾病预防与控制相关研究工作。
*通讯作者

行分离鉴定，并进行17种药物的敏感性试验。试验从15份乳样中分离到10株金黄色葡萄球菌、1株表皮葡萄球菌和2株化脓隐秘杆菌，其中，10株金黄色葡萄球菌均为荚膜多糖5型分离株；所分离金黄色葡萄球菌对青霉素G、红霉素、头孢氨苄和呋喃妥因耐药率分别为100%、90%、10%和10%；表皮葡萄球菌对青霉素G、复方新诺明和甲氧苄氨嘧啶耐药，而对其他14种抗菌药物敏感；化脓隐秘杆菌对四环素和甲氧苄氨嘧啶耐药，对强力霉素和复方新诺明中度敏感，对其他13种抗菌药物敏感。该牛场顽固性乳房炎是多种病原菌混合感染所致，但以荚膜多糖5型金黄色葡萄球菌为主。不同的病原菌的药物敏感性不同，故而治疗困难，建议使用恩诺沙星治疗病牛并加强隐性乳房炎的筛查与防治，病情得到控制。

关键词：奶牛；顽固性乳房炎；耐药性；金黄色葡萄球菌；荚膜多糖5型

（发表于《动物医学进展》）

“菌毒清”口服液对人工感染鸡传染性喉气管炎的治疗*

牛建荣**，张继瑜，周绪正，李剑勇，
李　冰，魏小娟，杨亚军，刘希望

（中国农业科学院兰州畜牧与兽药研究所/农业部兽用药物创制重点实验室/甘肃省新兽药工程重点实验室，兰州　730050）

摘　要：试验分为7组，“菌毒清”4组：饮水中加入2.0%（A组）、1.5%（B组）、1.0%（C组）和0.5%（D组）“菌毒清”；“瘟毒清”治疗组（E组）：饮水中加入1.0%“瘟毒清”；阳性空白对照组（F组）和阴性空白对照组（G组），让鸡自由饮用，连用7d为1疗程，比较“菌毒清”口服液对人工感染鸡传染性喉气管炎病的治愈率、有效率，与现有的治疗鸡呼吸道疾病的中药“瘟毒清”口服液的治疗效果．结果表明：经1个疗程的治疗，“菌毒清”4个组的治愈率分别为69.23%、81.48%、78.83%和59.26%，“瘟毒清”组的治愈率为77.78%；“菌毒清”4个组的有效率分别为84.61%、92.59%、93.43%和77.78%，“瘟毒清”组的有效率为88.89%．多重分析表明，D、F组的治愈率与A、B、C、E组差异显著（$P<0.05$），其余组间无差异；有效率除阳性对照组（F组）与“菌毒清”组（A、B、C、D组）、“瘟毒清”组（E组）差异显著（$P<0.05$）外，其他组之间差异不显著．结果说明，“菌毒清”是治疗鸡传染性喉气管炎病的有效药物之一，与现有的治疗鸡呼吸道疾病的中药“瘟毒清”治疗

* 基金项目：农业部公益性行业专项“防治畜禽卫气分证中兽药生产关键技术研究与应用”（201303040-09）。

** 牛建荣（1968—　），男，硕士，副研究员，长期从事兽医临床和新兽药开发研究．E-mail：lzn-jr1968@ 126. com.

效果相当。

关键词："菌毒清"口服液；人工感染鸡传染性喉气管炎；疗效

（发表于《甘肃农业大学学报》）

牦牛 *C1H21orf62* 基因的克隆、生物信息学及表达分析*

赵娟花**，梁春年，裴　杰，褚　敏，
吴晓云，张良斌，佘平昌，阎　萍***
（中国农业科学院兰州畜牧与兽药研究所/甘肃省牦牛繁育工程重点实验室，兰州　730050）

摘　要：应用 RT-PCR 方法克隆牦牛 *C1H21orf62* 基因并进行生物信息学分析，利用实时荧光定量 PCR 方法检测 *C1H21orf62* 基因在有角牦牛和无角牦牛角芽组织的 mRNA 表达。结果表明，*C1H21orf62* 基因编码区全长 662bp，编码 161 个氨基酸。*C1H21orf62* 是一种亲水性蛋白，含有 13 个潜在的磷酸化位点，但无跨膜域和信号肽。*C1H21orf62* 基因 mRNA 在无角牦牛角芽组织的表达水平极显著高于有角牦牛。

关键词：牦牛；无角；有角；*C1H21orf62*；克隆

（发表于《广东农业科学》）

---

* 基金项目：国家现代农业（肉牛牦牛）产业技术体系专项（CARS-38）；国家科技支撑计划重点项目（2012AD13B05）。

** 赵娟花（1989— ），女，在读硕士生，E-mail：zjh326303202@163.com；

*** 通讯作者：阎萍（1963— ），女，博士，研究员，E-mail：pingyanlz@163.com.

13 种大毛细皮动物毛纤维物理性能的研究*

李维红**，高雅琴，梁丽娜，熊　琳

（中国农业科学院兰州畜牧与兽药研究所/农业部动物毛皮及制品质量监督检验测试中心（兰州）/农业部畜产品质量安全风险评估实验室（兰州），兰州　730050）

摘　要：为了测定 13 种大毛细皮类动物毛纤维物理性能指标，试验采用测定其强力、伸长率、长度、细度及细度离散等方法，对毛纤维物理性能进行研究。结果表明：不同动物种类毛纤维之间的强力、伸长率、长度、细度及细度离散等物理性能指标差异极显著（$P<0.01$），同一动物种类不同动物品种毛纤维之间的强力、伸长率、长度、细度及细度离散等物理性能指标差异不显著（$P>0.05$），同一动物品种的针毛和绒毛之间的强力、伸长率、长度、细度及细度离散等物理性能指标差异极显著（$P<0.01$）。

关键词：大毛细皮动物；毛纤维；物理性能；针毛；绒毛

（发表于《黑龙江畜牧兽医》科技版）

B 超活体测定牦牛眼肌面积和背膘厚的研究***

郭　宪[1]****，裴　杰[1]，保善科[2]，吴晓云[1]，赵索南[2]，包鹏甲[1]，孔祥颖[2]，阎　萍[1]

（1. 中国农业科学院兰州畜牧与兽药研究所/甘肃省牦牛繁育工程重点实验室，兰州　730050；2. 青海省海北藏族自治州畜牧兽医科学研究所，海北　810200）

摘　要：为了提高牦牛肉用性能选育效率，试验采用 B 超测定技术活体测定牦牛背膘厚与眼肌面积。结果表明：成年牦牛（4 岁以上）背膘厚为（4.87 ± 1.09）mm、眼肌面积为（25.58 ± 1.75）cm^2，青年（1 岁）牦牛背膘厚为（1.93 ± 0.28）mm、眼肌

---

* 基金项目：中央级公益性科研院所基本科研业务费专项资金（中国农业科学院兰州畜牧与兽药研究所）资助项目（1610322013008）。

** 李维红（1978—　），女，助理研究员，博士，研究方向动物纤维、毛皮及制品质量检测及其质量标准，E-mail：lwh0923@ 163. com.

*** 基金项目："十二五"国家科技支撑计划项目（2013BAD16B09）；国家自然科学基金项目（31301976）；现代农业（肉牛牦牛）产业技术体系建设专项（CARS-38）。

**** 郭宪（1978—　），男，副研究员，博士，研究方向为动物繁育，E-mail：guoxian@ caas. cn.

通讯作者：阎萍（1963—　），女，研究员，博士，研究方向为动物遗传育种与繁殖，E-mail：pingyanlz@ 163. com.

面积为（12.93±2.78）cm^2。说明牦牛脂肪沉积晚，这可能与牦牛在高寒牧区生长发育慢、体成熟迟有关。

关键词： B超技术；背膘厚；眼肌面积；体重；牦牛

（发表于《黑龙江畜牧兽医》科技版）

阿维菌素类药物残留检测的研究进展

文　豪，周绪正，
李　冰，魏小娟，张继瑜
（中国农业科学院兰州畜牧与兽药研究所/
农业部兽用药物创制重点实验室，兰州　730050）

摘　要： 阿维菌素具有强大的抗虫药效，开创了新一代抗虫药物的里程碑，对螨类和昆虫具有胃毒和触杀作用。随着人们生活水平的不断提高，人们对于动物性食品中的药物残留量提出了更高的要求，欧美与我国都制定了阿维菌素类药物的残留标准，快速有效的检测阿维菌素的残留量具有较大的实用意义与安全意义。对近年来阿维菌素的残留检测研究及发展现状作了详细的综述。

关键词： 阿维菌素；残留；检测方法；研究进展；趋势；检测限

（发表于《黑龙江畜牧兽医》科技版）

不同来源抑制素抗原对牛羊繁殖力影响研究

孙晓萍[1]*，刘建斌[1]，张万龙[2]，杨博辉[1]，
郭　健[1]，冯瑞林[1]，岳耀敬[1]，郭婷婷[1]
（1. 中国农业科学院兰州畜牧与兽药研究所，兰州　730050；
2. 金昌市绵羊繁育技术推广站，永昌　737200）

摘　要： 抑制素是一种由性腺分泌的糖蛋白，由α、β两个亚基组成。研究表明，它具有选择性抑制促卵泡素（FSH）合成和分泌的作用。最初在研究抑制素生理作用时发现，用卵泡液提取物免疫动物后，可提高排卵率，增加产仔数。另一方面的研究发现，在多产

---

* 孙晓萍（1962—　），女，副研究员，本科，主要从事动物遗传育种研究，E-mail：xiaopingsun163@163.com.

Booroola 美利奴母羊卵巢中缺少具生物活性的抑制素，这说明这种母羊体内 FSH 浓度较高，这显然是由于该种母羊卵巢内缺少抑制素的结果。于是，上述现象促使人们想到在生产实践中以抑制素作免疫原免疫家畜，用以中和内源性抑制素，削弱它对 FSH 合成与分泌的抑制作用，提高血液中 FSH 的浓度，进而促使家畜提高排卵率及增加产仔数。

产复康中 5 种中药的薄层色谱鉴别*

王东升**，尚小飞，苗小楼，张世栋，
董书伟，魏立琴，邝晓娇，严作廷

（中国农业科学院兰州畜牧与兽药研究所/农业部兽用药物创制重点实验室/甘肃省中兽药工程技术研究中心/甘肃省新兽药工程重点实验室，兰州 730050）

摘　要：为了对产复康进行质量控制，试验采用薄层色谱法对产复康中丹参、当归、川芎、枳壳和大血藤等主要中药进行定性鉴别。结果表明：在供试品色谱中，在与对照品或对照药材色谱相应的位置上，显示出相同颜色的斑点，阴性对照无干扰。说明该鉴别方法操作简便，专属性强，重复性好，可用于产复康的定性质量控制。

关键词：产复康；薄层色谱；丹参；当归；川芎；枳壳；大血藤

（发表于《黑龙江畜牧兽医》科技版）

动物源性食品中性激素残留的危害及检测方法***

高旭东[1,2]****，黄　鑫[1]，郝宝成[1]，陈士恩[2]，梁剑平[1]

（1. 中国农业科学院兰州畜牧与兽药研究所/农业部兽用药物创制重点实验室/甘肃省新兽药工程重点实验室，兰州 730050；
2. 西北民族大学生命科学与工程学院，兰州 730030）

摘　要：我国畜牧业资源丰富，且食品开发及加工业也已跃居成为我国国民经济的主要

---

* 基金项目：“十二五”国家科技支撑计划项目（2012BAD12B03）。

** 王东升（1979—　），男，助理研究员，硕士，研究方向为中药药理与奶牛繁殖疾病，E-mail：lzmyswds@ 126. com.

通讯作者：严作廷（1962—　），男，研究员，博士，研究方向为中兽医临床和奶牛疾病防治，E-mail：yanzuoting@ caas. cn.

*** 基金项目：中国农业科学院科技创新工程兽用天然药物创新团队项目（CAAS-ASTIP-2014-LIHPS-04）。

**** 高旭东（1988—　），女（满族），硕士研究生，研究方向为食品安全与品质控制，E-mail：228168214@ qq. com. 通讯作者：陈士恩（1963—　），男，教授，硕士，硕士生导师，研究方向为食品安全与品质控制，E-mail：chshien @ 163. com；梁剑平（1962—　），男，研究员，博士，博士生导师，研究方向为兽医药理与毒理，E-mail：liangjp100@ sina. com.

支柱产业之一，工农业发展和人们生活水平的提高也必将受其影响。文章参考国内外有关激素残留相关文献，重点介绍目前动物源食品中性激素残留的危害及检测方法，并对前景进行了展望。

关键词：动物源性食品；性激素；危害；残留；检测

（发表于《黑龙江畜牧兽医》科技版）

甘南亚高山草甸草原植被群落结构分析及其生长量测定*

朱新书**[1]，王宏博[1]，包鹏甲[1]，李世红[2]，
陈胜红[2]，夏永祺[2]，汪海成[2]，裴　杰[1]

（1. 中国农业科学院兰州畜牧与兽药研究所/甘肃省牦牛繁育工程重点实验室，兰州　730050；2. 甘肃省甘南藏族自治州临潭县农牧局畜牧草原工作站，临潭　747500）

摘　要：为了阐述甘南亚高山草甸草原的牧草品质和产量，试验在甘南州临潭县选取有代表性的样地，对草原植被的群落结构进行了分析，对生物生长量进行了测定。结果表明：甘南亚高山草甸草原牧草种类繁多，群落结构复杂，主要以莎草科和禾本科牧草为优势种或亚优势种。植被总覆盖度比较好，在 75% ~ 97%。8 月牧草产量明显高于 6 月，8 月、6 月莎科草、禾本科干草产量分别为 91.42g /m^2和 50.05g /m^2，二者差异极显著（$P<0.01$）；8 月、6 月杂类草干草产量产量分别为 123.69 g /m^2 和 79.87g /m^2，差异不显著（$P> 0.05$）；8 月、6 月总干草产量分别为 215.11g /m^2和 129.91g /m^2，差异极显著（$P< 0.01$）。

关键词：亚高山草甸草原；植被；群落结构；盖度；生长量

（发表于《黑龙江畜牧兽医》科技版）

* 基金项目：公益性行业（农业）科研专项经费专项（201003061；201303062）；现代农业（肉牛牦牛）产业技术体系专项（CARS-38）。

** 朱新书（1957—　），男，副研究员，本科，主要从事牦牛、藏羊资源利用研究，E-mail：1151758706@ qq. com.

回流法和超声法提取常山碱的研究*

郭志廷[1]**，罗晓琴[2]，梁剑平[1]，郭文柱[1]，尚若锋[1]，杨　珍[1]

（1. 中国农业科学院兰州畜牧与兽药研究所/农业部兽用药物创制重点实验室/甘肃省新兽药工程重点实验室，兰州　730050；
2. 兰州市动物卫生监督所/畜产品质量安全检测实验室，兰州　730050）

摘　要：为了比较回流法和超声法从中药常山中提取常山碱的效果，试验采用紫外—可见分光光度计测定常山碱含量，并以浸膏得率和常山碱含量作为评价指标。结果表明：回流法和超声法的浸膏平均得率分别为11.31%和14.17%；浸膏中常山碱的平均含量分别为4.15%和4.79%。说明回流法和超声法的提取工艺均切实可行，具有省时、操作简便和提取率高等优点，且超声法相对更好一些。

关键词：常山；常山碱；回流法；超声法；提取；中药；浸膏；得率

（发表于《黑龙江畜牧兽医》综合版）

基于CNKI数据库的兽用抗寄生虫药伊维菌素的文献计量学分析***

文　豪****，周绪正，李　冰，牛建荣，魏小娟，张继瑜

（中国农业科学院兰州畜牧与兽药研究所/农业部兽用药物创制重点实验室，兰州　730050）

摘　要：为了实时了解我国抗寄生虫药伊维菌素研究的前沿动态与现状，文章对中国知识资源总库（CNKI）数据库近11年收录的相关科研文献进行了文献计量学比较和研究，分析了国内的研究现状。结果表明：兽用抗寄生虫药伊维菌素对各种寄生虫有着良

* 基金项目：中央级公益性科研院所专项（1610322011004）。

** 郭志廷（1979—　），男，助理研究员，硕士，主要从事中药免疫学和药理学方面的研究，E-mail：16114140@ qq. com.

*** 基金项目：国家现代农业肉牛牦牛产业技术体系建设专项（CARS-38）；科技部科技基础性工作专项（2013FY110600-8）。

**** 文豪（1990—　），男，硕士研究生，研究方向为兽医药理学与毒理学，E-mail：wh901226@ sina. com

通讯作者：张继瑜（1967—　），男，研究员，博士，博士生导师，研究方向为兽医药理与毒理学，E-mail：infzjy@ sina. com.

好的疗效，得到了畜牧业科技人员和技术人员的肯定。

关键词：伊维菌素；畜牧业；分析；文献计量学；研究现状；中国知识资源总库（CNKI）

（发表于《黑龙江畜牧兽医》科技版）

苦豆子提取物的急性毒性及抗炎试验研究*

辛任升[1,2]**，刘　宇[1]，梁剑平[1]***

（1. 中国农业科学院兰州畜牧与兽药研究所/农业部兽用药物创制重点实验室/甘肃省新兽药工程重点实验室，兰州　730050；
2. 甘肃农业大学动物医学院，兰州　730070）

摘　要：为了观察苦豆子提取物（AL）对昆明小鼠的急性毒性反应和抗炎效果，试验采用腹腔注射方式给小鼠注射 AL，观察小鼠毒副反应情况，采用改良寇式法计算半数致死量（LD_{50}）及95% 可信限，并测定了 AL 的抗炎效果。结果表明：AL 的 LD_{50} 为 101.95mg/kg，95% 可信限为 98.5 ~105.5mg/kg；AL 的高、低剂量均可使小鼠耳廓肿胀度明显降低，其抑制率分别为 65.6% 和 50.4%。说明 AL 对小鼠有一定毒性，并且抗炎效果良好。

关键词：苦豆子提取物（AL）；急性毒性；抗炎；半数致死量（LD_{50}）；耳廓肿胀度；腹腔注射

（发表于《黑龙江畜牧兽医》综合版）

* 基金项目："十二五"农村领域国家科技计划项目（2011AA10A214）。

** 辛任升（1987—　），男，硕士研究生，E-mail：renshengxin0824@ 163. com.

*** 梁剑平（1962—　），男，研究员，博士，主要从事中草药开发与利用方面的工作，E-mail：liangjp100@ sina. com.

兰州地区奶牛乳房炎流行病学及病原菌调查*

罗金印**，李宏胜，李新圃，杨　峰

（中国农业科学院兰州畜牧与兽药研究所，兰州　730050）

摘　要：为了开展兰州地区奶牛乳房炎流行病学及病原菌的调查研究，试验应用 LMT 诊断液对兰州地区 4 个中等规模奶牛场的 344 头泌乳牛的 1 237 个乳区进行了隐性乳房炎检测，并对临床型乳房炎乳样进行了病原菌的分离鉴定。结果表明：奶牛隐性乳房炎的头阳性率为 58.43%，乳区阳性率为 36.86%；临床型乳房炎的头发病率为 5.52%，乳区发病率为 2.91%；主要病原菌检出率为无乳链球菌 30.0%、大肠杆菌 27.1%、金黄色葡萄球菌 21.4%。说明兰州地区奶牛乳房炎发病率较高，病乳中无乳链球菌、大肠杆菌和金黄色葡萄球菌的检出率较高。

关键词：奶牛；乳房炎；LMT 诊断液；流行病学；病原菌；调查

（发表于《黑龙江畜牧兽医》科技版）

绵羊皮肤脆裂症遗传病的研究进展***

岳耀敬[1]****，王天翔[2]，郭婷婷[1]，刘建斌[1]，李桂英[2]，孙晓萍[1]，
李文辉[2]，冯瑞林[1]，牛春娥[1]，李范文[2]，郭健[1]，杨博辉[1]*****

（1. 中国农业科学院兰州畜牧与兽药研究所，兰州　730050；
2. 甘肃省绵羊繁育技术推广站，张掖　734031）

摘　要：随着特克赛尔羊、杜泊羊、澳洲美利奴羊等国外优良绵羊品种的大量引进，在显著提高绵羊生产性能的同时，使得携带的各种遗传病在全国范围内快速蔓延，给羊产

---

* 基金项目：“十二五”国家科技支撑计划项目（2012BAD12B03）；甘肃省自然科学基金项目（1308RJZA119）。

** 罗金印（1969—　），男，副研究员，本科，研究方向为奶牛产科疾病，E-mail：luojy818@163.com.

*** 基金项目：国家绒毛用羊产业技术体系分子育种岗位项目（CARS-40-03B）；中央级公益性科研院所基本科研业务费专项（BRF1610322014006）；甘肃省青年自然基金项目（1308RJYA037）。

**** 作者简介：岳耀敬（1980—　），男，助理研究员，博士，研究方向为绵羊分子育种，E-mail：yueyaojing@126.com.

***** 通讯作者：杨博辉（1964—　），男，研究员，博士，研究方向为绵羊分子育种，E-mail：yangbh2004@163.com.

业带来巨大的经济损失，严重威胁我国绵羊育种和羊产业的健康发展。文章对绵羊皮肤脆裂症遗传病的临床症状、分子遗传基础和检测方法进行了综述，以期为我国绵羊皮肤脆裂症遗传病的净化提供理论借鉴。

关键词：绵羊；遗传病；皮肤脆裂症；分子遗传机制；分子诊断；研究进展

（发表于《黑龙江畜牧兽医》科技版）

陶赛特、波德代与藏羊杂交1代羔羊肉用性能分析*

孙晓萍[1]**，刘建斌[1]，张万龙[2]，杨博辉[1]，郭　健[1]，冯瑞林[1]

（1. 中国农业科学院兰州畜牧与兽药研究所，兰州　730050；
2. 金昌市绵羊繁育技术推广站，金昌　737200）

摘　要：为了提高藏羊的产肉性能，试验采用引进的陶赛特羊、波德代羊分别与藏羊杂交，用随机抽样的方法对陶×藏、波×藏1代羊出生到4月龄生长发育指标、130日龄屠宰性能指标等进行测定。结果表明：在西北高原严酷的生态环境条件下，陶×藏、波×藏1代羔羊生长发育快于藏羊，4月龄活重极显著高于藏羊（$P<0.01$），体长、胸围、胸宽、胸深、十字部宽也极显著高于藏羊（$P<0.01$）；产肉性能和胴体品质明显优于藏羊，陶×藏、波×藏1代羊比藏羊胴体重分别高2.36kg、2.06kg，屠宰率分别高3.17个百分点、3.01个百分点（$P<0.05$），净肉重和净肉率分别高1.94kg、1.56kg和1.89个百分点、1.78个百分点（$P<0.05$），眼肌面积和GR值分别高2.60cm^2（$P<0.01$）、1.09cm^2（$P<0.05$）和0.60cm（$P<0.01$）、0.45cm（$P<0.05$）。

关键词：陶赛特羊；波德代羊；藏羊；杂交1代；生长发育；肉用性能

（发表于《黑龙江畜牧兽医》科技版）

* 基金项目：甘肃省科技支撑计划项目（1104NKCA083）。

** 孙晓萍（1962— ），女，副研究员，本科，硕士生导师，研究方向为动物遗传育种，E-mail：xiaopingsun163@163.com.

陶赛特羊与蒙古羊、小尾寒羊及滩羊杂交生长发育性状研究*

孙晓萍[1]**，刘建斌[1]，张万龙[2]，杨博辉[1]，郭 健[1]，冯瑞林[1]

（1. 中国农业科学院兰州畜牧与兽药研究所，兰州 7300502；
2. 金昌市绵羊繁育技术推广站，金昌 737200）

摘 要：为了提高蒙古羊、小尾寒羊、滩羊的产肉性能，采用引进的陶赛特羊分别与其杂交，采用随机抽样的方法，对不同阶段的杂交组合羊只的生长发育性状进行测定，探讨最佳产肉组合模式。结果表明：杂交三代生长发育速度高于对应杂交二代，二代高于对应杂交一代。杂交三代中，陶寒F3、陶蒙F3组合的生长发育较陶滩寒F3快；杂交二代、一代中，同样是陶寒、陶蒙组合生长发育快于陶滩寒组合。说明在饲养小尾寒羊、蒙古羊和滩羊的地区，以陶赛特羊为父本杂交进行羔羊肉生产，经济效益显著。

关键词：陶赛特羊；蒙古羊；小尾寒羊；滩羊；杂交组合；生长发育；研究

（发表于《黑龙江畜牧兽医》科技版）

伊维菌素微乳制剂的溶血性试验***

邢守叶[1,2]****，周绪正[2]，李 冰[2]，牛建荣[2]，魏小娟[2]，张继瑜[2]*

（1. 甘肃农业大学动物医学院，兰州 730070；
2. 中国农业科学院兰州畜牧与兽药研究所/农业部兽用药物创制重点实验室，兰州 730050）

摘 要：为了观察新型兽药制剂伊维菌素微乳是否具有溶血和红细胞凝聚等反应，试验以《化学药物刺激性、过敏性和溶血性研究技术指导原则》为指导，取兔血，采用常规体外试管法，观察伊维菌素微乳制剂是否出现溶血现象。结果表明：伊维菌素微乳制剂未出现溶血和红细胞凝聚现象，可供注射使用。说明伊维菌素微乳制剂无溶血性，安

---

* 基金项目：甘肃省“十二五”重大科技项目（2014XCNA003）。

** 作者简介：孙晓萍（1962— ），女，副研究员，本科，研究方向为动物遗传育种，E-mail：xiaopingsun163@ 163. com.

*** 基金项目：国家现代农业肉牛牦牛产业技术体系建设专项基金项目（CARS-38）；重点牧区生产生态生活保障技术集成与示范项目（2012BAD13BOO）。

**** 作者简介：邢守叶（1988— ），女，硕士研究生，主要从事兽医药理学与毒理学方面的研究。* 通讯作者

全可靠。

关键词： 伊维菌素微乳制剂；日本大耳白兔；溶血性试验；2%红细胞悬液；红细胞凝聚

（发表于《动物医学进展》）

中草药饲料添加剂对河西肉牛生产性能及食用品质影响的研究*

周学辉[1]**，李　伟[1]，杨世柱[1]，郭兆斌[2]

（1. 中国农业科学院兰州畜牧与兽药研究所，兰州　730050；
2. 甘肃农业大学食品科学与工程学院，兰州　730070）

摘　要： 为了研究中草药饲料添加剂“速肥绿药”对河西肉牛生产性能及食用品质的影响，试验采用该中草药饲料添加剂对河西肉牛进行饲喂试验和屠宰试验，测定河西肉牛生产性能（始重、宰前活重、胴体重、屠宰率、日增重、眼肌面积）及食用品质（色度值、pH 值、失水率、熟肉率和剪切力）。结果表明：饲喂高剂量“速肥绿药”的肉牛生产性能明显提高，即增重快，屠宰率高，产肉性能良好；“速肥绿药”对牛肉色度的影响不大；对屠宰后、排酸前牛肉及肝脏的酸碱度有一定影响；高剂量中草药组牛的肝脏嫩度比低剂量组高；肉牛肝脏相对较嫩，肉则较老；中草药饲料添加剂对牛肉、肝脏及心脏失水率和熟肉率均无明显影响。

关键词： 中草药饲料添加剂；河西肉牛；屠宰率；日增重；眼肌面积；食用品质

（发表于《黑龙江畜牧兽医》科技版）

* 基金项目：甘肃省科技重大专项（1002NKDA023）。

** 作者简介：周学辉（1964—　），男，助理研究员，本科，研究方向为中草药绿色环保新产品的研究、应用与开发，E-mail：zhouxuehui888@163.com.

复方板黄口服液对畜禽常见病原菌的体外抑菌作用研究*

魏小娟**，张继瑜，李宏胜，周绪正，牛建荣，李剑勇，李　冰

（中国农业科学院兰州畜牧与兽药研究所/农业部兽用药物创制重点实验室/甘肃省新兽药工程重点实验室，兰州　730050）

摘　要：为了研究板黄口服液对畜禽常见病原菌的体外抑菌作用，研究通过体外试验观察药物对禽巴氏杆菌、鸡葡萄球菌、鸡白痢沙门杆菌、肺炎克雷伯菌、牛巴氏杆菌、牛大肠杆菌的抑菌效果。结果表明：板黄口服液对鸡葡萄球菌的最小抑菌浓度（MIC）为125mg /mL，对牛巴氏杆菌、禽巴氏杆菌、鸡白痢沙门杆菌、肺炎克雷伯菌、牛大肠杆菌的MIC均为250mg /mL。说明板黄口服液对禽巴氏杆菌、鸡葡萄球菌、鸡白痢沙门杆菌、肺炎克雷伯菌、牛巴氏杆菌、牛大肠杆菌有很好的抑制作用。

关键词：板黄口服液；畜禽；常见病原菌；体外抑菌；最小抑菌浓度（MIC）

（发表于《黑龙江畜牧兽医》科技版）

酵母菌制剂保存的稳定性研究***

李春慧[1,2]****，蒲万霞[1]，吴　润[2]

（1. 中国农业科学院兰州畜牧与兽药研究所/农业部兽用药物创制重点实验室/ 甘肃省新兽药工程重点实验室，兰州　730050；
2. 甘肃农业大学动物医学院，兰州　730070）

摘　要：选取3种不同浓度的酵母菌制剂（1.67×10^{-2} g/mL、5.00×10^{-2} g/mL、8.00×10^{-2} g/mL）分为试验组1、2和3，在4 ℃下保存360 d，每30 d取样1次，采用稀释平板法进行活酵母菌计数，检测酵母菌制剂保存的稳定性。结果表明，随着保存时间延长，试验组

---

* 基金项目：国家现代农业肉牛牦牛技术体系项目（nycytx-38）。

** 作者简介：魏小娟（1976—　），女，助理研究员，硕士，研究方向为兽医微生物学，E-mail：w_7788@126.com.

通讯作者：张继瑜（1967—　），男，研究员，博士，研究方向为兽医病理学与毒理学。

*** 基金项目：公益性行业（农业）科研专项（201303038-4）。

**** 作者简介：李春慧（1989—　），女，河南新乡人，兽医师，硕士，主要从事生物制药研究工作，E-mail：chunhuili2013@163.com；通讯作者，蒲万霞，研究员，主要从事兽医微生物与微生物制药的研究，E-mail：puwanxia@caas.cn.

1、3 的活酵母菌数呈下降趋势，试验组 2 的活酵母菌数呈先下降后升高的趋势。在第 30 天、90 天、180 天和 360 天时，试验组 1 酵母菌存活率分别为 76.27%、28.64%、20.85% 和 11.10%，试验组 2 酵母菌存活率分别为 106.25%、13.63%、121.25%和 101.88%，试验组 3 酵母菌存活率分别为 126.37%、90.11%、69.78%和 56.04%。可见，酵母菌制剂试验组 3 保存的稳定性良好，4 ℃条件下保存期可达到 180d 以上。

关键词：酵母菌；制剂；保存；稳定性

（发表于《湖北农业科学》）

天祝白牦牛 *KAP*1.1 基因亚型 *B2A* 克隆及鉴定*

张良斌**，梁春年[2,3]，裴　杰[2,3]，褚　敏[2,3]，
吴晓云[2,3]，张建一[2,3]，潘和平[1]，阎　萍[2,3]

（1. 西北民族大学生命科学与工程学院，兰州　730030；
2. 中国农业科学院兰州畜牧与兽药研究所，兰州　730050；
3. 甘肃省牦牛繁育工程重点实验室，兰州　730050）

摘　要：通过对天祝白牦牛高硫角蛋白基因启动子 *B2A* 克隆得到其序列，将序列提交到 NCBI，登录号：KJ910005，并对其进行生物信息学分析，通过亚细胞定位，磷酸化分析，二级、三级结构和保守结构域预测结果比较，MEGA 5.10 软件进行 NJ 法进化树构建和 MegAlign 进行蛋白质相似分析来分析 *B2A* 基因与 *KAP*1.1（Keratin associated protein1.1）基因的差异性。*B2A* 基因的 CDS 区全长为 471 bp，共编码 156 个氨基酸，二级结构含有延伸链、β 转角和无规卷曲 3 种。B2A 蛋白和 *KAP*1.1 蛋白的亚细胞定位，磷酸化分析，保守结构域，二级结构、三级结构和同源性差异较大。但是从功能预测结果显示 2 个蛋白功能完全一致，并且使用 Clustal X 进行氨基酸序列比对分析显示 B2A 蛋白从第 33 位开始相对于 *KAP*1.1 蛋白有 10 个氨基酸的缺失以外，其余序列之间的保守性相当高。因此，可以证明 B2A 基因是 *KAP*1.1 基因的基因亚型。

关键词：牦牛；B2A 基因；*KAP*1.1 蛋白；基因亚型

（发表于《华北农学报》）

* 基金项目：科技支撑计划项目（2012BAD13B05）；行业科研项目（201003061z）；甘肃省农业生物技术研究与应用开发项目（CNSW-2011-23）；西北民族大学 2014 年度研究生科研创新项目（ycx14162）。

** 作者简介：张良斌（1988—　），男，内蒙古乌海人，在读硕士，主要从事动物遗传育种与繁殖研究。

通讯作者：潘和平（1962—　），男，甘肃西和人，教授，博士，主要从事动物遗传育种与繁殖研究；阎萍（1963—　），女，山西运城人，研究员，博士，主要从事动物遗传育种与繁殖研究。

11 种小毛细皮动物毛纤维的物理性能*

李维红**，熊　琳，高雅琴，梁丽娜

（中国农业科学院兰州畜牧与兽药研究所/农业部动物毛皮及制品质量监督检验测试中心/农业部畜产品质量安全风险评估实验室，兰州　730050）

摘　要：以 11 种小毛细皮类动物毛纤维为试验材料，测定其强力、伸长率、长度、细度及细度离散等物理性能指标。结果显示，不同动物种类毛纤维之间强力、伸长率、长度、细度及细度离散等物理性能指标差异极显著（$P<0.01$）；同一动物种类不同动物品种毛纤维之间强力、伸长率、长度、细度及细度离散等物理性能差异不显著（$P>0.05$）；同一动物品种的针毛和绒毛之间强力、伸长率、长度、细度及细度离散等物理性能差异极显著（$P<0.01$）。

关键词：小毛细皮动物；毛纤维；强力；伸长率；长度；细度

（发表于《江苏农业科学》）

畜禽产品风险评估过程探讨***

杨晓玲[1,2]****，高雅琴[1,2]，李维红[1,2]，熊　琳[1,2]，郭天芬[1,2]

（1. 中国农业科学院兰州畜牧与兽药研究所，兰州　730050；
2. 农业部畜产品质量安全风险评估实验室（兰州），兰州　730050）

摘　要：从风险评估的危害识别、危害特征描述、暴露评估、风险特征描述 4 个步骤，借鉴农产品、食品的风险评估经验，系统地对风险评估的过程加以梳理和阐述，从而为做好畜禽产品风险评估工作提供理论参考。

关键词：畜禽产品；风险评估；危害识别；质量安全

（发表于《江苏农业科学》）

---

* 基金项目：中央级公益性科研院所基本科研业务费专项资金（中国农业科学院兰州畜牧与兽药研究所）（编号：1610322013008）。

** 作者简介：李维红（1978—　），女，甘肃靖远人，博士，助理研究员，主要从事动物纤维、毛皮及制品质量检测及其质量标准的研究。E-mail：lwh0923@163.com.

*** 基金项目：2014 年畜禽产品质量安全风险评估专项（编号：GJFP2014007）；甘肃省科技计划（编号：145RJZA150）；中央级公益性科研院所基本科研业务费专项（编号：1610322013008）。

**** 作者简介：杨晓玲（1987—　），女，甘肃榆中人，硕士，实习研究员，主要从事畜产品质量安全与检测研究。E-mail：yangxl066@163.com.

饲料原料白酒糟基本成分测定及评价*

王晓力[1**]，孙尚琛[2]，王永刚[2]，朱新强[1]，
李秋剑[3]，王春梅[1]，张　茜[1]，刘锦民[4]

（1. 中国农业科学院兰州畜牧与兽药研究所，兰州　730050；
2. 兰州理工大学生命科学与工程学院，兰州　730050；
3. 深圳市农产品质量安全检测中心，深圳　518005；
4. 甘肃省粮食局，兰州　730050）

摘　要：为了解白酒糟的基本成分以确定其利用价值，测定了白酒糟（2012-WS-007）的营养成分及对其进行营养学评价，并与葡萄渣、甜菜渣和小麦秸秆等饲料原料进行了比较。结果表明：粗蛋白质、粗脂肪和粗纤维三大类物质占白酒糟干物质总量的64.99%，在所含的13中氨基酸中，必须氨基酸占总氨基酸的67.56%，每千克白酒糟中含维生素 B_1 15mg，维生素 B_2 1.8mg，此外还含有丰富的铜、铁和锌等矿物质。上述结果表明白酒糟中各类营养物质种类和含量均符合国标规定的饲料原料要求，以白酒糟为原料开发制备饲料等产品具有较好的经济效益。

关键词：白酒糟；营养；效益

（发表于《粮油加工》）

糟渣类物质干燥技术的研究

王晓力[1]，孙尚琛[2]，王永刚[2]，
朱新强[1]，李秋剑[3]，王春梅[1]，张　茜[1]

（1. 中国农业科学院兰州畜牧与曾 · 药研究所，兰州　730050；
2. 兰州理工大学生命科学与工程学院，兰州　730050；
3. 深圳市农产品质量安全检测中心，深圳　518005）

摘　要：将糟渣类物质干燥后制备饲料是充分利用工农业废弃物，减少环境污染，实现

---

* 基金项目：农业部兽用药物创制重点试验室和甘肃省新兽药工程重点试验室开放课题经费，公益性行业（农业）科研专项子课题：甘肃河西走廊荒漠灌区苜蓿高效种植关键技术研究与集成示范；工业副产品的优化利用技术研究与示范项目资助。

** 作者简介：王晓力（1965—　），女，副研究员，硕士，从事畜牧方面的研究工作，E-mail：412316788@163.com.

资源循环利用的有效途径本文就现阶段几种干燥理论和干燥技术进行介绍与分析，从干燥后饲料的品质、干燥工艺的节能环保和简便程度综合考虑，提出最优方案，以期对实际生产作出指导。
关键词：糟渣；干燥技术

（发表于《粮油加工》）

紫菀不同极性段提取物的药效比较*

任丽花，辛蕊华，彭文静，王贵波，罗永江，
罗超应，谢家声，李锦宇，郑继方**
（中国农业科学院兰州畜牧与兽药研究所/甘肃省新兽药
工程重点实验室/甘肃省中兽药工程技术研究中心，兰州　730050）

摘　要：［目的］对比分析紫菀不同极性段提取物的镇咳、平喘及祛痰效果，为进一步研究其药效机理提供理论基础，也为其开发利用提供科学参考。［方法］分别采用浓氨水致咳法、喷雾引喘法、毛细玻管法观察紫菀不同极性段提取物对小鼠的镇咳效果、对豚鼠的平喘效果及对麻醉大鼠的祛痰效果。［结果］石油醚组（Ⅰ）、最后母液组（Ⅳ）及75%乙醇组（Ⅴ）可极显著延长小鼠咳嗽的潜伏期和减少2 min内咳嗽次数（$P<0.01$，下同），正丁醇组（Ⅲ）有显著的镇咳作用（$P<0.05$，下同）；石油醚组（Ⅰ）、正丁醇组（Ⅲ）及75%乙醇组（Ⅴ）能极显著延长4%氯化乙酰胆碱引喘豚鼠的潜伏期；乙酸乙酯组（Ⅱ）、正丁醇组（Ⅲ）、最后母液组（Ⅳ）及75%乙醇组（Ⅴ）可极显著增加大鼠痰液分泌量，石油醚组（Ⅰ）和阳性对照氯化铵组有显著作用。［结论］紫菀75%乙醇提取物具有镇咳、平喘、祛痰作用，3种药效对应的主要有效极性部位分别为石油醚和最后母液部分、石油醚和正丁醇部分、正丁醇和最后母液部分。紫菀镇咳、平喘、祛痰各药效对应的有效极性部位不同，对其药效进行开发利用时应作针对性选择。
关键词：紫菀；极性段；提取物；镇咳；平喘；祛痰

（发表于《南方农业学报》）

---

* 基金项目：“十二五”国家科技支撑计划项目（2011BAD34B03）；公益性行业科研专项项目（201303040-18）；甘肃省科技支撑计划项目（1304NKCA155）。
** 作者简介：为通讯作者，郑继方（1958—　），研究员，主要从事中兽医学研究工作，E-mail：zhengjifang100@126.com；任丽花（1988—　），研究方向为中兽医药学，E-mail：726660797@qq.com.

牛羊焦虫病综合防控技术

张继瑜，周绪正

（中国农业科学院兰州畜牧与兽药研究所，兰州 730050）

摘 要： 牛羊焦虫病是由焦虫（巴贝斯焦虫、泰勒焦虫）寄生于牛羊的巨噬细胞、淋巴细胞和红细胞所引起的一类蜱传性的血液性原虫病。以高热、贫血、出血、消瘦和体表淋巴结肿大为主要临床症状，常导致牛羊死亡，给畜牧业带来了严重的经济损失。该病呈很强的地方流行性，多呈急性型，死亡率高达90%以上，国际兽疫局将其列为B类疫病，我国将其列为二类疫病。在我国主要疫区是北方的干旱和半干旱地区。近年来，随着经济的飞速发展，牛羊养殖产业也随之进一步发展，牛羊交易活动的频繁发生，使该病的发生和流行频度、流行区域也逐步扩大。采用特效药物和疫苗早期治疗和预防的综合防控措施，可有效地防治本病。

（发表于《农家顾问》）

牦牛 *CAV*-3 基因的克隆及其在牦牛和黄牛组织的表达分析*

赵娟花**，裴 杰，梁春年，
郭 宪，吴晓云，张良斌，阎 萍

（中国农业科学院兰州畜牧与兽药研究所/甘肃省牦牛繁育工程重点实验室，兰州 730050）

摘 要： 旨在对牦牛 *CAV*-3 基因进行克隆、生物信息学分析，并对其在牦牛组织中的表达规律进行初步研究。根据 GenBank 数据库中已知的黄牛 *CAV*-3 基因的 mRNA 序列并设计特异性引物，应用 RT-PCR 技术克隆牦牛 *CAV*-3 基因的编码区。运用生物信息学方法，分析并预测牦牛 Caveolin-3 蛋白的理化性质、疏水性、蛋白结构域以及蛋白质二级结构。通过半定量 PCR 技术检测 *CAV*-3 基因 mRNA 在牦牛和黄牛各组织中的表达；利用实时荧光定量 PCR 技术检测牦牛和黄牛肌肉组织中 *CAV*-3 基因 mRNA 表达水平。牦牛 *CAV*-3 的编码区全长 631 bp，共编码 151 个氨基酸。*CAV*-3 在牦牛肺、脾脏、肾脏、肝脏、卵巢组织中均不表达，仅在心脏和肌肉组织中表达，且在心脏组织的表达水平高于肌肉组织，*CAV*-3 基因在黄牛各组织中的表达结果与牦牛一致。*CAV*-3 基因在牦牛肌肉中的表达低于黄牛肌肉组织，但差异不显著（$P>0.05$）。

关键词： 牦牛；Caveolin-3；克隆；表达分析

---

* 基金项目：现代农业（肉牛牦牛）产业技术体系专项（CARS-38），甘肃甘南牧区“生产生态生活”保障技术集成与示范（2012AD13B05）。

** 作者简介：赵娟花，女，硕士研究生，研究方向：动物遗传育种；E-mail：zjh326303202@163.com. 通讯作者：阎萍，女，研究员，博士生导师，研究方向：动物遗传育种；E-mail：pingyanlz@163.com.

（发表于《生物技术通报》）

回流法和超声法从中药常山中提取常山乙素的比较研究*

郭志廷[1**]，罗晓琴[2]，梁剑平[1]，杨　珍[1]，艾　鑫[1]

（1. 中国农业科学院兰州畜牧与兽药研究所/农业部兽用药物创制重点实验室/甘肃省新兽药工程重点实验室，兰州　730050；
2. 兰州市动物卫生监督所，畜产品质量安全检测实验室，兰州　730050）

摘　要：目的比较回流法和超声法从中药常山中提取常山乙素的提取效果。方法采用高效液相色谱法测定常山乙素含量，以浸膏得率和常山乙素含量作为评价指标。结果回流法和超声提取法浸膏平均得率分别为10.11%和12.87%；浸膏中常山乙素平均含量分别为0.160%和0.181%。结论回流法和超声法提取工艺具有省时、操作简便和提取率高等优点，超声法相对更好一点。

关键词：常山；常山乙素；回流法；超声法；高效液相色谱法

（发表于《时珍国医国药》）

肉品中β-受体激动剂类药物残留检测技术研究进展***

熊　琳[1,2****]，李维红[1,2]，高雅琴[1,2]，郭天芬[1,2]，杨晓玲[1,2]

（1. 中国农业科学院兰州畜牧与兽药研究所，兰州　730050；
2. 农业部畜产品质量安全风险评估实验室（兰州），兰州　730050）

摘　要：肉品中β-受体激动剂类药物残留问题严重影响着人体健康。近几年来暴露出来的肉品中β-受体激动剂类药物残留事件层出不穷，为了加强监控，越来越多的检测

---

* 基金项目：中央级公益性科研院所专项资金项目（No. 1610322011004）。

** 作者简介：郭志廷（1979—　），男（汉族），内蒙古凉城人，中国农业科学院兰州畜牧与兽药研究所助理研究员，硕士学位，主要从事中药免疫学和药理学研究工作。

*** 基金项目：中央级科研院所基本科研业务费专项（1610322014014）、甘肃省科技计划项目（145RJZA150）。

**** 通讯作者：熊琳，硕士，助理研究员，主要研究方向为农产品质量安全。E-mail：xionglin807@sina.com.

方法被建立起来，极大地提高了肉品中β-受体激动剂类药物残留检测的效率。本文综述了检测肉品中β-受体激动剂类药物残留的色谱法、酶联免疫法、胶体金法和其他一些新方法的研究进展，以及我国现行的β-受体激动剂类药物残留标准检测方法。对现有的检测方法的优缺点和发展阶段作了归纳总结，并对β-受体激动剂类药物残留检测方法发展的方向提出了一些展望。未来的发展主要是一些特异性和灵敏度更高的快检方法和样品前处理更加简化的液质色谱联用法。

关键词：食品安全；肉品；β-受体激动剂类药物残留

（发表于《食品安全质量检测学报》）

响应曲面法优选苦豆子总生物碱的提取工艺研究*

刘　宇[1]**，尚若锋[1]，程富胜[1]，
郝宝成[1]，梁剑平[1]，王学红[1]，蒲秀瑛[2]
（1. 中国农业科学院兰州畜牧与兽药研究所/农业部兽用药物创制重点实验室/甘肃省新兽药工程重点实验室，兰州　730050；
2. 兰州理工大学，兰州　730050）

摘　要：以甘肃苦豆子为原料，利用超声波辅助提取苦豆子总生物碱，并采用 Box-Behnken 对超声波超声时间、乙醇浓度、料液比进行响应面优化。结果表明：最佳提取工艺参数为乙醇浓度65%，料液比1 ：8 g/mL，提取时间30 min。在此条件下，总生物碱提取率为29. 1mg/g，超声提取工艺合理、可行，为苦豆子总生物碱的新药开发提供了理论参考。

关键词：苦豆子；总生物碱；响应曲面法

（发表于《食品研究与开发》）

---

* 基金项目：甘肃省省自然科学研究基金计划项目（1010RJZA004）。

** 作者简介：刘宇（1981— ），男（汉），助理研究员，硕士，主要从事有机及天然药物化学等研究。

潜江市土壤养分空间分布及其水系对它的影响*

李润林[1,2]**，姚艳敏[3]

（1. 农业部兰州黄土高原生态环境重点野外科学观测实验站，兰州　730050；
2. 中国农业科学院兰州畜牧与兽药研究所，兰州　730050；
3. 中国农业科学院农业资源与农业区划研究所，北京　100081）

摘　要：结合地统计与GIS技术研究潜江市土壤养分的空间分布特征及其影响因素，结果表明：潜江市土壤养分呈西高东低的格局，按照全国第2次土壤普查的土壤养分分级标准评估，发现潜江市有机质低于20g/kg的土壤占21.4%，碱解氮低于90mg/kg的土壤占24.9%，有效磷低于10mg/kg土壤占18.5%，速效钾低于100mg/kg的土壤占0.3%。其余部分土壤处于中等丰富状况，其中速效钾大于100mg/kg的土壤占到99.7%，说明潜江市不缺速效钾。定量分析水系密集区和水系稀疏区土壤养分含量，发现水系密集区土壤养分平均含量高于水系稀疏区土壤养分平均含量，有机质高1.36g/kg、碱解氮高10.13mg/kg、有效磷高0.32mg/kg、速效钾6.62mg/kg。方差分析发现，水系密集区和水系稀疏区土壤养分含量存在显著差异，说明水系对土壤养分的空间分布具有重要影响，对解释水系密集区土壤养分沿水系成带状分布具有重要价值。

关键词：空间变异；土壤养分；地统计学；GIS

（发表于《土壤通报》）

---

* 基金项目：国家“973”计划课题（2010CB951502）、公益性行业（农业）科研专项（201203008）和中央级公益性科研院所基本科研业务费专项（1610322014016）资助。

** 作者简介：李润林（1982—　），男，甘肃省武威市人，硕士，助理研究员，主要从事农业资源遥感研究。E-mail：lirunlin555@163.com.

甘肃地区新生仔猪腹泻细菌性病原的分离鉴定及其耐药性分析*

黄美州**，刘永明***，王　慧，王胜义，崔东安，李胜坤

（中国农业科学院兰州畜牧与兽药研究所/农业部兽用药物创制重点实验室/甘肃省新兽药工程重点实验室/甘肃省中兽药工程技术研究中心，兰州 730050）

摘　要：为了解甘肃地区导致新生仔猪腹泻的主要细菌性病原。采集甘肃省部分地区新生仔猪腹泻病料 104 份，采用 16Sr DNA 扩增技术对细菌性病原进行鉴定，经标准诊断血清确定其血清型，并用药敏纸片法对细菌性病原进行耐药性分析。结果显示：在 35 份病料中分离出致病性大肠埃希菌，4 份病料中分离得到与致病性大肠埃希菌混合感染的沙门菌；肠道产毒性大肠埃希菌的血清型主要以 O78：K80（B）为主；该地区的致病性大肠埃希菌对多种抗生素产生耐药性。表明：甘肃地区新生仔猪腹泻的细菌性病原主要是大肠埃希菌，且多重耐药性严重。

关键词：新生仔猪；腹泻；大肠埃希菌；耐药性

（发表于《西北农业学报》）

基因本体论在福氏志贺菌转录组研究中的应用****

朱　阵*****，刘翠翠，王　婧，张继瑜******，魏小娟，周绪正，李　冰，郭　肖

（中国农业科学院兰州畜牧与兽药研究所/农业部兽用药物创制重点实验室，兰州　730050）

摘　要：利用 RNA-seq 技术进行环丙沙星敏感和耐药的福氏志贺菌 2457T 株的序列测

---

*　基金项目：公益性行业专项（201303040-17）；国家科技支撑计划（2012BAD12B03）。

**　第一作者：黄美州，男，硕士研究生，研究方向为仔畜腹泻病防治。E-mail：13141254071@163. com.

***　通讯作者：刘永明，男，研究员，研究方向为营养代谢病和中毒病。E-mail：myslym@ sina. com.

****　基金项目：国家自然科学基金（31272603、31101836）。

*****　第一作者：朱阵，男，在读硕士，从事兽医药理与毒理学的研究。E-mail：zhuzhen234@ yeah. net.

******　通讯作者：张继瑜，男，博士，研究员，主要从事新兽药的研究与开发工作。E-mail：infzjy@ sina. com.

定与组装，对其碱基序列借助基因本体（Gene Ontology，GO）在生物进程、细胞组分、分子功能三方面进行注释和生物学分析。研究经过耐药诱导的转录组差异表达情况，并探讨差异表达基因在耐药性产生过程中的作用及相互关系。结果表明，通过耐药诱导有3 641个基因差异表达，其中961个GO条目注释到39个功能类别。同时发现硫酸盐、硝酸盐氧化还原相关酶系，电子/H^+传导基因，膜转运蛋白，ABC外排泵家族4类耐药相关的功能基因组分。
关键词：RNA-seq；福氏志贺菌；转录组；基因本论

（发表于《西北农业学报》）

牛源鲍曼不动杆菌的鉴定及体外药敏试验*

王孝武**，王旭荣，李建喜，王学智，林　杰，杨志强***

（中国农业科学院兰州畜牧与兽药研究所/农业部兽用药物创制重点实验室/甘肃省新兽药工程重点实验室，兰州　730050）

摘　要：为鉴定山西省某奶牛场引起奶牛子宫内膜炎的相关病原菌。采用细菌分离培养、革兰染色镜检和16SrDNA测序分析法对采集的奶牛子宫分泌物进行细菌鉴定；通过体外药敏试验和小鼠致病性试验检测其耐药性和致病性，并与已知序列进行比对，了解其亲缘性。结果显示，共分离到1株鲍曼不动杆菌，该株菌与中国内地和美国的分离菌株属于同一分支，且与美国的分离菌株亲缘较近；该菌对青霉素、氨苄西林、阿莫西林、卡那霉素、链霉素、庆大霉素、红霉素、先锋霉素Ⅳ均耐药；对小白鼠有较强的致病性。
关键词：鲍曼不动杆菌；子宫内膜炎；奶牛

（发表于《西北农业学报》）

---

* 基金项目：国家奶牛产业技术体系科学家岗位项目（CARS37）；“948”项目（2014-Z9）。

** 第一作者：王孝武，男，硕士研究生，研究方向为奶牛疾病。E-mail：xiaowu2053@ 163. com；

*** 通讯作者：杨志强，男，研究员，博士生导师，从事中兽医药学、兽医药理毒理、动物营养代谢与中毒病等研究。E-mail：zhiqyang2006@ 163. com.

紫外线和亚硝基胍对益生菌 FGM 发酵提取黄芪多糖的影响[*]

陈　婕[**]，王旭荣，张景艳，秦　哲，
王　磊，孟嘉仁，尚利明，杨志强，王学智[***]，李建喜[****]

（中国农业科学院兰州畜牧与兽药研究所/农业部兽用药物创制
重点实验室/甘肃省中兽药工程技术研究中心，兰州　730050）

摘　要：旨在评价紫外线（UV）和亚硝基胍（NTG）诱变处理对菌株 FGM 发酵提取黄芪多糖的影响。分别用适量 UV、NTG、UV+NTG 诱变处理对数生长期的益生菌株 FGM，筛选生长性能较好的诱变株发酵中药黄芪，水提醇沉发酵产物中的粗多糖，苯酚—硫酸法测定多糖质量分数。结果显示，UV 处理 90s 得到的诱变菌株 U90 发酵黄芪后，产物中多糖质量分数比其他诱变剂量较高，为 235.51mg/g，与诱变前相比增加 58.13%；1.0g/L 的 NTG 处理菌株 10min 后，得到的诱变株 N1-1 发酵黄芪后，产物中多糖质量分数比其他剂量高，为 161.2mg/g，与诱变前比较增加 10.64%；UV 照射 10s 后，再用 1.0g/L 的 NTG 处理 10min，得到的诱变菌株 UN10-1 发酵黄芪后，产物中的多糖质量分数为 293.39mg/g，较诱变前增加了 94.99%；UV 照射 10~120s 所得的菌株发酵黄芪后产物中多糖质量分数均呈升高趋势，而 NTG 处理所得菌株发酵黄芪后产物中多糖质量分数变化无规律性。由此可见，UV 和 NTG 诱变均有改善益生菌株 FGM 发酵提取黄芪多糖的作用，但 NTG 改善菌株发酵性能不明显，适量 UV 照射结合一定浓度 NTG 诱变所得的菌株发酵提取黄芪多糖的性能优于 UV 和 NTG 单独作用。

关键词：紫外线；亚硝基胍；发酵黄芪；多糖质量分数

（发表于《西北农业学报》）

---

* 基金项目：国家自然科学基金（31072162）；公益性行业（农业）科研专项（201303040）；公益性研究所基本科研费（1610322011007）。

** 第一作者：陈婕，女，硕士研究生，从事益生菌发酵研究。E-mail：chenjielz@126.com；

*** 通讯作者：王学智，男，副研，博士，主要从事兽医临床及毒理学研究。E-mail：wangxuezhi@caas.cn；

**** 李建喜，男，研究员，博士，主要从事中兽药现代化和奶牛疾病研究。E-mail：lzjianxil@163.com.

西藏苜蓿种子繁育研究进展及其前景 *

杨　晓[1]，余成群[2]，李锦华[1] **，朱新强[1]，江　措[3]

（1. 中国农业科学院兰州畜牧与兽药研究所，兰州　730050；
2. 中国科学院地理科学与资源研究所，北京　100101；
3. 西藏自治区山南地区草原工作站，山南　856000）

摘　要：文章通过概述苜蓿种子西藏本土化生产中存在的主要问题，分析了影响当地苜蓿结实率的主要因素，介绍了近年来西藏苜蓿种子繁育研究的进展，探讨了进一步解决西藏苜蓿种子生产技术的途径，以期为西藏苜蓿种业发展提供参考。

关键词：苜蓿；种子生产；西藏

（发表于《西藏科技》）

医药认知模式创新与中医学发展***

罗超应[1]，罗磐真[2]，谢家声[1]，李锦宇[1]，
王贵波[1]，辛蕊华[1]，罗永江[1]，郑继方[1]

（1. 中国农业科学院兰州畜牧与兽药研究所/甘肃省中兽药工程技术研究中心，兰州　730050；2. 户县中医医院急诊科，户县　710300）

摘　要：在对中医传统认知模式与中西医结合困惑进行综合分析的基础上，论述了中医学既不能完全回归传统文化，搞独立发展，也不能再延续“中西医结合”或“中医学现代化”的做法，以传统科学的思路与方法来解释与衡量中医学；而是要进行中医认识模式创新——创立“复杂-整体-状态医学模式”，正确认知中西医学关系，促进中西医学和谐发展，以使中医学在新的历史条件下既可以保持传统优势与特点，又能吸收西医学乃至整个现代科学的新知识与新技术，从而促进其更快地发展。

关键词：医学模式；中西医结合；中医学现代化；复杂性科学

（发表于《医学与哲学》）

---

* 西藏自治区科技专项“西藏主要栽培豆科牧草繁育研究与示范”（Z2013C02N03-05）；“抗霜霉病苜蓿品种的示范与推广”（2012F4001002）。

** 通讯作者

*** 基金项目：科技部科研院所技术开发研究专项“新型中兽药射干地龙颗粒的研究与开发”，项目编号：2013EG134236；科技部科技基础性工作专项“华东区传统中兽医药资源抢救和整理”，项目编号：2013FY110600-8。

甘青乌头抗炎镇痛活性部位的半仿生-酶法提取工艺优化*

吴国泰[1,2]**，王瑞琼[2,3]，邵　晶[2]，杜丽东[2]，尚若峰[1]，景　琪[2]，梁剑平[1]***

（1. 中国农业科学院兰州畜牧与兽药研究所，兰州　730050；
2. 甘肃中医学院甘肃省中药药理与毒理学重点实验室，
兰州　730000；3. 甘肃农业大学，兰州　730070）

摘　要：用半仿生-酶法提取甘青乌头抗炎镇痛的活性部位，考察加酶量、酶解时间、半仿生提取（SBE）温度和总时间 4 个因素的影响。结果表明，优选工艺为：加酶量 36 U/g，酶解时间 90 min，SBE 温度 80℃，SBE 总时间 180 min。

关键词：甘青乌头；抗炎镇痛；响应面法；半仿生-酶法；提取工艺

（发表于《应用化工》）

牧草的航天诱变研究****

杨红善[1,2]*****，王彦荣[1]，常根柱[2]，周学辉[2]，包文生[3]

（1. 兰州大学草地农业科技学院，兰州　730020；
2. 中国农业科学院兰州畜牧与兽药研究所，兰州　730050；
3. 甘肃省航天育种工程技术中心，天水　741030）

摘　要：航天诱变育种是近十几年来快速发展的农业高科技新领域，将高科技的空间技术与常规农业育种相结合，是一项高创新性育种方法，其目的是利用空间诱变技术

---

* 基金项目：国家“十一五”科技支撑计划项目（2011AA10A214）；甘肃省中药药理学和毒理学重点实验室开放基金项目（ZDSYS-KJ-2012-004）。

** 作者简介：吴国泰（1978—　），男，甘肃清水人，甘肃中医学院讲师，在读博士，师从梁剑平研究员，从事天然药物化学与新药开发研究。E-mail：wgt@ gszy. edu. cn.

*** 通讯联系人：梁剑平（1977—　），男，研究员，研究方向为天然药物化学与新药开发。E-mail：ljping100@ sina. com. cn.

**** 基金项目：甘肃省省青年科技基金计划项目（145RJYA273）；甘肃省农业生物技术研究与应用开发项目（GNSW-2014-19）；中央级公益性科研院所基本科研业务费专项资金项目（中国农业科学院兰州畜牧与兽药研究所，1610322014022）资助。

***** 作者简介：杨红善（1981—　），男，甘肃兰州人，助研，主要从事牧草航天育种的工作，E-mail：Yanghsh123@ 126. com.

选育植物新种质、新材料、培育新品种。我国的航天育种技术走在世界前列，已在农作物、蔬菜及花卉上广泛应用，且已通过国家或省级审定、鉴定了70多个品种，牧草航天育种尚处于初始阶段，目前主要集中于紫花苜蓿（*Medicago sativa*）方面的相关研究。为此，简要阐述了航天诱变育种的定义、机理和特点，介绍了紫花苜蓿、燕麦、猫尾草、红三叶、白三叶、黄花矶松、红豆草和沙打旺等几种牧草航天诱变研究的自身特点，以及目前我国牧草航天诱变研究进展，并提出存在的问题和今后的研究内容。

关键词：航天诱变；牧草

（发表于《中国草地学报》）

一起小尾寒羊黄花棘豆中毒病例的诊断与治疗

郝宝成[1,2]，夏晨阳[3]，刘建枝[3]，
王保海[3]，胡永浩[2]，梁剑平[1,2]

（1. 中国农业科学院兰州畜牧与兽药研究所/农业部兽用药物创制重点实验室，甘肃省新兽药工程重点实验室，兰州 730050，2. 甘肃农业大学动物医学院，兰州 730000；3. 西藏自治区农牧科学院畜牧兽医研究所，拉萨 850000）

摘　要：本文报告了2014年10月3日甘肃省兰州市榆中县一羊养殖厂发生一起小尾寒羊集体疑似疯草中毒事件。疑似中毒羊为不同年龄、性别的多只小尾寒羊。经对询问饲养人员饲养情况，观察疑似中毒羊各种症状，如后肢无力、目光呆滞、不食、消瘦、犬坐式、头颈后仰、手提羊耳出现转圈运动，对放养草地采集植物观察、辨认、比较后，诊断为典型的小尾寒羊疯草中毒。

关键词：小尾寒羊；疯草中毒；黄花棘豆；诊断；治疗

（发表于《亚洲兽医病例研究》）

Improving the Local Sheep in Gansu via Crossing with Introduced Sheep Breeds Dorset and Borderdale*

SUN Xiao-ping[1**], LIU Jian-bin[1], ZHANG Wan-long[2],
LANG Xia[1], YANG Bo-hui[1], GUO Jian[1], FENG Rui-lin[1]

(1. Lanzhou Institute of Husbandry and Pharmaceutical Sciences, Chinese Academy of Agricultural Sciences, Lanzhou 730050, China;
2. Extension Station forSheep Breeding Technique in Jinchang City, Jinchang 737200, China)

Abstract: In order to improve the meat performance of local sheep in Gansu Province, Dorset and Borderdale were introduced to crossbreed with local sheep whichwere Tan sheep, Small-tail Han sheep and Mongolia sheep. The offspring under different crossbreeding combinations were sampled randomly at the different growingstage to measure their growth traits so as to select optimize the crossbreeding mode. The results indicated that, for the same crossbreeding mode, the growth rate ofprogeny was in order F3 > F2 > F1; for the F3 progeny, the combinations Dorset-Borderdale-Small tail Han sheep and Dorset-Borderdale-Mongolia sheep gavea higher growth rate, with a body weight of 1.57%, 3.17%, 8.23%, 1.15% higher in male and female individuals than the counterparts of Dorset and Tan sheepand Small tail Han sheep; for the F2 progeny, the combinations Dorset-Borderdale-Small tail Han sheep and Dorset-Borderdale-Mongolia sheep also gave ahigher growth rate, with a body weight of 2.15%, 4.53%, 9.21% and 2.75% higher in male and female individuals than the counterparts of Dorset and Tan sheepand Small tail Han sheep; for the F1 progeny, the combination Borderdale and Small tail Han sheep assumed a higher growth rate, with a body weight of 3.23%, 6.07%, 7.42% and 8.66% higher in male and female individuals than the counterparts of Borderdale - Mongolia sheep and Tan sheep - Small tail Han sheep, respectively. Therefore, in the Small-tail Han sheep and Mongolia sheep producing regions, the F2 or F3 progeny bred by using Dorset or Borderdale sheep as maleparent to cross with local breeds, or the hybrid lambs of Small-tail Han sheep and Borderdale sheep as highly qualified commodity, would produce significant economicbenefit. Moreover, the novel breeds obtained by crossing were the valuable genetic resource for breeding meat sheep.

Key words: Dorset; Borderdale; Tan sheep; Small tail Han sheep; Mongolia sheep; Improvement; Growth and development

(发表于《Animal Husbandry and Feed Science》)

---

* Supported by Science & Technology Pillar Program of Gansu Province (1104NKCA083).

** Corresponding author. E-mail: xiaopingsun163@ 163. com.

低致病性禽流感所致产蛋下降的中药防治效果*

谢家声**，王贵波，李锦宇，罗超应

（中国农业科学院兰州畜牧与兽药研究所，兰州 730050）

摘 要：低致病性禽流感不仅导致蛋鸡产蛋性能下降而且长时间难以恢复，已严重威胁蛋鸡养殖业健康发展。因此，研制恢复低致病性流感病后产蛋的有效中兽药，对蛋鸡养殖具有十分重要的意义。试验选用低致病性禽流感 H_9N_2 病愈后，产蛋不能回复的鸡群，连续7d在饲料中以日粮的1.0%添加中药复方（暂定名为红花益母增蛋散），观察其恢复产蛋性能的效果。试验随机分为3个处理，即红花益母增蛋散组、药物对照组和空白对照组。结果，红花益母增蛋散组平均净增产蛋率比空白对照组的平均净增产蛋率提高10.6%，差异显著（$P<0.05$）、比药物对照组的提高7.4%，差异显著（$P<0.05$）；红花益母增蛋散组破、软蛋分别比药物对照组和空白对照组的降低66.7%和71.4%。试验结果，红花益母增蛋散防治禽流感所致的降蛋，具有良好的效果。

关键词：红花益母增蛋散；低致病性禽流感；降蛋

（发表于《家禽科学》）

21种紫花苜蓿在西藏"一江两河"地区引种试验研究***

杨 晓[1****]，李锦华[1*****]，朱新强[1]，余成群[2]

（1. 中国农业科学院兰州畜牧与兽药研究所，兰州 730050；
2. 中国科学院地理科学与自然资源研究所，北京 100000）

摘 要：针对西藏自治区（全书简称西藏）"一江两河"高海拔河谷地区豆科牧草种植

* 基金项目：科技部科研院所技术开发研究专项（项目编号2013EG134236）；科技部科技基础性工作专项课题（项目编号2013FY110600-8）。

** 作者简介：谢家声，男，高级实验师，从事兽医临床研究。E-mail：13220469024@163.com.

*** 基金项目：西藏自治区科技专项"西藏主要栽培豆科牧草繁育研究与示范"和"抗霜霉病苜蓿品种的示范与推广"课题。

**** 作者简介：杨晓（1985— ），男，助理研究员，硕士。

***** 通讯作者：李锦华（1963— ），男，副研究员，博士，从事牧草栽培与育种研究。

的发展需求，试验研究了21个国内外紫花苜蓿（*Medicago sativa* L.）品种在该地区的适应性。结果表明：参试的21个品种均能完成生育期；以鲜草产量为评价指标，德宝、新疆大叶、甘农4号和中兰1号4个品种的适应性优于其他苜蓿品种；以种子产量为评价指标，阿尔冈金最优。通过聚类分析综合考虑，新疆大叶、甘农4号和德福在高海拔河谷地区具有较强的适应性，可以进行大面积种植以提供优质蛋白饲草来源。
关键词：紫花苜蓿；“一江两河”；越冬率；鲜草产量；种子产量

（发表于《草地与牧草》）

哺乳期藏绵羊冬春季放牧+补饲优化模式研究

朱新书[1]，王宏博[1]，包鹏甲[1]，李世红[2]，
陈胜红[2]，夏永祺[2]，汪海成[2]，张　功[2]

（1. 中国农业科学院兰州畜牧与兽药研究所/甘肃省牦牛繁育工程重点实验室，兰州　730050；2. 甘肃省甘南藏族自治州临潭县农牧局畜牧草原工作站，临潭）

摘　要：为研究哺乳期藏绵羊成年母羊冬春季放牧+补饲的最优化模式，采用单因子设计，对藏绵羊进行了补饲试验。结果表明：在放牧和补饲粗饲料相同的条件下，全价精饲料补饲效果明显好于玉米补饲料，其中以全价精料只均日补饲量200~250g效果最好，为最优化补饲模式。
关键词：藏绵羊；哺乳期；冬春季；放牧+补饲

（发表于《中国草食动物科学》）

不同哺乳方式对甘南牦犊牛生长发育影响的研究*

牟永娟[1,2,3]**，阎　萍[1,2]***，梁春年[1,2]，丁学智[1,2]，李明娜[1,2]，杨树猛[3]

（1. 中国农业科学院兰州畜牧与兽药研究所，兰州　730050；
2. 甘肃省牦牛繁育工程重点实验室，兰州　730050；
3. 甘肃省甘南州畜牧科学研究所，合作　747000）

摘　要：为探索甘南牦犊牛培育的有效方法，试验采用全哺乳和半哺乳方式对其生长发育情况进行了研究。结果表明：6月龄时试验组（全哺乳）犊牛体重达78.58kg，较对照组（日挤奶1次）高16.54kg；试验组1~6月龄犊牛腹泻病平均发病率9.98%，对照组16.47%；全哺乳牦犊牛生长发育快，抗病力强。说明在甘南牧区开展全哺乳培育对牦犊牛生长发育有显著的促进作用。

关键词：全哺乳；半哺乳；牦犊牛；生长发育

（发表于《中国草食动物科学》）

藏绵羊成年母羊四季放牧采食量研究****

朱新书[1]*****，王宏博[1]，包鹏甲[1]，李世红[2]，
陈胜红[2]，夏永祺[2]，汪海成[2]，张　功[2]

（1. 中国农业科学院兰州畜牧与兽药研究所/甘肃省牦牛繁育工程重点实验室，
兰州　730050；2. 甘肃省甘南藏族自治州临潭县农牧局畜牧草原工作站，临潭）

摘　要：为了研究藏绵羊成年母羊四季放牧干物质采食情况，试验采用饱和烷烃法对藏绵羊的放牧采食量进行了测定。结果表明：藏绵羊成年母羊春、夏、秋、冬四季放牧干物质日采食量分别为(1.09±0.29) kg、(1.77±0.34) kg、(1.38±0.24) kg 和 (0.23±0.04) kg；相对采

* 基金项目：现代农业产业技术体系专项（MATS－Beef Cattle System）；公益性行业科技专项（201203008）；国家科技支撑计划（2012BAD13B05）。

** 作者简介：牟永娟（1980—　），女，高级兽医师。

*** 通讯作者：阎萍（1963—　），女，研究员，博士生导师。

梁春年[1,2]（1973—　），男，副研究员，博士。

**** 基金项目：农业部公益性行业（农业）科研专项（201303062）；现代农业（肉牛牦牛）产业技术体系专项（CARS-38）。

***** 作者简介：朱新书（1957—　），男，副研究员，大学本科，主要从事牦牛、藏羊资源利用研究。

食量分别为2.97%、4.44%、3.07%和0.88%。藏绵羊成年母羊放牧干物质采食量随季节呈现明显变化，夏季采食量最高，显著高于秋季（$P<0.05$），极显著高于春季和冬季（$P<0.01$）；冬季采食量最低，与春、夏、秋三季采食量比较，差异均极显著（$P<0.01$）。
关键词：藏绵羊；成年母羊；放牧采食量；链烷烃

（发表于《中国草食动物科学》）

藏绵羊后备母羊四季放牧采食量研究*

朱新书[1**]，王宏博[1]，包鹏甲[1]，李世红[2]，
陈胜红[2]，夏永祺[2]，汪海成[2]，张　功[2]
（1. 中国农业科学院兰州畜牧与兽药研究所，甘肃省牦牛繁育工程重点实验室，兰州　730050；2. 甘肃省甘南藏族自治州临潭县农牧局畜牧草原工作站，临潭）

摘　要：为了探讨藏绵羊后备母羊四季放牧干物质采食情况，试验采用饱和烷烃法对藏绵羊后备母羊四季放牧采食量进行了测定。结果表明：藏绵羊后备母羊春、夏、秋、冬四季放牧干物质日采食量分别为1.27～0.29kg、1.09～0.18kg、1.24～0.24kg和0.20～0.03kg，相对放牧干物质采食量分别为4.13%、3.77%、3.41%和0.89%。其中，春季干物质采食量最高，显著高于夏季（$P<0.05$），极显著高于冬季（$P<0.01$），与秋季的采食量接近（$P>0.05$）；冬季干物质采食量最低，与春、夏、秋三季采食量比较，差异极显著（$P<0.01$）。提示冬季应该对羊群进行有效补饲。
关键词：藏绵羊；后备母羊；放牧采食量；链烷烃

（发表于《中国草食动物科学》）

---

* 基金项目：公益性行业（农业）科研专项（201303062）；现代农业（肉牛牦牛）产业技术体系专项（CARS-38）。

** 作者简介：朱新书（1957—　），男，副研究员，大学本科，主要从事牦牛、藏羊资源利用研究。

代乳料对甘南牦犊牛生长发育及母牦牛繁殖性能的影响*

牟永娟[1,2,3]**，阎　萍[1,2]，梁春年[1,2]，丁学智[1,2]，杨树猛[3]，马登录[3]

（1. 中国农业科学院兰州畜牧与兽药研究所，兰州　730050；
2. 甘肃省牦牛繁育工程重点实验室，兰州　730050；
3. 甘肃省甘南州畜牧科学研究所，合作　747000）

摘　要：试验用代乳料饲喂45头牦犊牛每头牦犊牛饲喂量0.4kg/d。经过160d试验，6月龄公、母牦犊牛体高、体长、胸围、管围比同期对照组公、母犊牛分别增加了10.34cm和6.76cm、5.32cm和4.94cm、12.34cm和10.00cm、0.3cm和0.2cm，体重分别增加了7.4kg和7.3kg，除管围外均差异极显著（$P<0.01$）。试验组18月龄公、母牦犊牛体高、体长、胸围和管围比对照组公、母牦犊牛分别增加了6.3cm和4.9cm、8.0cm和6.7cm、11.1cm和12.2cm、1.3cm和1.0cm，体重分别增加了15.2kg和14.7kg，差异均极显著（$P<0.01$）。试验组发情率26.7%，对照组8.9%，差异显著（$P<0.05$）。

关键词：代乳料；甘南牦犊牛；生长发育；繁殖性能

（发表于《中国草食动物科学》）

高效液相色谱法测定饲料中己烯雌酚含量的方法学考察

王东伟[1]***，陈建华[2]，周　磊[3]

（1. 公安部警犬技术学校，沈阳　110034；
2. 辽宁省沈阳市兽药饲料监察所，沈阳　110031
3. 中国农业科学院兰州畜牧与兽药研究所，兰州　730050）

摘　要：己烯雌酚是一种人工合成的雌性激素，被禁止添加在畜禽饲料中，人们长期食用这种含有雌激素的畜禽、鱼虾等，可以在人体内引起蓄积，引发一系列的健康问题。

---

* 基金项目：现代农业产业技术体系专项经费资助（MATS-Beef Cattle System）；公益性行业科技专项经费（201203008）。

** 作者简介：牟永娟（1980—　），女，高级兽医师；
通讯作者：梁春年（1973—　），男，副研究员，博士；
杨树猛（1976—　），男，高级畜牧师。

*** 作者简介：王东伟（1979—　），男，助理研究员，硕士；
通讯作者：周磊（1979—　），男，助理研究员，硕士，从事畜牧生产系统管理工作。

应用 BP028—2002 高效液相色谱法可以快速检测到饲料中己烯雌酚的微量含量。此方法经过大量的试验操作，方法摸索，方案优化，具有灵敏度高、检测时间短、回收率高等优点。

关键词：高效液相；色谱柱；色谱峰；己烯雌酚；回收率

（发表于《中国草食动物科学》）

两种多糖对奶牛乳房灌注的刺激性试验

张 哲[1]，李新圃[1]，杨 峰[1]，
罗金印[1]，李晓强[2]，刘龙海[1]，李宏胜[1]
（1. 中国农业科学院兰州畜牧与兽药研究所/农业部兽用药物创制重点实验室/
甘肃省新兽药工程重点实验室，兰州 730050；2. 西藏奇正藏药股份有限公司）

摘 要：研究低聚糖硫酸酯和右旋糖酐硫酸酯两种多糖对奶牛乳房灌注的刺激性，为其临床应用提供参考。将 0.025g/mL 的两种多糖溶液分别给奶牛乳房灌注 10~50mL，连续 3d，每天观察奶牛体温、食欲、过敏反应、乳房性状（红、肿、热、痛等症状）、乳汁外观性状，并以 LMT 法对乳汁进行检测。试验结果表明：奶牛乳房连续 3d 灌注 2.5%的两种多糖液体 10~15mL，奶牛乳房及乳汁未出现任何不良反应，但当乳房连续灌注 20~50mL 多糖溶液后，部分奶牛乳房出现红肿、产奶量下降及乳汁变性等不良反应。说明两种多糖可用于奶牛乳房灌注，但安全使用范围为每个乳区每次灌注量不应超过 0.45g。

关键词：奶牛；乳房炎；低聚糖硫酸酯；右旋糖酐硫酸酯；乳房灌注；刺激性试验

（发表于《中国草食动物科学》）

速康解毒口服溶液对兔皮肤刺激性和过敏性的实验研究*

黄　鑫**[1]，郝宝成[1]，高旭东[1]，刘建枝[2]，王保海[2]，梁剑平[1]***

（1. 中国农业科学院兰州畜牧与兽药研究所/农业部兽用药物创制重点实验室/甘肃省新兽药工程重点实验室，兰州　730050；
2. 西藏自治区农牧科学院畜牧兽医研究所，拉萨　850000）

摘　要：试验旨在为评价动物疯草中毒解毒治疗药物——速康解毒口服溶液的临床前应用安全性，为临床推广应用提供实验依据。用速康解毒口服溶液、赋形剂（蒸馏水）局部涂抹大白兔破损皮肤及完整皮肤，并逐日观察症状表现，统计观察结果。用速康解毒口服溶液或赋形剂、2，4-二硝基氯苯（DNCB）对大白兔多次致敏，并于2周后观察大白兔的反应。结果表明：速康解毒口服溶液对家兔完整皮肤刺激反应的平均分值为零，属于无刺激药物，对破损皮肤也无明显刺激；在过敏性实验中，给药组5只家兔中1只有轻微的过敏现象，结合致敏反应评价的分值法和致敏反应发生率，速康解毒口服溶液对家兔皮肤过敏反应的分值为0.2（1/5），致敏率为20%（1/5），具有弱致敏性。结论：速康解毒口服溶液对家兔皮肤无明显刺激性，有轻微过敏现象，属于临床安全药物。

关键词：速康解毒口服溶液；皮肤刺激性；皮肤过敏性；安全性

（发表于《中国草食动物科学》）

* 基金项目：国家高新技术研究发展计划（863计划）（2011AA10A214）；西藏横向委托项目“动物疯草中毒解毒新制剂的中试生产”（2013-2014）。

** 作者简介：黄鑫（1989—　），男，在读硕士研究生；

*** 通讯作者：梁剑平（1962—　），男，博士生导师。

西部某省羊养殖过程风险隐患调研及羊肉产品风险因子分析

高雅琴，杜天庆，李维红，
熊　琳，郭天芬，杨晓玲，
王宏博，周绪正，杨亚军，靳亚娥

（农业部畜产品质量安全风险评估实验室（兰州）/
中国农业科学院兰州畜牧与兽药研究所，兰州　730050）

摘　要：在西部某省羊主产区，以兽药和饲料预混料等投入品为研究重点，调查了不同季节、不同养殖方式、不同生长期羊疾病种类及治疗用药情况、饲料添加剂的使用情况以及养殖过程中其他的风险隐患情况，并采集40份羊肉样品，对其中可能存在的危害因子进行验证，以期探明羊养殖全过程中存在的质量安全风险隐患，为羊养殖场质量安全管控提供指南，为农产品质量安全监管提供参考。

关键词：羊肉；饲料及添加物；兽药残留；风险因子

（发表于《中国草食动物科学》）

乌锦颗粒剂的急性毒性与亚慢性毒性试验研究*

李胜坤**，王胜义，崔东安，王　慧，黄美州，齐志明，刘永明***

（中国农业科学院兰州畜牧与兽药研究所/甘肃省中兽药工程技术中心/农业部兽用药物创制重点实验室，兰州　730050）

摘　要：本研究旨在评价乌锦颗粒剂的毒性，为临床安全用药提供理论依据。试验分别以昆明小鼠和Wistar大鼠为研究对象，进行急性毒性试验和亚慢性毒性试验研究。急性毒性试验结果显示，乌锦颗粒剂的半数致死量（LD_{50}）>40g/kg体重，最大给药量为160g/kg体重，相当于临床用药量的80倍；在亚慢性毒性试验中，动物一般情

---

* 基金项目：公益性行业专项课题（20130304-17）；国家科技支撑计划项目（2012BAD12B03）。

** 作者简介：李胜坤（1987—　），男，河南开封人，硕士，研究方向：羔羊痢疾的防治，E-mail：lsk-99@163.com.

*** 通讯作者：刘永明（1957—　），男，甘肃景泰人，学士，研究员，主要从事动物营养与代谢病研究，E-mail：myslym@sina.com.

况正常，试验组 Wistar 大鼠增重和饲料消耗量与对照组相比无显著差异（$P>0.05$）；与对照组相比，高剂量组中雌鼠血清胆红素（T-BIL）、总蛋白（TP）、白蛋白（ALB）、尿素氮（BUN）和肌酐（CREA）及雄鼠 CREA 和肝脏指数均有显著差异（$P<0.05$），而其他指标与对照组相比差异均不显著（$P>0.05$）；低剂量组和中剂量组 Wistar 大鼠血常规、血液生化指标和脏器指数与对照组相比差异均不显著（$P>0.05$）；病理学检查发现高剂量组 Wistar 大鼠肝脏出现轻微颗粒变性，其他组 Wistar 大鼠组织结构清晰正常。结果表明，高剂量的乌锦颗粒剂能抑制肝脏对游离胆红素的摄入及蛋白质的合成；低剂量和中剂量乌锦颗粒剂此作用不明显。综合分析，乌锦颗粒剂临床用药是安全的。

关键词：乌锦颗粒剂；急性毒性；亚慢性毒性；安全评价

（发表于《中国畜牧兽医》）

牦牛卵泡液差异蛋白质组双向电泳图谱的构建与质谱分析*

郭　宪**，裴　杰，褚　敏，王宏博，丁学智，阎　萍***

（中国农业科学院兰州畜牧与兽药研究所/甘肃省牦牛繁育工程重点实验室，兰州　730050）

摘　要：从蛋白质水平了解牦牛季节性繁殖规律，利用双向电泳与质谱鉴定技术分析牦牛卵泡液与血浆蛋白质组分变化。以青海高原牦牛卵泡液与血浆为研究对象，采用双向电泳技术构建牦牛卵泡液与血浆蛋白质双向电泳图谱，银染后利用 Image Master 2DPlatinum 软件分析并采用 MALDI-TOF-MS 进行质谱鉴定。用试剂盒 ProteoExtractAlbumin/IgG Removal Kit 去除高丰度蛋白质后，利用 2-DE 技术获得了分辨率较高的卵泡液与血浆蛋白质电泳图谱，卵泡液与血浆蛋白质图谱对比分析共发现了 24 个差异表达蛋白质点，其中 2 个蛋白质点表达上调，22 个蛋白质点表达下调。经质谱分析，最终成功鉴定出 8 个蛋白质点、5 个未知蛋白质点。本研究成功构建了蛋白质图谱及分离鉴定的差异蛋白质，为从蛋白质水平揭示牦牛卵泡发育规律及了解卵母细胞发育的微环境提供了试验依据。

关键词：牦牛；卵泡液；血浆；差异蛋白质组；双向电泳；质谱鉴定

（发表于《中国畜牧兽医》）

---

* 基金项目：国家自然科学基金（31301976）；现代农业（肉牛牦牛）产业技术体系建设专项（CARS-38）；国家科技支撑计划项目子课题（2013BAD16B09）；中国农业科学院创新工程项目（CAAS-ASTIP-2014-LIHPS-01）。

** 作者简介：郭宪（1978—　），男，甘肃环县人，博士，副研究员，研究方向：动物遗传育种与繁殖，E-mail：guoxian@ caas. cn.

*** 通讯作者：阎萍（1963—　），女，山西临猗人，博士，研究员，博士生导师，研究方向：动物遗传育种与繁殖，E-mail：yanping@ caas.

世界奶牛乳房炎疫苗研究的知识图谱分析*

杨　峰**，王　玲，王旭荣，李新圃，罗金印，张世栋，李宏胜***

（中国农业科学院兰州畜牧与兽药研究所/农业部兽用药物创制
重点实验室/甘肃省新兽药工程重点实验室，兰州　730050）

摘　要：为了展示世界奶牛乳房炎疫苗领域的整体研究状况和研究热点，基于Citespace Ⅲ研究了1995—2015年Web of Science™核心合集数据库所收录的奶牛乳房炎疫苗相关的4 205条科技文献，分析了年度发文量，绘制了主要国家、机构、作者、被引期刊、被引文献及关键词的知识图谱。结果发现，奶牛乳房炎疫苗研究领域的年度发文量从1995至2012年持续上升，2013年发文量开始呈下滑的趋势；研究奶牛乳房炎疫苗的国家和机构主要分布在美洲、欧洲和亚洲，大学是该研究领域的主要研究机构；奶牛乳房炎疫苗研究领域的热点集中于：病原菌的分离鉴定、耐药性和致病机理；受染奶牛机体自身的免疫应答及相关的调控因子；预防和治疗乳房炎的最佳时期；有效抗原的筛选。该领域未来一段时期的研究方向将侧重于NF-κB和生物被膜的相关研究。

关键词：奶牛乳房炎；疫苗；知识图谱；Web of Science™核心合集；Citespace Ⅲ

（发表于《中国畜牧兽医》）

---

* 基金项目：甘肃省省青年科技基金计划项目（145RJYA311）；中央级公益性科研院所基本科研业务费专项资金项目（1610322015007）；“十二五”国家科技支撑项目（2012BAD12B03）。

** 作者简介：杨峰（1985—　），男，陕西榆林人，硕士，助理研究员，主要从事奶牛乳房炎的研究，E-mail：yangfeng@ caas. cn.

*** 通讯作者：李宏胜（1964—　），男，甘肃兰州人，博士，研究员，主要从事奶牛乳房炎的研究，E-mail：lihsheng@ sina. com.

伊维菌素微乳制剂的过敏性研究*

邢守叶[1,2]**，周绪正[1]，李　冰[1]，牛建荣[1]，魏小娟[1]，张继瑜[1]***

（1. 中国农业科学院兰州畜牧与兽药研究所/农业部兽用药物创制重点实验室，兰州　730070；2. 甘肃农业大学动物医学院，兰州　730050）

摘　要：本试验旨在观察伊维菌素微乳制剂对动物的过敏性反应。以《化学药物刺激性、过敏性和溶血性研究技术指导原则》为指导，选用豚鼠及Wistar大鼠为试验动物，采用全身主动过敏试验和被动皮肤过敏试验观察伊维菌素微乳制剂对动物的致敏作用。结果显示伊维菌素微乳制剂对豚鼠和Wistar大鼠的过敏性反应均为阴性。本试验为伊维菌素微乳制剂的安全性评价提供了试验依据。

关键词：伊维菌素微乳制剂；豚鼠；大鼠；过敏性试验

（发表于《中国畜牧兽医》）

家畜不孕证辨证施治

王华东****

（中国农业科学院兰州畜牧与兽药研究所，兰州　730050）

不孕是成年母畜不发情或发情后经多次配种难于受孕的一类繁殖障碍性疾病。临床上适龄母畜屡配不孕，或产1~2胎后不能再受孕者，均称为不孕症，也称为难孕症。目前以牛、猪不孕症对临床生产影响最严重，伴侣动物如犬不孕症也较常见。

（发表于《中兽医》）

---

* 基金项目：国家现代农业肉牛牦牛产业技术体系建设专项资金（CARS-38）；重点牧区生产生态生活保障技术集成与示范（2012BAD13BOO）。

** 作者简介：邢守叶（1988—　），女，山东人，硕士，研究方向：兽医药理学与毒理学。

*** 通讯作者：张继瑜（1967—　），男，甘肃人，研究员，博士生导师，从事兽医药理与毒理学的研究。E-mail：infzjy@ sina. com.

**** 作者简介：王华东（1979—　），男，硕士，助理研究员，E-mail：whd1979@ 126. com.

丹翘灌注液抗炎镇痛作用的研究*

魏立琴[1,2]**，王东升[2]，苗小楼[2]，张世栋[2]，
董书伟[2]，邝晓娇[2]，杨志强[2]，余四九[1]，严作廷[2]***

（1. 甘肃农业大学动物医学院，兰州 730070；
2. 中国农业科学院兰州畜牧与兽药研究所/农业部兽用药物创制
重点实验室/甘肃省中兽药工程技术研究中心，兰州 730050）

摘 要：研究丹翘灌注液的抗炎镇痛作用。采用二甲苯致小鼠耳廓肿胀、棉球致小鼠肉芽肿、小鼠醋酸扭体反应、小鼠福尔马林致痛模型观察丹翘灌注液的抗炎和镇痛作用。40g/kg、20g/kg、10g/kg 的丹翘灌注液能不同程度的抑制二甲苯引起的小鼠耳廓肿胀、棉球引起的肉芽肿增生；不同程度的减少醋酸引起的小鼠扭体次数和福尔马林引起的舔足时间。结果表明，丹翘灌注液具有抗炎、镇痛的作用。

关键词：丹翘灌注液；抗炎；镇痛；药理作用

（发表于《中国农学通报》）

---

* 基金项目：“十二五”国家科技支撑计划项目“奶牛健康养殖重要疾病防控关键技术研究”（2012BAD12B03）；中国农业科学院科技创新工程项目“奶牛疾病防控”；中央级公益性科研院所基本科研业务费项目“奶牛疾病诊断与防治”（1610322014028）。

** 第一作者简介：魏立琴，女，1986 年出生，甘肃兰州人，硕士。E-mail：24weiliqin@ 163. com.

*** 通讯作者：严作廷，男，1962 年出生，甘肃武威人，研究员，博士，主要从事中兽医临床和奶牛疾病防治的研究。E-mail：yanzuoting@ caas. cn.

藿芪灌注液毒性研究*

王东升**，张世栋，董书伟，苗小楼，
魏立琴，邝晓娇，那立冬，闫宝琪，严作廷***

（中国农业科学院兰州畜牧与兽药研究所/农业部兽用药物
创制重点实验室/甘肃省中兽药工程技术研究中心/甘肃省新兽药
工程重点实验室，兰州　730050）

摘　要：为了测定小鼠口服藿芪灌注液的急性毒性和长期毒性，评价其安全性，为临床安全用药提供依据，采用最大给药量法进行了急性毒性研究，选择健康小鼠20只，24h内按40g/kg体重灌胃给药3次，给药后连续观察10天，测定其1日最大给药量，确定藿芪灌注液的急性毒性。将80只健康大鼠随机分为高、中、低剂量组和对照组，每组雌雄各10只，高、中、低剂量组分别按20g/kg、10g/kg、3g/kg体重灌服藿芪灌注液，对照组按20mL/kg灌服生理盐水，每天一次，连续给药35天，通过临床观察、病理组织学检查、血液生理生化指标测定，研究其长期毒性。结果表明，小鼠口服藿芪灌注液的1日最大给药量为120g/kg，相当于临床奶牛日用量的600倍。藿芪灌注液不影响大鼠的采食、活动、饮水、增重、脏器指数、血液生理生化指标，不会引起大鼠发病和死亡。说明藿芪灌注液实际无毒，至少在20g/kg体重给药剂量下大鼠连续给药35天是安全的。

关键词：藿芪灌注液；急性毒性；长期毒性；最大给药量；小鼠；大鼠

（发表于《中国农学通报》）

---

* 基金项目："十二五"国家科技支撑计划项目"奶牛健康养殖重要疾病防控关键技术研究"（2012BAD12B03）；中国农业科学院科技创新工程项目"奶牛疾病"。

** 第一作者简介：王东升，男，1979年出生，甘肃临洮人，助理研究员，硕士，主要从事中药药理与奶牛繁殖疾病的研究。E-mail：lzmyswds@ 126. com.

*** 通讯作者：严作廷，男，1962年出生，甘肃武威人，研究员，博士，主要从事中兽医临床和奶牛疾病防治的研究。E-mail：yanzuoting@ caas. cn.

西藏“一江两河”地区紫花苜蓿生产性能灰色关联综合评价*

杨　晓[1**]，李锦华[1***]，朱新强[1]，余成群[2]

（1. 中国农业科学院兰州畜牧与兽药研究所，兰州　730050；
2. 中国科学院地理科学与自然资源研究所，北京　100101）

摘　要：针对西藏“一江两河”高海拔地区豆科牧草种植需求，应用灰色关联综合分析，用总生长高度、茎叶比、鲜干比、越冬率、鲜草产量等指标构建品种评价模型，对20个国内外紫花苜蓿品种在该地区的生产性能进行比较，结果表明参试品种中的综合表现较好的有“雷西斯”“苜蓿王”“中兰1号”“德宝”“赛特”等品种，而“阿尔冈金”“杜普梯”“皇后”“博勒维”等品种生产性能较低。评价指标中所占的权重顺序为茎叶比>总生长高度>鲜干比>越冬率>鲜草产量。

关键词：“一江两河”；紫花苜蓿；灰色关联度；综合评价；生产性能

（发表于《中国农学通报》）

* 基金项目：西藏自治区科技专项“西藏主要栽培豆科牧草繁育研究与示范”（Z2013C02N02），“抗霜霉病苜蓿品种的示范与推广”（2012F4001002）。

** 第一作者简介：杨晓，男，1985年出生，助理研究员，硕士，从事牧草栽培与育种研究。E-mail：yx760@sina.com.

*** 通讯作者：李锦华，男，1963年出生，副研究员，博士，从事牧草栽培与育种研究。E-mail：ljh63814@sina.com.

奶牛蹄病之疣性皮炎的防治*

董书伟[1**]，杨小伟[2]，张世栋[1]，王东升[1]，
王　慧[1]，苗小楼[1]，严作廷[1***]，杨志强[1****]
（1. 中国农业科学院兰州畜牧与兽药研究所/农业部兽用药物创制重点实验室/
甘肃省新兽药工程重点实验室；2. 宁夏回族自治区吴忠市利通动物卫生监督所）

摘　要：疣性皮炎是奶牛常见的一种蹄病，患牛生产性能下降，被迫过早淘汰，给奶牛养殖造成了巨大的经济损失。本文就奶牛疣性皮炎的病因、症状、治疗和预防方法进行了综述。

关键词：奶牛；疣性皮炎；蹄病；防治

（发表于《中国乳业》）

菌毒清口服液治疗人工感染鸡传染性鼻炎试验

牛建荣，张继瑜，周绪正，李剑勇，
李　冰，魏小娟，杨亚军，刘希望
（中国农业科学院兰州畜牧与兽药研究所/农业部兽用药物创制重点实验室/
甘肃省新兽药工程重点实验室，兰州　730050）

摘　要：为了观察“菌毒清”口服液对人工感染鸡传染性鼻炎的治疗效果与现有的治疗鸡呼吸道疾病的中药“瘟毒清”口服液疗效进行比较。试验分为7组，包括2.0%、1.5%、1.0%和0.5%“菌毒清”饮水治疗组，1.0%“瘟毒清”饮水治疗组，阳性空白对照组和阴性空白对照组，鸡只自由饮用，连用7d为1疗程。经1个疗程的治疗，“菌毒清”治疗组的治愈率分别为69.23%、77.78%、81.48%和62.96%，“瘟毒清”组的治愈率为78.57%；“菌毒清”治疗组的有效率分别为88.46%、88.8 0%、94.81%和8 1.48%，“瘟毒清”治疗组的有效率为89.29%；不同剂量“菌毒清”治疗组的有

---

* 基金项目：“十二五”国家科技支撑计划项目（2012BAD12B03）；国家自然科学青年基金项目（31302156）；中央级基本科研业务费预算增量项目（2014ZL012）。

** 作者简介：董书伟（1980—　），男，河南人，博士生，助理研究员，主要从事奶牛疾病研究；E-mail：dongshuwei@caas.com.

*** 严作廷（1962—　），男，博士，研究员；E-mail：yanzuoting@caas.cn.

**** 通讯作者：杨志强（1957—　），男，博导，研究员；E-mail：yangzhiqiang@caas.cn.

效率与“瘍毒清”治疗组差异不显著，但不同药物治疗组与阳性空白对照组差异显著($P<0.05$)。表明“菌毒清”是治疗鸡传染性鼻炎的有效药物之一。
关键词：“菌毒清”口服液；鸡传染性鼻炎；疗效

（发表于《中国兽医杂志》）

HPLC 法测定丹翘灌注液中丹参酮 II_A 的含量

王东升，苗小楼，张世栋，董书伟，
魏立琴，那立冬，闫宝琪，严作廷

（中国农业科学晓兰州畜牧与兽药研究所/农业部兽用药物创制重点实验室/甘肃省中兽药工程技术研究中心/甘肃省新兽药工程重点实验室，兰州 730050）

摘 要：测定丹翘灌注液中丹参酮Ⅱ$_A$含量。采用高效液相色谱法，色谱柱：Kromasil：100-5C_{18}（4.6mm×250mm，5μm）流动相：甲醇-水（75：25），流速：1.0mL/min，柱温：30℃，检测波长：270nm。结果：丹参酮Ⅱ$_A$在0.217 6～2.176 0μg范围内线性关系良好（r=0.979 7），丹参酮Ⅱ$_A$的平均回收率为100.03%，RSD为0.57%。丹翘灌注液中丹参酮Ⅱ$_A$含量为0.979 7mg/mL。结果表明采用高效液相色谱法测定丹翘灌注液中丹参酮Ⅱ$_A$含量简便、快速、灵敏、准确，可用于该制剂中丹参酮Ⅱ$_A$的含量测定。
关键词：丹翘灌注液；丹参酮Ⅱ$_A$；高效液相色谱法；含量测定

（发表于《中兽医医药杂志》）

板黄口服液的薄层鉴别研究*

刘希望**，史梓萱，杨亚军，周绪正，
李　冰，牛建荣，魏小娟，李剑勇，张继瑜***

（中国农业科学院兰州畜牧与兽药研究所/农业部兽用药物创制重点实验室/甘肃省新兽药工程重点实验室，兰州　730050）

摘　要：为了建立板黄口服液的质量标准，采用薄层色谱法对组方中黄连、黄芩、金银花、知母、牛蒡子药材进行定性鉴别。试验结果表明，黄连、黄芩、金银花、知母、牛蒡子的特征成分在与对照品和对照药材色谱相应位置上显示相同颜色的斑点，阴性对照无干扰。提示所建立的方法简便、专属性强、重复性好，可为板黄口服液的质量控制提供依据。

关键词：板黄口服液；薄层色谱法；定性鉴别；质量标准

（发表于《中兽医医药杂志》）

* 基金项目：公益性行业（农业）科研专项经费项目（201303040-12）；兰州市科技计划项目（2012-2-73）。

** 作者简介：刘希望（1985—　），男，助理研究员，主要从事新兽药研发，E-mail：xiwangliu@126.com.

*** 通讯作者：张继瑜（1967—　），男，研究员，主要从事新兽药研发与药理学研究。

樗白皮提取物对腹泻模型小鼠抗氧化能力的影响*

杨　欣[1,2]**，张　霞[2]，刘　宇[1]，
王华东[1]，张继瑜[1]，周绪正[1]，程富胜[1]***

（1. 中国农业科学院兰州畜牧与兽药研究所/农业部兽用药物创制重点实验室/甘肃省新兽药工程重点实验室，兰州　730050；
2. 甘肃农业大学生命科学学院，兰州　730070）

摘　要：为研究腹泻病理状态下樗白皮提取物对机体氧化还原状态的调节功能，采用番泻叶提取物建立小鼠药物性腹泻模型，经不同剂量樗白皮提取物治疗，对小鼠血清中MDA、SOD、GSH、GSSG、GSH-Px等抗氧化指标进行测定。结果显示，模型组MDA含量较正常组显著升高（$P<0.05$），樗白皮提取物不同剂量治疗组均低于模型组，与治疗对照组无显著差异；模型组SOD酶活性显著低于其他各组（$P<0.01$），正常对照组与其他实验组之间无显著差异；模型组GSH含量显著低于正常对照组和治疗对照组（$P<0.05$），高剂量组和中剂量组高于模型组（$P>0.05$），低剂量组与模型组无显著差异；正常对照组GSSG含量显著低于其他各组（$P<0.01$），模型组显著高于治疗对照组和高剂量组（$P<0.01$），高于中剂量组和低剂量组（$P>0.05$）；GSH-Px模型组均低于其他各组，除与低剂量组无显著差异外，与其他各组差异均极显著（$P<0.01$）。结果表明，樗白皮提取物能够降低腹泻模型小鼠体内氧化性物质含量，提高还原酶活性及促进还原性物质生成，对机体的氧化还原稳态具有一定的调节作用，并且与给药剂量之间存在一定的量效关系。

关键词：樗白皮提取物；腹泻，抗氧化；番泻叶；小鼠

（发表于《中兽医医药杂志》）

---

* 基金项目：农业部现代农业肉牛牦牛体系（CARS-38）；科技基础性工作专项（2013FY110600-04）；甘肃省科技支撑项目（1204NKCA088）；兰州市科技计划项目（2010-1-151）。

** 作者简介：杨欣（1988—　），女，在读硕士，研究方向：天然药物免疫药理学。

*** 通讯作者：程富胜，副研究员，E-mail：chengfusheng@126.com.

两种多糖对奶牛乳房炎常见病原菌体外抑菌效果研究

张　哲，李新圃，杨　峰，罗金印，刘龙海，李宏胜

（中国农业科学院兰州畜牧与兽药研究所/农业部兽用药物创制重点实验室/甘肃省新兽药工程重点实验室，兰州　730050）

摘　要：研究低聚糖硫酸酯和右旋糖酐硫酸酯两种多糖对奶牛乳房炎常见病原菌的体外抑菌效果。采用杯碟法对两种多糖及青霉素对于乳房炎常见的8种病原菌进行了体外抑菌活性研究。试验结果表明，1%的低聚糖硫酸酯及2.5%和5%的右旋糖酐硫酸酯对金黄色葡萄球菌表现为高度敏感；2.5%的低聚糖硫酸酯对金黄色葡萄球菌表现为中度敏感；5%和2.5%的两种多糖对无乳链球菌均表现为中度敏感；但两种多糖对其余6种病原菌均无抑菌作用。5%的青霉素对8种病原菌均表现为高度敏感。

关键词：低聚糖硫酸酯；右旋糖酐硫酸酯；奶牛乳房炎；病原菌；抑菌试验

（发表于《中兽医医药杂志》）

犬体针灸穴位图的26处订正探讨*

罗超应**，李锦宇，王贵波，谢家声，罗永江，辛蕊华，郑继方

（中国农业科学院兰州畜牧与兽药研究所，兰州　730050）

摘　要：鉴于目前犬体针灸穴位主要文献中的文字记述与图示标记有些不妥之处，尤其是某些教学设备公司生产的犬体针穴模型上的穴位标示错误众多，笔者在对《兽医针灸手册》《中国兽医针灸学》与《中国兽医针灸图谱》等兽医针灸学著作进行比较研究的基础上，结合自己的应用实践经验，对其中26处穴位标记错误进行订正分析与探讨，以为后学提供更加准确与直观的应用参考。

关键词：犬；针灸；穴位；图示/模型；订正

（发表于《中兽医医药杂志》）

* 基金项目：科技部科技基础工作专项（2013FY110600-8）。

** 作者简介：罗超应（1960—　），男，研究员，主要从事中兽医学研究。

血液流变学在奶牛蹄叶炎研究中的应用探讨*

董书伟[1]**，李　巍[2]，张世栋[1]，
王东升[1]，苗小楼[1]，严作廷[1]***，杨志强[1]***

（1. 中国农业科学院兰州畜牧与兽药研究所/农业部兽用药物
创制重点实验室/甘肃省新兽药工程重点实验，兰州　730050；
2. 河南牧业经济学院，郑州　450000）

摘　要：蹄叶炎是奶牛常见的营养代谢病之一，血液流变学是研究营养代谢病和微循环障碍的有效方法。文章总结了奶牛蹄叶炎和血液流变学的主要研究进展，并探讨了血液流变学在该病研究中的应用前景。

关键词：蹄叶炎；血液流变学；微循环障碍；奶牛

（发表于《中兽医医药杂志》）

仔猪腹泻研究现状浅析****

杨　欣[1,2]*****，张　霞[2]，刘　宇[1]，王华东[1]，程富胜[1]******

（1. 中国农业科学院兰州畜牧与兽药研究所/农业部兽用药物创制
重点实验室/甘肃省新兽药工程重点实验室，兰州　730050；
2. 甘肃农业大学生命科学学院，兰州　730070）

摘　要：本文就仔猪腹泻疾病和仔猪腹泻对养猪业带来的危害进行了阐述和总结，针对不同类型的腹泻和引起腹泻发生的不同原因进行详细的分析和归纳，在传统防治仔猪腹泻方法的基础上，提出和探讨了防治仔猪腹泻的新观点和新方法，为养猪业和相关仔猪腹泻的防治提供新思路。

---

* 基金项目：国家自然科学青年基金项目（31302156）；中央级公益性科研院所基本科研业务费专项资金项目（1610322014012）；“十二五”国家科技支撑计划项目（2012BAD12B03）。

** 作者简介：董书伟（1980—　），男，在读博士，助理研究员，主要从事动物疾病蛋白质组学研究，E-mail：dongshuwei@ caas. com.

*** 通讯作者：严作廷（1962—　），男，博士，研究员，E-mail：yanzuoting@ caas. cn；杨志强（1957—　），男，博导，研究员，E-mail：yangzhiqiang@ caas. cn.

**** 基金项目：甘肃省科技计划支撑项目（1204NKCA088）；兰州市科技计划项目（2010-1-151）。

***** 作者简介：杨欣（1988—　），女，在读硕士研究生，研究方向：天然药物免疫药理学。

****** 通讯作者：程富胜，副研究员，E-mail：chengfusheng@ 126. com.

关键词：仔猪；腹泻；现状

（发表于《中兽医医药杂志》）

正交设计优化常山碱的醇提工艺*

郭志廷**[1]，梁剑平[1]，刘　宏[2]，刘志国[2]，徐海城[2]，尚若锋[1]

（1. 中国农业科学院兰州畜牧与兽药研究所/农业部兽用药物创制
重点实验室/甘肃省新兽药工程重点实验室，兰州　730050；
2. 石家庄正道动物药业有限公司，石家庄　050200）

摘　要：［目的］优化常山碱的乙醇回流提取工艺。［方法］采用紫外-可见分光光度计测定常山碱含量，以浸膏得率和常山碱含量作为评价指标，应用正交设计法优化提取工艺参数。［结果］常山碱乙醇回流提取最佳工艺：70%乙醇、提取时间1.5 h、溶媒用量7.5倍，提取2次，在此条件下，浸膏得率为10.43%，浸膏中常山碱含量为4.26%。方差分析表明，对常山碱提取率影响因素的大小顺序：提取次数>溶媒用量>提取时间>乙醇浓度。工艺验证试验表明，3批常山的平均浸膏得率和平均常山碱含量分别为11.27%和4.11%。结论：常山碱采用乙醇回流提取工艺切实可行，具有省时、操作简便和提取率高等优点。

关键词：常山；常山碱；乙醇回流；正交设计

（发表于《中兽医医药杂志》）

重离子辐照截短侧耳素产生菌的诱变选育

王学红，郝宝成，赵晓斌，刘宇，梁剑平

（中国农业科学院兰州畜牧与兽翁研究所/农业部兽用兹物创制
重点实验室/甘肃省新兽药程重点实验室，兰州　730050）

摘　要：利用重离子辐照对截短侧耳素产生菌进行诱变，筛选高产菌株，采用$^{12}G^{6+}$重离子束，剂量分别为20 Gy、30Gy、4 0 Gy、50 Gy。结果显示重离子辐照的最适剂量为40Gy。筛选出一株截短侧耳素高产菌株K 40-3，摇瓶效价达到16120μg/mL，较出发菌株提高25.3%。

* 基金项目：中央级公益性科研院所专项资金项目（1610322011004）。
** 作者简介：郭志廷（1979—　），男，助理研究员，主要从事中药免疫学和药理学研究。

关键词：截短侧耳素产生菌；重离子；辐照；筛选

（发表于《中兽医医药杂志》）

藏药牦牛角有效成分的分离与解热镇痛物质基础的研究*

王鸣刚[1,2**]，李晓茹[2]，张新国[2]，相炎红[2]

（1. 中国农业科学院兰州畜牧与兽药研究所/甘肃省牦牛繁育工程重点实验室，兰州 730050；2. 兰州理工大学生命学院，兰州 730050）

摘　要：[目的] 寻找藏药材牦牛角中解热镇痛物质。[方法] 实验以牦牛角为原料，经回流提取得到粗提液，再经过透析和非变性电泳分离，得到不同分子量大小的多肽；采用皮下酵母注射法建立小鼠发热模型，热板法评价镇痛活性。[结果] 牦牛角回流粗提取液经透析、非变性电泳分离后，得到的小分子多肽在给药 30min 后能显著降低肛温，达到最大痛阈，并提高值到 67%。[结论] 研究证实牦牛角粗提取液有明显的解热和镇痛作用，为进一步获得新的多肽类解热镇痛候选药物奠定基础。

关键词：牦牛角；分离；凝胶电泳；小分子多肽；解热镇痛

（发表于《中医药学报》）

喷施硼肥对黄土高原紫花苜蓿产量的影响***

陈　璐[1****]，胡　宇[2]，路　远[2*****]，周　恒[2]，张　茜[2]，田福平[2]

（1. 甘肃农业大学草业学院，兰州 730070；2. 中国农业科学院兰州畜牧与兽药研究所/农业部兰州黄土高原生态环境重点野外科学观测试验站，兰州 730050）

摘　要：为研究硼肥对黄土高原紫花苜蓿(*Medicago Sotiva* L.)地上生物量和种子产量的影响,试验于 2014 年 5 月 28 日苜蓿开花前在中兰 1 号紫花苜蓿叶面分别施用硼含量为 20.5%和

---

* 基金项目：甘肃省牦牛繁育工程重点实验室开放课题资助。

** 作者简介：王鸣刚（1962— ），男，教授，主要从事分子生物学研究。

*** 基金项目：国家自然科学基金面上项目（31372368）；公益性行业（农业）科研专项（201203006）；“十二五”农村领域国家科技计划课题（2012BADl3807）；中央级公益性科研院所基本科研业务费专项资金项目（1610322014009）。

**** 作者简介：陈璐（1990— ），女，在读硕士研究生。

***** 通讯作者：路远（1980— ），女。助理研究员，硕士，从事草地生态研究。

15.0%的硼肥。两种硼肥分别设置4个施肥量梯度和2个对照，共10个处理，进行施肥试验。研究结果表明：喷施含硼量分别为20.5%和15.0%的两种硼肥，对8年生紫花苜蓿草产量及种子产量均有明显增产作用。当喷施肥量达到3 000g/hm^{2}时，硼含量为20.5%的硼肥对紫花苜蓿鲜草产量和干草产量增产效果显著；硼含量为15.0%的硼肥可使紫花苜蓿的种子产量显著增加。说明喷施硼肥可以增加紫花苜蓿的鲜草产量、干草产量及种子产量。
关键词： 紫花苜蓿；硼肥；草产量；种子产量

（发表于《中国草食动物科学》）

常山、常山碱及其衍生物防治鸡球虫病的研究进展*

郭志廷[1,2**]，梁剑平[1]，韦旭斌[2]

（1. 中国农业科学院兰州畜牧与兽药研究所/农业部兽用药物创制重点实验室/甘肃省新兽药工程重点实验室，兰州 730050；
2. 吉林大学 动物医学学院，长春 130062）

摘 要： 鸡球虫病（coccidiosis in chicken）是一种全球流行、无季节性、高发病率和高死亡率的肠道寄生性原虫病[1]。据报道，全世界每年因鸡球虫病造成高达30亿美元的经济损失，中国每年用于防制鸡球虫病的药物费用高达数亿元人民币，占鸡病全部防制费用的1/3[2]。目前，国内外对于鸡球虫病的防治主要依靠化学药物，如地克珠利、妥曲珠利和常山酮等，它们对现代养禽业的健康发展贡献巨大[3]。随着球虫耐药性的日趋严重、免疫预防尚未普遍应用和人们健康意识的不断加强，迫切需要安全、高效、环保的抗球虫新药研发。中兽药就是将中医药理论应用于动物身上，中药既可预防和治疗疾病、促进动物生长，又不会产生危害人体健康的药物残留，近年逐渐成为科研工作者研究的热点。
关键词： 常山；常山碱；常山酮；抗球虫

（发表于《中国兽医学报》）

* 基金项目：中央级公益性科研院所专项资金资助项目（1610322011004）。
** 作者简介：郭志廷（1979— ），男，助理研究员，硕士。

微量元素硒、铜、锌在饲料添加应用中存在的问题与对策*

王 慧[1**]，黄美州[1]，王胜义[1]，李胜坤[1]，
崔东安[1]，齐志明[1]，张 林[2]，刘永明[1***]

（1. 中国农业科学院兰州畜牧与兽药研究所/农业部兽用药物创制重点实验室/
甘肃省新兽药工程重点实验室/甘肃省中兽药工程技术研究中心，
兰州 730050；2. 西藏自治区山南地区畜牧兽医总站，山南 856000）

摘 要：微量元素是机体不可缺少的营养物质，饲粮中添加一定量的微量元素对畜禽的健康具有重要的作用，但是添加过多或过少都会对机体造成一定影响。微量元素的应用从无机盐发展到微量元素有机酸和氨基酸微量元素，由于无机微量元素吸收效果较差且易对环境造成污染，近年来有机微量元素已成为国内外动物营养学家的研究热点。本文就硒、铜、锌的生物学功能，缺乏或过量时对机体的影响，以及有机微量矿质元素的特性作一综述。

关键词：微量元素；生物学功能；饲料添加剂；毒性

（发表于《畜牧与兽医》）

藏药研究概况****

曹明泽*****，孔小军，王 磊，张景燕，
王旭荣，秦 哲，孟嘉仁，李建喜，王学智*

（中国农业科学院兰州畜牧与兽药研究所/农业部兽用药物创制重点
实验室/甘肃省新兽药工程重点实验室，兰州 730050）

摘 要：藏药作为我国丰富的民族药物资源，已有2 000多年的历史。由于生长环境的独

---

* 基金项目：中央级科研院所基本科研业务费（1610322013003）；农业科技成果转化资金项目（2010GB23260564）。

** 作者简介：王慧（1985— ），男，硕士，研究实习员。

*** 通讯作者：刘永明，研究员，硕士生导师，主要从事中兽药开发和微量元素代谢病研究，E-mail：myslym@ sina. com.

**** 基金项目：国家公益性行业专项课题（201303040-14）；国家公益性行业科研专项（201203008-2）。

***** 作者简介：曹明泽（1987— ），女，河北石家庄人，硕士研究生，主要从事临床兽医学研究。* 通讯作者

特性，赋予了藏药不可替代的疗效和医用价值。它以其传统的治疗方法，发挥着自身的特点和优势，尤其对某些特殊的疑难病证具有特定的疗效。近年来，随着人们对藏药材的不断开发研究，其产品越来越受到人们的认可，但藏药的药物剂型、化学成分及药理作用研究都还处于相对较低的水平，与现代化的要求还有很大差距，需要通过新技术不断的完善。论文概述了藏药的发展史和藏医药特点，并从藏药药材资源、化学成分、药理作用、药物开发4个方面综述了藏药的研究概况，为更好的研究藏药提供数据和资源。
关键词：藏药；藏药资源；化学成分；药理学；药物开发

（发表于《动物医学进展》）

牦牛角性状与相关基因的研究进展*

赵娟花[1,2]**，梁春年[1,2]，裴　杰[1,2]，
郭　宪[1,2]，褚　敏[1,2]，吴晓云[1,2]，阎　萍[1,2]***
（1. 中国农业科学院兰州畜牧与兽药研究所，兰州　730050；
2. 甘肃省牦牛繁育工程重点实验室，兰州　730050）

摘　要：牦牛是能在青藏高原恶劣条件下生活和繁衍的牛种。由于牦牛具有较强的野性，易发生争斗，且可能对饲养人员造成伤害，所以给牦牛的饲养管理带来诸多困难。牦牛的培育方向除了提高生产性能、品质性状外，无角性状也是其中一个重要的选育方向。无角牦牛有利于规模化饲养管理，也可避免由于意外伤害造成的经济损失。文章主要从微卫星标记、基因芯片、单核苷酸多态性(SNP)方法和角型性状基因的表达研究四个方面介绍了近年来牦牛角性状的研究进展。
关键词：牦牛；角性状；基因定位；多态性；基因表达；基因芯片；研究进展

（发表于《黑龙江畜牧兽医》科技版）

---

* 基金项目：甘肃甘南牧区“生产生态生活”保障技术集成与示范项目（2012AD13B05）；中国农业科学院牦牛种质资源创新与利用创新工程项目。

** 作者简介：赵娟花（1989—　），女，硕士研究生，研究方向为动物遗传育种，E-mail：zjh326303202@163.com。

*** 通讯作者：阎萍（1963—　），女，研究员，博士，研究方向为动物遗传育种，E-mail：pingyanlz@163.com.

丁香酚在畜牧业生产中的应用 *

李世宏 **，杨亚军，刘希望，孔晓军，秦　哲，李剑勇

（中国农业科学院兰州畜牧与兽药研究所/农业部兽用药物创制重点实验室/甘肃省新兽药工程重点实验室，兰州　730050）

摘　要：丁香是一种天然的食药两用植物，在食品、医药和畜牧等领域有广泛的应用。文章综述了丁香酚在动物疾病的防治、作为饲料添加剂及水产动物麻醉剂等方面的应用现状，并对其前景进行了展望，为将来在畜牧业上更加广泛地利用丁香酚提供有价值的参考。

关键词：丁香酚；防治；疾病；饲料添加剂；麻醉剂

（发表于《中国草食动物科学》）

奶牛乳房炎综合防控关键技术研究进展***

刘龙海[1]****，李新圃[1]，杨　峰[1]，

罗金印[1]，王旭荣[1]，张　哲[1,2]，李宏胜[1]*****

（1. 中国农业科学院兰州畜牧与兽药研究所/农业部兽用药物创制重点实验室/甘肃省新兽药工程重点实验室，兰州　730050；2. 甘肃农业大学动物医学院）

摘　要：奶牛乳房炎是一种给奶牛养殖业造成极大经济损失的常见多发病。目前的研究表明，采取综合防控技术是控制本病最有效的方法。文章综述了奶牛乳房炎发病情况、病因、发生规律及造成的危害、引起奶牛乳房炎的主要病原菌及其区系分布状况、乳房炎诊断方法、治疗药物及疫苗研制等最新研究进展，同时还对奶牛乳房炎综合防控技术的前景进行了展望，以期为更好地防治奶牛乳房炎提供借鉴和参考。

关键词：奶牛乳房炎；综合防控；病原菌；诊断治疗；疫苗

（发表于《中国草食动物科学》）

---

*　基金项目：中国农业科学院兽用化学药物创新工程基金。

**　作者简介：李世宏（1974—　），男，副研究员。研究方向：兽药药理和兽医临床。

***　基金项目："十二五"国家科技支撑计划（2012BAD12B03）；甘肃省科技支撑计划（144NKCA240）；甘肃省农业科技创新项目（GNCX-2013-59）；中国农业科学院创新工程项目（奶牛疾病研究团队）。

****　作者简介：刘龙海（1990—　），男，硕士研究生。

*****　通讯作者：李宏胜（1964—　），男，研究员，博士。研究方向：主要从事奶牛乳房炎免疫及预防研究。

奶牛乳腺上皮细胞原代培养、纯化及鉴定技术研究进展*

林　杰**，王旭荣，李建喜，王孝武，杨志强***

（中国农业科学院兰州畜牧与兽药研究所/农业部兽用药物创制重点实验室/甘肃省中兽药工程技术研究中心，兰州　730050）

摘　要：奶牛乳腺上皮细胞不仅具有合成和分泌乳汁的功能，而且在乳腺的先天免疫中扮演着重要角色，对泌乳机制、乳房炎发病机制的研究，以及药物筛选具有重要意义。原代培养的奶牛乳腺上皮细胞适宜建立细胞模型，可作为生理、病理、药理等方面研究的良好介质，解决体内试验周期长、成本高、个体差异大的难题。作者主要从奶牛乳腺上皮细胞原代培养的发展历程、培养技术、纯化技术及鉴定方法等方面的最新研究情况进行综述，以期为奶牛乳腺上皮细胞培养相关研究提供参考。

关键词：奶牛；乳腺上皮细胞；原代培养；鉴定

（发表于《中国畜牧兽医》）

奶牛乳腺炎疫苗免疫佐剂研究进展****

张　哲*****，李新圃，杨　峰，
罗金印，王旭荣，刘龙海，李宏胜******

（中国农业科学院兰州畜牧与兽药研究所/农业部兽用药物创制重点实验室/甘肃省新兽药工程重点实验室，兰州　730050）

摘　要：奶牛乳腺炎是一种多因素导致的奶牛常见疾病，严重影响奶牛业的健康发展。

---

* 基金项目：国家奶牛产业技术体系（CARS37）。

** 作者简介：林杰（1991—　），女，河北唐山人，硕士，研究方向：动物疾病预防与控制，E-mail：qielinjie2009@163.com.

*** 通讯作者：杨志强（1958—　），男，四川武胜人，研究员，博士生导师，主要从事奶牛疾病相关研究，E-mail：zhiqyang2006@163.com.

**** 基金项目：“十二五”国家科技支撑项目（2012BAD12B00），甘肃省科技支撑计划（144NKCA240）；甘肃省农业科技创新项目（GNCX-2013-59）；甘肃省农业生物技术研究与应用开发项目（GNSW-2013-28）；中国农业科学院奶牛疾病研究创新团队项目。

***** 作者简介：张哲（1990—　），男，河南人，在读硕士，研究方向为主要从事奶牛乳房炎疫苗方向。

****** 通讯作者：李宏胜（1964—　），男，陕西人，博士，研究员，研究方向为主要从事奶牛乳房炎免疫及预防研究。

接种疫苗防治奶牛乳腺炎，被认为是一种最为经典有效的方法。佐剂作为疫苗的重要组成部分，对疫苗免疫效果及安全性起着至关重要的作用。探明疫苗佐剂的作用机制是研制有效乳腺炎疫苗的前提。本文从多个角度对不同佐剂的作用机制和安全性以及目前用于奶牛乳腺炎疫苗的佐剂进行了综述，同时对奶牛乳腺炎疫苗佐剂的研发及应用前景进行了展望，对进一步深入研究和开发高效的奶牛乳腺炎疫苗具有一定的借鉴作用。
关键词：奶牛；乳腺炎疫苗；免疫佐剂；作用机制

（发表于《中国奶牛》）

作者简介

杨志强（1957—　），中共党员，学士，二级研究员，博士生导师。甘肃省优秀专家，甘肃省第一层次领军人才，中国农业科学院跨世纪学科带头人，甘肃省“555”创新人才，《中兽医医药杂志》主编。兼任中国毒理学会兽医毒理学分会会长，中国畜牧兽医学会常务理事，中国兽医协会常务理事，中国畜牧兽医学会动物药品学分会副会长，中国畜牧兽医学会毒物学分会副会长，中国畜牧兽医学会中兽医学分会副会长，中国畜牧兽医学会西北地区中兽医学会理事长，农业部兽药评审委员会委员。长期从事中兽医药学、兽医药理毒理、动物营养代谢与中毒病等研究工作，是该领域内的知名专家，先后主持和参加国家、省、部级科研课题 33 项，其中主持 20 项，获奖 9 项，自主和参与研发新产品 8 个，获授权专利 3 项。先后培养硕士研究生 20 名，培养博士研究生 10 名。在国内和国际学术刊物上共发表学术论文 100 余篇，其中主笔发表论文 80 篇。主编和参与编写《微量元素与动物疾病》等学术专著 13 部。

刘永明（1957—　），中共党员，大学文化程度，三级研究员，硕士研究生导师。先后担任中国农业科学院中兽医研究所党委办公室副主任、主任，中国农业科学院兰州畜牧与兽药研究所人事处处长、副所长、党委副书记和纪委书记等职务，2001 年 7 月至今任中国农业科学院兰州畜牧与兽药研究所党委书记、副所长、工会主席，兼任《中兽医医药杂志》和《中国草食动物科学》杂志编委会主任、中国农业科学院思想政治工作研究会理事、中国兽医协会会员和兰州市科学技术奖励委员会委员等职务。主要从事动物营养与代谢病研究工作。先后主持国家科技支撑计划、公益性行业专项、科技成果转化基金项目、948 项目以及省级科研课题或子专题 12 项，主持基本建设项目 4 项，获授权专利 8 项，取得新兽药证书 1 个、添加剂预混料生产文号 5 个；主编（主审）、副主编著作 6 部，参与编写著作 6 部。

张继瑜（1967—　），中共党员，博士研究生，三级研究员，博（硕）士生导师，中国农业科学院三级岗位杰出人才，中国农业科学院兽用药物研究创新团队首席专家，国家现代农业产业技术体系岗位科学家。兼任中国兽医协会中兽医分会副会长，中国畜牧兽医学会兽医药理毒理学分会副秘书长，农业部兽药评审委员会委员，农业部兽用药物创制重点实验室常务副主任，甘肃省新兽药工程重点实验室常务副主任，中国农业科学院学术委员会委员，黑龙江八一农垦大学和甘肃农业大学兼职博导。主要从事

兽用药物及相关基础研究工作，重点方向包括兽用化学药物的研制、药物作用机理与新药设计、细菌耐药性研究。带领的研究团队在动物寄生虫病、动物呼吸道综合症防治药物研究上取得了显著进展。在肠杆菌耐药机理、血液原虫药物作用靶标筛选的研究处于领先地位。先后主持完成国家、省部重点科研项目20多项，获得科技奖励6项，研制成功4个兽药新产品，其中国家一类新药一个，以第一完成人申报10项国家发明专利，取得专利授权5项。培养研究生21名，发表论文170余篇，主编出版《动物专用新化学药物》和《畜牧业科研优先序》等著作2部。

阎萍（1963— ）中共党员，博士，三级研究员，博士生导师。2012年享受国务院特殊津贴，是中国农业科学院三级岗位杰出人才，甘肃省优秀专家，甘肃省“555”创新人才，甘肃省领军人才。曾任畜牧研究室副主任、主任等职务，2013年3月任研究所副所长职务。兼任国家畜禽资源管理委员会牛马驼品种审定委员会委员，中国畜牧兽医学会牛业分会副理事长，全国牦牛育种协作组常务副理事长兼秘书长，中国畜牧兽医学会动物繁殖学分会常务理事和养牛学分会常务理事等。阎萍研究员主要从事动物遗传育种与繁殖研究，特别是在牦牛领域的研究成绩卓越，先后主持和参加完成了科技部支撑计划、科技部基础性研究项目、科技部“863”计划、“948”计划、农业部行业科技项目、国家肉牛产业技术体系岗位专家、人事部回国留学基金项目、科技部成果转化项目、甘肃省科技重大专项计划、甘肃省农业生物技术项目等20余项课题。现为国家肉牛牦牛产业技术体系牦牛选育岗位专家，甘肃省牦牛繁育工程重点实验室主任。作为高级访问学者多次到国外科研机构进行学术交流。培育国家牦牛新品种1个，填补了世界上牦牛没有培育品种的空白。获国家科技进步奖1项，省部级科技进步奖5项及其他科技奖励3项。培养研究生15名，发表论文180余篇，出版《反刍动物营养与饲料利用》、《现代动物繁殖技术》、《牦牛养殖实用技术问答》、《Recend Advances in Yak Reproduction》、《中国畜禽遗传资源志—牛志》等著作。

董书伟（1980— ），男，汉族，在读博士，助理研究员，毕业于西北农林科技大学。主要从事奶牛营养代谢病与中毒病的蛋白质组学研究。先后主持参加国家科技支撑计划、中央级公益性科研院所基本科研业务费专项资金项目、中国科学院西部之光项目、国家自然科学青年基金，发表学术论文20余篇，申请专利18项。

郭宪（1978— ），中国农业科学院硕士生导师，博士，副研究员。中国畜牧兽医学会养牛学分会理事，中国畜牧业协会牛业分会理事，全国牦牛育种协作组理事。主要从事牛羊繁育研究工作。先后主持国家自然科学基金、国家支撑计划子课题、中央级公益性科研院所基本科研业务费专项资金项目等5项。科研成果获奖3项，参与制定农业行业标准3项，授权专利3项。主编著作2部，副主编著作5部，参编著作2部。主笔发表论文30余篇，其中SCI收录5篇。

郭志廷（1979— ），男，内蒙古人，助理研究员，执业兽医师，九三学社兰州市青年委员会委员，中国畜牧兽医学会中兽医分会理事。2007年毕业于吉林大学，获中兽医硕士学位。近年主要从事中药抗球虫、免疫学和药理学研究。先后主持或参加国家、省部级科研项目5项，包括中央级公益性科研院所专项基金。作为参加人获得兰州市科技进步一等奖1项，兰州市技术发明一等奖1项，完成甘肃省科技成果鉴定4项，授权国家发明专利5项（1项为第一完成人），参编国家级著作2部。在国内核心期刊上发表学术论文80余篇（第一作者30篇）。

郝宝成（1983— ），甘肃古浪人，硕士研究生，助理研究员，研究方向为新型天然兽用药物研究与创制。先后主持和参与了中央级公益性科研院所基本科研业务费专项资金、国家支撑计划、863项目子课题等项目6项，以第一作者发表论文23篇（其中SCI论文2篇，一级学报2篇），参与编写著作3部《中兽药学》、《兽医中药学及试验技术》、《天然药用植物有效成分提取分离与纯化技术》，以第一发明人获得国家发明专利3项，荣获2012年度兰州市九三学社参政议政先进工作者。

李宏胜（1964— ），九三学社社员，博士，研究员，硕士生导师，甘肃省“555”创新人才。中国畜牧兽医学会家畜内科学分会常务理事。多年来主要从事兽医微生物及免疫学工作，尤其在奶牛乳房炎免疫及预防方面有比较深入的研究。先后主持和参加完成了国家自然基金、国家科技支撑计划、国际合作、甘肃省、兰州市及企业横向合作等20多个项目。先后获得农业部科技进步三等奖1项；甘肃省科技进步二等奖2项、三等奖2项；中国农业科学院技术成果二等奖3项；兰州市科技进步一等奖2项、二等奖2项。获得发明专利4项，实用新型专利14项，培养硕士研究生5名，在国内外核心期刊上发表论文160余篇，其中主笔论文60余篇。

李剑勇（1971— ），研究员，博士学位，硕士和博士研究生导师，国家百千万人才工程国家级人选，国家有突出贡献中青年专家。现任中国农业科学院科技创新工程兽用化学药物创新团队首席专家，农业部兽用药物创制重点实验室副主任，甘肃省新兽药工程重点实验室副主任，甘肃省新兽药工程研究中心副主任，农业部兽药评审专家，甘肃省化学会色谱专业委员会副主任委员，中国畜牧兽医学会兽医药理毒理学分会理事，国家自然基金项目同行评议专家，《黑龙江畜牧兽医杂志》常务编委，《PLOS ONE》、《Medicinal Chemistry Research》等SCI杂志审稿专家。多年来一直从事兽用药物创制及与之相关的基础和应用基础研究工作。曾先后完成药物研究项目40多项，主持16项。获省部级以上奖励10项，2011年度获第十二届中国青年科技奖；2011年度获第八届甘肃青年科技奖；获2009年度兰州市职工技术创新带头人称号。获国家一类新兽药证书，均为第2完成人。申请国家发明专利22项，获授权9项。发表科技论文200余篇，其中SCI收录22篇，第一作者和通讯作者15篇。出版著作4部，培养研究生15名。

王学智（1969— ），研究员、博士，硕士研究生导师，科技管理处处长。主要从事科技管理工作和兽医临床科研工作。主要在中兽药、动物营养代谢病等兽医临床学方面开展基础应用研究，先后主持“国家公益性行业专项中兽药生产关键技术研究与应用课题防治螨病和痢疾藏中兽药制剂制备”、“科技基础性工作专项-传统中兽医药资源抢救和整理”和“甘肃省科技重大专项——防治奶牛繁殖病中药研究与应用”、甘肃省科技支撑计划项目等科研课题10项。参与完成“948项目”、“国家自然基金项目”、“国家科技支撑计划”等各类科研项目16项。先后获得省部院级科技成果奖励7项：“奶牛乳房炎联合诊断和防控新技术研究及示范”获甘肃省农牧渔业丰收一等奖，“重金属镉/铅与喹乙醇抗原合成、单克隆抗体制备及ELLSA检测技术研究”和“新型中兽药饲料添加剂“参芪散”的研制与应用”获中国农业科学院科学技术成果二等奖，“归蒲方中草药饲料添加剂的研制与应用”、“禽用复合营养素的研制及其应用”和“非解乳糖链球菌发酵黄芪转化多糖的研究与应用”获甘肃省科技进步三等奖。在国家级学术刊物上先后发表学术论文15篇，其中SCI. 7篇，参编著作17部，主编5部，获得专利22个。

李维红（1978— ），博士，副研究员。主要从事畜产品质量评价技术体系、畜产品检测新方法及其产品开发利用研究等。先后主持和参加了中央级公益性科研院所基本科研业务费专项资金项目、甘肃省自然基金项目等。主笔发表论文20余篇。主编《动物纤维超微结构图谱》，副主编《绒山羊》，参加编写《动物纤维组织学彩色谱》和《甘肃高山细毛羊的育成和发展》著作2部。获得专利4项，作为参加人获甘肃省科技进步一等奖一项、三等奖一项。

李新圃（1962— ），博士，副研究员。主要从事兽医药理学研究工作。已主持完成省、部、市级科研项目7项。参加完成国际合作、省、部、市级科研项目三十余。参加项目“奶牛重大疾病防控新技术的研究与应用”获2010年甘肃省科技进步二等奖；“奶牛乳房炎主要病原菌免疫生物学特性的研究”在2008年获兰州市科技进步一等奖、中国农科院科学技术成果二等奖和甘肃科技进步三等奖；“绿色高效饲料添加剂多糖和寡糖的应用研究”获2005年兰州市科技进步一等奖；“奶牛乳房炎综合防治配套技术的研究及应用”获2004年甘肃省科技进步二等奖。已发表研究论文40余篇，其中5篇被SCI收录。获实用新型专利授权3个。

梁春年（1973— ），硕士生导师，博士，副研究员，副主任，兼任全国牦牛育种协作组常务理事，副秘书长，中国畜牧兽医学会牛学会理事，中国畜牧业协会牛业分会理事，中国畜牧业协会养羊学分会理事等职。主要从事动物遗传育种与繁殖方面的研究工作。现主持国家科技支撑计划子课题“甘肃甘南草原牧区牦牛选育改良及健康养殖集成与示范”和国家星火计划项目子课题“牦牛高效育肥技术集成示范”等课题4项；参加完成国家及省部级科研项目30余项，获得省部级科技奖励7项。参与制定农业

行业标准 5 项。参加国内各类学术会议 30 余次，国际学术会议 4 次。主编著作 2 本，副主编著作 5 本，参编著作 6 本，发表论文 90 余篇，其中 SCI 文章 5 篇。

孙晓萍（1962— ），硕士生导师，学士，副研究员。主要从事绵羊遗传育种工作，先后主持的项目有：甘肃省自然科学基金：绵羊毛生长机理研究；甘肃省农委推广项目：绵羊双高素推广应用研究；甘肃省支撑计划项目：奶牛产奶量的季节性变化规律研究 ；甘肃省星火项目：肉羊高效繁殖技术研究；甘肃省支撑计划项目：肉用绵羊高效饲养技术研究。先后发表学术论文 40 余篇，主编参编著作 11 部，实用新型专利 3 个。获甘肃省畜牧厅科技进步二等奖 1 项，甘肃省科技进步二等奖 1 项，中国农业科学院科技进步一等奖 1 项，中华农业科技二等奖 1 项。

王东升（1979— ），农学硕士，助理研究员。主要从事奶牛繁殖疾病的研究。主持“奶牛子宫内膜中天然抗菌肽的分离、鉴定及其生物学活性研究”和“狗经穴靶标通道及其生物效应的研究”2 个课题，参加“十二五”国家科技支撑计划项目“奶牛健康养殖重要疾病防控关键技术研究”和“十一五”国家科技支撑计划项目“奶牛主要繁殖障碍疾病防治药物研制”等 10 多个项目。参与申请并获得发明专利 4 项，实用性新专利 5 项，取得三类新兽药证书 1 个，参加的成果获甘肃省科技进步二等奖 2 项和兰州市科技进步二等奖 2 项。参编著作《兽医中药配伍技巧》、《兽医中药学》、《奶牛围产期饲养与管理》和《奶牛常见病综合防治技术》等 5 部，主笔发表论文 20 余篇。

王宏博（1977— ），博士，助理研究员，博士。主要从事动物营养与饲料科学。先后主持甘肃省科技厅项目 2 项，中央级公益性科研院所科研业务费专项资金项目 3 项，先后参加农业部公益性行业（农业）专项 3 项，“948”项目 1 项，获得国家发明专利 2 项，实用新型专利 1 项，参与完成国家发明专利和实用新型专利总计 10 余项。参与制定国家和农业部标准 10 余项。主编 1 部，副主编著作 1 部，参编著作 3 部。发表学术论文 30 余篇。

王磊（1985— ），硕士，研究实习员。主要从事中兽医药理学、奶牛疾病防治和药物残留研究。先后参加各类研究课题 8 项，主持课题 1 项。2013 年参加的“非解乳糖链球菌发酵黄芪转化多糖的研究与应用”获甘肃省科技进步三等奖、2014 年参加的“重金属镉/铅与喹乙醇抗原合成、单克隆抗体制备及 ELISA 检测技术研究”获中国农业科学院科学技术成果二等奖；主笔发表科技论文 4 篇、其中 SCI 收录 1 篇，获得授权专利 2 项。

王晓力（1965— ），副研究员。主要从事牧草资源开发利用方面的研究工作。近年来参加省部级课题多项，主持农业行业科研专项子课题2项，获全国农牧渔业丰收二等奖1项、内蒙古自治区农牧业丰收一等奖1项、2011年贵州省科学技术成果转化二等奖1项。发表论文30余篇，出版著作4部，参编4部，授权专利5项。

王旭荣（1980— ），博士，助理研究员。主要从事分子病毒学和分子细菌学方面的研究。主持完成“犬瘟热病毒”的基本科研业务费项目1项；现主持“奶牛乳房炎病原菌“方面的农业行业标准项目1项和甘肃省农业生物技术项目1项；参与国家奶牛产业技术体系项目、948项目、国家科技支撑项目、国家自然基金、科技基础性工作专项等10余项。参与的研究项目在2011—2014年期间获得奖项3个，其中甘肃省科技进步三等奖1项，甘肃省农牧渔丰收奖一等奖1项、中国农业科学院科学技术成果二等奖1项。发表科技论文40余篇，主笔15篇；参编著作1部；实用新型授权20余项，第一发明人授权5项；申请发明专利（第一发明人）3项。

王学红（1975— ），硕士，高级兽医师。主要从事天然药物提取研究。先后主持及参加国家级省部级药物研究项目20余项，包括“十二五”农村领域国家科技计划项目、国家支撑计划项目、甘肃省科技支撑计划项目及中央级公益性科研院所基本科研业务费专项资金项目等。获省部级科技进步奖4项，院厅级奖2项，国家发明专利10余个。在国内外学术刊物上发表论文20余篇，出版著作4部。

吴晓云（1986— ），博士，研究实习员，主要从事牦牛低氧适应机制和牦牛肉品质形成遗传机制的研究。目前以第一作者发表论文11篇，其中7篇被SCI收录。

严作廷（1962— ），硕导，博士，研究员。现为第五届农业部兽药评审专家、中国畜牧兽医学会家畜内科学分会常务理事、中国畜牧兽医学会中兽医学分会理事。主要从事工作奶牛疾病和中兽医学研究（先后主持国家自然科学基金、国家科技支撑计划等项目10多项，现主持“十二五”国家科技支撑计划课题“奶牛健康养殖重要疾病防控关键技术研究”课题）。主要成就（主编或副主编《家畜脉诊》、《奶牛围产期饲养与管理》等著作3部，参编《中兽医学》、《奶牛高效养殖技术及疾病防治》和《犬猫病诊疗技术及典型医案》等著作6部。发表论文70余篇，获省部级科技进步奖6项，院

厅级奖 7 项。获得国家新兽药证书 2 个，国家发明专利 10 个，实用新型专利 5 个）。

张世栋（1983— ），硕士，助理研究员。主要从事奶牛疾病与中兽医药研究工作。近年来，主持中央级基本科研业务费及其增量项目共 3 项，国家自然基金 1 项；参与国家“十一五”、“十二五”课题 3 项，甘肃省、兰州市科技项目多项。参与获得兰州市科技进步二等奖 1 项。发表论文 10 篇，获得专利 6 项，参与申请专利 10 多项。

李建喜（1971— ），研究员，博士学位，硕士研究生导师。现任中国农业科学院兰州畜牧与兽药研究所中兽医（兽医）研究室主任，中国农业科学院科技创新工程中兽医与临床创新团队首席专家，甘肃省中兽药工程技术研究中心副主任，农业部新兽药中药组评审专家，国家自然基金项目同行评议专家，国家现代农业（奶牛）产业技术体系后备人选等。从事兽医病理学、动物营养代谢病与中毒病、兽医药理与毒理、奶牛疾病防治、中兽医药现代化等研究工作。完成国家和省部级科研项目 40 余项，其中主持 24 项，包括国家自然基金面上项目、国家科技支撑计划课题、国家公益性行业（农业）科研专项课题、国家科技基础性工作专项课题、农业部 948 计划项目、甘肃省农业科技重大专项、甘肃省生物技术专项、中国西班牙科技合作项目、中泰科技合作项目等。先后获科技奖励 6 项，其中获省部级奖励 2 项，院厅级奖励 4 项。研发新产品 6 个，获授权发明专利 9 项，以第一导师培养硕士研究生 6 名，共同培养硕士研究生 17 名，参与培养博士研究生 8 名。在国内和国际学术刊物上共发表学术论文 99 篇，其中以第一作者和通讯作者发表论文 54 篇，SCI 收录 6 篇，编写著作 7 部。

梁剑平（1962— ），博士，三级研究员，博士生导师。2005 年获“农业部有突出贡献的中青年专家”称号。2005 年分别获中国科协“西部开发突出贡献奖”、中央统战部“为全面建设小康社会做出贡献的先进个人”、甘肃省“陇上骄子”、九三学社甘肃省委、“十佳青年”称号。现任兰州畜牧与兽药研究所兽药研究室副主任、农业部兽用药物创制重点实验室副主任、中国农业科学院新兽药工程重点开放实验室副主任。是中国农业科学院二级岗位杰出人才，甘肃“555”创新人才。兼任中国毒理学会兽医毒理学分会及中国兽医药理学分会理事，农业部新兽药评审委员会委员，农业部兽药残留委员会委员，中国兽药典委员会委员，中国农业科学院学术委员会委员，中国农科院研究生院教学委员会委员、政协兰州市委常委，九三学社兰州市七里河区主任委员，九三学社兰州市副主委。梁剑平研究员主要从事兽药化学合成和中草药的提取及药理研究，先后主持和参加国家和省部级重大科研项目 20 余项，获奖 8 项，其中获国家科技进步二、三等奖各 1 项，省（部）级二等奖 2 项、三等奖 2 项，发明专利 10 项。

蒲万霞（1964— ）博士后，四级研究员，硕士生导师，甘肃省微生物学会理事，政协兰州市七里河区第六届委员。从事兽医微生物与微生物制药研究，重点方向为兽用微生态制剂的研制及细菌耐药性研究，在金黄色葡萄球菌耐药性研究方面取得一定进展。先后主持农业部公益性行业科研专项“新型动物专用药物的研制与应用”子专题，国家科技支撑计划“奶牛主要疾病综合防控技术研究及开发”子专题，中央级公益性科研院所基本科研业务费专项资金项目、甘肃省科技成果重点推广计划项目、甘肃省农业科技创新项目、兰州市院地校企合作项目、兰州市科技局项目、农业部畜禽病毒学开放实验室基金项目等各级各类项目15项。获得各级政府奖7项。取得国家发明专利2项。主编《食品安全与质量控制技术》等著作5部，副主编著作1部，发表论文70多篇，培养硕士生11名。

朱新书（1957— ），学士，副研究员，主要从事牦牛藏羊品种资源开发和利用工作，主要主持和参加的项目有：公益性行业（农业）科研专项《放牧牛羊营养均衡需要研究与示范》；农业部重点攻关项目《大通牦牛选育与杂交利用》；甘肃省重大科技项目《甘南牦牛改良与选育技术研究示范》；国家肉牛牦牛产业技术体系《牦牛繁育技术研究与示范岗位科学家团队》等。发表学术论文20余篇，主编参编著作3部，荣获农业部科技进步三等奖1项。

岳耀敬（1980— ），在读博士。主要从事绵羊繁殖、羊毛和高原适应性等重要经济、抗逆性状的分子调控机制研究。国家绒毛用羊产业技术体系分子育种岗位团队成员，兼任中国畜牧兽医学会养羊学分会副秘书长、世界美利奴育种者联盟成员，曾先后到法国、澳大利亚、新西兰等国学习考察细毛羊育种工作。现主持国家自然青年基金、甘肃省青年基金等项目3项；申报发明专利5项，授权发明专利2项、实用新型2项；参编著作5部，发表学术论文38篇，其中SCI5篇。

谢家声（1956— ），大学专科学历，高级实验师。主要从事猪、禽、奶牛等动物疾病的中兽医临床防治研究，主持完成了甘肃省自然基金—“复方中草药防治畜禽病毒性传染病新制剂的研究”、甘肃省科技攻关项目—“治疗奶牛胎衣不下天然药物的研制”等项目。完成国家“七五”、“八五”攻关项目、国家“十五”及“十一五”奶业专项、国家“十一五”、“十二五”支撑计划以及省市各类研究项目30余项。发表论文60余篇，合编专著8部，获得国家发明专利3项。1990年以来先后获甘肃省科技进步三等奖1项；甘肃省科技进步二等奖3项；中国农业科学院科技进步二等奖4项；兰州市科技进步二等奖3项；甘肃省农牧厅二等奖1项；甘肃省天水市科技进步二等奖2项。

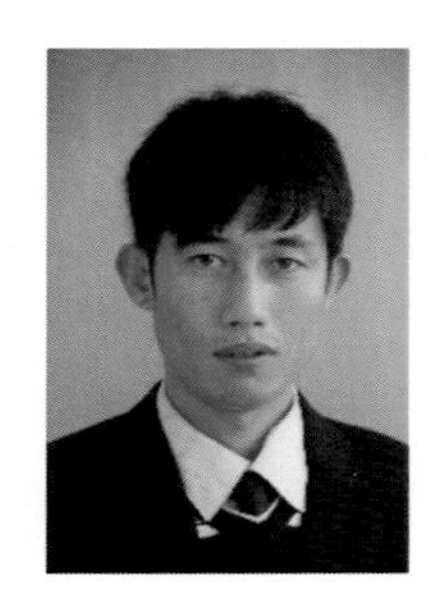

尚小飞（1986— ），硕士，助理研究员。主要从事藏兽医药的现代化研究。目前主持中央级公益性院所基本科研业务费项目一项，参与多项国家及省部级课题。在 *Journal of Ethnopharmacology*，*Veterinary Parasitology* 等 SCI 杂志发表文章 8 篇，参与编写著作一部。

杨红善（1981— ），硕士，助理研究员。主要从事牧草种质资源搜集与新品种选育研究工作，主持在研项目 3 项，其中甘肃省青年基金项目 1 项、甘肃省农业生物技术研究与应用开发项目 1 项、中央级公益性科研院所基本科研业务费专项资金项目 1 项，参加国家支撑计划子课题等各类项目共计 3 项。工作期间以第一完成人或参加人审定登记甘肃省牧草新品种 4 个。获中国农业科学院科技进步二等奖 1 项（第三完成人）。参加编写《高速公路绿化》著作 1 本（副主编）。在各类期刊发表论文 20 篇，其中主笔 13 篇。

刘宇（1981— ），硕士，助理研究员。2007 年 7 月来中国农业科学院兰州畜牧与兽药研究所工作至今，主要从事有机及天然药物化学研究。先后主持及参与国家级省部级药物研究项目 10 余项，包括“十二五”农村领域国家科技计划项目、国家支撑计划项目、甘肃省科技支撑计划项目、甘肃省自然科学基金项目及中央级公益性科研院所基本科研业务费专项资金项目等。获省部级科技进步奖 1 项，院厅级奖 2 项，国家发明专利 4 个。在国内外学术刊物上发表论文 20 余篇，出版著作 2 部。

程富胜（1971— ）博士，副研究员，硕士生导师。主要从事天然药物活性成分免疫药理学研究及其制剂开发研制。先后主持和参加国家、省、部级科研课题 20 项。参加完成的国家支撑项目“中草药饲料添加剂‘敌球灵’的研制”获农业部科技进步二等奖；国家自然基金项目“免疫增强剂-8301 多糖的研究与应用”成果获中国农业科学院科技进步二等奖；“蕨麻多糖免疫调节作用及其机理研究与临床应用”成果已分别获中国农科院、兰州市可进步二等奖。主持完成的“富含活性态微量元素酵母制剂的研究”获兰州市科技进步二等奖。在国内和国际学术刊物上共发表学术论文 50 余篇，其中主笔发表论文 30 余篇。

李冰（1981— ），女，硕士研究生、助理研究员，研究方向为兽药新制剂研制与安全评价。主要从事新兽药制剂质量标准制定、体内药代动力学研究与残留研究。中国畜牧兽医学会兽医药理毒理学分会理事。主持在研项目 2 项，其中公益性行业专项子课题 1 项、中央级公益性科研院所基本科研业务费专项资金项目 1 项；参加国家支撑计划项目、国家高技术研究发展计划（863 计划）、甘肃省科技重大专项、国家自然基金项目、国家基础专项等各类项目十余项。以主要完成人申报国家新兽药 1 个。前三

以前作者发表论文30余篇，主笔核心期刊3篇，主笔SCI收录2篇，参编著作2部。获2011年兰州市科学技术进步奖一等奖，第6完成人；2012年中国农业科学院科技进步二等奖，第3完成人；2013年甘肃省科学技术发明奖一等奖，第3完成人；2013年兰州市科学技术发明奖一等奖，第3完成人。

褚敏（1982— ），中共党员，在读博士，助理研究员。主要从事牦牛分子遗传与育种研究，先后参加国家科技支撑计划、农业部行业科技、甘肃省重大科研项目、“863”“948”项目、甘肃省生物技术和甘肃省科技支撑等重大课题十余项，现主持中央级公益性科研院所基本科研业务费专项资金项目1项。主笔发表SCI学术论文2篇，中文10多篇，参与编写著作《牦牛养殖实用技术问答》和《适度规模肉牛场养殖技术示范》2部，翻译并编辑第五届国际牦牛会议英文学术论文集1部，发明实用新型专利6项。

杨博辉（1964— ）博士，四级研究员，博士生导师。现为“国家绒毛用羊产业技术体系”分子育种岗位科学家，中国农科院兰州畜牧与兽药研究所“绵羊新品种（系）培育”科技创新团队首席专家。兼任中国畜牧兽医学会养羊学分会副理事长兼秘书长，中国农科院学位委员会委员，中国博士后基金评审专家，中国畜牧业协会羊业分会特聘专家，中国农业科学院三级岗位杰出人才。主要从事绵羊新品种（系）培育、分子育种及产业化研发。先后主持完成国家863计划、国家支撑计划、国家基础专项及国家（部）级标准等项目20余项。制定国颁标准6项，部颁标准5项。获国家发明专利2项，实用新型专利8项。已培育出“大通牦牛”新品种、西北肉羊新品群，基本培育出高山美利奴羊新品种。发表论文120篇，其中7篇SCI，获国际论文一等奖1篇。主编《甘肃高山细毛羊的育成和发展》与《中国野生偶奇蹄目动物遗传资源》著作2部，副主编著作4部，参编著作4部。培养博、硕士研究生21名，其中国际留学生1名。

王慧（1985— ），硕士，研究实习员。主要从事动物营养代谢病研究。现主持“中央级科研院所基本科研业务费专项资金项目”（NO. 1610322013003）。2013年获农业部全国农牧渔业丰收二等奖（第7完成人）；兰州市科学技术进步二等奖（第9完成人）。目前第一作者发表论文16篇，其中SCI收录8篇。申请专利2项。

尚若锋（1974— ），博士，副研究员。主要从事兽用药物研发工作。主持或参与国家支撑计划、“863”计划以及其他国家级和省部级的科研项目20余项。获得省部级科研奖励4项，国家发明专利12项，以第一作者或通讯作者发表文章30余篇，其中SCI文章12篇。

刘建斌（1977—　），甘肃农业大学动物生产系统与工程专业博士，副研究员，主要从事绵羊现代遗传学理论与育种，动物种质资源优良基因发掘和重要经济性状分子遗传机理研究。先后主持省部级科研项目5项。获甘肃省科技进步二等奖1项、中国农业科学院科技进步二等奖1项、中华神农科技三等奖1项。主编出版《现代肉羊生产实用技术》和《绒山羊》著作2部，副主编出版《甘肃高山细毛羊的育成和发展》、《中国野生偶奇蹄目动物遗传资源》、《羊繁殖与双羔免疫技术》、《适度规模肉羊场高效生产技术》和《优质羊毛生产技术》著作5部，参编出版著作5部；主笔发表专业论文30余篇，其中SCI论文7篇；申报国家发明专利6项，授权发明专利3项，合作授权国家实用新型专利30项。

辛蕊华（1981—　），硕士，助理研究员。主要从事中兽医药物学工作，先后主持和参与过中央级公益性科研院所基本科研业务费专项资金项目、公益性行业（农业）科研专项、国家科技支撑项目等；发表论文12篇，参与著作两部。参加的“富含活性态微量元素酵母制剂的研究”科研成果获得兰州市科学技术二等奖。

杨峰（1985—　），助理研究员，长期从事奶牛乳房炎的诊断、预防和治疗工作。主持2项中央级公益性科研院所基本科研业务费专项资金和1项甘肃省科技计划项目，先后参与国家自然科学基金、国家支撑计划、省部级等项目10多项。以第一完成人获得授权实用新型专利8项。

罗超应（1960—　），学士，研究员，中西兽医药学结合研究（主持科技部科研院所开发研究专项“新型中兽药射干地龙颗粒的研究与开发”、科技部基础工作专项“华东区传统中兽医学资源抢救与整理”与国家“十一五”科技支撑子项目“中兽药中试及其生产工艺研究”等）。主编、副主编与参编出版《牛病中西医结合治疗》等著作16部，共计679余万字；主笔发表“Variability of the Dosage, Effects and Toxicity of Fu Zi (Aconite) From a Complexity Science Perspective”、“以复杂科学理念指导中西医药学结合”、“奶牛乳房炎的复杂性及其对传统科学观念的挑战”等学术论文近100篇，其中英文期刊文章5篇；专利12项，成果奖励5项。

冯瑞林（1959—　），本科毕业，助理研究员，研究方向为羊繁殖育种。主要从事羊双羔免疫技术研究工作，参与甘肃省重大专项“甘肃超细毛羊新品种培育示范与推广”、中国农业科学院“羊绿色增产增效技术集成与示范”等十余项。工作期间参与申报细毛羊新品种1个。获国家科技进步三等奖1项（第九完成人），甘肃省科技进步一等奖1项（第九完成人）。主编出版著作《羊繁殖与双羔免疫技术》1部，参加编写《甘肃高

山细毛羊的育成与发展》（副主编）、《绒山羊》（副主编）等著作5部。在各类期刊发表论文50余篇，其中以第一作者发表论文12篇。

郑继方（1958— ），研究员，硕士生导师。现任甘肃省中兽药工程技术研究中心主任，中国农业科学院兰州畜牧与兽药研究所中兽医药创新团队首席科学家，中国农业科学院兰州畜牧与兽药研究所学术委员会委员，《中兽医医药杂志》编委，亚洲传统兽医学会常务理事，中国畜牧兽医学会中兽医分会理事，中国生理学会甘肃分会理事，西北地区中兽医学术研究会常务理事，中国畜牧兽医学会高级会员，农业部项目评审专家，农业部新兽药评审委员会委员，科技部国际合作计划评价专家，西南大学客座教授。从事中兽医药学的研究工作。先后主持国家自然科学基金、国家科技攻关、国家支撑计划、省部级等项目20多项。获省部级科技进步奖3项，院厅级奖4项，国家级新兽药证书两个，国家发明专利十余项。主编出版了《兽医中药学》、《中兽医诊疗手册》、《兽医药物临床配伍与禁忌》、《常用兽药临床新用》、《中草药饲料添加剂配制与应用》、《甲型H1N1流感防控100问》和《兽医中药临床配伍技巧》等专著10部，先后在国内外各学术刊物发表论文80余篇。

刘希望（1986— ），硕士，助理研究员。2007年毕业西北农林科技大学环境科学专业，2010年获西北农林科技大学化学生物学专业硕士学位，同年参加工作至今。曾主持农业部兽用药物创制重点实验室开放基金项目1项，现主持中央公益性科研院所基本科研业务费1项。以第一作者发表SCI论文4篇，主要从事新兽药研发，药物合成方面的研究工作。

杨晓（1985— ），硕士，助理研究员。主要从事牧草栽培和草原生态研究，先后参与国家科技支撑计划项目、公益行业专项项目等。发表代表性文章5篇，其中SCI文章1篇。

王华东（1979— ），硕士，助理研究员。主要从事科技期刊编辑工作，先后出版著作3部（副主编1部，参编2部），主笔发表学术论文5篇。编辑、出版的《中兽医医药杂志》发行范围涵盖中国、北美、澳洲、西欧、韩国、日本、新加坡等海内外15个国家和地区。随着中兽医药在畜禽疫病防治及公共卫生安全等方面的独特作用的日益凸显，作为中兽医领域唯一的国家级学术刊物，《中兽医医药杂志》全面、详实地报道了近年来该领域的最新研究成果，并力促其推广应用与转化，为继承和传播祖国兽医学诊疗技术、促进中西兽医结合、繁荣学术、服务“三农”等做出了重大贡献，取得了显著的经济效益、社会效益和生态效益。

熊琳（1984—　），硕士，助理研究。主要从事农产品质量安全的研究。发表相关科研论文 4 篇，其中 SCI 收录 1 篇，授权专利 8 项，参与制定国家标准 1 项。

魏小娟（1976 年），女，硕士研究生、助理研究员，研究方向为分子药理学。曾先后主持国家自然科学基金 1 项、甘肃省青年基金项目 1 项、中央级公益性科研院所基本科研业务费专项资金项目 1 项；先后参加国家自然科学基金项目，863 项目、国家支撑计划项目、公益性行业专项、甘肃省重大专项等项目 10 余项。工作期间获甘肃省科技进步一等奖 1 项，中国农业科学院科技进步二等奖 2 项，兰州市科技进步二等奖 2 项。参加编写著作 5 部。在各类期刊发表论文 30 篇，其中 SCI 5 篇。

崔东安（1981—　），博士，主要从事奶牛胎衣不下方证代谢组学研究（奶牛胎衣不下血瘀证的代谢组学研究 No. 1610322015006），发表文章 4 篇，其中 3 篇 SCI 文章。

路远（1980—　），硕士，副研究员。主要从事牧草新品种选育及植物组培研究。自毕业以来，主持院所长基金项目“美国杂交早熟禾引进驯化及种子繁育技术研究”、和“黄花矶松驯化栽培及园林绿化开发应用研究”，曾参与并完成了 973 合作子课题项目“气候变化对西北春小麦单季玉米区粮食生产资源要素的影响机理研究”，全球环境基金（GEF）项目“野生牧草种质资源应用研究”、“放牧利用与草原退化关系研究” 和国家科技基础条件平台工作项目子课题“牧草种质资源的实物共享及标志性数据采集”、“ 牧草种质资源的标准化整理和整合”、国家科技支撑计划项目子课题“西北优势和特色牧草生产加工关键技术研究与示范” 等项目 20 余项。发表论文 20 余篇，主编著作 1 部，或甘肃省科技进步二等奖 1 项，选育牧草新品种 1 个，获专利 5 项。

高雅琴（1964—　），研究员，硕士研究生导师，主要从事畜产品加工及动物毛皮质量评价研究工作。现主持农业部畜产品质量安全风险评估项目中牛羊肉质量安全风险评估子项目。曾主持的国家公益类科研项目“动物纤维毛皮产品质量评价技术的研究”，2009 年获甘肃省科技进步三等奖；主持国家标准制定项目“GB/T 25885—2010 羊毛纤维直径试验方法——激光扫描仪法”、“GB/T 25880—2010 毛皮掉毛测试方法——掉毛测试仪法”、“GB/T 26616—2011 裘皮獭兔皮” 均已颁布并实施；主持的农

业行业标准制定项目“NY 1164—2006 裘皮蓝狐皮”、“NY 2222—2012 动物纤维直径及成分分析方法——显微图像分析仪法”已颁布实施，“动物毛皮各类鉴别方法 显微镜法”上报农业部审定。参与多项国家标准和农业行业标准“GB/T25243—2010 甘肃高山细毛羊”、“河西绒山羊”和“大通牦牛”、“天祝白牦牛”、“NY 1173—2006 动物毛皮检验技术规范”等。主编的《动物毛纤维组织学彩色图谱》，《动物毛皮质量鉴定技术》；参与出版著作 5 部。发表科技论文 150 余篇，其中主笔发表 40 余篇。